BIOPHYSICS OF
COMPUTATION

COMPUTATIONAL NEUROSCIENCE

SERIES EDITOR
Michael Stryker

BIOPHYSICS OF COMPUTATION *Information Processing in Single Neurons*
Christof Koch

BIOPHYSICS OF COMPUTATION

Information Processing in Single Neurons

Christof Koch

New York   Oxford
OXFORD UNIVERSITY PRESS

Oxford University Press

Oxford   New York
Athens   Auckland   Bangkok   Bogotá   Buenos Aires
Calcutta   Cape Town   Chennai   Dar es Salaam
Delhi   Florence   Hong Kong   Istanbul   Karachi
Kuala Lumpur   Madrid   Melbourne   Mexico City
Mumbai   Nairobi   Paris   São Paulo   Singapore
Taipei   Tokyo   Toronto   Warsaw

and associated companies in
Berlin   Ibadan

Library of Congress Cataloging-in-Publication Data
Koch, Christof.
   Biophysics of computation: information processing in single
neurons / by Christof Koch.
      p.   cm.—(Computational neuroscience)
   Includes bibliographical references (p.   ) and index.
   ISBN 0-19-510491-9 (cl.); 0-19-518199-9 (pbk.)
   1. Computational neuroscience.   2. Neurons.   3. Neural networks
(Neurobiology)   4. Action potentials (Electrophysiology)   5. Neural
conduction.   I. Title   II. Series.
QP357.5.K63   1998
573.8'536—dc21   97-51390
                                                                          CIP

1 3 5 7 9 8 6 4 2

Printed in the United States of America
on acid-free paper

Dedicated to Hannele, Alexander, and Gabriele

Memento homo, quia pulvis es,
et in pulverem reverteris

Contents

Preface

The book that you are holding in your hands is aimed at trying to reconcile two disparate scientific fields: the biophysics of ionic channels, synapses, dendrites and neurons on the one hand and computation and information processing on the other. While excellent textbooks exist in both fields, not a single one adopts the perspective of linking specific biophysical mechanisms, say the nonlinear interaction among excitatory and inhibitory synapses, with specific computations, in this case, multiplication implemented in the dendritic tree of retinal ganglion cells. Analyzing such linear and nonlinear operations requires that we seek to understand the reliability with which information can be encoded at this level, the bandwidth of these signals, and so on.

Thinking about brain-style computation requires a certain frame of mind, related to but distinctly different from that of the biophysicist. For instance, how should we think of a chemical synapse? In terms of the complicated pre- and postsynaptic elements? Ionic channels? Calcium-binding proteins? Or as a non-reciprocal and stochastic switching device that transmits a binary signal rapidly between two neurons and remembers its history of usage? The answer is that we must be concerned with both aspects, with biophysics as well as computation.

This book evolved over the course of many years. It was first conceived when I was working on my doctoral thesis under Tomaso Poggio and Valentin Braitenberg at the Max Planck Institut for Biological Cybernetics in Tübingen in the early eighties. It germinated when I moved to the Massachusetts Institute of Technology and continued to work with Tomaso Poggio on a "Biophysics of Computation," driven by the hope that a small number of canonical, biophysical mechanisms is used throughout biology to implement all relevant computations. The ideas expounded here took further shape when I joined the new "Computation and Neural Systems" graduate program at the California Institute of Technology. In this intellectual hothouse the outline of this book slowly took shape, influenced by such colleagues as John Hopfield and Carver Mead, as well as Francis Crick and Terrence Sejnowski at the Salk Institute. Many of the chapters have their origin in lectures I gave at Caltech and during the "Methods in Computational Neuroscience" summer course James Bower and I directed for five years at the Marine Biological Laboratory in Woods Hole.

The first chapters were written on sabbatical while visiting Klaus Hepp at the Theoretical Physics Department at the Eidgenössische Technische Hochschule in Zürich in the fall of 1993. Between running a large laboratory with diverse interests and teaching and supervising students, I found it very difficult to spend much time on this book during the academic year. Thus, the predominant fraction was penned—actually typed—during the past five summers, while away from Caltech. I would like to thank the Aspen Institute of Physics, the Telluride Summer Research Center and, in particular, the Institute for Neuroinformatics in Zürich and its leaders, Rodney Douglas and Kevan Martin, for their kind hospitality.

A note pertaining to style. Science is a social endeavor; many, if not most, new ideas are born out of discussions with colleagues, reading books and papers, attending seminars and so on. We are often not even explicitly aware of these influences, but they are there none-the-less. It is to acknowledge this that I use the *pluralis modestati* form of the "we" throughout the book.

This is also the place to thank everybody who made this book possible. Foremost, a profound *Danke* to my wife Edith and my children, Alexander and Gabi, who had to put up with "vacations," that involved dad spending most of the time on his Macintosh powerbook, typing away. I owe you more than I can ever express in words.

Thanks to the members of my laboratory who helped me in ways small and large with simulations, figures and text, in particular Öjvind Bernander, Fabrizio Gabbiani, Reid Harrison, Candace Maechtlen, Dusan Misevic and Adam Strassberg. I particularly like to thank Gary Holt and Dave Flowers who gave so generously of their time to satisfy my demands for simulations, figures and instant fixes to the myriad of postscript problems that I encountered on the way. I would like to gratefully acknowledge Anthony Zador who co-authored Chapter 13, as well as Barak A. Pearlmutter and Zador who wrote Appendix C. Lyle Borg-Graham, David Heeger, Gabriel Kreiman and Mark Kvale provided detailed feedback on the entire oeuvre. I appreciate their hard work. Fabrizio Gabbiani and Larry Abbott helped me with their insights into coding and the stochastic properties of neurons. I would like to acknowledge the colleagues worldwide who inspired me, sent me their latest publications and plots, and who corrected various chapters: Larry Abbott, Hagai Agmon-Snir, Bill Bialek, Daniel Blank, Barry Connors, Francis Crick, Alain Destexhe, Rodney Douglas, Yuki Goda, Henry Lester, Virginia Meskenaite, Read Montague, Tomaso Poggio, Muki Rapp, John Rinzel, Rahul Sarpeshkar, Idan Segev, Terrence Sejnowski, Gordon Shepherd, Diana Smetters, Nelson Spruston, Chuck Stevens and Eric Vu. Finally, I would like to thank all of the authors and publishers who kindly gave me permission to use their figures throughout this book.

All of this was made possible by the generous financial support of three US institutions over the years, the National Science Foundation, the National Institute of Mental Health, and the Office of Naval Research. Their wisdom and the willingness of the American taxpayer to support modern scientific research has made this a wonderful time and place to search for the elusive Truth with a capital "T."

A final cautionary note. I have tried to be as careful as I could when ascribing specific ideas to specific publications. If I have missed anybody or if a reader feels that an important topic has been neglected, please let me know and I will rectify this in future editions. I am open to any thoughtful suggestion.

Professor Christof Koch
California Institute of Technology
Pasadena
April 20, 1998

List of Symbols

Following is a list of symbols used frequently in the book. Many symbols appearing in only one chapter are not included. Appendix A includes a detailed discussion of units. In this table, an entry of "1" in the "Dimension" column refers to a dimensionless variable. "Definition" refers to the equation number where the variable in question is defined.

Symbol	Description	Dimension	Definition
γ	Single-channel conductance	nS	.
λ	Steady-state electrotonic space constant	μm	2.13
$\tau_h(V)$	Time constant of sodium inactivation	msec	
τ_m (or τ)	Passive membrane time constant	msec	1.10
$\tau_m(V)$	Time constant of sodium activation	msec	
$\tau_n(V)$	Time constant of potassium activation	msec	6.8
$\tilde{A}_{ij}(f)$	Frequency-dependent voltage attenuation	1	3.35
$\tilde{A}_{ij}$	Steady-state voltage attentuation between i and j	1	
[B]	Concentration of free intracellular buffer	μM	11.36
[B · Ca]	Concentration of bound intracellular Calcium buffer	μM	11.36
c_m	Specific membrane capacitance per unit length	μF/cm	2.10
C	Total capacitance	nF	
C_m	Specific membrane capacitance	μF/cm^2	
C_V	Coefficient of variation	1	15.15
$[Ca^{2+}]_i$	Concentration of free intracellular calcium ions	μM	
d	Cable diameter	μm	
D	Diffusion coefficient	μm^2/msec	11.13
D_{ii} (or D_{xx})	Input or local delay	msec	2.46
D_{ij} (or D_{xy})	Transfer delay	msec	2.45
E_{syn}	Synaptic reversal potential or battery	mV	4.3
E_i	Ionic reversal potential	mV	
E_K	Potassium reversal potential	mV	
E_{Ca}	Calcium reversal potential	mV	

Symbol	Description	Dimension	Definition
E_{Na}	Sodium reversal potential	mV	
F	Ratio of EPSP to net EPSP/IPSP	1	5.14
$F(T)$	Fano factor	1	15.18
$f(t)$	Instantaneous firing rate	Hz	
$\langle f(t) \rangle$	Time-averaged firing rate	Hz	14.1
$\langle f_b \rangle$	Average synaptic background firing rate	Hz	
g_{leak}	Leak conductance	nS	
g_{peak}	Peak synaptic-induced conductance change	nS	4.5
$g_{\text{syn}}(t)$	Synaptic-induced conductance change	nS	4.5 and 4.6
G_m	Specific leak conductance	mS/cm^2	
$\overline{G}_{\text{i}}$	Maximal ionic conductance	mS/cm^2	
$h(t)$	Sodium inactivation	1	6.15
$h_\infty(V)$	Steady-state sodium inactivation	1	
$I_{\text{Ca}}(t)$	Inward calcium current[1]	nA/cm^2	9.1
$I_{\text{syn}}(t)$	Synaptic current[1]	nA/cm^2	4.4
$I_m(t)$	Membrane current[1]	nA/cm^2	
$I_{\text{K}}(t)$	Outward potassium current[1]	nA/cm^2	6.4
$I_{\text{Na}}(t)$	Inward sodium current[1]	nA/cm^2	6.13
I_{th}	Current threshold for spike initiation[1]	nA	
K_d	Dissociation constant	Molar	11.38
$\tilde{K}_{ii}(f)$	Complex input impedance as a function of frequency		
$\tilde{K}_{ii}$	Steady-state input resistance	MΩ	3.19
$\tilde{K}_{ij}(f)$	Complex transfer impedance as a function of frequency		
$\tilde{K}_{ij}$	Steady-state transfer resistance	MΩ	
$\tilde{K}_{ss}$	Steady-state somatic input resistance	MΩ	
ℓ	Length of a cable	μm	
L	Normalized electrotonic length ($1/\lambda$)	1	
$\tilde{L}_{ij}^{v}(f)$	Frequency-dependent log voltage attenuation	1	3.38
$m(t)$	Sodium activation	1	6.14
$m_\infty(V)$	Steady-state sodium activation	1	
$n(t)$	Potassium activation	1	6.6
$n_\infty(V)$	Steady-state potassium activation	1	6.9
n_{th}	Number of synaptic inputs for spike initiation	1	

[1] Note that current is sometimes expressed as absolute current and sometimes as current per unit area.

Symbol	Description	Dimension	Definition
$p(t)$	Probability of release of a single synaptic vesicle	1	
P_{ij} (or P_{xy})	Propagation delay	msec	2.51
Q_i	Charge transferred from location i to the soma	C	3.44
r_a	Intracellular resistance per unit length	Ω/cm	2.8
R_∞	Input resistance of a semi-infinite cable	MΩ	2.16
R_i	Intracellular resistivity	$\Omega \cdot$cm	
R_{in}	Input resistance	MΩ	
R_m	Specific leak or membrane resistance	$\Omega \cdot$cm^2	
r_m	Specific membrane resistance per unit length	$\Omega \cdot$cm	2.9
t	Time	msec	
t_{peak}	Time-to-peak for synaptic conductance change	msec	4.5
T	Normalized time (t/τ_m)	1	
T_{B}	Concentration of total buffer	μM	11.37
T_{th}	Time to trigger a spike	msec	
$V(x,t)$	Transmembrane potential relative to V_{rest}	mV	
$\langle V_m(t) \rangle$	Time-averaged membrane potential	mV	17.12
$V_m(x,t)$	Absolute transmembrane potential	mV	
V_{rest}	Resting potential	mV	
V_{th}	Voltage threshold for spike initiation	mV	
Δw_{ij}	Change in synaptic weight ij	1	13.7
x	Position	μm	
X	Normalized distance (x/λ)	1	

Introduction

The brain computes! This is accepted as a truism by the majority of neuroscientists engaged in discovering the principles employed in the design and operation of nervous systems. What is meant here is that any brain takes the incoming sensory data, encodes them into various biophysical variables, such as the membrane potential or neuronal firing rates, and subsequently performs a very large number of ill-specified operations, frequently termed computations, on these variables to extract relevant features from the input. The outcome of some of these computations can be stored for later access and will, ultimately, control the motor output of the animal in appropriate ways.

The present book is dedicated to understanding in detail the biophysical mechanisms responsible for these computations. Its scope is the type of information processing underlying perception and motor control, occurring at the millisecond to fraction of a second time scale. When you look at a pair of stereo images trying to fuse them into a binocular percept, your brain is busily computing away trying to find the "best" solution. What are the computational primitives at the neuronal and subneuronal levels underlying this impressive performance, unmatched by any machine? Naively put and using the language of the electronic circuit designer, the book asks: "What are the diodes and the transistors of the brain?" and "What sort of operations do these elementary circuit elements implement?"

Contrary to received opinion, nerve cells are considerably more complex than suggested by work in the neural network community. Like morons, they are reduced to computing nothing but a thresholded sum of their inputs. We know, for instance, that individual nerve cells in the locust perform an operation akin to a multiplication. Given synapses, ionic channels, and membranes, how is this actually carried out? How do neurons integrate, delay, or change their output gain? What are the relevant variables that carry information? The membrane potential? The concentration of intracellular Ca^{2+} ions? What is their temporal resolution? And how large is the variability of these signals that determines how accurately they can encode information? And what variables are used to store the intermediate results of these computations? And where does long-term memory reside?

Natural philosophers and scientists in the western world have always compared the brain to the most advanced technology of the day. In the mid-seventeenth century, René Descartes was inspired by the fountains and waterplays at Versailles to write about the brain in terms of hydraulics in his celebrated "Traité de l'homme." In the time between the two world wars, the brain was compared to a telephone switchboard. Today, much of our thinking about the brain is dominated by our favorite new artifact, the digital computer. Indeed, the present work frequently appeals to comparisons between computers and brains, and between electronic and neuronal circuits. This is useful to highlight commonalties or to point out profound differences in style between the two.

Before we proceed, let us briefly pause and consider what we mean by computation (Sejnowski, Koch, and Churchland, 1988). Any physical process that transforms variables can be thought of as a *computation* as long as it can be mapped onto one or more mathematical operations that perform some useful function. For instance, a marble running down a hill can be thought of as finding the local minimum of a two-dimensional energy function. Two pieces of marked wood sliding past each other can be mapped onto a multiplication, provided the markings have logarithmic spacings. A silicon retina, where at each pixel a photocurrent is injected into a two-dimensional linear resistive network, is equivalent to low-pass filtering the image (Mead, 1989). Likewise as we will see a few chapters down the road, excitatory and inhibitory synapses in a dendritic tree can be mapped onto a low-order polynomial (multiplication) operation.

Of course we must determine the extent to which such a mapping is actually being exploited by the organism. Many physical systems do not lend themselves in a natural way to perform a large range of computations. A case in point is applied minimization problems. Since they usually involve hundreds of dimensions rather than two, landscaping the real world will not be a terribly practical way to solve them. Contrariwise, slide rules have been immensely useful to generations of students for multiplication, taking the square root, and related operations. In the brain we must in each instance seek experimental evidence that any particular mechanism actually implements some specific operation of ultimate relevance to the behavior of the animal.

Any complex information processing system can be analyzed at several levels (Marr and Poggio, 1976; Marr, 1982). The *computational level of description* involves a description of the mathematical function the system performs. Thus, we can think of some synapses as switching elements that signal changes in the mean firing rate, rather than the absolute firing rate. Or, in the example cited above, groups of dendritic synapses can instantiate a so-called sigma-pi neuron (summing product terms). In order to make this case, we need to understand these events at the physical or *biophysical level of description*. Such a description is also necessary to characterize the bandwidth, variability, and temporal resolution of the associated variables, in our example from above, the membrane potential.

The interplay between both levels is what the present book is all about. It characterizes a *Biophysics of Computation*, the biophysical mechanisms implementing specific operations that have been postulated to underlie computation. The text deals in depth with the spatio-temporal dynamics of the various biophysical variables thought to encode information in single neurons, in particular the membrane potential and the intracellular calcium concentration. It also deals with the stochastic character of spiking trains, the principal communication channel among neurons. In the process of doing so, this monograph covers experiments and models of computations from a wide variety of animals, ranging from sea slugs and insects to electric fish, frogs, rodents, cats, and monkeys and from all sorts of sensory and motor systems.

Understanding the detailed biophysics of neurons and their components is important for another reason. In electronic circuit design, work on the physics of computation has characterized the physical mechanisms that are exploited to perform elementary information processing operations in digital computers (Mead and Conway, 1980).[1] These mechanisms constrain the types of operations that can easily and efficiently be performed. An example is the complementary metal-oxide-semiconductor (CMOS) technology used to design the microprocessors at the heart of today's digital computers (Fig. 0.1). Because of the physics of the MOS transistor and the way the transistor switches, the canonical logic circuit is the

---

1. Indeed, the term "Biophysics of Computation," first used in Koch and Poggio (1987), was adopted from the title of the closing chapter "Physics of Computational Systems" of this seminal text.

A)

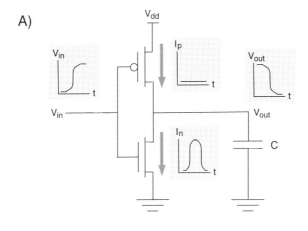

B)

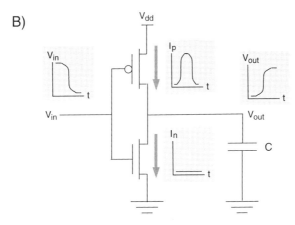

Fig. 0.1 **Basic Logic Circuit Underlying Most Digital Computers** The microprocessors at the heart of today's personal computers and workstations are built using complementary metal-oxide-semiconductor (CMOS) technology. Their basic circuit building block is the inverter shown here. The input and output voltages V_{in} and V_{out} can take on values of V_{dd} (corresponding to a logic 1) or V_{ground} (corresponding to a logic 0). The inverter consists of a p-type field effect transistor (pFET) (transistor at the top of the circuit) and an nFET (lower transistor). Conceptually, the output of every gate can be treated as charging up a capacitance C. (Switching transients are neglected in this figure.) **(A)** The input node charges up from ground to V_{dd}. The pFET fails to conduct while the charge on the gate of the nFET allows a current to flow through its channel. This connects V_{out} with ground, a current I_n flows, and V_{out} goes low. **(B)** In the converse case V_{in} goes from high to low. This induces the pFET to open up, allowing its source-to-drain current I_p to charge up the output node V_{out}, which goes high. In other words, $V_{out} = \overline{V_{in}}$. All other necessary circuit elements such as NAND and NOR can be crafted from this basic circuit by adding additional transistors in series or parallel. We conclude that the physics of the FETs render this the most natural and efficient implementation.

inverter. NAND and NOR logic circuits can be constructed as simple expansions of the basic inverter circuit. It is not that OR, AND, and other logic gates cannot be designed, but just that they require more circuitry and are therefore less suited to this technology (Keyes, 1985).

We must ask ourselves to what extent the biophysics of membranes, ionic channels, and synapses dictate the types of operations implemented in the nervous system. In order to understand the style of computation practiced by nervous systems, we must study their hardware. Can one list a small set of elemental biophysical operations that are efficient and common to most neural computation (Koch and Poggio, 1987)?

In the following, we will address these questions. We emphasize the temporal dynamics of the electrical potential $V_m(t)$ across the neuronal membrane at synapses, in passive and active dendrites, and in the axon. This is because we understand the membrane potential relatively well, it is readily accessible to our instruments, and only voltage signals—in the form of asynchronous binary pulses—propagate across large distances in relatively short times. V_m appears to be the primary physical substrate of rapid information processing and propagation in the nervous system.

The first ten chapters discuss how V_m evolves and is distributed in passive dendritic cables, how synapses work, how the membrane potential responds to synaptic input, and how the action potential comes about and describes the various ionic currents that control and shape V_m.

Other, less well explored biophysical variables, in particular the intracellular calcium concentration $[Ca^{2+}]_i$, appear to be relevant for many computational processes, in particular those involving short- and long-term adaptation and synaptic plasticity. Chapters 11 through 13 are dedicated to those.

Chapter 14 provides a link to the neural network literature by focusing on two families of simple and popular models of single neurons (or units), one emphasizing the all-or-none nature of the neuronal output, action potentials, while the second one treats the output as a continuous variable. We here also introduce the notion of a continuous firing rate code. Much evidence has accumulated for the view that spiking neurons communicate on the basis of such a code that is modulated at the 5–10 msec level, possibly supplemented by coincidence detection across neurons. Chapters 15 and 16 deal with the stochastic nature of spike trains, presenting models that can account for these. Chapters 17 through 19 synthesize the previously learned lessons into a complete account of the events occurring in realistic dendritic trees with all of their attendant nonlinearities. We will see that dendrites can indeed be very powerful, nontraditional computational devices, implementing a number of continuous operations.

Chapter 20 treats sundry unconventional biophysical mechanisms that have been hypothesized to underlie specific operations. They range from molecular computing at the submicrometer scale to the effects of neuromodulators on large numbers of neurons.

We summarize what we have learned about the style of neuronal computations in Chap. 21, the way brain operations differ from computation of the type taught in computer science and electrical engineering departments around the world. We conclude with a list of strategic questions that need to be answered in the near future.

This book addresses itself to advanced undergraduate students as well as to doctoral students and more senior researchers. It assumes that the reader has some familiarity with elemental notions of calculus and differential equations. Only Chap. 15 requires a bit more advanced mathematical knowledge. A familiarity with the basic elements of electrical circuit theory, such as possessed by any experimentalist, comes in handy. Little prior background in biophysics or systems neuroscience is assumed, although such knowledge would certainly be helpful to understanding the larger context.

Let us now start with the simplest of all models, a patch of neuronal membrane. It might be surprising to some readers what this system can already accomplish.

1

THE MEMBRANE EQUATION

Any physical or biophysical mechanism instantiating an information processing system that needs to survive in the real world must obey several constraints: (1) it must operate at high speeds, (2) it must have a rich repertoire of computational primitives, with the ability to implement a variety of linear and nonlinear, high-gain, operations, and (3) it must interface with the physical world—in the sense of being able to represent sensory input patterns accurately and translate the result of the computations into action, that is motor output (Keyes, 1985).

The membrane potential is the one physical variable within the nervous system that fulfills these three requirements: it can vary rapidly over large distances (e.g., an action potential changes the potential by 100 mV within 1 msec, propagating up to 1 cm or more down an axon within that time), and the membrane potential controls a vast number of nonlinear gates—ionic channels—that provide a very rich substrate for implementing nonlinear operations. These channels transduce visual, tactile, auditory, and olfactory stimuli into changes of the membrane potential, and such voltage changes back into the release of neurotransmitters or the contraction of muscles.

This is not to deny that ionic fluxes, or chemical interactions of various substances with each other, are not crucial to the working of the brain. They are, and we will study some of these mechanisms in Chap. 11. Yet the membrane potential is the incisive variable that serves as primary vehicle for the neuronal operations underlying rapid computations—at the fraction of a second time scale—in the brain.

We will introduce the reader in a very gentle manner to the electrical properties of nerve cells by starting off with the very simplest of all neuronal models, consisting of nothing more than a resistance and a capacitance (a so-called *RC circuit*). Yet endowed with synaptic input, this model can already implement a critical nonlinear operation, *divisive normalization* and *gain control*.

1.1 Structure of the Passive Neuronal Membrane

As a starting point, we choose a so-called *point* representation of a neuron. Here, the spatial dependency of the neuron is reduced to a single point or compartment. Such an approximation would be valid, for instance, if we were investigating a small, spherical cell without a significant dendritic tree.

1.1.1 Resting Potential

The first thing we notice once we managed to penetrate into this cell with a wire from which we can record (termed an intracellular *microelectrode*) is the existence of an electrical potential across this membrane. Such experiments, carried out in the late 1930s by Cole and Curtis (1936) in Woods Hole, Massachusetts, and by Hodgkin and Huxley (1939) on the other side of the Atlantic, demonstrated that almost always, the membrane potential, defined as the difference between the intracellular and the extracellular potentials, or

$$V_m(t) = V_i(t) - V_e(t), \tag{1.1}$$

is negative. Here t stands for time. In particular, at rest, all cells, whether neurons, glia or muscle cells, have a negative resting potential, symbolized throughout the book as V_{rest}. Depending on the circumstances, it can be as high as -30 mV or as low as -90 mV. Note that when we say the cell is at "rest," it is actually in a state of dynamic equilibrium. Ionic currents are flowing across the membrane, but they balance each other, in such a manner that the net current flowing across the membrane is zero. Maintaining this equilibrium is a major power expenditure for the nervous system. Half of the metabolic energy consumed by a mammalian brain has been estimated to be due to the membrane-bound pumps that are responsible for the upkeep of the underlying ionic gradients (Ames, 1997).

The origin of V_{rest} lies in the differential distribution of ions across the membrane, which we do not further describe here (see Sec. 4.4 and Hille, 1992). V_{rest} need not necessarily be fixed. Indeed, we will discuss in Sec. 18.3 conditions under which a network of cortical cells can dynamically adjust their resting potentials.

1.1.2 Membrane Capacity

What is the nature of the membrane separating the intracellular cytoplasm from the extracellular milieu (Fig. 1.1)? The two basic constitutive elements of biological membranes, whether from the nervous system or from nonneuronal tissues such as muscle or red blood cells, whether prokaryotic or eukaryotic, are *proteins* and *lipids* (Gennis, 1989).

The backbone of the membrane is made of two layers of phospholipid molecules, with their polar heads facing the intracellular cytoplasm and the extracellular space, thereby separating the internal and external conducting solutions by a 30–50-Å-thin insulating layer. We know that whenever a thin insulator is keeping charges apart, it will act like a *capacitance*. The capacitance C is a measure of how much charge Q needs to be distributed across the membrane in order for a certain potential V_m to build up. Or, conversely, the membrane potential V_m allows the capacitance to build up a charge Q on both sides of the membrane, with

$$Q = CV_m. \tag{1.2}$$

In membrane biophysics, the capacitance is usually specified in terms of the *specific membrane capacitance* C_m, in units of microfarads per square centimeter of membrane area ($\mu F/cm^2$). The actual value of C can be obtained by multiplying C_m by the total membrane area. The thickness and the dielectric constant of the bilipid layer determine the numerical value of C_m. For the simplest type of capacitance formed by two parallel plates, C_m scales

A)   B)

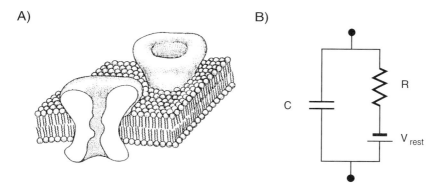

Fig. 1.1 NATURE OF THE PASSIVE NEURONAL MEMBRANE (A) Schematic representation of a small patch of membrane of the types enclosing all cells. The 30–50 Å thin bilayer of lipids isolates the extracellular side from the intracellular one. From an electrical point of view, the resultant separation of charge across the membrane acts akin to a capacitance. Proteins inserted into the membrane, here ionic channels, provide a conduit through the membrane. Reprinted by permission from Hille (1992). (B) Associated lumped electrical circuit for this patch, consisting of a capacitance and a resistance in series with a battery. The resistance mimics the behavior of voltage-independent ionic channels inserted throughout the membrane and the battery accounts for the cell's resting potential V_{rest}.

inversely with the thickness separating the charges (the thinner the distance between the two plates, the stronger the mutual attraction of the charges across the insulating material). As discussed in Appendix A, the specific capacitance per unit area of biological membranes is between 0.7 and 1 μF/cm^2. For the sake of convenience, we adopt the latter, simple to remember, value. This implies that a spherical cell of 5-μm radius with a resting potential of -70 mV stores about -0.22×10^{-12} coulomb of charge just below the membrane and an equal but opposite amount of charge outside.

When the voltage across the capacitance changes, a current will flow. This *capacitive current*, which moves on or off the capacitance, is obtained by differentiating Eq. 1.2 with respect to time (remember that current is the amount of charge flowing per time),

$$I_C = C \frac{dV_m(t)}{dt} .$$
(1.3)

For a fixed current, the existence of the membrane capacitance imposes a constraint on how rapidly V_m can change in response to this current; the larger the capacitance, the slower the resultant voltage change.

It is important to realize that there is never any actual movement of charge across the insulating membrane. When the voltage changes with time, the charge changes and a current will flow, in accordance with Eq. 1.3, but never directly across the capacitance. The charge merely redistributes itself across the two sides by way of the rest of the circuit.

Can any current flow directly across the bilipid layers? As detailed in Appendix A, the extremely high resistivity of the lipids prevents passages of any significant amount of charge across the membrane. Indeed, the specific resistivity of the membrane is approximately one billion times higher than that of the intracellular cytoplasm. Thus, from an electrical point of view, the properties of the membrane can be satisfactorily described by a sole element: a capacitance.

1.1.3 Membrane Resistance

With no other components around, life would indeed be dull. What endows a large collection of squishy cells with the ability to move and to think are the all-important *proteins* embedded within the membrane. Indeed, they frequently penetrate the membrane, allowing ions to pass from one side to the other (Fig. 1.1). Protein molecules, making up anywhere from 20 to 80% (dry weight) of the membrane, subserve an enormous range of specific cellular functions, including ionic channels, enzymes, pumps, and receptors. They act as doors or gates in the lipid barrier through which particular information or substances can be transferred from one side to the other. As we shall see later on, a great variety of such "gates" exists, with different keys to open them. For now, we are interested in those membrane proteins that act as ionic *channels* or *pores*, enabling ions to travel from one side of the membrane to the other. We will discuss the molecular nature of these channels in more detail in Chap. 8.

For now, we will summarily describe the current flow through these channels by a simple linear resistance R. Since we also have to account for the resting potential of the cell, the simplest electrical description of a small piece of membrane includes three elements, C, R, and V_{rest} (Fig. 1.1). Such a circuit describes a *passive* membrane in contrast to *quasi-active* and *active* membranes, which contain, respectively, linear, inductance-like, and nonlinear voltage-dependent membrane components. For obvious reasons, it is also sometimes known as an *RC circuit*. Fortuitously, the membranes of quite a few cells can be mimicked by such *RC* circuits, at least under some limited conditions.

The membrane resistance is usually specified in terms of the *specific membrane resistance* R_m, expressed in terms of resistance times unit area (in units of $\Omega \cdot cm^2$). R is obtained by dividing R_m by the area of the membrane being considered. The inverse of R_m is known as the passive conductance per unit area of dendritic membrane or, for short, as the *specific leak conductance* $G_m = 1/R_m$ and is measured in units of siemens per square centimeter (S/cm^2).

1.2 A Simple *RC* Circuit

Let us now carry out a virtual electrophysiological experiment. Assume that we have identified a small spherical neuron of diameter d and have managed to insert a small electrode into the cell without breaking it up. Under the conditions of our experiments, we have reasons to believe that its membrane acts passively. We would like to know what happens if we inject current $I_{inj}(t)$ through the microelectrode directly into the cell. This electrode can be thought of as an ideal *current source* (in contrast to an ideal voltage source, such as a battery).

How can we describe the dynamics of the membrane potential $V_m(t)$ in response to this current? The cell membrane can be conceptualized as being made up from many small *RC* circuits (Fig. 1.2A). Because the dimensions of the cell are so small, the electrical potential across the membrane is everywhere the same and we can neglect any spatial dependencies; physiologists will say the cell is *isopotential*. This implies that the electrical behavior of the cell can be adequately described by a single *RC* compartment with a current source (Fig. 1.2B). The net resistance R is determined by the specific membrane resistance R_m divided by the total membrane area πd^2 (since the current can flow out through any one part of the membrane) while the total capacitance C is given by C_m times the membrane area.

A)   B)

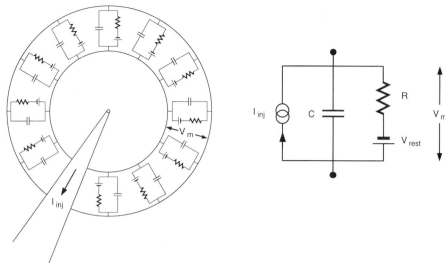

Fig. 1.2 ELECTRICAL STRUCTURE OF A SMALL PASSIVE NEURON (A) Equivalent electrical model of a spherical cell with passive membrane. An intracellular electrode delivers current to the cell. By convention, an outward current is positive; thus, the arrow. We assume that the dimensions of the cell are small enough so that spatial variations in the membrane potential can be neglected. (B) Under these conditions, the cell can be reduced to a single *RC* compartment in series with an ideal current source I_{inj}.

It is straightforward to describe the dynamics of this circuit by applying *Kirchhoff's current law*, which states that the sum of all currents flowing into or out of any electrical node must be zero (the current cannot disappear, it has to go somewhere). The current across the capacitance is given by expression 1.3. The current through the resistance is given by *Ohm's law*,

$$I_R = \frac{V_m - V_{\mathrm{rest}}}{R} . \tag{1.4}$$

Note that the potential across the resistance is not equal to V_m, but to the difference between the membrane potential and the fictive battery V_{rest}, which accounts for the resting potential. Due to conservation of current, the capacitive and resistive currents must be equal to the external one, or

$$C\frac{d V_m(t)}{dt} + \frac{V_m(t) - V_{\mathrm{rest}}}{R} = I_{\mathrm{inj}}(t) . \tag{1.5}$$

With $\tau = RC$, with units of $\Omega \cdot F =$ sec, we can rewrite this as

$$\tau \frac{d V_m(t)}{dt} = -V_m(t) + V_{\mathrm{rest}} + R I_{\mathrm{inj}}(t) . \tag{1.6}$$

A minor, but important detail is the sign of the external current (after all, we could have replaced $+I_{\mathrm{inj}}$ by $-I_{\mathrm{inj}}$ in Eq. 1.6). By convention, an outward current, that is positive charge flowing from inside the neuron to the outside, is represented as a positive current. An outward going current that is delivered through an intracellular electrode will make the inside of the cell more positive; the physiologist says that the cell is *depolarized*. Conversely,

an inward directed current supplied by the same electrode, plotted by convention in the
negative direction, will make the inside more negative, that is, it will *hyperpolarize* the
cell. If the current is not applied from an external source but is generated by a membrane
conductance, the situation is different (see Chap. 5).

Due to the existence of the battery V_{rest}, the electrical diagram in Fig. 1.2B does *not*,
formally speaking, constitute a *passive* circuit, since its current-voltage $(I-V)$ relationship
is not restricted to the first and third quadrants of the $I-V$ plane. This implies that
power is needed to maintain this $I-V$ relationship, ultimately supplied by the differential
distributions of ions across the membrane. Because the $I-V$ relationship has a nonzero,
positive derivative for every value of V_m, it is known as an *incrementally passive* device.
This point is not without interest, since it relates to the stability of circuits built using such
components (Wyatt, 1992). We here do not take a purist point of view, and we will continue
to refer to a membrane whose equivalent circuit diagram is similar to that of Figs. 1.1B and
1.2B as passive.

Equation 1.6 is known as the *membrane equation* and constitutes a first-order, ordinary
differential equation. With the proper initial conditions, it specifies an unique voltage
trajectory. Let us assume that the membrane potential starts off at $V_m(t = 0) = V_{rest}$.
We can replace this into Eq. 1.6 and see that in the absence of any input $(I_{inj} = 0)$
this assumption yields $dV_m/dt = 0$, that is, once at V_{rest}, the system will remain at
V_{rest} in the absence of any input. This makes perfect sense. So now let us switch on, at
$t = 0$, a step of current of constant amplitude I_0. We should remember from the theory of
ordinary differential equations that the most general form of the solution of Eq. 1.6 can be
expressed as

$$V_m(t) = v_0 e^{-t/\tau} + v_1 \tag{1.7}$$

where v_0 and v_1 depend on the initial conditions. Replacing this into Eq. 1.6 and canceling
identical variables on both sides leaves us with

$$v_1 = V_{rest} + R I_0 . \tag{1.8}$$

We obtain the value of v_0 by imposing the initial condition $V_m(t = 0) = v_0 + v_1 = V_{rest}$.
Defining the steady-state potential in response to the current as $V_\infty = R I_0$, we have solved
for the dynamics of V_m for this cell,

$$V_m(t) = V_\infty(1 - e^{-t/\tau}) + V_{rest} . \tag{1.9}$$

This equation tells us that the time course of the deviation of the membrane potential from
its resting state, that is, $V_m(t) - V_{rest}$, is an exponential function in time, with a time constant
equal to τ. Even though the current changed instantaneously from zero to I_0, the membrane
potential cannot follow but plays catch up. This is demonstrated graphically in Fig. 1.3.
How slowly V_m changes is determined by the product of the membrane resistance and the
capacitance; the larger the capacitance, the larger the current that goes toward charging up
C. Note that τ is independent of the size of the cell,

$$\tau = RC = R_m C_m . \tag{1.10}$$

As we will discuss in considerable detail in later chapters, passive time constants range
from 1 to 2 msec in neurons that are specialized in processing high-fidelity temporal
information to 100 msec or longer for cortical neurons recorded under slice conditions. A

A)

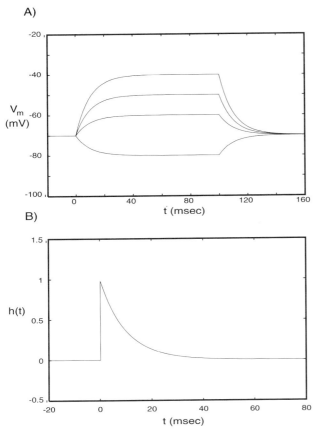

B)

Fig. 1.3 BEHAVIOR OF AN *RC* CIRCUIT **(A)** Evolution of the membrane potential $V_m(t)$ in the single *RC* compartment of Fig. 1.2B when a current step of different amplitudes I_0 (see Eq. 1.9) is switched on at $t = 0$ and turned off at 100 msec. Initially, the membrane potential is at $V_{\text{rest}} = -70$ mV. We here assume $R = 100$ MΩ, $C = 100$ pF, $\tau = 10$ msec, and four different current amplitudes, $I_0 = -0.1, 0.1, 0.2,$ and 0.3 nA. **(B)** Normalized *impulse response* or *Green's function* (Eq. 1.17) associated with the *RC* circuit of Fig. 1.2B. The voltage $V_m(t)$ in response to any current input $I_{\text{inj}}(t)$ can be obtained by convolving this function with the input.

typical range for τ recorded from cortical pyramidal cells in the living animal[1] is between 10 and 20 msec.

Remember the origin of the membrane capacitance in the molecular dimensions of the bilipid membrane. A thicker membrane would lead to a smaller value for C_m and faster temporal responses.[2]

The final voltage level in response to the current step is $RI_0 + V_{\text{rest}}$ (from Ohm's law). If $I_0 > 0$, the cell will depolarize (that is, $V_\infty > V_{\text{rest}}$), whereas for $I_0 < 0$, the converse occurs. The resistance R is also termed the *input resistance* of the cell; the larger R, the larger the voltage change in response to a fixed current. The input resistance at the cell bodies of neurons, obtained by dividing the steady-state voltage change by the current causing it, ranges from a few megaohms for the very large motoneurons in the spinal cord to hundreds of megaohms for cortical spiny stellate cells or cerebellar granule cells.

---

1. This is called *in vivo*. Such experiments need to be distinguished from the cases in which a very thin slice is taken from an animal's brain, placed in a dish, and perfused with a nutrient solution. This would be termed an *in vitro* experiment.

2. As an aside to the neuromorphic engineers among us designing analog integrated electronic circuits, $C_m = 1$ μF/cm^2 is about 20 times higher than the specific capacitance obtained by sandwiching a thin layer of silicon dioxide between two layers of poly silicon using a standard 2.0 or 1.2 μm CMOS process (Mead, 1989).

What happens if, after the membrane potential reaches its steady-state value V_∞, the current is switched off at time t_{off}? An analysis similar to the above shows that the membrane potential returns to V_{rest} with an exponential time course; that is,

$$V_m(t) = V_\infty e^{-(t-t_{off})/\tau} + V_{rest} \tag{1.11}$$

for $t \geq t_{off}$. (This can be confirmed by placing this solution into Eq. 1.6; see also Fig. 1.3A.)

Now that we know the evolution of the membrane potential for a current step, we would like to know the solution in the general case of some time-dependent current input $I_{inj}(t)$. Are we condemned to solve Eq. 1.6 explicitly for every new function $I_{inj}(t)$ that we use? Fortunately not; because the RC circuit we have been treating here is a shift-invariant, linear system, we can do much better.

1.3 *RC* Circuits as Linear Systems

Linearity is an important property of certain systems that allows us—in combination with shift invariance—to completely characterize their behavior to *any* input in terms of the system's *impulse response* or *Green's function* (named after a British mathematician living at the beginning of the nineteenth century). Since the issue of linear and nonlinear systems runs like a thread through this monograph, we urge the reader who has forgotten these concepts to quickly skim through Appendix B, which summarizes the most relevant points.

1.3.1 *Filtering by* RC *Circuits*

Let us compute the voltage response of the RC circuit of Fig. 1.2B in response to a current impulse $\delta(t)$. We will simplify matters by only considering the deviation of the membrane potential from its resting state V_{rest}. Here and throughout the book we use $V(t) = V_m(t) - V_{rest}$ when we are dealing with the potential relative to rest and reserve $V_m(t)$ for the absolute potential. This transforms Eq. 1.6 into

$$\tau \frac{dV(t)}{dt} = -V(t) + R\delta(t). \tag{1.12}$$

We can transform this equation into Fourier space, where $\tilde{V}(f)$ corresponds to the Fourier transform of the membrane potential (for a definition, see Appendix B). Remembering that the $dV(t)/dt$ term metamorphoses into $i2\pi \tilde{V}(f)\hat{V}(P)$, where $i^2 = -1$, we have

$$\tilde{V}(f) = \frac{R}{1 + i2\pi f \tau}. \tag{1.13}$$

A simple way to conceptualize this is to think of the input as a sinusoidal current of frequency f; $I_{inj}(t) = \sin(2\pi f t)$. Since the system is linear, it responds by a sinusoidal change of potential at the same frequency f, but of different amplitude and shifted in time: $V(t) = \tilde{A}(f)\sin(2\pi f t + \tilde{\phi}(f))$. The amplitude of the voltage response at this frequency, termed $\tilde{A}(f)$, is given by

$$|\tilde{A}(f)| = \frac{R}{\sqrt{1 + (2\pi f \tau)^2}} \tag{1.14}$$

and its phase by

$$\tilde{\phi}(f) = -\arctan(2\pi f \tau).\tag{1.15}$$

In the general case of an arbitrary input current, one can define the complex function $\tilde{A}(f)$ as the ratio of the Fourier transform of the voltage transform to the Fourier transform of the injected current,

$$\tilde{A}(f) = \frac{\tilde{V}(f)}{\tilde{I}_{\text{inj}}(f)}.\tag{1.16}$$

$\tilde{A}(f)$ is usually referred to as the *input impedance* of the system. Its value for a sustained or dc current input, $\tilde{A}(f = 0) = R$, is known as the *input resistance* and is a real number. It is standard engineering practice to refer to the inverse of the input impedance as the *input admittance* and to the inverse of the input resistance as the *input conductance* in units of siemens (S).

Does this definition of $\tilde{A}$ make sense? Let us look at two extreme cases. If we subject the system to a sustained current injection, the change in voltage in response to a sustained input current I is proportional to R, Ohm's law. Conversely, what happens if we use a sinusoidal that has a very high frequency f? The amplitude of the voltage change becomes less and less since at high frequencies the capacitance essentially acts like a short circuit. In the limit of $f \to \infty$, the impedance goes to zero.

For intermediate values of f, the amplitude smoothly interpolates between R and 0. In other words, our circuit acts like a *low-pass* filter, preferentially responding to slower changing inputs and severely attenuating faster ones: $|\tilde{A}(f)|$ is a strictly monotonically decreasing function of the frequency f.

Experimentally, the impedance can be obtained by injecting a sinusoidal current of frequency f and measuring the induced voltage at the same frequency. The ratio of the voltage to the current corresponds to $|\tilde{A}(f)|$. The use of impedances to describe the electrical behavior of neurons and, in particular, of muscle cells has a long tradition going back to the 1930s (Cole and Curtis, 1936; Falk and Fatt, 1964; Cole, 1972).

The result of such a procedure, carried out in a regular firing cell in a slice taken from the visual cortex of the guinea pig, is shown in Fig. 1.4. Carandini and his colleagues (1996) injected either sinusoidal currents or a noise stimulus into these cells and recorded the resultant somatic membrane potential (in the presence of spikes). Given their very fast time scale, somatic action potentials do not contribute appreciably to the total power of the voltage signal. Indeed, when stimulating with a sine wave at frequency f, the power of the voltage response at all higher frequencies was only 3.8% (median) of the power of the fundamental f. This implies that when judged by the membrane potential and not by the firing rate, and only considering input and output at the soma, at least some cortical cells can be quite well approximated by a linear filter (Carandini et al., 1996).

This is surprising, given the presence of numerous voltage-dependent conductances at the soma and in the dendritic tree. It is, however, not uncommon in neurobiology to find that despite of—or, possibly because of—a host of concatenated nonlinearities, the overall system behaves quite linearly (see Sec. 21.1.3). Sometimes one has the distinct impression that evolution wanted to come up with some overall linear mechanism, despite all the existing nonlinearities.

We will study later how adding a simple, absolute voltage threshold to the *RC* compartment that gives rise to an output spike accounts surprisingly well for the spiking behavior

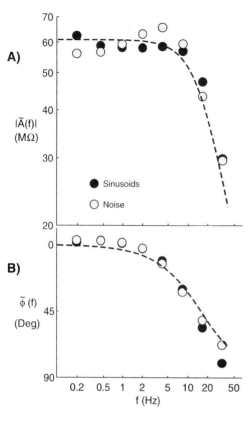

Fig. 1.4 CORTICAL CELLS BEHAVE LIKE AN RC CIRCUIT When either noise or sinusoidal currents are injected into the cell body of regularly firing cells in guinea pig visual cortex, the membrane potential can be adequately modeled as resulting from convolving the current input by a low-pass filter of the sort described in Eqs. 1.14 and 1.15 (dashed lines; here with $R = 58.3$ MΩ and $\tau = 9.3$ msec; $V_{\text{rest}} = -70.7$ mV; Carandini et al., 1996). **(A)** The amplitude of the filter and **(B)** its phase. The noise current curve reveals a shallow peak at around 8 Hz. We conclude that from the point of view of somatic input-output, these cells can be reasonably well described by a single RC compartment. The responses were obtained by computing the first harmonic of the membrane potential response and dividing by the current. The power of the first harmonic was between 9 and 141 times the power of the higher harmonics. Reprinted by permission from Carandini et al., (1996).

of such cells. This simplest model of a spiking neuron, known as a *leaky integrate-and-fire unit*, is so important that it deserves its own detailed treatment in Chap. 14.

We can recover the Green's function $h(t)$ of the RC compartment by applying the inverse Fourier transform to Eq. 1.13, which results in

$$h(t) = \frac{1}{C} e^{-t/\tau} \tag{1.17}$$

for $t \geq 0$ and 0 for negative times (the units of the Green's function are ohms per second (Ω/sec)). Conceptually, the extent of this filter, that is, the temporal duration over which this filter is significantly different from zero, indicates to what extent the distant past influences the present behavior of the system. For a decaying exponential as in an RC circuit, an event that happened three time constants ago (at $t = -3\tau$) will have roughly 1/20 the effect of something that just occurred (Fig. 1.3B). This is expected in a circuit that implements a low-pass operation. Input is integrated in time, with long ago events having exponentially less impact than more recent ones.

1.4 Synaptic Input

So far, we have not considered how the output of one neuron provides input to the next one. Fast communication among two neurons occurs at specialized contact zones, termed *synapses*. Synapses are the elementary structural and functional units for the construction of neuronal circuits. Conventional point-to-point synaptic interactions come in two different flavors: *electrical synapses*—also referred to as *gap junctions*—and the much more common *chemical synapses*. At about 1 billion chemical synapses per cubic millimeter of cortical grey matter, there are lots of synapses in the nervous system (on the order of 10^{15} for a human brain). In order to give the reader an appreciation of this, Fig. 1.5 is a photomicrograph of a small patch of the monkey retina at the electron-microscopic level, with a large number of synapses visible. Synapses are very complex pieces of machinery that can keep track of their history of usage over considerable time scales. In this chapter, we introduce fast, voltage independent chemical synapses from the point of view of the postsynaptic cell, deferring a more detailed account of synaptic biophysics, as well as voltage dependent synapses and electrical synapses, to Chap. 4, and an account of their adapting and plastic properties to Chap. 13.

Upon activation of a fast, chemical synapse, one can observe a rapid and transient change in the *postsynaptic* potential. Here, postsynaptic simply means that we are observing this signal on the "far" or "output" side of the synapse; the "input" part of a synapse is referred to as the *presynaptic* terminal. When the synapse is an excitatory one, the membrane potential rapidly depolarizes, returning more slowly to its resting state: an *excitatory postsynaptic potential* (EPSP) has occurred. Conversely, at an inhibitory synapse, the membrane will typically be transiently hyperpolarized, resulting in an *inhibitory postsynaptic potential* (IPSP). These EPSPs and IPSPs are caused by so-called *excitatory* and *inhibitory postsynaptic currents* (EPSCs and IPSCs), triggered by the spiking activity in the presynaptic cell.

Figure 1.6 illustrates some of the properties of a population of depolarizing synapses between the axons of granule cells, also called *mossy fibers*, and a CA3 hippocampal

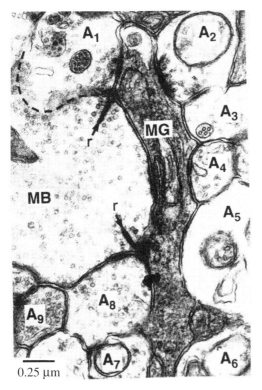

0.25 μm

Fig. 1.5 SYNAPSES AMONG RETINAL NEU-RONS Electron microscopic photograph of a few square micrometers of tissue in the central portion of the retina in the monkey. Here a midget bipolar cell (**MB**) makes two *ribbon synapses* onto a midget ganglion cell (**MG**). It is surrounded by nine processes belonging to amacrine cells (A_1 to A_9). Some of these feed back onto the bipolar cell (e.g., A_8), some feed forward onto the ganglion cell (e.g., A_1), some do both, and some also contact each other (e.g., $A_2 \rightarrow A_3$). Since neither the bipolar cell nor the amacrine cell processes have been shown to generate action potentials, these synapses are all of the analog variety, in distinction to synapses in the more central part of the nervous system that typically transform an action potential into a graded, postsynaptic signal. Reprinted by permission from Calkins and Sterling (1996).

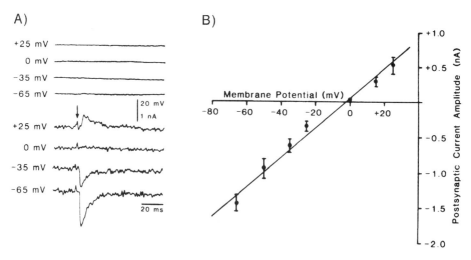

Fig. 1.6 A FAST EXCITATORY SYNAPTIC INPUT Excitatory postsynaptic current (EPSC) caused by the simultaneous activation of synapses (arrow) made by the mossy fibers onto CA3 pyramidal cells in the rodent hippocampus (Brown and Johnston, 1983). This classical experiment showed how a central synapse can be successfully voltage clamped. **(A)** The voltage-clamp setup stabilizes—via electronic feedback control—the membrane potential at a fixed value. Here four experiments are shown, carried out at the holding potentials indicated at the left. The current that is drawn to keep the membrane potential constant, termed the clamp current, corresponds to the negative EPSC. It is maximal at negative potentials and reverses sign around zero. The synaptic current rises within 1 msec to its peak value, decaying to baseline over 20–30 msec. The experiments were carried out in the presence of pharmacological agents that blocked synaptic inhibition. **(B)** When the peak EPSC is plotted against the holding potential, an approximately linear relationship emerges; the regression line yields an x-axis intercept of -1.9 mV and a slope of 20.6 nS. Thus, once the synaptic reversal potential is accounted for, Ohm's law appears to be reasonably well obeyed. We conclude that synaptic input is caused by a transient increase in the conductance of the membrane to certain ions. Reprinted by permission from Brown and Johnston (1983).

pyramidal cell.[3] The figure is taken from an experiment by Brown and Johnston (1983), which demonstrated for the first time how a synapse within the central nervous system could be voltage clamped. The *voltage-clamp* technique was previously used on the very large synapse made between the axonal terminals of motoneurons and the muscle, the so-called *neuromuscular junction* (Katz, 1966; Johnston and Wu, 1995). It allows the experimentalist to stabilize the membrane potential (via a feedback loop) at some fixed value, irrespective of the currents that are flowing across the membrane in response to some stimulus. This allows the measurement of EPSCs at various fixed potentials (as in Fig. 1.6). The EPSC has its largest value at a holding potential of -65 mV, becoming progressively smaller and vanishing around 0 mV. If the membrane potential is clamped to values more positive than zero, the EPSC reverses sign (Fig. 1.6A). When the relationship between the peak current and the holding potential is plotted (Fig. 1.6B), the data tend to fall on a straight line that goes through zero around -1.9 mV and that has a slope of 20.6 nS.

What we can infer from such a plot is that the postsynaptic event is caused by a temporary increase in the membrane conductance, here by a maximal increase of about 20 nS

---

3. It should be pointed out that we are here looking at a population of synapses, made very close to the soma of the pyramidal cell, thereby minimizing space-clamp problems.

(due to simultaneous activation of numerous synapses) in series with a so-called *synaptic reversal battery* or potential, $E_{syn} = -1.9$ mV (since the conductance change is specific for a particular class of ions). Spiking activity in the presynaptic cell triggers, through a complicated cascade of biophysical events (further discussed in Chap. 4), a conductance change in the membrane of the postsynaptic cell. Typically, the conductance $g_{syn}(t)$ transiently increases within less than 1 msec, before this increase subsides within 5 msec. The equivalent electrical circuit diagram of a synapse embedded into a patch of neuronal membrane is shown in Fig. 1.7A. It is important to understand that from a biophysical, postsynaptic point of view, a synapse does not correspond to a fixed current source—in that case the slope of the I–V curve in Fig. 1.6 should have been zero—but to a genuine increase in the membrane conductance. As we will reemphasize throughout the book, this basic feature of the neuronal hardware has a number of important functional consequences.

Because of the existence of the synaptic battery, the *driving potential* across the synapse is the difference between E_{syn} and the membrane potential. The postsynaptic current due to a single such synapse is given by Ohm's law

$$I_{syn} = g_{syn}(t)(V_m(t) - E_{syn}) . \tag{1.18}$$

Inserting this synapse into a patch of membrane (Fig. 1.7A) gives rise to the following ordinary differential equation (on the basis of Kirchhoff's current law):

$$C\frac{dV_m}{dt} + g_{syn}(t)(V_m - E_{syn}) + \frac{V_m - V_{rest}}{R} = 0 \tag{1.19}$$

or, with $\tau = CR$, the passive membrane time constant in the absence of any synaptic input, we can transform this into

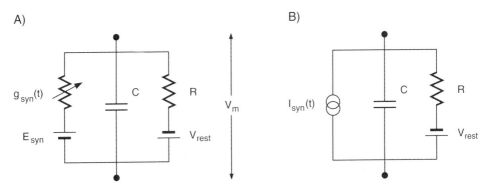

Fig. 1.7 Equivalent Electrical Circuit of a Fast Chemical Synapse (A) Electrical model of a fast voltage-independent chemical synapse. This circuit was put forth to explain events occurring at the neuromuscular junction by Katz (1969). Remarkably, all fast chemical synapses in the central nervous system, with the exception of the voltage-dependent NMDA receptor-synaptic complex, operate on the same principle. Activation of the synapse leads to the transient opening of ionic channels, selective to certain ions. This corresponds to a transient increase in the membrane conductance $g_{syn}(t)$ in series with the synaptic reversal potential E_{syn}, shown here in parallel with a passive membrane patch. (B) If the evoked potential change is small relative to the synaptic reversal potential, the synapse can be approximated by a current source of amplitude $g_{syn}(t)E_{syn}$. In general, however, this will not be the case and synaptic input must be treated as a conductance change, a fact that has important functional consequences.

$$\tau \frac{dV_m}{dt} = -(1 + Rg_{syn}(t))V_m + Rg_{syn}(t)E_{syn} + V_{rest} \,. \tag{1.20}$$

Frequently the time course of synaptic input is approximated by a so-called α function. It describes the transient behavior of synaptic input for a number of preparations, such as nicotinic input to vertebrate sympathetic ganglion cells or the synaptic input mediated by the mossy fibers (Brown and Johnston, 1983; Williams and Johnston, 1991; Yamada, Koch, and Adams, 1998), reasonably well:[4]

$$g_{syn}(t) = const \cdot t e^{-t/t_{peak}} \,. \tag{1.21}$$

We can now integrate Eq. 1.20 using the same values for t_{peak} and g_{peak}, but three different values for the synaptic reversal potential (Fig. 1.8).

If $E_{syn} > V_{rest}$, the synaptic current is inward and—by convention—negative and will act to depolarize the membrane. This is the hallmark of an EPSP, as observed at the most

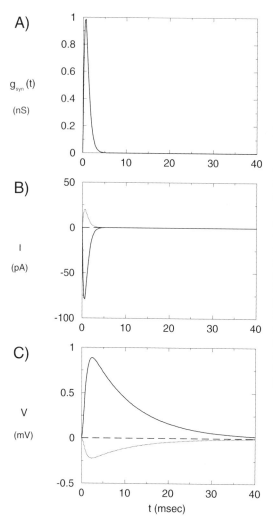

A)

$g_{syn}(t)$

(nS)

B)

I

(pA)

C)

V

(mV)

t (msec)

Fig. 1.8 ACTION OF A SINGLE SYNAPSE INSERTED INTO A MEMBRANE Three different types of synaptic inputs and their differential effect on the membrane potential. (A) Time course of the synaptic-induced conductance increase, here with $t_{peak}=0.5$ msec and $g_{peak} = 1$ nS (Eq. 1.21). The synapse is inserted into a patch of membrane (Fig. 1.7A) with $R = 100$ MΩ, $C = 100$ pF, and $\tau = 10$ msec. (B) Postsynaptic current in response to the conductance increase if the synaptic reversal potential is positive ($E_{syn} = 80$ mV relative to rest; solid line), negative ($E_{syn} = -20$ mV; dotted line), and zero (so-called shunting inhibition; dashed line). By convention, an inward current that depolarizes the cell is plotted as a negative current. (C) Associated EPSP (solid line) and IPSP (lower dashed line), relative to V_{rest}, solved by numerical integration of Eq. 1.20. Notice that the time course of the postsynaptic potential is much longer than the time course of the corresponding postsynaptic current due to the low-pass nature of the membrane. Shunting inhibition by itself does not give rise to any change in potential (center dashed line).

4. g_{peak}, its peak value, attained at $t = t_{peak}$, defines const $= g_{peak} \, e^1 / t_{peak}$.

common fast, excitatory synapse in the brain, the so-called non-NMDA synapse (named after its insensitivity to N-methyl-D-aspartic acid) that uses the neurotransmitter glutamate with a reversal potential about 80–100 mV above the resting potential of the cell (for more details, see Sec. 4.6).

If the converse occurs, that is, $E_{syn} < V_{rest}$, the current is outward and the membrane is hyperpolarized, away from the threshold for spike generation. In the central nervous system, this is typically caused by a slower form of inhibition due to γ-aminobutyric acid B receptors (GABA$_B$) at synapses that release the neurotransmitter GABA and that let potassium ions out of the cell and have a reversal potential 10–30 mV below the resting potential (Fig. 1.8).

What happens if a synapse is activated whose battery potential is close to the membrane potential, that is, $E_{syn} \approx V_{rest}$? If the membrane is at rest, no driving potential exists across the synaptic conductance, since $V_m - E_{syn} \approx 0$, and the membrane potential remains unperturbed. But the total conductance of the membrane increases by $g_{syn}(t)$.

If this system is now depolarized by excitatory input, activation of this *silent* or *shunting*[5] inhibition causes a reduction in the EPSP amplitude.

Activation of a GABA$_A$ synapse, one of the most common forms of fast inhibition in cortex and associated structures, increases the membrane conductance for chloride ions and has a reversal potential in the neighborhood of many cells' resting potential, thereby implementing a form of shunting inhibition.

How do we deal with multiple synaptic inputs? Since currents add, we can extend Eq. 1.19 in a straightforward manner by placing the synapses in parallel with the RC circuit,

$$C\frac{dV_m}{dt} = \sum_{i=0}^{n} g_{syn,i}(t)(E_{syn,i} - V_m) + \frac{V_{rest} - V_m}{R} \qquad (1.22)$$

where the sum is taken over all synapses (each of which can have its own reversal potential). Of course, this is very reminiscent of the linear summation of inputs in the "units" of standard neural network theory (see Sec. 14.4).

1.5 Synaptic Input Is Nonlinear

What is not immediately apparent from Eq. 1.22 is the fact that synaptic input as conductance change is of necessity nonlinear; that is, the change in membrane potential is a nonlinear function of the synaptic input. Yet this turns out to be crucial. From the point of view of information processing, a linear noiseless system cannot create or destroy information. Whatever information is fed into the system is available at the output. Of course, any system existing in the real world has to deal with noise, which places restrictions on the amplitude of signals that can be discriminated. Therefore in a noisy linear system, information can be destroyed. But what is needed in a system that processes information are nonlinearities that can perform discriminations and decisions. Similarly, in order for a digital system to be Turing universal, a nonlinearity such as negation and logical ANDing is required. As we will see later on, one ever-present nonlinearity is the voltage threshold for spike initiation. As we will see now, another nonlinearity that comes for free with synaptic hardware is saturation.

---

5. Because both excitatory and inhibitory fast synapses act to increase a postsynaptic membrane conductance, all of them can properly be said to be shunting. However, in this book we follow widespread usage and only refer to shunting inhibition as a conductance increase with a reversal potential in the neighborhood of the cell's resting potential.

1.5.1 Synaptic Input, Saturation, and the Membrane Time Constant

The nature of this effect can be perfectly well understood for a single synaptic input. If we consider the change in membrane potential relative to rest in response to a slowly varying synaptic input (that is, we can reduce $g_{syn}(t)$ to g_{syn}), we can express the dynamics of V as

$$\tau' \frac{dV}{dt} = -V + \frac{g_{syn} E_{syn}}{G_{in}} \tag{1.23}$$

where the new value of the input conductance is

$$G_{in} = g_{syn} + \frac{1}{R} \tag{1.24}$$

and the new value of the time constant in the presence of synaptic input is

$$\tau' = \frac{C}{G_{in}} = \frac{RC}{1 + Rg_{syn}}. \tag{1.25}$$

In other words, each synaptic input, whether excitatory, shunting, or inhibitory, increases the synaptic input conductance, thereby decreasing the membrane time constant (Fig. 1.9). This is, of course, equally true when one considers the effect of numerous simultaneous synaptic inputs. As we shall see further along in this book, under physiological conditions neurons can be bombarded with massive synaptic input, which will lower the membrane time constant significantly, as compared to the value of τ measured under slice conditions in the absence of normal synaptic input.

How does the membrane potential behave as a function of g_{syn}? Solving for Eq. 1.23 yields for the steady-state potential

$$V_\infty = \frac{Rg_{syn} E_{syn}}{1 + Rg_{syn}}. \tag{1.26}$$

If the synaptic input is small, that is, if $Rg_{syn} \ll 1$ or $g_{syn} \ll 1/R$, the denominator is roughly equal to 1, and the EPSP is

$$V \approx Rg_{syn} E_{syn}. \tag{1.27}$$

Here, the input can be approximated to a fair degree by a constant current source of amplitude $g_{syn} E_{syn}$. Doubling the input under these conditions leads to a doubling of the voltage change.

As the EPSP becomes larger and larger, the driving potential across the synaptic conductance $V_m - E_{syn}$ becomes smaller and smaller, disappearing eventually at $V_m = E_{syn}$. No matter how large the conductance increase is made, there is no more potential to drive ions across the membrane. Here, $Rg_{syn} \gg 1$ (or equivalently, $g_{syn} \gg 1/R$; that is, the synaptic input is considerably larger than the input conductance), and we have

$$V \approx \frac{Rg_{syn} E_{syn}}{Rg_{syn}} = E_{syn}. \tag{1.28}$$

The membrane potential has saturated at the synaptic reversal potential (Fig. 1.9).

1.5.2 Synaptic Interactions among Excitation and Shunting Inhibition

As we are devoting the entire Chap. 5 to the topic of synaptic interaction, we focus here on the specific interaction between excitatory input and shunting inhibition occurring within a single RC compartment (Fig. 1.10A). For the sake of simplicity, we assume that at $t = 0$, both excitation (of constant amplitude g_e and battery E_e) and shunting inhibition (of constant

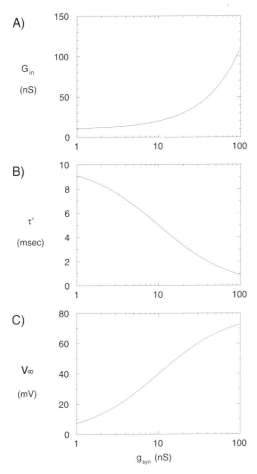

A)

B)

C)

g_{syn} (nS)

Fig. 1.9 SYNAPTIC INPUT, SATURATION, AND THE TIME CONSTANT   The effect of varying the synaptic input conductance on **(A)** the input conductance G_{in} (Eq. 1.24), **(B)** the membrane time constant τ' (Eq. 1.25), and **(C)** the steady-state change in membrane potential V_∞ (Eq. 1.26) in a single-compartment model (Fig. 1.7A) as a function of g_{syn}. Conceptually, if we assume an excitatory synapse (with $E_{syn} = 80$ mV) and a fixed peak amplitude of $g_e = 1$ nS, the x axis is logarithmic in the number of synapses involved in the overall synaptic event. Note that τ' as well as G_{in} will increase irrespective of whether the synapses are depolarizing, shunting, or hyperpolarizing. The fact that the input to neurons comes in the form of a change in the membrane conductance implies that the very structure of the neuronal hardware changes with the input, since the dynamics of the cell speeds up in the presence of strong synaptic input.

amplitude g_i and battery $E_i = V_{rest}$) are turned on and remain on. Using Kirchhoff's current law, we can express the change in membrane potential relative to V_{rest} in this circuit as

$$C\frac{dV}{dt} = g_e(E_e - V) - g_i V - \frac{V}{R}. \tag{1.29}$$

As in Eq. 1.23, we can transform this into

$$\tau'\frac{dV}{dt} = -V + \frac{g_e E_e}{G_{in}} \tag{1.30}$$

where the input conductance in the presence of the two synaptic inputs is

$$G_{in} = g_e + g_i + \frac{1}{R} \tag{1.31}$$

and the time constant is

$$\tau' = \frac{C}{G_{in}}. \tag{1.32}$$

The solution to this is a low-pass filter function multiplied with some constant, or

A)

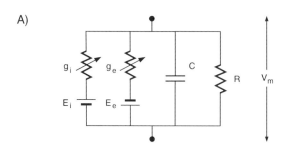

B)

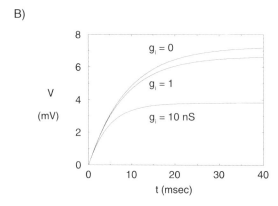

Fig. 1.10 NONLINEAR INTERACTION BETWEEN EXCITATION AND SHUNTING INHIBITION Inhibitory synaptic input of the shunting type, that is, whose reversal potential is close to the cell's resting potential, can implement a form of division. (A) This is demonstrated for an RC circuit ($R = 100\,M\Omega$, $C = 100\,pF$) in the presence of both excitation (with battery $E_e = 80\,mV$) and shunting inhibition (with $E_i = 0$). We are here only considering the change in membrane potential relative to V_{rest}. (B) Time course of the membrane depolarization in response to a step onset of both excitation (of amplitude $g_e = 1\,nS$) and shunting inhibition (for three values of $g_i = 0, 1$, and $10\,nS$). One effect of increasing g_i is an almost proportional reduction in EPSP amplitude. A further consequence of increasing the amount of shunting inhibition is to decrease the time constant τ', from its original 10 msec in the absence of any synaptic input to 9 msec in the presence of only excitation to 4.8 msec in the presence of excitation and the 10 times larger shunting inhibition.

$$V(t) = \frac{g_e E_e}{G_{in}}(1 - e^{-t/\tau'}).\tag{1.33}$$

(This can be checked by replacement into Eq. 1.30.) The steady-state potential for $t \to \infty$ converges to

$$V_{\infty} = \frac{g_e R E_e}{1 + g_e R + g_i R}.\tag{1.34}$$

What is important to realize is that the numerator does not include any contribution from the shunting inhibition (since the synaptic reversal potential is equal to the resting potential); g_i only appears in the denominator. Increasing g_i therefore reduces the EPSP from its peak value in a *division*-like manner. This is the reason shunting inhibition is frequently also referred to as *divisive inhibition* (Bloomfield, 1974; Torre and Poggio, 1978). Of course, due to the offset in the denominator, g_i only implements a true division in the limit of $g_i \gg g_e + 1/R$. Under these conditions,

$$V_{\infty} \approx \frac{g_e E_e}{g_i}.\tag{1.35}$$

Increasing g_i also affects the speed with which the cell converges to its steady-state, since the time constant decreases with increasing shunting inhibition, as illustrated in Fig. 1.10B.

Finally, let us consider the *voltage gain*, that is, the sensitivity of the output to variations in the excitatory input: by how much does the amplitude of the EPSP vary if g_e varies? This amounts to computing

$$\frac{dV_\infty}{dg_e} = \frac{RE_e(1 + g_i R)}{(1 + g_e R + g_i R)^2} .$$ (1.36)

We see that the gain is maximal in the absence of any shunting inhibition, becoming progressively smaller as g_i increases. In the limit of $g_i \to \infty$, the gain becomes zero; in the presence of massive amounts of inhibition, the excitatory input becomes swamped and is completely dominated by inhibition.

1.5.3 Gain Normalization in Visual Cortex and Synaptic Input

One example demonstrating the use of synaptic conductance changes to implement a nonlinear operation crucial to the behavior of neurons in visual cortex has been proposed by Carandini and Heeger (1994).

The standard model of a simple cell in primary visual cortex (V1) is that its firing rate is a linear function of the visual input, using a Gabor filter for its spatial receptive field and some low-pass or band-pass filter to account for its temporal behavior (Wandell, 1995). While much evidence has accumulated in favor of this position, many V1 neurons do show a number of nonlinearities: (1) the response saturates with increasing visual contrast, (2) at higher contrast level, the response occurs earlier, and (3) superposition does not hold; that is, the response of the cell to a bar at its optimal orientation superimposed onto a bar orthogonal to its optimal orientation is not equal to the sum of the response to the two bars when presented by themselves. Carandini and Heeger (see also Nelson, 1994) account for this behavior by using a simple RC model for a V1 cell, as in Fig. 1.10A, augmented by an input from a hyperpolarizing synapse. Their intriguing idea is to have the amplitude of the shunting inhibition depend on the response rate of the entire population of cells at this particular location in space, summed over all orientations and direction of motions (Fig. 1.11). At high contrasts when g_e is large, the network is very active and g_i is also very large; this leads to the divisive normalization witnessed in Fig. 1.10B as well as to a reduction in the time constant, explaining the advance of the response at high contrast levels. Heeger, Simoncelli, and Movshon (1996) have argued that the very same normalization mechanism is also operating in other cortical areas. Note that a more physiological implementation of this idea needs to take the additional conductance due to massive, excitatory recurrent feedback among cortical cells into account (Ahmed et al., 1994; Douglas et al., 1995).

While the Carandini and Heeger (1994) model is elegant, leads to a simple mathematical model of a cortical cell, and can account for much of the data (Fig. 1.11), it does have one serious flaw that we can only allude to here. As will become apparent in Sec. 18.5 (Kernell, 1969, 1971; Holt and Koch, 1997), shunting inhibition acts in a linear, subtractive manner when treated within the context of a spiking neuron, rather than in the saturating manner we are accustomed to from a passive membrane. This is a natural consequence of the biophysics of spike generation and throws doubt onto the hypothesis that contrast normalization is implemented using the natural properties of conductance-increasing synapses.

1.6 Recapitulation

In this introductory chapter, we meet some of the key actors underlying neuronal information processing. Basic to all cells is the capacitance inherent in the bilipid layer, limiting how quickly the membrane potential can respond to a fixed input current. The simplest of all neuronal models is that of a single compartment that includes a resistance, in series with a

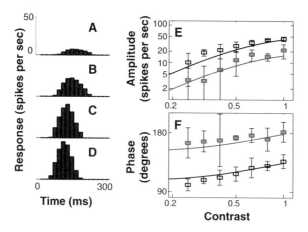

Fig. 1.11 GAIN NORMALIZATION IN NEURONS IN VISUAL CORTEX Response properties of a simple cell in primary visual cortex of the monkey in response to drifting sinusoidal gratings (Carandini and Heeger, 1994). (**A**) through (**D**) One cycle of the response to gratings of contrast 0.125, 0.25, 0.5, and 1.0. The cell saturates with contrast (doubling the contrast doubles the neuronal response when going from **A** to **B**, but not when going from **C** to **D**) and advances its response (a shift of about 50 msec occurs between **A** and **D**). (**E**) Amplitude and (**F**) Phase of the fundamental Fourier component to sinusoidal gratings drifting at 6 Hz. Shown are the responses of the cell at its preferred orientation (open symbols) and 20° away from the preferred orientation (solid symbols). Error bars represent ±1 standard deviation and the solid lines correspond to the best fit of the model equation that uses shunting inhibition, activated via massive feedback, to carry out this gain normalization. Reprinted by permission from Carandini and Heeger (1994).

battery, and the capacitance. It can be completely described by a linear, low-pass filter. As we will see in a later chapter, such an *RC* circuit, augmented by a simple voltage threshold, constitutes one of the simplest yet also most powerful models of a spiking neuron: the *leaky integrate-and-fire unit*.

We introduced fast chemical synapses, the stuff out of which computations arise. Chemical synapses convert the presynaptic voltage signal—via a chemical process—into a postsynaptic electrical signal, via a change in the membrane conductance specific to certain ions. Such a synapse can be described by a time-dependent synaptic conductance $g_{syn}(t)$ and a synaptic battery E_{syn}. In general, synapses cannot be treated as constant current sources.

Chemical synapses, similar to an operational amplifier wired up as a follower, isolate the electrical properties of the postsynaptic site from the presynaptic one. This allows synapses to link neurons with very different electrical impedances. Furthermore, the amplitude, duration, and sign of the postsynaptic signal can be quite different from those of the presynaptic one. Electrical synapses, discussed in Sec. 4.10, share none of these properties.

The fact that synapses act by changing, usually increasing, the postsynaptic membrane conductance has a number of important consequences. It allows for the natural expression of several nonlinear operations, in particular saturation and gain normalization. As an example, we saw how shunting inhibition, mediated by a type of synapse whose synaptic reversal potential is close to the cell's resting potential, acts similar to division. We also studied how synaptic input that increases the postsynaptic membrane conductance for some combinations of ions, no matter what its reversal potential, acts to decrease the cell's input resistance and thus its membrane time constant. As postulated by Carandini and Heeger (1994), the effect of massive feedback synaptic input might, in a very physiological manner, implement *gain normalization* in cortical areas.

2

LINEAR CABLE THEORY

In the previous chapter, we briefly met some of the key actors of this book. In particular, we introduced the *RC* model of a patch of neuronal membrane and showed an instance where such a "trivial" model accounts reasonably well for the input-output properties of a neuron, as measured at its cell body (Fig. 1.4). However, almost none of the excitatory synapses are made onto the cell body, contacting instead the very extensive dendritic arbor. As we will discuss in detail in Chap. 3 (see Fig. 3.1), dendritic trees can be quite large, containing up to 98% of the entire neuronal surface area. We therefore need to understand the behavior of these extended systems having a cablelike structure (Fig. 2.1).

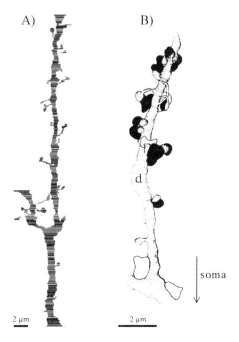

A)

B)

d

soma

2 µm   2 µm

Fig. 2.1 CLOSEUP VIEW OF DENDRITES Two reconstructed dendrites of a spiny stellate cell in the visual cortex of the cat. The reconstructions were carried out by a very laborious serial electron microscopic procedure. Notice the thin elongated, thornlike structures, *dendritic spines*. The vast majority of neuronal processes, whether axons or dendrites, possess such an elongated, cylindrical geometry. Studying the spread of electrical current in these structures is the subject of cable theory. **(A)** Cross section of a branching dendrite. **(B)** Three-dimensional view of another dendrite. The black blobs are excitatory synapses and the three clear blobs are inhibitory synapses. Reprinted by permission from Anderson et al. (1994).

The basic equation governing the dynamics of the membrane potential in thin and elongated neuronal processes, such as axons or dendrites, is the *cable equation*. It originated in the middle of the last century in the context of calculations carried out by Lord Kelvin, who described the spread of potential along the submarine telegraph cable linking Great Britain and America. Around the turn of the century, Herman and others formulated the concept of *Kernleitermodel*, or core conductor model, to understand the flow of current in nerve axons. Such a core conductor can be visualized as a thin membrane or sheath surrounding a cylindrical and electrically conducting core of constant cross section placed in a solution of electrolytes (see Fig. 2.2).

The study of the partial differential equations describing the evolution of the electrical potential in these structures gave rise to a body of theoretical knowledge termed *cable theory*. In the 1930s and 1940s concepts from cable theory were being applied to axonal fibers, in particular to the giant axon of the squid (Hodgkin and Rushton, 1946; Davis and Lorente de No, 1947).[1] The application of cable theory to passive, spatially extended dendrites started in the late 1950s and blossomed in the 1960s and 1970s, primarily due to the work of Rall (1989). In an appropriate gesture acknowledging his role in the genesis of quantitative modeling of single neurons, Segev, Rinzel, and Shepherd (1995) edited an annotated collection of his papers, to which we refer the interested reader. It also contains personal recollections from many of Rall's colleagues as well as historical accounts of the early history of this field.

We restrict ourselves in this chapter to studying *linear cable theory*, involving neuronal processes that only contain voltage-independent components. In particular, we assume that the membrane can be adequately described by resistances and capacitances (*passive membrane*). Given the widespread existence of dendritic nonlinearities, it could be argued that studying neurons under such constraints will fail to reveal their true nature. However, it is also true that one cannot run before one can walk, and one cannot walk before one can crawl. In order to understand the subtlety of massive synaptic input in spatially extended passive and active cables, one first needs to study the concepts and limitations of linear cable theory before advancing to nonlinear phenomena.

Cable theory, whether linear or nonlinear, is based on a number of assumptions concerning the nature and geometry of neuronal tissue. Let us discuss these assumptions prior to studying the behavior of the membrane potential in a single, unbranched, passive cable.

2.1 Basic Assumptions Underlying One-Dimensional Cable Theory

In a standard copper wire, electrons drift along the gradient of the electrical potential. In axons or dendrites the charge carriers are not electrons but, in the main, one of two ionic species, sodium and potassium, and, to a lesser extent, calcium and chloride. How can this current be quantified?

1. Starting point for any complete description of electrical currents and fields must be Maxwell's equations governing the dynamics of the electric field $\mathbf{E}(x, y, z, t)$ and the magnetic field $\mathbf{B}(x, y, z, t)$,[2] supplemented by the principle of conservation of charge

---

1. For a detailed account of all the twists and turns of this story, see Cole (1972) and Hodgkin (1976). When reading these down-to-earth monographs, one becomes painfully aware of the very limited amount of real knowledge and insight gained during decades of intensive experimental and theoretical research. Most of one's effort is usually spent on pursuing details that turn out to be irrelevant and in constructing and developing incorrect models.

2. We follow standard convention in using boldface variables for all vector quantities.

A)

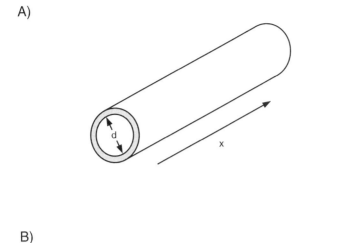

B)

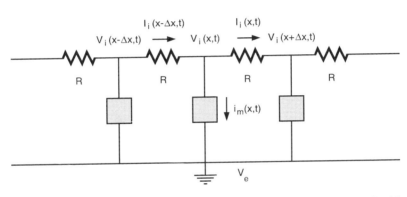

Fig. 2.2 ELECTRICAL STRUCTURE OF A CABLE (A) Idealized cylindrical axon or dendrite at the heart of one-dimensional cable theory. Almost all of the current inside the cylinder is longitudinal due to geometrical (the radius is much smaller than the length of the cable) and electrical factors (the membrane covering the axon or dendrite possesses a very high resistivity compared to the intracellular cytoplasm). As a consequence, the radial and angular components of the current can be neglected, and the problem of determining the potential in these structures can be reduced from three spatial dimensions to a single one. On the basis of the bidomain approximation, gradients in the extracellular potentials are neglected and the cable problem is expressed in terms of the transmembrane potential $V_m(x, t) = V_i(x, t) - V_e$. (B) Equivalent electrical structure of an arbitrary neuronal process. The intracellular cytoplasm is modeled by the purely ohmic resistance R. This tacitly assumes that movement of carriers is exclusively due to drift along the voltage gradient and not to diffusion. Here and in the following the extracellular resistance is assumed to be negligible and V_e is set to zero. The current per unit length across the membrane, whether it is passive or contains voltage-dependent elements, is described by i_m and the system is characterized by the second-order differential equation, Eq. 2.5.

(Feynman, Leighton, and Sands, 1964). As detailed in Rosenfalck's thesis (1969), the magnetic vector potential associated with the movement of charges during an action potential in biological tissues only has a negligible effect (10^{-9}) on the electric field and can therefore safely be neglected. Indeed, it took the technological development of very sensitive quantum devices (SQUIDs) to be able to measure the magnetic field associated with massive electrical activity in the brain. So the first simplification involves neglecting the magnetic field.

2. This leaves us with three fundamental relationships governing electrodynamics in neuronal structures.

    a. Gauss's law, stating that the divergence of the field **E** is identical to the charge density normalized by the electrical permittivity ϵ. Equivalently, Poisson's equation, which links the Laplacian of the electrical potential to the negative charge density normalized by ϵ, serves as well.

    b. Charge conservation, that is, the sum of the flux of current through any closed surface and the change of the charge over time inside this surface must be zero.

    c. An equation linking the electrical current to the electric field. In general, charged carriers can move either by drift along an electric field or by diffusion, from a volume of high carrier concentration into one of lower concentration, and the total current flow is the sum of these two independent components. The mathematical expression of this fact constitutes the Nernst-Planck electrodiffusion equation, treated in Sec. 11.3. As discussed there, for almost all cases of interest the changes in concentration of the various ions (Na^+, K^+, Ca^{2+} and Cl^-) are too small to measurably contribute to current flow. Only in very thin fibers of less than 1 μm diameter does longitudinal current flow due to concentration differences begin to play any role. In other words, Ohm's law is perfectly adequate to describe the electrical current moving within an axon or dendrite.[3]

3. This is the starting point for most derivations of cable theory (Lorente de Nó, 1947; Clark and Plonsey, 1966, 1968; Plonsey, 1969; Rall, 1969b; Eisenberg and Johnson, 1970).

    a. The dominant fraction of current inside a neuronal process, such as a dendrite or axon, flows parallel to its longitudinal axis. Only a very small fraction of the current flows across the neuronal membrane. This is true both for geometrical reasons—the diameter of axons and dendrites being much smaller than their longitudinal extent—as well as for electrical ones. As detailed in Appendix A, the neuronal membrane is all but impermeable to current flow. Charged carriers can only cross the membrane through the ionic channels. The high transmembrane resisitivity stands in contrast to the relatively small intracellular resistivity.

    A major implication is that instead of having to solve for the voltage in three dimensions, our problem is reduced to one of describing the voltage along a single spatial dimension. In a careful comparison between the membrane potential derived as the solution of Laplace's equation in a three-dimensional cylindrical coordinate system and the solution of the one-dimensional cable equation, Rall (1969b) showed that the radial and angular membrane potential terms typically decay 10^4 times faster than the components of the membrane potential along the axis. Fortuitously, we can safely neglect two out of three dimensions for all of the cases considered in this book.

    b. Electrical charge in the cytoplasm, no matter whether inside or outside the cell, relaxes in a matter of microseconds or less. In other words, any capacitive effects of the cytoplasm itself can be totally ignored on the millisecond or longer time scale (inductive effects can be completely neglected; Scott, 1971). Thus, from an electrical point of view, the extracellular as well as the intracellular cytoplasm can be approximated by ohmic resistances.

---

3. This is analogous to the situation prevalent in a copper wire, where the current flow due to drift down the gradient of the electrical potential exceeds by many orders of magnitude the current flow due to differences in the local densities of electrons.

c. The solution of the equation for the electrical potential is still extremely complicated if all the neuronal structures and membranes outside the dendrite or axon under investigation are explicitly included. Fortuitously for modelers (but less so for the electrophysiologist, who has to infer the neuronal activity of a cell from its extracellular signature), the extracellular potential (1) usually is small (since the small amount of current making it through the membrane encounters a relatively large extracellular volume), and (2) decays over distances which are usually much larger than the diameters of the fiber itself. This implies that the extracellular space can be treated as a homogeneous dielectric, averaging over local inhomogeneities. The problem of computing the membrane potential is therefore reduced to two homogeneous domains, the extracellular and the intracellular ones.

The extracellular resistivity is often defined in the case when the external medium is a shell of conducting cytoplasm surrounding the cable, a shell that can be characterized by a resistance per unit length of cylinder r_e. For large external volumes (think of the case of a single neuronal fiber placed in a bath solution) r_e is assumed to be zero. In this case, no extracellular voltage gradients exist and the entire extracellular space is isopotential, $V_e(x, t) = $ const, which we set to zero. Including a uniform extracellular resistivity complicates matters only slightly, and the solution of the cable equation is qualitatively similar to the solution for $r_e = 0$. Therefore, the *membrane potential* $V_m(x, t)$, defined as the intracellular potential minus the extracellular potential (Eq. 1.1), is identical to the intracellular potential. Indeed, throughout the book, we use these two variables interchangeably. Yet it should always be kept in mind that the membrane potential corresponds to the difference in voltage across the membrane separating the inside from the outside.

A timely research topic of considerable interest is a detailed investigation of electrical coupling of realistically modeled neurons via the extracellular potential. Lengthy experimental and theoretical studies have been carried out for the case of two parallel axons. For this geometry, any direct electrical coupling is slight (the extracellular potential due to a spike is in the 10 μV range; Clark and Plonsey, 1968, 1971; Marks and Loeb, 1976; Scott and Luzader, 1979; Barr and Plonsey, 1992; Bose and Jones, 1995; Struijk, 1997). However, extracellular potentials recorded close to dendritic trees can be much larger (up to a few mV) than those next to axons. Given the extremely tight packing among neurons, this type of ephaptic[4] coupling could be of functional relevance, yet almost no theoretical work has been carried out on this subject (Lorente de Nó, 1953; Hubbard, Llinás and Quastel, 1969; Holt, 1998).

At this stage, we represent the neuronal tissue with the help of a series of discrete electrical circuits of the type shown in Fig. 2.2B. Without making any specific assumption concerning the detailed nature of the neuronal membrane, we express the current per unit length flowing through the membrane at location x as $i_m(x, t)$. We can write down Ohm's law for the discrete circuit illustrated in Fig. 2.2B,

$$V_i(x, t) - V_i(x + \Delta x, t) = R I_i(x, t) \tag{2.1}$$

or, in the limit of an infinitesimal small interval Δx, and with $V_m = V_i$,

$$\frac{\partial V_m}{\partial x}(x, t) = -r_a \cdot I_i(x, t), \tag{2.2}$$

---

4. Greek for "touching onto," rather than *synaptic*, "touching together."

where $r_a = R/\Delta x$ is the intracellular resistance per unit length of cable with dimensions of ohms per centimeter. I_i is the intracellular core current flowing along the cable, assumed to be positive when flowing toward the right, in the direction of increasing values of x. Kirchhoff's law of current conservation stipulates that the sum of all currents flowing into and out of any particular node must equal zero. Applied to the node at x in Fig. 2.2B, we have

$$i_m(x, t)\Delta x + I_i(x, t) - I_i(x - \Delta x, t) = 0 \tag{2.3}$$

or, in differential form in the limit that $\Delta x \to 0$,

$$i_m(x, t) = -\frac{\partial I_i}{\partial x}(x, t). \tag{2.4}$$

Inserting the spatial derivative of Eq. 2.2 into Eq. 2.4 leads to

$$\frac{1}{r_a}\frac{\partial^2 V_m}{\partial x^2}(x, t) = i_m(x, t). \tag{2.5}$$

This second-order ordinary differential equation, together with appropriate boundary conditions, describes the membrane potential in an extended one-dimensional cable structure with an ohmic intracellular cytoplasm, regardless of the exact nature of the neuronal membrane.

2.1.1 Linear Cable Equation
In Sec. 1.1, we discussed the nature of a patch of passive membrane and assumed that the membrane current includes a capacitive (Eq. 1.3) and a resistive (Eq. 1.4) component (Figs. 1.1 and 1.2). Including an external current term $I_{inj}(x, t)$, the membrane current per unit length of the cable, i_m, is given by

$$i_m(x, t) = \frac{V_m(x, t) - V_{rest}}{r_m} + c_m\frac{\partial V_m(x, t)}{\partial t} - I_{inj}(x, t), \tag{2.6}$$

where r_m is the membrane resistance of a unit length of fiber, measured in units of ohms-centimeter. If the electrical nature of the membrane is constant along the length of the passive fiber under investigation (Fig. 2.3), we can replace $i_m(x, t)$ on the right-hand side of Eq. 2.5 with Eq. 2.6 and multiply both sides with r_m to arrive at

$$\lambda^2\frac{\partial^2 V_m(x, t)}{\partial x^2} = \tau_m\frac{\partial V_m(x, t)}{\partial t} + (V_m(x, t) - V_{rest}) - r_m I_{inj}(x, t), \tag{2.7}$$

with the membrane time constant $\tau_m = r_m c_m$ and the *steady-state space constant* $\lambda = (r_m/r_a)^{1/2}$. We will discuss their significance forthwith.

Equation 2.7 is the *linear cable equation*, a partial differential equation, first order in time and second order in space. This type of parabolic differential equation is quite similar to the heat and diffusion equations. The behavior of all three is characterized by dissipation and the absence of any wavelike solution with constant velocity. Parabolic differential equations have a well-specified and unique solution if appropriate initial conditions, such as the voltage throughout the cable at $t = 0$ should be zero, or boundary conditions, such as no current should leak out at either end of the cable, are specified. The cable equation is fundamental to understanding the behavior of the membrane potential, the principal state variable used for rapid intracellular communication in neurons. We will discuss its behavior in both this chapter and the next.

As expressed in Eq. 2.7, a simple, unbranched cable has a nonzero resting potential V_{rest} which does not vary with the position along the cable. For a homogeneous cable in the

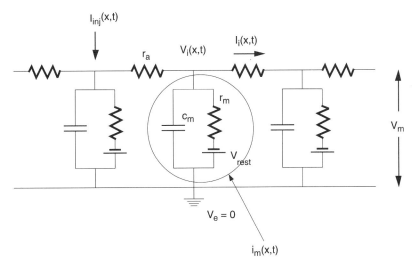

Fig. 2.3 A SINGLE PASSIVE CABLE Equivalent lumped electrical circuit of an elongated neuronal fiber with passive membrane. The intracellular cytoplasm is described by an ohmic resistance per unit length r_a and the membrane by a capacitance c_m in parallel with a passive membrane resistance r_m and a battery V_{rest}. The latter two components are frequently referred to as *leak resistance* and *leak battery*. An external current $I_{\text{inj}}(x, t)$ is injected into the cable. The associated linear cable equation (Eq. 2.7) describes the dynamics of the electrical potential $V_m = V_i - V_e$ along the cable.

absence of any input $I_{\text{inj}}(x, t)$, the membrane potential throughout the cable will be equal to a constant. The amplitude of V_{rest} varies between -50 and -90 mV, depending on cell type and other circumstances, with the inside of the neuron being at the negative potential. V_{rest} need not always be constant throughout the dendritic tree (see Sec. 18.3.4).

Because the resting potential is simply an offset, it is often set to zero. This can be thought of as defining the membrane potential $V_m(x, t)$ as relative to this resting potential. Very often the equations will be somewhat simplified when the potential is defined as relative to V_{rest}. We use the convention that $V_m(x, t)$ refers to the absolute membrane potential, while $V(x, t)$ refers to the potential relative to V_{rest}.

We should here also allude to the vexing question of units. The three voltage-independent components of a passive cable are commonly specified in one of two ways. If they are expressed as quantities per unit length, they are conventionally labeled

$$r_a = \frac{4R_i}{\pi d^2} \tag{2.8}$$

in units of Ω/cm,

$$r_m = \frac{R_m}{\pi d} \tag{2.9}$$

in units of $\Omega \cdot \text{cm}$ and

$$c_m = C_m \cdot \pi d \tag{2.10}$$

in units of F/cm. Using these variables has the advantage that the cable equation contains no explicit terms depending on the diameter d of the cable.

The more common way, and the one we adopt throughout the book, is to specify these quantities in units that are independent of the diameter of the fiber, using capital letters: the *intracellular resistivity* R_i, the *specific membrane resistance* R_m and the *specific membrane capacitance* C_m, with dimensions of $\Omega\cdot\text{cm}$, $\Omega \cdot \text{cm}^2$ and F/cm^2, respectively. For more details, consult Appendix A.

2.2 Steady-State Solutions

Let us investigate the behavior of the cable equation in response to a current $I_{\text{inj}}(x)$ injected at location x via an intracellular microelectrode or a synapse. We assume that the current is switched on at $t = 0$ and remains on. One frequently encounters this situation in experiments to investigate the cable properties of neurons and axons. After some initial transients, the voltage will reach a steady-state value. To compute the steady-state membrane potential, we set $\partial V/\partial t = 0$ and write the cable equation as

$$\lambda^2 \frac{d^2 V(x)}{dx^2} = V(x) - r_m I_{\text{inj}}(x) .\tag{2.11}$$

This reduces the original partial differential Eq. 2.7 to an ordinary second-order differential equation depending solely on space. We now study its solutions for different neuronal geometries.

2.2.1 Infinite Cable

We begin by assuming that a current I_{inj} of constant amplitude is injected at the origin, $x = 0$, of an infinite cable of diameter d. Mathematically, we describe this by setting $I_{\text{inj}}(x)$ to $I_0 \delta(x)$, where $\delta(x)$ is the Dirac *delta* or *impulse* distribution in space. As boundary condition we assume that the voltage at the two infinitely distant terminals goes to zero as $|x| \rightarrow \infty$. Using the theory of Fourier transforms (see Appendix B) we arrive at the solution

$$V(x) = V_0 e^{-|x|/\lambda} ,\tag{2.12}$$

with $V_0 = I_0 r_m/(2\lambda)$. This solution can easily be verified by placing it into Eq. 2.11. The stationary voltage distribution, sometimes referred to as the *electrotonus*, in the infinite cable decays exponentially away from the site of injection. The parameter controlling this decay is the *space constant* λ. The voltage decreases to e^{-1}, that is, to 37% of its original value, at $x = \lambda$ and to e^{-2}, or 13% of its original value, at $x = 2\lambda$. In the derivation of the cable equation, the *steady-state space constant* is defined as

$$\lambda = \left(\frac{r_m}{r_a}\right)^{1/2} = \left(\frac{R_m}{R_i} \cdot \frac{d}{4}\right)^{1/2} .\tag{2.13}$$

The larger the membrane resistance R_m, the less current leaks across the membrane and the larger the space constant λ. Furthermore, a thick dendrite has a larger space constant than a thin one, reflecting the fact that the spread of current is enhanced by a larger diameter. Another way of deriving λ involves computing the distance l over which the total resistance to current flowing across the membrane is identical to the total longitudinal resistance. Paying careful attention to the relevant units, we have $r_m/l = r_a l$, or $l = \sqrt{r_m/r_a} = \lambda$. For a typical apical dendrite of a cortical cell with a 4 μm diameter, $R_i = 200 \ \Omega\cdot\text{cm}$ and $R_m = 20,000 \ \Omega \cdot \text{cm}^2$, the space constant λ comes out to be 1 mm. This large distance, compared to the diameter of the dendrite, is the reason why we can neglect the radial components of voltage along these cables.

Given the importance of λ for the electrotonic spread of the potential in a neuron, we frequently normalize the spatial coordinate x with respect to λ, expressing it in dimensionless units: $X = x/\lambda$. Any particular distance ℓ can likewise be expressed in terms of the associated dimensionless *electrotonic distance* $L = \ell/\lambda$.

What is the input resistance of the infinite cable? Operationally, it is measured by inserting an electrode that passes current and, at a distance that is small compared to λ, an electrode to record the voltage. In the limit that this distance shrinks to zero, we can write

$$R_{\text{in}} = \frac{V(x)}{I_i(x)} = \frac{V(x = 0)}{I_0}. \tag{2.14}$$

The last equality holds because the input resistance at the location of the injecting electrode is, by definition, equal to the ratio of the evoked potential to the injected current causing this change. It follows that

$$R_{\text{in}} = \frac{r_m}{2\lambda} = \frac{r_a \lambda}{2} = \frac{(r_a r_m)^{1/2}}{2}. \tag{2.15}$$

The input resistance is—as expected—constant throughout the infinite and homogeneous cable. Confirming our intuition, increasing either the membrane resistance or the intracellular resistivity will increase R_{in}.

Conceptually, we can think of an infinite cable as two semi-infinite cables, one going off to the left and one to the right. The input resistance associated with a single semi-infinite cable R_∞ must therefore be twice the resistance associated with the infinite cable (since current can only flow in one direction); or

$$R_\infty = (r_a \cdot r_m)^{1/2} = r_a \lambda = \frac{r_m}{\lambda} = (R_m R_i)^{1/2} \frac{2}{\pi d^{3/2}}. \tag{2.16}$$

This variable, rather than the resistance associated with an infinite cable, is called R_∞, since it corresponds to the situation of a soma with a single dendrite extending into infinity (Rall, 1959).

The input conductance of a semi-infinite cylinder is given by the inverse of Eq. 2.16,

$$G_\infty = \frac{1}{R_\infty} = \left(\frac{1}{R_m R_i}\right)^{1/2} \frac{\pi d^{3/2}}{2}. \tag{2.17}$$

The input conductance decreases as the square root of the membrane resistance R_m and increases as the $\frac{3}{2}$ power of the diameter of the fiber, a relationship that will be important later on.

The input resistance of a patch of membrane is linearly related to the membrane resistance R_m (with the constant of proportionality given by the total membrane area). In general, as the dimensionality of the space increases, the dependency of the input resistance on R_m lessens. Thus, R_{in} in an infinite cable is proportional to the square root of R_m. For a two-dimensional resistive sheet, $R_{\text{in}} \propto \log(R_m)$. In a three dimensional syncytium (such as muscle tissue) $R_{\text{in}} \propto e^{-1/R_m^{1/2}}$ (see Chap. 3 in Jack, Noble, and Tsien, 1975; Eisenberg and Johnson, 1970). Given the area- or volume-filling geometry of the dendritic tree, the dependency of its input resistance on R_m falls somewhere between that of an infinite cable and that of the resistive sheet.

2.2.2 Finite Cable

Real neurons certainly do not possess infinitely long dendrites, so we need to consider a finite piece of cable of total electrotonic length $L = \ell/\lambda$. The general solution to the linear second-order ordinary differential cable equation can be expressed in normalized electrotonic units as

$$V(X) = \alpha \cosh(L - X) + \beta \sinh(L - X),$$ (2.18)

with $\cosh(x) = (e^x + e^{-x})/2$ and $\sinh(x) = (e^x - e^{-x})/2$. The values of α and β depend on the type of boundary conditions imposed at the two terminals. (What happens at the end of the finite cable influences the voltage throughout the fiber.) We distinguish three different boundary conditions.

Sealed-End Boundary Condition

This is the boundary condition of most relevance to neurons embedded in the living tissue. It assumes that the end of the fiber is covered with neuronal membrane with resistance R_m. It follows that the resistance terminating the equivalent circuit in Fig. 2.3 has the value $4R_m/\pi d^2$. For $d = 2\ \mu\text{m}$ and $R_m = 10^5\ \Omega \cdot \text{cm}^2$ this is about 3000 GΩ, a value so high that for all intents and purposes we can consider it to be infinite. If the terminating resistance is infinite, no axial current $I_i(X = L)$ will flow. And since the axial current is given by the derivative of the voltage along the cable, this implies that

$$\left.\frac{dV(X)}{dX}\right|_{X=L} = 0$$ (2.19)

at the terminal. This *zero-slope* or *von Neumann* boundary condition is referred to as a *sealed-end* boundary condition and is the one commonly adopted to model the terminals of dendrites or other neuronal processes. Applying Eq. 2.19 to Eq. 2.18 leads to

$$V(X) = V_0 \frac{\cosh(L - X)}{\cosh(L)}.$$ (2.20)

Figure 2.4 illustrates the voltage profile in a short and a long cable with such a sealed-end boundary condition. As expected from Eq. 2.19, the slope of both curves flattens out as the terminal is approached. Furthermore, both curves lie above the voltage decay in a semi-infinite cable. In other words, the voltage in a cable with a sealed end—regardless of its length—decays less rapidly than the voltage in a semi-infinite cable.

We compute the input resistance R_{in} at the origin of the cable, looking into the cable toward its terminal, using the same strategy as in the previous section,

$$R_{\text{in}} = R_\infty \coth(L),$$ (2.21)

with $\coth(x) = \cosh(x)/\sinh(x)$. This is plotted in Fig. 2.5 (upper curve). This input resistance is always higher than that of the semi-infinite cable, since the intra-axial current I_i is prevented from leaving the cable at the endpoint of the cable.

Killed-End Boundary Condition

Another type of boundary condition is of relevance when the dendrite or axon is physically cut open or otherwise short-circuited. Under these conditions the intracellular potential at the terminal is identical to the extracellular potential, that is, the effective potential is set to zero,

$$V(X)|_{X=L} = V_L = 0.$$ (2.22)

This *Dirichlet* type of boundary condition is also known as *open-* or *killed-end* boundary

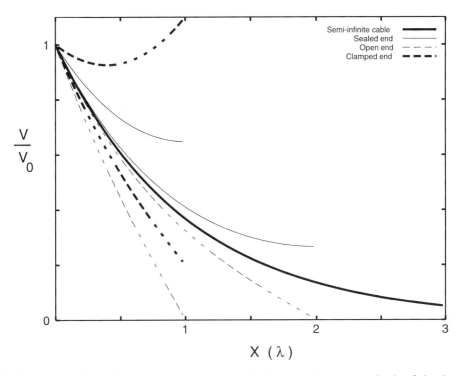

$$\frac{V}{V_0}$$

$$X\ (\lambda)$$

Fig. 2.4 STEADY-STATE VOLTAGE ATTENUATION Steady-state voltage attenuation in a finite piece of cable as a function of the normalized electrotonic distance $X = x/\lambda$ from the left terminal. The potential at the left terminal is always held fixed at $V = V_0$, while the normalized potential throughout the cable varies with the boundary condition at the right terminal. The bold continuous line corresponds to the voltage in a semi-infinite cable, showing a pure exponential decay. The thin continuous lines show the voltage decay for two cables that terminate in a sealed end (Eq. 2.20) at $X = 1$ or $X = 2$. This is the type of boundary condition used most commonly in simulations. The two thin dashed curves show the same two cables, but now terminating in a short circuit (killed-end boundary condition; Eq. 2.23). Note that either the spatial derivative of voltage (sealed-end) or the voltage itself (killed-end) is zero at the rightmost terminal. That the spatial voltage profile can be nonmonotonic in a passive cable is witnessed by the topmost bold dashed curve, where the voltage at $X = 1$ is clamped to 1.1 times the voltage at the origin. For the lower bold dashed curve, the voltage at the terminal is clamped to $0.2V_0$. Reprinted in modified form by permission from Rall (1989).

condition and corresponds to setting the terminating resistance to zero. It follows that the voltage along the cable is

$$V(X) = \frac{V_0 \sinh(L - X)}{\sinh(L)}, \tag{2.23}$$

and the input resistance is

$$R_{\text{in}} = R_\infty \tanh(L), \tag{2.24}$$

with $\tanh(x) = \sinh(x)/\cosh(x)$. The two thin dashed curves in Fig. 2.4 are the voltage profiles along two cables of electrotonic length $L = 1$ and 2 with a killed-end boundary condition. Their values are always less than the voltage at the corresponding location in a semi-infinite cable. Correspondingly, the input resistance of these cables is always less than that of the semi-infinite cable (Fig. 2.5). The input resistance at the origin $X = 0$ of the

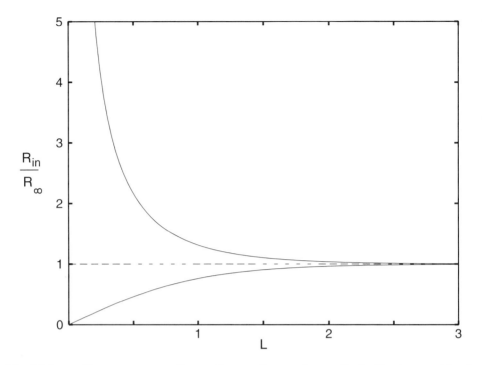

Fig. 2.5 INPUT RESISTANCE OF A FINITE CABLE   Input resistance R_{in} looking into a cable of electrotonic length L toward the right terminal. The ordinate is normalized in terms of the input resistance R_{in} of a semi-infinite cylinder (Eq. 2.16). The normalized input resistance for a sealed-end boundary condition (upper curve) is always larger than R_∞, while the input resistance of a cable with killed-end boundary condition (lower curve) is always less. In the former case, the current is prevented from leaving the cable at the endpoint, while the voltage is "shorted to ground" in the latter case. For cables longer than two space constants, $R_{in} \approx R_\infty$.

cable is inversely proportional (see Eq. 2.2) to the slope of V. The actual input resistance as a function of the electrotonic length of a killed-end cable is shown in Fig. 2.5 (lower curve).

Arbitrary Boundary Condition

In general, the terminal has neither infinite (sealed-end) nor zero (killed-end) resistance, but some finite value R_L. This, for instance, is the case if the cable is connected to some other cable or even to an entire dendritic tree. If we know the value of the voltage at this boundary, that is, V_L, we can express the voltage as

$$V(X) = \frac{V_0 \sinh(L - X) + V_L \sinh(X)}{\sinh(L)}. \qquad (2.25)$$

Notice how this expression takes on the value V_0 at $X = 0$ and V_L at $X = L$. In Fig. 2.4 we show two such cases in which V_L is either clamped to $0.2V_0$ or to $1.1V_0$ (causing the non-monotonic appearance). Note that this sagged appearance is a direct consequence of the unusual boundary condition.

The leak current through the terminal follows from Ohm's law and Eq. 2.2 as

$$i_L = \left.\frac{V_L}{R_L}\right|_{X=L} = \left.\frac{-1}{r_a}\frac{dV(X)}{dX}\right|_{X=L}. \qquad (2.26)$$

We can now rewrite Eq. 2.25 as

$$V(X) = V_0 \cdot \frac{\cosh(L - X) + (R_\infty/R_L)\sinh(L - X)}{\cosh(L) + (R_\infty/R_L)\sinh(L)}, \tag{2.27}$$

resulting in a general expression for the voltage in a finite piece of cable.

With the help of Eq. 2.3 and the above equation, we can derive an expression for the input resistance of a cable of length L with a terminating resistance R_L,

$$R_{\text{in}} = R_\infty \frac{R_L + R_\infty \tanh(L)}{R_\infty + R_L \tanh(L)}. \tag{2.28}$$

The previous two equations allow us to obtain the values for the voltage and the input resistance for the sealed-end and the killed-end boundary conditions by setting R_L to either ∞ or 0. Furthermore we recover $R_{\text{in}} = R_\infty$ for an infinite cable (since $\tanh(L)$ goes to 1 as $L \to \infty$).

2.3 Time-Dependent Solutions

So far, we have only been concerned with the behavior of the voltage in a cable in response to a stationary current injection, a situation where the voltage settles to a constant value. In general though, we need to consider the voltage trajectory in response to some time-varying current input. Since the time-dependent solution of the cable equation is substantially more complex than the steady-state solution treated above, we will only discuss the solution to two special cases. The interested reader is referred to the monographs by Jack, Noble, and Tsien (1975) and by Tuckwell (1988a) for a treatment of many more cases of interest.

Before we do so, we will introduce a normalized version of the cable equation. Recalling the definition of the neuronal time constant from Chap. 1 as

$$\tau_m = r_m c_m = R_m C_m \tag{2.29}$$

allows us to introduce dimensionless variables for both time, $T = t/\tau_m$, and space, $X = x/\lambda$. Written in these units and taking care to properly transfrom the input current (Sec. 4.4 in Tuckwell, 1988a), the cable equation becomes

$$\frac{\partial^2 V(X, T)}{\partial X^2} = \frac{\partial V(X, T)}{\partial T} + V(X, T) - \frac{I_{\text{inj}}(X, T)}{\lambda c_m}, \tag{2.30}$$

with $I_{\text{inj}}(X, T) = \lambda \tau_m I_{\text{inj}}(x, t)$ and $I_{\text{inj}}(x, t)$ corresponds to the stimulus current density.

2.3.1 Infinite Cable

In order to compute the dynamic behavior of the infinite cable in response to current injections we will once again exploit the linearity of Eq. 2.30, that is, the fact that if the response of the membrane to the current $I(X, T)$ is $V(X, T)$, the polarization in response to the current $\alpha I(X, T)$ is $\alpha V(X, T)$.

Voltage Response to a Current Pulse

As discussed in the first chapter (and summarized in Appendix B), we can completely characterize the system by computing the *impulse response* or *Green's function* associated with Eq. 2.30, which we do by transforming to the Fourier domain, assuming that $V(X) \to 0$ as $|X| \to \infty$, and transferring back to the time domain (Jack, Noble, and Tsien, 1975).

Assuming that a fixed amount of charge Q_0 is applied at $X = 0$ as an infinitely brief pulse of current $I_0 = Q_0/\tau_m$, the resultant voltage is

$$V_\delta(X, T) = \frac{I_0 r_m}{2\lambda(\pi T)^{1/2}} e^{\frac{-X^2}{4T}} e^{-T} . \tag{2.31}$$

In order to gain a better intuitive understanding of cable theory, let us review various special cases. If we record the voltage at the same location at which we injected the current, corresponding to $X = 0$, the Green's function is proportional to $e^{-T}/\sqrt{T}$. $V_\delta(0, T)$ diverges at the origin and decays a bit faster than exponentially for large times (Figs. 2.6 and 2.7B). The singularity at the origin comes about due to the infinitesimal amount of capacitance for $X = 0$ between the site of current injection and that of the measuring device. Using L'Hôpital's rule, we can see that the limit of $V_\delta(X, T)$ for any value of X other than the origin is zero. No singularity exists for the membrane patch model, which possesses a simple exponentially decaying Green's function with a fixed amount of capacitance (Fig. 2.6).

If one waits long enough, the voltage decay throughout the cable will be identical, approaching more and more to the decay seen at the spatial origin. This can be observed

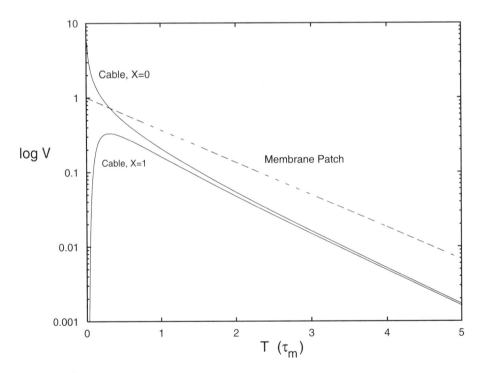

Fig. 2.6 Impulse Response at the Origin of an Infinite Cable Comparison of the impulse response or Green's function for an infinite cable $V_\delta(X, T)$ (Eq. 2.31) for $X = 0$ and $X = 1$ and the normalized Green's function for a patch of passive membrane (e^{-T}; Eq. 1.17; dashed line) on a logarithmic voltage scale. Time is expressed in units of τ_m. The voltage in the infinite cable (solid lines) is measured at the location where the δ pulse of current is applied or one space constant λ away. The Green's function diverges at $X = 0$, since the amount of membrane capacitance between the current injection and the voltage recording electrode is infinitely small, but has a constant value of C for the membrane patch case.

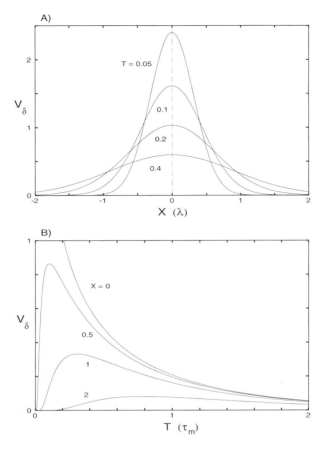

Fig. 2.7 IMPULSE RESPONSE OF THE INFINITE CABLE Impulse response or Green's function $V_\delta(X, T)$ (Eq. 2.31) at $X = 0$ and $T = 0$ as a function of normalized space **(A)** and time **(B)** using different linear scales. At any point in time, the spatial profile of the voltage along the cable in the upper panel can be described by a Gaussian. One of the consequences of the low-pass nature of the membrane is evident in the bottom panel: more distant locations respond with a delay.

very well using a logarithmic scale, as in Fig. 2.6, where the decay ultimately scales as e^{-t/τ_m}.

In other words, immediately following the current injection, the voltage is sharply peaked around the injection site. As time goes by, the spatial voltage profile becomes broader, smearing out more and more (Fig. 2.7A). At any particular instant T, the spatial distribution of the voltage along the cable is proportional to e^{-X^2}, corresponding to a Gaussian function centered at the injection site. If the time course of voltage is recorded at increasing distances from the site of the charge application, the voltage response takes longer to reach its peak due to the low-pass nature of the membrane capacitance (Fig. 2.7B).

Since the nature of the filtering carried out by the membrane does not depend on the applied potential across it, the system is a linear one. It follows that the voltage in response to an arbitrary current stimulus $I_{inj}(T)$ injected at the origin is given by the superposition of the impulse response function with the input current. Conceptually, one can think of the input $I_{inj}(T)$ as a series of delta pulses staggered in time. The final voltage is the sum of the impulse functions associated (and weighted) with the individual pulses. This is concisely expressed by the convolution integral

$$V(X, T) = \frac{\tau_m}{Q_0} V_\delta(X, T) * I_{\text{inj}}(T) = \frac{\tau_m}{Q_0} \int_0^T V_\delta(X, T') I_{\text{inj}}(T - T') dT' , \quad (2.32)$$

where $*$ denotes convolution and the τ_m/Q_0 factor is responsible for the correct normalization (converting the voltage V_δ into an impedance).

Voltage Response to a Current Step

If a rectangular step of current is injected into the cable, such that $I_{\text{inj}}(T) = I_0$ for $T \geq 0$ and 0 otherwise, the previous integral evaluates to $I_0(\tau_m/Q_0) \int_0^T V_\delta(X, T') dT'$. Due to the presence of the Gaussian term in Eq. 2.31, this integral has no closed-form solution and must be expressed in terms of the error function $\text{erf}(x) = \frac{2}{\sqrt{\pi}} \int_0^x e^{-y^2} dy$ (as first carried out by Hodgkin and Rushton, 1946). While we will not discuss the full solution in all of its glory (see Eq. 3.24 in Jack, Noble, and Tsien, 1975), we will consider several cases of particular interest to us. The transient voltage response at the site of current injection is

$$V_{\text{Step}}(0, T) = \frac{I_0 R_\infty}{2} \text{erf}(\sqrt{T}) . \quad (2.33)$$

Since $\text{erf}(1) = 0.84$, the voltage in the cable at the site of the current step rises to 84% of its steady-state value in one time constant, compared to 63% of its peak value for the exponential charging in the case of a patch of membrane (see Eq. 1.9). Conceptually, this latter case can be thought of as resulting from injecting a current into a cable, only that in the membrane patch case the entire cable has been "space clamped" by introducing an imaginary wire along its length. Let us plot the normalized membrane potential for both cases (Fig. 2.8A), that is, the potential relative to its steady-state value,

$$W(X, T) = \frac{V(X, T)}{\lim_{T \to \infty} V(X, T)} , \quad (2.34)$$

and plotted in Fig. 2.8A.

When considering why the potential in the cable reaches its steady-state value faster than the potential across a patch of membrane with the same input impedance, it is helpful to consider the closely related problem (for linear cables) of why voltage decays faster in a cable than across a membrane patch. In the former, current can flow longitudinally and therefore escapes more rapidly than when it must all flow across the membrane. The same argument holds for an electrode injecting current into the soma of a neuron with an extended dendritic tree. A significant fraction of the injected current flows onto the extensive dendritic membrane surface, and as a result, the buildup and decay of the somatic voltage is faster compared to the isopotential membrane patch. This effect was first recognized by Rall (1957). Coombs, Eccles, and Fatt (1955) had fitted the experimentally observed membrane transients at the motoneuron soma with a single exponential with $\tau_m = 2$ msec. Rall argued that the cable properties of the dendrites needed to be accounted for and—on the basis of Eq. 2.33—estimated a τ_m of 4 msec.

Frequency-Dependent Space Constant

As expressed by Eq. 2.12, the steady-state space constant λ is defined as the distance in an infinite cable over which a *steady-state* voltage decays by $1/e$. This variable can be generalized to a function depending on frequency f. If a sinusoidal current $I(t) = I_0 \sin(2\pi f t)$ is injected into an infinite cable, the theory of Fourier transforms can be

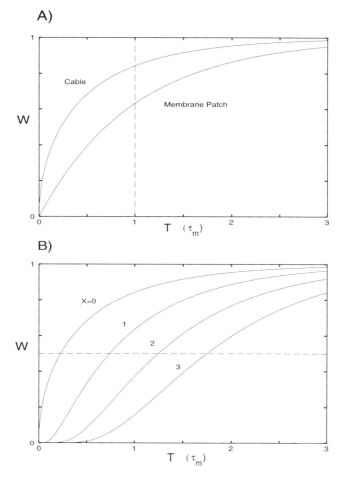

Fig. 2.8 VOLTAGE IN AN INFINITE CABLE IN RESPONSE TO A CURRENT STEP (**A**) Normalized voltage $W(X, T)$, that is, $V(X, T)$ divided by its steady-state value at location X (Eq. 2.34), in an infinite cable in response to a current step injected at the same location is compared to the voltage increase in response to the same current injected into a patch of membrane. In the cable, the voltage rises faster to its final value than in a patch of membrane. (**B**) The normalized voltage in response to a current step in an infinite cylinder at different distances X from the site of current injection. As X increases, the response becomes more smeared out. The point at which the voltage at any one location X reaches half of its final value moves—in the limit for long times—with constant velocity $2\lambda/\tau_m$ along the cable (see Sec. 2.4).

used to introduce a generalized *frequency-dependent* or *transient* space constant $\lambda(f)$ (see Appendix B).

It is relatively straightforward to formulate the cable equation in a generalized linear system by representing a finite or an infinite cable as a ladder network with arbitrary intracellular impedance $z_a(f)$ and membrane impedance $z_m(f)$ (Koch and Poggio, 1985a). In this case, the membrane can contain any collection of linear elements, including inductances, capacitances, and resistances. Under certain conditions, explored in more detail in Chap. 10, a nonlinear membrane can be linearized and can be expressed by a combination of these linear circuit elements.

If the current $I_0 \sin(2\pi f t)$ is injected into an infinite cable, the voltage at any point in the cable will be proportional to $\sin(2\pi f t + \phi)$, with a phase shift $\phi(f)$. The constant of proportionality is given by $e^{-\gamma(f)x}$, where x is the distance between the site of current injection and the recording electrode, and the *propagation constant* is

$$\gamma(f) = \sqrt{\frac{z_a(f)}{z_m(f)}} . \tag{2.35}$$

For a passive membrane, $z_a(f) = r_a$ and $z_m(f) = r_m/(1 + i2\pi f \tau_m)$. By extracting the real part of this function, we can define the frequency-dependent space constant $\lambda(f)$ as

$$\lambda(f) = \frac{\lambda(0)}{\text{Re}\{\sqrt{1 + i2\pi f \tau_m}\}} , \tag{2.36}$$

with $\lambda(0)$ the sustained or steady-state space constant (Eisenberg and Johnson, 1970). As seen in Fig. 2.9, $\lambda(f)$ decays steeply with increasing frequency, becoming proportional to $1/\sqrt{f \tau_m}$ in the limit of $2\pi f \tau_m \gg 1$. This decay is due to the distributed membrane capacitance that soaks up more and more of the current as the frequency increases. For instance, at 1000 Hz (roughly corresponding to the inverse of the width of a typical action potential), λ has decreased to 8% of its steady-state value (assuming $\tau_m = 50$ msec). This emphasizes once again the low-pass nature of the passive neuronal membrane: high frequencies are preferentially filtered out.

The frequency-dependent space constant also informs us about the limit of one-dimensional cable theory. It is clear that as $\lambda(f) \approx d$, the diameter of the fiber, one can no longer neglect the radial dimensions of the cable and has to treat the full three-dimensional problem. In general, this is not expected to occur until very high frequencies. For instance, in an apical dendrite of 4-μm diameter with $R_m = 50,000 \ 2\pi f$ cm^2, $R_i = 200 \ 2\pi f$ cm, and $C_m = 1 \ \mu F/cm^2$. The steady-state space constant λ equals 1.581 mm. Under these conditions, one has to go to frequencies close to 1 MHz in order for $\lambda(f) \approx d$. We conclude that under most circumstances, one-dimensional cable theory will hold.

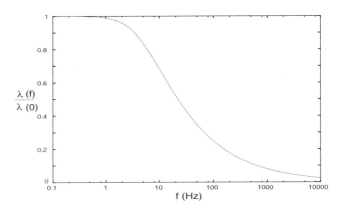

Fig. 2.9 FREQUENCY-DEPENDENT SPACE CONSTANT   If a sinusoidal current of frequency f is injected at $x = 0$ into an infinite passive cable, the voltage at location x will also be a sinusoid of frequency f, but attenuated by $e^{-x/\lambda(f)}$ and phase shifted. We here plot $\lambda(f)$, normalized by the steady-state space constant $\lambda(0)$, for $\tau_m = 50$ msec. At 1 kHz, the space constant has decayed to 8% of its original value.

2.3.2 Finite Cable

A number of different techniques are available for computing the Green's function in finite cables (Tuckwell, 1988a). One classical method, known as separation of variables, assumes that the Green's function can be written as the product of two functions, one depending only on X while the other one depends solely on T. Rall (1969a) used this technique to derive the voltage in a finite cable with sealed-end boundary conditions at $X = 0$ and $X = L$. The voltage in response to an arbitrary current input anywhere in the cable can be expressed as an infinite series,

$$V(X, t) = \sum_{n=0}^{\infty} B_n \cos \frac{n\pi X}{L} \cdot e^{-\alpha_n t/\tau_m} . \tag{2.37}$$

The B_n depend on the initial conditions chosen, such as injection of a delta pulse current or a current step. The coefficients α_n are the ratios of the membrane time constant τ_m to the *equalizing time constants* that are associated with the redistribution of charge and with the reduction of voltage differences between different regions of the cable. They are defined as

$$\alpha_n = 1 + \left(\frac{n\pi}{L}\right)^2 . \tag{2.38}$$

The physical intuition behind Eq. 2.37 is that each term results from a "reflection" of the voltage at one of the terminals. Each term becomes progressively smaller as it is reflected back and forth an infinite number of times (Fig. 2.10). Another way to understand Eq. 2.37 is to note that the $n = 0$ term, relating to the slowest decay, is constant throughout the cable and corresponds to an exponential decay away from the average voltage along the finite cable. The $n = 1$ term is associated with decay and rapid equalization of charge between two half-lengths of the cylinder ($V(X, T)$ is positive for $0 \leq X \leq L/2$ and negative for the other half of the cylinder). Higher order terms lead to an even more rapid equalization of charge over shorter lengths of the cable. The sum in Eq. 2.37 can also be expressed as

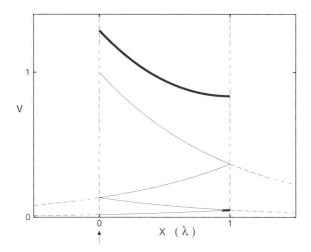

Fig. 2.10 VOLTAGE RESPONSE IN A FINITE CABLE In a finite cable with sealed end boundary conditions at $X = 0$ and 1, the voltage V in response to any current input can be described as the sum (bold line) of infinitely many "reflection" terms (thin lines), each term becoming progressively smaller. This leads to the convergent series in Eq. 2.41. Here, the current is injected at $X = 0$.

$$V(X, t) = V_\infty(X) + C_0 e^{-t/\tau_0} + C_1 e^{-t/\tau_1} + C_2 e^{-t/\tau_2} + \cdots, \qquad (2.39)$$

where the C_i's depend both on the initial conditions as well as on X, and V_∞ captures the steady-state components of the voltage. τ_0 equals the passive membrane time constant τ_m, while all other values of τ_n, called *equalizing time constants*, are smaller than τ_0, with

$$\tau_n = \frac{\tau_m}{1 + \left(\dfrac{n\pi}{L}\right)^2}. \qquad (2.40)$$

The voltage decay for large t is always dominated by the largest time constant $\tau_0 = \tau_m$. In other words, if the logarithm of the membrane potential in response to, say, a current step, is plotted as a function of time, the linear slope at the tail end of the pulse is identical to the membrane time constant τ_m (see also Fig. 2.6 and the next chapter).

A different way to express the voltage in a finite piece of cable with sealed-end boundary condition at $X = 0$ and $X = L$ in response to a current pulse (with $I_0 = Q/\tau_m$) at $X = 0$ uses the theory of Laplace transforms (Jack, Noble, and Tsien, 1975; Tuckwell, 1988a),

$$V(X, T) = \frac{R_\infty I_0}{2} \frac{e^{-T}}{(\pi T)^{1/2}} \sum_{n=-\infty}^{+\infty} e^{-(-\frac{X}{2} + nL)^2/T}. \qquad (2.41)$$

As $L \to \infty$, we recover the impulse response function for the infinite cable (Eq. 2.31). This expression has a simple graphical interpretation (as expressed in Fig. 2.10) in terms of ever reflecting and decreasing contributions, corresponding to an infinite number of virtual electrodes that inject charge at $\pm L, \pm 2L, \cdots$

While the time constants τ_n depend on L, they are independent of the site of the current pulse, the site of recording, or the initial conditions. Thus, they provide a convenient way to calculate the electrotonic length of a cable. In particular, the ratio of the first two time constants provides a measure of the cable's electrotonic length,

$$L = \frac{\pi}{\sqrt{\dfrac{\tau_0}{\tau_1} - 1}}. \qquad (2.42)$$

This expression has frequently been applied to experimental data from different preparations by measuring the two slowest time constants, obtained by *peeling* the slopes (time constants) of the logarithm of the transient voltage response (see Fig. 3.12). The outcome of this procedure is quite dependent on the neuronal geometry of the cell recorded from and the amount of noise in the measured voltage transient and needs to be used with great care. For an overview of the advantages and limitations of this method see Rall et al. (1992) and Holmes, Segev, and Rall, (1992).

A related, but more complicated, expression for the voltage can be derived if one end of the finite cable is terminated with an equipotential "soma," consisting of a somatic leak and capacitance (Rall, 1962, 1969a). We refer the interested reader to the monograph by Tuckwell (1988a) that lists the Green's functions associated with a host of other neuronal geometries and initial conditions.

2.4 Neuronal Delays and Propagation Velocity

How fast does the potential induced by the current step propagate along the cable? Figure 2.8B shows the relative voltage change along an infinite cable in response to a current step. The potential is normalized at each location by its steady-state value (Eq. 2.34). This normalization accounts for the effect of the exponential attenuation of V along the cable (Eq. 2.12). Because the membrane capacitance preferentially "soaks" up electrical charge associated with high temporal frequencies, the farther any particular point X is away from the site of current injection, the longer it takes $W(X, T)$ to rise to a particular value, say 0.5 (that is, half of its maximum). Is this delay proportional to the distance or, in other words, can one define a propagation velocity?

The linear cable equation does not admit any wave solution due to the dissipation of energy through the passive membrane. As long as no inductive elements are present in the neuronal membrane across which the current cannot change instantaneously, the voltage will, in principle, respond infinitely fast to a change in the current input an arbitrary distance away. In other words, the answer to the above question is, "No, in general one *cannot* define a velocity in a passive cable." This can change if voltage-dependent nonlinear components are incorporated in the membrane, as witnessed by the propagation of spikes at constant velocities along axons.

Yet all is not lost. Because one cannot readily define the delay of voltages in passive cables, Agmon-Snir and Segev (1993) used the trick of computing the propagation delay of the *centroid* or the *first moment* of the voltage or the current in a passive cable. Following the nomenclature of Zador, Agmon-Snir, and Segev (1995), we define the centroid of the signal $h(x, t)$ at location x as

$$\hat{t}_x^h = \frac{\int_{-\infty}^{+\infty} t h(x, t) dt}{\int_{-\infty}^{+\infty} h(x, t) dt} . \tag{2.43}$$

Here h can be either a current or a voltage with a single peak or with multiple peaks. This measure is frequently also called the *center of mass* if t is thought of as distance and $h(x, t)$ as the mass distribution, with

$$\int_{-\infty}^{+\infty} (t - \hat{t}_x^h) h(x, t) dt = 0 . \tag{2.44}$$

We define the *transfer delay* as the difference between the centroid of the induced voltage measured at location y and the centroid of current that was injected at location x,

$$D_{x \to y} = D_{xy} = \hat{t}_y^V - \hat{t}_x^I . \tag{2.45}$$

We define the *input* or the *local delay* in the same spirit as the difference between the centroids of the voltage response and the current that gave rise to it,

$$D_{xx} = \hat{t}_x^V - \hat{t}_x^I . \tag{2.46}$$

It is possible to prove a number of useful properties of these delays by multiplying the cable equation by t and integrating over t. This results in an ordinary linear differential equation, similar to the steady-state cable equation, which can be analyzed by very similar techniques. Most importantly, Agmon-Snir and Segev (1993) prove that the transfer delay D_{xy} is always positive and is independent of the shape of the transient input current. In other words, D_{xy} is a property of the passive cable and not of the input. Furthermore, no matter what the electrical structure of the cable under consideration, the transfer delay is symmetric, that is,

$$D_{xy} = D_{yx} , \tag{2.47}$$

and it does not depend on the direction of travel.

In the simple case of an isopotential neuron,

$$D_{xx} = \tau_m , \qquad (2.48)$$

due to the capacitive nature of the neuronal membrane (Fig. 2.11A). In other words, the centroid for a depolarizing potential occurs exactly one time constant later than the centroid for the current underlying this potential.

In an infinite (or semi-infinite) cable, potentials rise and decay faster, as we saw already in Fig. 2.8A. Indeed, because the charge injected into the cable not only flows onto

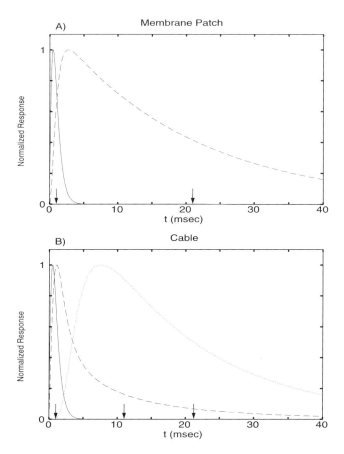

Fig. 2.11 NEURONAL INPUT AND PROPAGATION DELAYS An elegant way to define propagation delays in passive cables involves tracking the *centroid* or *center of mass* of voltages or currents in passive cable (Agmon-Snir and Segev, 1993). This is illustrated in **(A)** for an isopotential patch of membrane with $\tau_m = 20$ msec. A brief current pulse (solid profile with $t_{\text{peak}} = 0.5$ msec) gives rise to a rapidly rising but very slowly decaying depolarizing potential (shown dashed using normalized units). The centroids of the two signals (see arrows at 1 and 21 msec) are displaced by one time constant. In **(B)**, the same current is injected into a very long cable, and the normalized potential at the same location (dashed) and at a location one space constant displaced (dotted) are plotted. In an infinite cable, the *transfer delay* D_{xy} between the centroid of the current at x and the centroid of the voltage at y is $(1 + |x - y|/\lambda)\tau_m/2$ (see the arrows at 1, 11, and 21 msec). As witnessed already in Fig. 2.8A, the potential decays faster in a cable than in an isopotential patch of membrane.

the capacitance at the location of the electrode but via the intracellular resistance onto neighboring capacitances, the input delay between any current input and the associated potential at the same location is only delayed by half a time constant,

$$D_{xx} = \frac{\tau_m}{2}, \tag{2.49}$$

as compared to τ_m for an isopotential patch of membrane. The faster local response time of an infinite cable compared with that of a patch of membrane is obvious in Fig. 2.8A. If the current is injected at x and the voltage recorded at y, the two centroids are displaced by

$$D_{xy} = \left(1 + \frac{|x - y|}{\lambda}\right) \frac{\tau_m}{2}. \tag{2.50}$$

This is clearly evident in Fig. 2.11B. Again, this delay does not depend on the particular form of the current input but holds for any input. This dependency on distance allows us to define a *propagation delay* P_{xy} as the difference between the centroids of the voltage at x and at y,

$$P_{xy} = \hat{t}_x^V - \hat{t}_y^V = D_{xy} - D_{xx}. \tag{2.51}$$

For an infinite cable,

$$P_{xy} = \left(\frac{|x - y|}{\lambda}\right) \frac{\tau_m}{2}. \tag{2.52}$$

This linear relationship between space and time is equivalent to the notion of a *propagation velocity* in an infinite cable,

$$v = 2\frac{\lambda}{\tau_m} = \left(\frac{d}{R_m R_i C_m^2}\right)^{1/2}. \tag{2.53}$$

It is important to emphasize that v is a "pseudovelocity" rather than the physical velocity of a constant wave moving along the cable for which $V(x, t) = V(x - vt)$ should hold. Yet, despite the fact that any potential will decay and become smeared out as it moves along a cable, its center of mass travels with a fixed velocity.

We feel obliged to point out a serious drawback when using D_{xy} or P_{xy} in active structures. The D_{xy} measure of a dendritic input giving rise to a somatic EPSP with an undershoot, that is, a hyperpolarization due to potassium current activation (as in Fig. 18.1B), will seriously underestimate the delay due to the small, but long-lasting negative contribution to the centroid, rendering it less useful for real cells than for purely passive structures (Bernander, 1993). Under these conditions, D_{xy} can be negative.

2.5 Recapitulation

One-dimensional cable theory is based on several approximations. (1) The magnetic field due to the movement of charge can be neglected. (2) Changes in the concentration of the charged carriers, Na^+, K^+, and other ions, is slight so that the current can be expressed by Ohm's law, and the intracellular cytoplasm can be mimicked by an ohmic resistance. (3) Due to the wirelike geometry of dendrites and axons and the high resistivity of the neuronal membrane, the radial and angular components of voltage can be neglected, reducing the complexity of the solution from three spatial dimensions to a single one. (4) The extracellular space is reduced to a homogeneous resistive milieu whose resisitivity is usually set to zero.

This allows us to solve for the potential $V(x, t)$ across the neuronal membrane on the basis of a single equation.

Linear cable theory further assumes that for a limited range of voltage excursions around the resting potential, the membrane properties are independent of the membrane potential, reducing the electrical description of the membrane to resistances and capacitances, greatly simplifying analysis.

Starting in the late 1950s and early 1960s, the linear cable equation was solved by Rall and others to study the dynamics of the membrane potential in dendritic trees. Several key concepts associated with the linear cable equation for a single finite or infinite cylinder are the *space constant* λ, determining the distance over which a steady-state potential in an infinite cylinder decays e-fold, the neuronal *time constant* τ_m, determining the charging and discharging times of $V(x, t)$ in response to current steps, and the *input resistance* R_{in}, determining the amplitude of the voltage in response to slowly varying current injections.

The voltage in response to a current input, whether delivered by an electrode or by synapses, can be expressed by convolving the input with an appropriate Green's function. For passive cables, this always amounts to filtering the input by a low-pass filter function.

While the class of parabolic differential equations (to which the cable equation belongs) does not admit to any wave solutions but only shows dissipative behavior, one can define input, transfer, and propagation delays by computing and tracking the centroid or center of mass of $V(x, t)$ relative to the centroid of the current input. In the following chapter, we will apply these concepts to realistic dendritic trees.

3

PASSIVE DENDRITIC TREES

The previous chapter dealt with the solution of the cable equation in response to current pulses and steps within a single unbranched cable. However, real nerve cells possess highly branched and extended dendritic trees with quite distinct morphologies. Figure 3.1 illustrates the fantastic variety of dendritic trees found throughout the animal kingdom, ranging from neurons in the locust to human brain cells and cells from many different parts of the nervous system. Some of these cells are spatially compact, such as retinal amacrine cells, which are barely one-fifth of a millimeter across (Fig. 3.1E), while some cells have immense dendritic trees, such as α motoneurones in the spinal cord extending across several millimeters (Fig. 3.1A). Yet, in all cases, neurons are very tightly packed: in vertebrates, peak density appears to be reached in the granule cell layer of the human cerebellum with around 5 million cells per cubic millimeter (Braitenberg and Atwood, 1958) while the packing density of the cells filling the 0.25 mm^3 nervous system of the housefly *Musca domestica* is around 1.2 million cells per cubic millimeter (Strausfeld, 1976). The dendritic arbor of some cell types encompasses a spherical volume, such as for thalamic relay cells (Fig. 3.1G), while other cells, such as the cerebellar Purkinje cell (Fig. 3.1F and K), fill a thin slablike volume with a width less than one-tenth of their extent.

Different cell types do not appear at random in the brain but are unique to specific parts of the brain. By far the majority of excitatory cells in the cortex are the *pyramidal cells* (Fig. 3.1C). Yet even within this class, considerable diversity exists. But why this diversity of shapes? To what extent do these quite distinct dendritic architectures reflect differences in their roles in information processing and computation? What influence does the dendritic morphology have on the electrical properties of the cell, or, in other words, what is the relationship between the morphological structure of a cell and its electrical function?

One of the few cases where a quantitative relationship between form and some aspect of neuronal function has been established is the retinal neurons. It seems obvious that the larger their dendritic tree, the larger the visual area over which they can receive input.[1] That this is true for large cells was shown by Peichl and Wässle (1983). They correlated the electrophysiologically defined receptive field of one type of retinal ganglion cell (*brisk transient* or Y cell) with its dendritic tree size. It is well known that the physiologically

---

1. The area in visual space in which an appropriate visual input will excite a cell is termed its *classical receptive field*.

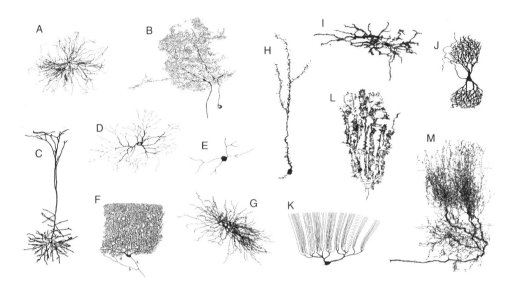

Fig. 3.1 DENDRITIC TREES OF THE WORLD Great variety of dendritic trees (in addition to a glia cell and an axonal tree) observed in the nervous systems of animals. The cells are not drawn to scale. (**A**) α motoneuron in spinal cord of cat (2.6 mm). Reprinted by permission from Cullheim, Fleshman, and Burke (1987). (**B**) Spiking interneuron in mesothoracic ganglion of locust (0.54 mm). Unpublished data from G. Laurent, with permission. (**C**) Layer 5 neocortical pyramidal cell in rat (1.03 mm). Reprinted by permission from Amitai et al., (1993). (**D**) Retinal ganglion cell in postnatal cat (0.39 mm). Reprinted by permission from Maslim, Webster, and Stone (1986). (**E**) Amacrine cell in retina of larval tiger salamander (0.16 mm). Reprinted by permission from Yang and Yazulla (1986). (**F**) Cerebellar Purkinje cell in human. Reprinted by permission from Ramón y Cajal (1909). (**G**) Relay neuron in rat ventrobasal thalamus (0.35 mm). Reprinted by permission from Harris (1986). (**H**) Granule cell from olfactory bulb of mouse (0.26 mm). Reprinted by permission from Greer (1987). (**I**) Spiny projection neuron in rat striatum (0.37 mm). Reprinted by permission from Penny, Wilson, and Kitai (1988). (**J**) Nerve cell in the nucleus of Burdach in human fetus. Reprinted by permission from Ramón y Cajal (1909). (**K**) Purkinje cell in mormyrid fish (0.42 mm). Reprinted by permission from Meek and Nieuwenhuys (1991). (**L**) Golgi epithelial (glia) cell in cerebellum of normal-reeler mouse chimera (0.15 mm). Reprinted by permission from Terashima et al., (1986). (**M**) Axonal arborization of isthmotectal neurons in turtle (0.46 mm). Reprinted by permission from Sereno and Ulinski (1987). The lengths given are approximate and correspond to the maximal extent. Reprinted by permission from Mel (1994).

defined Y cells correspond to the anatomically defined α ganglion cells while x corresponds to β (Saito, 1983; Stanford and Sherman, 1984). Figure 3.2 shows the dendritic tree and the cell body of an β cell, including its presynaptic input from bipolar cells, reconstructed from serial electron micrographs (Freed and Sterling, 1988). Because a cell can only respond to light stimulation in that part of the retina where it has functional connections with the photoreceptors (via horizontal and bipolar) cells, it is not surprising that the position, shape, and size of the receptive field are determined by the position, shape, and size of the underlying dendritic tree (Fig. 3.3). Indeed, a compartmental model of the type discussed in this chapter has shown that the passive electrotonic properties of the dendritic tree itself contribute very little to the spatial profile (Freed, Smith, and Sterling, 1992). Because of the horizontal spread of the neuronal elements interposed between the photoreceptors, the receptive field is on average 1.4 times larger than the dendritic tree.

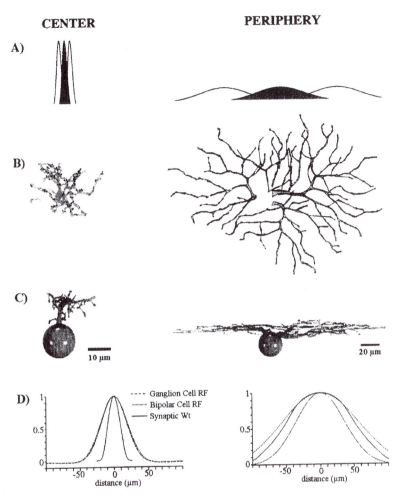

CENTER   **PERIPHERY**

A)

B)

C)

10 μm   20 μm

D)

---- Ganglion Cell RF
— · — Bipolar Cell RF
——— Synaptic Wt

distance (μm)

distance (μm)

Fig. 3.2 STRUCTURE-FUNCTION RELATIONSHIP IN RETINAL GANGLION CELLS The relationship between the extent of the receptive field of visual neurons and the size of their dendritic tree differs depending on the size of the receptive field. Shown here is the functional architecture of a central (1°; left column) and a peripheral (20°; right column) β ganglion cell in the cat retina. (**A**) Receptive field of these cells, determined by fitting a Gaussian sensitivity curve through their center (dark) and their surround fields. (**B**) Their dendritic tree, reconstructed at the electron-microscopic (left) and the light-microscopic (right) levels. The small cell receives excitatory input from about 170 bipolar ribbon synapses, while the large one is excited by 1700 bipolar ribbon synapses (see Fig. 1.5). (**C**) Compartmental modeling of these two cells, assuming a passive dendritic tree, shows that voltage attenuation is negligible. (**D**) The receptive field of the two ganglion cells, obtained from the literature, is superimposed onto the receptive field of the presynaptic neuron, a bipolar cell. The effect of synaptic input from the bipolar cell onto the ganglion cell—expressed by the dendritic tree itself—is summarized by the "synaptic weight" curve. For a central β cell, it contributes little to the ganglion cell receptive field, since the input receptive field is substantially broader than the contribution from synaptic input onto the dendritic tree (solid inner curve in the left panel). Since the dendritic tree contribution for a cell in the periphery (solid middle curve in the right panel) is larger than the extent of the receptive field of the input, it gives rise to the observed linear scaling of the receptive field with the dendritic tree size for large cells. Unpublished data from P. Sterling and R. Smith, printed with permission.

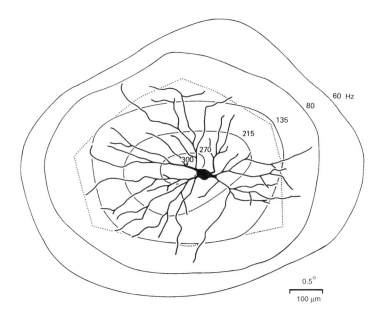

Fig. 3.3 RELATIONSHIP BETWEEN RECEPTIVE-FIELD SIZE AND DENDRITIC TREE   Comparison of the dendritic tree and the receptive field center, shape, and extent of a retinal ganglion cell in the cat. Morphologically, this cell type is known as α; its axon (not shown) projects to the geniculate. The continuous contour lines correspond to equal response contours and represent peak firing rates of the indicated amount in response to a flashing spot of light. The maximum discharge of 300 spikes per second is marked by a spot. The maintained background activity of the cell was 32 Hz. The dotted line corresponds to the cell's receptive field. The physiologically defined receptive field extends roughly 100–150 μm beyond the dendritic tree. Reprinted by permission from Peichl and Wässle (1983).

A very similar structure-function relationship also holds for the anatomically defined β class of cat retinal ganglion cells in the periphery, where the dendritic tree of these cells is relatively large (Fig. 3.2). This cell type corresponds to the physiologically defined X type (Saito, 1983; Stanford and Sherman, 1984). The proportionality between receptive field size and the extent of the dendritic tree fails for small β type ganglion cells near the center of the retina. Here, the receptive field of the presynaptic neurons providing the driving input, bipolar cells, is already substantially larger in extent than the dendritic tree of the ganglion cell, obscuring any contribution the passive dendritic tree might make.

Besides this simple size relationship, what other functional aspects might the different retinal ganglion cell morphologies reflect? And what about possible structure-function relationships of other cell types? In the periphery, one can expect specific structural elements of the nervous system to reflect specific attributes of the physical world they are designed to extract. This is unlikely to be the case for more central structures. Can one still say something meaningful about how information processing is related to different dendritic trees (Borst and Egelhaaf, 1994)?

In this chapter, central to the book, we discuss how the previously introduced concepts from linear cable theory can be applied to study the electrical properties of complex, spatially extended dendritic trees. Why should the reader study passive dendritic trees when evidence from calcium imaging experiments, coupled with intradendritic recordings, shows that most, if not all, dendritic trees include substantial voltage-dependent, that is, nonlinear membrane components? This challenge can be answered by citing the truism "You can't run before you can walk." An understanding of information processing in active trees can only be based upon knowing the properties of passives ones. Furthermore, under certain well-specified conditions, for instance, for "small" synaptic inputs, voltage-dependent nonlinearities do not come into play and the dendrite acts essentially like a passive one. Let us therefore press on, delaying the confrontation with active trees until Chap. 19.

3.1 Branched Cables

An extended dendritic tree can be treated as a branched cable structure, where each cable or cylinder can, in principle, have different diameters and membrane properties. Since there exists no evidence for loops within a given dendritic tree structure, that is, one dendritic branch connecting—either directly or with the aid of a synapse—to another dendrite of the same cell, dendritic arbors can be considered to be true *trees* in the graph theoretic sense, with a unique path between any two points in the tree.

3.1.1 What Happens at Branch Points?

The cable equation (Eq. 2.7) as applied to a finite cable holds for any particular unbranched cylindrical segment with constant membrane properties and a constant diameter. How can we use this equation to study the potential in two or more such segments connected together? The way in which finite cable segments are linked resides in the appropriate choice of boundary conditions for each segment. In order to understand this better, let us focus on the highly simplified dendritic tree illustrated in Fig. 3.4A. We assume that the three branches have electrotonic lengths L_0, L_1, and L_2, with infinite input resistances $R_{\infty,0}$, $R_{\infty,1}$, and $R_{\infty,2}$. As discussed in Sec. 2.2.2, the *infinite input resistance* of a finite cable is equal to the input resistance of a semi-infinite cable of the same diameter and having the same R_m and R_i values, looking toward the infinitely far away terminal, and is computed according to Eq. 2.16.

We will assume, for the sake of simplicity, that the two rightmost branches terminate with a sealed-end boundary condition. The input resistances of the two rightmost daughter branches at the branch point, looking toward their terminals, are specified by Eq. 2.21 as

$$R_{\text{in},1} = R_{\infty,1} \coth(L_1) \tag{3.1}$$

and

$$R_{\text{in},2} = R_{\infty,2} \coth(L_2). \tag{3.2}$$

For an arbitrary boundary condition at the rightmost terminal of the daughter branches (rather than the sealed end assumed here) we replace these expressions by Eq. 2.28.

The right-hand side of the main cable ends at the branch point with the two daughter cables. Its terminal resistance $R_{L,0}$ is determined by the input resistances of the daughter branches, by making use of the fact that the net conductance of two parallel conductances is simply given by their sum. In other words, we compute the terminal resistance by effectively replacing the two conductances of the daughter branches by a single one. The effective terminal resistance of the main cable is given by

$$\frac{1}{R_{L,0}} = \frac{1}{R_{\text{in},1}} + \frac{1}{R_{\text{in},2}}. \tag{3.3}$$

A)

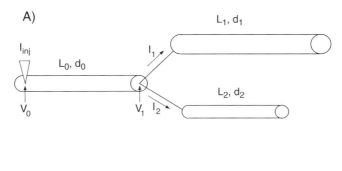

B)

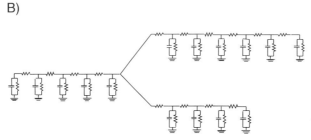

Fig. 3.4 PASSIVE BRANCHING DENDRITE (A) Schematic drawing of a passive cylindrical dendrite, with diameter d_0 and electrotonic length $L_0 = l_0/\lambda_0$, with two daughter branches, each with its distinct diameter and electrotonic length $L_i = l_i/\lambda_i$. A simple recursive scheme exists to compute exactly the voltage in such tree structures in response to current input (Rall, 1959). All terminals are assumed to be sealed. (B) Compartmental-model representation of this passive dendritic tree. The voltage dynamics in this circuit approximate the solution to the cable equation of the continuous cable in **A** in the sense that the cable equation (Eq. 2.7) describes the dynamics of the membrane potential in the limit that the grid size becomes infinitely fine. For the sake of simplicity, we set V_{rest} to zero (see Fig. 2.3). Standard software packages, such as NEURON or GENESIS, automatically solve for the voltage in these circuits.

In a final step, we apply Eq. 2.28 to arrive at the input resistance at the left terminal of the main branch (looking toward the daughter branches),

$$R_{\text{in},0} = R_{\infty,0} \frac{R_{L,0} + R_{\infty,0} \tanh(L_0)}{R_{\infty,0} + R_{L,0} \tanh(L_0)}. \tag{3.4}$$

The same algorithm can, of course, be applied in a recursive manner to evaluate the input resistance of arbitrary passive trees. Moving from the outermost dendritic tips and using either Eq. 2.21 or Eq. 2.24 (depending on the choice of boundary conditions), we compute the input resistances of the daughter branches. Adding the inverse of these input resistances gives the inverse of the terminal resistance of the branch of the parent generation. We now use Eq. 2.28 to compute the input resistance of this branch and its sibling, add them in parallel, and so on. Of course, this "folding" algorithm needs to be carried out on both ends of the cylinder. (In our pedagogical example in Fig. 3.4, we did not consider the effect of one or more cables connecting to the left terminal of the parent cable.) In this manner, the input resistance at any point in the tree can be evaluated exactly, a procedure proposed by Rall (1959) and described in considerable detail in Chap. 7 of Jack, Noble, and Tsien (1975).

   Once we compute the input resistance at any particular location in the dendritic tree, we can derive the steady-state voltage at that location in response to a current injection I_{inj} using Ohm's law. For instance, the voltage at $x = 0$ in the parent cable is

$$V_0 = R_{\text{in},0} \cdot I_{\text{inj}}. \tag{3.5}$$

Given V_0, we can use Eq. 2.27 to compute the sustained voltage change anywhere along the main branch in response to the sustained current injection I_{inj}. Furthermore, recursive evaluation of this equation allows us to compute the voltage at any location in the tree of Fig. 3.4. Altogether, four linear equations (Eqs. 2.27, 2.28, 3.3, and 3.5) need to be applied iteratively to solve for the input resistance, and therefore for the sustained voltage following a sustained current input, in any dendritic tree structure in response to a current injection at any location (Rall, 1959).

Assuming that the parent cable delivers the current I_0 to the two daughter branches, what fraction I_1 of this current will flow into one and what fraction I_2 into the other (Fig. 3.4A)? Using Eqs. 2.2 and 2.20 and the definition in Eq. 2.16, we compute the current flowing in one of the daughter cables, assuming that its terminal is sealed (this is not necessary),

$$I_1(X) = \frac{V_1}{R_{\infty,1}} \frac{\sinh(L_1 - X)}{\cosh(L_1)} \tag{3.6}$$

where V_1 is the voltage at the branch point. At this point, $X = 0$ (for the daughter branch), and the current flowing into this branch is given by

$$I_1(X = 0) = \frac{V_1}{R_{\text{in},1}} \tag{3.7}$$

where we made use of Eq. 3.1. Since the same principle applies to the second daughter branch and also holds true for an arbitrary boundary condition (in this case Eq. 2.20 must be replaced by Eq. 2.27), it follows that

$$I_0 = I_1 + I_2 = V_0(G_{\text{in},1} + G_{\text{in},2}) \tag{3.8}$$

where $G_{\text{in},1} = 1/R_{\text{in},1}$ and likewise for $G_{\text{in},2}$. In other words, the current divides among these branches according to their input conductances.

3.2 Equivalent Cylinder

Under certain conditions one does not need to solve numerous equations to derive the potential in a branched tree, a great simplification discovered and exploited by Rall (1962, 1964). Let us compute the voltage distribution in the simple tree shown in Fig. 3.4, assuming that the two daughter branches are identical (that is, $L_1 = L_2$ and $d_1 = d_2$ with identical membrane parameters) and that a sustained current is injected into the left terminal of the parent cylinder. The normalized voltage profile throughout this minitree is illustrated in Fig. 3.5 for three different combinations of parent and daughter branch diameters: (1) $d_0 = 2d_1$, (2) $d_0^{3/2} = 2 \cdot d_1^{3/2}$, and (3) $d_0 = d_1$. The upper and lower curves indicate the two limiting cases in which the parent cable is terminated at the right end either with a sealed-end boundary condition or in a short circuit. In the cases where the parent diameter has either twice the diameter or the same size as the daughter branches, close inspection of the curves in Fig. 3.5 reveals a discontinuity in the derivative of the voltage at the branch point.

In the third case, $d_0^{3/2} = 2d_1^{3/2}$, all derivatives of the voltage profile are continuous and a rather interesting phenomenon occurs. In contrast to the other two cases, the voltage decay along both sets of cables can be described by a single expression. To see why, let us return

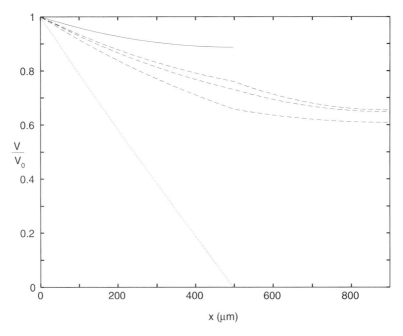

Fig. 3.5 VOLTAGE DECAY ALONG A SIMPLE DENDRITIC TREE Steady-state voltage decay along the tree shown in Fig. 3.4A, assuming that the two daughter branches are identical, with $d_1 = d_2$ and $L_1 = L_2 = 0.5$. The three dashed curves correspond to (1) $d_0 = 2d_1$, (2) $d_0^{3/2} = 2d_1^{3/2}$, and (3) $d_0 = d_1$. For the dotted curve, the potential at the right-hand terminal of the main branch is set to zero (short circuit). For the upper, continuous curve, the membrane is sealed. At the branchpoint $x = 500\,\mu$m, the voltage profile has a discontinuous derivative for $d_0 = 2d_1$ and for $d_0 = d_1$. If the input resistance of the parent at this point is matched to that of the daughter branches (as is the case for the second condition, $d_0^{3/2} = 2d_1^{3/2}$), the voltage decay across the cables can be described by a single, simple expression. This trick is exploited by Rall in his concept of the *equivalent tree*.

to Eq. 2.27, expressing the voltage in a finite cable of length L_0 with an arbitrary terminal condition specified via the terminal resistance $R_{L,0}$,

$$V(X) = V_0 \cdot \frac{R_{L,0}\cosh(L_0 - X) + R_{\infty,0}\sinh(L_0 - X)}{R_{L,0}\cosh(L_0) + R_{\infty,0}\sinh(L_0)}. \tag{3.9}$$

Since both daughter cables are identical and therefore have identical input resistance, $R_{in,1} = R_{in,2}$ (Eqs. 3.1 and 3.2), it follows from Eq. 3.3 that the combined input resistance of both cables is

$$R_{L,0} = \frac{R_{\infty,1}\coth(L_1)}{2}. \tag{3.10}$$

We now return to Eq. 2.16 for the input resistance of a semi-infinite cable, noticing its dependency on the inverse of the cable diameter to the $\frac{3}{2}$ power

$$R_{\infty,0} = \left(\frac{R_m \cdot R_i}{\pi^2}\right)^{1/2} \frac{2}{d_0^{3/2}}. \tag{3.11}$$

If $d_0^{3/2} = 2d_1^{3/2}$, then

$$R_{\infty,0} = \frac{R_{\infty,1}}{2}. \tag{3.12}$$

That is, the infinite input resistance of the parent cable is matched to the infinite input resistances of the two daughter branches in parallel. Combining Eqs. 3.10 and 3.12 leads to

$$R_{L,0} = R_{\infty,0} \coth(L_1). \tag{3.13}$$

If we place this back into Eq. 3.9, we have

$$\begin{aligned} V(X) &= V_0 \cdot \frac{\cosh(L_0 - X) + \sinh(L_0 - X)/\coth(L_1)}{\cosh(L_0) + \sinh(L_0)/\coth(L_1)} \\ &= V_0 \frac{\cosh(L_0 + L_1 - X)}{\cosh(L_0 + L_1)}. \end{aligned} \tag{3.14}$$

Comparing this to Eq. 2.20, we see that the above equation describes the voltage in a single, unbranched cylinder of length $L_0 + L_1$ with a sealed-end boundary condition. Instead of having to model this structure as three interconnected cylinders, we can reduce it to a single *equivalent cylinder*, whose electrotonic length is given by $L_0 + L_1$ and whose diameter is identical to the diameter of the main branch. The reason we are able to carry out this simplification is that the infinite input resistance of the main branch is "matched" to the infinite input resistance of the two daughter branches in parallel, a principle known as *impedance matching*. In particular, the voltage and the input resistance now have continuous derivatives across the branch point, as witnessed in Fig. 3.5. Note that, in general, impedance matching does not imply that the input resistance R_{in} of the daughter cables is matched to that of the main cable (since the input resistance of the main cable at the right terminal is proportional to $\coth(L_0)$ while the input resistance of the daughter branches is proportional to $\coth(L_1)$).

As Rall (1962, 1964) first showed in the early 1960s, an entire class of dendritic trees can be reduced or collapsed into a single equivalent cylinder provided the following four conditions are met. (1) The values of R_m and R_i are the same in all branches. (2) All terminals end in the same boundary condition. (3) All terminal branches end at the same electrotonic distance L from the origin in the main branch, where L is the sum of the L_i values from the origin to the distal end of every terminal; L corresponds to the total electrotonic length of the equivalent cylinder. (Note that this requirement does not necessarily include only trees with perfect symmetric branching.) (4) At every branch point, infinite input resistances must be matched. If all cables possess the same membrane resistance and intracellular resistivity, this implies

$$d_0^{3/2} = d_1^{3/2} + d_2^{3/2}. \tag{3.15}$$

This last condition is known as the $d^{3/2}$ law. If these four conditions are met, the equivalent cylinder mimics perfectly the behavior of the entire tree. In other words, injecting current into the leftmost terminal of the tree will yield exactly the same voltage response as injecting the identical current into its equivalent cylinder.

If input to any of the daughter branches is considered, an additional constraint must be obeyed. (5) Identical synaptic inputs, whether current injection or conductance change, must be delivered to all corresponding dendritic locations. In the case of Figs. 3.4 and 3.5, injecting a current I_0 into both daughter branches at the same electrotonic distance X_1 from the branch point corresponds to injecting the current $2I_0$ into the equivalent cable at location $X = L_0 + X_1$. If we are only interested in describing the voltage in the primary branch, because this is close to the cell body where the intracellular electrode is usually located, injecting the current $2I_0$ into only one of the two daughter branches will lead to the same voltage change in the primary branch as injecting the current $2I_0$ at the distance $L_0 + X_1$ in the equivalent tree (Rall and Rinzel, 1973). This ceases to be true when considering the voltage profile in other parts of the equivalent tree, outside of the region $X \leq L_0$.

The transformation from a branching cable structure to an equivalent cylinder preserves membrane area. In other words, the branching tree has the same membrane surface area as the unbranched cylinder that has the same diameter as the trunk of the tree and the same electrotonic length as the whole dendritic tree. This implies that all variables that are expressed per unit membrane area, such as channel densities, R_m, and C_m, are conserved in the transformation to an equivalent cylinder.

The principle of impedance matching is a powerful one and can be generalized to many situations. While here we have only considered the case of stationary inputs, the reduction of a tree to a cylinder can also be carried out for transient inputs, as long as the above four conditions are satisfied and C_m is constant throughout the entire tree. In other words, the dynamic behavior of the voltage is identical in the two cases. Furthermore, impedance matching can be generalized to tapering cables (Rall, 1962) and even holds in the presence of an active membrane containing voltage-dependent elements. In particular, axonal trees fulfilling the above conditions can similarly be reduced to a single equivalent axon.

Rall exploited this technique to great advantage in his extensive research on the electrical behavior of cat α motoneurons (Rall, 1977; Rall et al., 1992; see also Fleshman, Segev, and Burke, 1988), where the somatic potential of the cell in response to synaptic input in the dendritic tree is modeled by a single equivalent cylinder with the cell body being represented by an RC element tacked onto one end of the cylinder. Under these conditions, the electrical behavior of an entire neuron is characterized by a simple model with six free parameters (the R and C values of the isopotential soma, the R_m, C_m, and R_i of the equivalent cylinder, as well as its electrotonic length L).

Real dendritic trees rarely conform to all of the assumptions underlying the equivalent tree reduction. In some cases the $d^{3/2}$ rule is observed to a remarkable extent at branch points (e.g., relay cells in the cat's lateral geniculate; Bloomfield, Hamos, and Sherman, 1987). In general, neither this local rule dictating the diameter of branching processes nor the assumption that all dendrites terminate at the same distance from the cell body hold (Hillman, 1979; Burke, 1998; Larkman et al., 1992; Turner et al., 1995).

With a few exceptions, then, most neurons violate one or more of the constraints required for reducing a tree to an equivalent tree. In fact, one can ask why should dendritic trees obey the $d^{3/2}$ law at all? What functional advantages can be realized for a neuron if its dendritic tree can be reduced to a single equivalent cylinder? Voltage "reflections" at branch points will be absent if the $d^{3/2}$ law is obeyed, but it is unclear why this should be relevant to neurons.[2]

3.3 Solving the Linear Cable Equation for Branched Structures

Historically, much effort was spent on deriving solutions to linear differential equations like the cable equation in terms of series expansions of specialized functions (such as Bessel functions) and we will briefly touch upon these methods. Yet, given the widespread usage of digital computers and the complicated geometries of cells as well as their non-linear membrane characteristics, the majority of research today is being carried out using numerical methods.

---

2. Concepts from the *matched transmission lines* literature in electrical engineering are germane to some of these questions. We will not pursue any of the interesting analogies; see Davidson (1978) or Brühl, Jansen, and Vogt (1979).

3.3.1 *Exact Methods*

A number of iterative techniques have been developed to solve exactly for the voltage transient in treelike structures in response to an arbitrary current input $I(t)$. All of them rely on the superposition principle, that is, on the linearity of the membrane (although a linear membrane need not necessarily be passive; see Chap. 10).

Linearity of the membrane implies that the injection of a sinusoidal current of frequency f only gives rise to a voltage response at the same frequency. The above equations can be reformulated in the Laplace (or Fourier) domain to compute the potential $\tilde{V}(f)$ anywhere in the tree in response to a sinusoidal current $I(t) = I_0 \sin(2\pi f t)$ applied anywhere else. One set of methods (Butz and Cowan, 1974; Horwitz, 1981; Koch, 1982; Koch and Poggio, 1985a; Holmes, 1986) uses variants of this approach to derive $\tilde{V}_m(f)$ in arbitrary dendritic trees with the help of a small number of rules. Usually, these equations have simple graphical interpretations and can be implemented recursively, leading to very efficient and small programs.

For instance, Koch and Poggio's (1985a) four rules specify the impedance of a single cylinder (Eq. 2.28), the effective impedance at a branch point (Eq. 3.3), and the voltage at any point in a cylinder as a function of either the injected current (by combining Eqs. 2.25 and 2.26) or the voltage at one terminal (Eq. 2.27). One advantage of this method is that it can be applied to evaluate the potential in dendritic trees with arbitrary linear membranes, such as those containing inductances as can be obtained by linearizing certain active types of membranes (see Chap. 10). The number of equations that need to be evaluated in these methods is proportional to the number of cylindrical segments in the dendritic tree considered. Note that these methods require the inverse Laplace (or Fourier) transform to obtain $V(t)$.

A completely different approach was pioneered by Abbott and his group (Abbott, Farhi, and Gutmann, 1991; Abbott, 1992; Cao and Abbott, 1993), based on a path integral approach of the type used so successfully in quantum mechanics and quantum field theory (Feynman and Hibbs, 1965). The voltage at x in response to current input at y in an arbitrary tree (including trees with loops) is obtained by evaluating the Green's function of the cable (for example, Eq. 2.31) along all possible "paths" between x and y. The number of paths is potentially infinite, since a path, starting out at point x and heading toward y, may change direction at every node or terminal it encounters on its way and may pass through x and y an arbitrary number of times but must begin at x and end at y. Of course, the longer the path, the smaller its final contribution toward the potential. (The Green's function in Eq. 2.31 suppresses long paths of length L_{path} exponentially as $e^{-L_{\mathrm{path}}^2/4T}$.) We met such an approach while evaluating the potential in a single finite cylinder by summating over infinitely many reflecting terms (Eq. 2.41 and Fig. 2.10). The primary advantage of this method lies in computing $V(t)$ explicitly for short times, since under these conditions only as few as four paths need to be evaluated (Cao and Abbott, 1993). Furthermore, distinct from the methods working in the Laplace/Fourier domain, the path integral method yields the membrane potential directly as a function of time.

3.3.2 *Compartmental Modeling*

All the algorithms discussed so far solve for $\tilde{V}_m(f)$ and $V_m(t)$ in response to an arbitrary current input $\tilde{I}(f)$ and $I(t)$, respectively. They come, though, with serious drawbacks. (1) In the presence of n current inputs—rather than a single one—on the order of n^2 additional computations have to be performed (see Sec. 3.3); that is, these methods do not at all scale well to massive synaptic input. (2) While these methods can be adapted to treat synaptic input as conductance changes, it is not computationally efficient to do so. (3) Finally, and fatally, they assume linearity and fail in the presence of voltage-dependent membrane components.

The method of choice for most work in the field is to solve the partial differential equations numerically by discretizing the underlying equations. This approach, introduced by Rall in his landmark (1964) paper (see also Segev, Rinzel, and Shepherd, 1995), is known as *compartmental modeling* and leads to the discrete electrical circuits illustrated in Fig. 3.4B. Instead of solving the continuous, linear or nonlinear, partial differential cable equation, it is discretized into a system of ordinary differential equations, corresponding to small patches of neuronal membrane that are isopotential. These compartments are then coupled by sparse matrices.

The most fundamental requirement of any numerical method is *convergence*, that is, that the error between the solution using the numerical method and the exact solution be made as small as possible. Furthermore, the method must be stable. These requirements place constraints on how fine time and space must be discretized in order to arrive at an—approximately—correct solution. A conservative rule of thumb is to divide the cable into segments equal to or smaller than one-tenth of the associated effective length constant λ (Segev et al., 1985). How to achieve all of this properly and in an as efficient manner as possible is a well-established subject within applied mathematics. Appendix C, written by Barak Pearlmutter and Anthony Zador, treats modern numerical techniques—using the matrix formalism—for approximating the solution of these equations.

A number of public-domain single-cell simulators with graphical interfaces, in particular NEURON and GENESIS, implement such methods in an efficient manner. (For an overview of about half a dozen of them see DeSchutter, 1992; for more comprehensive references see Bower and Beeman, 1998 and Koch and Segev, 1998.) Indeed, the field has advanced to such a point that a set of benchmarks has been introduced to compare simulators in a more quantitative manner (Bhalla, Bilitch, and Bower, 1992).

The power of compartmental methods derives from their flexibility in permitting arbitrary levels of functional resolution to be included into the model. As long as the mechanism at hand, say a particular membrane conductance or pump, can be described by a steady-state or a differential equation, it can be incorporated into the model. Indeed, the numerical simulations discussed in this book range from the submillisecond to the second scale and from the submicrometer regime to several millimeters. Ultimately, the limiting factor in erecting ever more complex models is the exploding number of functions and parameters that need to be specified and measured as well as the speed of the computer implementing the algorithm.

3.4 Transfer Resistances

Let us now introduce a different, and much more intuitive, manner of calculating the voltage change in response to current input in passive dendritic trees of arbitrary geometry. Rather than expressing the properties of some electrical network in terms of an analytical expression figuring the various cellular parameters (e.g., the Green's function Eq. 2.31), the system is "observed" at one, two, or more discrete locations, and its properties are summarized in terms of a handful of functions called *transfer functions*. Their usage is based on multiport theory, in which a linear or nonlinear system is characterized by n ports, or independent pairs of current and voltage at each port (Oster, Perelson, and Katchalsky, 1971; Chua, Desoer and Kuh, 1987; Wyatt, 1992). Resistance, diode, capacitance, or inductance are all instances of one-port devices, since a pair of numbers, the voltage across the device and the current through it, completely captures their behavior, while the simplest model of a bipolar transistor is a three-terminal element (collector, emitter and base) with two pairs of current-voltage relationships. As we will see in Chap. 4, a chemical synapse can also be approximated as a two-port device (at a short time scale).

The two-port formulation of cable theory, in which voltage and current are manipulated at two locations in an arbitrary tree, has proven to be of great intuitive value. In particular, it allows the attenuation experienced by synaptic input on its way to the soma to be easily expressed. In the version popularized by Butz and Cowan (1974) (see also Fig. 3.6 and Koch, Poggio, and Torre, 1982; Carnevale and Johnston, 1982) it assumes that the dendritic tree is linear (although n-port theory does not require this straight-jacket).

3.4.1 General Definition

As long as one is dealing with a linear system, the voltage change $V_j(t)$ at location j in response to an arbitrary current input $I_i(t)$ at location i can always be expressed as

$$V_j(t) = K_{ij}(t) * I_i(t)$$

$$= \int_{-\infty}^{+\infty} K_{ij}(t')I_i(t - t')dt' = \int_{0}^{t} K_{ij}(t')I_i(t - t')dt', \qquad (3.16)$$

where $K_{ij}(t)$ is the *impulse response* or *Green's function* of the system (Fig. 3.6). The last relationship assumes that the impulse function is *causal*, implying that $K_{ij}(t) = 0$ for $t \leq 0$ (since no effect can precede its cause) and that no current is injected prior to $t = 0$.

The subscripts ij signify that the current is injected at i and the voltage recorded at j, a convenient mnemonic short form for summarizing the input-output relationship. Equation 3.16 holds for any cable structure and simply expresses the linearity of the system. The Green's function can be obtained by injecting a delta pulse of current into the cable, that is, $I_i(t) = I_0\delta(t)$, and computing $V_j(t)$ by solving the appropriate cable equation. This yields

$$K_{ij}(t) = \frac{V_j(t)}{I_0} \qquad (3.17)$$

with the dimension of a resistance. If location i coincides with location j, the equation reduces to

$$V_i(t) = K_{ii}(t) * I_i(t). \qquad (3.18)$$

One way to visualize the meaning of $K_{ij}(t)$ is to compute its Fourier transform (see Appendix B) $\tilde{K}_{ij}(f)$, where f is the temporal frequency in units of hertz. It is a complex

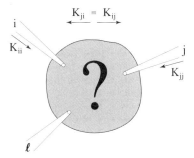

Fig. 3.6 TRANSFER FUNCTIONS Neuronal transfer functions are used throughout the book to characterize an arbitrarily complex, but linear neuron. The unknown system is observed at two points, or ports, i and j. The behavior at these two points can be fully described by knowledge of the three transfer functions K_{ii}, $K_{ij} = K_{ji}$, and K_{jj}. In particular, the attenuation of the input at i propagating to j can be easily expressed in terms of these functions. Its chief drawback is that these functions are not useful for expressing the voltage at some other location ℓ. This requires knowledge of $K_{i\ell}$ and $K_{\ell\ell}$.

function whose amplitude is measured in ohms. Any particular value of $\tilde{K}_{ij}(f)$ can be measured by injecting a sinusoidal current of frequency f at location i, that is, $I_i = \sin(2\pi f t)$, and recording the resultant voltage amplitude and phase at location j at the same frequency f.

The steady-state voltage change in response to a stationary current (that is, for $f = 0$) can be recovered with the aid of the real number

$$\tilde{K}_{ij}(f = 0) = \int_0^{+\infty} K_{ij}(t)dt . \tag{3.19}$$

We follow the practice of referring to the complex function $\tilde{K}_{ij}(f)$ as the *transfer impedance*, to its stationary or dc value $\tilde{K}_{ij}(f = 0) > 0$ as the *transfer resistance*, and to its inverse as the *transfer conductance*. Unless otherwise noted, $\tilde{K}_{ij}$ refers to $\tilde{K}_{ij}(f = 0) > 0$. Similarly, the amplitude of $\tilde{K}_{ii}(f)$ is called the *input impedance* and $\tilde{K}_{ii}(f = 0) > 0$ the *input resistance*.

3.4.2 An Example

In order to relate transfer resistances to the previously computed Green's functions, let us consider events in an infinite, passive cable of diameter d. Following Eq. 2.31 we can directly write

$$K_{ij}(T) = \frac{R_m}{2\pi d\lambda(\pi T)^{1/2}} e^{\frac{-(|i-j|/\lambda)^2}{4T}} e^{-T} \tag{3.20}$$

where $|i - j|$ is a shorthand notation to represents the distance between i and j, that is, between the sites where the current is injected and the voltage is recorded, and $T = t/\tau_m$ is the normalized time. To compute K_{ii}, we set $|i - j|$ to zero,

$$K_{ii}(T) = \frac{R_m}{2\pi d\lambda(\pi T)^{1/2}} e^{-T} . \tag{3.21}$$

The Fourier transform of $K_{ij}(T)$ yields

$$\tilde{K}_{ij}(f) = \frac{R_\infty}{2} \frac{1}{\sqrt{1 + i2\pi f \tau_m}} e^{-\sqrt{1+i2\pi f \tau_m}|i-j|/\lambda} \tag{3.22}$$

where R_∞ is the input resistance of a semi-infinite cylinder (Eq. 2.16). For the transfer resistance at $f = 0$ we arrive at

$$\tilde{K}_{ij} = \frac{R_\infty}{2} e^{-|i-j|/\lambda} \tag{3.23}$$

and for the input resistance,

$$\tilde{K}_{ii} = \frac{R_\infty}{2} . \tag{3.24}$$

The transfer resistance decays exponentially with the distance between the two sites and indicates the degree of coupling between the two sites i and j, which decreases with increasing distance.

3.4.3 Properties of $\tilde{K}_{ij}$

We will now assume that we are dealing with current inputs that either are stationary or change much more slowly than the membrane time constant τ_m. Since this corresponds to the solution of Eq. 3.16 with $t \to \infty$, the convolution is replaced by multiplication and the function $K_{ij}(t)$ is replaced by the dc component of its Fourier transform; that is,

$$V_j = \tilde{K}_{ij}(0) I_i = \tilde{K}_{ij} I_i \tag{3.25}$$

for the transfer resistance between i and j and

$$V_i = \tilde{K}_{ii}(0) I_i = \tilde{K}_{ii} I_i \tag{3.26}$$

for the input resistance at location i. Transfer and input resistances have several noteworthy properties.

Symmetry

The most important property is *symmetry*,

$$\tilde{K}_{ij} = \tilde{K}_{ji} . \tag{3.27}$$

If the current I_0 is injected at location i somewhere in the dendritic tree, and the voltage $V_0 = \tilde{K}_{ij} \cdot I_0$ is measured at location j, for instance, at the cell body, we obtain the same change in voltage V_0 at location i if the identical current I_0 is injected at location j: $V_i = \tilde{K}_{ji} \cdot I_j = \tilde{K}_{ij} \cdot I_0 = V_0$. This is true for any reciprocal linear system, irrespective of the positions of i and j. It is important to point out that this does not mean that the attenuation from location i to location j is the same as the other way around. Most emphatically not (see Sec. 3.5.2)!

Symmetry, in the electrical engineering literature also known as *reciprocity*, is a general property of n-ports created by interconnecting linear, two-terminal resistances. Its proof requires the use of Tellegen's theorem (Penfield, Spence, and Duinker, 1970; Reza, 1971) and is beyond the scope of this book.

Positivity

A second property is

$$\tilde{K}_{ij} \leq \tilde{K}_{ii} \quad \text{and} \quad \tilde{K}_{ij} \leq \tilde{K}_{jj} \tag{3.28}$$

from which it follows that the input resistance $\tilde{K}_{ii}$ is always larger than the transfer resistance between location i and any other location. Furthermore, the more removed i and j are, the smaller the transfer resistance $\tilde{K}_{ij}$. Thus, $\tilde{K}_{ij}$ is related to the degree of coupling of site i to site j.

Transitivity

A third property of any dendritic tree structure without loops is

$$\tilde{K}_{ij} = \frac{\tilde{K}_{i\ell} \tilde{K}_{\ell j}}{\tilde{K}_{\ell\ell}} \tag{3.29}$$

where ℓ corresponds to any location on the direct path between locations i and j. This relationship does not hold if ℓ is located on a branch off the direct path between i and j (Koch, 1982). For instance, if we consider the transfer resistance $\tilde{K}_{a3,s}$ between $a3$ and the soma in Fig. 3.7, Eq. 3.29 holds for points $a2$ and $a1$ but not for $a4$, $a11$, or $b1$. It can be viewed as a form of transitivity (as long as all three points are located along a direct path).

All three properties remain in force not only for the dc value of the transfer impedance but for any one frequency (e.g., $\tilde{K}_{ij}(f) = \tilde{K}_{ji}(f)$).

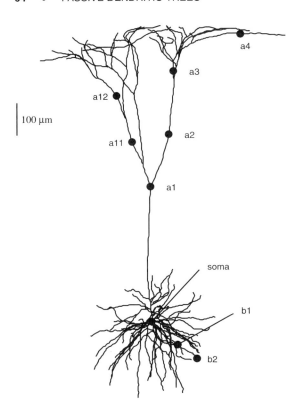

100 μm

Fig. 3.7 A LAYER 5 NEOCORTICAL PYRAMIDAL CELL Morphology of the neocortical pyramidal cell we use throughout this book. The cell was stained and reconstructed by Douglas, Martin, and Whitteridge (1991) from the visual cortex of an adult cat. The cell body is located in layer 5 while the most distal dendrites lie in layer 1. A few locations that are used as synaptic input sites are indicated by letters.

While these properties are true for any cable system with a linear membrane, an additional property of cables with passive membranes is that the membrane becomes increasingly more conductive to current flow as the temporal frequency increases due to the presence of the capacitive element. Or,

$$|\tilde{K}_{ij}(f)| < |\tilde{K}_{ij}(f')| \tag{3.30}$$

for any $|f| > |f'|$. In particular, $\tilde{K}_{ij}(0) > |\tilde{K}_{ij}(f)|$. In other words, the transfer and input impedances show low-pass behavior. If the membrane contains inductive-like elements, caused, for instance, by the small signal behavior of a voltage-dependent potassium conductance activated by depolarization, $|\tilde{K}_{ij}(f)|$ can have a maximum at some nonzero frequency, with interesting consequences for information processing (see Chap. 10).

3.4.4 Transfer Resistances in a Pyramidal Cell

In order for the reader to develop his or her intuition about these transfer resistances and other related concepts in the following section, we will make use of our "canonical" model of a nerve cell, a pyramidal cell from the mammalian neocortex. We choose this particular neuron both for idiosyncratic reasons as well as for the general interest many scientists have in understanding the information processing operations occurring in cortex. Most of the concepts discussed in this book apply equally to other cell types.

The morphology of this neuron (Fig. 3.7) is derived from a layer 5 pyramidal cell in primary visual cortex filled with the intracellular dye HRP during experiments in the anesthetized adult cat (Douglas, Martin, and Whitteridge, 1991) and is translated into several hundreds of compartments, similar to the ones shown in Fig. 3.4. The dendrites are passive while eight voltage-dependent membrane conductances at the cell body enable the cell to generate action potentials and shape the interspike interval between them accordingly (for details consult Bernander, 1993; Bernander et al., 1991, 1992, 1994). In this chapter we consider a linear model of the cell, obtained by replacing the active somatic membrane conductances by an appropriately chosen passive conductance, such that the input resistance and the resting potential throughout the cell are identical to the ones obtained in the full model. In other words, for small steady-state inputs this "passified" model behaves as the full, active model.

Unless otherwise noted, the ever-present spontaneous synaptic background activity has been incorporated into the passive membrane parameters. Because each transient opening of a synapse briefly increases the local membrane conductance, the random background activity will affect the "passive" membrane resistance R_m, with the effective membrane resistance varying from around 12,000 $\Omega \cdot cm^2$ at the soma to 52,000 $\Omega \cdot cm^2$ for distal dendrites (see Sec. 18.3). The membrane capacitance is set to 1 $\mu F/cm^2$ and the intracellular resistivity to 200 $\Omega \cdot cm$.

A number of the analytical techniques presented in Sec. 3.3 can be used to compute $\tilde{K}_{ij}$ in trees of arbitrary geometry, in particular the recursive techniques of Rall (1959), Butz and Cowan (1974), and Koch and Poggio (1985a). We refer the reader to these and other papers discussed in Sec. 3.3 and to Appendix C for more details. Table 3.1 shows representative values for $\tilde{K}_{ii}$ and $\tilde{K}_{is}$ for sustained and ac inputs. The steady-state input resistance $\tilde{K}_{ii}$ increases dramatically as one moves along the apical tree toward more distal sites, from 16.5 MΩ at the soma to several gigaohms in the far periphery. This is not unexpected, as we can see from Eq. 2.21 that the input resistance of a finite cable of electrotonic length L with a sealed end increases as $1/L$ as $L \to 0$, somewhat analogous to moving toward the distal dendritic tips. Using rapidly varying sinusoidal input of 10 or 1000 Hz reduces the input resistance moderately.

The transfer resistance to the soma $\tilde{K}_{is}$ (by Eq. 3.27 identical to the transfer resistance $\tilde{K}_{si}$ from the soma to the dendritic site i), on the other hand, only varies by about a factor

TABLE 3.1
Transfer and Input Resistances

Location	$\tilde{K}_{ii}$	$\tilde{K}_{is}$	$\|\tilde{K}_{ii}(10)\|$	$\|\tilde{K}_{is}(10)\|$	$\|\tilde{K}_{ii}(1000)\|$	$\|\tilde{K}_{is}(1000)\|$	L
soma	16.5	16.5	14.6	14.6	1.71	1.71	0
b1	388	15.5	383	13.6	71.4	0.18	0.24
b2	751	15.2	667	13.3	99.3	0.053	0.43
a1	63.4	13.9	50.7	11.1	6.48	0.052	0.33
a2	194	12.1	156	8.43	23.4	3.3×10^{-3}	0.52
a3	342	10.5	258	6.62	30.8	0.11×10^{-3}	0.75
a4	3298	7.8	2609	4.10	216	0.053×10^{-6}	1.31
a11	178	12.4	148	8.94	30.9	5.2×10^{-3}	0.49
a12	404	11.2	335	7.53	45.2	0.30×10^{-3}	0.69

Input and transfer resistances (in megaohms) between dendritic locations and the soma of the layer 5 pyramidal cell shown in Fig. 3.7. $|\tilde{K}_{is}|$ is computed for sustained as well as for 10 and 1000 Hz sinusoidal inputs. The voltage-dependent conductances at the soma were replaced by their slope conductances, rendering the cell passive but with the same steady-state input resistance and resting potential as the active cell. This is reflected in the low-pass behavior of $|\tilde{K}_{ij}|$ which decreases with frequency. The last column lists the normalized electrotonic distance between the synapse and the soma (computed according to Eq. 3.33). Notice the dramatic uncoupling between dendritic sites and the soma for rapidly varying input. For a discussion of the parameters used see Sec. 3.4.4.

of 2 between the most proximal and the most distal sites. Remember that the somatic potential induced by the current injection is given by the product of the injected current and the transfer resistance $\tilde{K}_{is}$. This implies that the coupling between dendritic sites and the soma depends only weakly on distance for dc or slowly varying current inputs. For rapidly varying inputs, on the other hand, the coupling will effectively be zero since the distributed capacitance throughout the tree will absorb the charge before it reaches the soma. We will develop this theme further in the following section.

To derive transfer resistances between dendritic sites we can use the third property of $\tilde{K}_{ij}$ (Eq. 3.29). For instance, to obtain the transfer resistance between two points in the apical tree, say $a3$ and $a1$, we use

$$\tilde{K}_{a3,s} = \frac{\tilde{K}_{a3,a1}\tilde{K}_{a1,s}}{\tilde{K}_{a1,a1}} \qquad (3.31)$$

yielding $\tilde{K}_{a3,a1} = 47.65\ M\Omega$.

3.5 Measures of Synaptic Efficiency

The introduction of the transfer impedance $\tilde{K}_{ij}(f)$, or, for stationary current input, the transfer resistance $\tilde{K}_{ij}(0) = \tilde{K}_{ij}$, allows us to characterize the voltage as well as the current attenuation between the site i of current injection and any other location in the dendritic tree, for instance, the soma s. As we will see, these are but two of a number of different measures characterizing the *efficiency*, or the *degree of coupling*, between any particular synaptic input site and the cell body where the initiation of the action potential occurs.

3.5.1 Electrotonic Distance
The simplest of all measures, yet a potentially misleading one, is an estimate of the electrotonic distance between any one location i and the soma. In an infinite cylinder or in a single finite cylinder of diameter d, the electrotonic distance between two points i and j is

$$L_{ij} = \frac{|i - j|}{\lambda} = |i - j|\left(\frac{dR_m}{4R_i}\right)^{-1/2}, \qquad (3.32)$$

where $|i - j|$ is a shorthand notation to represent the physical distance between the two points and λ the space constant. This expression immediately generalizes to the frequency-dependent electrotonic distance by replacing λ with $\lambda(f)$, as in Eq. 2.36. In as far as the cell under investigation can be approximated by a single equivalent cylinder (see Sec. 3.2; Rall et al., 1992), any point along this cylinder, corresponding to one or more equivalent locations in the original tree, can be assigned a meaningful electrotonic distance, say, relative to one end of the equivalent tree. However, as discussed in Sec. 3.2, real cells rarely satisfy the conditions necessary for them to be reduced to an equivalent cylinder.

One method to estimate the "distance" between i and the soma is to compute the electrotonic lengths of all the cylinders between i and the soma and add them up. In the continuous case, we have

$$L_{is} = \int_s^i \frac{dx}{\lambda(x)} \qquad (3.33)$$

where integration takes place along the direct path between the cell body and location i. In an infinite cylinder, there exists a simple exponential relationship between the electrotonic length and the voltage attenuation (Sec. 2.2.1). In all other structures, L_{is} does *not* provide a measure of the efficacy of signal transfer, since the underlying space constant λ assumes the convenient fiction of an infinite and constant cylinder. This can be seen graphically in Fig. 2.4 for the case of a single, finite cylinder. Depending on the type of boundary condition imposed, the actual voltage attenuation in a cable can be stronger (for a killed-end) or weaker (for a sealed-end boundary condition) than in an infinite cylinder with the same diameter and membrane parameters. This difference can be substantial in real trees, given the heavy load imposed by all the additional branching. Under these conditions, L_{is} seriously underestimates the attenuation experienced by dc or transient inputs on their way to the soma. For instance, the electrotonic distance between point $a4$ in the distal apical tree and the soma is 1.31 (Table 3.1), giving rise to the expectation that the attenuation from that point to the soma (assuming an infinite cylinder) should be $e^{L_{a4,s}} = 3.7$, yet the true voltage attenuation is around 423 for sustained current inputs and higher for transient inputs. Thus, for almost any dendritic tree, L_{is} fails to even be in the ballpark of the true voltage attenuation.

3.5.2 Voltage Attenuation

One of the most common measures is the *voltage attenuation* among two sites (Rall and Rinzel, 1973), that is, the ratio of the voltage at location i to the voltage at location j,

$$A_{ij} = \frac{V_i}{V_j}. \tag{3.34}$$

Since we are dealing with passive structures without a voltage-dependent membrane that can amplify the signal, the voltage at the input site i will always be larger than the voltage anywhere else in the tree: $A_{ij} > 1$. For the general case of a sinusoidal input current $I_i(f)$ of frequency f, we have (Koch, Poggio, and Torre, 1982)

$$\tilde{A}_{ij}(f) = \frac{\tilde{K}_{ii}(f)I_i(f)}{\tilde{K}_{ij}(f)I_i(f)} = \frac{\tilde{K}_{ii}(f)}{\tilde{K}_{ij}(f)}. \tag{3.35}$$

In general $\tilde{A}_{ij}(f)$ is a complex number. As usual, if we do not explicitly express the dependency of $\tilde{A}_{ij}$ on f, we refer to the dc component for $f = 0$.

This definition does not depend on whether this current injection was caused by a synapse or by an intracellular injection. The magnitude of $\tilde{A}_{ij}(f)$ describes by how much the voltage attenuates between i and j, with large values indicating large decrements. For a fixed input site i, $\tilde{A}_{ij}$ depends only on the transfer resistance $\tilde{K}_{ij}$. Figure 3.8 illustrates this dependency for the voltage attenuation between site $a3$ in the apical tuft (Fig. 3.7) and the rest of the cell as well as the attenuation from the soma to other points. Because the ordinate is specified in units of $\tilde{K}_{ij}$, this axis is inversely proportional to $\tilde{A}_{ij}$. This type of graph is quite instructive and was first used in Rall and Rinzel (1973) to demonstrate a number of important points.

1. The attenuation from any one point toward more distal points is small. Witness this when considering the attenuation away from the soma (lower curve) or the voltage decrement from $a3$ backward, that is, toward $a4$, or in a branch away from the soma (e.g., $a11$ and $a12$). This is not surprising, since the injected current has a much easier job depolarizing high-impedance sites than low-impedance ones.

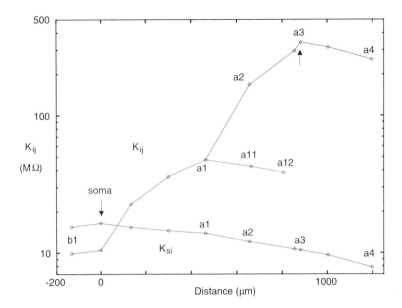

Fig. 3.8 **VOLTAGE AND CHARGE ATTENUATION IN A PYRAMIDAL CELL**   Two curves showing the steady-state transfer resistance $\tilde{K}_{ij}$ between a site in the apical tuft (at $a3$ in Fig. 3.7) and other parts of the tree (upper trace) as well as from the soma to dendritic sites for the layer 5 pyramidal cell. The upper curve is inversely proportional to the voltage attenuation from $a3$ to other sites in the tree (Eq. 3.34), while the lower curve is inversely proportional to the charge attenuation between sites in the tree and the soma (Eq. 3.47). Note the pronounced asymmetry between large voltage attenuation but much smaller charge attenuation. The input resistances $\tilde{K}_{a3,a3}$ and $\tilde{K}_{s,s}$ are indicated by arrows.

2. On the other hand, the voltage decrement at sites on the way to the soma, a very low impedance site, is substantial. For instance, $\tilde{A}_{a3,s} = 341.68/10.46 \approx 33$: if a synaptic input to $a3$ causes a local sustained EPSP of 10 mV, the sustained somatic EPSP will be only 0.3 mV.

3. For rapid synaptic input, the actual attenuation will be larger due to the distributed capacitance that "soaks" up the high-frequency components on their way to the cell body. When the input consists of a rapid and transient current (peaking at 0.5 msec) into the dendrite at $a3$, the effective $\tilde{A}_{a3,s}$, defined as the peak of the local potential (here equal to 4.86 mV) divided by the peak somatic potential (0.051 mV), is 94, about three times larger than for sustained inputs.

4. Both sustained as well as transient voltage attenuation can be considerably larger. For the distal site $a4$, the dc attenuation alone is already 423.

5. The attenuation away from the soma is considerably smaller (Rinzel and Rall, 1974). For instance, $\tilde{A}_{s,a3}$ is only 1.6, about twentyfold less than the voltage attenuation in the reverse direction (see also the lower curve in Fig. 3.8). In fact, except for an infinite cylinder,

$$\tilde{A}_{ij} \neq \tilde{A}_{ji} . \tag{3.36}$$

This relationship appears to run counter to the symmetry of the transfer impedance, but upon closer inspection this discrepancy is resolved: Eq. 3.27 expresses symmetry between current input and voltage output, while Eq. 3.36 expresses lack of symmetry between voltages at two different sites. The symmetry of $\tilde{K}_{a3,s}$ is expressed in the fact that the location marked *soma* on the upper curve in Fig. 3.8 has the same numerical value as the location marked $a3$ on the lower one (Rall, 1989).

The physical reason for the pronounced asymmetry between $\tilde{A}_{is}$ and $\tilde{A}_{si}$ is the large difference in input impedance between dendritic sites and the soma. While most of the current flows toward the low-impedance (or high-conductance) cell body (as expressed by Eq. 3.8), this current causes a much smaller change in the membrane potential at the soma than at high-impedance locations. This fact also explains why the voltage attenuation between $a3$ and the most distal part of the apical tree ($a4$) is comparatively minor. Given thin branches and sealed-end boundary conditions, only a very small voltage gradient exists. This is important to keep in mind when trying to understand the antidromic spread of the somatic action potential to dendritic sites.

By exploiting Eq. 3.29, we can express the voltage attenuation from i to j using any point ℓ lying on the direct path between i and j as a function of the attenuation between i and ℓ and between ℓ and j,

$$\tilde{A}_{ij} = \frac{\tilde{K}_{ii}}{\tilde{K}_{ij}} = \frac{\tilde{K}_{ii}\tilde{K}_{\ell\ell}}{\tilde{K}_{i\ell}\tilde{K}_{\ell j}} = \left(\frac{\tilde{K}_{ii}}{\tilde{K}_{i\ell}}\right)\left(\frac{\tilde{K}_{\ell\ell}}{\tilde{K}_{\ell j}}\right) = \tilde{A}_{i\ell}\tilde{A}_{\ell j}. \tag{3.37}$$

This property allows us to introduce a sort of pseudo *attenuation metric*. Following Zador (1993), Zador, Agmon-Snir, and Segev (1995) and Brown et al., (1992), we define the *log attenuation* as

$$\tilde{L}_{ij}^{v}(f) = \ln|\tilde{A}_{ij}(f)| \tag{3.38}$$

using the natural logarithm. Note that to differentiate this dimensionless variable from the electrotonic distance, it carries a superscript. Because $|\tilde{A}_{ij}| \geq 1$,

$$\tilde{L}_{ij}^{v} \geq 0 \tag{3.39}$$

as expected for a distance. We now make use of Eq. 3.37 and have for any point ℓ on the direct path from i to j,

$$\tilde{L}_{ij}^{v} = \tilde{L}_{i\ell}^{v} + \tilde{L}_{\ell j}^{v}. \tag{3.40}$$

Additivity constitutes an elegant feature of $\tilde{L}_{ij}^{v}$, rendering it somewhat similar to a distance measure. Note that $\tilde{L}_{ij}^{v}$ is not a true metric, since symmetry does not hold due to the asymmetry in the voltage attenuation,

$$\tilde{L}_{ij}^{v} \neq \tilde{L}_{ji}^{v}. \tag{3.41}$$

In the special case of an infinite cylinder, it follows from Eq. 3.23 that

$$\tilde{A}_{ij} = \tilde{A}_{ji} = e^{+|i-j|/\lambda} \tag{3.42}$$

and, therefore,

$$\tilde{L}_{ij}^{v} = \tilde{L}_{ji}^{v} = \frac{|i-j|}{\lambda}. \tag{3.43}$$

Because the logarithm of the voltage attenuation between two points in an infinite cable equals the electrotonic distance between them, $\tilde{L}_{ij}^{v}$ can be viewed as a generalization of the traditional notion of electrotonic distance to arbitrary cable structures (Zador, Agmon-Snir, and Segev, 1995).

While voltage attenuation has been and continues to be a popular measure of synaptic efficiency, it should not be forgotten that it represents a relative measure. And whether or not the cell spikes by exceeding a threshold voltage depends on the absolute amplitude of the somatic EPSP and not on the relative ratio of voltages.

3.5.3 Charge Attenuation

Another measure of synaptic efficiency is the ratio of the charge transferred across the membrane at the location of the input i to the charge reaching location j (Koch, Poggio, and Torre, 1982). The total charge transferred at i due to the current $I_i(t)$ is

$$Q_i = \int_{-\infty}^{+\infty} I_i(t)dt \ . \tag{3.44}$$

This equals the steady-state value of the Fourier transform of the current

$$\tilde{I}_i(f = 0) = \int_{-\infty}^{+\infty} I_i(t)dt = Q_i \ . \tag{3.45}$$

The charge transferred across the membrane at j is

$$Q_j = \int_{-\infty}^{+\infty} I_j(t)dt = \tilde{I}_j = \frac{\tilde{K}_{ij}}{\tilde{K}_{jj}} \tilde{I}_i \ . \tag{3.46}$$

The *charge attenuation* from i to j, defined by analogy to the voltage attenuation, is

$$\frac{Q_i}{Q_j} = \frac{\tilde{K}_{jj}}{\tilde{K}_{ij}} = \tilde{A}_{ji} \ . \tag{3.47}$$

Or, the charge attenuation from i to j is identical to the voltage attenuation in the reverse direction (for this reason, Zador, Agmon-Snir, and Segev, 1995, also refer to the charge attenuation as *outward going attenuation*). This allows us to use the lower trace in Fig. 3.8, which is proportional to the voltage decrement $\tilde{A}_{si}$ from the soma to dendritic sites, as an indicator for the charge attenuation between these sites and the soma. For the parameters used in this model, $\tilde{A}_{si}$ is usually below 2, even for the most distal sites in the layer 1 dendrites. It should not be forgotten that $\tilde{A}_{si}$ is a relative measure and does not tell us anything about the absolute amount of charge transferred from the synapse i to the cell body. Due to the equivalence of charge attenuation and inverse voltage attenuation, the logarithmic transform of Eq. 3.38 can be used to define a pseudodistance.

The charge attenuation has, in contrast to the voltage attenuation, the distinction that it is independent of the time course of the current $I_i(t)$. It simply provides an index of the relative amount of charge reaching the cell body. Of course, this measure also stipulates that the period of integration extends to infinity, which means in practice several time constants, until all ions have charged and discharged the neuronal membrane between i and the soma.

It is also possible to define the *current attenuation* as the ratio of the input current flowing at location i and the current flowing across the neuronal membrane at j,

$$\frac{I_i(f)}{I_j(f)} = \frac{\tilde{K}_{jj}(f)}{\tilde{K}_{ij}(f)} = \tilde{A}_{ji}(f) \ . \tag{3.48}$$

This implies that the steady-state current attenuation between i and j is identical to the charge attenuation between these two sites. It is important to understand here that the current I_j does *not* correspond to the intracellular current that flows from i to j. Rather, I_j is the current that needs to be injected across the membrane at location j in order to change the potential at j by the same amount as the current i does (namely, by $\tilde{K}_{ij} I_i$).

3.5.4 Graphical Morphoelectrotonic Transforms

It is more and more common that the detailed, three-dimensional morphology of reconstructed neurons is available in computer readable form. It therefore becomes possible to reexpress the geometry of a specific tree using different measures, for instance, the electrotonic length, the log attenuation pseudodistance measure, or the dendritic delay. As a group, these transforms, introduced by Zador (1993; Zador and co-workers 1992, 1995; see also Bernander, Douglas, and Koch, 1992), have been called *morphoelectrotonic transforms* (METs). In a MET, the anatomical length of each dendritic cable segment is replaced by one of these various measures, while the diameter and the orientation of the segment are preserved. In this way, different aspects of the electrotonic structure of the cell can be directly visualized. Color can be used as a further visual aid to identify corresponding points between different METs and the original anatomy.

Figure 3.9 illustrates these procedures without the use of color. The original anatomy, derived from projecting the three-dimensional digitized coordinates of the tree onto the image plane, is shown in (A). In (B), the anatomical length of each compartment is replaced by its electrotonic length. Therefore, the distance between a point i and the soma corresponds to L_{is} (Eq. 3.33). In particular, the thin basal and distal apical dendrites are accentuated at the cost of the thick apical trunk, since thinner cables have shorter space constants and are therefore electrotonically longer. The electrotonic length does *not* correspond to the strength of coupling between i and the cell body, since the use of L is conditioned upon the space constant λ that is defined for an infinite cylinder. With the total electrotonic length of the cell being on the order of one λ, we would expect that stationary signals from the periphery decay by a factor of $1/e$, while their actual decay is more substantial (Table 3.1).

The log transform of the voltage attenuation from the dendritic tree to the soma ($\tilde{L}_{is}^v$; Fig. 3.9C) and from the soma to dendritic sites ($\tilde{L}_{si}^v$; Fig. 3.9D) emphasizes the large differences (about a factor of 10) between incoming and outgoing voltage decrement. The outgoing voltage attenuation MET in Fig. 3.9D corresponds to the logarithmic transform of the charge attenuation from sites in the dendritic tree to the soma. The transforms are functional, in the sense that the voltage transfer can be directly read off from the figure. (This is a consequence of the additivity of the $\tilde{L}^v$ measure, as expressed in Eq. 3.40.) For instance, the voltage attenuation from $b2$ at the tip of a basal dendrite is $e^{3.9} \approx 50$, while the charge injected at the same site will only attenuate by $e^{0.08} \approx 1.09$. This also means that a long-lasting hyperpolarization at the cell body of 10 mV will hyperpolarize the entire basal tree, an area that accounts for 62% of the cell's membrane area, by at least 9.5 mV. Since the most distal sites in layer 1 of the apical tree have an $\tilde{A}_{si}$ value of about 2.1, even these parts will be hyperpolarized by about 4.8 mV.

Zador, Agmon-Snir, and Segev (1995) offer another interpretation of $\tilde{L}_{si}^v$. A frequently asked question relates to the *cost* of moving a synapse from the soma to a particular location i in the tree. Assuming that the synaptic input can be treated as a current and is identical to I_0 at both locations, the voltage induced by the somatic location is given by $\tilde{K}_{ss} I_0$, while the somatic potential in response to the dendritic input is $\tilde{K}_{is} I_0$. The ratio of the

A) Morphology

B) L

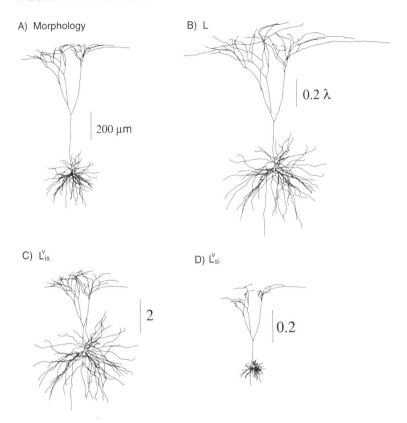

200 μm

0.2 λ

C) $\overset{v}{L}_{is}$

D) $\overset{v}{L}_{si}$

2

0.2

Fig. 3.9 MORPHOELECTROTONIC TRANSFORMS Graphical means of visualizing different measures of synaptic efficiency using morphoelectrotonic transforms (METs) (Zador, 1993). **(A)** Two-dimensional projection of the layer 5 pyramidal cell (see also Fig. 3.7). **(B)** The physical length ℓ of each dendritic compartment is replaced by the electrotonic length $L = \ell/\lambda$, while the orientation of the compartment remains constant. In an infinite cable, the voltage attenuation between two sites a distance L apart is e^{-L}. But because the electrical structure of the dendritic tree is quite distinct from that of an infinite cable, the voltage decay from dendritic sites to the soma is much larger than indicated by this measure. **(C)**, **(D)** METs corresponding to the logarithmic voltage attenuation from sites i in the tree toward the soma, $\tilde{L}_{is}^{v} = \ln|\tilde{A}_{is}|$, and from the soma to the tree, $\tilde{L}_{si}^{v} = \ln|\tilde{A}_{si}|$, for stationary input. The latter measure is equivalent to the charge attenuation, that is, to the ratio of the charge injected at i to the charge reaching the soma (after integrating for a time that is long compared to τ_m). Notice the very different scales (corresponding to the distance over which the signal attenuates by a factor of $e^{2.0} = 7.39$ and $e^{0.2} = 1.22$, respectively), attesting to the fact that neurons are very compact from the point of view of charge but very dispersed from the point of view of voltage attenuation. These METs provide a snapshot of the cell's electroanatomy for a particular configuration of electrical parameters.

membrane potential induced by the somatic input to the EPSP induced by the dendritic input is $\tilde{K}_{ss}/\tilde{K}_{is} = \tilde{A}_{si}$. As an example, moving a sustained current injection from the soma to site $a4$ in the distal apical tree, reduces the effect of this input by only a factor of 2.1. Of course, for rapidly varying input, the cost of moving to the dendrites can be substantially higher (for the same synapse, a factor of about 4 for input changing at the 10-Hz scale and a factor of 20 million for input varying at the kilohertz scale).

If transient input is used at the soma, the outgoing voltage attenuation is much larger. This is demonstrated in Fig. 3.10 for a sinusoidal input of 1000 Hz. This frequency corresponds approximately to the inverse of the width of a typical action potential and provides us with some intuition for how far the very rapid portion of a somatic action potential will propagate into a passive tree (a phenomenon known as *antidromic spike invasion*). While the somatic potential in response to current input at a distal, apical site (*a*4 in Fig. 3.7) is 423 times smaller than the dendrite EPSP for stationary inputs (but only losing half of the injected charge in the process) and 630 times smaller for a 10-Hz sinusoidal, it is attenuated by a staggering factor of 4 billion for a 1000-Hz sinusoidal input. In other words, without dramatic amplification by active processes, a distal high-frequency input has no chance of ever influencing anything occurring close to the soma (see Chap. 19).

A further feature of METs is that they reflect the momentary state of the cell, a state that can be modulated by any number of different events in the brain. For instance, as discussed in Sec. 18.3.3, varying the synaptic background—corresponding to effectively varying the membrane resistance R_m—modulates the electrotonic geometry of the cell by changing the L_{is} or $\tilde{L}_{is}^{v}$ (see Fig. 18.6). Or if some brainstem afferent releases a neuromodulator, such as noradrenaline or acetylcholine, onto part of the cell, its electroanatomy can shrink or grow, reflecting the corresponding change in the efficiency of synaptic inputs.

Figure 3.11, from Zador, Agmon-Snir and Segev (1995), provides a graphical comparison of $\tilde{L}_{si}^{v}$ for four different neurons with identical, albeit passive, membrane properties. The differences between them are dramatic. Given the large diameter of the Purkinje cell dendrites, almost no charge attenuation occurs for sustained current inputs. The electrotonic behavior of the layer 2/3 pyramidal cell is very similar to the behavior of the basal tree of the layer 5 cell. Finally, the apical tree of the hippocampal pyramidal cell sprouts into a long, central stalk across which substantial charge attenuation occurs.

A) $\overset{v}{\tilde{L}}_{is}$ (1000)   B) $\overset{v}{L}_{si}$ (1000)

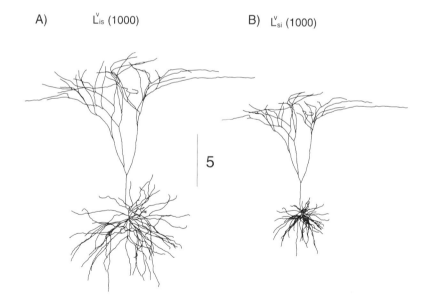

5

Fig. 3.10 Dynamic Morphoelectrotonic Transforms (A) Logarithm of the voltage attenuation, $\ln |\tilde{A}_{is}(f)| = \tilde{L}_{is}^{v}(f)$ for a 1000 Hz input. For such transient input, the most distal site is 27 units away from the cell body (for stationary inputs, the corresponding distance is 6.7; see Fig. 3.9C). (B) Logarithm of the voltage attention $\tilde{L}_{si}^{v}(f)$ from the soma to dendritic sites for the same 1000-Hz input. This frequency has been choosen to demonstrate how far the fast components in a somatic action potential (with an approximate width of 1 msec) can propagate back into the dendritic tree.

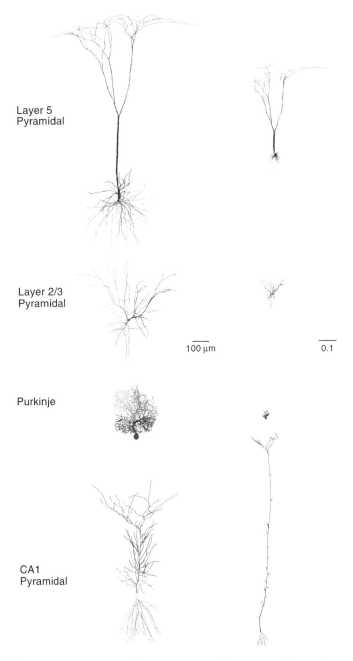

Layer 5
Pyramidal

Layer 2/3
Pyramidal

100 μm   0.1

Purkinje

CA1
Pyramidal

Fig. 3.11 MORPHOELECTROTONIC TRANSFORMS FOR DIFFERENT CELL TYPES Comparison of the logarithm of the charge attenutation—corresponding to the log-voltage attentuation from the soma to dendritic sites for stationary inputs—for four different neurons: the layer 5 neocortical pyramidal cell used throughout, a layer 2/3 neocortical pyramidal cell, a Purkinje cell from the cerebellum of a guinea pig, and a CA1 pyramidal cell from rat hippocampus (Zador, Agmon-Snir, and Segev, 1995). The left column corresponds to the morphology and the right one to $\hat{L}^{v}_{si}(0)$. The scale bar, indicating the distance over which the voltage attenuates by $e^{0.1} \approx 1.1$, applies to all cells. To facilitate comparison, all neurons are passive with identical membrane parameters ($R_m = 100,000\ \Omega\cdot\text{cm}^2$, $R_i = 100\ \Omega\cdot\text{cm}$). Reprinted in modified form by permission from Zador, Agmon-Snir, and Segev (1995).

3.6 Signal Delays in Dendritic Trees

We discussed in Sec. 2.3 transient inputs and their effect on the voltage in single cables. The most important parameter regulating how fast the membrane potential equilibrates is the neuronal *time constant* τ_m. But how is τ_m measured experimentally in dendritic trees? Following Agmon-Snir and Segev (1993), we also introduced a new measure of neuronal delay in terms of the difference between the centroid or center of mass of the current input and the voltage signal. How do these definitions fare in spatially extended structures?

3.6.1 Experimental Determination of τ_m

A general property of the voltage response to current input into an arbitrary complex, passive cable structure (whether a single finite cylinder or a large dendritic tree) with uniform electrical parameters throughout the tree is that it can always be expressed as an infinite sum of exponentials (Rall, 1969a; for more details, consult Appendix C),

$$V(x, t) = V_\infty(x) + B_0 e^{-t/\tau_0} + B_1 e^{-t/\tau_1} + B_2 e^{-t/\tau_2} + \cdots, \tag{3.49}$$

where the B_i's depend both on the initial conditions and on x, and V_∞ captures the steady-state components of the voltage. The τ_i's can be thought of as *equalizing constants* that govern the rapid flow of current (and the reduction of voltage differences) between different regions of the cable. They are independent of the site of the current input, the site of recording, or the initial voltage distribution in the tree. In general, solving for these τ_i's involves extracting the root of a recursively defined transcendental equation. (For more recent work on this, see Holmes, Segev, and Rall, 1992; Major, Evans, and Jack, 1993a.)

Several important points need to be mentioned here. First, in a tree with uniform membrane properties with sealed ends (and without a shunt or a voltage clamp) the slowest time constant τ_0 is always equal to the membrane time constant, $\tau_0 = \tau_m = R_m C_m$. Second, all time constants scale with the membrane capacitance C_m. Third, introducing a shunt conductance at the soma—mimicking the effect of an intracellular electrode—reduces all the time constants, including τ_0. In other words, under realistic experimental conditions, the measured slowest τ_0 represents a lower bound on the actual value of τ_m. As evident in Eq. 2.40, these τ_i's have a particularly pleasing and simple interpretation for a single finite passive cable with sealed-end boundaries.

Direct measurements of τ_m in neurons depended on the development of the intracellular electrode, introduced by Graham and Gerard (1946) for muscles and first applied to central mammalian neurons (spinal motoneurons) by Woodbury and Patton (1952) and Brock, Coombs, and Eccles (1952). These early studies neglected the cable properties of the dendrites and estimated $\tau_m = 2$ msec. A common numerical technique for extracting time constants involves the *peeling* of low-order exponentials from a semilogarithmic plot of the voltage decay (Rall, 1969a,b; see Fig. 3.12). It was first applied to the problem of extracting time constants in α motoneurons by Burke and Ten Bruggencate (1971).

They estimated that τ_m ranges from 5 to 7 msec. Averaging over 50 voltage transients allowed them to determine not only $\tau_0 = \tau_m$ but also the next slowest time constant τ_1, which they calculated to be around 1 msec. Assuming that the dendritic tree of the motoneuron can be reduced to a single, equivalent cable, allowed Burke and Ten Bruggencate (1971) to exploit Eq. 2.42 to estimate the electrotonic length at around 1.5.

Experimentally, τ_m can be determined by injecting a brief current pulse into the soma and recording the voltage response at the same point. Long current steps tend to activate

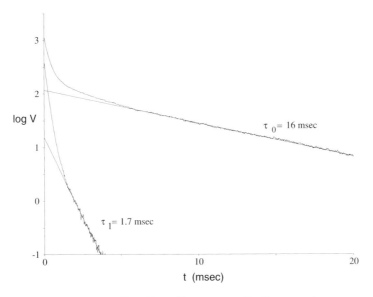

Fig. 3.12 ESTIMATING THE FIRST TWO TIME CONSTANTS   Peeling procedure to assess the time constants of voltage decay in response to a small and short current pulse. The data shown here were obtained by recording from a pyramidal cell located in the upper layers in the frontal cortex of the guinea pig. The voltage response is shown as a logarithmic function of time. Due to the nonisopotential nature of the dendrites, the voltage decays initially rapidly; the slowest time constant $\tau_0 = \tau_m$ is only evident toward the tail of the response, with $\tau_0 = 16$ msec. Peeling it away reveals the second slowest time constant $\tau_1 = 1.7$ msec. The recordings were carried out at the cell's resting potential of -65 mV in the laboratory of Yosef Yarom. Reprinted by permission from Koch, Rapp, and Segev (1996).

voltage-dependent components (e.g., the over- and undershoot dealt with in detail by Ito and Oshima, 1965) that severely interfere with the measurement of the passive time constant. In a linear system, the response to a current pulse is given by the temporal derivative of the response to a step. Because the derivative of Eq. 3.49 still leaves an infinite number of exponential terms with the slowest term decaying as e^{-t/τ_m}, the peeling method is also applicable to the derivative of $V(t)$ in Eq. 3.49 with the advantage that the steady-state component of the voltage disappears. The slope of the tail of the decaying phase of V_m, when plotted on a semilogarithmic scale, is $-1/\tau_m$. Figure 3.12 illustrates this procedure for a voltage transient in a guinea pig neocortical pyramidal cell, showing how the voltage gradients over the cell surface equalize after about 6–8 msec; from this time on, the voltage decays exponentially. The slowest time constant $\tau_0 = 16$ msec corresponds to the inverse of the slope at the tail end of the voltage distribution (see also Kim and Connors, 1993). Subtracting this component from the original voltage curve gives rise to a new transient (noisy curve at bottom) whose tail can now again be fitted with a straight line whose slope is $-1/\tau_1$. In this case $\tau_1 = 1.7$ msec. If we assume that the membrane of this cell is uniform and that the intracellular electrode did not cause a significant shunt (injury), $\tau_0 = \tau_m$.

It is important to emphasize that the quality of the electrical recording depends to a large extent on the tightness of the resistive *seal* around the electrode. During the impalement of the neuronal membrane, the intracellular electrode frequently rips a large hole in the membrane, allowing ions to flow through, thereby seriously compromising the quality of the recording. The introduction of the whole-cell recording of *in vitro* cells and the perforated patch clamp technique (Edwards et al., 1989; Spruston and Johnston, 1992) has dramatically improved the situation with an effective electrode shunt on the order of 0.1 nS. Indeed, over the last four decades the estimate for τ_m in central neurons has grown significantly. In the 1950s it was assumed to be only a few milliseconds. With improved averaging and recording techniques, the recent estimates of τ_m from intracellular recordings range from 10 to 40 msec for the major types of central neurons. (See Fleshman, Segev, and Burke, 1988 and Clements and Redman, 1989, for α motoneurons, Brown, Fricke, and Perkel, 1981, for hippocampal neurons, Nitzan, Segev, and Yarom, 1990, for vagal motoneurons and Rapp, Segev, and Yarom, 1994, for cerebellar Purkinje cells.) With tight-seal whole-cell recordings, the estimates are growing even further and are approaching 100 msec in slice preparations (e.g., Andersen, Raastad and Storm 1990 found time constants ranging from 50 to 140 msec in the hippocampus; Major et al., 1994 report an average value of 93 msec in CA3 pyramidal cells; see also Spruston, Jaffe, and Johnston, 1994). A notable exception to such high values is the 2-msec time constant in slices in the avian cochlear nucleus responsible for processing high-fidelity sound information, measured by using whole-cell recording with tight seal (Reyes, Rubel, and Spain, 1994). We conclude this section by noting that τ_m estimates depend heavily on the composition of the physiological solution during the experiment. Adding specific blockers to the solution, as is frequently done (e.g., caesium ions to block K^+-dependent rectification or blockers of NMDA-dependent channels), directly affects (typically increases) estimates of τ_m.

3.6.2 Local and Propagation Delays in Dendritic Trees

The problem with the standard definition of τ_m is twofold. Firstly, because τ_m is a property of the membrane it does not take into consideration the effect of the neuronal structure on the dynamics of the membrane potential at the input site (which can be the soma, a thin dendrite, or an elongated spine). Secondly, τ does not inform us about how rapidly the membrane can respond to more physiological input than a current pulse or step. Both problems were addressed by Agmon-Snir and Segev (1993) (see also Zador, Agmon-Snir, and Segev, 1995). In Chap. 2, we defined the *transfer delay* D_{ij} to be the difference in the centroid of the induced voltage at location j and the centroid of the current flowing at i (Eq. 2.45) and the *input delay* D_{ii} as the difference between the centroids of the local voltage and the injected current at the same location (Eq. 2.46). For an infinite cable, the transfer delay $D_{ij} = (1 + |i - j|/\lambda)\tau_m/2$ (Eq. 3.50).

The center of mass method can be applied to calculate analytically the local delay at any point in an arbitrary dendritic tree. Figure 3.13 illustrates a few delays arising in the pyramidal cell for fast current injections of the form expected during activation of fast voltage-independent synaptic input. The time course follows that of the α function introduced in Eq. 1.21,

$$I_{\text{syn}}(t) = \frac{I_{\text{peak}}}{t_{\text{peak}}} t e^{1-t/t_{\text{peak}}} . \tag{3.50}$$

(We here approximate the synaptic input by a current source.) The current attains its peak value I_{peak} at $t = t_{\text{peak}}$. Values for the latter are in the 0.25–1-msec range, depending on the exact circumstances (for more details see Sec. 4.6). The centroid of the current is located at $2t_{\text{peak}}$.

The local delay D_{ii} reflects the RC properties of the local membrane as well as the electrotonic structure of the cell as seen from location i. It is maximal (and equal to τ_m) for the membrane patch case (Eq. 2.48), since here the input current can only discharge through

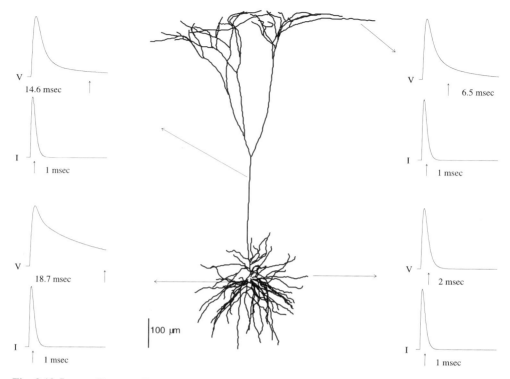

Fig. 3.13 LOCAL DELAY Four examples of the local delay D_{ii} defined as the difference between the centroid of the local voltage (upper curves in the four panels) and the centroid of the local current injection that gave rise to this depolarization (lower curves in the four panels) for the layer 5 pyramidal cell from neocortex. D_{ii} reflects the contribution of the geometry of the cable structure to the voltage dynamics and is independent of the waveform of the current input. The input current is given by Eq. 3.50 with $t_{peak} = 0.5$ msec and a centroid at 1 msec (the centroids are marked with arrows). At the soma (lower left) $D_{ii} = 17.7$ msec $\approx \tau_m$, while at distal sites, such as a layer 1 apical dendrite (upper right) or the tip of a layer 5 basal dendrite (lower right), D_{ii} can be much faster than τ_m. $R_m = 20,000 \ \Omega \cdot cm^2$ and $C_m = 1 \ \mu F/cm^2$ throughout the cell. The amplitude of all responses have been normalized and shifted by 0.5 msec to the right. Reprinted by permission from Koch, Rapp, and Segev (1996).

one pathway. In an infinite cable, $D_{ii} = \tau_m/2$, since current can flow through membrane resistance as well as through the intracellular resistivity (Fig. 2.11). Adding additional current sinks, in the form of a complex and highly branched dendritic tree, further reduces D_{ii}.

Because the soma is relatively large (with a radius of 20 μm), D_{ii} at the soma is not very far from that of an isopotential membrane patch, $D_{ii} = 17.7$ msec $\approx \tau_m$ (lower left in Fig. 3.13). For distal sites in the layer 5 pyramidal cell, $D_{ii} \approx \tau_m/10$ or less (upper and lower right-hand panels in Fig. 3.13), while for dendritic spines $D_{ii} \approx \tau_m/20$ (not shown). Such small values of D_{ii} go hand in hand with a narrow (brief) voltage transient at the input sites (Rinzel and Rall, 1974). Indeed, distal dendritic arbors, in particular the many fine branches of the basal tree, provide for multiple sites with a very brief (compared to τ_m) and local voltage response, provided the input is also brief. This will be taken up in more detail further below as well as in Chap. 19.

The transfer delay between i and j really includes two components, the time it takes for the current input to depolarize (or hyperpolarize) the local membrane (that is, D_{ii}) and the time it takes the centroid of the local potential to propagate to j. Accordingly, Agmon-Snir and Segev (1993) define the *propagation delay* as

$$P_{ij} = \hat{t}_j^V - \hat{t}_i^V = D_{ij} - D_{ii} . \qquad (3.51)$$

This measure has a number of similarities with the log attenuation $\tilde{L}_{ij}^v$ measure defined in Sec. 3.5.2. In particular, it is not symmetric,

$$P_{ij} \neq P_{ji}, \qquad (3.52)$$

but it is additive for all locations ℓ on the direct path between locations i and j:

$$P_{ij} = P_{i\ell} + P_{\ell j}. \qquad (3.53)$$

This last property allows us to compute the morphoelectrotonic transforms for P_{ij} just like we did for the log voltage attenuation measure (Zador, Agmon-Snir, and Segev, 1995). Before we discuss the functional significance of these different delays in realistic dendritic trees, we would like to note the analogies between input and transfer resistances and input and transfer delays. Both sets of variables are defined as quantities relating current at location i to voltage at location i or j and both obey the same symmetry relationships.

The propagation delay P_{is} from a dendritic site i to the soma s is obtained by subtracting the local delay at i from the transfer delay from i to the soma, reflecting the actual propagation time of the voltage centroid from the input site to the soma. We can, of course, also define a propagation delay from the soma to dendritic sites P_{si} (Agmon-Snir and Segev call this the *net dendritic delay*, to indicate that P_{si} measures the "cost" in terms of delay that results from placing the input at the dendritic site i rather than at the cell body). This measure turns out to be the most useful of all. Reverting back to the definitions of P_{ij} and D_{ij} given in Sec. 2.4, we have

$$P_{si} = D_{si} - D_{ss} = D_{is} - D_{ss} \qquad (3.54)$$

that is, we can think of P_{si} as the difference between the voltage response at i to a somatic input and the voltage response at the soma to a somatic input. Because the decay phases of both EPSPs are relatively similar (being primarily dictated by the properties of the large cell body), P_{si} reflects differences in the rising phases of the EPSPs. In this sense, P_{si} has similar values and can be interpreted as the time delay between the peak of the local EPSP and the peak in the somatic EPSP. Figure 3.14 illustrates the MET of the propagation delay P_{si} for the layer 5 pyramidal cell devoid of synaptic background activity. An extreme case is shown for which R_m is reduced tenfold. In an infinite cable, this leads to a $\sqrt{10}$ reduction in the propagation delay (Eq. 2.52). In fact, the scaling is more complex for a distributed dendritic tree. It takes the centroid of the somatic EPSP between 1 and 2 msec to propagate to the tips of the basal dendrites (and proximal apical dendrites) independent of R_m, while P_{si} to the tip of a typical apical dendrite decreases by about a factor of 2 to about 10–15 msec.

The *outward going propagation delay* P_{si} has another interesting interpretation. Using Eq. 2.51, we have

$$P_{si} = D_{is} - D_{ss} = (\hat{t}_i^V - \hat{t}_i^I) - (\hat{t}_s^V - \hat{t}_s^I) = \hat{t}_s^I - \hat{t}_i^I . \qquad (3.55)$$

In other words, it corresponds to the difference between the centroids of the current at the soma and at site i.

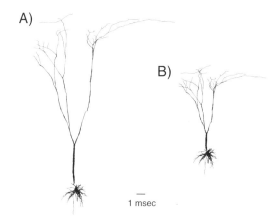

Fig. 3.14 Propagation Delay Morphoelectrotonic transform (MET) for the outward going propagation delay P_{si} from the soma to dendritic sites (Agmon-Snir and Segev, 1993) computed for the layer 5 pyramidal cell with (**A**) a very high value of $R_m = 100,000\ \Omega \cdot \text{cm}^2$ or (**B**) a 10 times reduced value of $10,000\ \Omega \cdot \text{cm}^2$. P_{si} corresponds to the effective delay between the peak of the local EPSP at i and the somatic EPSP. Changing R_m by a factor of 10 should reduce the propagation by $\sqrt{10}$ for an infinite cable, while the effect is much less in a distributed tree. P_{si} to the main branch point of the apical tree is 6 msec for the high and 2.5 msec for the low R_m value.

3.6.3 Dependence of Fast Synaptic Inputs on Cable Parameters

While the dynamics of the membrane potential in response to a slowly varying input current is dictated by τ_m, this is most emphatically not the case for rapid inputs (Segev and Parnas, 1983). For fast synapses onto thin dendrites, the rise time as well as the initial decay of the associated postsynaptic potential is almost completely independent of R_m and is primarily limited by the dynamics of the synaptic current, that is, ultimately by how fast the underlying ionic channels open and close (e.g., D_{ii} for distal dendritic sites in Fig. 3.13).

Let us consider how three aspects of the excitatory postsynaptic potential depend on the three cable parameters, R_m, C_m, and R_i: (1) its rise time or rate of depolarization, (2) its peak potential, and (3) its rate of repolarization.

For a fast synaptic input, the rate of depolarization in an infinite cable of diameter d is limited by the temporal derivative of the membrane potential in response to a current step of amplitude I_0. Using Eq. 2.33 and recalling that the derivative of the error function is $2e^{-x^2}/\sqrt{\pi}$ yields

$$\frac{dV_{\text{step}}(0,t)}{dt} = \frac{R_i^{1/2} I_0}{(\pi d)^{3/2} C_m^{1/2}} \frac{e^{-t/\tau_m}}{t^{1/2}} . \qquad (3.56)$$

At very short times when the exponential term can be developed in a Taylor series, the derivative behaves as $\text{const}/\sqrt{t}$, where const depends on R_i, I_0, and C_m, but *not* on R_m. What happens is that initially the injected current primarily charges up the membrane capacitance at the site of current injection as well as at neighboring locations (via the axial resistance R_i). The larger the R_i or the smaller the C_m, the faster the voltage will change. It is only the slow components of the decay that are affected by low-pass filtering, as demonstrated graphically in Fig. 3.15. Here the dendritic potential in response to a single depolarizing current input (Eq. 3.50) in the pyramidal cell is computed for

$R_m = 10,000 \, \Omega \cdot cm^2$ and $R = 100,000 \, \Omega \cdot cm^2$. While the dc input and transfer resistances are affected appropriately, almost no effect on the rise time is evident.

Surprisingly, both the peak depolarization as well as the initial phase of the repolarization of the voltage change are not affected by the tenfold change in R_m. This surprising behavior is only seen in thin dendrites and not at the soma: because of the limited membrane area, all of the injected current charges up the associated capacitance. The rapid repolarization

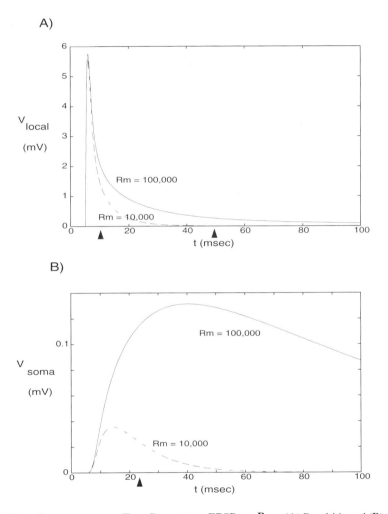

Fig. 3.15 WEAK DEPENDENCE OF FAST DENDRITIC EPSP ON R_m (A) Dendritic and (B) somatic synaptic potentials in response to a single fast excitatory synapse activated 5 msec into the simulation and located at $a3$ in the distal apical tree of the layer 5 pyramidal cell (see Fig. 3.7) for two different values of R_m (in units of $\Omega \cdot cm^2$). The current peaks 0.5 msec after onset and is over within 2 msec. This is so fast compared to the membrane time constant of either 10 or 100 msec that the rising phase as well as the peak dendritic response is more or less independent of R_m. Indeed, because the input is so fast compared to the time constant of the dendrite, the local voltage response approximates the impulse response. Due to the low-pass filtering between the synapse and the soma, the somatic potential is strongly affected by the decrease in electrotonic distance and the increase in τ_m. This shows the fallacy of using τ_m as an indicator for how rapidly the membrane can respond to synaptic input. Arrows point to the location of the centroids.

occurs because the capacitances outside the immediate neighborhood of the site of current injection (which includes the distant cell body) soak up the charge. The synaptic current is so rapid that all of this takes place with only little involvement of the membrane resistance. Because of the much larger capacitance present at the soma, the dendrites branching from the cell body fail to provide any substantive capacitive sink and repolarization is much slower (witness the decay of the somatic and the dendritic voltages in Fig. 3.13).

Rall (1964) first observed that this capacitive effect allows for faster than exponential decay, in particular in dendrites (reviewed extensivley in Jack, Noble, and Tsien, 1975). Softky (1994) derives an approximate expression for the peak potential and the rate of repolarization. By neglecting any leak conductances (that is, working in the limit of $R_m \rightarrow \infty$), the situation is formally equivalent to treating diffusion of charge inside a cylinder (see Fig. 11.3). This allows Softky to find the following expression for the peak depolarization in response to a very fast current pulse,

$$V_{\text{peak}} \approx 1.5 \frac{R_i^{1/2} I_{\text{peak}}}{(\pi d)^{3/2} C_m^{1/2}} \sqrt{t_{\text{peak}}} , \qquad (3.57)$$

(see also Eq. 3.56). Note that R_m never appears here and that the peak potential scales with the square root of t_{peak}. Softky quantifies the rate of repolarization by estimating the time $\tau_{1/2}$ it takes for the potential to decay from V_{peak} halfway to zero,

$$\tau_{1/2} \approx 6 t_{\text{peak}} . \qquad (3.58)$$

Figure 3.16 illustrates how well this approximation works for very rapid current inputs. It does badly on the slower components of the voltage decay since it completely neglects the effect of the membrane resistance.

As enthusiastically argued by Softky (1994, 1995), such rapid membrane dynamics, outpacing τ_m by far, give rise to the possibility of submillisecond coincidence detection in the thin basal and apical dendrites, in particular if Hodgkin-Huxley-like nonlinearities are present. We will pick up this story in Chap. 19.

3.7 Recapitulation

While Chap. 2 is concerned with solutions to the cable equation in single cables, this one concentrates on studying the membrane potential in realistic dendritic trees with the aim of understanding the relationship between dendritic architecture and function. A number of exact, recursive techniques exist to derive the membrane potential at any point in the dendritic tree in response to an arbitrary current injection at some other point. Because these techniques do not generalize well to n simultaneous inputs and do not work in the presence of membrane nonlinearities, today's method of choice is the numerical solution of the appropriately spatio-temporal discretized cable equation (compartmental method). Several well-documented, graphics-oriented, public software packages are available that implement the appropriate algorithms.

An alternative approach for gaining intuition about the events occurring in a dendritic tree is to "observe" the system at two points i and j, where the input is applied at i and the output recorded at j. The entire system can then be characterized in terms of three frequency-dependent functions that, in general, take on complex values: the input impedances $\tilde{K}_{ii}(f)$ and $\tilde{K}_{jj}(f)$ and the transfer impedance $\tilde{K}_{ij}(f) = \tilde{K}_{ji}(f)$. For sustained current input, these reduce to three real numbers, corresponding to the conventional input resistances $\tilde{K}_{ii}$

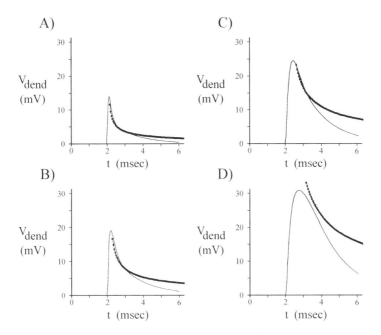

Fig. 3.16 SUBMILLISECOND RISE AND DECAY OF POTENTIALS Rapid membrane depolarization at the center of a thin basal dendrite of the pyramidal cell (close to $b1$ in Fig. 3.7) in response to local current injections with different rise times. The current took the form of Eq. 3.50 with t_{peak} of **(A)** 0.05, **(B)** 0.1, **(C)** 0.2, and **(D)** 0.4 msec. The approximations derived by Softky (1994) are indicated in bold. They describe well the peak membrane potential (Eq. 3.57) and the early part of the rapid repolarization. It is only during the slow, late part of the membrane depolarization, where the approximation diverges from the correct solution, that the membrane resistance plays any role. The somatic depolarizations in response to these inputs are minute, typically less than 0.5 mV. The morale is that brief and rapid synaptic inputs can give rise to changes in the membrane potential in the submillisecond range. Reprinted in modified form by permission from Softky (1994).

and $\tilde{K}_{jj}$ and transfer resistance $\tilde{K}_{ij}$ (with the dimensions of a resistance). This approach allows us to define in a straightforward manner the voltage attenuation $\tilde{A}_{is}$ between any dendritic site i and the soma s and the outgoing voltage attenuation $\tilde{A}_{si}$ from the soma to sites in the dendritic tree. This last measure is identical to the charge attenuation, the ratio of the total charge injected at the synaptic site i to the total charge reaching the soma.

The chapter also introduces the logarithmic transform of the attenuation, $\tilde{L}_{ij}^{v} = \ln |\tilde{A}_{ij}|$. As compared to a true distance metric, this measure is not symmetric: the logarithm of the voltage attenuation between i and j is different from the logarithm of the voltage decrement between j and i. Indeed, for passive trees, the voltage attenuation from any dendritic site to the soma is always considerably larger than in the reverse direction. The reason for this profound asymmetry is that dendritic sites have high impedance while the cell body has a low impedance. A further consequence is that, if integrated over sufficiently long times (in practice several time constants), about half of the total charge injected into distal dendritic sites can reach the soma. We discussed a frequently used measure of synaptic efficiency, the total electrotonic distance L_{is}. Due to the fact that real trees are quite distinct from infinite cables, this measure sharply underestimates the true voltage attenuation. All of this can be visualized graphically, using the morphoelectrotonic transforms. METs are a useful dynamic tool for portraying different features of the electroanatomy of a cell.

We also considered delays in the dendritic tree and between dendritic sites and the soma. Using the definition of delays in terms of differences in the centroids of voltages and currents, we can show that τ_m represents an upper bound on how slowly the voltage decays in response to fast inputs. At the soma, the local delay is on the order of the membrane time constant, while for distal sites the local delay can be much faster, as small as 5 or 10% of τ_m, implying that events in the dendritic tree can be very rapid. This is particularly true for rapid synaptic events in thin dendrites, where the rise and early decay times of the dendritic membrane potential are independent of R_m, allowing for the possibility of carrying out precise timing relationships in these structures that are not limited by τ_m.

The take-home message is that one has to be careful in applying the concepts that characterize an infinite cable (λ, τ_m, and R_{in}) to realistic neurons. These have finite arbors, with multiple dendrites terminating at different distances and branching patterns that do not obey the necessary geometrical constraints for reduction to an equivalent cylinder. These considerations have significant effects on how the voltage spreads and attenuates in dendrites and hence in how synaptic inputs are integrated.

4

SYNAPTIC INPUT

How information circulates in the brain was the subject of a heated debate lasting a decade or more among anatomists in the closing years of the last century. One camp argued that neural tissue consisted of a continuum, a *syncytium*, with no discernible functional units while the opposing view held that the brain consisted of discrete units, the nerve cells, communicating through point-to-point contacts that Sherrington dubbed *synapses*. Although in principle both views can be supported, in practice the majority of rapid communication occurs via specific point-to-point contacts, at either chemical and electrical synapses. *Ephaptic transmission* refers to nonsynaptic, electrical interactions between neurons. While such interactions do occur, for instance, among adjacent, parallel axons across the extracellular space, they are, by their very nature, neither very strong among any one pair of processes nor very specific. Their functional significance—if any—is currently not known, and we will not discuss them here (Traub and Miles, 1991; Jefferys, 1995).

In the beginning chapter, we introduced the action of fast, chemical synapses. Given their importance, we will now return to this topic in greater depth. We first overview the pertinent biophysical events underlying chemical synaptic transmission and some of the vital statistics of synapses before we come to the mathematical treatment of synaptic input. In the last section, we will summarize our knowledge of electrical synapses and their computational role.

Most typically, a synapse consists of a *presynaptic* axonal terminal and a *postsynaptic* process that can be located on a dendritic spine, on the trunk of a dendrite, or on the cell body. Figure 4.1 shows some examples of synapses among cortical cells as seen through a high-powered electron microscope. It is not easy at first to identify the synapses amid all the curved, irregular, and densely packed structures making up the neuronal tissue. In a number of locations, such as the retina or the thalamus, a synaptic connection is made between two dendrites, rather than between an axon and a dendrite. These synapses are called *dendro-dendritic* synapses; they are believed to be relatively rare in the adult cortex (see Sec. 5.3). Most synapses are small and highly specialized features of the nervous system. As we will see, a chemical synapse converts a presynaptic electrical signal into a chemical signal and back into a postsynaptic electrical signal.

Occasionally, an axon makes a synaptic connection, called an *autapse*, onto its own dendritic tree (van der Loos and Glaser, 1972; Karabelas and Purpura, 1980; Lübke et al.,

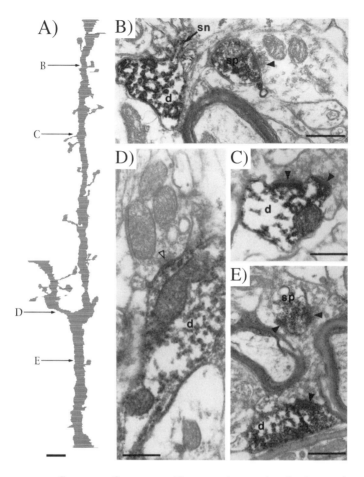

Fig. 4.1 **ANATOMY OF CORTICAL SYNAPSES**   Electron micrographs of spines and synapses on a dendrite of a *spiny stellate* neuron in layer 4 of the visual cortex of an adult cat. Around 300 million synapses are packed into a single cubic millimeter of this tissue, providing the substrate for an extraordinary degree of neuronal interconnectivity. **(A)** 625 ultrathin sections of part of a proximal dendrite of the neuron are shown here in this computer-assisted view. The lower end is close to the soma. The letters and arrows correspond to the electron-microscopic views shown in the adjacent five photographic panels. **(B)** The dendrite (**d**) gives rise to a spine neck (**sn**) and spine head (**sp**). The latter receives an asymmetric, that is, excitatory synaptic profile at the solid arrow. These arrows always point from the unstained presynaptic toward the postsynaptic site. **(C)** Two asymmetric synapses on the dendritic shaft. **(D)** A symmetric, that is, inhibitory synapse (open arrow) at the branch point shown in A. **(E)** A single presynaptic element makes two asymmetric synapses with a spine and a dendrite of the cell. **(F)** An asymmetric synapse with the dendritic shaft. Scale bars equal 2 μm in **A** and 0.5 μm in **B–F**. Reprinted by permission from Anderson et al., (1994).

1996). Autapses appear to be identical to normal chemical synapses. While autapses are rare on pyramidal cells, they are found more frequently on two subclasses of cortical inhibitory interneurons, raising the possibility that they might have a functional role and are not just the unfortunate consequence of an imprecise developmental rule (Tamás, Buhl, and Somogyi, 1997; Bekkers, 1998).

   The literature on the properties of synaptic transmission and its molecular nature is gargantuan and keeps growing (Kandel, Schwartz, and Jessell, 1991; Jessell and Kandel,

1993; Stevens, 1993; for a historical overview, see Shepherd, and Erulkar, 1997). We recommend Hille (1992) and Johnston and Wu (1995).

4.1 Neuronal and Synaptic Packing Densities

From the anatomical point of view, synapses in the central nervous system can be conveniently classified according to the detailed morphology of the synaptic profiles in electron-microscopic images into one of two classes, *Gray type I* and *Gray type II* synapses (Gray, 1959). Using a combination of electrophysiological, pharmacological, and anatomical criteria, type I synapses, also known as *asymmetrical synapses*, have been found to be excitatory, while type II synapses, also known as *symmetrical synapses*, act in an inhibitory manner (White, 1989; Braitenberg and Schüz, 1991; Douglas and Martin, 1998; see Fig. 4.1).

Synapses are small. The area of contact between pre- and postsynaptic processes has a diameter of 0.5–1.0 μm, while the presynaptic terminal is only slightly larger. This implies very high synaptic packing densities of around 7.2×10^8 synapses per cubic millimeter in the mouse (Braitenberg and Schüz, 1991). In the cat, the density has been estimated around 3×10^8 per cubic millimeter (Beaulieu and Colonnier, 1985). If the synapses were located on a three-dimensional lattice, they would be spaced a mere 1.1 μm apart. Braitenberg and Schüz (1991) estimate the extent of neuronal processes in a 1-mm^3-cube of cortical tissue, coming up with the staggering amount of 4.1 km total length of axonal processes (at an average diameter of 0.3 μm) and 456 m total length of dendrites (with a diameter of 0.9 μm) in this volume. In other words, the "average" nerve cell in the mouse cortex receives input from 7800 synapses along its 4 mm of dendrites and is connected with 4 cm of "axonal wire" to other cells.

In higher mammals (e.g., mouse, cat, monkey, humans) the total number of neurons in a column of unit area that reaches across all cortical layers is constant, irrespective of what cortical area this column is taken from (the sole exception is the primary visual cortex with 2.5 times more cells). Thus, while the cortical sheet thickened by a factor of 2–3 in its evolution from mouse to human, the number of cells below 1 mm^2 of neocortex remained fixed, in the neighborhood of around 100,000 cells below a square millimeter of cortex (Rockel, Hiorns and Powell, 1980).

Given an approximate density of 100,000 cells per mm^3 in the primate, a synaptic density of 6×10^8 per mm^3, a total surface area of about 100,000 mm^2 for one hemisphere, and an average thickness of about 2 mm, the average human cortex contains on the order of 20 billion neurons and 240 trillion synapses (2.4×10^{14}), quite impressive numbers given the current count of about 10^{11} transistors in the memory and central processing unit (CPU) of a modern parallel supercomputer.[1]

4.2 Synaptic Transmission Is Stochastic

Our view of the synapse is based upon the influential work of Katz and his collaborators (Katz, 1969). Working on the frog neuromuscular junction, they described the basic features

---

1. Of course, the connectivity among the components in the CPU is three to four orders of magnitude lower than the connectivity of a cortical cell. On the other hand, the cycle time of such machines, in the low nanosecond range, compares very favorably with neuronal time constants in the millisecond range.

of synaptic transmission upon which every further model has been based (Salpeter, 1987). This canonical view is outlined in Fig. 4.2.

The neurotransmitter is prepackaged in numerous small (30–40-nm-diameter) spheres or *vesicles* which reside in the presynaptic terminal. Invasion of the presynaptic axonal terminal by an action potential causes a inrush of calcium ions (via voltage-dependent calcium channels in the presynaptic terminal). These calcium ions—through a complex chain of events—cause one or more synaptic vesicle to fuse with the membrane at special sites. Here, the vesicle now releases its "quantum" of transmitter into the narrow synaptic cleft between the pre- and postsynaptic cell membranes (Fig. 4.2), a process termed *exocytosis*. Calcium is necessary for synaptic transmission. Reducing the extracellular calcium concentration can drastically lower the efficiency of the synapse (Mintz, Sabatini, and Regehr, 1995; Borst and Sakmann, 1996). This might have important functional consequences (Sec. 20.3).

The neurotransmitter rapidly diffuses across the 20-nm-wide *synaptic cleft* and binds to postsynaptic receptors, usually ionic channels. These receptors are then responsible for the great diversity of postsynaptic events that cause the postsynaptic membrane potential to change. Each quantum of transmitter is released probabilistically and independently of the others, and their postsynaptic effects add linearly. One reason for the success of Katz's theory is that its predictions have been evaluated quantitatively (Katz, 1969). If n sites exist, and each such site independently releases either no or only a single packet or quantum of transmitter with an associated amplitude q and with probability p, then the probability of k quanta being released at the entire synapse is given by the binominal distribution

$$p(n, k) = \frac{n!}{(n - k)!k!} p^k (1 - p)^{n-k} . \tag{4.1}$$

At the frog's neuromuscular junction, where the theory was developed and the critical experiments were performed, the probability of release can be quite low. Yet because there exist on the order of 100 to 1000 release sites, overall synaptic transmission from the

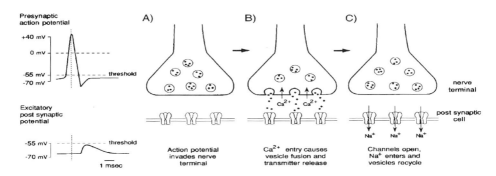

Fig. 4.2 ELEMENTS OF SYNAPTIC TRANSMISSION Standard model of synaptic transmission at a chemical synapse, as elucidated by Katz for the neuromuscular junction. An action potential (far upper left) invades the presynaptic terminal (**A**) and—mediated by the resultant influx of calcium ions—causes the vesicles to fuse with the membrane and release the neurotransmitter inside the vesicle into the synaptic cleft separating the presynaptic membrane from the postsynaptic one (**B**). The neurotransmitter molecules rapidly diffuse across the cleft and bind to the receptors there. In this illustration, they cause the opening of Na$^+$ selective channels (**C**), leading to an excitatory postsynaptic potential (far bottom left). At central synapses, a similar chain of events occurs, except that the entire process of synaptic transmission appears to be highly stochastic. Reprinted by permission from Jessell and Kandel (1993).

nerve onto the muscle is very reliable. In a somewhat more general version of this model, spontaneous release of vesicles, that is, the spontaneous postsynaptic potentials, can be accounted for by making the probability of release time dependent such that it is very small, but nonzero, in the absence of any input and much larger following a nerve impulse arrival (Katz and Miledi, 1965; Barrett and Stevens, 1972).

While many of the same principles elucidated by Katz also appear to apply to more central synapses, several crucial differences have emerged (Walmsley, Edwards, and Tracey, 1987; Redman, 1990; Stevens, 1993). The most important one concerns the reliability and variability of synaptic transmission. At central synapses, it appears that each synaptic bouton contains only one or a few active release zones, rather than the hundreds found at the neuromuscular junction (Korn and Faber, 1991). In other words, upon arrival of an action potential at an individual anatomically identified synapse, typically none or only one vesicle is released (*one-vesicle hypothesis*; Korn and Faber, 1993). Because only one vesicle is released at each synapse, the probability of release p looms large in whether or not a presynaptic action potential results in a postsynaptic signal.

We know from *in vitro* studies that the probability of release at an individual synapse in vertebrates as well as in invertebrate systems can be highly variable and is usually in the range of between 0.1 to 0.9 (Korn, Faber, and Triller, 1986; Bekkers and Stevens, 1989; Redman, 1990; Edwards, Konnerth, and Sakmann, 1990; Raastad, Storm, and Andersen, 1992; Laurent and Sivaramakrishnan, 1992; Hessler, Shirke and Malinow, 1993; Gulyás et al., 1994).

The molecular origin of the probabilistic release is not yet clear. However, it is likely to relate to the fact that exocytosis depends on a very high local calcium concentration inside the synaptic terminal and that this depends on the vagaries of the exact spatial relationship between calcium channels in the membrane and the location of the vesicle (Borst and Sakmann, 1996; Bennett, 1997).

4.2.1 Probability of Synaptic Release p

Failure of synaptic transmission is vividly demonstrated in Fig. 4.3 (see also Bekkers, Richerson, and Stevens, 1990). Using a so-called minimal stimulation paradigm, only a single Schaffer collateral axon is stimulated. These axons typically make only a single excitatory synapse onto CA1 pyramidal cells (Sorra and Harris, 1993). If the presynaptic-postsynaptic connection were secure, each presynaptic stimulus generated by the electrode should evoke a postsynaptic event. In this case, the membrane potential in the pyramidal cells is clamped to its resting potential and the resultant excitatory postsynaptic current (EPSC) is recorded. The five records in the left column and the four top records in the right column in Fig. 4.3 show nine trials, of which only three lead to a postsynaptic event. The bottom right record corresponds to the average postsynaptic event (averaged over the nine records).

As we will discuss in Chap. 13, the probability of synaptic release p is itself subject to change, depending on the recent history of the presynaptic terminal. Depending on the number and timing of presynaptic action potentials, p can either decrease, as in long-term depression, or increase, as in long-term potentiation. p also varies from one synapse to the next. This behavior might well be different from one synaptic type to the next (e.g., thalamo-cortical versus intracortical synapses; Stratford et al., 1996). Understanding the dynamics of p and its relationship to synaptic plasticity has been a hot research topic of the last few years.

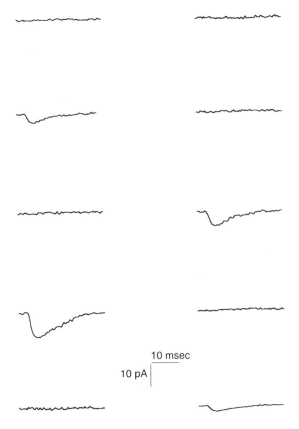

10 msec

10 pA

Fig. 4.3 SYNAPTIC TRANSMISSION IS VERY UNRELIABLE Synaptic transmission among central neurons can be highly unreliable and should be thought of as a binary event with a probability of success p, which can be as low as 0.1. This is exemplified by this recording of the excitatory inward current caused by a single synapse, made by a Schaffer collateral axon onto a CA1 pyramidal cell in the mammalian hippocampal slice. The five records on the left and the top four records on the right (30 msec long) show the postsynaptic current in response to nine stimulations of the presynaptic axon using an electrode. Only in three trials does the presynaptic spike lead to the release of a vesicle filled with neurotransmitter that causes a postsynaptic current to flow. The amplitude of this postsynaptic event, if it occurs, is also highly variable. The average postsynaptic EPSC (computed over these nine trials) is indicated on the lower right. It is not clear how the brain deals with the unreliability present at its elementary computational units. Unpublished data from Y. Wang and C. Stevens, printed with permission.

In the example of Fig. 4.3, the probability of release is about 0.3. This is quite remarkable, since it implies that the connecting elements among neurons are binary but highly unreliable circuit elements, transmitting only one out of three events. Imagine a transistor that only conducts, that is, switches, one out of three times that a charge is dumped onto its gate. Synapses made onto pyramidal cells in slices from sensory and motor cortex appear to behave in an identical fashion (Smetters and Nelson, 1993; Thompson, Deuchars, and West, 1993a,b). An important *caveat* is that these results were not obtained under physiological conditions in intact animals but in slices or other isolated preparations. It is possible that in a behaving animal, various neuromodulators can significantly boost or reduce p.

The lack of reliability of synapses should be compared to a reliability of transmission among electronic circuits, which is many, many orders of magnitude higher.[2] In principle, the nervous system could compensate for such unreliable components by exploiting redundancy (Moore and Shannon, 1956; von Neumann, 1956), that is, by using many parallel synapses. Yet, it is common for a cortical axon to make only one or two synaptic contacts onto its target cell (Gulyás et al., 1993; Sorra and Harris, 1993). The brain must have found a different way in which it deals with such unreliable synaptic components.

The reliability of synaptic transmission in the central nervous system is further compromised by the fact that the variability in the amplitude of the postsynaptic response to the release of a single packet of neurotransmitters is much larger than at peripheral synapses. Thus, even if a packet of neurotransmitters is released, a number of factors, such as the variation in the size of the vesicle and the number of postsynaptic receptors, cause the postsynaptic response to this presynaptic event to vary from one trial to the next (see the three successful instances of transmission in Fig. 4.3; Bekkers, Richerson, and Stevens, 1990; Larkman, Stratford, and Jack, 1991). In the study by Mason, Nicoll, and Stratford (1991), the variance in the size of the EPSP—evoked in one rat neocortical pyramidal cell by electrically activating a nearby pyramidal cell —is as large as its mean (Fig. 4.4), with the largest EPSP (2.08 mV) about 40 times larger than the smallest (0.05 mV). In these experiments, the number of synapses between the pair of cells recorded from is not known (see also Stratford et al., 1996).

We defer a discussion of the possible functional role of the stochastic nature of synaptic transmission to Sec. 13.5.5.

4.2.2 What Is the Synaptic Weight?

Let us briefly address the thorny issue of what exactly constitutes *synaptic weight* or *synaptic strength*. In the neural network literature, this is a single scalar, usually labeled w_{ij} or T_{ij} (see Sec. 14.4). However, synapses are very complicated devices, which are characterized by numerous parameters. Traditionally—as expressed by Eq. 4.1—biophysicists have used three scalar variables to describe a synapse: the number of quantal release sites n, the probability of synaptic release per site p, some measure of the postsynaptic effect of the synapse q. Depending on the experiment, q can be the peak conductance, the maximal synaptic current at some potential, or the peak EPSP. Of course, in real life, n is drawn from some probability distribution (usually binominal or Poisson), p depends on the previous spiking history, and q itself is a function of time (such as the time-dependent conductance change of Eq. 1.21).

One possible relationship between these state-dependent functions and the synaptic weight is the expected or mean postsynaptic action, computed by averaging over a Poisson distributed presynaptic spike distribution with

$$R = npq .$$
(4.2)

We shall return to this in Chap. 13. However, one should never forget that the averaging assumptions used to compute this weight might not be relevant to the brain under its normal operating conditions (for instance, since the input is oscillatory or tends to fire in bursts or because its stochastic nature is critical).

---

2. Indeed, in digital CMOS technology, where the signal gain at an inverter, the elementary computational unit, is in the neighborhood of 15 to 20, the thermal fluctuations—on the order of 0.65 mV at room temperatures—are so much less than the switching voltage, that the probability of switching is about $1 - 10^{-14}$. Given that each inverter restores the input signal, the effect of thermal noise does not accumulate along a cascade of inverters.

A)

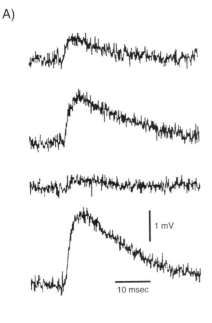

Fig. 4.4 Postsynaptic Amplitude Is Highly Variable Fluctuations in the amplitude of EPSPs observed in a pyramidal cell by evoking an action potential in a nearby pyramidal cell. Both pairs of neurons are located in layer 2/3 of brain slices taken from rat visual cortex. **(A)** Four individual sweeps from the same synaptic connection. The EPSP amplitude over the entire population of cell pairs is 0.55 ± 0.49 mV. **(B)** Average of 2008 such sweeps, together with the presynaptic record, showing that the EPSP in the postsynaptic cell is caused by the presynaptic spike. Due to technical reasons, no events less than 0.03 mV—and, in particular, no failures as in Fig. 4.3—can be recorded. Reprinted by permission from Mason, Nicoll, and Stratford (1991).

B)

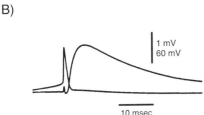

In summary, because of the highly probabilistic nature of the release of neurotransmitter, compounded by the variability in the size of the postsynaptic responses to this neurotransmitter, a basic and unavoidable feature of interneuronal communications using chemical synapses is their lack of reliability and consistency. As pointed out by Stevens (1994), it is crucial that any theory of the brain account for this fundamental property of neuronal hardware. We will return to the theme of stochastic computations in the brain in Chap. 15.

4.3 Neurotransmitters

The plethora of time courses, amplitudes, and types of actions seen in synaptic transmission is the result of the interplay between different *neurotransmitters*, stored in vesicles at the presynaptic terminal, and different synaptic *receptors*, inserted in the postsynaptic membrane.

There exist three major chemical classes of neurotransmitters, *amino acids, biogenic amines,* and *neuropeptides* (Table 4.1). Fast synaptic transmission in the central nervous system of vertebrates is mainly mediated by amino acids, the major excitatory neuro-transmitters being *glutamate* and *aspartate*, and the major inhibitory neurotransmitters being γ-amino-butyric acid (GABA) and *glycine*. The onset of the associated excitatory or inhibitory postsynaptic currents (EPSCs or IPSCs), forming the principal substrate of neural computation, is rapid (<1 msec) and their durations are relatively short (<20 msec). All these neurotransmitters also have slow effects.

The list of biogenic amines that can modulate the response of the cell to synaptic input includes *acetylcholine* (ACh), *norepinephrine* (also referred to as *noradrenaline*), *dopamine, serotonin* and *histamine*. Usually, their actions have a much slower onset than the action of amino acids and may persist hundreds of milliseconds to seconds. However, some neuroactive substances, such as acetylcholine, cause fast (e.g., nicotinic-receptor-mediated fast synaptic transmission at the vertebrate neuromuscular junction) as well as slow (e.g., muscarinic-receptor-mediated changes in potassium conductances) effects, depending on which receptor is present in the postsynaptic membrane.

A long and ever-growing list of *peptides*, that is, short chains of amino acids, modulates the response of neurons over very long time scales (that is, minutes). *Hormones*, that is substances that are transported via the bloodstream (in vertebrates) or the hemolymph (in invertebrates), also affect neuronal responses, with the distinction between the actions of neuropeptides and hormones being a gradual one. These neuroactive substances, frequently also termed *neuromodulators*, are usually *colocalized* with conventional fast neurotransmit-ters in individual neurons. Indeed, sometimes two or more neuropeptides can be colocalized within a conventional terminal in different vesicles (Kupfermann, 1991). Release can be differential, in the sense that the release of the vesicles containing the modulators requires a higher presynaptic firing rate than release of the vesicles harboring the fast neurotransmitter. Dozens of peptides and hormones have been identified in small invertebrate ganglia that are accessible to neurochemical methods (Marder, Christie, and Kilman, 1995; see also Fig. 20.5), and it is unlikely that the situation in the cortex will be much simpler.

The concentration of these substances can be thought of as the closest equivalent to a *global variable* in the brain. While a global variable within a computer program is defined for the entire program (rather than for any particular procedure), similarly, release of a long-lasting neuromodulatory substance will affect all neurons within a given distance from the site of release. As we will see in the next section, neurons possess receptors for a variety of different neurotransmitters and modulators, giving rise to a very complex and possibly redundant system for modulating the electrical responses of neurons over a great variety

TABLE 4.1
Principal Neurotransmitters

Amino acids	Biogenic amines	Neuropeptides
Glutamate	ACh	Substance P
Aspartate	Dopamine	Somatostatin
GABA	Noradrenaline	Proctolin
Glycine	Serotonin	Neurotensin
	Histamine	Luteinizing-hormone-releasing hormone (LHRH)

Three main chemical classes of neurotransmitters found in the nervous systems of all animals. Amino acids are usually involved in fast (glutamate/aspartate) excitatory or inhibitory (GABA/glycine) synaptic traffic. Many substances, such as glutamate or acetylcholine (ACh), have fast and transient as well as slow and long-lasting effects, depending on the postsynaptic receptor. The very large number of *neuropeptides*, of which only a handful are listed, act on the time scale of seconds to minutes.

of different spatial and temporal scales (for reviews see Nicoll, 1988; Kupfermann, 1991; Hille, 1992; Bourne and Nicoll, 1993; McCormick, 1998). This implies that cells can be "addressed" using a unique bar-code-like combination of neuromodulators and receptors (see Sec. 20.5).

4.4 Synaptic Receptors

Postsynaptic receptors come in two different flavors (Fig. 4.5). *Ionotropic* receptors are directly coupled to ionic channels, which open and permit certain types of ions to cross the postsynaptic membrane. Such channels, also called *ligand-gated* channels, include the *nicotinic* acetylcholine receptor in the peripheral nervous system of vertebrates and many invertebrates, the $GABA_A$ receptor complex, and the ubiquitous glutamate *N*-methyl-D-aspartate (NMDA) and non-NMDA channel complexes (Table 4.2). Given the close point-to-point connection between the presynaptic release zone and the postsynaptic receptor (Fig. 4.1) and the direct coupling between the receptor and the channel (Figs. 4.2

A) Channel Using Intrinsic Sensor

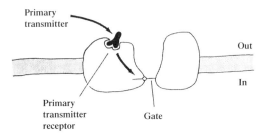

B) Channel Using Remote Sensor

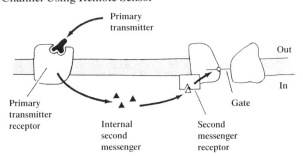

Fig. 4.5 IONOTROPIC AND METABOTROPIC SYNAPTIC ACTION (A) Fast excitatory and inhibitory input mediated by a tightly linked *ionotropic* receptor-channel complex. Binding of the neurotransmitter leads to a rapid opening of the associated ionic channel. (**B**) In the case of a *metabotropic* receptor, binding of the neurotransmitter leads to activation of a second messenger substance (such as Ca^{2+} ions). This messenger molecule, possibly after diffusing to its site of action, binds to a particular ionic channel and will modulate its properties. While the action of the ionotropic receptor is point to point and rapid, both the onset and the duration of the metabotropic mediated synaptic input are usually slow and its action can extend over larger distances. Both receptor types can be colocalized. Reprinted by permission from Hille (1992).

TABLE 4.2
Synaptic Receptor Types

Neurotransmitter	Receptor	E_{syn}	Type	Comments
Glutamate	Non-NMDA	0	I	Very fast
Glutamate	NMDA	0	I	Voltage dependent
GABA	$GABA_A$	-70	I	Fast inhibition
GABA	$GABA_B$	-100	M	Slow inhibition
ACh	Nicotinic	-5	I	Neuromuscular junction
ACh	Muscarinic	-90	M	Decreases K conductance
Noradrenaline	α_2	-100	M	Increases K conductance
Noradrenaline	β_1	-100	M	Decreases K conductance

List of major types of synaptic receptors and the associated neurotransmitters. The four top listings are the dominant transmitters used for fast communication in the vertebrate central nervous system. The synaptic reversal potential E_{syn} is specified in millivolts absolute potential. The type corresponds to either ionotropic (I) or metabotropic (M) receptors.

and 4.5A), the action of ionotropic receptors is rapid and transient. They implement the computations underlying rapid perception and motor control.

Binding of neurotransmitter to *metabotropic* receptors, on the other hand, leads to the activation of some intracellular molecules, termed *second messengers*, which in turn may induce a conformational change in some ionic channel and therefore in its kinetic behavior.

It has been estimated that the mammalian genome has devoted about 1000 of its 10^5 genes to these receptors. They could encode for a truly staggering number of different metabotropic receptors. At the molecular level, all of these receptors span the cell membrane in a snakelike fashion, crossing the bilipid membrane seven times (Chap. 8). These receptors connect with so-called *G proteins* just inside the cell membrane. G proteins, named because they bind guanosine triphosphate (GTP), include at least 20 different proteins and are among the most versatile nanomachines in biology (Clapham, 1996). Acting on perhaps 100 different receptors throughout the brain, the muscles, the glands, and other organs, they recognize photons (rhodopsin) and odor molecules as well as conventional neurotransmitters, such as glutamate, GABA, and ACh (Ross, 1989; Hille, 1992). Activation of these receptors is linked via a cascade of biochemical reactions to the ionic channel that is modulated (Fig. 4.5B). Functionally, these multiple intracellular steps can greatly amplify the signal. Thus, a single occupied receptor might activate many G proteins, each one of which can, in turn, activate many other proteins, and so on. Second messengers can act to increase or to decrease the postsynaptic membrane conductance, in particular for potassium.

An important distinction between ionotropic and metabotropic receptors is their time scale. While members of the former class act rapidly, terminating within a very small fraction of a second, the speed of the latter class is limited by diffusion. Biochemical reactions can happen nearly instantaneously at the neuronal time scale. However, if a synaptic input to a metabotropic receptor induces the release of some messenger, such as calcium ions, which have to diffuse to the cell body in order to "do their thing," the time scale is extended to seconds or longer. Section 9.3 details the involvement of one such metabotropic receptor type in the control of firing frequency adaptation.

It is difficult to overemphasize the importance of modulatory effects involving complex intracellular biochemical pathways. The sound of stealthy footsteps at night can set our heart to pound, sweat to be released, and all our senses to be at a maximum level of alertness, all actions that are caused by second messengers. They underlie the difference in sleep-wake behavior, in affective moods, and in arousal, and they mediate the induction of long-term memories. It is difficult to conceptualize what this amazing adaptability of neuronal hardware implies in terms of the dominant Turing machine paradigm of computation.

Because of the indirect coupling between receptor and effector—in particular if the second messengers involved have to diffuse within the postsynaptic cell—the onset of their effects is usually slow but long-lasting and is not restricted to a single site. They modulate the properties of groups of neurons over long time scales, that is, hundreds of milliseconds to seconds, minutes, and longer. Yet, this view does not do justice to the overlapping time scales of ionotropic and metabotropic receptors. If receptor and effector are spatially close to each other, both can act at the same time scale. For instance, a ligand-gated event such as a glutamate-induced NMDA depolarization lasts for 50–100 msec, approximately the same duration as a second-messenger-mediated hyperpolarization caused by activation of $GABA_B$ receptors.

4.5 Synaptic Input as Conductance Change

Activation of an ionotropic receptor permits the associated ionic channel in the postsynaptic channel to change its configuration, thereby allowing the passage of ions across the membrane. Given the small diameter of the open channel, about 3-7 Å, channels are highly selective for certain ions that diffuse through the open channel down their electrochemical gradient, leading to a change in postsynaptic potential. Chapter 8 will treat ionic channels in more detail. At this point, all we need to know is that individual channels act as binary elements, having zero conductance in their closed state and a fixed, nonzero conductance value in their open state (ranging between 5 and 50 pS, depending on the channel type).[3] The graded nature of the observed postsynaptic conductance change comes from the simultaneous openings of tens to hundreds of these binary elements. How long these channels stay open—the determinant of the duration of synaptic input—depends on two factors: (1) the presence of active uptake systems in the synaptic cleft that can remove the neurotransmitter and degrade or recycle it, and (2) the internal kinetics of the channel.

4.5.1 Synaptic Reversal Potential in Series with an Increase in Conductance

Given the distribution of ions in the intracellular and extracellular cytoplasm and the specificity of the ionic channels, each type of synaptic input has an associated *Nernst potential*, also referred to as the *ionic battery*, or *ionic reversal potential*, E_{syn}. The origin of this potential—crucial to neuronal excitability— relates to the equilibrium established between the concentration gradient of the different ions across the neuronal membrane and the electrical force opposing this gradient. Using the Boltzmann equation of classical statistical mechanics, the value of the synaptic potential at body temperature for a channel permeable to a single ionic species is given by the Nernst equation

$$E_{syn} = \frac{61.5}{z} \log_{10} \frac{[S]_o}{[S]_i} \tag{4.3}$$

where E_{syn} is measured in millivolts, z is the valence of the ions involved (positive for Ca^{2+}, Na^+ and K^+ ions, negative for Cl^- ions) and $[S]_o$ is the extra- and $[S]_i$ the

---

3. We here neglect the fact that many channels have more than one open state; this is usually described by saying that a channel has conductance *sublevels*.

intracellular concentration of the ionic species considered. The meaning of E_{syn} becomes clearer if one considers that in the absence of any other ionic species, the membrane potential across the open channel would stabilize at E_{syn}. At this potential, the ions on both sides of the membrane are in a dynamic equilibrium, with no net flux of ions through the channel. Many channels are permeable to a mixture of two or more ionic species, requiring a more complex expression, such as the *Goldman-Hodgkin-Katz* (GHK) *voltage equation* (Goldman, 1943; Hodgkin and Katz, 1949; Hille, 1992).

As we saw in the first chapter, the opening of a synaptic channel corresponds, from an electrical point of view, to an increase in the membrane conductance in series with the ionic reversal battery E_{syn} (Fig. 1.7A). The key insight, namely that the binding of neurotransmitters causes an increase in the *conductance* of the membrane, came from research carried out by Fatt, Katz, and Eccles in the 1950s as they were working on the neuromuscular junction (Eccles, 1964, 1990).

The lumped effect of many ionic channels opening in response to the binding of the transmitter molecules to the receptor is treated as a time-varying change in the membrane conductance $g_{syn}(t)$ in series with the synaptic reversal potential E_{syn}. This conductance change gives rise to a transient ionic current through the open channels, transporting charge across the membrane. How does the synaptic current relate to the conductance of the channel and to the membrane potential?

In general, this relationship is a complex one. One such description is referred to as the *Goldman-Hodgkin-Katz current equation* (Goldman, 1943; Hodgkin and Katz, 1949; Hille, 1992; see Eq. 9.1). Assuming that ions cross the membrane without interacting with each other and that the potential drops linearly across the membrane, the GHK current equation expresses a nonlinear dependency between the ionic current and the membrane voltage. This nonlinearity, referred to as *rectification* since current passes more easily in one direction than in the other, is caused by the nonhomogeneous distribution of ions across the membrane and increases with increasing concentration gradient.

However, in general, a linear relationship is observed between synaptic current and the membrane potential (with the important exception of the NMDA receptor discussed below). In fact, once the reversal potential has been accounted for, *Ohm's law* appears as a satisfactory first-order description of the synaptic current, and this is the one we adopted at the beginning of this book (e.g., Eq. 1.18),

$$I_{syn}(t) = g_{syn}(t)(V_m(t) - E_{syn}) \qquad (4.4)$$

It is important to realize that the synaptic current depends on V_m. Only under certain conditions can a synaptic input be approximated as a constant current source (Fig. 1.7B). This seemingly trivial observation has a number of important consequences, outlined further below as well as in Chaps. 5 and 18. Equation 4.4 is a purely phenomenological description of the current as a function of the membrane potential and is not derived from first principles. Experimentally, linearity is not necessarily assured and needs to be experimentally confirmed (as, for instance, in Hestrin et al., 1990a, for the fast, non-NMDA input onto hippocampal pyramidal cells).

A synapse is excitatory if the synaptic current I_{syn} depolarizes the postsynaptic membrane by giving rise to an EPSP. Activation of an inhibitory synapse, on the other hand, can either clamp the potential to remain around V_{rest} or cause an outward current to flow, thereby giving rise to an IPSP that hyperpolarizes the cell. In the language of the electrophysiologist, an excitatory synapse injects current into the cell; such inward currents are represented, by convention, as negative currents. For synapses with $g_{syn} \geq 0$, an excitatory synapse has a reversal potential more positive than the resting potential, while an inhibitory synapse has a reversal potential close to or negative to the resting potential.[4]

Figure 4.6 illustrates the typical sequence of a fast EPSP followed by a fast and a slow IPSP seen in a cortical pyramidal cell upon electrical stimulation of the fibers projecting into cortex. The hyperpolarizing potentials are caused by inhibition acting on two separate inhibitory receptors ($GABA_A$ and $GABA_B$) discussed further below.

4.5.2 Conductance Decreasing Synapses

In general, the release of neurotransmitters at an ionotropic synapse increases the postsynaptic membrane conductance, that is, $g_{syn}(t) > 0$. At metabotropic receptors, this need not be the case. A well-known example is the loss of firing rate adaptation seen in hippocampal neurons following the stimulation of fibers releasing noradrenaline. This input leads to the reduction of a calcium-dependent potassium conductance (Fig. 9.9). Another example is the change occurring in the responsivity of thalamic relay cells when mammals wake

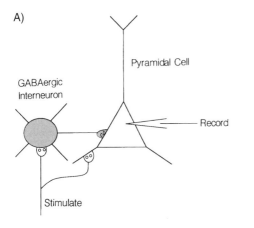

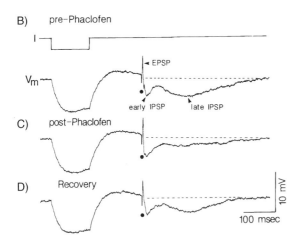

Fig. 4.6 SYNAPTIC POTENTIALS GENERATED IN A CORTICAL CELL (A) Electrical stimulation of axons ascending into the cerebral cortex generates a very rapid EPSP, terminating within 10 msec, followed by two IPSPs, an early fast one and a late slow one. This sequence of events is common to most cortical cells upon stimulation of their extracortical afferents (Douglas, Martin, and Whitteridge, 1991; Tseng and Haberly, 1988). (B) and (D) The inhibition is caused by recruitment of cortical interneurons, which release GABA acting on two distinct synaptic receptors. The early IPSP is due to activation of the ionotropic $GABA_A$ receptors, causing an increase in a chloride membrane conductance, while the late IPSP is due to $GABA_B$ receptor activation. This metabotropic synapse causes a hyperpolarizing potassium conductance to open. (C) Applying the substance phaclofen to the slice blocks the $GABA_B$ receptors while not affecting the $GABA_A$-mediated IPSP. The constant current pulse at the beginning of the trace (see B) assesses the extent to which the input resistance changes during application of phaclofen (it does not). Reprinted by permission from McCormick (1998).

---

4. However, under certain circumstances inhibitory input can act functionally as if it were exciting a cell and vice versa (Lytton and Sejnowski, 1991).

up (that is, the switch from slow-wave sleep to awake, attentive state). Activation of a variety of different metabotropic receptors (e.g., muscarinic ACh receptors, receptors for adrenaline and histamine; McCormick, 1992) causes a 10-mV or greater increase in the resting potential. This sustained depolarization, triggered by the *reduction* of a potassium "leak" conductance (that is, $g < 0$), is sufficient to switch the cell from a bursting into a single action potential firing mode with important consequences for information processing (Steriade and McCarley, 1990; McCormick, 1992).

In summary, the postsynaptic effect of a chemical synapse involves a transient change $g_{syn}(t)$ in the membrane conductance in series with a synaptic battery E_{syn}. Before we discuss the biophysical consequences of this further, let us summarize the key properties of the most common forms of excitatory and inhibitory synaptic input.

4.6 Excitatory NMDA and Non-NMDA Synaptic Input

The predominant fast, excitatory neurotransmitter of the vertebrate central nervous system is the amino acid *glutamate*, activating synaptic receptor channels on nearly every nerve cell as well as on many of the supporting glia cells. In the peripheral nervous system of vertebrates, glutamate synapses are nearly unknown. Here the dominant fast neurotransmitter is ACh. Invertebrates use ACh as well as glutamate for fast transmission of information throughout their nervous systems.

Applying glutamate or its close relative *aspartate* onto central neurons causes a fast depolarizing event, providing the substrate for the fast excitatory traffic in the central nervous system. However, receptors sensitive to glutamate turn out to be a diverse lot and their various subtypes are now being expressed using molecular techniques (Sommer and Seeburg, 1992; Westbrook, 1994). About two dozen glutamate receptor subunits have been cloned from the rat brain, some of which are ionotropic and some metabotropic receptors (Dingledine and Bennett, 1995). These are expressed in different locations throughout the brain and differ in their kinetics, degree of voltage dependency, and so on. Thus, quite different from commercial analog and digital CMOS highly integrated silicon circuits, where the degree of specialization of transistors and circuit types across the chip is minimal, each circuit in an animal may use a slightly different synaptic subtype (possibly also depending on the history of the animal as well as on its developmental stage).

The study of glutamate receptors and their associated synaptic properties is currently in a very active phase. We summarize here the pertinent facts of this unfolding story (Ascher and Nowak, 1988; Thomson, Girdlestone, and West, 1988; Bekkers and Stevens, 1989; Hestrin et al., 1990a,b; Mason, Nicoll, and Stratford, 1991; Williams and Johnston, 1991; Stern, Edwards, and Sakmann, 1992; Jonas and Spruston, 1994; Destexhe, Mainen, and Sejnowski, 1994a).

The most important distinction among excitatory glutamate synapses is based on applying various pharmacological substances to the receptor and measuring their action. One set of these agents, called generically *agonists*, activates one subclass of glutamate receptors, with different chemical substances binding to different receptors, like keys fitting into locks. The different receptors are usually identified by the name of their agonists.

In the case of the glutamate-sensitive receptor, two major functional subclasses exist (as mentioned above, many more molecular subtypes have been cloned, but their specific functions are not yet known). One receptor binds kainate, quisqualate, and α-amino-3-hydroxy-5-methyl-4-isoxalone propionic acid (AMPA). The other class of glutamate receptors can be selectively activated by NMDA. In other words, dumping NMDA onto the synapse causes it to bind to the NMDA receptor and the associated synapse to open. It is important to realize that NMDA, AMPA, and the other agonists do not exist within the nervous system but are pharmacological substances used by biophysicists to identify the receptors. In the brain, all glutamate synapses, regardless of their receptor types, are activated by the presynaptic release of glutamate. Because we are not interested in the various subtypes and the proper nomenclature is still being debated, we simply refer throughout this book to NMDA and non-NMDA synapses.

Other pharmacological agents, so-called *antagonists*, very specifically block the various subclasses of glutamate receptors, allowing us to isolate the contributions of the non-NMDA and the NMDA receptors to EPSPs. For instance, the substance 6-cyano-7-nitroquinoxaline-2,3-dione (CNQX) blocks the fast, non-NMDA synapse without interfering with the NMDA synapse, while DL-2-amino-5-phosphono-valeric acid (APV) blocks the slower NMDA component without blocking the non-NMDA component. Again, antagonists do not occur naturally in the brain but are used as a tool to investigate synaptic transmission.

Once glutamate binds to the AMPA receptor the associated channel opens, allowing monovalent cations, mainly Na^+ and K^+, to flow across the membrane. The synaptic reversal potential of this synapse is about 0 mV (absolute potential). At a non-NMDA receptor, the postsynaptic channels activate very rapidly. The synaptic current peaks within a few hundred microseconds, with an exponential decay whose time constant varies between 0.5 and 3 msec (Hestrin, Sah, and Nicoll, 1990b; Hestrin, 1992; Trussel, Zhang, and Raman, 1993; Jonas, Major, and Sakmann, 1993).

The time course of the synaptic conductance increase can in general be described by an nth-state Markov process with $n(n-1)$ time constants (Destexhe, Mainen, and Sejnowski, 1994a; Johnston and Wu, 1995). Although efficient implementations for these schemes are known (Destexhe, Mainen and Sejnowksi, 1994b), in general simplified expressions are used, the most common being the α function of Eq. 1.21 (Rall, 1967; Jack, Noble, and Tsien, 1975),

$$g_{\text{syn}}(t) = \text{const} \cdot t e^{-t/t_{\text{peak}}} . \tag{4.5}$$

The constant is chosen such that $g_{\text{syn}}(t_{\text{peak}}) = g_{\text{peak}}$, that is, $\text{const} = g_{\text{peak}} e/t_{\text{peak}}$. The function $g(t)$ decays to 1% of its peak value at about $t = 7.64 t_{\text{peak}}$. Using $t_{\text{peak}} = 0.5$ msec with a conductance change of between 0.25 and 1 nS reproduces well the observed time course of the non-NMDA input (Fig. 4.7).

Different from the non-NMDA receptor, with whom it usually appears to be colocalized, the amplitude of the conductance change associated with the NMDA synapse depends on the membrane potential. Figures 4.8A and 4.9A illustrate the nonlinear instantaneous I–V relationship. What is apparent is the negative slope-conductance region between -70 and -40 mV. This negative slope is conferred upon the NMDA current by the action of Mg^{2+} ions, which are naturally present in the extracellular environment. If the postsynaptic potential is at rest and glutamate is bound to the NMDA receptor, the channel opens but is physically obstructed by Mg^{2+} ions. As the membrane is depolarized, the Mg^{2+} ions move out, and the channel becomes permeable for a mixture of Na^+, K^+ and a small number of Ca^{2+} ions. The fraction of the current carried by Ca^{2+} ions through the NMDA channel is about 7% (at negative potentials), five times the fractional calcium current through the voltage-independent glutamate channel (Schneggenburger et al., 1993). If the Mg^{2+} ions are removed from the saline solution bathing the neurons in an experimental slice setup, the synapse looses its nonlinearity and becomes purely ohmic (solid line in Fig. 4.8A). The synapse has a reversal potential close to zero. Another difference compared to the

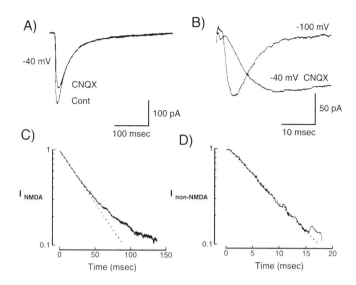

Fig. 4.7 TIME COURSE OF EXCITATORY SYNAPTIC INPUT   Time course of the non-NMDA and NMDA excitatory inputs recorded from cells in hippocampal slices. **(A)** The excitatory postsynaptic current, obtained by clamping the cell to −40 mV and applying glutamate, is shown in the absence (Cont) and presence of the non-NMDA channel blocker CNQX. The remaining current is mediated via NMDA receptors. By convention, depolarizing inward currents are represented as negative currents. **(B)** Closeup of the time course of the NMDA current (labeled CNQX) and the non-NMDA current, obtained by clamping the EPSC to −100 mV and thereby blocking all NMDA channels. The NMDA component continues to rise as the non-NMDA component has started to decay already. **(C)** A semilogarithmic plot of the NMDA-mediated current demonstrates that the decay of the NMDA current is not well fitted by a single exponential (at 31° C). **(D)** The decay of the non-NMDA current (from the curve labeled "−100 mV" in B) is well fitted by a single exponential with a time constant of 7.2 msec (at 24° C). Reprinted by permission from Hestrin, Sah, and Nicoll (1990b)

non-NMDA input is the much slower time course of the NMDA-mediated conductance change due to the intrinsic kinetics of the receptor. Both rise and decay times are at least a factor of 10 times slower, with rise times on the order of 10 msec. Indeed, maximum NMDA receptor activation occurs when the non-NMDA conductance change has almost subsided (Fig. 4.7C).

The NMDA conductance can be derived from a model in which the binding rate constant of Mg^{2+} varies as an exponential function of voltage (Ascher and Nowak, 1988; Jahr and Stevens, 1990). Modeling the time dependency by the difference between two exponentials, we have

$$g_{syn}(t) = g_n \frac{e^{-t/\tau_1} - e^{-t/\tau_2}}{1 + \eta[Mg^{2+}]e^{-\gamma V_m}}, \tag{4.6}$$

with $\tau_1 = 80$ and $\tau_2 = 0.67$ msec, $\eta = 0.33/mM$, $\gamma = 0.06/mV$ and $g_n \approx 0.2$ to 0.4 nS (at 35° C). The physiological concentration of Mg^{2+} ions is around 1 mM.

The synaptic current is obtained by multiplying this conductance by the membrane potential (since the associated $E_{syn} \approx 0$). Figure 4.8B shows the time course of an experimentally measured NMDA current as compared with this model.[5] Better fits can be obtained by using more complex models (Clements and Westbrook, 1991).

5. The temperature at which electrophysiological experiments occur has a significant impact on the dynamics of neuronal processes. Experiments on brain slices and cultured neurons are generally carried out around 22–25° C, significantly below the

A)

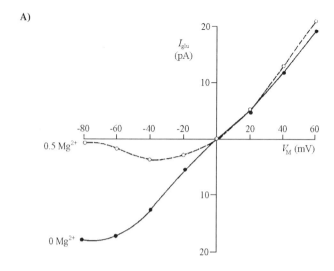

B)

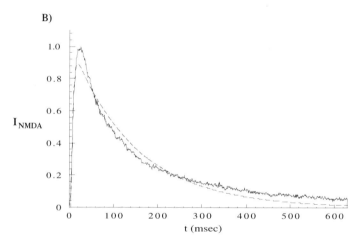

Fig. 4.8 VOLTAGE DEPENDENCY OF THE NMDA CURRENT Current-voltage relationship associated with the NMDA receptor. Because of the action of Mg^{2+} ions in blocking the underlying channel, the associated postsynaptic conductance increase is voltage dependent, different from other fast synaptic inputs. **(A)** Current-voltage relationship of the NMDA current in the absence and presence of 0.5 mM magnesium. While the current behaves relatively ohmic in the absence of magnesium, under physiological concentrations of Mg^{2+} a strong voltage dependency is revealed. Reprinted by permission from Nowak et al. (1984). **(B)** Experimentally recorded normalized NMDA current in a brain slice kept at room temperature (Hessler, Shirke, and Malinow, 1993; heavy jagged line). Superimposed is the current computed from Eq. 4.6, with $\tau_1 = 145.5$ and $\tau_2 = 4.1$ msec. Unpublished data from A. Destexhe, printed with permission.

(continued) $37°$ C baseline in mammals. In general, the absolute conductances associated with ionic channels vary little with the temperature. However, the rates with which these channels change conformation (Chap. 8) speed up considerably at higher temperatures. The extent of this sensitivity is characterized by the *temperature coefficient* Q_{10}, defined as the increase in rate as the temperature changes by $10°$ C. Hestrin, Sah, and Nicoll (1990b) report an average temperature coefficient associated with the decay time constant of the fast non-NMDA current of $Q_{10} = 2.7 \pm 0.8$, with a decay equal to 7.9 msec at $26.5°$ C. This implies a much faster decay of $7.9 \, Q_{10}^{(26.5-37.0)/10} = 2.78$ msec at $37°$ C in the animal, which explains why the NMDA current plotted in Fig. 4.9B, obtained at $23°$ C, has larger values of τ_1 and τ_2 than those indicated here.

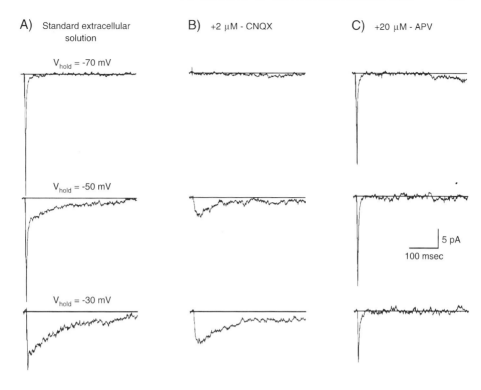

A) Standard extracellular
        solution

V_{hold} = -70 mV

V_{hold} = -50 mV

V_{hold} = -30 mV

B) +2 μM - CNQX

C) +20 μM - APV

5 pA

100 msec

Fig. 4.9 EXCITATORY POSTSYNAPTIC CURRENTS Excitatory postsynaptic current recorded at the soma of inhibitory stellate cells in slices of rat visual cortex in response to the synaptic input, recorded by clamping the membrane potential to three different values (Stern, Edwards, and Sakmann, 1992). The EPSCs are measured at three different clamp potentials. **(A)** As discussed in the text, the EPSC mainly reflects the time course of the underlying synaptic channels, while the time course of the EPSP is dictated by the electrotonic structure of the postsynaptic site. **(B)** Pharmacological dissection of the two components of a stimulus-evoked EPSC. Here, the fast component is blocked by application of the non-NMDA antagonist CNQX, while in **(C)** the slow NMDA-mediated component is blocked by APV. Because the experiments were performed at room temperatures, EPSCs in the living animal will be faster. Reprinted by permission from Stern, Edwards, and Sakmann (1992).

As discussed in Sec. 1.4, the *voltage-clamp* paradigm allows the physiologist to measure the current needed to keep the membrane at a fixed potential (the clamp potential V_{clamp}). This technique was developed by Marmont (1949), Cole (1949), and Hodgkin, Huxley, and Katz (1949) to record accurately the membrane current flowing across the axonal membrane. Using a high-gain feedback amplifier, enough current is supplied to the membrane to stabilize the potential at V_{clamp}, even if the membrane conductance is changing rapidly (as during an action potential; see Chap. 6). The basic circuit is presented in an highly idealized form in Fig. 4.10. For a primer on the theory of voltage clamping, consult Appendix A in Johnston and Wu (1995).

When voltage clamping is applied to a cell receiving synaptic input, the clamp current I_{clamp} needed to keep the membrane at V_{clamp} is usually termed *excitatory postsynaptic current* (EPSC) or *inhibitory postsynaptic current* (IPSC), depending on whether the current is negative or positive. If the synaptic input is at, or close to, the location of the electrode, cable properties can be neglected and the synaptic current can be estimated via the use of the voltage-clamp technique by fixing the somatic potential to the resting potential,

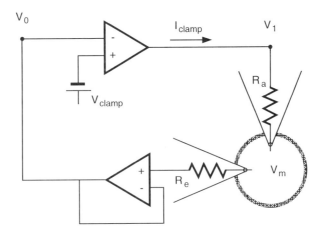

Fig. 4.10 VOLTAGE-CLAMP SETUP The technique of *clamping* the membrane potential to a particular value and measuring the resultant current constitutes a key technological advance in cellular biophysics. We here illustrate the vanilla flavored two-electrode voltage-clamp circuit schematically. The voltage recording electrode at the bottom connects to a high impedance follower circuit that acts as a buffer, drawing only minimal current from the cell. Its output voltage $V_0 \approx V_m$. This output is connected to the negative input of an amplifier with gain A. Its output voltage $V_1 = A(V_{\mathrm{clamp}} - V_0)$. The associated current I_{clamp} flows across the access resistance R_a into or out of the cell. In the limit of a very large gain A, the membrane potential across the cell V_m is "clamped" to V_{clamp}.

$$I_{\mathrm{clamp}}(t) = g_{\mathrm{syn}}(t)(V_{\mathrm{clamp}} - E_{\mathrm{syn}}) \tag{4.7}$$

with $V_{\mathrm{clamp}} = V_{\mathrm{rest}}$. For synaptic inputs that only give rise to a small postsynaptic potential at the soma, the driving potential is not much changed and I_{clamp} constitutes a fair approximation to the actual postsynaptic current, defined via Eq. 4.4.

   If the site of the synapse does not coincide with the site at which the membrane potential is clamped—common to most situations when one is recording from the cell body and the synaptic input is located somewhere in the dendritic tree—cable properties intervene and one no longer measures the pure synaptic current (the synapse is said to be improperly space clamped). Establishing whether or not voltage clamping is complete is a challenging problem with no cut and dry answers (see Smith et al., 1985).

   In many inhibitory interneurons, excitatory input makes frequent contacts directly onto the cell bodies, allowing them to be easily voltage clamped. Stern, Edwards, and Sakmann (1992) exploit this fact to record excitatory postsynaptic currents in inhibitory stellate cells in rat visual cortex (given their small size, the associated high input resistances are very high, between 0.5 and 2 GΩ). Stimulating neighboring cells leads to the small unitary EPSCs shown in Fig. 4.9 (unitary in the sense that below a certain stimulus threshold no EPSC can be seen, while the peak amplitude of the EPSC remains constant once the threshold has been exceeded). Making judicious use of CNQX and APV, the specific blockers of the non-NMDA and the NMDA receptors, Stern and his colleagues separate the two components. Clamping the membrane to near its resting potential, it can be seen that the fast and large component is due to non-NMDA input, while the slow and late component is NMDA mediated. Clamping at progressively more depolarizing currents increases the component remaining after application of CNQX, that is, the NMDA component, while the APV-insensitive component is decreased. Because the synaptic driving potential is progressively reduced as V_{clamp} is increased, the non-NMDA-mediated current decreases, while the NMDA current increases, the decrease in driving potential being compensated for by the increase in the conductance in the negative slope-conductance region. Glutamate-mediated synaptic input usually has both NMDA and non-NMDA components, arguing for colocalization of both receptors in the postsynaptic membrane (as in Fig. 4.9).

NMDA channels are in a sense a theoretician's dream come true, since they implement within a single protein complex a molecular AND gate: a significant depolarizing current can only be obtained in the presence of presynaptic neurotransmitters *and* postsynaptic de-polarization. This action is only possible because of the fact that, different from non-NMDA excitation and GABAergic inhibition, the amplitude of the postsynaptic conductance change depends on the postsynaptic potential. NMDA receptors have been experimentally impli-cated as contributing toward the conjunctive nonlinearities required for Hebb's learning rule (Chap. 13).

Before finishing this section, we need to point out a crucial difference between the EPSC and the resultant EPSP (the same applies for the relationship between IPSC and IPSP): the postsynaptic current always *leads* the postsynaptic voltage. Since the current flowing across a pure capacitance is proportional to the derivative of the voltage, the voltage response to a sinuosidal current will also be sinusoidal but with a *phase shift* of $90°$; the voltage lags the current by this phase angle. No such phase shift is observed for a pure resistance. For any distributed, mixed resistive and capacitive system, the current will always lead the voltage by a phase angle somewhere between $0°$ and $90°$. While the EPSC is not equal to the current flowing during an EPSP (since in the former case the voltage is clamped to some value while it changes in the latter) it usually is similar if the voltage is clamped to the resting potential of the cell (since the voltage excursion away from V_{rest} is usually small relative to E_{syn}).

The dynamics of the current flowing at the synapse is only limited by how fast the un-derlying channels can switch—in the submillisecond range for fast excitatory or inhibitory input—while the evoked potential change is constrained by postsynaptic parameters (such as the intracellular resistance, the membrane capacity, and, to a much lesser extent, the membrane resistance) as discussed at length in Sec. 3.6.3.

One functional reason to stress EPSCs over EPSPs is that the *current* flowing at the soma in response to synaptic input is a much more meaningful measure of synaptic efficiency than the peak postsynaptic potential (see Chaps. 17 and 18).

4.7 Inhibitory GABAergic Synaptic Input

The most common inhibitory neurotransmitter in the central nervous system of both invertebrates and vertebrates appears to be GABA. In the thalamus and cortex, about a quarter of all cells utilize GABA. (For the recent physiological literature on GABA see McCormick, 1990; Edwards, Konnerth, and Sakmann, 1990; LaCaille, 1991; Berman, Douglas, and Martin, 1992.) The second important inhibitory neurotransmitter, glycine, appears to mediate IPSP in the spinal cord onto motoneurons (Young and MacDonald, 1983) and in the brainstem.

There are two major forms of postsynaptic receptors associated with GABA releasing terminals, termed A and B receptors. They are quite distinct from each other, with their main commonality being that they both bind GABA. As in the case for the excitatory neurotransmitters, our knowledge about these different receptor classes derives from the existence of specific blockers (antagonists): the pharmacological agents *bicuculline* and *picrotoxin* reversibly block $GABA_A$ receptors while *phaclofen* blocks the B type (see Fig. 4.6).

The $GABA_A$ receptor is an ionotropic one, with binding of GABA leading to the direct opening of channels selective to chloride ions (as with the glutamate receptor, more than a dozen subreceptor types are known). The current through these channels reverses at the equilibrium potential for chloride ions, in the neighborhood of -70 mV. In many cells, only chloride conductances are open at rest, implying that their resting potential is close to E_{syn} for $GABA_A$ synapses, an important fact to which we will return. The associated postsynaptic conductance change rises very rapidly, that is, within 1 msec or less, and decays within 10–20 msec.

GABA can also bind to a metabotropic receptor, the $GABA_B$ receptor. With the help of a second messenger (a G protein), activation of this receptor type leads to the opening of channels selective to potassium ions. This makes the associated synaptic battery considerably more hyperpolarized than the reversal potential of the $GABA_A$ receptor: it is between -90 and -100 mV. Due to the indirect coupling between the release of GABA and the opening of the associated K^+ channels, the onset and the duration of the postsynaptic conductance change are much slower, with the peak not occurring until 10 msec or more after transmitter release, the total duration being on the order of 100 msec or longer.

Different from the non-NMDA and the NMDA receptors, $GABA_A$ and $GABA_B$ do not appear to be colocalized. In fact, pharmacological evidence argues for a segregation of these two receptor classes, with $GABA_A$ occurring at or close to the soma and the hyperpolarizing $GABA_B$ type farther away, out on the dendrites (Tseng and Haberly, 1988; Douglas and Martin, 1998). As we will see in the following chapter, $GABA_A$ synapses can implement a multiplication, albeit a "dirty" one, while the hyperpolarizing action of a $GABA_B$ synapse has more similarity to a linear subtraction.

4.8 Postsynaptic Potential

Section 1.4 dealt with a synapse embedded within a simple RC circuit. Let us now treat the more general case of synaptic input to some postsynaptic site i, characterized by the time-dependent input impedance $K_{ii}(t)$. The voltage change $V_i(t)$ in response to a synaptic current (Eq. 4.4) is given by Ohm's law (with all potentials relative to the resting potential and the sign of the synaptic current—following the convention of physiologists— inverted to account for the flow of electric charge),

$$V_i(t) = K_{ii}(t) * (-I_{syn}(t)) = K_{ii}(t) * \{g_{syn}(t)(E_{syn} - V_i(t))\}, \qquad (4.8)$$

where $*$ represents convolution. Equation 4.8 is just a shorthand way of writing

$$V_i(t) = \int_{-\infty}^{+\infty} K_{ii}(t - t')g_{syn}(t')(E_{syn} - V_i(t'))dt'$$

$$\qquad (4.9)$$

$$= \int_0^t K_{ii}(t - t')g_{syn}(t')(E_{syn} - V_i(t'))dt'$$

The last transformation follows from our assumption that $K_{ii}(t)$ and $g_{syn}(t)$ are zero for negative times. Equation 4.9 is a *Volterra integral equation* of the second type, characterized by the fact that the membrane potential appears both on the left-hand and on the right-hand side of the equation (Poggio and Torre, 1978).

Following Eq. 3.16, we can express the voltage at location j caused by the synaptic input at i in terms of the transfer impedance $K_{ij}(t)$,

$$V_j(t) = K_{ij}(t) * (-I_{\text{syn}}(t)) = K_{ij}(t) * \{g_{\text{syn}}(t)(E_{\text{syn}} - V_i(t))\}. \qquad (4.10)$$

Because the expression for V_j depends on V_i, finding V_j necessitates the solution of a system of coupled integral equations (Eqs. 4.9 and 4.10).

4.8.1 Stationary Synaptic Input

In order to develop some intuitions about these equations, let us first treat the stationary case, that is, the case in which a fixed increase in synaptic conductance g_{syn} leads to some fixed change in the voltage. The convolution in Eqs. 4.9 and 4.10 is then reduced to a simple multiplication, and the time-dependent input impedance $K_{ii}(t)$ is replaced by the dc part of the Fourier transform, that is, by the input resistance $\tilde{K}_{ii}(f = 0) = \tilde{K}_{ii}$,

$$V_i = \tilde{K}_{ii} \cdot g_{\text{syn}}(E_{\text{syn}} - V_i) \qquad (4.11)$$

or,

$$V_i = \frac{\tilde{K}_{ii} E_{\text{syn}} g_{\text{syn}}}{1 + \tilde{K}_{ii} g_{\text{syn}}}. \qquad (4.12)$$

This nonlinear equation implies that the postsynaptic potential V_i does not increase indefinitely with g_{syn}, but saturates. Figure 4.11 demonstrates this in the case of transient synaptic input. The reason for the sublinear behavior is the fact that the synaptic current pushes the membrane potential toward E_{syn}, reducing the *driving potential* $E_{\text{syn}} - V_i$. As this becomes smaller, the synaptic current will also become smaller. At $E_{\text{syn}} = V_i$, the current ceases, no matter how large the conductance change, since no potential difference exists across the synaptic conductance g_{syn}.

The determinant of whether the synaptic input excites or inhibits the cell is primarily its reversal potential E_{syn}. If the battery is above the resting potential, an increase in $g_{\text{syn}}(t)$ causes an EPSP; if the converse is true, a hyperpolarizing IPSP results. In the remainder of this chapter, we only treat excitatory inputs. However, the same principles apply for inhibitory synaptic inputs.

Let us consider the membrane potential for a small synaptic input (Rinzel and Rall, 1974). "Small" is defined here relative to the input resistance. If the dimensionless product of the dc input resistance and the conductance change is much less than one, that is, $g_{\text{syn}} \cdot \tilde{K}_{ii} \ll 1$, the input is considered small. For instance, relative to a 500 MΩ dendritic input site, a 0.2 nS synaptic input is small, because $5 \times 10^8 \times 2 \times 10^{-10} = 0.1$. Under these circumstances we can approximate Eq. 4.12 by

$$V_i \approx \tilde{K}_{ii} E_{\text{syn}} g_{\text{syn}} = \tilde{K}_{ii} I_{\text{syn}}. \qquad (4.13)$$

In other words, if the synaptic input is small enough, the membrane potential changes only little (that is, $E_{\text{syn}} - V_i \approx E_{\text{syn}}$), and the synaptic input can be approximated by a constant current source, $I_{\text{syn}} = E_{\text{syn}} g_{\text{syn}}$, independent of the membrane potential. If synaptic input can be treated as a current source, then doubling the amplitude of the current source will simply double the postsynaptic potential change, no bothersome saturating behavior exists, and the output is linear in the input.

We discussed in Secs. 4.6 and 4.7 evidence for conductance increases at single glutamate and GABA synapses of 1 nS or less. This implies that as long as the postsynaptic site has

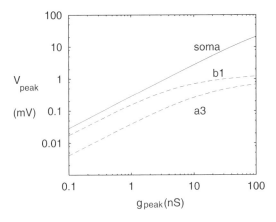

Fig. 4.11 SYNAPTIC SATURATION Relationship between the postsynaptic conductance change and the peak somatic EPSP on a log-log plot. A single fast ($t_{peak} = 0.5$ msec) synapse of variable amplitude g_{peak} is activated at one of three different locations—the soma and at two dendritic sites ($a3$ and $b1$; Fig. 3.7)—and the peak somatic EPSP is shown. Given the relatively high input conductance at the cell body, the somatic input does not saturate within the range shown here, while the two dendritic synapses are saturating for inputs above 10 nS. The simultaneous or near-simultaneous activation of many synapses can easily increase the membrane conductance by this amount or more. Synaptic saturation has important consequences for the cell during massive synaptic input. $R_m = 20,000 \ \Omega \cdot cm^2$, $V_{rest} = -65$ mV and $E_{syn} = 0$ mV.

an input resistance of less than several hundred megaohms, individual inputs can be treated as current injection. This is frequently the case for cortical pyramidal cells, where a single synaptic input can often be treated as a current source. Only in the distal dendrites and thin and elongated spines will the "current" approximation be invalid. However, in the presence of massive synaptic input, as would occur under physiological conditions when hundreds of inputs can be active simultaneously, saturation occurs and has important functional consequences (Chap. 18).

If the product of g_{syn} and $\tilde{K}_{ii}$ becomes larger, but still less than 1, we can use the Taylor series expansion of $1/(1+x) = 1-x+x^2+O(x^3)$ and express the postsynaptic potential as

$$V_i = g_{syn}\tilde{K}_{ii}E_{syn}\left[1 - g_{syn}\tilde{K}_{ii} + (g_{syn}\tilde{K}_{ii})^2 + O((g_{syn}\tilde{K}_{ii})^3)\right] \qquad (4.14)$$

where $O((\ldots)^3)$ is a shorthand notation for cubic and higher order terms. Intuitively, we can think of this series expansion in terms of adding higher order correction terms to the linear current term: first the synapse opens and the current $I_{syn} = E_{syn}g_{syn}$ flows into the cell, causing a change in membrane potential (given by $\tilde{K}_{ii}E_{syn}g_{syn}$; see Eq. 4.13). This variation in membrane potential changes in turn the driving potential from E_{syn} to $E_{syn}(1 - g_{syn}\tilde{K}_{ii})$, which will decrease the potential by $E_{syn}(g_{syn}\tilde{K}_{ii})^2$, which again leads to a change in the driving potential, and so on. Adding up all these terms leads to the above equation. Effectively, we approximate the conductance input g_{syn} by a current input plus a number of "correction" terms: $V_i = \tilde{K}_{ii}(I_{syn}^{(1)} + I_{syn}^{(2)} + I_{syn}^{(3)} + \cdots)$.

In the other extreme case, when the product of the synaptic conductance change and the input resistance is very large, that is, $g_{syn}\tilde{K}_{ii} \gg 1$, we have

$$V_i \approx E_{syn} . \qquad (4.15)$$

The synapse is saturated (Fig. 4.11). As we have seen already in Fig. 1.11 and will discuss further on in more detail, this nonlinear behavior of synapses can be exploited to implement a number of nonlinear neuronal operations, in particular multiplication. The nonlinear transduction process between conductance change and membrane potential was fully recognized by Rall (1964), although usually disregarded by him and other early modelers, both because the nonlinearity involved usually precludes analytical treatment and because the input resistance in the dendritic tree was thought to be relatively small.

The steady-state potential at any other location j in response to a constant synaptic input at location i (Eq. 4.10) is given by

$$V_j = \tilde{K}_{ij} I_{\text{syn}} = \tilde{K}_{ij} \frac{V_i}{\tilde{K}_{ii}} = \frac{\tilde{K}_{ij} E_{\text{syn}} g_{\text{syn}}}{1 + \tilde{K}_{ii} g_{\text{syn}}}. \tag{4.16}$$

The denominator of this expression is identical to the one describing the potential at location i (Eq. 4.12). Therefore, the saturation behavior (e.g., whether or not the synaptic input can be treated as a current) only depends on the electrical properties at the synapse i and not on the transfer resistance $\tilde{K}_{ij}$. Indeed, the potential at j is given by the voltage at i divided by the voltage attenuation,

$$V_j = \frac{V_i}{A^v_{ij}} \tag{4.17}$$

4.8.2 Transient Synaptic Input

In the previous section, we assumed for the sake of mathematical convenience that the synaptic input is so slow that it can be treated as a sustained dc input. Under most physiological conditions, though, fast excitatory synaptic input is anything but stationary but can be over in a time interval less than τ_m.

Solving analytically for the voltage in response to transient conductance inputs is difficult. Numerical evaluations can be carried out in one of two different manners. Either one can first compute the appropriate input and transfer impedances using the cable equation and then solve Eqs. 4.8 and 4.10, or one can directly integrate the cable equation. The first method suffers from the disadvantage that computing the potential in response to n spatially distributed inputs requires the evaluation of about $n^2/2$ input and transfer impedances and only works for a linear membrane, while the second method does not depend on the number of inputs and can be extended in a straightforward manner to deal with membrane nonlinearities.

For a single synaptic input $g_{\text{syn}}(t)$ at location x_0 onto a single passive dendrite, the modified cable equation (Eq. 2.7) has the form

$$\lambda \frac{\partial^2 V_m(x, t)}{\partial x^2} = \tau_m \frac{\partial V_m(x, t)}{\partial t} + V(x, t) + \delta(x_0) r_m g_{\text{syn}}(t)(V_m(x, t) - E_{\text{syn}}) \tag{4.18}$$

with the appropriate boundary conditions.

Historically, it was Rall and his collaborators who first studied solutions to this equation for a variety of different dendritic tree geometries (Rall and Rinzel, 1973; Rinzel and Rall, 1974; Segev et al., 1985). This equation is difficult to solve in closed form and we will not attempt to do so. Rather, the course we follow throughout the book is to integrate the cable equation numerically. Because of their generality, simple numerical integration algorithms are the methods of choice today for solving these partial differential equations (see Appendix C). We made use of the very efficient and freely distributed single-cell simulator program package called NEURON for generating most of the figures in this book. NEURON was developed by Hines (1984, 1989, 1998) and is widely used in the computational neuroscience community. An even more powerful tool is the single cell and neuronal network simulation package GENESIS, developed by Bower and his group (DeSchutter, 1992; Bower and Beeman, 1998).

Figure 4.12 shows the dendritic and somatic EPSPs due to a single non-NMDA excitatory synaptic input located in the basal tree of our layer 5 pyramidal cell. Notice the small size of the somatic EPSP, requiring summation of many excitatory inputs before the membrane potential reaches the firing threshold. If the same input is located at more distal locations (not shown; see, however, Fig. 18.1), the somatic EPSP is even smaller and its peak occurs later in time. Rall has derived in the case of a single equivalent cylinder a relationship between the time of peak potential and the location of synaptic input (Rall et al., 1967), which has been used to infer synaptic location in the case of Ia synaptic input to α motoneurons (Smith, Wuerker, and Frank, 1967). Because pyramidal cells do not fulfill the conditions for reduction to an equivalent cable (Sec. 3.2), no simple analytical relationship between time to peak and location exists. Furthermore, because of the low-pass action of the dendritic tree, the peak potential becomes more and more smeared out as the synaptic input moves away from the soma, making the time at which the peak potential occurs more and more difficult to define.

4.8.3 Infinitely Fast Synaptic Input

What occurs when we consider the other extreme, that of a conductance change that occurs infinitely fast or, in practice, much faster than the neuronal time constant τ_m? Because the conductance change is so rapid, it does not have time to change the membrane potential,

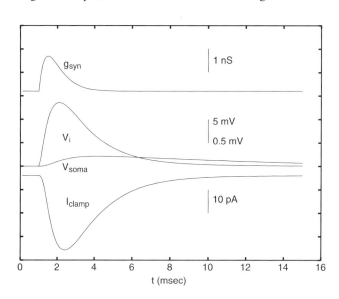

Fig. 4.12 EXCITATORY SYNAPTIC INPUT Simulated synaptic input in response to activation of a single, fast non-NMDA excitatory input in the basal tree of our pyramidal cell (at location $b1$ in Fig. 3.7). We here show the synaptic conductance change $g_{syn}(t)$ (activated at 1 msec and peaking 0.5 msec later) and the local as well as the somatic EPSP (the 5-mV scale bar applies to V_i, while the 0.5-mV scale applies to V_{soma}). Notice the small amplitude of the somatic EPSP, in agreement with experiments (Fig. 4.4). The bottom trace shows the clamp current in response to this input, when the soma is clamped to -65 mV. This approximates the current at the soma when the synaptic input is activated.

and therefore the driving potential, appreciably before it is over. Thus, we expect a very fast conductance input to act like a current input. This can be shown quite easily by assuming $g_{\text{syn}}(t) = g_{\text{peak}}\delta(t)$. Replacing this into the convolution Eq. 4.9 yields

$$V_i(t) = g_{\text{peak}} \int_{-\infty}^{+\infty} K_{ii}(t - t')\delta(t')(E_{\text{syn}} - V_i(t'))dt' = g_{\text{peak}}K_{ii}(t)E_{\text{syn}} \quad (4.19)$$

assuming that $V_i(t = 0) = 0$ (relative to V_{rest}). The resulting EPSP faithfully follows the impulse response function $K_{ii}(t)$. From this we expect that very fast synaptic inputs will show little evidence of conductance changes, even if they are large.

4.9 Visibility of Synaptic Inputs

A question of experimental interest is to what extent a synaptic conductance change can be detected at the cell body using an intracellular electrode. This issue is of paramount concern when trying to resolve the question of the existence and strength of conductance changes, in particular those associated with inhibition, during visual stimulation of neurons in both retinal ganglion cells (Marchiafava, 1979; Watanabe and Murakami, 1984) and cortical neurons (Douglas, Martin, and Whitteridge, 1988; Pei et al., 1991; Ferster and Jagadeesh, 1992; Borg-Graham, Monier, and Frégnac, 1998).

It has been proposed that shunting or silent inhibition vetoes the response of the cell when the visual stimulus, usually a spot or bar of light, moves in the cell's null direction (Torre and Poggio, 1978; Koch, Poggio, and Torre, 1982; Koch and Poggio, 1985b; Sec. 5.1). These theoretical ideas can be tested experimentally by inferring the existence of synaptically mediated shunting inhibition, raising the general issue of the conditions under which a synaptic input somewhere in the dendritic tree can be seen by an electrode at the cell body (Rall, 1967).

One problem is that an intracellular electrode cannot record conductance inputs directly; rather they have to be inferred from their shunting effect on the voltage. A sustained conductance change is measured by injecting a steady-state current I_s at the soma and recording the resultant stationary voltage V_s; the ratio of these two corresponds to the input conductance $\tilde{G}_{ss} = 1/\tilde{K}_{ss}$. The same procedure is repeated during synaptic stimulation. The difference between the old and the new values of the somatic input conductance then corresponds to the change in input conductance $\Delta \tilde{G}_{ss}$ due to the synaptic input g_{syn}. The voltage at the postsynaptic site i is the sum of the synaptic contribution and the current spread from the somatic current injection,

$$V_i = g_{\text{syn}}\tilde{K}_{ii}(E_{\text{syn}} - V_i) + \tilde{K}_{is}I_s . \quad (4.20)$$

The somatic EPSP in the presence of both synaptic input and injected current is

$$V_s' = g_{\text{syn}}\tilde{K}_{is}(E_{\text{syn}} - V_i) + \tilde{K}_{ss}I_s . \quad (4.21)$$

It is now straightforward to show (Koch, Douglas, and Wehmeier, 1990) that the change in somatic input resistance ΔK_{ss} is given by

$$\Delta\tilde{K}_{ss} = \frac{-g_{\text{syn}}\tilde{K}_{is}^2}{1 + g_{\text{syn}}\tilde{K}_{ii}} . \quad (4.22)$$

As expected for any input with $g_{\text{syn}} \geq 0$, $\Delta\tilde{K}_{ss}$ is always negative; physically, this corresponds to the input resistance decreasing in response to an increase in the membrane conductance. Furthermore, it can be proven that the change in input conductance $\Delta \tilde{G}_{ss}$ is always bound by the total synaptic change, that is, $0 \le \Delta \tilde{G}_{ss} \le g_{syn}$.

The equation itself is simple to understand. The current injected at the soma I_s induces a voltage change $\tilde{K}_{is} I_s$ at the location of the synapse. This voltage provides a synaptic driving force, which converts the synaptic conductance input g_{syn} into a voltage change at location i, that is, $(\tilde{K}_{is} I_s) \tilde{K}_{ii} g_{syn} / (1 + g_{syn} \tilde{K}_{ii})$. This is propagated to the soma, causing a voltage change $g_{syn} \tilde{K}_{is}^2 I_s / (1 + g_{syn} \tilde{K}_{ii})$. Dividing this voltage by the current I_s leaves the resistance change.

The higher the input resistance $\tilde{K}_{ii}$ at the synapse, or the further removed the synapse is from the cell body ($\tilde{K}_{is} \to 0$), the less visible the synaptic input becomes. Even if the synaptic conductance change is very large, its effect on the soma will always be limited (as $g_{syn} \to \infty$, the change in input resistance will converge to $-\tilde{K}_{is}^2 / \tilde{K}_{ii}$), except if the input is located directly at the soma, since in that case $\Delta \tilde{K}_{ss} \to -\tilde{K}_{ss}$ and the new input conductance goes to zero.

Notice that $\Delta \tilde{K}_{ss}$ does not depend on the synaptic battery. In principle, this implies that hyperpolarizing, shunting, or excitatory synaptic inputs are all equally visible from the recording electrode. Due to the capacitive nature of the neuronal membrane, transient conductance changes are more difficult to see from the soma than sustained ones (since additional charge between the synapse and the soma flows onto the distributed capacitances). Therefore, Eq. 4.22 represents an upper bound on what can be seen of a conductance change at the cell body.

4.9.1 Input Impedance in the Presence of Synaptic Input

In the previous chapter, we defined the input impedance in a quiescent system without any synaptic input. If a sustained synaptic input occurs at i, the new value of the somatic input resistance

$$\tilde{K}'_{ss} = \tilde{K}_{ss} + \Delta \tilde{K}_{ss} \tag{4.23}$$

follows simply from Eq. 4.22.

In the presence of a time-dependent conductance change, the system is a nonstationary one and the new input conductance is not simply the sum of the time-varying conductance change $g_{syn}(t)$ and the old conductance change. In the case of a patch of membrane without spatial extent characterized by the input impedance $K(t)$, the new input impedance $K'(t, t')$ is a two-dimensional function depending on the time the synaptic input arrived,

$$\frac{1}{K'(t, t')} = g_{syn}(t) + \frac{1}{K(t' - t)}. \tag{4.24}$$

The exact value of the conductance change depends on the relative timing between the onset of the synaptic input and the onset of the current pulse used to measure the input impedance. For stationary input, we obtain the familiar $1/K' = g_{syn} + 1/K$.

4.10 Electrical Gap Junctions

As we mentioned, neurons are also coupled using direct electrical connections (Dermietzel and Spray, 1993). This point-to-point coupling occurs at specialized channels spanning the pre- and postsynaptic membranes called *gap junctions*. Such a junction provides a direct, high-conductance pathway between between neurons and eliminates the possibility that the gap junction current would be shunted by the extracellular space. The strength of the coupling can be modulated by a number of factors, such as high internal calcium concentration (a feature thought to protect the surrounding tissue from the death of any one cell) or intracellular pII (Bennett and Spray, 1987).

Many cells are coupled by gap junctions, giving rise to a *syncytium* of two- or three-dimensional cellular networks. The most dramatic example is the cardiac muscle. Gap junctions allow single action potentials, originating in a group of pacemaker cells, to sweep through all cells in a wavelike manner, generating the rhythmic squeeze and relaxation that is the stuff of life (Noble, 1979).

Glial cells are another example. These cells, thought to play mainly a supporting, metabolic role in the nervous system—such as maintaining the ionic distributions and gradients necessary to homeostasis—lack conventional chemical synapses. They communicate instead via an extensive grid of electrical gap junctions with each other (Giaume and McCarthy, 1996).

One way to study gap junctions is to record from two coupled cells using pipettes, and measuring the current flow between the two as a function of the voltage gradient (Fig. 4.13A). In general, one observes a linear I–V curve, whose slope depends on the number of gap junction channels among the pair of cells (Fig. 4.13B). Typically, each channel contributes about 100 pS of conductance. Gap junctions are usually symmetrical, with a depolarization in one cell leading to a depolarization in the other one, albeit of less value, and a hyperpolarization leading to a less pronounced hyperpolarization in the other cell. From an electrical point of view, this situation can be mimicked by postulating a coupling conductance g_c among the two cells (Fig. 4.13C).

When current is injected into cell 1, as shown in the figure, the application of Kirchhoff's current law to the two nodes and the neglect of all transient behavior leads to the following expression for the transfer resistance:

$$\tilde{K}_{12} = \frac{g_c}{g_1 g_2 + g_1 g_c + g_2 g_c} = \tilde{K}_{21}, \tag{4.25}$$

where g_1 and g_2 correspond to the input conductance of each cell. The inverse of the slope of the I–V curve in Fig. 4.13B corresponds to this coupling resistance. The input resistance of cell 1 is

$$\tilde{K}_{11} = \frac{g_c + g_2}{g_1 g_2 + g_1 g_c + g_2 g_c}. \tag{4.26}$$

Equation 4.25 implies that injecting some current into cell 1 induces the same voltage in cell 2 as injecting the current in cell 2 and measuring the voltage in cell 1. In order to see the asymmetry inherent in direct electrical coupling, we compute the voltage attenuation experienced by going from cell 1 to cell 2,

$$A_{12}^v = \frac{g_2 + g_c}{g_c} \tag{4.27}$$

and in the opposite direction,

$$A_{21}^v = \frac{g_1 + g_c}{g_c} \tag{4.28}$$

Let us assume a coupling conductance of $g_c = 1$ nS and an input conductance of $g_2 = 10$ nS for the second cell. If the first cell has the same input conductance, $A_{12}^v = A_{21}^v = 1.1$, that

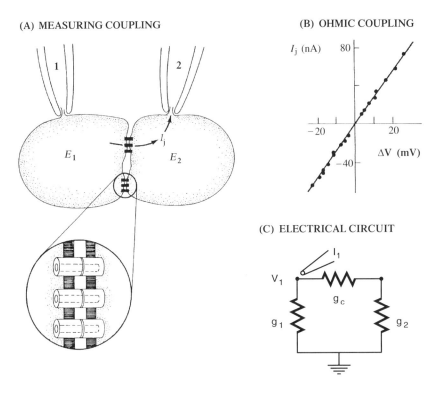

(A) MEASURING COUPLING

(B) OHMIC COUPLING

(C) ELECTRICAL CIRCUIT

Fig. 4.13 ELECTRICAL COUPLING A second type of specific cell-to-cell coupling occurs via *gap junctions* or *electrical synapses*. **(A)** They can be studied by applying a membrane potential across two adjacent cells and measuring the junctional current. Different from a chemical synapse, the current flows instantaneously, with no delay. The inset provides a graphical rendition of the molecular nature of the voltage-independent channels underlying this coupling. Reprinted by permission from Hille (1992). **(B)** The resultant $I–V$ curve, here taken from the gap junction among axons in the crayfish motor system (Watanabe and Grundfest, 1961). The inverse of the slope corresponds to the transfer resistance $\tilde{K}_{12}$ of Eq. 4.25. **(C)** The majority of such junctions can be modeled by a coupling conductance g_c. Current flows from one cell to the next without being shunted by the extracellular cytoplasm. The capacitive nature of the pre- and postsynaptic membranes has been neglected.

is, an EPSP would be attenuated by about 10% in either direction. However, if cell 1 is much smaller, with a tenfold lower conductance of $g_1 = 1$ nS, then the voltage attenuates by a factor of 2 when going from the low-input to the high-input conductance cell, while it still only attenuates by 10% in the opposite direction. Attenuation is symmetrical when the two input conductances g_1 and g_2 are identical. It should be pointed out that attenuation across an electrical synapse is in marked contrast to the high degree of amplification possible at chemical synapses.

Electrical coupling has been studied in detail in the outer layers of the vertebrate retina. Gap junctions among rods as well as among horizontal cells give rise to extended two-dimensional sheets of cells (Kaneko, 1976; Copenhagen and Owen, 1976; Attwell and Wilson, 1980). Theoretical calculations show that a spatio-temporal filter can be ascribed to this syncytium (Torre, Owen, and Sandini, 1983; Torre and Owen, 1983). Indeed, the experimental data suggest that the high-pass filtering properties of the rod network serve to optimize the signal-to-noise ratio by integrating visually evoked signals over a large area for rapid signals and over a small area for slowly changing ones (Detwiler, Hodgkin, and McNaughton, 1980).

In the adult thalamus and cortex, gap junctions have only occasionally been reported. At this point in time it is safe to say that the mere existence as well as any possible functional role of electrical synapses in cortex and associated structures remain unknown.

Electrical synapses do not have any intrinsic delay, unlike chemical ones. Thus, one frequently finds them in time-critical pathways. The classical example is electrical synapses made by presynaptic fibers onto axons of giant motor neurons that mediate the emergency tail flips in the crayfish, the basis of its escape reaction during danger (Furshpan and Potter, 1959). Interestingly, here the connection is rectifying, with a diode-like relationship between the voltage across and the current flowing through this junction.

A further example is the part of the electrosensory system in weakly electric fish that conveys information about the time of occurrence of the zero crossing of the electrical wave signal that the fish can send out and also sense (Heiligenberg, 1991). At multiple locations in this pathway, electric synapses mediate cell-to-cell coupling. For instance, spikes in the so-called T afferents from the periphery carry temporal information in action potentials with a time jitter of 30 ± 25 μsec. A number of these afferents make gap junctions with a spherical cell in the *electrosensory lateral line lobe*, such that a randomly occurring input in any one input fiber does not bring the cell to threshold. Using this conjunctive logic together with synapses that do not introduce any additional temporal smear, reduces the temporal jitter of the output spikes to 11 ± 3 μsec (Carr, Heiligenberg, and Rose, 1986).

One functional role of gap junctions most certainly has to do with the fact that small molecules, such as cAMP (or intracellular dye in an experiment), can pass through these channels, enabling them to diffuse across many cells. This might explain the more wide-spread distribution of gap junctions during early development (DeHaan and Chen, 1990).

4.11 Recapitulation

Fast communication among nerve cells occurs at specialized junctions called synapses. They are very compact: between several hundred million and one billion synapses can be packed into one cubic millimeter of neuronal tissue. Of the two types, chemical and electrical, we focus on the former since they are much more frequent and make very specific point-to-point connections.

It is useful to distinguish fast ionotropic chemical synapses, acting on a millisecond time scale, from metabotropic chemical synapses, acting on a time scale of a fraction of a second to minutes. Conceptually and *cum grano salis*, ionotropic synapses are of the essence in the rapid forms of neuronal communication and computations underlying perception and motor control.

In response to a presynaptic change in membrane potential at a synapse, neurotransmitters are released and diffuses within a fraction of a millisecond across the cleft separating pre- and postsynaptic terminal. At the postsynaptic terminal, neurotransmitter molecules bind to specific receptors, which usually—either directly or indirectly, via involvement of a second-messenger system—open specific ionic channels. Depending on the neurotransmitter-receptor kinetics, these channels remain open for some time and a synaptic current flows across the membrane. Synaptic transmission at central synapses appears to be stochastic. A presynaptic action potential has a probability p of causing a release of a vesicle and a postsynaptic response, where p can be as small as a few percent and depends on the spiking history of the synapse. The amplitude of the postsynaptic signal is variable as well. These factors need to be taken into consideration when thinking about neural computation.

An electrical engineer would be justified in treating a chemical synapse (at the time scale of tens of milliseconds) as a nonreciprocal, two-port device (see the introduction to Sec. 3.4). A two-port description is necessary, since a pair of equations (for both the pre- and the postsynaptic current and voltage changes) is required to completely characterize its behavior; it is non-reciprocal since changes at the postsynaptic side have no (fast) effect on the presynaptic side. Synapses serve to decouple neuronal elements that can each have very different electrical impedances, rather like a follower-amplifier circuit.

At the macroscopic and phenomenological level, a fast synaptic input induces a time-dependent increase in conductance $g_{syn}(t)$ in series with a battery E_{syn}. The sign of E_{syn} relative to the membrane potential at the postsynaptic terminal determines whether synaptic input causes an EPSP (excitatory synapse), an IPSP (inhibitory synapse), or no change in membrane potential (silent or shunting inhibition). The five dominant types of fast synaptic inputs are (1) non-NMDA or AMPA voltage-independent excitation; (2) ACh-mediated excitation; (3) the voltage-dependent and slower NMDA excitatory input that is thought to be crucially involved in synaptic plasticity; (4) the $GABA_A$ type of silent inhibition; and (5) the slower $GABA_B$ hyperpolarizing inhibition. The fact that synaptic input increases the postsynaptic membrane conductance in series with a battery has important consequences. In particular, synaptic inputs can saturate, will influence the input conductance of the cell, and can interact with each other nonlinearly. Only if the amplitude of the synaptic input conductance change is small relative to the local input impedance can synaptic input be treated as a constant current source.

The immense variety of neurotransmitters and postsynaptic receptors gives rise to a staggering combination of possible pairings that act on all possible time scales, and across different spatial scales, from a single synapse to a single ganglion or an entire neural system, such as the thalamus. Synapses are responsible for the most salient difference between nervous systems and even our most advanced digital computers: while the former adapt and learn—a subject we will cover in depth in Chap. 13—the latter do not.

Electrical synapses allow for direct current flow among adjacent neurons. A gap junction can usually be modeled by a fixed conductance. Different from chemical synapses where amplification between the pre- and postsynaptic sites can occur, here the signal is always attenuated. One advantage of this mode of cellular communication is speed, since no synaptic delay occurs. Thus, electrical synapses are frequently found in neuronal pathways, which subserve information that needs to be communicated very rapidly and faithfully. In the retina, gap junctions among photoreceptors and horizontal cells create vast, electrically interconnected networks that filter the incoming visual signal.

5

SYNAPTIC INTERACTIONS IN
A PASSIVE DENDRITIC TREE

Nerve cells are the targets of many thousands of excitatory and inhibitory synapses. An extreme case are the Purkinje cells in the primate cerebellum, which receive between one and two hundred thousand synapses onto dendritic spines from an equal number of parallel fibers (Braitenberg and Atwood, 1958; Llinás and Walton, 1998). In fact, this structure has a crystalline-like quality to it, with each parallel fiber making exactly one synapse onto a spine of a Purkinje cell. For neocortical pyramidal cells, the total number of afferent synapses is about an order of magnitude lower (Larkman, 1991). These numbers need to be compared against the connectivity in the central processing unit (CPU) of modern computers, where the gate of a typical transistor usually receives input from one, two, or three other transistors or connects to one, two, or three other transistor gates.[1] The large number of synapses converging onto a single cell provide the nervous system with a rich substratum for implementing a very large class of linear and nonlinear neuronal operations. As we discussed in the introductory chapter, it is only these latter ones, such as multiplication or a threshold operation, which are responsible for "computing" in the nontrivial sense of information processing.

It therefore becomes crucial to study the nature of the interaction among two or more synaptic inputs located in the dendritic tree. Here, we restrict ourselves to passive dendritic trees, that is, to dendrites that do not contain voltage-dependent membrane conductances. While such an assumption seemed reasonable 20 or even 10 years ago, we now know that the dendritic trees of many, if not most, cells contain significant nonlinearities, including the ability to generate fast or slow all-or-none electrical events, so-called *dendritic spikes*. Indeed, truly passive dendrites may be the exception rather than the rule in the nervous

---

1. The reason for this very small *fan-in* and *fan-out* is that each additional gate whose parasitic capacitance needs to be charged up increases the signal propagation delay between consecutive stages, slowing down how fast the clock can operate. Since the clock speed is one of the most important determinants of performance of the final chip, convergence as well as divergence are kept to a minimum. The situation is quite different for random access memory circuits. For example, in a 64-Mbit DRAM (dynamic random access memory), the fan-in and fan-out associated with the switching transistor at each memory location is approximately 8192. The time taken to charge the large interconnect capacitance that dominates these circuits is reduced through the use of regenerative amplifiers and buffers.

system. We will take up the theme of active dendrites and outline their putative role in computation in Chap. 19.

The voltage change in a passive tree in response to two or more current injections is given by the sum of the voltages induced by the individual current inputs. This linearity between current and voltage is expressed in the cable equation and is the basis for using input and transfer impedances to completely charter the behavior of this system. In the absence of any nonlinearity, no true information processing operations can occur. However, as we keep on emphasizing, a synaptic input acts to change the membrane conductance (in series with a battery), implying that the change in membrane voltage caused by two or more synaptic inputs is *not* the sum of the voltages induced by the individual synaptic inputs. It has not escaped the attention of theoreticians that this nonlinearity could be used to implement a type of multiplication (Blomfield, 1974; Srinivasan and Bernard, 1976; Poggio and Torre, 1978, 1981; Torre and Poggio, 1978).

In the following section, we describe the nonlinear interaction between excitation and inhibition, which has received particular scrutiny in the literature. In the second part of this chapter, we focus on the multiplicative-like interaction among large numbers of voltage-dependent NMDA synaptic inputs. The example discussed here is the most plausible instance of a higher level operation, in this case storing and discriminating complex patterns, implemented using a biophysically very detailed model of a nerve cell. The last section focuses on the important topic of *synaptic microcircuits* (Shepherd, 1972, 1978), that is, very small and specific arrangements of synapses between particular neurons. Such microcircuits usually involve so-called dendro-dendritic synapses among two dendrites and are thought to subserve very specific computations.

5.1 Nonlinear Interaction among Excitation and Inhibition

In Sec. 1.5, we studied this interaction for the membrane patch model. With the addition of the dendritic tree, the nervous system has many more degrees of freedom to make use of, and the strength of the interaction depends on the relative spatial positioning, as we will see now. That this can be put to good use by the nervous system is shown by the following experimental observation and simple model.

5.1.1 Absolute versus Relative Suppression

For an animal, it is often necessary to suppress certain behaviors completely, while under other conditions the threshold for initiating a behavior should only be elevated but not prevented altogether. Vu and Krasne (1992) study how these two operations can be implemented in the same neuron for the tail-flip escape response in the crayfish. Here, as indeed in any other animal, during the execution of the escape reflex propelling the animal away from a potential dangerous situation, no additional escape behavior should be initiated. The neuronal influence responsible for this absolute suppression is called recurrent inhibition. While the crayfish is feeding or otherwise engaged, the escape reflex becomes more difficult to initiate. This suppressive influence, termed tonic inhibition, can be overridden in the presence of a strong enough stimulus.

The escape response is mediated by a pair of *lateral giant command* (LG) neurons. On the basis of intracellular recordings, Vu and Krasne (1992; see also Vu, Lee, and Krasne, 1993) correlate these two types of suppression with distal (tonic inhibition) and proximal (recurrent inhibition) inhibitory inputs onto the dendritic tree of the LG cell (Fig. 5.1). Since it is known that the excitation mediating the escape reflex is distal from the spike-initiating zone, Vu and his colleagues use the two compartment circuits shown in Fig. 5.1A and B to model the two forms of inhibition (here by assuming that the GABA-mediated inhibition is of the silent or shunting type with $E_i = 0$). In both cases, excitation is spatially removed from the output of the circuit. For "proximal inhibition," the inhibitory conductance change is located at the output (close to the soma), while for "distal inhibition," excitation and inhibition are both colocalized in the distal compartment. Vu and Krasne point out that in the former case, inhibition always reduces the EPSP by some amount, no matter how strong excitation is. This, they argue, is the manifestation of *absolute suppression* in the sense that no matter the level of excitation, the EPSP—and therefore the behavior—will be suppressed. Quite a different behavior is observed for distal inhibition: any amount of inhibition can always be overcome by more excitation, that is, the system demonstrates *relative suppression*. They argue that any behavior that needs to be suppressed no matter what the circumstances should use synaptic inhibition that is close to the soma (*absolute suppression*). Conversely, if the threshold for initiating a behavior needs to be elevated—without abolishing it altogether—distal synaptic inhibition is required.

Let us follow Vu and Krasne (1992) and reduce the neuron to but two compartments, one proximal and one distal to the spike initiation zone. To understand the principle of the spatial interaction we will further assume that the time course of synaptic input is slow compared to the membrane time constant and that inhibition is of the shunting type (that is, $E_i = 0$). Let us first analyze the case when inhibition is proximal, here in the somatic compartment (Fig. 5.1A). Following Kirchhoff's current law, which stipulates that the sum of all currents flowing into a node must be zero, we can write for the compartment proximal or close to the spike triggering zone,

$$-V_p g_i + \frac{-V_p}{R_p} + \frac{V_d - V_p}{R_c} = 0 \qquad (5.1)$$

and for the dendritic compartment,

$$(E_e - V_d)g_e + \frac{-V_d}{R_d} + \frac{V_p - V_d}{R_c} = 0. \qquad (5.2)$$

After a few algebraic manipulations, we arrive at an expression for the proximal depolarization,

$$V_p = \frac{E_e g_e R_d R_p}{R_c + R_d + R_p + g_e(R_d R_p + R_d R_c + g_i R_d R_p R_c) + g_i(R_d R_p + R_p R_c)} \qquad (5.3)$$

The amplitude of the EPSP as a function of g_e for various settings of inhibition is plotted in Fig. 5.1A. What is apparent is that a fixed amout of inhibition at the output, that is, the soma, can reduce the EPSP amplitude, no matter what the level of excitation g_e. This absolute dependency on g_i readily becomes apparent in the limit of large excitatory inputs,

$$\lim_{g_e \to \infty} V_p = \frac{E_e R_p}{R_p + R_c + g_i R_c R_p}. \qquad (5.4)$$

Repeating the same analysis for the case that excitation and inhibition are colocalized in the distal compartment leads to

$$V_p = \frac{E_e g_e R_p R_d}{R_c + R_d + R_p + g_e(R_p R_d + R_d R_c) + g_i(R_p R_d + R_d R_c)}. \qquad (5.5)$$

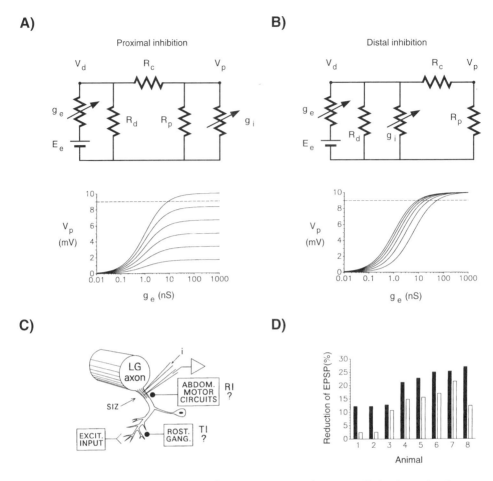

Fig. 5.1 RELATIVE VERSUS ABSOLUTE SUPPRESSION An important distinction exists between inhibition, which can partially suppress an excitatory input (*relative suppression*), and inhibitory input, which cannot be overridden by excitation, no matter what its amplitude (*absolute suppression*). These different types of inhibition were demonstrated by Vu and Krasne (1992) in the case of the tail-flip escape behavior in the crayfish. Two circuits emphasize the relative placement of excitation (with an associated conductance change g_e in series with $E_e = 100\,\text{mV}$) in the periphery and shunting inhibition (of amplitude g_i and $E_i = 0$) in either the proximal (**A**) or the distal (**B**) compartment (see Eqs. 5.3 and 5.5). The curves were generated using a variable amount of inhibition (0, 0.2, 0.5, 1, 2, and 5 from top to bottom in units of $g_e R_d$). No matter how large the excitatory input, proximal inhibition can reduce the peak potential in a graded manner (*absolute suppression*). This is not true if excitation and inhibition are colocalized, where excitation can always overcome the inhibition, providing a substrate for *relative suppression*. (**C**) Part of the neurobiological circuitry underlying the escape reflex. Excitatory inputs onto the dendrites of the lateral giant (LG) command neuron mediate the reflex. It can be partially suppressed by so-called tonic inhibition (TI), which is inferred to be in the distal dendritic tree. The escape reflex can be completely abolished by recurrent inhibition (RI), which is thought to synapse close to the spike trigger zone. SIZ demarcates the spike initiating zone. (**D**) Another prediction concerns the relative amount of attenuation (the F factor of Eq. 5.14) for small (that is, small g_e value; solid bars) and large (large g_e value; open bars) EPSPs for distal inhibition. The model predicts that smaller EPSPs should be associated with a larger reduction (larger F values) than larger EPSPs, as borne out by this data from eight different animals. For proximal inhibition, the F factor should depend only weakly on g_e (see upper left panel). Reprinted in modified form by permission from Vu and Krasne (1992).

Although similar to Eq. 5.3, a crucial difference emerges in the limit,

$$\lim_{g_e \to \infty} V_p = \frac{E_e R_p}{R_p + R_c} \tag{5.6}$$

For any fixed level of inhibition, excitation can always override its effect (relative inhibition; Fig. 5.1B). If excitation and inhibition are not located close to each other, things are different since the local potential saturates at E_e, placing a cap on the amount of current the excitatory input can deliver to the soma. Because of this saturation, inhibition will always be able to reduce the total amount of current delivered to the soma. This would not be true if the synaptic input were to behave as a constant current source, since for $g_e \to \infty$, an infinite amount of depolarizing current would flow to the soma, dominating any loss of current via inhibitory synapses.

5.1.2 General Analysis of Synaptic Interaction in a Passive Tree

We have seen how treating a neuron as more than a single spatial compartment enhances its computational power. Let us now analyze the general theory of nonlinear synaptic interactions in a dendritic tree using two-port analysis developed in Chap. 3 (Koch, Poggio, and Torre, 1982).

A word of warning. The theory developed here assumes that the cell operates in its subthreshold domain, where no action potentials are generated. As discussed more fully in Chap. 18, modeling the effect of synaptic interaction on a cell's firing rate can yield surprising results quite different from those in the subthreshold domain (Holt and Koch, 1997).

We assume a constant excitatory input at location e (of amplitude $g_e > 0$ and battery $E_e > 0$) and a constant inhibitory conductance change at location i (of amplitude $g_i > 0$ and battery $E_i \leq 0$; see Fig. 5.2). We neglect the time-dependent aspects of this problem, a simplification that allows us to use the input and transfer resistances described in Sec. 3.4, expressing the change in membrane potential as the sum of the currents contributed by the two synapses. At synapse e the synaptic current $I_e = g_e(E_e - V_e)$ will flow, and at location i the current $I_i = g_i(E_i - V_i)$. The composite postsynaptic potential at location e is given by the synaptic current at e multiplied by the input resistance $\tilde{K}_{ee}$ plus the synaptic current at i times the transfer resistance between i and e, $\tilde{K}_{ei}$,

$$V_e = \tilde{K}_{ee} I_e + \tilde{K}_{ei} I_i \tag{5.7}$$

or,

$$V_e = \tilde{K}_{ee} g_e (E_e - V_e) + \tilde{K}_{ei} g_i (E_i - V_i). \tag{5.8}$$

Likewise, for the voltage change at location i we have

$$V_i = \tilde{K}_{ei} g_e (E_e - V_e) + \tilde{K}_{ii} g_i (E_i - V_i) \tag{5.9}$$

where we exploit the symmetry property $\tilde{K}_{ie} = \tilde{K}_{ei}$. Finally, the potential at the soma is given by

$$V_s = \tilde{K}_{es} g_e (E_e - V_e) + \tilde{K}_{is} g_i (E_i - V_i). \tag{5.10}$$

Notice that the voltage is always specified as the product of the appropriate input or transfer resistance and the synaptic current and that the input and transfer resistances are computed in the absence of any synaptic input. Solving these three algebraic equations is straightforward, resulting in the following expression for the somatic potential:

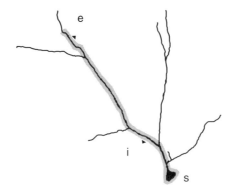

Fig. 5.2 INTERACTION AMONG AN EXCITATORY AND AN INHIBITORY SYNAPSE How does the interaction between an excitatory synapse (at location e) and an inhibitory synapse (at i) in a passive dendritic tree depend on their spatial positions? And what role do the synaptic architecture and the dendritic morphology play? In general, the potential at the soma s is not simply the sum of the individual IPSP and EPSP but can be much less. If the inhibition is of the *shunting* type, with a reversal potential close to the resting potential of the cell, inhibition by itself leads to no significant potential change while still being able to veto the EPSP, as long as the inhibitory synapse is either close to the excitatory one or "on the direct path" between excitation and the soma s (shaded area). The effectiveness of shunting inhibition drops substantially outside this zone.

$$V_s = \frac{g_e E_e (\tilde{K}_{es} + g_i \tilde{K}_e^+) + g_i E_i (\tilde{K}_{is} + g_e \tilde{K}_i^+)}{1 + g_e \tilde{K}_{ee} + g_i \tilde{K}_{ii} + g_e g_i \tilde{K}^*} \tag{5.11}$$

with

$$\tilde{K}_i^+ = \tilde{K}_{is} \tilde{K}_{ee} - \tilde{K}_{es} \tilde{K}_{ie}$$

$$\tilde{K}_e^+ = \tilde{K}_{es} \tilde{K}_{ii} - \tilde{K}_{is} \tilde{K}_{ie}$$

$$\tilde{K}^* = \tilde{K}_{ee} \tilde{K}_{ii} - \tilde{K}_{ie}^2$$

Whether the somatic potential is positive or negative, that is, whether it corresponds to an EPSP or to an IPSP, depends on the relative magnitude of the two contributions.

In order to arrive at a qualitative picture of the behavior of the system, let us assume that the conductance inputs g_e and g_i are small, such that $g_e \tilde{K}_{ee} \ll 1$, $g_i \tilde{K}_{ii} \ll 1$, and $g_e g_i \tilde{K}^* \ll 1$, and that higher than second-order terms in g_e and g_i can be neglected (e.g., $g_e^3 \approx 0$ and $g_e g_i^2 \approx 0$, etc.). Similarly to Eq. 4.14, we can use the Taylor series expansion of Eq. 5.11 to arrive at

$$V_s \approx \tilde{K}_{es} \left[g_e E_e - g_e \tilde{K}_{ee} (g_e E_e) - g_e \tilde{K}_{ie} (g_i E_i) \right]$$

$$+ \tilde{K}_{is} \left[g_i E_i - g_i \tilde{K}_{ii} (g_i E_i) - g_i \tilde{K}_{ie} (g_e E_e) \right] \tag{5.12}$$

$$\approx \tilde{K}_{es} [I_e + I_{ee} + I_{ei}] + \tilde{K}_{is} [I_i + I_{ii} + I_{ie}]$$

The three components in the first bracket are a shorthand version of the currents flowing at locations e that are propagated to the cell body, and the three currents in the second bracket are propagated from the site of inhibition to the soma. I_e is the current generated under the assumption that the driving potential remains unperturbed (that is, $E_e - V_e \approx E_e$). I_{ee} "corrects" the first term by approximating the driving potential as the difference between

E_e and the voltage change $\tilde{K}_{ee}g_eE_e$ associated with the first term. I_{ei} results from the interaction between the two synapses. It can be interpreted as follows: the first-order approximation to the inhibitory current is g_iE_i, causing an IPSP at e of amplitude $\tilde{K}_{ie}g_iE_i$, which will affect the driving potential at e. The appropriate correction current is given by the IPSP multiplied by the local conductance change $-g_e\tilde{K}_{ie}g_iE_i$. By analogy, the currents at the site of the inhibition can likewise be explained by the superposition of a zero-order current plus the first two correction terms.

5.1.3 Location of the Inhibitory Synapse

So far we did not discuss any specific spatial arrangements between excitation and inhibition. We are interested in finding the location where synaptic inhibition can be maximally effective in reducing the amplitude of the excitatory input. Specifically, given an excitatory synapse (of amplitude g_e and battery E_e) at location e, where should the inhibitory synapse (of amplitude g_i and battery E_i) be located such that it maximally reduces the EPSP? This question was the subject of an investigation into the relationship between synaptic architecture and dendritic morphology (Koch, Poggio, and Torre, 1982, 1983). It is tedious but straightforward (Koch, 1982) to prove the following:

On-the-path theorem. *For arbitrary values of $g_e > 0$, $g_i > 0$, $E_e > 0$, and $E_i \leq 0$, the location where inhibition is maximally effective is always on the direct path from the location of the excitatory synapse to the soma (Fig. 5.2).*

Where exactly the optimal location is on this path depends on the details of the system and can shift from the location of the excitatory synapse to somewhere on the path to the cell body. As inhibition is moved toward the soma, its specificity decreases, since the same inhibition now reduces not only the EPSP coming from synapse e but also those from other locations. This is most pronounced at the soma, where an excitatory input anywhere in the dendritic tree will be attenuated. Thus, mapping synaptic architecture onto the dendritic morphology can give rise to different classes of computations.

Three useful properties concerning the optimal location of inhibition (see also Jack, Noble, and Tsien, 1975; Rall, 1967, 1970) that are valid for stationary conductance inputs in arbitrary, passive dendritic trees are:

1. For small synaptic inputs g_e and g_i, the most relevant parameter is the distance between the two synapses. It makes only little difference whether inhibition is behind (with respect to the cell body) excitation or on the path. The strength of the interaction among synapses decreases as the amplitudes of the conductance changes are decreased.

2. As g_e increases while g_i remains constant, the optimal location of inhibition moves along the direct path toward the soma.

3. For very large excitatory inputs ($g_e \to \infty$), all inhibitory synapses located behind the excitatory synapse are completely ineffective.

These properties are a direct consequence of Eq. 5.11, the fact that there exists a unique path between any two points in a dendritic tree, and properties of the transfer resistances (Sec. 3.4).

Experimental verification of the specific nature of synaptic inhibition comes from a study by Skydsgaard and Hounsgaard (1994) carried out on the large dendritic arbor of motoneurons. Using three independent electrodes that can release either glutamate or GABA (so-called *iontophoresis* electrodes) as well as a fourth recording electrode at the cell body, they showed in 10 out of 12 experiments a spatially specific reduction in the glutamate-induced excitatory response. In other words, the shunting action of GABA was primarily effective in reducing the excitatory action of glutamate when the two iontophoresis electrodes were close to each other. No or little effect was observed on the response due to a more distal glutamate-releasing electrode. This constitutes *prima facie* evidence for the spatial selectivity of shunting inhibition.

5.1.4 Shunting Inhibition Implements a "Dirty" Multiplication

Let us consider the specific case when inhibition has a reversal potential close or equal to the resting potential of the cell. The GABA$_A$ mediated increase in chloride conductance approximates such a *shunting* inhibition. In this case, the nonlinear interaction between excitation and inhibition is most pronounced, because activation of the inhibition by itself does not cause any change in potential (hence its proper—but rarely used—name of *silent* inhibition), while inhibition during synaptic excitation can greatly reduce the EPSP. For small enough inputs Eq. 5.11 reduces to

$$V_s \approx E_e(g_e \tilde{K}_{es} - g_e^2 \tilde{K}_{ee} \tilde{K}_{es} - g_e g_i \tilde{K}_{ei} \tilde{K}_{is}) . \tag{5.13}$$

Here the somatic potential is given by two terms in g_e in addition to a crossterm involving a multiplicative interaction between g_e and g_i. Conceptually, one can think of this as a *dirty* or *approximate multiplication* between two synaptic inputs g_e and g_i, with the value of the offset $(g_e \tilde{K}_{es} - g_e^2 \tilde{K}_{ee} \tilde{K}_{es})$ depending in a nonlinear manner on one of the inputs (Poggio and Torre, 1978).

In order to quantify the effectiveness of shunting inhibition, Koch, Poggio, and Torre (1982) introduced the F factor as the ratio of the somatic EPSP in the absence of any inhibition to the somatic EPSP in the presence of inhibition; the larger this number, the stronger the effect of inhibition. On the basis of Eq. 5.11 this can be expressed as

$$F = \frac{g_e \tilde{K}_{es}}{1 + g_e \tilde{K}_{ee}} \cdot \frac{1 + g_e \tilde{K}_{ee} + g_i \tilde{K}_{ii} + g_e g_i \tilde{K}^*}{g_e \tilde{K}_{es} + g_e g_i \tilde{K}_e^+} \tag{5.14}$$

where $\tilde{K}_e^+ = 0$ for inhibition located on the path between e and s. A large F value describes an effective inhibition, while $1/F$ indicates the relative decrease of the somatic EPSP by inhibition. As expected, $F \geq 1$ for all cases. Table 5.1 illustrates some typical F values obtained in a numerical simulation of the effect of inhibition in a retinal ganglion cell responding to a single excitatory input at location 1 (Fig. 5.3A). Let us summarize these results.

Effectiveness of Shunting Inhibition

1. For small synaptic inputs the on-the-path effect is weak and the strength of the interaction depends mainly on the distance between excitation and inhibition. Under these conditions, all locations close to the excitatory synapse are equally effective in reducing excitation.

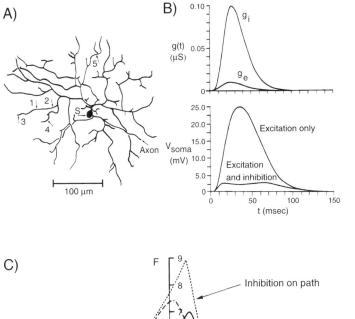

Fig. 5.3 Spatio-Temporal Specificity of Synaptic Interaction Effectiveness of shunting inhibition ($E_i = V_{rest}$) in vetoing an EPSP in a retinal ganglion cell. (**A**) Excitation is always at location 1, while the position of the inhibitory synapse varies. The cell's morphology was reconstructed based on Golgi material from Wässle, Illing, and Peichl (1979). (**B**) The time course of the excitatory and the inhibitory conductance changes reflects the slow retinal dynamics (Baylor and Fettiplace, 1979), with identical time courses but different peak conductance changes (10 nS for excitation and 100 nS for inhibition). (**C**) The F factor, that is, the reduction in the peak somatic EPSP due to inhibition, as a function of the delay between the onset of $g_e(t)$ and $g_i(t)$. The strength of the interaction (see also Table 5.1) is specific in space and time. This result holds over a large parameter range. In particular, R_m (here set to 14,000 $\Omega \cdot cm^2$) can be increased over several orders of magnitude. Reprinted by permission from Koch, Poggio, and Torre (1983).

TABLE 5.1
Effectiveness of Shunting Inhibition

Inhibition g_e	1	1	1	10
g_i	1	10	100	100
3	1.16	2.06	3.43	1.96
1	1.17	2.70	1.80	7.72
2	1.14	2.40	15.00	8.72
4	1.12	1.72	2.47	1.87
5	1.08	1.46	1.84	1.62
s	1.11	2.06	11.57	9.18

F values, describing the effectiveness of shunting inhibition in reducing a somatic EPSP, for stationary synaptic inputs in the retinal ganglion cell model of Fig. 5.3A. The excitatory synaptic input g_e (with $E_e = 80$ mV relative to V_{rest}) is always at location 1, while inhibition g_i is either located behind excitation (3), at the same location as excitation (1), between excitation and the soma (2), in a branch just off the path (4), in a very different part of the tree (5), or at the soma (s). The amplitudes of g_e and g_i are in nanosiemens. The somatic EPSP in the absence of any inhibition for $g_e = 1$ nS is 6.56 mV. $R_m = 14,000\ \Omega \cdot cm^2$ and $R_i = 70\ \Omega \cdot cm$. From Koch (1982).

2. If the two synapses coincide anywhere in the dendritic tree ($i = e$), F reduces to

$$F = 1 + \frac{g_i \tilde{K}_{ii}}{1 + g_e \tilde{K}_{ii}}. \qquad (5.15)$$

This is also the F factor obtained when dealing with a patch of passive membrane in the presence of two synapses (if we identify R with $\tilde{K}_{ii}$). The difference between Eqs. 5.14 and 1.34 (or 5.15) is due to the spatially distributed tree. As discussed in Sec. 5.1.1, excitation can always overcome the effect of fixed but large inhibition.

3. If the amplitude of inhibition is above a critical value (about 50 nS in the case of the retinal ganglion cell simulation; Koch, Poggio, and Torre, 1983), F values can be quite high (2–8), even if the excitatory input is much larger than the inhibitory one, as long as the inhibition is between excitation and the soma. Comparing this against the F factor obtained in a lumped circuit model (Eq. 5.15), we ascertain that these high F values are due to the cable properties of the tree. Inhibition behind excitation or on a neighboring branch (about 10 or 20 μm for the retinal ganglion cell considered here) off the direct path is ineffective in reducing excitation significantly.

4. The specificity for high values of g_i persists even for very large values of the membrane resistivity, when the cell is relatively compact (in terms of electrotonic distance or of the logarithm of the voltage attenuation L^v). For instance, for $R_m = 1$ M$\Omega \cdot cm^2$, the electrotonic size of the ganglion cell in Fig. 5.3A shrinks to 0.03λ and its input resistances increase to over 1 GΩ. Yet for $g_e = 10$ nS and $g_i = 100$ nS, we obtain $F = 10.5, 14.5, 21.0, 2.4, 2.7$, and 2.7 for inhibition at the location of excitation (1), on the path (2), at the cell body (s), behind excitation (3), on a neighboring branch (4), and in a different part of the tree (5; Fig. 5.3A). These effects can be visualized with the aid of a loose analogy between dendrites and water pipes and reservoirs. Shunting inhibition corresponds to opening a hole in a pipe which is coupled to a water reservoir with the same pressure as in the quiescent pipe system. If this hole is between the site where water is being injected and the central pool, it is clear that more water will leave the hole on its way to the large somatic sink than if the hole were to be opened upstream from excitation.

We conclude that if g_i is large enough, the on-the-path condition becomes very specific such that blocking an EPSP is only possible if inhibition is either in the close neighborhood of the excitatory synapse or between excitation and the cell body. Rall (1964) observed earlier that in a single unbranched cable, shunting inhibition effectively vetoes more distal but not more proximal excitation.

Temporal Specificity

So far we only considered the case of stationary synaptic inputs. In the general time-dependent case, the algebraic Eqs. 5.8–5.10 are replaced by integral equations,

$$V_e(t) = (g_e(t)[E_e - V_e(t)]) * K_{ee}(t) + (g_i(t)[E_i - V_i(t)]) * K_{ei}(t)$$

$$V_i(t) = (g_e(t)[E_e - V_e(t)]) * K_{ei}(t) + (g_i(t)[E_i - V_i(t)]) * K_{ii}(t) \qquad (5.16)$$

$$V_s(t) = (g_e(t)[E_e - V_e(t)]) * K_{es}(t) + (g_i(t)[E_i - V_i(t)]) * K_{is}(t)$$

Typically, these equations are solved numerically (Segev and Parnas, 1983; Koch, Poggio, and Torre, 1983). The efficacy of inhibition is characterized by a slight generalization of the F factor of Eq. 5.14 (F is defined as the ratio of the peak amplitude of the somatic EPSP without inhibition to the peak of the somatic EPSP in the presence of inhibition).

Figure 5.3 shows the dependency of F on the relative timing between the onset of excitation and inhibition. Negative delays correspond to inhibition preceding excitation, while positive delays are associated with inhibition following excitation. As can be seen, the specificity of shunting inhibition in vetoing EPSPs generalizes from space to time: on-the-path inhibition can effectively veto excitation if it occurs within about 10 msec of the onset of excitation, about two-thirds of the membrane time constant. As we discussed in Sec. 3.6, the local delay D_{ii} (Agmon-Snir and Segev, 1993) specifies the window during which synaptic inputs can interact with each other. Different locations have different temporal windows associated with them (e.g., Fig. 3.13), allowing for an operationalized definition of what is meant by two *simultaneous* inputs.

The delay between the onset of excitation and optimal inhibition increases with increasing distance between the two synapses, as expected from the definition of the propagation delay in Eq. 2.51.

Decreasing the duration of inhibition below that of excitation decreases its effectiveness, in some cases dramatically. In no case did the F factor for transient inputs exceed the F factor for the corresponding stationary case. That is, inhibition appears to be at its most effective if temporal effects can be discounted.

Figure 5.3 demonstrates the on-the-path effect very vividly. Locating inhibition between the site of the excitatory synapse and the soma can shunt the somatic EPSP by a maximal factor of 9, while moving inhibition beyond excitation (approximately conserving the distance between the two synapses) renders inhibition much less effective, with the F factor dropping to 1.3 (a reduction of about 7). Yet this specificity only occurs in the presence of large conductance changes, possibly outside the physiological range (see Table 5.1), and becomes much less pronounced for smaller values of g_i, when the interaction depends more on the distance between excitation and inhibition.

In summary, the veto effect of inhibition can be strong and specific with respect to spatial location and relative timing provided the following requirements are fulfilled: (1) inhibition must have a reversal potential close to the resting potential of the cell, (2) inhibition should be close to excitation or on the direct path between excitation and the cell body, (3) peak inhibition must be large enough, and (4) inhibition must last (at least) as long as excitation and their time courses should overlap substantially.

5.1.5 Hyperpolarizing Inhibition Acts Like a Linear Subtraction

As we saw, in the limit of small synaptic inputs ($g_e \tilde{K}_{ee} \ll 1$ and $g_i \tilde{K}_{ii} \ll 1$), the action of shunting inhibition on excitation can be characterized as a multiplication with an offset. Even though the interaction between hyperpolarizing inhibition, such as the GABA$_B$ complex, which increases the postsynaptic potassium conductance (reversing close to -100 mV), and excitation also contains such multiplicative effects, it is more linear due to the direct contribution the hyperpolarizing battery makes toward the membrane potential. Inspection of Eq. 5.12 reveals that the first two terms in $\tilde{K}_{is}$ are proportional to the reversal potential of the inhibitory synapses. The more negative the inhibitory reversal potential (relative to V_{rest}), the more these terms will dominate and the more the overall operation will resemble a linear subtraction.

A physiological instance of such a subtractive inhibition that can be linked to a specific computation occurs in the fly's *motion detection system*. Hassenstein and Reichardt (1956; Reichardt, 1961) first suggested that the key mechanism underlying the optomotor response of insects to moving stimuli is mediated by a correlation-like operation. In this scheme, the output from one receptor is multiplied by the temporally delayed (e.g., low-pass filtered) output from a neighboring receptor and then temporally averaged. The component that is direction selective can be extracted from the component that is independent of the direction of motion by subtracting the output of two mirror-symmetric pairs of Reichardt detectors from each other.

A computational analysis (Egelhaaf, Borst, and Reichardt, 1989) of this system predicts that as the subtraction stage between oppositely oriented motion detectors is blocked, the power of the second-order component of the membrane potential in the cellular output stage increases. Experiments with picrotoxin, a blocker of GABA$_A$ receptors, confirm this (Egelhaaf, Borst, and Pilz, 1990). Intracellular recordings from large tangential cells in the lobula plate in the blowfly in combination with pharmacological blockers show that the excitatory input is mediated by fast cholinergic synapses while the subtraction relies on GABAergic synapses that reverse between -70 mV and -80 mV (Borst, Egelhaaf, and Haag, 1995; Brotz and Borst, 1996). A stable resting potential around -50 mV (Fig. 5.4) gives the membrane enough maneuvering room in both directions to implement a linear operation of importance to optomotor behavior in flies.

5.1.6 Functional Interpretation of the Synaptic Architecture and Dendritic Morphology: AND-NOT *Gates*

The specificity of the interaction between excitation and shunting or silent inhibition provides *in principle* the substrate for implementing different classes of analog computations in different dendritic trees. This situation is portrayed in a highly idealized but suggestive manner in Fig. 5.5 using a logical metaphor (Koch, Poggio, and Torre, 1982), bearing in mind that these interactions are continuous and not all or none.

In an unbranched cable (Fig. 5.5A), input i_1 can inhibit more distal excitatory input e_1, e_2 and e_3, while inhibition i_3 can only significantly affect e_3. If we adopt the logic AND-NOT all-or-none operation to characterize this interaction (an output only occurs in the presence of excitation *and no* inhibition), such a cable implements the logic expression

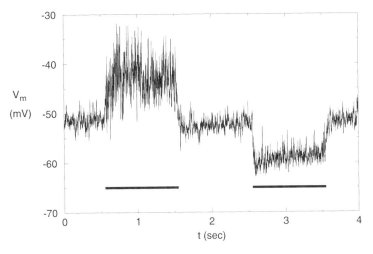

Fig. 5.4 **LINEAR SUBTRACTION IN THE MOTION PATHWAY OF THE FLY** Intracellular recording from a VS cell in the third optic ganglion of the blowfly *Calliphora* during motion of a whole-field grating. This stimulus is moved for 1 sec in the preferred (first black bar) and for 1 sec in the cell's opposite, null direction (second black bar) past the fly. The resting potential is stable around -50 mV, explaining the more or less symmetric response pattern driven by opposing pairs of Reichardt correlation detectors (Borst, Egelhaaf, and Haag, 1995). Excitation is mediated by fast cholinergic and inhibition by GABAergic synaptic input onto this cell. Unpublished data from J. Haag and A. Borst, printed with permission.

$$[e_3 \text{ AND-NOT } (i_1 \text{ OR } i_2 \text{ OR } i_3)] \text{ OR } [e_2 \text{ AND-NOT } (i_1 \text{ OR } i_2)] \text{ OR } (e_1 \text{ AND-NOT } i_1) \quad (5.17)$$

In other words, the soma receives a significant signal only if e_3 is high and not blocked by i_1, i_2, or i_3, or if e_2 is high and not vetoed by either i_1 or i_2, or if e_1 is high and not vetoed by i_1. Due to the on-the-path effect, the more branched dendritic tree in Fig. 5.5B implements a different expression,

$$(e_1 \text{ AND-NOT } i_1) \text{ OR } (e_2 \text{ AND-NOT } i_2) \text{ OR } \{[(\ e_3 \text{ AND-NOT } i_3) \text{ OR }$$
$$(e_4 \text{ AND-NOT } i_4) \text{ OR } (e_5 \text{ AND-NOT } i_5) \text{ OR } (e_6 \text{ AND-NOT } i_6)] \text{ AND-NOT } i_7\} \quad (5.18)$$

Thus, the combination of dendritic morphology coupled with specific synaptic circuits conspires to create a rich class of nonlinear operations, with different dendritic trees carrying out different computations. These interactions would be based on voltage-independent AMPA excitatory synapses and GABA_A inhibitory synapses.

The "dendritic tree as logic network" metaphor has been a popular concept, which has been extended to a number of other biophysical situations and logical operators (Segev and Rall, 1988; Shepherd and Brayton, 1987; Shepherd, 1992; Zador, Claiborne, and Brown, 1992; see Chaps. 12 and 19).

Yet as succinctly summarized by Mel (1994), this simile can be criticized on several grounds. The principal argument against postulating that specific arrangements of individual synapses act as logical or analog gates is the great demand that this places on developmental processes. In order to achieve the precise type of wiring idealized in Fig. 5.5, some learning mechanism has to guide individual synapses to individual branches in the dendritic tree to achieve the precise spatial arrangements required. For instance, moving i_7 in Fig. 5.5B to the adjacent branch significantly changes the associated logical expression. Thus, the great specificity of this interaction represents at the same time its own Achilles heel (Mel, 1994).

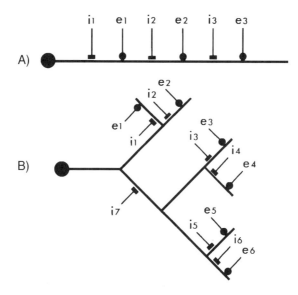

Fig. 5.5 IMPLEMENTING AND-NOT LOGIC IN A DENDRITIC TREE   Idealized binary view of the continuous nonlinear interactions occurring between excitation and shunting inhibition. Inhibitory inputs (rectangles) veto more distal excitatory inputs (circles), but have only a marginal effect on inputs more proximal to the soma, thereby approximating a logic AND-NOT gate. The two distinct dendritic architectures, coupled with specific synaptic wiring, implement very different logical expressions (see Eqs. 5.17 and 5.18). Reprinted by permission from Koch, Poggio, and Torre (1982).

It is possible, of course, that on-the-path effects are exerted not by individual synaptic inputs but by groups of synapses, as is the case in the distinction between absolute and relative suppression (Vu and Krasne, 1992). This spatially less precise degree of spatial interactions places a much reduced burden on developmental processes, since it only specifies that inhibitory synapses should be either adjacent to the excitatory ones or in a different part of the dendritic tree.

On biophysical grounds, it is unclear how specific the interaction between excitatory and inhibitory synapses actually is. As illustrated in Table 5.1, if the amplitude of the inhibitory conductance change is small, little specificity results. Furthermore, during physiological conditions, the cell potential may never be at V_{rest} because the cell receives tonic excitatory input. Under these conditions, the distinction between hyperpolarizing and shunting inhibition is much smaller than we assume here, since both reverse negative to the "effective" resting potential.

5.1.7 Retinal Directional Selectivity and Synaptic Logic

Directional selectivity of retinal ganglion cells is one example of a complex nonlinear operation that appears to use synaptic logic (Koch, Poggio, and Torre, 1986). A subset of ganglion cells in the vertebrate retina fire vigorously in response to the motion of a spot of light in one direction, the preferred one, but are silent to motion in the opposite, null, direction. The classical experiments of Barlow and Levick (1965) inferred that synaptic inhibition—now known to be GABAergic—plays a critical role by preventing the cell from responding in the null direction. More recent work on rabbit ganglion cells reveals an additional facilitatory direction-selective component (Grzywacz and Amthor, 1993).

A popular model for the biophysical basis of directional selectivity is nonlinear synaptic interaction between cholinergic excitation and GABAergic inhibition mediated by $GABA_A$ receptors in the dendritic tree of ganglion cells (Torre and Poggio, 1978; Koch, Poggio, and Torre, 1982, 1983, 1986). Intracellular recordings from turtle and frog direction-selective ganglion cells support activation of shunting inhibition in the null direction (Marchi-afava, 1979; Watanabe and Murakami, 1984). The very complex dendritic morphology of direction-selective ganglion cells (Oyster, Amthor, and Takahashi, 1993) provides an ideal substrate for numerous local and nonlinear veto operations of the type schematized in Fig. 5.5B.

Yet in some ganglion cells, direction selectivity is maintained or even reversed in the presence of pharmacological blockers of GABA. Patch clamping the dendrites of turtle ganglion cells revealed an excitatory input that is already direction selective (Borg-Graham, 1991b). Furthermore, motion in the preferred direction causes a much larger increase in the input conductance than motion in the null direction, in conflict with the models presented here, in which the exact temporal relationship between excitation and inhibition shapes the postsynaptic response but the total conductance change remains invariant to the direction of motion (e.g., Fig. 5.3).

It thus appears that at least some fraction of the direction-selective response is computed presynaptically, most likely in cholinergic neurons possessing a unique dendritic branching pattern called *starburst amacrine* cells (Masland, Mills, and Cassidy, 1984). In the rabbit, starburst amacrine cells are well situated to provide synaptic input from their distal dendrites to directionally selective ganglion cells via graded, dendro-dendritic synapses (Vaney, Collin and Young, 1989). Their distinctive dendritic branching pattern is especially suited for local generation of directionally selective outputs, such that directional selectivity could be implemented using a combination of nonlinear synaptic interaction and inhibitory kinetics that are slower than those for excitation (Borg-Graham and Grzywacz, 1992). In the second scheme, excitation is always trailed by a slower inhibitory wave—if the activated synaptic input is moving *toward* a site of synaptic output (in this case a dendritic tip), excitation reaches the output before the inhibition catches up. On the other hand, if the input is moving *away* from the output, the inhibition, interposed between the excitation and the output, effectively shunts the excitatory current, preventing it from depolarizing the output zone (see also Rall, 1964). Yet, biophysical simulations of amacrine cells suggest that the nonlinear interaction of excitation and inhibition probably is the dominant mechanism generating a direction-selective response under normal conditions (Borg-Graham and Grzywacz, 1992).

The final verdict is not in. It appears likely that even such a simple operation as distinguishing the direction of a moving stimulus is implemented using a plurality of biophysical mechanisms acting at several sites, some requiring inhibition and some not (Amthor and Grzywacz, 1993). Such redundancy might be necessary in the face of demands that the circuitry wire itself up during development and retain its specificity in the face of a constantly varying environment.

5.2 Nonlinear Interaction among Excitatory Synapses

What about the interaction expected between two excitatory voltage-independent synaptic inputs (with synaptic reversal potential E_e and stationary conductance increases g_1 and g_2)? If both synapses are colocalized, the somatic EPSP will be of amplitude

$$\frac{(g_1 + g_2)E_e \tilde{K}_{es}}{1 + (g_1 + g_2)\tilde{K}_{ee}} = \frac{g_1 E_e \tilde{K}_{es}}{1 + (g_1 + g_2)\tilde{K}_{ee}} + \frac{g_2 E_e \tilde{K}_{es}}{1 + (g_1 + g_2)\tilde{K}_{ee}}. \qquad (5.19)$$

In the opposite case, when the two sites 1 and 2 are electrotonically decoupled, that is, $\tilde{K}_{12} \to 0$, the somatic potential is given by

$$\frac{g_1 E_e \tilde{K}_{1s}}{1 + g_1 \tilde{K}_{11}} + \frac{g_2 E_e \tilde{K}_{2s}}{1 + g_2 \tilde{K}_{22}}. \qquad (5.20)$$

Assuming that $\tilde{K}_{ee} \approx \tilde{K}_{11} \approx \tilde{K}_{22}$ and $\tilde{K}_{es} \approx \tilde{K}_{1s} \approx \tilde{K}_{2s}$ (that is, 1 and 2 have the same electrotonic properties), it is clear that the somatic EPSP is smaller if both synapses coincide (Eq. 5.19) than if both are far apart (Eq. 5.20). The same results holds true even if the coupling term $\tilde{K}_{12}$ is nonzero and transient inputs are considered (Koch, Poggio, and Torre, 1982). The reason for this *sublinear addition* is simply the fact that as the postsynaptic potential increases, the amount of current flowing through voltage-independent excitatory synaptic channels decreases (Rall, 1977).

Experimental studies of these predictions have been scant, with the large motoneurons providing a favorite preparation. Two earlier studies (Burke, 1967; Kuno and Miyahara, 1969) had concluded, based on an indirect inference, that excitatory inputs interact sublinearly when located somewhere in the distal dendritic tree. In the Skydsgaard and Hounsgaard (1994) experiment mentioned, glutamate was applied locally to different parts of the dendritic tree of turtle motoneurons from two independent iontophoresis electrodes. At the cell's resting potential they observed linear summation at the soma (see also Langmoen and Andersen, 1983), which turned supralinear as the membrane was depolarized by the recording electrode (most likely due to amplifying voltage-dependent membrane conductances). Yet basic biophysics implies nonlinear saturation due to the conductance-increasing nature of synaptic input. Is it possible that synaptic saturation is always compensated for by voltage-dependent outward currents distributed in the dendritic tree? And if so, why?

5.2.1 Sensitivity of Synaptic Input to Spatial Clustering

As emphasized in Chap. 4, in the central nervous system many, if not most, neurons receive synaptic input from a mixture of voltage-independent AMPA and voltage-dependent NMDA ionotrophic synapses.

For instance, a major role for NMDA input in cat visual cortex is supported by experiments that blocked NMDA responses using the pharmacological agent APV. This procedure caused the visual response of neurons whose cell bodies are located in the upper three layers to be either strongly attenuated or eliminated altogether, while neurons in input layer 4 could still be visually activated (Miller, Chapman, and Stryker, 1989; Fox, Sato, and Daw, 1990; Daw, Stein, and Fox, 1993). Thus, NMDA distribution is most likely not homogeneous but layer specific.

Such a voltage-dependent excitatory input can implement a multiplicative-like interaction: while the EPSP produced by a single NMDA synapse might not be strong enough to allow current to flow through the associated channels, two or more simultaneously active

NMDA inputs could depolarize the potential sufficiently to relieve the Mg^{2+}-mediated blockage and lead to a much larger depolarization (Fig. 4.8; Koch, 1987). Motivated by the possibility that such interactions can instantiate a specific class of computations underlying certain forms of learning, Mel (1992, 1993, 1994; Mel and Koch, 1990; see also Brown et al., 1992) initiated a detailed biophysical investigation of synaptic integration in the dendritic tree of cortical pyramidal cells on the basis of three different mechanisms: NMDA receptors and calcium- and sodium-mediated active dendritic membrane conductances. In this chapter we focus on the interaction occurring in a passive dendritic tree among NMDA synapses, deferring the more complex set of events that can occur in a dendritic tree with voltage-dependent conductances to Sec. 19.3.4.

The basic question Mel addresses is the following: what is the sensitivity of excitatory input to spatial clustering in the postsynaptic neuron? In particular, are spatially adjacent synapses in a passive dendritic tree more or less effective than the same number of synapses spread throughout the tree?

To answer these questions, Mel (1992) models the layer 5 pyramidal cell (Fig. 3.7) from cat visual cortex with a passive dendritic tree and a cell body that contains the two basic Hodgkin and Huxley currents I_{Na} and I_K necessary for spike initiation (and which are described exhaustively in the following chapter). Mel randomly distributes 100 synapses in clusters of k over the entire dendritic tree, like sprinkling salt over food. Figure 5.6 illustrates this procedure for cluster sizes $k = 1$ and 7. In the first case 100 synapses are placed at random within the dendritic tree, while in the second case clusters of seven synapses each (plus one cluster of two synapses) are assigned randomly to 14 locations in the tree. In the absence of any NMDA input, each synapse is placed onto a dendritic spine and treated as a fast, glutamergic input of the AMPA type. In the alternative case, 90% of the postsynaptic conductance increase is of the NMDA type, using the membrane potential and time dependence of Eq. 4.6 (Fig. 5.7). The NMDA-mediated synaptic current, the product of the driving potential (which decreases for increasing levels of membrane depolarization) and the conductance change (which first increases and subsequently saturates with increasing V_m), increases to about -30 mV potential, after which it decreases and reverses sign at the synaptic reversal potential of 0 mV. The remaining 10% of the synaptic conductance is of the AMPA type.

All synapses (whether belonging to the same cluster or not) are activated independently of each other at a mean rate of 100 Hz. As a measure of cell response, Mel considers the number of spikes triggered over a 100-msec period. Using the peak somatic potential or the time integral of the somatic potential (in the absence of a somatic spiking mechanism) gives results that are qualitatively very similar. In the passive cell with voltage-independent inputs, the somatic response following activation of 100 synapses randomly located throughout the dendritic arbor is much bigger than the response following activation of 10 clusters of 10 neighboring synapses (Fig. 5.7). Indeed, the former situation triggers four spikes, while the latter none (even though here the local response is larger; but, of course, fewer clusters contribute toward the somatic excitability).

If the cell's membrane is endowed with NMDA channels, it will respond selectively to patterns of stimulation in which activated synapses are spatially clustered rather than uniformly distributed across the tree (Fig. 5.7). The cell fails to spike in response to random activation of 100 NMDA "isolated" synapses (firing at 100 Hz), while it fires at an effective rate of 20 Hz if the same 100 synapses are clustered in 10 locations (Mel, 1993).

This preference for spatial clustering as compared to a more uniform synaptic distribution for NMDA input is confirmed statistically by generating a large number (either 50 or 100)

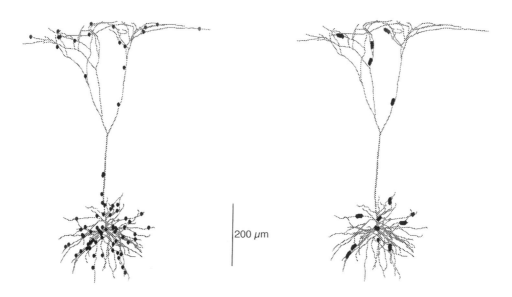

200 μm

Fig. 5.6 CELLULAR RESPONSE TO CLUSTERED AND NONCLUSTERED INPUT In the study by Mel (1992), the layer 5 pyramidal cell is festooned with 100 fast excitatory synapses that are either spread out over 100 randomly selected individual locations (left) or clustered at 14 randomly chosen locations of 7 synapses each (right). In a passive dendritic tree and in the absence of NMDA input, clustered synaptic activity leads to a reduced somatic response compared to the case when synapses are dispersed. For synaptic input of the NMDA type, cooperativity exists, such that spatial clustering causes an enhanced somatic response. Unpublished data from B. Mel, printed with permission.

of randomized synaptic distributions of the type shown in Fig. 5.6 and averaging the results for cluster sizes ranging between 1 and 15. The effect is clear and unambiguous (Fig. 5.8). Due to synaptic saturation, non-NMDA input will always be more effective when spread out in space, while the converse is true for NMDA input: as the size of the cluster is increased (and even though the number of cluster sites goes down since the total number of synapses remains fixed) the cellular response increases. For very large cluster sizes the attenuating effects of synaptic saturation begin to offset the excitatory effects of the NMDA voltage dependence.

While the exact shape of the curve in Fig. 5.8B, such as the most effective cluster size, depends on the details of the biophysical setting, the overall shape does not. Its convex shape is due to cooperativity for small cluster sizes and to synaptic saturation for large cluster sizes (that is, the subsynaptic potential approaches the synaptic reversal potential) and is extremely robust to parameter variations (Mel, 1992).

We conclude that due to the voltage dependency of the NMDA receptor complex, cooperativity pays off, in the sense that a cluster of adjacent synapses will lead to a larger postsynaptic response than if the synapses are randomly distributed throughout the tree. As opposed to the more demanding spatial accuracy in positioning excitation and shunting inhibition to obtain the AND-NOT effect discussed in Sec. 5.1, cluster sensitivity does not depend on the exact positioning of synapses within a cluster. Thus, it places much less premium on development mechanisms to carefully "wire" up the synaptic architecture of the cell. Mel (1992, 1993) made no special assumptions about temporal synchronization among synaptic inputs: if input to a cluster arrives as a highly synchronized wave of excitation, it

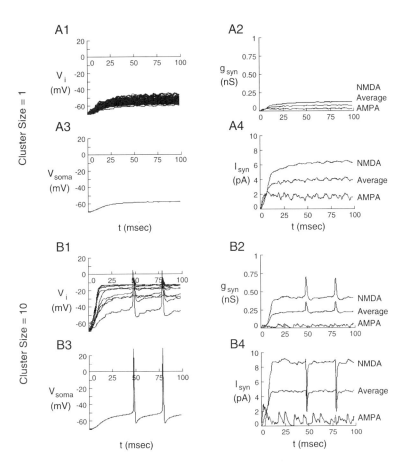

Fig. 5.7 **DETAILS OF THE CELL'S RESPONSE TO NMDA SYNAPTIC INPUT** Membrane potential at the synapses (panels A_1, B_1) and at the soma (panels A_3, B_3) in response to 100 uniformly distributed synapses (upper panels) and to 10 clusters of 10 synapses. 90% of the synaptic conductance change is of the NMDA type, the remainder being voltage independent. Each synapse, whether within a cluster or not, is independently activated by a Poisson process of 100-Hz rate. Because in the first case, synapses act "alone," the average synaptic conductance (A_2) and synaptic current (A_4) remain modest. If, however, synapses can "cooperate" due to spatial proximity, the effective synaptic conductance (B_2) and synaptic current flowing (B_4) can be much larger, inducing the cell to fire at 20 Hz. The synaptic conductance changes scale inversely with the local input resistance, with high g_{peak} values at the soma and low ones in the distal dendritic tree. Reprinted by permission from Mel (1993).

can be much more powerful than if spread out over time. In this sense, his analysis represents a worst-case scenario.

The sensitivity to spatial clumping of synaptic input is present under a broad parameter regime, in particular in the presence of calcium or sodium conductances in the dendritic tree (as we will discuss in more detail in Chap. 19). Cluster sensitivity appears to represent a neurobiologically plausible mechanism to confer nonlinear, multiplicative properties to local *subunits* in the dendritic tree. Such a subunit could be more rigorously defined as a region of the dendritic tree within which synaptic interaction is strong and nonlinear, while interaction between two or more subunits would be linear (Koch, Poggio and Torre, 1982).

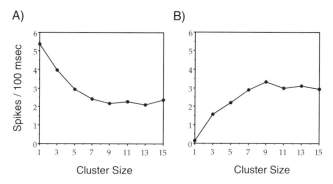

Fig. 5.8 **CLUSTER SENSITIVITY IN A PASSIVE DENDRITIC TREE**   Average number of spikes within a 100-msec period using **(A)** zero NMDA and **(B)** high NMDA conditions as a function of cluster size. In each case, 100 synapses are placed on spines in $100/k$ clusters randomly distributed throughout the passive tree (with uniform probability density over the dendritic length). This procedure is repeated either **(A)** 100 or **(B)** 50 times for cluster sizes between 1 and 15, and the cellular responses are averaged. Due to the cooperativity of NMDA synapses, spatial clustering increases the postsynaptic response. This basic effect is very robust to parameter variations. Reprinted in modified form by permission from Mel (1993).

5.2.2 Cluster Sensitivity for Pattern Discrimination

Mel (1992) provides an illustrative example of how the nonlinear cooperative interaction of NMDA synapses can implement a high-level computation, discriminating one pattern among many others.

The basic idea is straightforward: if a set of synaptic afferents coding for pattern A terminates in a synaptic cluster on a particular neuron, and the synapses expressing some pattern B are randomly spread throughout the cell, the cell can be said to recognize or discriminate pattern A, since activating its associated synaptic cluster makes the cell fire, while the cellular response to activation of the randomly distributed synapses will be negligible. This is illustrated by showing how a neuron can discriminate among a set of 100 black-and-white photos from a summer vacation (Mel, 1992).

The coding of the gray-scale pictures is accomplished in the following manner. First 50 randomly chosen images, designated as training set, are convolved with four oriented Gabor filters on a 64- by 64-pixel grid, resulting in a population of 16,384 orientation-selective visual units. Each one of these oriented units is assumed to make a single all-or-none output synapse on the layer 5 pyramidal cell in Fig. 5.6. The activity of one such unit would represent the presence of a particular oriented edge at that location in the image. Out of this large population, the 80 most active units are selected—the output of all other units being suppressed—and are mapped onto 10 cluster of eight neighboring synapses known to trigger at least one spike. This process is repeated for each of the 50 training images, except that a visual unit, once mapped onto the cell, is not remapped if it occurs in a subsequent training image. Following this "learning" procedure, the remaining unused orientation-selective units (about 12,000) are mapped randomly onto individual synaptic sites. Although a caricature of what is expected to occur in visual cortex, this procedure assures that each of the 50 images is mapped onto at most 80 clustered synapses (because of the potential overlap among images).

As seen in Fig. 5.9, the cell responds to any one of the 50 training images with an effective spike frequency of 12.5 Hz, while the 50 randomly assigned test images fail to bring the cell above threshold. Mel (1992) also uses three sets of partially corrupted training images as input to the pyramidal cell. In the "half/half" stimulation paradigm, half of the vector components coding for one training image were swapped with half of the components encoding another training image. In a linear perceptron, the superposition of two training patterns cannot be distinguished from a single training pattern. In this case the cell's response is significantly reduced compared to its response to a full training pattern (by about 33%; Fig. 5.9). The same performance is seen if 20% of a training image is randomly corrupted. If much more noise is added, the cell only responds weakly.

The performance of the pyramidal cell endowed with NMDA synapses is quite remarkable. The cell can be trained to respond to any one of 50 high-dimensional vectors, discriminating them from random input or vectors not beloging to the training set. The probability of misclassification is 14%, with the majority being false negatives, that is, training images not recognized as such. Given the highly dispersed three-dimensional geometry of dendritic and axonal arbors and the rich combinatorics possible in choosing particular spatial arrangements among 10,000 afferent synapses, the true discrimination capacity of even a single pyramidal cell could be much greater than demonstrated here (Mel, 1993, 1994; see also Brown et al., 1991a). The thread of this story is picked up in Sec. 14.4.2 in the context of understanding the computational capabilities of neural networks.

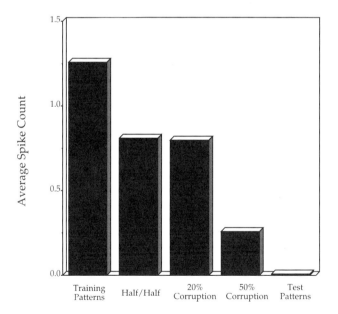

Fig. 5.9 NMDA Endowed Pyramidal Cell as Pattern Discriminator Average response (over a 100-msec period) of the pyramidal cell to five types of images mapped onto its synapses. The largest response is to the 50 training images and the smallest to the remaining 50 test images. In the intermediate cases, training images are partially corrupted. In the "half/half" case, half of the features of one image are combined with half of the features of another image. Note the significantly reduced response to the superposition of two training patterns, direct evidence of the nonlinear pattern discrimination ability of a neuron endowed with NMDA input. In the two other cases, either 20% or 50% of the image is replaced by random features. Reprinted by permission from Mel (1992).

To what extent pyramidal cells in cortex implement such a scheme is difficult to ascertain since the detailed microanatomy of the origin of all synapses would, by itself, not be sufficient to answer the question. Yet this mechanism predicts the existence of a learning rule that favors the placement of simultaneously active synapses in clusters on the dendritic tree, while uncorrelated synapses should have no priviliged spatial relationship to each other. Biophysical experiments involving synaptic plasticity should uncover such a clustering rule.

The degree of specificity required is less than that of the AND-NOT type of interaction discussed at the beginning of this chapter. Yet similar to the interaction between excitation and inhibition, clustering effectively fractionates the dendritic tree into many small subunits, within which nonlinear interactions take place.

5.2.3 Detecting Coincident Input from the Two Ears

One situation where sublinear addition is exploited is in aiding individual neurons to detect coincident inputs. In both birds and mammals, sound localization involves comparing the travel time of inputs from the two ears (Konishi, 1992). If the sound source is located straight ahead, there will be no delay between inputs from the two ears, while shifting the source to one side will cause the auditory input from one side to be delayed relative to input from the opposite side (Carr and Konishi, 1990). Individual cells can perform quite a remarkable job in detecting the resultant small interaural time differences. For instance, a neuron specialized to sounds in the 5 kilohertz auditory range reduces its firing range when the delay between the two ears changes by a mere 20 μsec (Moiseff and Konishi, 1981). This remarkable sensitivity goes hand in hand with a very low passive time constant of less than 2 msec (Reyes, Rubel, and Spain, 1994).[2]

Auditory brainstem cells receive binaural inputs segregated onto their two dendrites (Fig. 5.10), a mechanism that enhances the discriminability of sounds that arrive in phase (straight ahead) relative to those arriving 180° out of phase. Numerous synaptic inputs converge onto each dendrite and the associated synaptic amplitude has a significant stochastic component (Sec. 4.2). Suppose that six simultaneous synaptic events are just enough to trigger a spike in a point neuron without a dendrite. Since twice as many presynaptic inputs are active simultaneously when both inputs arrive in a coincident manner, the neuron responds more strongly to in-phase than to out-of-phase input. In the bipolar neuron with segregated inputs from the two ears to its two dendrites, the amplitude of the synaptic conductance is adjusted. The resulting three synaptic inputs on one dendrite, coupled with three synaptic events on the other, give rise to the same firing rate as in the point neuron. If—due to the stochastic nature of how many inputs arrive per sound cycle—four, five, or more synaptic events were to occur on one dendrite (and not on the other), the neuron would still not fire due to the synaptic saturation that limits how much current each dendrite by itself can deliver to the soma (see also Sec. 18.4.1). As long as the voltage threshold for spike initiation is set sufficiently high, synaptic current from two independent dendrites is required to trigger a spike and temporal discrimination is enhanced compared to a point neuron with no dendrites (Agmon-Snir, Carr, and Rinzel, 1998).

5.3 Synaptic Microcircuits

Excitation and inhibition by individual synapses usually have little computational significance by themselves. It is the assembly of synapses into specific patterns of connectivity

---

2. Since these cells fire at high rates, their membrane is rarely at rest and the effective time constant is probably in the submillisecond range (Gerstner et al., 1996).

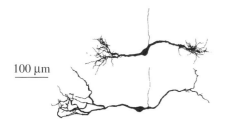

100 μm

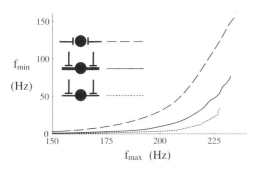

Fig. 5.10 COINCIDENCE DETECTION IN BIPOLAR NEURONS Two auditory brain stem neurons from the guinea pig that receive segregated inputs from the two ears onto their two bipolar dendrites. In a model of these cells (Agmon-Snir, Carr, and Rinzel, 1998), the average firing rate evoked by the two inputs arriving synchronously (f_{max}) or maximally out of phase (f_{min}) is plotted as the amplitude of the synaptic conductance is varied. The least amount of discrimination (smallest difference between f_{max} and f_{min}) is obtained in a point neuron (dashed curve). Synaptic saturation in the long and thin dendrites (dotted curve) prevents randomness in the number of inputs to each side from firing the cell when inputs arrive only to one side without coincident inputs from the other ear. Thicker dendrites (solid curve) perform at an intermediate level. Thus, dendrites enhance the ability of an individual cell to detect coincident input. Reprinted by permission from Agmon-Snir, Carr, and Rinzel, (1998).

which gives rise to identifiable building blocks, which can be found throughout the nervous system. Shepherd (1972, 1978) has termed such specific synaptic arrangements, which have a spatial extent measured in micrometers, *synaptic microcircuits* (Fig. 5.11). Any particular neuron may have dozens or hundreds of such microcircuits. Very often they will include *dendro-dendritic* synapses, that is, fast excitatory or inhibitory chemical synapses from one dendrite of the presynaptic cell onto a dendrite from a different, postsynaptic, cell. Examples of such circuits abound in the central nervous systems of both vertebrates and invertebrates. (For an excellent and detailed account of microcircuits in the mammalian brain see the monograph by Shepherd, 1998.)

Well-known examples include dendro-dendritic interactions among spiking and nonspiking interneurons in the sensory-motor system of the locust (Laurent and Burrows, 1989; Laurent, 1990), synaptic arrangements among the nonspiking neurons in the vertebrate retina (Dowling, 1979, 1987; Sterling, 1998; Fig. 1.5), the reciprocal dendro-dendritic inhibition occurring among the dendrites of the mitral cells and the spines of the granule cells in the mammalian olfactory bulb (Rall et al., 1966; Rall and Shepherd, 1968; Woolf, Shepherd, and Greer, 1991a,b) and the spine-triad circuit in the thalamus (Hamos et al., 1985; Koch, 1985; Fig. 12.8). What characterizes these synaptic arrangements is that they involve at least one excitatory and one inhibitory synapse on a very compact spatial scale, operate at the millisecond time scale, and exist in very large numbers (Shepherd, 1972, 1978; Shepherd and Koch, 1998).

Curiously, synaptic microcircuits do not appear to be numerous in mature cortical structures. To a good first approximation, fast excitatory and inhibitory traffic is mediated from the presynaptic axon of one cell to the postsynaptic dendrites of another cell without the involvement of higher order synaptic arrangements nor the dendro-dendritic synapses.[3]

---

3. It could be argued that an exception to this rule are the small fraction of dendritic spines that carry both an excitatory and an inhibitory synaptic profile; it is doubtful, though, that such arrangements have a specific function (Sec. 12.3.3).

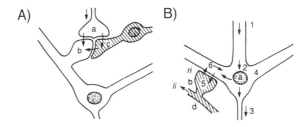

Fig. 5.11 SYNAPTIC MICROCIRCUITS MEDIATING DIFFERENT TYPES OF INHIBITION   Two types of synaptic microcircuits that mediate two types of inhibition in the vertebrate nervous system. (**A**) The *spine triade* consists of an afferent *a* making an excitatory synapse onto a dendrite or spine *b* as well as onto an interneuron *c*. The interneuron in turn inhibits, via a dendro-dendritic synapse, the postsynaptic neuron *b*. This feedforward inhibition is common in the retina and the thalamus (see Sec. 12.3.4 and Fig. 12.8). (**B**) Recurrent inhibition is implemented by a relay cell *a* that excites an interneuron *b*. This interneuron inhibits, via a dendro-dendritic synapse, the original relay cell. Dendro-dendritic reciprocal inhibition occurs between mitral and granule cells in the olfactory bulb. Such microcircuits, which are specified at the micrometer level, are common in structures outside the cortex and in invertebrates. Reprinted by permission from Shepherd and Koch (1998).

5.4 Recapitulation

The fact that a synaptic input changes the conductance in the postsynaptic membrane in series with a synaptic battery and does *not* correspond to a constant current source ultimately implies that synaptic inputs interact with each other via the membrane potential. In particular, the somatic potential in a passive tree in response to two or more inputs is *not* equal to the sum of the individual synaptic components. We explored the computational consequences of this in two cases.

The interaction between voltage-independent non-NMDA excitatory input and shunting inhibition (that is, when E_i reverses close to the cell's resting membrane potential) in the subthreshold domain can mediate a veto operation that is specific in space and time. If inhibition is adjacent to the excitatory synapse or on the path between excitation and the cell body and if their time courses overlap, inhibition can effectively suppress the effect of excitation. This implies that specific synaptic arrangements in a dendritic tree can implement logic-like AND-NOT operations, possibly one of the crucial nonlinearities underlying direction selectivity in retinal ganglion cells.

The specificity of synaptic interaction represents at the same time also its greatest weakness, in the sense that it places great demands on developmental mechanisms to precisely guide synapses and dendrites during development. A more plausible synaptic arrangement could be implemented at the level of synaptic populations: if excitatory and inhibitory synapses are colocalized onto the same part of the dendritic tree, excitatory input can always override any inhibitory influence. Conversely, if inhibition is at a different site, for instance, close to the spike initiating zone, and excitation at more distal sites, then any excitatory input can always be vetoed by inhibition. These two types of synaptic placements might instantiate different kinds of suppressive behaviors. Under certain conditions the threshold for initiating a reflex should be elevated (relative suppression) while under others the behavior needs to be totally abolished (absolute suppression).

If GABAergic inhibition has a reversal potential much below the membrane resting potential, as in the fly tangential interneurons, inhibition tends to act akin to a linear subtraction.

A more plausible mechanism to implement multiplicative behavior involves NMDA synapses clustered over the dendritic tree. When clusters of adjacent non-NMDA synapses are randomly sprinkled around the dendritic tree of a pyramidal cell, their activation causes a smaller cellular response than when the synapses are isolated from each other, a consequence of synaptic saturation. A very different behavior is obtained with clusters of voltage-dependent NMDA synapses. Because of their cooperative nature, clusters of 6 to 10 adjacent NMDA synapses are much more effective than the same number of synapses by themselves. A dendritic tree endowed with such synapses can be used to implement a very efficient nonlinear pattern discriminator that is robust to the presence of dendritic nonlinearities, while also serving as a plausible biophysical mechanism for multiplication, which is required for a host of computations, such as motion, binocular disparity, and tabular look-up storage.

Both case studies imply that the dendritic tree, rather than just performing a filtering operation onto the synaptic input, as suggested by linear cable theory presented in Chaps. 2 and 3, can be partitioned into numerous spatial subunits. Within each such subunit, synaptic inputs interact nonlinearly, while the interaction between two or more subunits is approximately linear.

Finally, we mentioned the concept of a *synaptic microcircuit*, pioneered by Shepherd (1978, 1998). These usually involve a combination of one or several excitatory and inhibitory synapses with a predilection for a specific arrangement among two or three neurons and are common in extracortical structures such as the retina, the olfactory bulb, and the thalamus and in invertebrates.

6

THE HODGKIN–HUXLEY MODEL
OF ACTION
POTENTIAL GENERATION

The vast majority of nerve cells generate a series of brief voltage pulses in response to vigorous input. These pulses, also referred to as *action potentials* or *spikes*, originate at or close to the cell body, and propagate down the axon at constant velocity and amplitude. Fig. 6.1 shows the shape of the action potential from a number of different neuronal and nonneuronal preparations. Action potentials come in a variety of shapes; common to all is the all-or-none depolarization of the membrane beyond 0. That is, if the voltage fails to exceed a particular threshold value, no spike is initiated and the potential returns to its baseline level. If the voltage threshold is exceeded, the membrane executes a stereotyped voltage trajectory that reflects membrane properties and not the input. As evident in Fig. 6.1, the shape of the action potential can vary enormously from cell type to cell type.

When inserting an electrode into a brain, the small all-or-none electrical events one observes extracellularly are usually due to spikes that are initiated close to the cell body and that propagate along the axons. When measuring the electrical potential across the membrane, these spikes peak between +10 and +30 mV and are over (depending on the temperature) within 1 or 2 msec. Other all-or-none events, such as the complex spikes in cerebellar Purkinje cells (Fig. 6.1G) or bursting pyramidal cells in cortex (Fig. 6.1H and Fig. 16.1C), show a more complex wave form with one or more fast spikes superimposed onto an underlying, much slower depolarization. Finally, under certain conditions, the dendritic membrane can also generate all-or-none events (Fig. 6.1H) that are much slower than somatic spikes, usually on the order to 50–100 msec or longer. We will treat these events and their possible significance in Chap. 19.

Only a small fraction of all neurons is unable—under physiological conditions—to generate action potentials, making exclusive use of graded signals. Examples of such *nonspiking cells*, usually spatially compact, can be found in the distal retina (e.g., bipolar, horizontal, and certain types of amacrine cells) and many neurons in the sensory-motor pathway of invertebrates (Roberts and Bush, 1981). They appear to be absent from cortex,

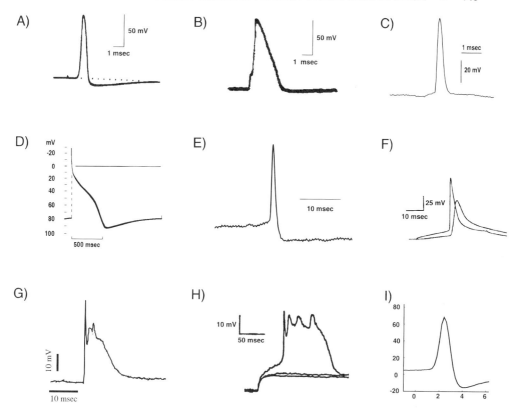

Fig. 6.1 ACTION POTENTIALS OF THE WORLD   Action potentials in different invertebrate and vertebrate preparations. Common to all is a threshold below which no impulse is initiated, and a stereotypical shape that depends only on intrinsic membrane properties and not on the type or the duration of the input. (**A**) Giant squid axon at 16° C. Reprinted by permission from Baker, Hodgkin, and Shaw (1962). (**B**) Axonal spike from the node of Ranvier in a myelinated frog fiber at 22° C. Reprinted by permission from Dodge (1963). (**C**) Cat visual cortex at 37° C. Unpublished data from J. Allison, printed with permission. (**D**) Sheep heart Purkinje fiber at 10° C. Reprinted by permission from Weidmann (1956). (**E**) Patch-clamp recording from a rabbit retinal ganglion cell at 37° C. Unpublished data from F. Amthor, printed with permission. (**F**) Layer 5 pyramidal cell in the rat at room temperatures. Simultaneous recordings from the soma and the apical trunk. Reprinted by permission from Stuart and Sakmann (1994). (**G**) A complex spike—consisting of a large EPSP superimposed onto a slow dendritic calcium spike and several fast somatic sodium spikes—from a Purkinje cell body in the rat cerebellum at 36° C. Unpublished data from D. Jaeger, printed with permission. (**H**) Layer 5 pyramidal cell in the rat at room temperature. Three dendritic voltage traces in response to three current steps of different amplitudes reveal the all-or-none character of this slow event. Notice the fast superimposed spikes. Reprinted by permission from Kim and Connors (1993). (**I**) Cell body of a projection neuron in the antennal lobe in the locust at 23° C. Unpublished data from G. Laurent, printed with permission.

thalamus, cerebellum, and associated structures (although it is difficult, on *a priori* grounds, to completely rule out their existence).

Action potentials are such a dominant feature of the nervous system that for a considerable amount of time it was widely held—and still is in parts of the theoretical community— that all neuronal computations only involve these all-or-none events. This belief provided much of the impetus behind the neural network models that originated in the late 1930s and early 1940s (Rashevsky, 1938; McCullough and Pitts, 1943).

The ionic mechanisms underlying the initiation and propagation of action potentials in nervous tissue were elucidated in the squid giant axon by a number of workers, most notably by Hodgkin and Huxley in Cambridge, England (1952a,b,c,d). Together with Eccles, they shared the 1963 Nobel prize in physiology and medicine. (For a historical account see Hodgkin, 1976.) Their quantitative model (Hodgkin and Huxley, 1952d) represents one of the high points of cellular biophysics and has been extremely influential in terms of enabling a large class of quite diverse membrane phenomena to be analyzed and modeled in terms of simple underlying variables. This is all the more surprising since the kinetic description of membrane permeability changes within the framework of the Hodgkin–Huxely model was achieved without any knowledge of the underlying ionic channels.

A large number of excellent papers and books describing in great detail various aspects of the Hodgkin–Huxley model are available today. Nothing matches the monograph by Jack, Noble, and Tsien (1975) for its 200-page extended coverage of various analytical and numerical approaches to understand all relevant aspects of initiation and conduction of action potentials. Cronin (1987) presents a mathematical account of the more formal aspects of Hodgkin and Huxley's model as well as related models, while Scott (1975) pays particular attention to questions of interest to physicists and applied mathematicians. The books by Hille (1992), Johnston and Wu (1995) and Weiss (1996) provide up-to-date and very readable accounts of the biophysical mechanisms underlying action potentials in neuronal tissues. The edited volume by Waxman, Kocsis, and Stys (1995) concentrates on the morphology and the pathophysiology of myelinated and unmyelinated axons.

Because the biophysical mechanisms underlying action potential generation in the cell body and axons of both invertebrates and vertebrates can be understood and modeled by the formalism Hodgkin and Huxley introduced 40 years ago, it becomes imperative to understand their model and its underlying assumptions. We will strive in this chapter to give an account of those properties of the Hodgkin–Huxley model that are of greatest relevance to understanding the initiation of the action potential. We will also discuss the propagation of spikes along unmyelinated and myelinated fibers. Chapter 9 extends the Hodgkin–Huxley framework to the plethora of other currents described since their days.

6.1 Basic Assumptions

Hodgkin and Huxley carried out their analysis in the giant axon of the squid. With its half-millimeter diameter, this fiber is a leviathan among axons. (The typical axon in cortex has a diameter more than 1000 times smaller; Braitenberg and Schüz, 1991.) In order to eliminate the complexity introduced by the distributed nature of the cable, a highly conductive axial wire was inserted inside the wire. This so-called *space clamp* keeps the potential along the entire axon spatially uniform, similar to the situation occurring in a patch of membrane. This, together with voltage clamping the membrane and the usage of pharmacological agents to block various currents, enabled Hodgkin and Huxley to dissect the membrane current into its constitutive components. The total membrane current is the sum of the ionic currents and the capacitive current,

$$I_m(t) = I_{\text{ionic}}(t) + C_m \frac{dV(t)}{dt}. \tag{6.1}$$

With the help of these tools, Hodgkin and Huxley (1952a,b,c) carried out a large number of experiments, which lead them to postulate the following *phenomenological* model of the

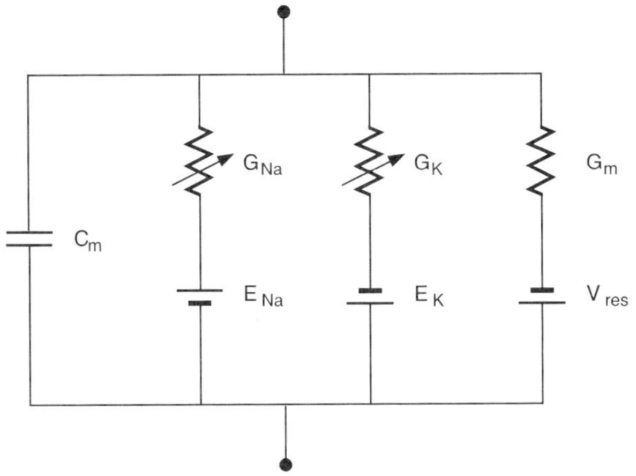

Fig. 6.2 Electrical Circuit for a Patch of Squid Axon Hodgkin and Huxley modeled the membrane of the squid axon using four parallel branches: two passive ones (membrane capacitance C_m and the leak conductance $G_m = 1/R_m$) and two time- and voltage-dependent ones representing the sodium and potassium conductances.

events underlying the generation of the action potential in the squid giant axon (Fig. 6.2; Hodgkin and Huxley, 1952d).

1. The action potential involves two major voltage-dependent ionic conductances, a sodium conductance G_{Na} and a potassium conductance G_K. They are independent from each other. A third, smaller so-called "leak" conductance (which we term G_m) does not depend on the membrane potential. The total ionic current flowing is the sum of a sodium current, a potassium current, and the leak current:

$$I_{ionic} = I_{Na} + I_K + I_{leak} . \tag{6.2}$$

2. The individual ionic currents $I_i(t)$ are linearly related to the driving potential via Ohm's law,

$$I_i(t) = G_i(V(t), t)(V(t) - E_i) \tag{6.3}$$

where the ionic reversal potential E_i is given by Nernst's equation for the appropriate ionic species. Depending on the balance between the concentration difference of the ions and the electrical field across the membrane separating the intracellular cytoplasm from the extracellular milieu, each ionic species has such an associated *ionic battery* (see Eq. 4.3). Conceptually, we can use the equivalent circuit shown in Fig. 6.2 to describe the axonal membrane.

3. Each of the two ionic conductances is expressed as a maximum conductance, $\overline{G}_{Na}$ and $\overline{G}_K$, multiplied by a numerical coefficient representing the fraction of the maximum conductance actually open. These numbers are functions of one or more fictive *gating particles* Hodgkin and Huxley introduced to describe the dynamics of the conductances. In their original model, they talked about *activating* and *inactivating* gating particles. Each gating particle can be in one of two possible states, open or close, depending on time and on the membrane potential. In order for the conductance to open, all of these gating particles must be open simultaneously. The entire kinetic properties of their model are contained in these variables. We will consider the physical and molecular interpretation of these gating particles in terms of numerous all-or-none microscopic *ionic channels* in Chap. 8.

6.2 Activation and Inactivation States

Let us specify how these activation and inactivation states determine the two ionic currents. This is important, since the vast majority of state-of-the-art ionic models is formulated in terms of such particles.

6.2.1 Potassium Current I_K

Hodgkin and Huxley (1952d) model the potassium current as

$$I_K = \overline{G}_K n^4 (V - E_K) \tag{6.4}$$

where the maximal conductance $\overline{G}_K = 36$ mS/cm^2 and the potassium battery is $E_K = -12$ mV relative to the resting potential of the axon. n describes the state of a fictional *activation particle* and is a dimensionless number between 0 and 1. Note that with today's physiological conventions, I_K as outward current is always positive (for $V \geq E_K$; see Fig. 6.5).

Chapter 8 treats the underlying microscopic and stochastic nature associated with the macroscopic and deterministic current. Let us for now develop our intuition by assuming that the probability of finding one activation particle in its *permissive* or open state is n (and it will be with probability $1 - n$ in its *nonpermissive* or closed state where no current flows through the conductance). Equation 6.4 states that in order for the channel to be open, the four gating particles must simultaneously be in their open state. We can also think of n as the proportion of particles in their *permissive* state; potassium current can only flow if four particles are in their permissive state.

If we assume that only these two states exist (for a single particle) and that the transition from one to the other is governed by first-order kinetics, we can write the following reaction scheme:

$$n \underset{\alpha_n}{\overset{\beta_n}{\rightleftharpoons}} 1 - n \tag{6.5}$$

where α_n is a voltage-dependent rate constant (in units of 1/sec), specifying how many transitions occur between the closed and the open states and β_n expresses the number of transitions from the open to the closed states (in units of 1/sec). Mathematically, this scheme corresponds to a first-order differential equation,

$$\frac{dn}{dt} = \alpha_n(V)(1 - n) - \beta_n(V)n . \tag{6.6}$$

The key to Hodgkin and Huxley's model—as well as the most demanding part of their investigation—was the quantitative description of the voltage dependency of the rate constants. Instead of using rate constants α_n and β_n, we can reexpress Eq. 6.6 in terms of a voltage-dependent time constant $\tau_n(V)$ and a steady-state value $n_\infty(V)$ with

$$\frac{dn}{dt} = \frac{n_\infty - n}{\tau_n} \tag{6.7}$$

where

$$\tau_n = \frac{1}{\alpha_n + \beta_n} \tag{6.8}$$

and

$$n_\infty = \frac{\alpha_n}{\alpha_n + \beta_n}. \qquad (6.9)$$

Both descriptions, either in terms of rate constants α_n and β_n or in terms of a time constant τ_n and a steady-state variable n_∞, are equivalent. While Hodgkin and Huxely used the former, we will use the latter, due to its simpler physical interpretation.

One of the most striking properties of the squid membrane is the steepness of the relation between conductance and membrane potential. Below about 20 mV, the steady-state potassium membrane conductance G_K increases e-fold by varying V by 4.8 mV, while the voltage sensitivity of the sodium conductance is even higher (an e-fold change for every 3.9 mV). For higher levels of depolarization, saturation in the membrane conductance sets in (Hodgkin and Huxley, 1952a). This steep relationship must be reflected in the voltage dependency of the rate constants. Hodgkin and Huxley (1952d) approximated the voltage dependencies of the rate constants by

$$\alpha_n(V) = \frac{10 - V}{100(e^{(10-V)/10} - 1)} \qquad (6.10)$$

and

$$\beta_n(V) = 0.125e^{-V/80} \qquad (6.11)$$

where V is the membrane potential relative to the axon's resting potential in units of millivolt. Figure 6.3 shows the voltage dependency of the associated time constant and the steady-state value of the potassium activation variable. While τ_n has a bell-shaped dependency, n_∞ is a monotonically increasing function of the membrane potential. The curve relating the steady-state potassium conductance to the membrane potential is an even steeper function, given the fourth-power relationship between G_K and n. This is a hallmark of almost all ionic conductances: depolarizing the membrane potential increases its effective conductance.[1] One of the few exceptions is the appropriately named *anomalous rectifier* current I_{AR} (frequently also termed *inward rectifier*), which turns on with increasing membrane hyperpolarization (Spain, Schwindt, and Crill, 1987).

The fraction of the steady-state potassium conductance open at any particular voltage $\overline{V}$, that is, for $t \to \infty$, is identical to $n_\infty(\overline{V})^4$. At V_{rest} this number is very small, $n_\infty(0)^4 = 0.01$, that is, only about 1% of the total potassium conductance is activated. Using the voltage-clamp setup, we now move the membrane potential as rapidly as possible to $\overline{V}$ and clamp it there. The evolution of the potassium conductance is dictated by the differential Eq. 6.7,

$$n(t)^4 = \left(n_\infty - (n_\infty - n_0)e^{-t/\tau_n(\overline{V})}\right)^4, \qquad (6.12)$$

where n_0 is the initial value of the potassium activation, $n_0 = n_\infty(0) = 0.32$, and n_∞ its final value, $n_\infty = n_\infty(\overline{V})$. The time course of any one activation variable follows an exponential, a reflection of the underlying assumption of a first-order kinetic scheme. The time course of the fourth power of $n(t)$ is plotted on the right-hand side of Fig. 6.4, following a sudden shift in the membrane potential, from rest to the various voltage values indicated. Superimposed are the experimentally measured values of the potassium conductance. It is remarkable how well the points fall onto the curve. Upon stepping back to the original membrane potential, n slowly relaxes back to its original low value.

These rate constants have a probabilistic interpretation, covered in more depth in Chap. 8.

---

1. Whether or not the associated ionic current also increases depends on the relevant ionic reversal potential (Eq. 6.3).

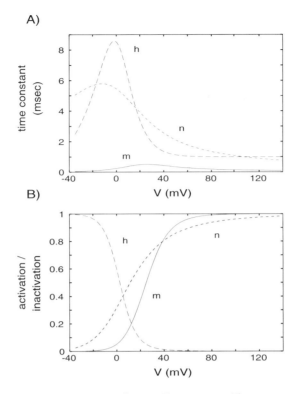

Fig. 6.3 **Voltage Dependency of the Gating Particles** Time constants (**A**) and steady-state activation and inactivation (**B**) as a function of the relative membrane potential V for sodium activation m (solid line) and inactivation h (long dashed line) and potassium activation n (short, dashed line). The steady-state sodium inactivation h_∞ is a monotonically decreasing function of V, while the activation variables n_∞ and m_∞ increase with the membrane voltage. Activation of the sodium and potassium conductances is a much steeper function of the voltage, due to the power-law relationship between the activation variables and the conductances. Around rest, G_{Na} increases e-fold for every 3.9 mV and G_K for every 4.8 mV. Activating the sodium conductance occurs approximately 10 times faster than inactivating sodium or activating the potassium conductance. The time constants are slowest around the resting potential.

6.2.2 Sodium Current I_{Na}

As can be seen on the left hand side of Fig. 6.4, the dynamics of the sodium conductance that we will explore now are substantially more complex.

In order to fit the kinetic behavior of the sodium current, Hodgkin and Huxley had to postulate the existence of a sodium activation particle m as well as an *inactivation particle h*,

$$I_{Na} = \overline{G}_{Na}m^3h(V - E_{Na}) \tag{6.13}$$

where $\overline{G}_{Na}$ is the maximal sodium conductance, $\overline{G}_{Na} = 120\,\text{mS/cm}^2$, and E_{Na} is the sodium reversal potential, $E_{Na} = 115$ mV, relative to the axon's resting potential. m and h are dimensionless numbers, with $0 \leq m, h \leq 1$. By convention the sodium current is negative, that is, inward, throughout the physiological voltage range (for $V < E_{Na}$; see Fig. 6.5).

The amplitude of the sodium current is contingent on four hypothetical gating particles making independent first-order transitions between an open and a closed state. Since these

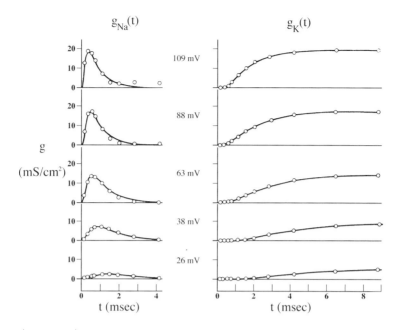

Fig. 6.4 K$^+$ and Na$^+$ Conductances during a Voltage Step Experimentally recorded (circles) and theoretically calculated (smooth curves) changes in G_{Na} and G_K in the squid giant axon at 6.3° C during depolarizing voltage steps away from the resting potential (which here, as throughout this chapter, is set to zero). For large voltage changes, G_{Na} briefly increases before it decays back to zero (due to *inactivation*), while G_K remains activated. Reprinted by permission from Hodgkin (1958).

particles are independent, the probability for the three m and the one h particle to exist in this state is m^3h. Notice that h is the probability that the inactivating particle is *not* in its inactivating state. Formally, the temporal change of these particles is described by two first-order differential equations,

$$\frac{dm}{dt} = \alpha_m(V)(1 - m) - \beta_m(V)m \tag{6.14}$$

and

$$\frac{dh}{dt} = \alpha_h(V)(1 - h) - \beta_h(V)h. \tag{6.15}$$

Empirically, Hodgkin and Huxley derived the following equations for the rate constants:

$$\alpha_m(V) = \frac{25 - V}{10(e^{(25-V)/10} - 1)} \tag{6.16}$$

$$\beta_m(V) = 4e^{-V/18} \tag{6.17}$$

$$\alpha_h(V) = 0.07e^{-V/20} \tag{6.18}$$

$$\beta_h(V) = \frac{1}{e^{(30-V)/10} + 1}. \tag{6.19}$$

The associated time constants and steady-state variables are plotted in Fig. 6.3 as a function

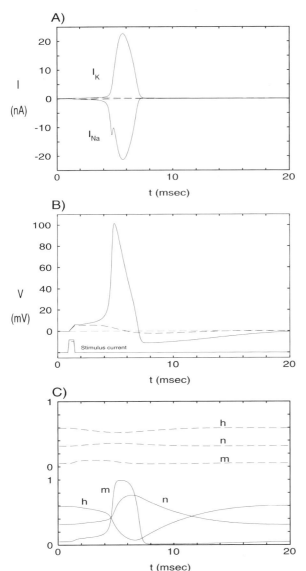

Fig. 6.5 HODGKIN-HUXLEY AC-TION POTENTIAL Computed action potential in response to a 0.5-msec current pulse of 0.4-nA amplitude (solid lines) compared to a subthreshold response following a 0.35-nA current pulse (dashed lines). **(A)** Time course of the two ionic currents. Note their large sizes compared to the stimulating current. **(B)** Membrane potential in response to threshold and subthreshold stimuli. The injected current charges up the membrane capacity (with an effective membrane time constant $\tau = 0.85$ msec), enabling sufficient I_{Na} to be recruited to outweigh the increase in I_K (due to the increase in driving potential). The smaller current pulse fails to trigger an action potential, but causes a depolarization followed by a small hyperpolarization due to activation of I_K. **(C)** Dynamics of the gating particles. Sodium activation m changes much more rapidly than either h or n. The long time course of potassium activation n explains why the membrane potential takes 12 msec after the potential has first dipped below the resting potential to return to baseline level.

of voltage. Similar to before, both τ_m and τ_h are bell-shaped curves,[2] but with a tenfold difference in duration. While m_∞ is a monotonically increasing function of V, as expected of an activation variable, h_∞ decreases with increasing membrane depolarization, the defining feature of an inactivating particle. Without inactivation, the sodium conductance would remain at its maximum value in response to a depolarizing voltage step.

The fraction of the steady-state sodium conductance open at rest is less than 1% of the peak sodium conductance. Inspection of Fig. 6.3 immediately reveals the reason: for voltages below or close to the resting potential of the axon, the activation variable m is close to zero while at positive potentials the inactivation variable h is almost zero. Thus,

---

2. Note that the voltage-dependent membrane time constant for the activation variable τ_m has the same symbol as the passive membrane time constant. When in doubt, we will refer to the latter simply as τ.

the steady-state sodium current $\overline{G}_{Na}m_\infty^3 h_\infty(V - E_{Na})$, also known as the *window current*, is always very small. The secret to obtaining the large sodium current needed to rapidly depolarize the membrane lies in the temporal dynamics of m and h. At values of the membrane close to the resting potential, h takes on a value close to 1. When a sudden depolarizing voltage step is imposed onto the membrane as in Fig. 6.4, m changes within a fraction of a millisecond to its new value close to 1, while h requires 5 msec or longer to relax from its previous high value to its new and much smaller value. In other words, two processes control the sodium conductance: activation is the rapid process that increases G_{Na} upon depolarization and outpaces inactivation, the much slower process that reduces G_{Na} upon depolarization.

6.2.3 Complete Model

Similar to most other biological membranes, the axonal membrane contains a voltage-independent "leak" conductance G_m, which does not depend on the applied voltage and remains constant over time. The value measured by Hodgkin and Huxley, $G_m = 0.3$ mS/cm^2, corresponds to a passive membrane resistivity of $R_m = 3333\ \Omega \cdot$ cm^2. The passive component also has a reversal potential associated with it. Hodgkin and Huxley did not explicitly measure V_{rest}, but adjusted it so that the total membrane current at the resting potential $V = 0$ was zero. In other words, V_{rest} was defined via the equation $G_{Na}(0)E_{Na} + G_K(0)E_K + G_m V_{rest} = 0$, and came out to be +10.613 mV. The membrane capacity $C_m = 1\ \mu$F/cm^2. At the resting potential, the effective membrane resistance due to the presence of the sum of the leak, the potassium, and the (tiny) sodium conductances amounts to 857 $\Omega \cdot$ cm^2, equivalent to an effective "passive" membrane time constant of about 0.85 msec.

We can now write down a single equation for all the currents flowing across a patch of axonal membrane,

$$C_m \frac{dV}{dt} = \overline{G}_{Na}m^3 h(E_{Na} - V) + \overline{G}_K n^4(E_K - V) + G_m(V_{rest} - V) + I_{inj}(t) \quad (6.20)$$

where I_{inj} is the current that is injected via an intracellular electrode. This nonlinear differential equation, in addition to the three ordinary linear first-order differential equations specifying the evolution of the rate constants (as well as their voltage dependencies), constitutes the four-dimensional Hodgkin and Huxley model for the space-clamped axon or for a small patch of membrane. Throughout the book, we shall refer to Eq. 6.20, in combination with the rate constants (Eqs. 6.7, 6.14, and 6.15) at 6.3° C as the *standard Hodgkin-Huxley membrane patch model*. In our simulations of these equations, we solve Eq. 6.20 for an equipotential $30 \times 30 \times \pi\ \mu$m^2 patch of squid axonal membrane and therefore express I_m in units of nanoamperes (nA), and not as current density.

We will explain in the following sections how this model reproduces the stereotyped sequence of membrane events that give rise to the initiation and propagation of all-or-none action potentials.

6.3 Generation of Action Potentials

One of the most remarkable aspect of the axonal membrane is its propensity to respond in either of two ways to brief pulses of depolarizing inward current. If the amplitude of the pulse is below a given threshold, the membrane will depolarize slightly but will return to the membrane's resting potential, while larger currents will induce a pulse-like action potential, whose overall shape is relatively independent of the stimulus required to trigger it.

Consider the effect of delivering a short (0.5-msec) inward current pulse $I_{inj}(t)$ of 0.35-nA amplitude to the membrane (Fig. 6.5). The injected current charges up the membrane capacitance, depolarizing the membrane in the process. The smaller this capacitance, the faster the potential will rise. The depolarization has the effect of slightly increasing m and n, in other words, increasing both sodium and potassium activation, but decreasing h, that is, decreasing potassium inactivation. Because the time constant of sodium activation is more than one order of magnitude faster than τ_n and τ_h at these voltages, we can consider the latter two for the moment to be stationary. But the sodium conductance G_{Na} will increase somewhat. Because the membrane is depolarized from rest, the driving potential for the potassium current, $V - E_K$, has also increased. The concomitant increase in I_K outweighs the increase in I_{Na} due to the increase in G_{Na} and the overall current is outward, driving the axon's potential back toward the resting potential. The membrane potential will slightly undershoot and then overshoot until it finally returns to V_{rest}. The oscillatory response around the resting potential can be attributed to the small-signal behavior of the potassium conductance acting phenomenologically similar to an inductance (see Chap. 10 for further discussion).

If the amplitude of the current pulse is increased slightly to 0.4 nA, the depolarization due to the voltage-independent membrane components will reach a point where the amount of I_{Na} generated exceeds the amount of I_K. At this point, the membrane voltage undergoes a runaway reaction: the additional I_{Na} depolarizes the membrane, further increasing m, which increases I_{Na}, causing further membrane depolarization. Given the almost instantaneous dynamics of sodium activation ($\tau_m \approx 0.1$–0.2 msec at these potentials), the inrushing sodium current moves the membrane potential within a fraction of a millisecond to 0 mV and beyond. In the absence of sodium inactivation and potassium activation, this positive feedback process would continue until the membrane would come to rest at E_{Na}. As we saw already in Fig. 6.4, after a delay both the slower sodium inactivation variable h as well as the potassium activation n will turn on (explaining why I_K is also called the *delayed rectifier* current I_{DR}). Sodium inactivation acts to directly decrease the amount of sodium conductance available, while the activation of the potassium conductance tends to try to bring the axon's membrane potential toward E_K by increasing I_K. Thus, both processes cause the membrane potential to dip down from its peak. Because the total sodium current quickly falls to zero after 1 msec, but I_K persists longer at small amplitudes (not readily visible in Fig. 6.5), the membrane potential is depressed to below its resting level, that is, the axon *hyperpolarizes*. At these low potentials, eventually potassium activation switches off, returning the system to its initial configuration as V approaches the resting potential.

6.3.1 Voltage Threshold for Spike Initiation

What are the exact conditions under which a spike is initiated? Does the voltage have to exceed a particular threshold value V_{th}, or does a minimal amount of current I_{th} have to be injected, or does a certain amount of electrical charge Q_{th} have to be delivered to the membrane in order to initiate spiking? These possibilities and more have been discussed in the literature and experimental evidence exists to support all of these views under different circumstances (Hodgkin and Rushton, 1946; Cooley, Dodge, and Cohen, 1965; Noble and Stein, 1966; Cole, 1972; Rinzel, 1978; for a thorough discussion see Jack, Noble, and Tsien, 1975). Because the squid axon is not a good model for spike encoding in central neurons, we will defer a more detailed discussion of this issue to Secs. 17.3 and 19.2. We here limit ourselves to considering spike initiation in an idealized nonlinear membrane, without dealing with the complications of cable structures (such as the axon).

To answer this question, we need to consider the $I–V$ relationship of the squid axonal membrane. Because we are interested in rapid synaptic inputs, we assume that the risetime of the synaptic current is faster than the effective passive time constant, $\tau = 0.85$ msec, and make use of the observation that the dynamics of sodium activation m is very rapid (the associated time constant is always less than 0.5 msec) and at least a factor of 10 faster than sodium inactivation h and potassium activation n (see Fig. 6.3). With these observations in mind, we ask what happens if the input depolarizes the membrane very rapidly to a new value V? Let us estimate the current that will flow with the help of the *instantaneous $I–V$* relationship $I_0(V)$ (Fig. 6.6).

I_0 is given by the sum of the ionic and the leak currents. We approximate the associated sodium and potassium conductances by assuming that h and n have not had time to change from the value they had at the resting potential $V = 0$, while m adjusts instantaneously to its new value at V. In other words,

$$I_0(V) = \overline{G}_{\mathrm{Na}}m(V)^3 h(0)(V - E_{\mathrm{Na}}) + \overline{G}_{\mathrm{K}}n(0)^4(V - E_{\mathrm{K}}) + G_m(V - V_{\mathrm{rest}}) \quad (6.21)$$

Figure 6.6 shows the inverted U-form shape of I_0 in the neighborhood of the resting potential, as well as its three ionic components I_{Na}, I_{K}, and I_{leak}.

In the absence of any input, the system rests at $V = 0$. If a small depolarizing voltage step is applied, the system is displaced to the right, generating a small, positive current. This current is outward since the increase in m (increasing the amplitude of I_{Na}) is outweighed by the increase in the driving potential $V - E_{\mathrm{K}}$ (increasing I_{K}) and the decrease in I_{leak}.

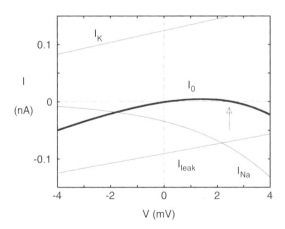

Fig. 6.6 Current-Voltage Relationship around Rest Instantaneous $I–V$ relationship, I_0, associated with the standard patch of squid axon membrane and its three components: $I_0 = I_{\mathrm{Na}} + I_{\mathrm{K}} + I_{\mathrm{leak}}$ (Eq. 6.21). Because m changes much faster than either h or n for rapid inputs, we computed G_{Na} and G_{K} under the assumption that m adapts instantaneously to its new value at V, while h and n remain at their resting values. I_0 crosses the voltage axis at two points: a stable point at $V = 0$ and an unstable one at $V_{\mathrm{th}} \approx 2.5$ mV. Under these idealized conditions, any input that exceeds V_{th} will lead to a spike. For the "real" equations, m does not change instantaneously and nor do n and h remain stationary; thus, I_0 only crudely predicts the voltage threshold which is, in fact, 6.85 mV for rapid synaptic input. Note that I_0 is specified in absolute terms and scales with the size of the membrane patch.

This forces the membrane potential back down toward the resting potential: the voltage trajectory corresponds to a subthreshold input. Similarly, if a hyperpolarizing current step is injected, moving the system to below $V = 0$, a negative inward current is generated, pulling the membrane back up toward V_{rest}. The slope of the $I–V$ curve around the resting potential $\partial I / \partial V$, termed the *membrane slope conductance*, is positive (for a substantial discussion of this concept, see Sec. 17.1.2). That is, the point $V = 0$ is a *stable attractor*. (For these and related notions, we defer the reader to the following chapter.)

$I_0(V)$ has a second zero crossing at $V = V_{th} \approx 2.5$ mV. If an input moves the membrane potential to exactly V_{th}, no current flows and the system remains at V_{th} (Fig. 6.6). Because the slope conductance is negative, the point is unstable, and an arbitrarily small perturbation will carry the system away from the zero crossing. A negative perturbation will carry the system back to V_{rest}. Conversely, a positive voltage displacement, no matter how minute, causes a small inward current to flow, which further depolarizes the membrane (due to the negative slope conductance), leading in turn to a larger inward current, and so on. The membrane potential rapidly increases to above absolute zero, that is, an action potential is triggered. During this phase, very large inward currents are generated, far exceeding the amplitude of the modest stimulus current. (Recall that around these potentials, I_{Na} increases e-fold every 3.9 mV.) For the patch of squid membrane simulated here (where the current scales linearly with the area of the patch), the peak of I_{Na} is about 23 nA.

This qualitative account of the origin of the voltage threshold for an active patch of membrane argues that in order for an action potential to be initiated, the net inward current must be negative. For rapid input, this first occurs at $V = V_{th}$. This analysis was based on the rather restrictive assumption that m changes instantaneously, while h and n remain fixed. In practice, neither assumption is perfect. Indeed, while our argument predicts $V_{th} = 2.5$ mV, the voltage threshold for spike initiation for rapid EPSPs for the Hodgkin–Huxley equation is, in fact, equal to 6.85 mV (Noble and Stein, 1966). As discussed in Sec. 17.3, reaching a particular value of the voltage for a rapid input in a single compartment is equivalent to rapidly dumping a threshold amount of charge Q_{th} into the system.

Applying a current step that increases very slowly in amplitude—allowing the system to relax always to its stationary state—prevents any substantial sodium current from flowing and will therefore not cause spiking. Thus, not only does a given voltage level have to be reached and exceeded but also within a given time window. We take up this issue in Sec. 17.3 in the context of our full pyramidal cell model and in Sec. 19.2 to explore how V_{th} is affected by the cable structure.

6.3.2 Refractory Period

Once the rapid upstroke exceeding 0 has been generated, the membrane potential should be pulled back to its resting potential, that is, repolarized, as rapidly as possible in order to enable the system to generate the next impulse.[3]

This is accomplished by inactivating G_{Na} and by increasing a potassium conductance, G_K. This conductance remains activated even subsequent to spike polarization (for up to 12 msec following the peak of the action potential in Fig. 6.5), causing the membrane to undergo a hyperpolarization. During this period, it is more difficult to initiate an action potential than before; the membrane remains in a *refractory* state. The reason for the reduced ability of the membrane to discharge again is the inactivation of I_{Na} (that is, h is small) and the continuing activation of I_K (n only decays slowly).

---

3. Given a specific membrane capacitance of 1 $\mu F/cm^2$, the 100 mV spike depolarization amounts to transferring about 6,000 positively charged ions per μm^2 of membrane area.

This *refractory period* can be documented by the use of a second current pulse (Fig. 6.7). At $t = 1$ msec a 0.5-msec current pulse is injected into our standard patch of squid axonal membrane. The amplitude of this pulse, $I_1 = 3.95$ nA, is close to the minimal one needed to generate an action potential. The input causes a spike to be triggered that peaks at around 5 msec and repolarizes to $V = 0$ at $t = 7$ msec. This time, at which the membrane potential starts to dip below the resting potential (Fig. 6.5), is somewhat arbitrarily assigned to $\Delta t = 0$. Following this point, a second 0.5-msec-long current pulse of amplitude I_2 is applied Δt msec later. The amplitude of I_2 is increased until a second action potential is generated. This first occurs at $\Delta t = 2$ msec (that is, 2 msec after the membrane potential has repolarized to zero). At this time, $I_2/I_1 = 23.7$, that is, the amplitude of the second pulse must be 23.7 times larger than the amplitude of the first pulse in order to trigger a spike. Since such large current amplitudes are unphysiological, the membrane is *de facto* not excitable during this period, which is frequently referred to as the *absolute refractory period*. The threshold for initiation of the second spike is elevated up to 11 msec after repolarization of the membrane due to the first spike (*relative refractory period*; Fig. 6.7). This is followed by a brief period of mild hyperexcitability, when a spike can be elicited by a slightly (15%) smaller current than under resting conditions.

From a computational point of view, it is important to realize that the threshold behavior of the Hodgkin–Huxley model depends on the previous spiking history of the membrane. In the squid axon, as in most axons, the threshold rises only briefly, returning to baseline levels after 20 msec or less. As warming the axon to body temperatures speeds up the rates of gating two- to fourfold,[4] the minimal separation time is expected to be only 1–2 msec

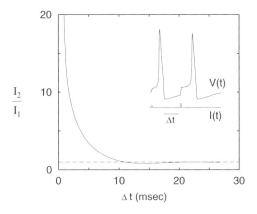

Fig. 6.7 REFRACTORY PERIOD A 0.5-msec brief current pulse of $I_1 = 0.4$ nA amplitude causes an action potential (Fig. 6.5). A second, equally brief pulse of amplitude I_2 is injected Δt msec after the membrane potential due to the first spike having reached $V = 0$ and is about to hyperpolarize the membrane. For each value of Δt, I_2 is increased until a second spike is generated (see the inset for $\Delta t = 10$ msec). The ratio I_2/I_1 of the two pulses is here plotted as a function of Δt. For several milliseconds following repolarization, the membrane is practically inexcitable since such large currents are unphysiological (*absolute refractory period*). Subsequently, a spike can be generated, but it requires a larger current input (*relative refractory period*). This is followed by a brief period of reduced threshold (hyperexcitability). No more interactions are observed beyond about $\Delta t = 18$ msec.

---

4. A crucial parameter in determining the dynamics of the action potential is the temperature. As first mentioned in footnote 5 in Sec. 4.6, if the temperature is reduced, the rate at which the ionic channels underlying the action potential open or close slows down, while the peak conductance remains unchanged. Hodgkin and Huxley recorded most of their data at 6.3° C and the rate constants are expressed at these temperatures (Eqs. 6.10, 6.11 and 6.16–6.19). To obtain the action potential at any other temperature T, all α's and β's need to be corrected by $Q_{10}^{(T-6.3)/10}$, with Q_{10} between 2 and 4 (Hodgkin, Huxley, and Katz, 1952;

for axons in warm-blooded animals. Nerve cells—as compared to axons—often display a much longer increase in their effective spiking threshold, depending on the number of action potentials generated within the last 100 msec or longer (Raymond, 1979). Section 9.3 will treat the biophysical mechanism underlying this short-term *firing frequency adaptation* in more detail.

6.4 Relating Firing Frequency to Sustained Current Input

What happens if a long-lasting current step of constant amplitude is injected into the space-clamped axon (Agin, 1964; Cooley, Dodge, and Cohen, 1965; Stein, 1967a)? If the current is too small, it will give rise to a persistent subthreshold depolarization (Fig. 6.8). Plotting the steady-state membrane depolarization as a function of the applied membrane current (Fig. 6.9A) reveals the linear relationship between the two. If the input is of sufficient amplitude to exceed the threshold, the membrane will generate a single action potential (Fig. 6.8). The minimal amount of sustained current needed to generate at least one action potential (but not necessarily an infinite train of spikes) is called *rheobase* (Cole, 1972). For our standard membrane patch, rheobase corresponds to 0.065 nA. (This current is obviously far less than the amplitude of the brief current pulse used previously.) After the spike has been trigged and following the afterhyperpolarization, $V(t)$ stabilizes at around 2 mV positive to the resting potential, limiting the removal of sodium inactivation as well as enhancing I_K. As the current amplitude is increased, the offset depolarization following the action potential and its hyperpolarization increases until, when the amplitude of the current step is about three times rheobase (0.175 nA), a second action potential is initiated. At around 0.18 nA (I_1 in Fig. 6.9A), the Hodgkin–Huxley equations will start to generate an indefinite train of spikes, that is, the membrane fires repetitively. After the membrane potential goes through its gyrations following each action potential, V creeps past V_{th} and the cycle begins anew: the system travels on a stable limit cycle. In a noiseless situation, the interval between consecutive spikes is constant and the cell behaves as a periodic oscillator with constant frequency.

Figure 6.9A shows the associated steady-state I–V relationship. Experimentally, it can be obtained by clamping the membrane potential to a particular value V and measuring the resultant clamp current I. The equations generate infinite trains of action potentials for $I \geq I_1$ (dashed line in Fig. 6.9A).

A mathematical curiosity of uncertain relevance is the observation that for a range of amplitudes of the current step, the Hodgkin–Huxley equations can display several solutions, including the stable and periodic oscillation emphasized here, a stable but steady-state depolarizing solution, and an unstable periodic solution (Troy, 1978; Rinzel and Miller, 1980). Which one is actually realized depends on the initial conditions.

If the current amplitude is further increased, the interspike intervals begin to decrease and the spiking frequency increases. Figure 6.9B shows the relationship between the amplitude of the injected current and the spiking frequency around threshold, and Fig. 6.10A over a larger current range. It is referred to as the *frequency–current* or f–I curve. Overall,

---

(continued) Beam and Donaldson, 1983; for a definition of Q_{10} see footnote 5 in Sec. 4.6). The Q_{10} for the peak conductances is a modest 1.3. As the temperature is increased, the *upstroke*, that is, the rate at which the voltage rises during the rapid depolarizing phase of the action potential, increases, because the speed at which I_{Na} is activated increases. At the same time, both sodium inactivation and potassium activation increase. Altogether, the total duration of the spike decreases. At temperatures above $33°$ C no spike is generated (Hodgkin and Katz, 1949; of course, the squid axon lives in far more frigid waters than these balmy temperatures).

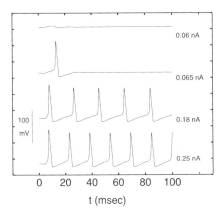

Fig. 6.8 REPETITIVE SPIKING Voltage trajectories in response to current steps of various amplitudes in the standard patch of squid axonal membrane. The minimum sustained current necessary to initiate a spike, termed *rheobase*, is 0.065 nA. In order for the membrane to spike indefinitely, larger currents must be used. Experimentally, the squid axon usually stops firing after a few seconds due to secondary inactivation processes not modeled by the Hodgkin–Huxley equations (1952d).

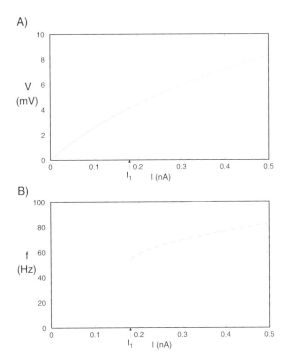

Fig. 6.9 SUSTAINED SPIKING IN THE HODGKIN-HUXLEY EQUATIONS (A) Steady-state $I-V$ relationship and **(B)** $f-I$ or discharge curve as a function of the amplitude of the sustained current I associated with the Hodgkin-Huxley equations for a patch of squid axonal membrane. For currents less than 0.18 nA, the membrane responds by a sustained depolarization (solid curve). At I_1, the system loses its stability and generates an infinite train of action potentials: it moves along a stable limit cycle (dashed line). A characteristic feature of the squid membrane is its abrupt onset of firing with nonzero oscillation frequency. The steady-state $I-V$ curve can also be viewed as the sum of all steady-state ionic currents flowing at any particular membrane potential V_m.

there is a fairly limited range of frequencies at which the membrane fires, between 53 and 138 Hz. If a current at the upper amplitude range is injected in the axon, the membrane fails to repolarize sufficiently between spikes to relieve sodium inactivation. Thus, although

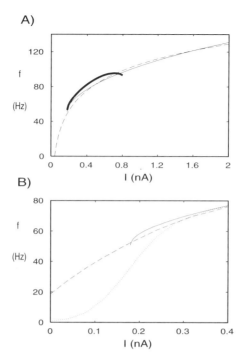

Fig. 6.10 HODGKIN-HUXLEY f–I CURVE AND NOISE (A) Relationship between the amplitude of an injected current step and the frequency of the resultant sustained discharge of action potentials (f–I curve) for a membrane patch of squid axon at 6.3° C (solid line) and its numerical fit (dashed line) by $f = 33.2 \log I + 106$. Superimposed in bold is the f–I curve for the standard squid axon cable (using normalized current). Notice the very limited bandwidth of axonal firing. **(B)** f–I curve for the membrane patch case around its threshold (rheobase) in the presence of noise. White (2000-Hz band-limited) current noise whose amplitude is Gaussian distributed with zero mean current is added to the current stimulus. In the absence of any noise (solid line) the f–I curve shows abrupt onset of spiking. The effect of noise (dotted curve—standard deviation of 0.05 nA; dashed curve—0.1 nA) is to linearize the threshold behavior and to increase the bandwidth of transmission (stochastic linearization). Linear f–I curves are also obtained when replacing the continuous and deterministic Hodgkin-Huxley currents by discrete and stochastic channels (see Sec. 8.3).

the membrane potential does show oscillatory behavior, no true action potentials are generated.

In the laboratory, maintained firing in the squid axon is not that common (Hagiwara and Oomura, 1958; see, however, Chapman, 1963). This is most likely due to secondary inactivation mechanisms which are not incorporated into the Hodgkin–Huxley equations. Yet for short times, the theoretical model of Hodgkin and Huxley makes reasonably satisfactory predictions of the behavior of the space-clamped axon (for a detailed comparison between experimental observations and theoretical predictions see Guttman and Barnhill, 1970 as well as Chap. 11 in the ever trustworthy Jack, Noble, and Tsien, 1975), in particular with respect to the small dynamic range of firing frequencies supported by the axonal membrane and the abrupt onset of spiking at a high firing frequency. The f–I curve can be well approximated by either a square root or a logarithmic relationship between frequency and injected current (Agin, 1964; see Fig. 6.10).

The f–I curves of most neurons, and not just the squid axon, bend over and saturate for large input currents. This justifies the introduction of a smooth, sigmoidal nonlinearity mimicking the neuronal input-output transduction process in continuously valued neural network models (Hopfield, 1984). It is important to keep in mind that the paradigm under which the f–I curves are obtained, sustained current input, represents only a very crude approximation to the dynamic events occurring during synaptic bombardment of a cell leading to very complex spike discharge patterns (see Chap. 14).

An important feature of the Hodgkin–Huxley model is that the frequency at the onset of repetitive activity has a well-defined nonzero minimum (about 53 Hz at 6.3° C; Fig. 6.10B). The membrane is not able to sustain oscillations at lower frequencies. This behavior, generated by a so-called *Hopf bifurcation* mechanism, is generic to a large class of oscillators occurring in nonlinear differential equations (Cronin, 1987; Rinzel and Ermentrout, 1998) and will be treated in more detail in the following chapter.

As first explicitly simulated by Stein (1967b), adding random variability to the input can increase the bandwidth of the axon by effectively increasing the range within which the membrane can generate action potentials. If the input current is made to vary around its mean with some variance, reflecting for instance the spontaneous release of synaptic vesicles, the sharp discontinuity in the firing frequency at low current amplitudes is eliminated, since even with an input current that is on average below threshold, the stimulus will become strong enough to generate an impulse with a finite, though small, average frequency. Depending on the level of noise, the effective minimal firing frequency can be reduced to close to zero (Fig. 6.10B). A similar linearization behavior can be obtained if the continuous, deterministic, and macroscopic currents inherent in the Hodgkin–Huxley equations are approximated by the underlying discrete, stochastic, and microscopic channels (Skaugen and Walloe, 1979; see Sec. 8.3).

Adding noise to a quantized signal to reduce the effect of this discretization is a standard technique in engineering known as *dithering* or *stochastic linearization* (Gammaitoni, 1995; Stemmler, 1996).

A large number of neurons can generate repetitive spike trains with arbitrarily small frequencies. As first shown by Connor and Stevens (1971c) in their Hodgkin–Huxley-like model of a gastropod nerve cell, the addition of a transient, inactivating potassium current (termed the I_A current) enables the cell to respond to very small sustained input currents with a maintained discharge of very low frequency. (This topic will be further pursued in Sec. 7.2.2.) Such low firing frequencies are also supported by pyramidal cells (Fig. 9.7).

6.5 Action Potential Propagation along the Axon

Once the threshold for excitation has been exceeded, the all-or-none action potential can propagate from the stimulus site to other areas of the axon. The hypothesis that this propagation is mediated by cable currents flowing from excited to neighboring, nonexcited regions was suggested already around the turn of the century by Hermann (1899). It was not until Hodgkin (1937) that direct experimental proof became available. A quantitative theory of this propagation had to await Hodgkin and Huxley's 1952 study. Because this has been a very well explored chapter in the history of biophysics, we will be brief here, only summarizing the salient points. Chapter 10 in Jack, Noble, and Tsien (1975) provides a deep and thorough coverage of nonlinear cable theory as applied to the conduction of action potentials. Section 19.2 will deal with how cable structures, such as an infinite cylinder, affect the voltage threshold for spike initiation.

6.5.1 Empirical Determination of the Propagation Velocity

The equivalent electrical circuit replicates the patch of sodium, potassium, and leak conductances and batteries (Fig. 6.2) along the cable in a fashion we are already familiar with from the passive cable (Fig. 6.11). Equation 2.5 specifies the relationship between the membrane current (per unit length) and the voltage along the cable,

$$i_m = \frac{1}{r_a}\frac{\partial^2 V}{\partial x^2}.$$ (6.22)

In Eq. 6.20, we derived the membrane current (per unit area) flowing in a patch of axonal membrane. Combining the two with the appropriate attention to scaling factors leads to an equation relating the potential along the axon to the electrical property of the active membrane,

$$\frac{d}{4R_i}\frac{\partial^2 V}{\partial x^2} = C_m\frac{\partial V}{\partial t} + \overline{G}_{\text{Na}}m^3 h(V - E_{\text{Na}}) + \overline{G}_{\text{K}}n^4(V - E_{\text{K}}) + G_m(V - V_{\text{rest}})$$ (6.23)

where d is the diameter of the axon. Hodgkin and Huxley (1952d) used a $d = 0.476$ mm thick axon in their calculations and a value of $R_i = 35.4$ Ω·cm. This nonlinear partial differential equation, in conjunction with the three equations describing the dynamics of m, h, and n and the appropriate initial and boundary conditions, constitutes the complete Hodgkin–Huxley model.

This type of second-order equation, for which no general analytical solution is known, is called a *reaction-diffusion* equation, because it can be put into the form of

$$\frac{\partial V}{\partial t} = D\frac{\partial^2 V}{\partial x^2} + F(V)$$ (6.24)

with $D > 0$ constant. We will meet this type of equation again when considering the dynamics of intracellular calcium (see Chap. 11). Under certain conditions, it has wave-like solutions.

Because Hodgkin and Huxley only had access to a very primitive hand calculator, they could not directly solve Eq. 6.23. Instead, they considered a particular solution to these equations. Since they observed that the action potential propagated along the axon without changing its shape, they postulated the existence of a wave solution to this equation, in which the action potential travels with constant velocity u along the axon, that is, $V(x, t) = V(x - ut)$. Taking the second spatial and temporal derivative of this

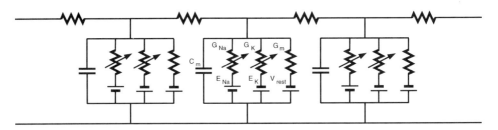

Fig. 6.11 ELECTRICAL CIRCUIT OF THE SQUID GIANT AXON   One-dimensional cable model of the squid giant axon. The structure of the cable is as in the passive case (Fig. 2.2B), with the *RC* membrane components augmented with circuit elements modeling the sodium and potassium currents (Fig. 6.2).

expression and using the chain rule leads to a second-order hyperbolic partial differential equation,

$$\frac{\partial^2 V}{\partial x^2} = \frac{1}{u^2} \frac{\partial^2 V}{\partial t^2} .$$

(6.25)

Replacing the second spatial derivative term in Eq. 6.23 with this expression yields

$$\frac{1}{K} \frac{d^2 V}{dt^2} = \frac{dV}{dt} + \frac{I_{ionic}}{C_m}$$

(6.26)

with $K = 4 R_i u^2 C_m / d$ and I_{ionic} defined in Eq. 6.2. Equation 6.26 is an ordinary second-order differential equation, whose solution is much easier to compute than the solution to the full-blown partial differential equation. It does require, though, a value for u. By a laborious trial-and-error procedure, Hodgkin and Huxley iteratively solved this equation until they found a value of u leading to a stable propagating wave solution. In a truly remarkable test of the power of their model, they estimated 18.8 m/sec (at 18.3° C) for the velocity at which the spike propagates along the squid giant axon, a value within 10% of the experimental value of 21.2 m/sec. This is all the more remarkable, given that their model is based on voltage- and space-clamped data, and represents one of the rare instances in which a neurobiological model makes a successful quantitative prediction.

We can establish the dependency of the velocity on the diameter of the fiber using the following argumentation. Because both I_i and C_m are expressed as current and capacitance per unit membrane area, their ratio is independent of the fiber diameter. The voltage across the membrane and its temporal derivatives must also be independent of d. This implies that the constant K in Eq. 6.26 must remain invariant to changes in diameter. Assuming that C_m and R_i do not depend on d, we are lead to the conclusion that the velocity u must be

$$u \propto \sqrt{d} .$$

(6.27)

In other words, the propagation velocity in unmyelinated fibers is expected to be proportional to the square root of the axonal diameter.[5] Indeed, this predicted relationship is roughly followed in real neurons (see Fig. 6.16; Ritchie, 1995).

This implies that if the delay between spike initiation at the cell body and the arrival of the spike at the termination of an axon needs to be cut in half, the diameter of the axon needs to increase by a factor of 4, a heavy price to pay for rapid communication. The premium put on minimizing propagation delay in long cable structures is most likely the reason the squid evolved such thick axons. As we will see further below, many axons in vertebrates use a particular form of electrical insulation, termed *myelination*, to greatly speed up spike propagation without a concomitant increase in fiber diameter.

It was more than 10 years later that Cooley, Dodge, and Cohen (1965; see also Cooley and Dodge, 1966) solved the full partial differential equation (Eq. 6.23) numerically using an iterative technique. Figure 6.12 displays the voltage trajectory at three different locations along the axon; at $x = 0$ a short suprathreshold current pulse charges up the local membrane capacitance. This activates the sodium conductance and Na^+ ions rush in, initiating the full-blown action potential (not shown). The local circuit current generated by this spike leads to an exponential rise in the membrane potential in the neighboring region, known as the "foot" of the action potential. This capacitive current in turn activates the local sodium conductance, which will increase rapidly, bringing this region above threshold:

---

5. Notice that we derived a similar square-root relationship between diameter and "pseudovelocity" for the decremental wave in the case of a passive cable (Eq. 2.53).

the spike propagates along the axon. Different from the space-clamped axon, where the capacitive current is always equal and opposite to the ionic currents once the stimulus current has stopped flowing (Eq. 6.20), the time course of current is more complex during the propagated action potential due to the local circuit currents. Because some fraction of the local membrane current depolarizes neighboring segments of the axonal cable (the so-called *local circuit* currents; see Fig. 6.13), the current amplitude required to trigger at least one action potential is larger than the current amplitude in the space-clamped case

If the voltage applied to the squid membrane is small enough, one can *linearize* the membrane, describing its behavior in terms of voltage-independent resistances, capacitances, and inductances. This procedure was first carried out by Hodgkin and Huxley (1952d) and will be discussed in detail in Chap. 10. Under these circumstances, a space constant λ can be associated with the "linearized" cable, describing how very small currents are attenuated along the axon. At rest, the dc space constant for the squid axon is $\lambda = 5.4$ mm, about 10 times larger than its diameter.

When long current steps of varying amplitudes are injected into the axon, the squid axon responds with regular, periodic spikes. However, the already small dynamic range of the f–I curve of the space-clamped axon (Fig. 6.10A) becomes further reduced to a factor of less than 1.7 when the sustained firing activity in the full axon is considered (from 58 to 96 Hz at 6.3° C). Thus, while the Hodgkin–Huxley model describes to a remarkable degree the behavior of the squid's giant axon, the equations do not serve as an adequate model for impulse transduction in nerve cells, most of which have a dynamic range that extends over two orders of magnitude.

As predicted by Huxley (1959), Cooley and Dodge (1966) found a second solution to the Hodgkin–Huxley equations. When the amplitude of the injected current step is very close to the threshold for spike initiation, they observed a decremental wave propagating away

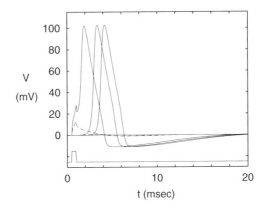

Fig. 6.12 PROPAGATING ACTION POTENTIAL   Solution to the complete Hodgkin–Huxley model for a 5-cm-long piece of squid axon for a brief suprathreshold current pulse delivered to one end of the axon. This pulse generates an action potential that travels down the cable and is shown here at the origin as well as 2 and 3 cm away from the stimulating electrode (solid lines). Notice that the shape of the action potential remains invariant due to the nonlinear membrane. The effective velocity of the spike is 12.3 m/sec (at 6.3° C). If the amplitude of the current pulse is halved, only a local depolarization is generated (dashed curve) that depolarizes the membrane 2 cm away by a mere 0.5 mV (not shown). This illustrates the dramatic difference between active and passive voltage propagation.

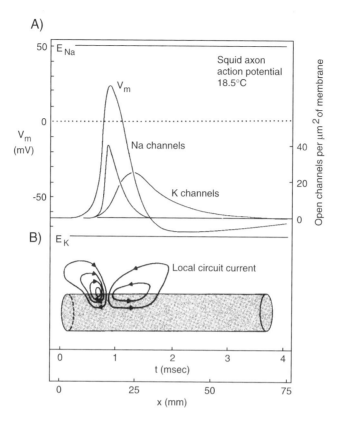

Fig. 6.13 LOCAL CIRCUIT CURRENT IN THE SQUID AXON   Illustration of the events occurring in the squid axon during the propagation of an action potential. Since the spike behaves like a wave traveling at constant velocity, these two panels can be thought of either as showing the voltages and currents in time at one location or as providing a snapshot of the state of the axon at one particular instant (see the space/time axes at the bottom). **(A)** Distribution of the voltage (left scale) or the number of open channels (right scale) as inferred from the Hodgkin–Huxley model at 18.5° C. **(B)** *Local circuit currents* that spread from an excited patch of the axon to neighboring regions bringing them above threshold, thereby propagating the action potential. The diameter of the axon (0.476 mm) is not drawn to scale. Reprinted by permission from Hille (1992).

from the source. This solution quickly dies away to zero potential as x increases and is only observed if the amplitude of the current step is within 0.1% of the threshold current needed to obtain at least one spike. This phenomenon reveals the fact that the Hodgkin–Huxley model does not possess a strict threshold in the true sense of the word. In other words, there exists a continuous transformation between the subthreshold and the threshold voltage response. Yet in order to reveal these intermediate solutions the excitation must be adjusted with a degree of accuracy impossible to achieve physiologically. Practically speaking, given unavoidable noise in any neuronal system, only the propagating wave solution (with its associated threshold) plays a significant role in propagating information along the axon.

Cooley and Dodge (1966) also considered what happens if the density of voltage-dependent channels underlying G_{Na} and G_K is attenuated by a factor of η (with $0 \leq \eta \leq 1$; the value of V_{rest} and G_{leak} were adjusted so that the resting potential and resting conductance were held constant). Reducing these conductances is somewhat analogous to the action of certain local anesthetics, such as lidocaine or procaine as used by dentists, in blocking action potential propagation. As η is reduced below 1, the velocity of propagation as well as the peak amplitude of the spike are reduced. For $\eta < 0.26$, no uniform wave solution is possible and the "action potential" decrements with distance.

6.5.2 Nonlinear Wave Propagation

Spikes moving down an axon are but one instance of a *nonlinear propagating wave*—nonlinear since in a linear dispersive medium, such as a passive cable, the different Fourier components associated with any particular voltage disturbance will propagate at a different velocity and the disturbance will lose its shape. This is why propagating spikes and the like are frequently referred to by mathematicians as resulting from *nonlinear diffusion*. Other examples include sonic shock waves or the digital pulses in an optical cable.

Scott (1975) argues for a broad classification of such phenomena into (1) those systems for which energy is conserved and which obey a conservation law and (2) those for which solitary traveling waves imply a balance between the rate of energy release by some nonlinearity and its consumption.

Waves associated with the first type of systems are known as *solitons* and are always based on energy conservation (Scott, Chu, and McLaughlin, 1973). Solitons emerge from a balance between the effects of nonlinearity, which tend to draw the wave together, fighting dispersion, which tends to spread the pulse out. This implies that solitons can propagate over a range of speeds. Furthermore, they can propagate through each other without any interference. Solitons have been observed in ocean waves and play a major role in high-speed optical fibers.

Action potentials are an example of the second type of propagating wave, similar to an ordinary burning candle (Scott, 1975). Diffusion of heat down the candle releases wax which burns to supply the heat. If P is the power (in joules per second) necesssary to feed the flame and E the chemical energy stored per unit length of the candle (joules per meter), the nonlinear wave in the form of the flame moves down the candle at a fixed velocity u given by

$$P = uE. \tag{6.28}$$

In other words, the velocity is fixed by the properties of the medium and does not depend on the initial conditions. Were we to light flames at both ends of a candle, the flames would move toward each other and annihilate themselves. This is also true if action potentials are initiated at the opposite ends of an axon. When they meet, they run into each other's refractory period and destroy each other. Thus, spikes are not solitons.

6.6 Action Potential Propagation in Myelinated Fibers

The successful culmination of the research effort by Hodgkin and Huxley heralded the coming of age of neurobiology. While we will deal in Chap. 9 with their methodology as applied in the past decades to the ionic currents found at the cell body of nerve cells, let us here briefly summarize spike propagation in *myelinated* axons (for more details, see Waxman, Kocsis, and Stys, 1995; Ritchie, 1995; Weiss, 1996).

Axons come in two flavors, those covered by layers of the lipid *myelin* and those that are not. The squid axon is an unmyelinated fiber, common to invertebrates. In vertebrates, many fibers are wrapped dozens or even hundreds of times with myelin, the actual diameter of the axon itself being only 60% or 70% of the total diameter (Fig. 6.14). This insulating material is formed by special supporting cells, called *Schwann* cells in the peripheral nervous system and *oligodendrocytes* in the central nervous system.

A second specialization of myelinated fibers is that the myelin sheet is interrupted at regular intervals along the axon by nodes, named for their discoverer *nodes of Ranvier*. Here, the extracellular space gains direct access to the axonal membrane. Typically, the length of a node is very small (0.1%) compared to the length of the *internodal* segment (Fig. 6.15). In the vertebrate, single myelinated fibers range in diameter from 0.2 to 20 μm, while unmyelinated fibers range between 0.1 and 1 μm. In stark contrast, the diameters of unmyelinated invertebrate fibers range from under 1 μm to 1 mm.

In myelinated axons, conduction does not proceed continuously along the cable, but jumps in a discontinuous manner from one node to the next. This *saltatory conduction* (from the Latin *saltus*, to leap) was clearly demonstrated by Huxley and Stämpfli (1949) and Tasaki (1953). What these and similar experiments on frog, rabbit, and rat myelinated fibers made clear is that ionic currents are strikingly inhomogeneously distributed across the axonal membrane (Fig. 6.15; FitzHugh, 1962; Frankenhaeuser and Huxley, 1964; Stämpfli and Hille, 1976; Rogart and Ritchie, 1977; Chiu et al., 1979; Chiu and Ritchie, 1980). Spike generation essentially only takes place at the small nodes of Ranvier, which are loaded with fast sodium channels (between 700 and 2000 per square micrometer). In mammalian myelinated nerves, the repolarization of the spike is not driven by a large outward potassium current, as in the squid axon[6] but is achieved using a rapid sodium inactivation in combination with a large effective leak conductance. Indeed, action potentials do not show any hyperpolarization (Fig. 6.1B), unlike those in the squid giant axon. The origin of the large voltage-independent leak might involve an extracellular pathway beneath the myelin that connects the nodal and internodal regions (Barrett and Barrett, 1982; Ritchie,

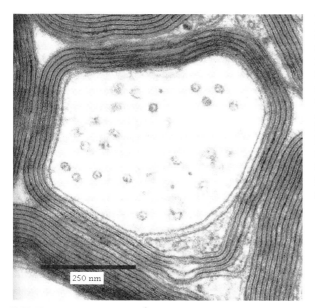

Fig. 6.14 Myelinated Axons
Electron micrograph of a cross section through a portion of the optic fiber in an adult rat. The complete transverse section through a single myelinated axon is shown in close neighborhood to other axons. About four wrappings of myelin insulation are visible. The circular structures inside the axonal cytoplasm are transverse sections through microtubules. Reprinted by permission from Peters, Palay, and Webster (1976).

250 nm

---

6. Pharmacological blockage of potassium channels has no effect on the shape of the action potential in the rabbit fibers (Ritchie, Rang, and Pellegrino, 1981).

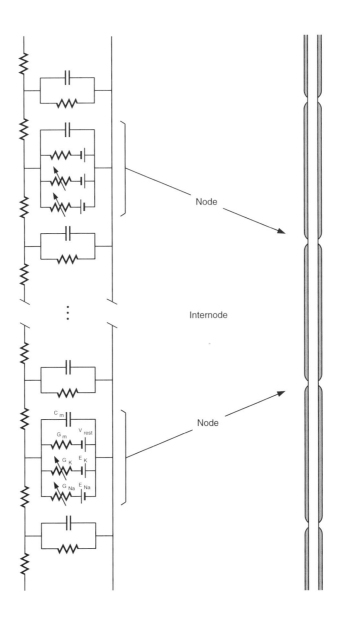

Fig. 6.15 Electrical Circuit for a Myelinated Axon   Geometrical and electrical layout of the myelinated axon from the frog sciatic nerve (Frankenhaeuser and Huxley, 1964; Rogart and Ritchie, 1977). The diameter of the axon and its myelin sheath is 15 μm, the diameter of the axon itself 10.5 μm, the difference being made up by 250 wrappings of myelin. The myelin is interrupted every 1.38 mm by a *node of Ranvier* that is 2.5 μm wide. The total distributed capacitance for the internode (2.2 pF) is only slightly larger than the capacitance of the much smaller node (1.6 pF). The same is also true of the distributed resistance. At each node, the spike is reamplified by a fast sodium current and is repolarized by a potassium current. Little or no potassium current is found at the nodes of Ranvier in mammalian myelinated axons. There, repolarization is accomplished by rapid sodium inactivation in conjunction with a large effective "leak" current.

1995). Potassium channels are present under the myelin sheet along the internodal section, although their functional role is unclear (Waxman and Ritchie, 1985).

The function of the numerous tightly drawn layers of myelin around the internodal segments is to reduce the huge capacitive load imposed by this very large cable segment, as well as to reduce the amount of longitudinal current that leaks out across the membrane. The effective membrane capacitance of the entire myelin sheath, which in the case of the frog axon illustrated in Fig. 6.15, is made up of 250 myelin layers, is $C_m/250$, with C_m the specific capacitance of one layer of myelin (similar to that of the axonal membrane), while the effective resistance is 250 times higher than the R_m of one layer of myelin. Even though the length of the interaxial node is typically 1000 times larger than the node, its total capacitance has the same order of magnitude (Fig. 6.15). This allows the action potential to spread rapidly from one node to the next, "jumping" across the intervening internodal areas and reducing metabolic cost (since less energy must be expended to restore the sodium concentration gradient following action potential generation). There is a safety factor built into the system, since blocking one node via a local anesthetic agent does not prevent blockage of the impulse across the node (Tasaki, 1953). Detailed computer simulations of the appropriately modified Hodgkin–Huxely equations (based on the circuit shown in Fig. 6.15) have confirmed all of this (Frankenhaeuser and Huxley, 1964; Rogart and Ritchie, 1977).

Single axons can extend over 1 m or more,[7] making conduction velocity of the electrical impulses something that evolution must have tried to maximize at all cost. Measurements (Huxley and Stämpfli, 1949) and computations indicate that the time it takes for the currents at one node to charge up the membrane potential at the next node is limited by the time it takes to charge up the intervening internodal membrane. This is determined by the time constant of the membrane τ, which is independent of the geometry of the axon. In this time the spike will have moved across the internodal distance, making the propagation velocity proportional to this distance divided by τ. Since anatomically the internodal distance is linearly related to the diameter of the axon, the velocity of spike propagation will be proportional to the fiber diameter,

$$u \propto d \tag{6.29}$$

rather than the square-root dependency found for unmyelinated fibers (Eq. 6.27). Rushton (1951) gave this argument a precise form using the principle of *dimensional scaling*. If, he argued, axons had the same specific membrane properties, then in order for points along two axons with different diameters to be in "corresponding states," certain scaling relationships must hold. In particular, space should increase in units of the internodal length and velocity should be roughly proportional to the fiber diameter (for more details, see Weiss, 1996). The latter is actually the case (Fig. 6.16). When comparing the fiber diameter against the propagation velocity for myelinated cat axons, a roughly linear relationship can be observed (Hursh, 1939; Rushton, 1951; Ritchie, 1982). With the exception of a 1.1-μm-thick unmyelinated mammalian fiber that propagates action potentials at 2.3 mm/msec (Gasser, 1950), spike velocity in very small fibers has, so far, been difficult to record.

The functional importance of myelinated fibers is clear. They provide a reliable and rapid means of communicating impulses at a much reduced cost compared to unmyelinated fibers (at the same conduction velocity, a myelinated fiber can be up to 50 times thinner than an

---

7. Think about the spinal nerve axons of an elephant or of the extinct Brontosaurus.

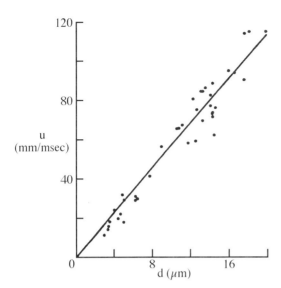

Fig. 6.16 **Diameter and Propagation Velocity** Relationship between (internal) diameter d of adult cat peripheral myelinated fibers and propagation velocity u of the action potential. The data are shown as dots (Hursh, 1939) and the least-square fit as a line. Peripheral myelinated fibers are bigger than 1 μm, while myelinated fibers in the central nervous system can be as thin as 0.2 μm, with an expected velocity in the 1-mm/msec range. Reprinted by permission from Ritchie (1982).

unmyelinated fiber). This large ($\times 2500$) factor in packing allows the brain to squeeze more than a million axons into a single nerve that supplies the brain with visual information. The primary cost of this insulation is the added developmental complexity and the possibility that demyelinating diseases, such as *multiple sclerosis*, can incapacitate the organism.

6.7 Branching Axons

The all-or-none nature of action potentials has lead to the idea that the axon serves mainly as a reliable transmission line, making a highly secure, one-way point-to-point connection among two processing devices. Furthermore, because of its high propagation velocity, the action potential is thought to arrive almost simultaneously to all of its output sites. Indeed, both properties have been used to infer that spikes propagating parallel fibers in the cerebellum serve to implement a very precise timing circuit (Braitenberg and Atwood, 1958; Braitenberg, 1967).

It will not come as a surprise that the axon-as-a-wire concept is not quite true and needs to be revised. Experimentally, it is known that trains of action potentials show failure at certain regions along the axon, most likely at the branch points. In other words, the train of spikes generated at the soma may have lost some of its members by the time it reaches the presynaptic terminals, with individual spikes "deleted" (Barron and Matthews, 1935; Tauc and Hughes, 1963; Chung, Raymond, and Lettvin, 1970; Parnas, 1972; Smith, 1983). For instance, conduction across a branching point in a lobster axon fails at frequencies above 30 Hz (Grossman, Parnas, and Spira, 1979a). This conduction block first appears in the thicker daughter branch and only later in the thinner branch, most likely due to a differential buildup of potassium ions (Grossman, Parnas, and Spira, 1979b). Physiological evidence indicates that such a switching mechanism might subserve a specific function in the case of the motor axon innervating the muscle used for opening the claw in the crayfish (Bittner, 1968). Depending on the firing frequencies, spikes are routed differentially into two branches of the axon going to separate muscle fibers.

These experimental studies have shown that action potentials may fail to invade the daughter branches of a bifurcating axon successfully. As the theoretical analysis by Goldstein and Rall (1974) pointed out, the single most important parameter upon which propagation past the bifurcation depends is its associated geometric ratio

$$\text{GR} = \frac{d_{\text{daughter},1}^{3/2} + d_{\text{daughter},2}^{3/2}}{d_{\text{parent}}^{3/2}} \tag{6.30}$$

where the d's are the fiber diameters and it is assumed that the specific membrane properties are constant in all three branches. This should remind us, of course, of the analysis of the branching passive cables (Secs. 3.1, 3.2, and Eq. 3.15) and, indeed, the reasoning is identical. GR equals the ratio of the input impedances if all cables are semi-infinite.

Goldstein and Rall's (1974) and subsequent analytical and modeling investigations (Khodorov and Timin, 1975; Parnas and Segev, 1979; Moore, Stockbridge, and Westerfield, 1983; Lüscher and Shiner, 1990a,b; Manor, Koch, and Segev, 1991) established the following principles. For GR $= 1$, *impedances match* perfectly and the spike propagates without any perturbation past the branch point (indeed, electrically speaking, for GR $= 1$ the branching configuration can be reduced to a single *equivalent cable*, albeit an active one; see Sec. 3.2). If GR < 1, the action potential behaves as if the axon tapers and it will slightly speed up. The far more common situation is GR > 1, that is, the combined electrical load of the daughters exceeds the load of the main branch. As long as GR is approximately < 10, propagation past the branch point is assured, although with some delay (that can be substantial for large values of GR). If GR> 10, propagation into both branches fails simultaneously, since the electrical load of the daughters has increased beyond the capacity of the electrical current from the parent branch to initiate a spike in the daughter branches. Parnas and Segev (1979) emphasize that for each constant geometric ratio, changes in the diameter ratio between the daughter branches never yields differential conduction into one of the daughters if the specific membrane properties are identical in both. This implies that the experimentally observed differential conduction (Bittner, 1968; Grossman, Parnas, and Spira, 1979a) must be due to other factors, such as a rundown in the ionic concentration across the membrane or saturation of the ionic pumps that are sensitive to the ratio of area to volume (and would thus be expected to occur earlier in larger fibers). Note that all of these modeling studies have assumed unmyelinated fibers and that axonal branch points appear to be devoid of myelin.

Up to 10–12 bifurcations (see the heavily branched axonal terminal arbor in Fig. 3.1M) can occur before the action potential reaches its presynaptic terminal where it initiates vesicular release. The delay at branch points with GR >1, in conjunction with other geometrical inhomogeneities, such as the short swellings at sites of synaptic terminals called *varicosities*, might add up to a considerable number, leading to a substantial broadening of spike arrival times at their postsynaptic targets.

The degree of temporal dispersion was simulated in the case of an axon from the somatosensory cortex of the cat (Manor, Koch, and Segev, 1991). Since it is almost entirely confined to cortical gray matter, it was taken to be unmyelinated (Fig. 6.17). In the absence of better data, Hodgkin–Huxley dynamics (at 20° C) were assumed. About 1000 boutons were added to the axon and the propagation time between spike initiation just beyond the cell body and these boutons is shown in the histogram of Fig. 6.17. The first peak (with a mean of 3.8 ± 0.5 msec) is contributed from terminals along the branches in cortical areas 3a and 3b (see the inset in Fig. 6.17A) and the more delayed one from those in area

4 (mean of 5.8 ± 0.4 msec). Of the total delay, about 22–33% is due to the branch points and geometrical inhomogeneities; the majority is simple propagation delay. Manor, Koch, and Segev (1991) conclude that temporal dispersion in the axonal tree will be minor, on the order of 0.5–1 msec.

Let us conclude with one observation. Computer simulations of branching axons by the author have shown that a strategically located inhibitory synapse of the shunting type onto one branch of the axon, following the on-the-path theorem (Sec. 5.1.3), can selectively veto an action potential from invading this branch while not affecting spike invasion into the second branch. This would allow for very fast synaptic switching or *routing* of information in an axonal tree (similar to a telephone network). While inhibitory synapses can be found directly at the axon initial segment (Kosaka, 1983; Soriano and Frotscher, 1989), no synapses, whether excitatory or inhibitory, have been observed on or around axonal branching points. It is anybody's guess why the nervous system did not avail itself of this opportunity to precisely (in space and time) filter or gate action potentials.

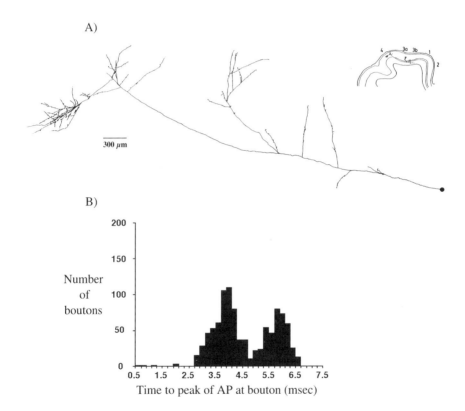

Fig. 6.17 PROPAGATION DELAYS ALONG A BRANCHING AXON   Delays that action potentials incur as they propagate through a highly branching axonal tree (**A**), simulated in the case of an HRP labeled axon originated in layer 5 of the somatosensory cortex of the adult cat (see inset). Reprinted by permission from Schwark and Jones (1989). (**B**) Histogram of the delay incurred between action potential initiation just beyond the cell body and the 977 terminals distributed in the terminal branches of the axon. The two humps correspond to locations in the proximal and distal parts of the axonal tree. Over the 3.5 mm of the tree, the total delay is 6.5 msec and temporal dispersion is minimal. Reprinted by permission from Manor, Koch, and Segev (1991).

6.8 Recapitulation

The Hodgkin–Huxley 1952 model of action potential generation and propagation is the single most successful quantitative model in neuroscience. At its heart is the depiction of the time- and voltage-dependent sodium and potassium conductances G_{Na} and G_K in terms of a number of gating particles. The state of G_{Na} is governed by three activation particles m and one inactivating particle h, while the fate of the potassium conductance is regulated by four activating particles n. The dynamics of these particles are governed by first-order differential equations with two voltage-dependent terms, the steady-state activation (or inactivation), and the time constant. The key feature of activating particles is that their amplitude increases with increasing depolarization, while the converse is true for inactivating particles. For rapid input to a patch of squid axonal membrane, spike initiation is exceeded whenever the net inward current becomes negative, that is, when a particular voltage threshold V_{th} is exceeded.

Inclusion of the cable term leads to a four-dimensional system of coupled, nonlinear differential equations with a wave solution that propagates at a constant velocity down the axon. This wave, the action potential, is due to the balance between dispersion and restoration caused by the voltage-dependent membrane. When injecting sustained currents into the axon, the equations predict two important aspects of the squid axon: the abrupt onset of sustained firing with a high spiking frequency and the very limited bandwidth of the firing frequency.

The Hodgkin–Huxley formalism represents the cornerstone of quantitative models of nerve cell excitability, and constitutes a remarkable testimony to the brilliance of these researchers. It should be remembered that their model was formulated at a time when the existence of ionic channels, the binary, microscopic, and stochastic elements underlying the continuous macroscopic, and deterministic ionic currents, was not known.

Wrapping axons in insulating material, such as the many layers of myelin observed in myelinated fibers that are found in all vertebrates, leads to a dramatic speedup over unmyelinated fibers. Conversely, at the same spike propagation speed, myelinated fibers can be up to 50 times thinner than unmyelinated fibers. In mammals, axons above 1 μm are usually myelinated, with speeds in the 5-mm/msec range, and rarely exceed 20 μm. When axons reach their target zone, they branch profusely, enabling them to make thousands of contacts on postsynaptic processes. As trains of spikes attempt to propagate past these points, they can be slowed down, depending on the exact geometry of the junction. In the more extreme cases, individual spikes can fail to propagate past branch points.

We conclude that pulses can communicate along axons reliably, rapidly (at speeds between 1 and 100 mm/sec) and with little temporal dispersion. The main exception to this appears to be the propagation of trains of spikes past branching points. Here, due to a variety of phenomena, conduction block can occur, which will differentially route information into one of the daughter branches or prevent conduction alltogether.

7

PHASE SPACE ANALYSIS OF
NEURONAL EXCITABILITY

The previous chapter provided a detailed description of the currents underlying the generation and propagation of action potentials in the squid giant axon. The Hodgkin–Huxley (1952d) model captures these events in terms of the dynamical behavior of four variables: the membrane potential and three state variables determining the state of the fast sodium and the delayed potassium conductances. This quantitative, conductance-based formalism reproduces the physiological data remarkably well and has been extremely fertile in terms of providing a mathematical framework for modeling neuronal excitability throughout the animal kingdom (for the current state of the art, see McKenna, Davis, and Zornetzer, 1992; Bower and Beeman, 1998; Koch and Segev, 1998). Collectively, these models express the complex dynamical behaviors observed experimentally, including pulse generation and threshold behavior, adaptation, bursting, bistability, plateau potentials, hysteresis, and many more.

However, these models are difficult to construct and require detailed knowledge of the kinetics of the individual ionic currents. The large number of associated activation and inactivation functions and other parameters[1] usually obscures the contributions of particular features (e.g., the activation range of the sodium activation particle) toward the observed dynamic phenomena. Even after many years of experience in recording from neurons or modeling them, it is a dicey business predicting the effect that varying one parameter, say, the amplitude of the calcium-dependent slow potassium current (Chap. 9), has on the overall behavior of the model. This precludes the development of insight and intuition, since the numerical complexity of these models prevents one from understanding which important features in the model are responsible for a particular phenomenon and which are irrelevant.

Qualitative models of neuronal excitability, capturing some of the *topological aspects* of neuronal dynamics but at a much reduced complexity, can be very helpful in this regard, since they highlight the crucial features responsible for a particular behavior. By topological

---

1. The electrical model of the spiking behavior of the bullfrog sympathetic ganglion cells, illustrated in Figs. 9.10–9.12, includes eight ionic currents, which are described with the help of two dozen voltage and/or calcium-dependent functions and an equal number of additional parameters. And this is for a spherical neuron devoid of any dendrites.

aspects we mean those properties that remain unchanged in spite of quantitative changes in the underlying system. These typically include the existence of stable solutions and their basins of attraction, limit cycles, bistability, and the existence of strange attractors. In this chapter we discuss two-dimensional models of spiking systems, a qualitative as well as quantitative, biophysical model, that will help us to understand a number of important features of the Hodgkin–Huxley (and related) membrane patch models using the theory of dynamical systems and phase space analysis. In particular, the FitzHugh–Nagumo model explains the origin and onset of neuronal oscillations, that is, periodic spike trains, in response to a sustained current input. It is considerably easier—from both a numerical and a conceptual point of view—to study the dynamics of populations of neuronal "units" described by such simplified dynamics rather than simulating the behavior of a network of biophysically complex neurons.

For a general introduction to the theory of dynamical systems, see the classical treatise of Hirsch and Smale (1974) or the informal and very friendly account by Strogatz (1994) that stresses applications. Two eminently readable introductions to physiological applications of this theory are the monograph by Glass and Mackey (1988) and the chapter by Rinzel and Ermentrout (1998).

Following the ancient wisdom that "there's no such thing as a free lunch," we can well ask what is the price that one pays for this elegance? The answer is that simple models usually lack numerically accurate predictions. As we will see below, we can understand why oscillations in one system will be initiated with a nonzero firing frequency, as in the squid giant axon, while other systems, such as neocortical pyramidal cells, can respond with a train of spikes having very long interspike intervals. Yet, in general, these models do not allow precise predictions to be made concerning the exact threshold for the onset of spikes and so on. This drawback is less serious than in other fields of science, however, since in almost all situations of interest, precise knowledge of the relevant biophysical parameters is lacking anyway. Given the fantastic diversity and variability among neurons, simple generic models are frequently much superior at capturing key aspects of neuronal function, in particular at the network level, than detailed ones.

7.1 The FitzHugh–Nagumo Model

The simplified model of spiking is justified by the observation that in the Hodgkin–Huxley equations, both the membrane potential $V(t)$ as well as the sodium activation $m(t)$ evolve on similar time scales during an action potential, while sodium inactivation $h(t)$ and potassium activation $n(t)$ change on similar, albeit much slower time scales. This can be visualized by plotting V and m, representing the "excitability" of the system, using normalized coordinates in response to a current step (Fig. 7.1A). Given the great similarity between both variables, it makes sense to lump them into a single "activation" variable V (Fig. 7.1C). The same observation can be made if one plots the degree of potassium activation n together with the amount of sodium inactivation $1 - h$ for the identical current stimulus (Fig. 7.1B). Again, because both variables show the same changes of almost the same amplitude over time, we combine both into a single variable W, characterizing the degree of "accommodation" or "refractoriness" of the system (Fig. 7.1D). The behavior of such a two-dimensional system with constant parameters is qualitatively very similar to the behavior of the four-dimensional Hodgkin–Huxley model with voltage-dependent functions $m_\infty(V)$, $\tau_m(V)$ and so on.

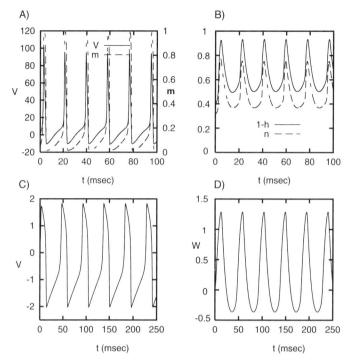

Fig. 7.1 REDUCING THE HODGKIN–HUXLEY MODEL TO THE FITZHUGH–NAGUMO SYSTEM
Evolution of the space-clamped Hodgkin–Huxley and the FitzHugh-Nagumo equations in response to
a current step of amplitude 0.18 nA in A and B and of amplitude $I = 0.35$ in C and D. **(A)** Membrane
potential $V(t)$ and sodium activation $m(t)$ (see also Fig. 6.8). Sodium activation closely follows
the dynamics of the membrane potential. **(B)** Sodium inactivation $1 - h$ and potassium activation n
of the Hodgkin–Huxley system. **(C)** "Excitability" $V(t)$ of the two-dimensional FitzHugh–Nagumo
equations (Eqs. 7.1) with constant parameters has a very similar time course to V and m of the squid
axon (notice the different scaling). **(D)** The "accommodation" variable W shows modulations similar
to $1 - h$ and n of the Hodgkin–Huxley equations.

We will let Fig. 7.1 (and others to follow) be sufficient justification for this reduction.
We refer those of our readers who are not satisfied by the somewhat *ad hoc* nature of
this procedure to Kepler, Abbott, and Marder (1992), who describe a general method for
reducing Hodgkin-Huxley-like systems using "equivalent potentials."

Historically, the equations underlying the FitzHugh–Nagumo model have their origin
in the work of van der Pol (1926), who formulated a nonlinear oscillator model (termed
a relaxation oscillator) and applied it to the cardiac pacemaker (van der Pol and van der
Mark, 1928). Using the phase space method explored by Bonhoeffer (1948) in the context
of a chemical reaction bearing some similarities to nerve cell excitation, FitzHugh (1961,
1969) and, independently, Nagumo, Arimoto, and Yoshizawa (1962) derived the following
two equations to qualitatively describe the events occurring in an excitable neuron:

$$\dot{V} = V - \frac{V^3}{3} - W + I$$

$$\dot{W} = \phi(V + a - bW) \tag{7.1}$$

where we use the conventional shorthand of writing $\dot{V}$ for the temporal derivative of

$V, dV/dt$. Equations 7.1 describe the so-called *FitzHugh–Nagumo* model or, in deference to earlier work by van der Pol (1926) and Bonhoeffer (1948), the *Bonhoeffer–van der Pol* model. The parameters a, b, and ϕ are dimensionless and positive, with a number of different versions of these equations in usage. In this chapter, we assume $a = 0.7$, $b = 0.8$, and $\phi = 0.08$ (Cronin, 1987). The amplitude of ϕ, corresponding to the inverse of a time constant, determines how fast the variable W changes relative to V. With $\phi = 0.08$, V changes substantially more rapidly than W (since $\dot{V}$ is typically $1/0.08 = 12.5$ times larger than $\dot{W}$).

Equations 7.1 represent an instance of a *singularly perturbed system*, where one variable (or set of variables) evolves much faster than the other variable (or set of variables). This has the consequence that the evolution of the system consists of "rapid" segments, in which the fast variable V evolves so rapidly that the slow variable W can be considered stationary, interconnected with "slow" segments, during which the slow variable will not remain constant, as during the fast segments, but will be an instantaneous function of the fast variable $W = W(V(t))$ (for more details, see Cronin, 1987).

Because of the nonlinear nature of these differential equations, closed-form solutions are difficult to derive and we are forced to integrate the equations numerically. We can, however, deduce the qualitative topological properties of these equations *without* explicitly solving them. The key concept here is to consider the evolution of the system, specified by the vector $r(t) = (V(t), W(t))$, in the associated two-dimensional *phase space*, spanned by V and W (Fig. 7.2). For every state or phase point $\mathbf{r}$ in this plane, Eqs. 7.1 assign a vector $\dot{\mathbf{r}} = (\dot{V}, \dot{W})$ to this point, specifying how the system will evolve in time. Indeed, these equations can be thought of as a vector field in phase space, akin to the motion of an imaginary laminar fluid on the plane. While only a few representative vectors are drawn in Fig. 7.2, they completely fill the plane. Such plots are known as *phase plane portraits* and provide an intuitive and geometrical way to understand certain qualitative aspects of the solution of the associated ordinary differential equations.

In cases such as Eqs. 7.1, where the derivatives on the right-hand side do not depend explicitly on time, different trajectories can never cross, since if two would intersect, there would be two different solutions starting from the same (crossing) point. This is ruled out by the existence and uniqueness theorem associated with a set of coupled differential equations, such as these, that are smooth enough. Thus, from any point in the phase space, the system can only evolve in one way, giving phase portraits their orderly looks.

7.1.1 Nullclines

In order to understand how the system evolves in time, let us consider two of its isoclines. An *isocline* is a curve in the (V, W) plane along which one of the derivatives is constant. In particular, the null isocline, or *nullcline*, is the curve along which either $\dot{V}$ or $\dot{W}$ is zero. The nullcline associated with the fast variable V, defined by $\dot{V} = 0$, is the cubic function $W = V - V^3/3 + I$. If the system currently is located on the V nullcline, its imminent future trajectory must be vertical, pointing either upward (for $\dot{W} > 0$) or downward (for $\dot{W} < 0$). Furthermore, for all points in the plane above this cubic polynomial, $\dot{V} < 0$, while the converse is true for all points below the curve $W = V - V^3/3 + I$.

The nullcline associated with the slow variable, $\dot{W} = 0$, is specified by the linear equation $W = (V + a)/b$. Thus, if the evolution of the system brings it onto the W nullcline, its trajectory in the immediate future must be horizontal, for only V, but not W, can change. Furthermore, all phase points in the half-plane to the left of $W = (V + a)/b$ have a

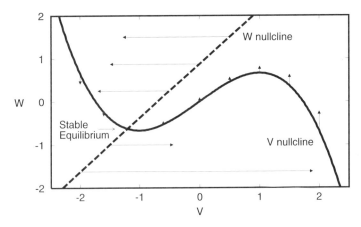

Fig. 7.2 PHASE PLANE PORTRAIT OF THE FITZHUGH–NAGUMO MODEL   Phase plane associated with the FitzHugh–Nagumo Eqs. 7.1 for $I = 0$. The fast variable V corresponds to membrane excitability while the slower variable W can be visualized as the state of membrane accommodation. The nullcline for the V variable, that is, all points with $\dot{V} = 0$, is a cubic polynomial, and the W nullcline (all points with $\dot{W} = 0$) is a straight line. The system can only exist in equilibrium at the intersection of these curves. For our choice of parameters and for $I = 0$, a single equilibrium point exists: $(\overline{V}, \overline{W}) = (-1.20, -0.625)$. The arrows are proportional to $(\dot{V}, \dot{W})$ and indicate the direction and rate of change of the system: V usually changes much more rapidly than W.

negative W derivative and therefore move downward, while points in the right half-plane are associated with increasing W values.

As mentioned above, the FitzHugh–Nagumo equations are an example of an *autonomous* system whose derivatives do not explicitly depend on time (assuming that I has some constant value for $t \geq 0$). If the right-hand side of Eqs. 7.1 would contain terms that explicitly depended on time t (such as a time-dependent forcing term $I(t)$), the nullclines would change with time. The trajectory $(V(t), W(t))$ of the system in phase space can now cross itself and our simple analysis no longer applies.

7.1.2 Stability of the Equilibrium Points

Let us first analyze the resting states of the system and their stability. The *fixed, critical,* or *singular points* of the system r^* are all those points (V_i, W_i) at which both derivatives are zero, that is, for which $(\dot{V}(V_i, W_i), \dot{W}(V_i, W_i)) = (0, 0)$. For Eqs. 7.1 with our choice of parameter values and $I = 0$, the two nullclines meet at a single point $r^* = (\overline{V}, \overline{W}) = (-1.20, -0.625)$. If the system started out exactly at this point, it would not move to any other point in phase space. This is why it is known as an equilibrium point. However, in the real world, the system will never be precisely at the critical point but in some neighborhood around this point. Or, if the system started exactly at the critical point $(\overline{V}, \overline{W})$, any noise would perturb it and carry it to a point in the immediate neighborhood $(\overline{V} + \delta V, \overline{W} + \delta W)$. The subsequent fate of the perturbed system depends on whether the fixed point is stable or not. In the former case, the system will eventually return to $(\overline{V}, \overline{W})$. If the equilibrium point is not stable, the system will diverge away from the singular point.

The stability of the equilibrium point can be evaluated by linearizing the system around the singular point and computing its eigenvalues. Conceptually, this can be thought of as zooming into the immediate neighborhood of this point. The linearization procedure corresponds to moving the origin of the coordinate system to the singular point and considering the fate of points in the immediate neighborhood of the origin (Fig. 7.3A). The associated eigenvalues of the linear system completely characterize the behavior in this neighborhood. Following Eqs. 7.1, we can write for any perturbation δ **r** around the fixed point **r***,

$$\dot{\overline{V}} + \delta\dot{V} = (\overline{V} + \delta V) - (\overline{V} + \delta V)^3/3 - (\overline{W} + \delta W) + I$$

$$\dot{\overline{W}} + \delta\dot{W} = \phi((\overline{V} + \delta V) + a - b(\overline{W} + \delta W)) \tag{7.2}$$

Remembering that $(\overline{V} - \overline{V}^3/3 - \overline{W} + I) = 0$ and $\phi(\overline{V} + a - b\overline{W}) = 0$ (the definition of the critical point) as well as $\dot{\overline{V}} = \dot{\overline{W}} = 0$ and neglecting higher order terms in δV (since the perturbation is assumed to be small), we arrive at

$$\delta\dot{V} = (1 - \overline{V}^2)\delta V - \delta W$$

$$\delta\dot{W} = \phi(\delta V - b\delta W) \tag{7.3}$$

or, in obvious vector notation,

$$\delta\dot{\mathbf{r}} = M\delta\mathbf{r} \tag{7.4}$$

with the matrix M given by

$$M = \begin{bmatrix} (1 - \overline{V}^2) & -1 \\ \phi & -b\phi \end{bmatrix} \tag{7.5}$$

This corresponds to the matrix of partial derivatives associated with Eqs. 7.1 and evaluated at the critical point.

Equation 7.4 describes how any perturbation $\delta\mathbf{r}$ evolves in the neighborhood of the singular point **r***. We can characterize the behavior around this point by finding the eigenvalues of M. The associated characteristic equation is

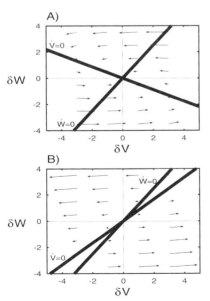

Fig. 7.3 **Behavior around the Equilibrium Point**   The behavior of the FitzHugh–Nagumo equations in a small neighborhood around the fixed point is determined by linearizing these equations around their fixed point and computing the associated pair of eigenvalues. **(A)** Evolution of $(\delta V, \delta W)$ in a coordinate system centered at the equilibrium point **r*** for $I = 0$. Because the real part of both eigenvalues is negative, a small perturbation away from the fixed point will decay to zero, rendering this point asymptotically stable. Any point in this plane will ultimately converge to the fixed point at the origin. **(B)** Similar analysis for the equilibrium point **r**$^{*\prime}$ for a sustained input with $I = 1$ (Fig. 7.5A). The fixed point is unstable.

$$\lambda^2 + (\overline{V}^2 - 1 + b\phi)\lambda + (\overline{V}^2 - 1)b\phi + \phi = 0. \tag{7.6}$$

The eigenvalues correspond to the solutions of this equation,

$$\lambda_{1,2} = \frac{-(\overline{V}^2 - 1 + b\phi) \pm \sqrt{(\overline{V}^2 - 1 - b\phi)^2 - 4\phi}}{2}. \tag{7.7}$$

The evolution of the system therefore takes the following form:

$$\delta\mathbf{r}(t) = c_1\mathbf{r}_1 e^{\lambda_1 t} + c_2\mathbf{r}_2 e^{\lambda_2 t} \tag{7.8}$$

with $\mathbf{r}_1$ and $\mathbf{r}_2$ the two eigenvectors associated with the eigenvalues λ_1 and λ_2 and c_1 and c_2 depending on the initial conditions of the system.

In the two-dimensional case considered here, the different types of solutions of Eqs. 7.1 are easy to classify and understand geometrically in terms of the eigenvalues. In order to study the ultimate fate of the fixed points, the exact values of λ_1 and λ_2 do not matter. What does matter is whether or not the eigenvalues are real or imaginary and the sign of their real parts. (For a thorough discussion of this see Hirsch and Smale, 1974, and Strogatz, 1994.)

1. If λ_1 and λ_2 are both real and at least one of them is positive, the perturbation $\delta r(t)$ will grow without bounds and the solution will diverge from the fixed point: the critical point is unstable. If both eigenvalues are positive, the point is called a *source*; if one of them is positive and the other negative, the point is called a *saddle*.

2. When the eigenvalues are real and both negative, $\delta\mathbf{r}(t) \to 0$ and the system will decay back to its resting state: the critical point is stable and termed a *sink*.

3. If the expression under the square root in Eq. 7.7 becomes negative, the two eigenvalues form a complex conjugate pair with $\lambda_1 = \alpha + i\omega$ and $\lambda_2 = \alpha - i\omega$, and we have

$$\delta\mathbf{r}(t) = c_1\mathbf{r}_1 e^{\alpha t} e^{i\omega t} + c_2\mathbf{r}_2 e^{\alpha t} e^{-i\omega t}. \tag{7.9}$$

Note that the eigenvectors $\mathbf{r}$ are complex, since the λ's are. Exploiting Euler's formula, $e^{i\omega t} = \cos\omega t + i\sin\omega t$, we can reexpress the evolution of the perturbation as a combination of terms involving $e^{\alpha t}\cos(\omega t)$ and $e^{\alpha t}\sin(\omega t)$.

This evolution equation has two parts; the first one, $e^{\alpha t}$, specifies whether the system will converge back to its resting state or whether it will diverge, depending on the sign of α, that is, on the sign of the real part of the eigenvalues. The second part can be thought of as a rotation around the origin, with the oscillation frequency given by $f = \omega/(2\pi)$. For $\omega > 0$, the trajectories will be traversed counterclockwise and clockwise if $\omega < 0$. If $\alpha = 0$, the system will always remain at a fixed distance from the center and will neither approach nor recede from this periodic trajectory. For $\alpha < 0$, the system will display damped oscillations and converge toward the equilibrium state. Here, the fixed point $(\overline{V}, \overline{W})$ is termed a stable *spiral* or *attractor*. Conversely, if $\alpha > 0$, the oscillations will grow exponentially in amplitude and the equilibrium point is an *unstable spiral* or focus.

Returning to the FitzHugh–Nagumo model in the case of no current input, the two eigenvalues associated with the equilibrium point at rest ($I = 0$) are $-0.50 \pm 0.42i$. According to our argumentation, this implies that any perturbation δr decays in an exponential manner back to zero, that is, back to the equilibrium point, since the real part of the eigenvalues is negative (see Fig. 7.3A). Because the solution will also oscillate, the equilibrium point is a stable spiral or attractor.

How large is the neighborhood around the equilibrium point within which any perturbation will ultimately die out? The size of the largest such neighborhood is called the *basin of attraction* associated with this stable point, that is, the set of initial conditions $\mathbf{r}_0 = (V_0, W_0)$ in phase space that will ultimately "fall into" this stable point. In our case, the basin is the entire plane. In other words, no matter what the initial conditions, in the absence of any stimulus current I the system will ultimately always come to rest at $\mathbf{r}^*$.

This illustrates the power of phase space analysis. The stability of any nonlinear autonomous system of differential equations can be analyzed in terms of the stability of the linearized system; the signs of the real part of the eigenvalues then determine the qualitative behavior of the system in some neighborhood around the equilibrium point (Hirsch and Smale, 1974). By looking for the intersections of the nullclines when changing the differential equations characterizing the system, one can immediately understand in an intuitive manner the steady-state properties of the system (see Strogatz, 1994, for many biologically inspired examples of this).

7.1.3 Instantaneous Current Pulses: Action Potentials

How will the FitzHugh–Nagumo model respond if an instantaneous current pulse $I(t) = Q\delta(t)$ (with $Q > 0$) is applied? The initial value of V will jump by Q, thereby moving the system in phase space a certain horizontal distance away from the equilibrium point (dashed line in Fig. 7.4A). The direction of the movement is determined by the sign of the current input.

If the amplitude of the current input is small, the system will almost immediately return to rest following a tight trajectory around the equilibrium point. In the process, V will over- and undershoot the resting state $\overline{V} = -1.2$ (since the eigenvalues are complex), typical of a system with an inductance-like behavior (see Chap. 10). However, in order to see these oscillations around $\mathbf{r}^*$, Fig. 7.4B would have to be enlarged. The trajectory of V reminds us of the subthreshold response of the Hodgkin–Huxley membrane model (Fig. 6.5B). There, a short current pulse also depolarizes the membrane, activating some potassium current and slightly hyperpolarizing the membrane, before finally settling down at the resting potential. Both in the Hodgkin–Huxley and in the FitzHugh–Nagumo systems, graded "potentials" can be obtained by varying the amplitude of the initial displacement over a small range of values.

What happens if the amplitude of the current pulse is made larger? If V is moved instantaneously past -0.64, the evolution of the system sharply veers away from the V nullcline, with V rapidly increasing while W remains more or less stationary until the rightmost branch of the V nullcline is encountered, defining the maximal value of V. The reason for this rapid change is that $\dot{V}$ is large, while $\dot{W}$, due to the small size of ϕ, is small (the slope dW/dV of the trajectory is on the order of ϕ; Eqs. 7.1). During the next phase, the system slowly "crawls" upward along the V nullcline until it reaches the knee at the top of this curve. This slow phase is dictated by the fact that $\dot{V}$ is very small in the neighborhood of this curve, causing W to be an instantaneous function of V (the slope of the trajectory dW/dV is large here). A third "fast" phase follows, where V is rapidly reduced, undershooting the resting value of $\overline{V}$ in the process. Finally, the system slowly loses its accommodation W, crawling back to the resting state along the left branch of the nullcline. This trajectory is very much reminiscent of a Hodgkin–Huxley action potential

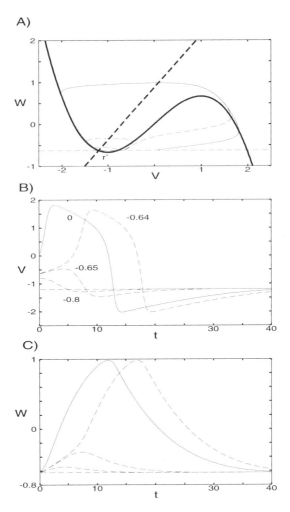

Fig. 7.4 RESPONSE OF THE FITZHUGH–NAGUMO MODEL TO CURRENT PULSES (A) The quiescent system is excited by a current pulse of different amplitudes $I = Q\delta(t)$, displacing the system from its resting state $\mathbf{r}^*$ along the dashed horizontal line. Following Eqs. 7.1, this briefly increases $\dot{V}$, in agreement with physical intuition, since a brief current pulse will cause a transient capacitive current $C\, dV/dt$ to flow. The evolution of the voltage V and of the adaptation variable W is plotted in (B) and (C). Changing V from its initial value of -1.2 to -0.8 or -0.65 only causes quick excursions of the voltage around the equilibrium point with the system rapidly returning to rest (the oscillatory manner in which the system does so is not readily apparent at the scale of these panels). If the current pulse is large enough so that V exceeds -0.64, a stereotyped "all-or-none" sequence is triggered: V rapidly increases to positive values but then dives below its resting value $\overline{V}$ before finally coming to rest again at $\mathbf{r}^*$. Notice that the trajectory in this case consists of "fast" segments, where V changes rapidly but W remains essentially constant (upper and lower segments), interconnected by "slow" segments, where the system changes so slowly that V is always in equilibrium (the "slow" segments closely coincide with the V nullcline).

(Fig. 6.5), with V and sodium activation m changing rapidly, while sodium inactivation h and potassium activation n vary at a more leisurely pace.

While the system is "hyperpolarized," that is, traveling along the leftmost branch of the V nullcline, it is clear that the inactivation variable W must be brought down below its resting value of about -0.625 before the system can cross over the W nullcline and once again execute its standard trajectory. This is equivalent to a *refractory* period, during which much larger current pulses are needed in order to trigger another spike.

As emphasized above, this behavior, with "fast" segments along which accommodation varies little but V changes rapidly, connected by "slower" segments where V is in equilibrium as W changes slowly, is typical of singularly perturbed systems (Cronin, 1987). Injecting larger current pulses will not change this basic trajectory, even though each value of I is associated with its own unique trajectory in phase space.

Although we stated above that the system goes through its stereotypical trajectory once V exceeds -0.64, the FitzHugh–Nagumo system shows no true threshold behavior. From a numerical point of view, the amplitude Q of the current pulse can be varied by minute amounts between -0.65 and -0.64, giving rise to all possible intermediate trajectories between sub- and suprathreshold (filling in the available phase space between the two associated trajectories in Fig. 7.4A). Indeed, it is known both computationally (Cooley and Dodge, 1966) and experimentally (at higher temperature; Cole, Guttman, and Bezanilla, 1970) that the squid action potential does not behave in a true all-or-none manner. In both systems, no true *saddle* exists in the sense that all points to one side of the saddle will remain below and all points beyond the saddle will be above threshold. However, if ϕ is small, points close to the middle part of the V nullcline act approximately as a *separatrix*, from which neighboring paths diverge sharply to the left or to the right. The "width" of the separatrix over which intermediate forms between sub- and suprathreshold responses exist is so small that from a physiological point of view (given the ambient level of variability in the membrane potential), the system does show threshold behavior, with the horizontal distance between the resting point and the separatrix corresponding to the amplitude of the threshold.

7.1.4 Sustained Current Injection: A Limit Cycle Appears

Let us stimulate the quiescent system with a sustained current step of amplitude I at $t = 0$. This input will, different from a current pulse, change the associated phase space diagram. We can visualize the new phase diagram by noticing that for a positive current step the V nullcline is shifted upward, while the W nullcline does not change (Fig. 7.5A). This, of course, changes the position of the equilibrium point. In the presence of a positive current up to three fixed points now coexist, since the cubic V nullcline can intersect the linear W nullcline at up to three different locations. However, both geometrical intuition (the slope of the W nullcline is too steep to intersect the V nullcline more than once) as well as an algebraic criterion (Cardan's solution) tell us that for $b < 1$ only a single solution exists. The solution is stable as long as the real part of the two eigenvalues is negative. Once the real part becomes zero and then positive, even infinitesimal small perturbations will become amplified and diverge away from the equilibrium point. Following Eq. 7.7, the real part changes sign at the two locations where

$$\overline{V}_\pm = \pm\sqrt{1 - b\phi}. \tag{7.10}$$

Thus, the equilibrium point is stable whenever the W nullcline meets the cubic nullcline along its right- and leftmost branches. Here the slope is negative and $|\overline{V}| \geq 1$. However, along the central part of the V nullcline, $|\overline{V}| < \sqrt{1 - b\phi}$, and the eigenvalues will acquire a positive real part, rendering the fixed point unstable.

Assuming that the system is in its quiescent state $\mathbf{r}^* = (\overline{V}, \overline{W}) = (-1.20, -0.625)$, then as long as the amplitude of the current step is small enough, the real part of the

A)

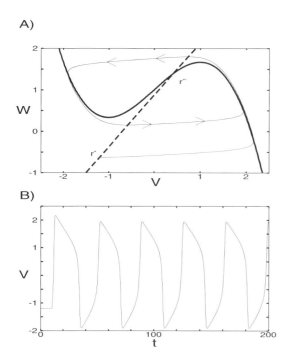

B)

Fig. 7.5 RESPONSE OF THE FITZ-HUGH–NAGUMO MODEL TO CURRENT STEPS (A) The quiescent system is subject to a current step of amplitude I. For $I < 0.32$, the new equilibrium $\mathbf{r}^{*\prime}$ is stable: the system depolarizes but remains subthreshold. For larger steps (here $I = 1$), the new equilibrium point lies along the middle portion of the V nullcline and is unstable. Because the system has a stable limit cycle, it will not diverge; rather, it generates a train of "action potentials" whose time course is shown in (B). Regardless of the initial state of the system, it will always converge rapidly onto this limit cycle.

eigenvalues associated with the new equilibrium point $\mathbf{r}^{*\prime}$ remains negative and $\mathbf{r}^{*\prime}$ is stable. Because of the upward shift of the V nullcline, the new resting state of the membrane potential $\overline{V}'$ is more positive than $\overline{V}$, that is, the system is depolarized.

For a critical value of the input current I_-, the nullclines intersect along the central part of the V nullcline at $\overline{V}_-$ (defined by Eq. 7.10) and local stability breaks down (as in Fig. 7.5B). Due to the cubic nonlinearity in Eqs. 7.1, the system does not diverge[2] but follows a stereotypical trajectory around $\mathbf{r}^{*\prime}$. As demonstrated in Fig. 7.5, the system moves from its starting point rapidly to the right ("the membrane depolarizes") until it meets the rightmost branch of the V nullcline where it will slowly creep upward (adaptation builds up). Subsequently, V decays rapidly by moving horizontally toward the right and down the leftmost branch of the V nullcline. So far, the behavior of the equations is very similar to that seen earlier, with fast phases alternating with phases during which the system crawls along the V nullclines (the arrows in Fig. 7.2 give a good feeling for the relative speed of these processes).

Different from before, the system does not return to its new equilibrium point $\mathbf{r}^{*\prime}$, since this is unstable. Rather the trajectory of the system undershoots $\mathbf{r}^{*\prime}$ and traces a path almost parallel to the V axis until it meets its previous trajectory, which it then follows. As long as the input I_0 persists, the system continues to evolve along the same trajectory, producing a constant and infinite stream of action potentials (Figs. 7.1C and 7.5B). Since these oscillations are stable, but the term "steady-state" usually refers to an unchanging and not to a changing state, another label is needed. The French mathematician Poincaré, to whom we owe phase space analysis of differential equations with two variables, called such steady oscillations a *stable limit cycle*.

2. As it would in a purely linear system once the system looses stability.

If the system starts out at a location in phase space inside the limit cycle, its trajectory will spiral outward until it meets the limit cycle. Likewise, for any point outside the limit cycle, the system will spiral into the limit cycle. Thus, the *basin of attraction* of the limit cycle is the entire phase plane (with the exception of the equilibrium point). The limit cycle is stable, since small perturbations away from the cycle will be quickly suppressed and the system will return to the limit cycle. In order for a closed, that is, oscillatory, solution to be a limit cycle is must have an *isolated trajectory*. This implies that any neighboring trajectories must either spiral toward (as they do here) or away from the limit cycle (unstable limit cycle). This is where the sense of the word "limit" in limit cycles comes from.

Limit cycles are inherently nonlinear phenomena; they cannot occur in a linear system. Obviously, linear systems can oscillate, as shown by the harmonic oscillator, that is they have closed orbits. Yet these trajectories are not isolated. If $\mathbf{r}(t)$ is an oscillatory solution, so will $c\mathbf{r}(t)$, for any value of c. For such a system, the initial conditions will always dictate the exact orbit chosen by the system. The system never forgets this, no matter how long one waits. This is quite distinct from a limit cycle in a nonlinear system, such as the Hodgkin–Huxley or the FitzHugh-Nagumo equations, where, given a strong enough input, the system will ultimately converge to the limit cycle. The cubic nonlinearity in Eqs. 7.1 is essential for the overall behavior of the system.

Limit cycles cannot exist in systems with only one degree of freedom that are governed by a first-order differential equation, no matter whether linear or nonlinear. Such systems either converge toward their fixed points or they diverge to infinity. No overshooting or damped oscillations can occur.

7.1.5 Onset of Nonzero Frequency Oscillations: The Hopf Bifurcation

What governs the onset of oscillations? Is it possible to obtain spiking with arbitrarily low frequencies? The coupled nonlinear differential equations that arise within a biophysical context can display two fundamentally different types of onset of oscillations, one related to a *Hopf bifurcation* and one related to the creation of a *saddle*. Both behave in very different ways, which are of functional relevance. We will study the first type of bifurcation here, deferring the second one to the following section.

As discussed in the previous section, increasing the amplitude of the injected current I from zero to more positive values shifts the intersection of the two nullclines (Fig. 7.5). As I increases, so does the real part α of the eigenvalue associated with the equilibrium point at its resting value. Breakdown of stability occurs when α changes sign, since for $\alpha > 0$, small perturbations in the neighborhood of the equilibrium point will fail to die out. Based on Eq. 7.7, $\alpha = 0$ occurs if and only if $\overline{V} = \pm\sqrt{1 - b\phi}$. These two voltages are associated with two different values of the input current, $I_- = 0.33$ and $I_+ = 1.42$.

At these two values of the input current, the equilibrium point ceases to be stable and develops into an unstable source. The system starts to oscillate, with a frequency determined by the imaginary part of the associated eigenvalues (Fig. 7.6B). Whether or not these oscillations develop into a stable limit cycle, as here or as in the case of the Hodgkin–Huxley equations, depends on the global character of the equations and cannot be determined based on a local criterion.

This type of dynamical phenomenon, in which stable large-amplitude oscillations arise abruptly as one particular parameter, known as the *bifurcation parameter* (here I), varies smoothly, is called *hard excitation*, or *subcritical Hopf bifurcation*. Hard excitation is feared in engineering applications, since one of its salient properties is that the amplitudes of the

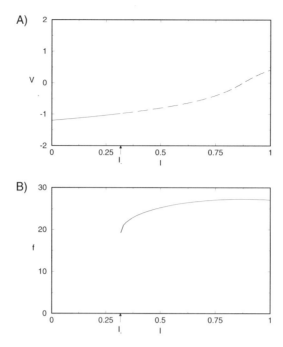

Fig. 7.6 DISCHARGE CURVE OF THE FITZHUGH–NAGUMO MODEL (A) Steady-state membrane potential and **(B)** oscillation frequency of the limit cycle for the FitzHugh–Nagumo equations as a function of a sustained current I. As the membrane is depolarized in response to increasing injection of current, it loses stability at $I_- = 0.33$ (arrow) and moves on a stable limit cycle. An important property of this type of bifurcation phenomenon, known as hard excitation or subcritical Hopf bifurcation, is that oscillations occur with nonzero frequency. This behavior is also characteristic for the Hodgkin–Huxley equations. Between I_- and $I_+ = 1.42$, the system moves along the limit cycle. The frequency of the oscillation is a function of I (dashed line in panel A). Beyond I_+, the equilibrium point becomes stable again, and the system remains "locked" at a depolarized level (not shown).

oscillations are nonzero and can be fairly large. A *soft excitation*—also known as *supercritical Hopf bifurcation*—occurs when oscillations arise with arbitrarily small amplitudes and gradually increase as the bifurcation parameter is ramped up. Hard excitation constitutes one of the generic mechanisms for the onset of oscillations in coupled nonlinear differential equations (Glass and Mackey, 1988; Strogatz, 1994). A subcritical Hopf bifurcation requires that a stable spiral or attractor change into an unstable spiral and is surrounded by a stable limit cycle.

A key feature of a Hopf bifurcation is the onset of a stable limit cycle with a nonzero oscillation frequency (since ω is finite). It is a well-known feature of the Hodgkin–Huxley model that the minimal stable spiking frequency is about 50 Hz (at 6.3°; see Fig. 6.9B). Experimentally, this is true for most axonal membranes as well as for the generation of spike trains at the cell body of certain neuronal types, for instance, the nonadapting cortical interneurons. As we keep on increasing I, the frequency of these stable oscillations first increases before the f–I curve peaks and decreases back to 22.8 Hz at I_+.

Also plotted in Fig. 7.6 is the steady-state voltage V as a function of the sustained current I_{ss} flowing across the membrane. I_{ss} can be obtained solving for I in Eqs. 7.1 under steady-state conditions. Similar to the steady-state I–V relationship for the squid axon membrane (Fig. 6.9A), this curve is a monotonically increasing function. Operationally, it can be obtained by injecting a current step of amplitude I into the system and plotting the resulting membrane potential.

7.2 The Morris–Lecar Model

It has been argued that the FitzHugh–Nagumo equations do not faithfully represent any particular neuronal membrane and that phase space analysis therefore is not a particular useful tool to understand the dynamics of "real" neurons. Because we think otherwise, we will apply phase space analysis to a second set of equations, which is meant to capture the dynamics of the membrane potential in the muscle of barnacles. Here all terms can be directly identified with various voltages and conductances. These equations also allow us to demonstrate a very different way in which the onset of oscillations can occur. In the so-called *saddle* bifurcation, oscillations first emerge with an arbitrarily small frequency. As the amplitude of the injected current is smoothly increased, the frequency of the spike trains also increases.

Barnacle muscle fibers that are subject to constant current inputs respond with a variety of oscillatory voltage patterns. In order to describe these, Morris and Lecar (1981) postulated a set of coupled, ordinary differential equations (neglecting any spatial dependencies) incorporating two ionic currents: an outward going, noninactivating potassium current and an inward going, noninactivating calcium current. Because the Ca^{2+} current responds much faster than the K^+ current, we assume that $I_{Ca}(t)$ is always in equilibrium for the time scales we are considering. This allows us to neglect the dynamics of the calcium current and we only need to treat the two-dimensional reduced version of the equations. Different variants of the Morris-Lecar equations are in usage. We will follow the presentation in Rinzel and Ermentrout (1998) which demonstrates best the two distinct manners in which a system can start to oscillate.

7.2.1 Abrupt Onset of Oscillations
In their reduced form, the equations showing this generic type of behavior are

$$C_m \frac{dV_m}{dt} = -I_{ionic}(V_m, w) + I(t)$$

$$\frac{dw}{dt} = \frac{w_\infty(V_m) - w}{\tau_w(V_m)} \tag{7.11}$$

where V_m is the absolute membrane potential (in millivolts), $C_m = 1\mu F/cm^2$, w is the activation variable for potassium, all currents are in units of microamperes per square centimeter ($\mu A/cm^2$) and time t is measured in milliseconds. The ionic current has three components,

$$I_{ionic}(V_m, w) = \overline{G}_{Ca} m_\infty(V_m)(V_m - E_{Ca}) + \overline{G}_K w(V_m - E_K) + G_m(V_m - V_{rest}). \tag{7.12}$$

*As stated above, we assume that the calcium current is always at equilibrium, with its activation curve given by

$$m_\infty(V_m) = 0.5 \left(1 + \tanh \frac{V_m + 1}{15} \right). \tag{7.13}$$

The potassium activation variable follows the standard first-order Eq. 6.7 with steady-state activation

$$w_\infty(V_m) = 0.5 \left(1 + \tanh \frac{V_m}{30} \right) \tag{7.14}$$

and time constant

$$\tau_w(V_m) = \frac{5}{\cosh V_m/60} . \tag{7.15}$$

The other parameters are $\overline{G}_{Ca} = 1.1, \overline{G}_K = 2.0, G_m = 0.5, E_{Ca} = 100, E_K = -70$, and $V_{rest} = -50$ (all conductances are in units of millisiemens per square centimeter and the reversal potentials in millivolts).

Following the methods described in the first part of this chapter, we characterize the phase space associated with these equations by the two nullclines (Fig 7.7A). For the parameters used here, the nullclines intersect at a single location, no matter what the amplitude of the injected current I. We obtain a physiological picture of events in a muscle when we depolarize the membrane by injecting a sustained current $I = 15 \ \mu A/cm^2$. Under these conditions, the equilibrium point is stable and the resting membrane potential is -31.7 mV.

Applying current pulses that briefly depolarize the system from its resting state demonstrates the existence of a threshold and a limit cycle. As long as the membrane is depolarized to less than -14.8 mV, the system quickly returns to its resting state (after a brief hyperpolarization; see Fig. 7.7). Depolarizing the membrane past -14.8 mV triggers a stereotypical behavior, following a very similar limit cycle to the one shown by the FitzHugh–Nagumo model. Indeed, different initial conditions (in the case of Fig. 7.7 depolarizing the membrane to either -14.7 or to -12 mV) has little effect on the phase space trajectory (besides speeding up initiation of the spike).

Figure 7.8 summarizes the behavior of the system to constant current steps. Panel 7.8A is a plot of the steady-state membrane potential as a function of the applied current. Following Eqs. 7.11, this is identical to the steady-state ionic current $I_{ionic}(V_m, w_\infty(V_m))$. In this system, the sustained I–V relationship is strictly monotonic. As more and more current is injected, the membrane is depolarized from rest at -31.7 mV (in the presence of the stabilizing 15-$\mu A/cm^2$ current injection), and the w nullcline intersects the V_m nullcline along its middle portion (Fig. 7.7A). Similar to the FitzHugh–Nagumo equations which experience the onset of oscillations via a subcritical Hopf bifurcation, the real part of the two conjugate eigenvalues goes to zero (at $I_- = 24.9 \ \mu A/cm^2$) and then becomes positive, indicating loss of stability. At the same time, a stable limit cycle appears: the system generates an infinite train of action potentials. For a limited region of current amplitudes, the equilibrium point remains unstable (as indicated by dashes in Fig. 7.8A) and the system spikes. All of this is more or less identical to what happens in the FitzHugh–Nagumo equations. Let us now change these equations in a minor way to obtain quite a different behavior.

7.2.2 Oscillations with Arbitrarily Small Frequencies

A key property of a Hopf bifurcation is that if the system oscillates, it will oscillate at a well-defined minimum frequency, as the squid axonal membrane and indeed most axons. However, numerous cell types, such as pyramidal cells in cortex, can generate trains of spikes with large interspike intervals, that is, with very low oscillation frequencies upon current injection (see Fig. 9.7; Connor and Stevens, 1971c).

Here, the onset of oscillations is governed by a dynamical mechanism different from the Hopf bifurcation, which can be observed in a slightly modified version of the Morris–Lecar equations. What has to occur is that the nullclines must intersect several times. We

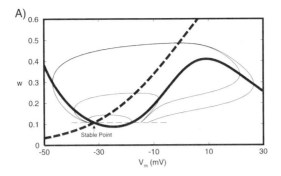

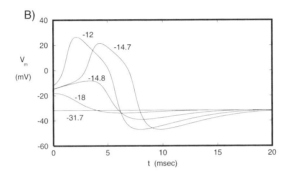

Fig. 7.7 RESPONSE OF THE MORRIS–LECAR MODEL TO CURRENT PULSES Equations 7.11, describing electrical events in the muscle cells of barnacles (Morris and Lecar, 1981), show a qualitatively similar behavior to the squid axon membrane and to the FitzHugh–Nagumo equations. **(A)** Phase plane portrait (here membrane potential V_m versus potassium activation w) for different stimulus conditions. The nullclines are plotted in bold. In the presence of a stabilizing current injection, the resting potential is -31.7 mV. From this state, the system is stimulated by brief current pulses, instantly depolarizing the membrane (thin lines). If the stimulus depolarizes the muscle to either -18 or to -14.8 mV, the system responds with a subthreshold excursion around the resting potential (two innermost loops around the resting potential). For larger values, the system moves along a limit cycle: it spikes. **(B)** Evolution of the membrane potential V_m for this stimulus paradigm.

can achieve this by assuming, for example, that the potassium activation is a much steeper function of voltage than before. Following Rinzel and Ermentrout (1998), we modify the potassium activation as well as its dynamics,

$$w_\infty(V_m) = 0.5 \left(1 + \tanh \frac{V_m - 10}{14.5} \right) \tag{7.16}$$

and

$$\tau_w(V_m) = \frac{3}{\cosh \dfrac{V_m - 10}{29}}. \tag{7.17}$$

We also decrease the amount of calcium conductance by 10% to $\overline{G}_{Ca} = 1$. All other parameters remain unchanged.

As witnessed by Fig. 7.9, this fundamentally changes the character of the phase portrait. Now, three equilibrium points exist. If we were to carry out a linear stability analysis, we would find that the lower point is a stable sink (that is, both eigenvalues are real and negative),

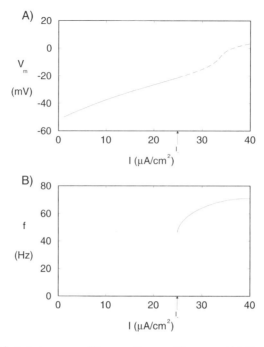

Fig. 7.8 SUSTAINED SPIKING IN THE MORRIS–LECAR MODEL (A) Steady-state voltage and (B) oscillation frequency as a function of the amplitude of the sustained current I for the reduced Morris–Lecar model (Eqs. 7.11). The steady-state I–V curve can also be viewed as the cumultative, steady-state ionic current flowing at any particular membrane potential V_m. At $I_- = 24.9 \ \mu\text{A/cm}^2$ (arrow), the single equilibrium point (Fig. 7.7A) loses stability via hard excitation (a subcritical Hopf bifurcation) when the real part of the two conjugate eigenvalues goes through zero and becomes positive. As in the case for the Hodgkin–Huxley and FitzHugh–Nagumo equations, this type of onset of oscillations implies a nonzero oscillation frequency (here a minimum of 50 Hz).

the middle one is an unstable saddle (one eigenvalue is positive and the other negative), while the upper equilibrium point is an unstable spiral (the eigenvalues are complex, with their real part being positive; Fig. 7.9A). Indeed, the lower resting state at $V_m = -29.2 \ \text{mV}$ and $w = 0.0045$ is a globally attracting rest state. A local perturbation in its neighborhood leads to a prompt decay back to rest, while a large input pulse will cause the system to move on its usual trajectory around the phase space, a trajectory that ultimately ends at rest again: the system responds to a brief current pulse with a spike before settling down again. However, different from the previous systems we analyzed, this one shows a true *threshold* due to the presence of the unstable saddle.

The middle equilibrium point, associated with the negative and the positive real eigenvalues, is called a *saddle* because it comes with a curve, called the *separatrix* curve, which sharply distinguishes between sub- and suprathresholds: any phase point to the left of this curve will veer away from equilibrium toward the stable sink, while any point to the right of the separatrix will move in that direction and generate a spike (Fig. 7.9B). If the initial condition places the system near the threshold separatrix, the system will display a long latency before decaying back or spiking, because being close to both nullclines implies that both variables V_m and w change only slowly.

What happens if the amplitude of the injected current is increased? This has the effect of lifting the V_m nullcline closer to the w nullcline and moving the two lower equilibrium points

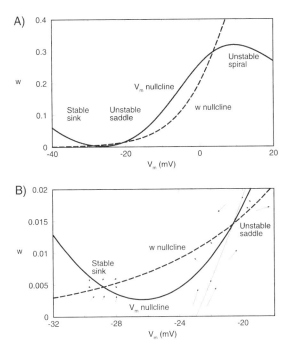

Fig. 7.9 SYSTEM WITH MULTIPLE EQUILIBRIUM POINTS The Morris–Lecar equations were modified by changing the dynamics of potassium activation and making its voltage dependency steeper (Eqs. 7.16 and 7.17). (**A**) Under these conditions, the two nullclines can intersect up to three times. Here, the lower equilibrium point is a globally attracting, *stable sink*, the middle one an *unstable saddle*, and the upper one an *unstable spiral*. (**B**) Phase space portrait around the two lower equilibrium points. The separatrix curve (thin line) at the saddle strictly separates sub- and suprathreshold regions of phase space. Points to the left of this curve will decay back to the stable sink, while points to the right will lead to a spike. If the current injected into the system is further increased, these two equilibrium points will move toward each other, coalesce, and disappear. If the initial state of the system lies in the neighborhood of these two curves, it will evolve very slowly since, by definition, $\dot{V}_m = 0$ along the V_m nullcline and $\dot{w} = 0$ along the w nullcline, explaining why the system is able to spike at very low frequencies.

closer and closer to each other. For a critical value $I_1 = 8.326 \ \mu A/cm^2$, the stable sink and the unstable saddle meet and coalesce and—upon further increases of I— disappear entirely. A *saddle-node bifurcation* has just occurred (Rinzel and Ermentrout, 1998), with the system losing two of its equilibrium points. At the point when the two critical points meet, the two eigenvalues are identical to zero.

If the state of the system is in the neighborhood of these two points (as in Fig. 7.9B), it will take a long time to move into a different region of phase space. (After all, by definition, the change in one or the other variable along the nullclines is zero.) Indeed, the closer the system is to the nullclines, the slower it will move. At I_1, that is, when the two equilibrium points have coalesced, the period of the oscillations is infinite. It can be shown under some fairly generic conditions (Strogatz, 1994) that the oscillation frequency is proportional to $\sqrt{I - I_1}$ (Figs. 7.10 and 7.11B).

A property of systems with a saddle-node bifurcation is an N-shaped relationship between the membrane potential and the injected current, or between the potential and the sustained current (Fig. 7.11A). In other words, for a range of membrane currents three different steady

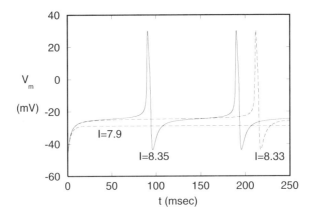

Fig. 7.10 Spiking at Low Frequencies Response of the modified Morris–Lecar equations (Eqs. 7.16 and 7.17) using current steps (starting at $t = 0$) of variable amplitude. In response to a current step of 7.9 μA/cm^2 amplitude, the membrane depolarizes. Close to $I_1 = 8.326$ μA/cm^2, the onset of spiking can be delayed almost indefinitely, similar to the delay observed in pyramidal cells due to the presence of the A-like current (Fig. 9.7). This is not caused by the very slow time constant of any one ionic current, but it is due to the structure of the nullclines in phase space (Fig. 7.9B).

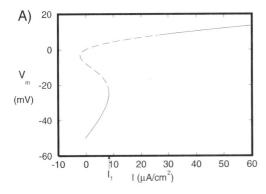

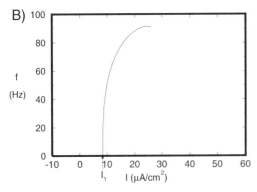

Fig. 7.11 Sustained Spiking in the Modified Morris–Lecar Model (A) Steady-state voltage and (B) oscillation frequency as a function of I for the modified Morris–Lecar model (Eqs. 7.16 and 7.17). Different from Fig. 7.8, the I–V relationship is N–shaped. Spiking first occurs at I_1 (arrow) when the slope of this curve is infinite and then becomes negative. This happens when the two lower equilibrium points in the phase portrait of Fig. 7.9 merge, creating oscillations with arbitrarily long interspike intervals via a saddle-node bifurcation.

state voltages exist, of which only one, however, is stable (solid portion of the curve). When the injected current is slowly increased, oscillations first occur when the slope of the I–V curve becomes infinite (for $I = I_1$) and then negative. If very large stimulus currents are used, the system locks up at very depolarized levels and spiking ceases.

In this model, the transition from a stable equilibrium point to a stable limit cycle is marked by arbitrarily low firing rates *without* having to rely on arbitrarily slow activation rates. Having a saddle is a necessary condition for such a continuous $f-I$ relationship. In our case, this is achieved via a steep nonlinear relationship between the membrane potential and activation of the potassium conductance. As we will see in Chap. 9, this is the characteristic action of a transient potassium current (the I_A current of Connor and Stevens, 1971c), which has a steep voltage dependency in the neighborhood of the resting state of the cell.

7.3 More Elaborate Phase Space Models

So far, we reviewed models of spiking that are constrained to a two-dimensional phase space. However, a variety of higher dimensional models have been studied to characterize more complex behavior, in particular *bursting* and intrinsic rhythmic behavior.

Thalamic cells can respond to input by either firing a few fast action potentials or generating a slow calcium, all-or-none event, on top of which ride a series of fast, conventional sodium spikes. Which response type occurs depends on the level of membrane depolarization (Jahnsen and Llinás, 1984a,b; Fig. 9.4). Researchers have generated detailed biophysical models of the various ionic currents that are responsible for this behavior (McCormick and Huguenard, 1992). Given the potential significance of bursting as a special means of communicating a privileged symbol, we dedicated Chap. 16 to this topic.

Rose and Hindmarsh (1985) use phase space models with three variables: one for the membrane potential, one for the degree of accommodation, and one to mimic slow adaptation. They proved that it is this slow variable that is responsible for determining whether the neuron fires in a tonic or in a burst mode in response to sustained current input. Indeed, a classification of bursting based on this type of phase-space analysis is beginning to emerge (Wang and Rinzel, 1995).

As reviewed extensively by Llinás (1988), many neurons do not behave at all like the squid axon membrane, but have a great deal of diversity of electrical behaviors. For instance, subsets of thalamic cells express complex intrinsic oscillations of the membrane potential in very different frequency bands (from a few to 40 Hz or more) depending on the behavioral state of the animal (sleep, arousal, etc.; see Steriade, McCormick, and Sejnowski, 1993). In order to understand the interplay between intrinsic cellular properties and network properties that generate these period phenomena, networks of thalamic cells with more than half a dozen ionic conductances have been simulated (Destexhe, McCormick, and Sejnowski, 1993; Wang, 1994; Contreras et al., 1997). Although the underlying phase spaces are high-dimensional, their projection onto suitable two-dimensional subspaces has shown the usefulness of analyzing the behavior of these systems in terms of the concepts introduced in this chapter.

7.4 Recapitulation

The theory of nonlinear dynamics represents a powerful tool to characterize the generic mechanisms giving rise to the threshold response, the stereotypical shape of action potentials, or the onset of oscillations. It allows us to understand why these phenomena occur even when we have insufficient information concerning the detailed kinetics or the exact shape of the various rate and time constants. In this chapter, we apply the theory of dynamical systems to analyze two-dimensional systems with the help of the phase portrait. In particular, we focus on the FitzHugh–Nagumo (FitzHugh, 1961, 1969; Nagumo, Arimoto, and Yoshizawa, 1962) and the Morris–Lecar (1981) models of spiking.

The FitzHugh–Nagumo model was based on the observation that the membrane potential and the sodium activation in the four-dimensional Hodgkin–Huxley equations evolve on a similar time scale, while the sodium inactivation and the potassium activation also share similar behavior, albeit on a slower scale. This feature can be exploited by expressing the membrane excitability via the two-dimensional FitzHugh–Nagumo equations with constant coefficients, with V corresponding to the excitability of the system and W its degree of accommodation. Linear stability analysis allows us to understand why the resting state is stable and when spiking first occurs. It is also helpful to explain the all-or-none shape of the action potential as an example of a limit cycle, upon which the system will rapidly converge once threshold is exceeded.

We also acquainted the reader with a qualitativly similar system of equations that describes muscle fiber excitability in terms of a leak, a calcium, and a potassium current (Morris and Lecar, 1981). Following the lead of Rinzel and Ermentrout (1998), we assume that the calcium current is so rapid that it is always in the steady-state with respect to the membrane potential and the potassium activation, and we discuss two variants of these equations. The first behaves similar to the squid axon and to the FitzHugh–Nagumo equations, generating oscillations with a nonzero oscillation frequency (via a subcritical Hopf bifurcation). Modifying the potassium conductance to be a steeper function of V_m allows the system to spike at very low frequencies via a saddle-node bifurcation, similar to what occurs in cells that have a marked delay between the onset of current injection and the first spike (due to the presence of I_A).

The advantage of using models based on a very small number of variables—rather than relying on biophysical very detailed models with an excuberance of variables—is that they offer us a qualitative, geometrical way to understand why a cell spikes and why it switches between two modes of firing without necessarily having to know every single detail of the system.

Given the enormous numerical load involved in simulating the dynamics of hundreds or thousands of neurons, such a simplified single-cell model will allow us to study neural networks containing realistic numbers of neurons. The price one pays for this reduced complexity is a lack of quantitative predictions.

8

IONIC CHANNELS

In the previous chapters, we studied the spread of the membrane potential in passive or active neuronal structures and the interaction among two or more synaptic inputs. We have yet to give a full account of *ionic channels*, the elementary units underlying all of this dizzying variety of electrical signaling both within and between neurons. Ionic channels are individual proteins anchored within the bilipid membrane of neurons, glia, or other cells, and can be thought of as water-filled macromolecular pores that are permeable to particular ions. They can be exquisitely voltage sensitive, as the fast sodium channel responsible for the sodium spike in the squid giant axon, or they can be relatively independent of voltage but dependent on the binding of some neurotransmitter, as is the case for most synaptic receptors, such as the acetylcholine receptor at the vertebrate neuromuscular junction or the AMPA and GABA synaptic receptors mediating excitation and inhibition in the central nervous system. Ionic channels are ubiquitous and provide the substratum for all biophysical phenomena underlying information processing, including mediating synaptic transmission, determining the membrane voltage, supporting action potential initiation and propagation, and, ultimately, linking changes in the membrane potential to effective output, such as the secretion of a neurotransmitter or hormone or the contraction of a muscle fiber.

Individual ionic channels are amazingly specific. A typical potassium channel can distinguish a K^+ ion with a 1.33 Å radius from a Na^+ ion of 0.95 Å radius, selecting the former over the latter by a factor of 10,000. This single protein can do this selection at a rate of up to 100 million ions each second (Doyle et al., 1998).

At the time of Hodgkin and Huxley's seminal study in the early 1950s, two broad classes of transport mechanisms were competing as plausible ways for carrying ionic fluxes across the membrane: *carrier molecules* and *pores*. At the time, no direct evidence for either one existed. It was not until the early 1970s that the fast ACh synaptic receptor and the Na channel were chemically isolated and purified and identified as proteins. One reason ionic channels were difficult to detect was the tiny amount of current they carry. In a technical breakthrough, Neher and Sakmann (1976) at the Max Planck Institute in Göttingen, Germany, reported the first direct measurement of the electrical current flowing through a single ionic channel using the *patch-clamp* technique. This method, subsequently developed into the *tight* or *giga-seal* recording technique (described in detail in Sakmann and Neher, 1983), allows one to record from very small patches of membrane containing one or a few ionic channels, using a glass pipette tightly sealed against the neuronal membrane. With modern amplifiers, this recording technology is limited by Johnson thermal noise and can measure currents in the 0.1 pA range. In recognition of the rapid progress in understanding ionic channels and their fundamental role in cellular and neurobiology, Neher and Sakmann were awarded the Nobel prize in Medicine in 1991.

The second technology that has allowed access to the detailed molecular structure of ionic channels is recombinant DNA and the associated instrumentarium of molecular biology. Channels are now routinely cloned, and the effect of replacing individual amino acids on the function of the channel can be assessed. This has allowed scientists to understand the family relationships between different ionic channels. Combined with crystallographic analyses, this has lead to an explosion of studies within the last decade that have characterized the nature of ionic channels, their three-dimensional structure and their biophysical properties, as well as their associated amino acid sequences.

Because of the tiny electrical current carried by individual ionic channels, it can be argued that there is no need to understand their properties in a monograph dedicated to studying macroscopic currents and how these currents underlie information processing. Yet we clearly need to discern the relationship between the macroscopic currents that are responsible for triggering action potentials and initiating synaptic transmission and the microscopic ionic fluxes through channels that underlie them. Different from digital computers, where the engineer strives to isolate the function of the system from its detailed physical properties (e.g., the function of core magnetic memory, bubble memory, CD-ROM, or optical memory is to store bits in certain patterns, regardless of the physical substrate), no such convenient isolation exists in biological systems. Here, *structure reflects function*. Thus, it behooves us to understand the components responsible for electrical signaling.

In the next section, we summarize some of the pertinent properties of ionic channels. We will be brief here; for *the* classical account of the biophysics of ionic channels see Hille (1992); for a more molecular treatment, see Hall (1992). In the section that follows, we introduce the reader to kinetic models of ionic channels. In the third and most important section of this chapter, we detail how microscopic, stochastic channels give rise to macroscopic, deterministic currents, and we provide a justification for why, in this book, we deal exclusively with the latter.

8.1 Properties of Ionic Channels

A trace of the tiny electrical current flowing through a single ionic channel as a function of time immediately reveals some of the unique aspects of channels. In the record shown in Fig. 8.1, acetylcholine (ACh) is applied to a single, fast nicotinic channel while the membrane potential across the channel is clamped to -80 mV. Thousands of these channels make up the much-studied neuromuscular endplate, the giant synapse transmitting the output of motoneurons to the muscle. Although the input is constant, the ionic current switches among two states, one defined by zero and the other by 3 nA current flow. These fluctuations are probabilistic in character, with the channel randomly alternating between the opened and the closed state. For ligand-gated synaptic ionic channels, the rates of opening and closing vary when one or more neurotransmitter molecules bind with the channel complex. For voltage-dependent channels, the opening and closing rates vary systematically with the applied membrane potential.

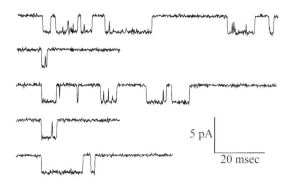

Fig. 8.1 CURRENT FLOWING THROUGH A SINGLE IONIC CHANNEL Several excerpts from a patch-clamp recording of a single acetylcholine-activated channel on a cultured muscle cell. The patch was held at a constant membrane potential of −80 mV. The openings of the channel (downward events) cause a unitary 3-nA current to flow, occasionally interrupted by a brief closing. These closings sometimes fail to reach the baseline due to the limited temporal resolution of the apparatus. Notice the random openings and closings of the channel, characteristic of all ionic channels. Fluctuations in the baseline current are due to thermal noise. Reprinted in modified form by permission from Sigworth (1983).

8.1.1 Biophysics of Channels

The single channel whose behavior is observed in Fig. 8.1 appears to act as a binary switching element: either the channel is closed and no current flows or the channel is open and—for a constant applied voltage level—a fixed amount of current flows. As the effective voltage gradient across a voltage-dependent channel is increased, two things change: the probability of opening as well as the amplitude of the current flowing through the channel in its open phase vary. As detailed in Sec. 9.1.1, the current across the membrane is frequently described using a nonlinear model, the Goldman-Hodgkin-Katz constant field equation (Eq. 9.1), as well as Ohm's law. If the Nernst reversal potential of the channel is properly taken into account, Ohm's law appears adequate for many channel types, in particular those permeable to Cl^-, Na^+, or K^+. In other words, the ionic current flowing through a single channel in the open state is given by

$$I_{ionic} = \gamma (V_m - E_{ionic}) \qquad (8.1)$$

with each channel having a characteristic *unitary channel* conductance γ (Fig. 8.2). This value ranges around 10 pS for the fast sodium channels in the squid axon to the very large calcium-dependent potassium channels in mammals with $\gamma \approx 200$–300 pS (see Table 8.1).[1]

Upon closer inspection, the I–V relationship of the calcium-dependent potassium channel in Fig. 8.2B is not really linear: for large de- or hyperpolarizing voltage excursions, the current through the channel starts to *saturate*. A number of factors are responsible for this. When the concentration gradient between inside and outside is very steep, current passes more easily in one direction than in the other (*rectification*). This is particularly true for calcium channels, with their five orders of magnitude concentration differential across the membrane. A second source of nonlinearities are ions in the intra- or extracellular cytoplasm that move into the channel and block it. Indeed, this is the reason for the nonlinearity of the

---

1. The value of γ depends weakly on temperature, with $Q_{10} \approx 1.3$–1.6.

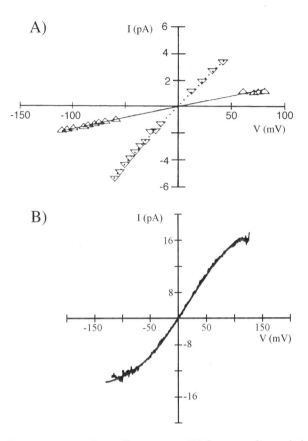

Fig. 8.2 **VOLTAGE DEPENDENCY OF IONIC CHANNELS** (**A**) Current-voltage relationship of a single nicotinic ACh-activated ionic channel. Since the concentration of the permeant ion on both sides of the membrane is identical (here set to either 150 mM NH$_4^+$ (downward pointing arrowhead) or Li$^+$ (upward pointing arrowhead)), the reversal potential is zero. As the voltage gradient across the channel increases, so does the current through the channel (up to a limit), in accordance with Ohm's law. The single-channel conductance γ is either 17 pS (in Li$^+$) or 79 pS (in NH$_4^+$). Reprinted in modified form by permission from Dani (1989). (**B**) $I–V$ relationship for a single voltage- and calcium-dependent potassium channel in a symmetrical 160-mM potassium solution. The slope of this curve around the origin is 265 pS. Deviations from Ohm's law are apparent for large voltage gradients. Reprinted in modified form by permission from Yellen (1984).

NMDA channel (Fig. 4.9A). Another difficulty is that many channels show subconductance levels; that is, they have two or more conductive states, each with its own characteristic conductance and with random voltage-dependent transitions among these sublevels. Indeed, in the long time series from which Fig. 8.1 was excerpted, occasionally the channel opens to a 4.3 pA level (Sigworth, 1983). Many, if not all, channels show evidence of some sublevels; this appears to be particularly true for GABA and glycine channels (Bormann, Hamill, and Sakmann, 1987). Yet another complication is that the current through an open channel sometimes briefly and transiently goes to zero. This rapid closing and opening of the channel is called *flickering* and is thought to be caused by the transient blocking of the pore by some ion or molecule. In some cases this happens so frequently that the current record has the appearance of a comb rather than a rectangular pulse. If the flickering is

TABLE 8.1
Properties of Different Types of Ionic Channels

Channel type	Preparation	γ	η
Fast Na$^+$	Squid giant axon[1]	14	330
Fast Na$^+$	Rat axonal node of Ranvier[2]	14.5	700
Fast Na$^+$	Pyramidal cell body[3]	14.5	4–5
Delayed rectifier K$^+$	Squid giant axon[4]	20	18
Ca^{2+}-dependent K$^+$	Mammalian preparation[5]	130–240	—
Transient A current	Insect, snail, mammal[5]	5–23	—
Nicotinic ACh receptor	Mammalian motor endplate[7]	20–40	10,000
GABA$_A$ Cl$^-$ receptor	Hippocampal granule cells[8]	14, 23	—

Single channel conductance γ (in pS) and average channel density η if known (per μm^2) for some ionic channels. The exact value of γ depends on many variables, in particular the temperature and the composition of the extracellular fluid.
[1] In the absence of external divalent ions at 5° C; Bezanilla (1987).
[2] At 20° C; Neumcke and Stämpfli (1982).
[3] Hippocampal pyramidal cell body and initial segment at 24° C; Colbert and Johnston (1996).
[4] Between 13–25° C; less frequent are channels with $\gamma = 10$ and 40 nS; Llano, Webb, and Bezanilla (1988).
[5] At 22° C; Latorre and Miller (1983).
[6] At 22° C; Adams and Nonner (1989).
[7] Peak density; the density decreases with distance from active zone; for a review, see Hille (1992).
[8] At 22° C; Edwards, Konnerth, and Sakmann (1990).

too rapid to be adequately resolved by the recording apparatus, it effectively lowers the apparent conductance of the channel. Nevertheless, to a good first approximation one can treat ionic channels as conductances (with the appropriate reversal potential depending on the concentration gradient across the membrane; see Eq. 4.2 and Fig. 8.2A; for more details, see Hille, 1992).[2]

Estimates of the single-channel conductance γ based on Ohm's macroscopic law and the geometry of the pore put the upper limit at about 300 pS and no channel with larger conductance has been reported. The current through single channels is usually larger than 1 pA, corresponding to about 6000 monovalent ions passing through the channel each millisecond; indeed most pass more current, with the record somewhere around 27 pA. The diffusion equation places an upper limit of around 10^8 ions per second diffusing through these small pores, whose width is only one or two atomic diameters at the most restricted part of the channel (Hille, 1992; Doyle et al., 1998).

The density of these channels can vary widely depending on the exact preparation used (Table 8.1). In the squid axon, there are on the order of 300 sodium and 18 potassium channels per square micrometer. Similar numbers hold for other rapidly conducting systems without myelin insulation. In the frog or the rat node of Ranvier, a membrane specialization in myelinated axons where all the sodium channels appear to be concentrated (Sec. 6.6), the density of Na channels can be as high as 2000 per square micrometer. Even at these densities, channel proteins do not constitute major chemical components of most neuronal membranes. In principle, between two and four orders of magnitude additional channels could be packed into the membrane before the maximum molecular crowding sustainable by a membrane is achieved (Hille, 1992). Given ionic fluxes of between 10^7 and 10^8 ions per second per channel and a membrane capacitance of around 1 $\mu F/cm^2$, these densities

---

2. Modern patch-clamp amplifiers have such high resolution that they are limited by the thermal noise of the recorded system, rather than by instrument noise. As is evident in Fig. 8.1, the current measurements fluctuate in both opened and closed states. The standard deviation of these fluctuations—dictated by statistical physics—is typically less than 0.1 pA.

are more than sufficient to lead to the rapid and pronounced changes in membrane potential evidenced in action potential initiation and propagation and other electrical events.

It should be emphasized that the properties of ionic channels discussed here, in particular their microscopic, all-or-none and stochastic nature, hold true regardless of whether we are dealing with ligand-gated NMDA or GABA receptors, voltage-dependent sodium, calcium, or potassium channels or ionic channels in nonneuronal systems, such as in T lymphocytes, cardiac pacemaker cells, muscle cells, mechanosensitive calcium channels in vertebrate hair cells or channels in the unicellular organism *Paramecium*. Furthermore, there appears to be very little systematic difference between channels recorded from molluscs, flies, squids, or mammals. All these channels are to a good, first approximation activated by the same stimuli, respond with the same kinetics, and are blocked by the same agents. In particular, ionic channels in more highly developed animals, such as vertebrates, are no more complex than channels in less complex animals, such as molluscs. Evidence from molecular cloning studies suggests that the major channel families evolved into their present form with the appearance of multicellular organisms about 700 million years ago (Hille, 1992). If these speculations concerning the evolution of channels bear out, nature seems to have converged on a small number of basic circuit elements—aqueous pores built using protein technology—a long time ago and has remained faithful to this basic design across all species.

8.1.2 Molecular Structure of Channels

Figure 8.3A shows an electron-microscopic image of an actual ACh-gated channel within its natural habitat, while Fig. 8.3B summarizes in a graphic manner the current working hypothesis for the structure of a voltage-dependent ionic channel. The entire macromolecule

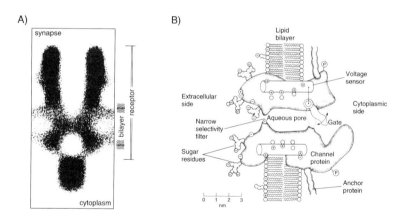

Fig. 8.3 MOLECULAR STRUCTURE OF AN IONIC CHANNEL   (A) Electron-microscopic image using crystallographic methods of the axial section of one of the nicotinic acetylcholine receptors in the electric fish *Torpedo*. This channel has the shape of a tall hour glass, with an overall length of 110 Å. Most of the protein lies outside the bilipid membrane. The complex extends about 20 Å into the cytoplasmic medium and about 60 Å into the extracellular environment. The most narrow portion of the channel is about 8 Å wide. The blob at the bottom of the image is not part of the channel. Reprinted by permission from Toyoshima and Unwin (1988). **(B)** Schematic view of a generic voltage-dependent ionic channel. The molecular details of such channels are now being charted by structural studies (e.g. Doyle et al., 1998). Reprinted by permission from Hille (1992).

making up the channel is sitting in the lipid bilayer of the membrane, anchored to elements of the intracellular cytoskeleton. It is quite large, consisting of several thousand amino acids (and associated sugar residues), with a molecular weight in excess of 300,000 daltons. When the channel is open, it forms a water-filled pore extending across the membrane. The selectivity for certain ions, that is, the fact that the sodium channel is primarily permeable to Na^+ ions and not to K^+ ions, is conveyed to the channel by the most narrow portion of the channel (termed the *selectivity filter*) with a diameter of around 3 Å(Hille, 1973).

The voltage-dependent gating of the channel is achieved by a conformational change in the three-dimensional shape of part of the molecule, such that the channel is either open or blocked. As discussed already by Hodgkin and Huxley (1952d), this requires the movement of hypothetical *gating particles* able to sense the electric field across the membrane. In fact, at a resting potential of -80 mV an electric field of about 200,000 V/cm is set up across the 40 Åthin membrane. Changes in this field move these gating charges through the membrane and change the configuration of the channel. The steepness of the voltage dependency of the sodium channel (that is, the steepness of m_∞^3 in Sec. 6.2.2) yields an estimate of six elementary gating charges that need to move across the entire membrane in order for a single sodium channel to open. This *gating current*, caused by the channel changing its configuration so that it can conduct current, is tiny compared to the 10^7 or so ions moving through the open channel each second but has been measured (Armstrong and Bezanilla, 1973). Qualitatively, the gating current can be thought of as a nonlinear capacitive current, proportional to $Q_m dm/dt$, where Q_m is the total charge associated with the transition of the m particle from close to open.

Recombinant DNA technology—in parallel with classical methods such as protein chemistry, X-ray diffraction, and electron microscopy—has given us the opportunity to image channels as well as to derive their amino acid sequence and to clone, express, and manipulate them. (For overviews of this very active research area see Catterall, 1995, 1996 and Doyle et al., 1998.) These methods have allowed the activation and inactivation processes of the sodium channel to be characterized at the molecular level.

The key culprit in voltage-dependent activation are four homologous segments (termed S4) of the sodium channel protein that span the membrane in an α helix. These S4 regions are highly conserved between Na^+ channels from different species and are also homologous to specific regions of the voltage-dependent K^+ and Ca^{2+} channels. Even before the molecular sequence of these channels was known, Armstrong proposed that two helices, one bearing multiple positive charges and the other multiple negative charges, move relative to each other, perhaps by rotating, so that relative translation by one turn would be equivalent to moving one charge across the membrane. Subsequently, Catterall (1988, 1992) outlined the *sliding helix* model for the action of each S4 segment in activating the channel, a model for which much experimental support is now available (Fig. 8.4). Upon membrane depolarization, each S4 segment rotates by 60°, moving outward by 5 Å. Every third amino acid residue making up this helix is positively charged and is about 5 Å away from the next charged residue. In other words, the rotating movement of the helix effectively moves one charge across the membrane. All four S4 segments, acting in concert, cause the conformational change that leads to the opening of the channel. A similar mechanism, acting also via S4, is thought to confer voltage-dependency onto certain potassium channels (Larsson et al., 1996).

Inactivation is thought to arise by a different process, usually explained by invoking a mechanistic analogy, the *ball-and-chain* model (Armstrong and Bezanilla, 1977; Hoshi, Zagotta, and Aldrich, 1990). Here, a part of the channel molecule on the inner, cytoplasmic

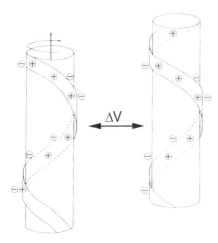

Fig. 8.4 VOLTAGE-DEPENDENT ACTIVATION OF THE Na$^+$ CHANNEL *Sliding helix* model postulated by Catterall (1988, 1992) to account for the voltage dependency of activation of the fast sodium conductance underlying action potential generation. The S4 segment of the channel protein that is extended as an α helix throughout the membrane is electrically neutral at rest. In the presence of sufficient depolarization, the helix rotates by 60° in a screw-like outward movement, until it moves into a new stable position, where locally the charges are balanced. The movements of four such S4 segments that are part of the channel molecule are thought to induce a conformational change that leaves the channel in its open state. Reprinted by permission from Catterall (1992).

side of the membrane is thought to act like a tethered gate. Under the right conditions, it swings shut, physically occluding the pore. Zagotta, Hoshi, and Aldrich (1990) showed in a series of elegant site-directed mutagenesis experiments that deletion of the appropriate cytoplasmic part of the inactivating, transient A-type K$^+$ channel molecule does, indeed, eliminate, or greatly reduce, inactivation. Furthermore, if this cytoplasmic region was added as a synthetic soluble peptide, inactivation was restored.

Because of the existence of four S4 segments in the Na$^+$ channel, it can be argued that a better model of sodium current activation would be m^4, rather than the m^3 term favored by Hodgkin and Huxley (see Eq. 6.13). Yet it needs to be remembered that the Hodgkin–Huxley equations are purely phenomenological fits of the data; the cubic m expression was introduced in order to make a first-order differential equation account for the sigmoidal rise of I_{Na}.

Once the detailed molecular structure of the sodium and other channels is worked out, we will be in a better position to derive the relevant equation relating the voltage across the channel to the time-varying current through it. It remains unclear, though, whether more complex models will increase our qualitative understanding of the dynamics of the macroscopic current.

One last comment on the molecular nature of ionic channels. These proteins are so amazingly specific that substituting one amino acid at one of two positions in the several thousands long amino acid sequence coding for the sodium channel with another amino acid will alter the ion-selection properties of the sodium channel to resemble those of calcium channels (Heinemann et al., 1992).

8.2 Kinetic Model of the Sodium Channel

In Chap. 6, we saw that Hodgkin and Huxley postulated a number of fictive "gating particles" to satisfactorily describe their macroscopic current records. For any current to flow, several (four in the case of I_{Na}) such particles have to come together simultaneously. Given the evidence alluded to above concerning internal repeats in the channel protein, m and h can be reinterpreted in molecular terms as the conformation of these internal repeats (Fig. 8.4). Similar activation and inactivation variables have been postulated for the myriads of sodium, potassium, calcium, and chloride channels described over the last two decades (see the following chapter).

Classical kinetic theory specifies the manner in which one can go from a description of these particles and their underlying states to the final, overall gating scheme. Two key assumptions underlie classical kinetic theory: (1) Gating is a *Markov process*, implying that the rate constants, regulating the transitions from one state to the next, do not depend on the previous history of the system. A sufficient description of the system is its present state, without knowing its previous state. (2) All transitions are assumed to be characterized by a first-order differential equation, with a single associated time constant. The applications of these methods to probabilistic ionic channels are summarized by Neher and Stevens (1977), Colquhoun and Hawkes (1981) and DeFelice (1981).

These two simplifying assumptions are used to construct the overall state diagram of the system, as well as its dynamic behavior, as we will now do for the fast sodium channel. Figure 8.5 illustrates the simplest symmetrical eight-state model of this channel. Here three identical m particles can be either in an "open" or in a "closed" position. α_m can be thought of as proportional to the probability for an m particle to make the transition from closed to open (and β_m proportional to the probability for the opposite transition). The fourth h particle can be either in an "inactivated" (top row of Fig. 8.5) or in a "not inactivated" (bottom row) position, with α_h the probability that the h particle will switch from inactivated to open. Alternatively, m can represent the fraction of m gates in their open state and h the fraction of h gates in their noninactivating state. Now, the α's and β's need to be interpreted as *rate constants*, governing the rate transitions from one state to the other.

If we assume that the system is in a state where no particle is in its open position, any one of the three identical m particles can open. Since any of these three particles can open, the rate constant is $3\alpha_m$ if we assume that the m particles act independent of each other. For this one open particle to switch back to its closed position, the rate constant is β_m. To make the next transition from one to two open particles, any one of the two remaining closed particles can switch and the appropriate rate constant is $2\alpha_m$, and so on. For the system to be in its

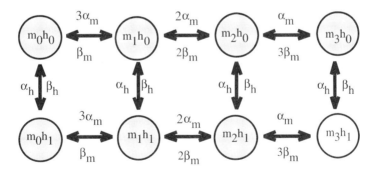

Fig. 8.5 KINETIC DIAGRAM OF THE FAST SODIUM CHANNEL Simplest kinetic diagram associated with the fast voltage-dependent Na⁺ channel. In order for the channel to be in its conductive state, all three m particles as well as the inactivating h particle need to be in their open position. Because we here assume that all three m particles are identical, eight different states exist. α_m is the rate constant at which the m particles switch from their closed to their open positions (that is, moving toward the right in this diagram), while β_m is the rate constant for the reverse operation (transition toward the left). Similarly, α_h is the rate constant governing the transition from inactivating to noninactivating state (downward) and β_h for the reverse operation. Many more complex gating schemes have been discussed in the literature.

conductive state, the h particle has to switch into the appropriate configuration; thus, of the eight possible different states (from zero m particles open and h closed to all three m as well as h open), only a single one—in the lower right corner in Fig. 8.5—corresponds to the open state.

Assuming the Markov property and first-order transitions among states, leads us to the calculus used by Hodgkin and Huxley in 1952. Following the logic laid out in Chap. 6, if the membrane potential is stepped from its initial value V_0 to V_1, the fraction of m particles in their activated state follows a simple exponential law,

$$m(t) = m_\infty(V_1) - (m_\infty(V_1) - m_\infty(V_0))e^{-t/\tau_m(V_1)} \tag{8.2}$$

with $m_\infty(V) = \alpha_m(V)/(\alpha_m(V) + \beta_m(V))$ and $\tau_m(V) = 1/(\alpha_m(V) + \beta_m(V))$. With our usual assumption of independence of the gating particles, the fraction of channels populating the conductive m_3h_1 state is

$$m^3h = \left(m_\infty(V_1) - (m_\infty(V_1) - m_\infty(V_0))e^{-t/\tau_m(V_1)}\right)^3$$

$$\times \left(h_\infty(V_1) - (h_\infty(V_1) - h_\infty(V_0))e^{-t/\tau_h(V_1)}\right) \tag{8.3}$$

By multiplying out the exponentials, this equation can be expressed as a sum of seven exponentially decaying terms,

$$m^3h = C_0 + \sum_{i=1}^{7} C_i e^{-t/\tau_i} \tag{8.4}$$

with seven associated time constants $\tau_i = \tau_m$, $2\tau_m$, $3\tau_m$, τ_h, $\tau_m\tau_h/(\tau_m + \tau_h)$, $\tau_m\tau_h/(\tau_m + 2\tau_h)$, and $\tau_m\tau_h/(\tau_m + 3\tau_h)$. For a system with eight independent states and first-order transitions among these, seven time constants are expected.

Experimentally, most of these time constants are difficult to resolve, since only the total channel current can be recorded, making the identification of these states difficult. If the three m subunits are assumed to differ from each other in some minor ways, the Na^+ channel needs to be characterized by 15 closed and a single open state and by 15 time constants, further compounding the problem. The rate constants between the different states can also be more complex than in the simple scheme illustrated in Fig. 8.5. If *cooperativity* exists, the transitions between states are not independent of each other. For instance, it has been postulated that the rate of inactivation depends on the number of open subunits and is much faster than in the standard Hodgkin–Huxley model (so-called *coupled models*; Armstrong and Bezanilla, 1977; Aldrich, Corey, and Stevens, 1983). A great deal of ingenuity continues to be spent deciphering the correct kinetic gating scheme with the help of highly refined measuring techniques. It remains an open question whether working out all the details of the underlying microscopic transitions will lead to a description of the ionic current at the macroscopic level substantially different from that in use today. Furthermore, as we will illustrate now, the large number of channels underlying most signaling in neurons allows us to average over individual stochastic channels and to only consider deterministic ionic currents.

8.3 From Stochastic Channels to Deterministic Currents

In Chap. 6 we learned how Hodgkin and Huxley successfully described the dynamics of the membrane voltage in terms of deterministic and graded ionic currents. Yet as detailed above, we now know this to be caused by the current through stochastic all-or-none molecular pores. Undoubtedly, the macroscopic current is constituted by the summation of the microscopic currents sweeping through many, many channels. But what is their exact relationship? Figure 8.6 expresses the general idea: although the precise time each channel switches into its open state is a random occurrence, the average duration of channel opening has a well-described temporal onset, depending systematically on the membrane voltage. On each of the 10 trials in Fig. 8.6, the exact sequence of openings and closings of a single delayed-rectifier potassium channel differs, yet averaging over a sufficient number of trials leads to the smoothly varying current expected from the continuous Hodgkin–Huxley formalism. Indeed, since h_∞ is close to 1 and m_∞ is close to 0 at -80 mV (absolute potential; Fig. 6.4), the time course of Fig. 8.6B can be approximated by $(1 - e^{-t/\tau_m})^3 e^{-t/\tau_h}$.

A number of researchers have explored the relationship between microscopic channels and macroscopic currents (Verveen and Derksen, 1968; Lecar and Nossal, 1971a,b; Skaugen and Walloe, 1979; Skaugen, 1980a, b; Clay and DeFelice, 1983; DeFelice and Clay, 1983; Strassberg and DeFelice, 1993; Fox and Lu, 1994; Chow and White, 1996). Let us delve into this topic in a bit more detail.

8.3.1 Probabilistic Interpretation

Let us go back to the idea expressed in the m^3h formalism of Hodgkin and Huxley. Ions are only free to drift down the potential gradient of the channel if all four gating particles are

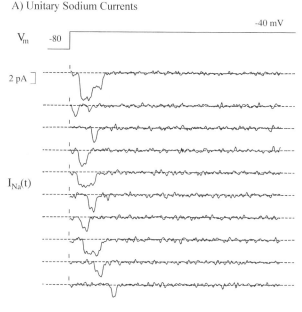

A) Unitary Sodium Currents

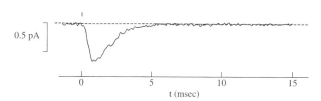

B) Ensemble Average

Fig. 8.6 STOCHASTIC OPEN-INGS OF INDIVIDUAL SODIUM CHANNELS (A) Random opening and closing of a handful of fast sodium channels in a mouse muscle cell. The membrane potential was stepped from -80 to -40 mV; the first trial reveals the simultaneous opening of two Na^+ channels, while on all other trials, only a single channel was open. **(B)** Averaging over 352 such trials leads to a smoothly varying current in accordance with the m^3h model of Hodgkin and Huxley. Experiment carried out at 15° C. Reprinted by permission from Patlak and Ortiz (1986).

in their open position. Let $m(t)$ be the fraction of activation particles in their open position and $\beta_m(V_m)$ the rate at which transitions occur from open to closed. In the case of the Hodgkin-Huxley model (Eq. 6.17), each 18-mV depolarization increases this rate e-fold. In the absence of any new open states being created, the fraction of open states must decay according to

$$\frac{dm(t)}{dt} = -\beta_m(V_m)m(t) . \tag{8.5}$$

If at the beginning of the experiment at $t = 0$, the fraction of particles in their open position was m_0, the temporal evolution of m is given by

$$m(t) = m_0 e^{-\beta_m(V_m)t} . \tag{8.6}$$

Since we assume that the gating particles have no memory, in other words, that the probability of a given particle to change its state does not depend on its previous history (Markovian property), we can express the probability that a single particle will remain in its open position after an observation period T as

$$m(T) = e^{-\beta_m(V_m)T} . \tag{8.7}$$

If the m variable is in its closed position it can switch into its open state at the voltage-dependent rate $\alpha_m(V_m)$. The total change in the fraction of particles open is the difference between the fraction of closed particles making the transition into open states and the fraction of open states decaying away into closed states,

$$\frac{dm(t)}{dt} = \alpha_m(V_m)(1 - m(t)) - \beta_m(V_m)m(t) . \tag{8.8}$$

Note that this is the equation we encountered in Chap. 6 in the context of the deterministic and macroscopic Hodgkin–Huxley formalism, while here we interpret $m(t)$ in a probabilistic manner.

Let us apply this interpretation to the fast sodium channels of the squid giant axon. Adopting the simplest kinetic scheme illustrated in Fig. 8.5, we need to compute the probabilities for each sodium channel to be in one of its eight distinct states. For instance, the probability of the m_2h_1 state (Fig. 8.5; two activation particles and the inactivation particle all open) is given by the product (since all four particles are assumed to act independently) of the probability of finding two open activation particles (m^2), one closed particle ($1 - m$) and one open inactivating particle (h). Because any two of the three m particles can be open, the final probability for this state is

$$p_1 = 3 \left(\frac{\alpha_m(V_0)}{\alpha_m(V_0) + \beta_m(V_0)} \right)^2 \left(1 - \frac{\alpha_m(V_0)}{\alpha_m(V_0) + \beta_m(V_0)} \right) \frac{\alpha_h(V_0)}{\alpha_h(V_0) + \beta_h(V_0)} \tag{8.9}$$

where V_0 is the voltage at rest. Similar expressions can be derived for all other seven states. In a 'Monte-Carlo computer simulation of the probabilistic openings and closings of single channels (as first carried out by Skaugen and Walloe, 1979, and Skaugen, 1980a,b), the actual initial state is computed by assigning each of the probabilities for the eight states a portion of the real line between 0 and 1. The length of this partitioning is given by the associated probability (with the sum of all probabilities equal to 1, of course) and by generating a random number r_1 (drawn from a uniform probability distribution between 0 and 1). The part of the line on which r_1 falls is the initial state.

Assume now that the membrane potential changes instantaneously from V_0 to V_1. How long will the channel remain in its initial state? The probability that the channel remains in this state after a period T is

$$p_2 = e^{-(\alpha_m(V_1)+\alpha_h(V_1)+2\beta_m(V_1))T} . \tag{8.10}$$

The three constants in the exponential account for the fact that the third subunit could open (with rate α_m), or either one of the two subunits could close again ($2\beta_m$), or the channel could inactivate (α_h). To determine the actual lifetime, another random number r_2 is generated; the duration of state m_2h_1 is then $-(\alpha_m + \alpha_h + 2\beta_m)^{-1}\ln(r_2)$. The probability for the channel to make a transition from m_2h_1 to the state where all four subunits are in their open state and the channel conducts (state m_3h_1) is given by

$$p_3 = \frac{\alpha_m(V_1)}{\alpha_m(V_1) + \alpha_h(V_1) + 2\beta_m(V_1)} . \tag{8.11}$$

Likewise, the probability for the channel to switch from m_2h_1 into m_2h_0 is

$$p_4 = \frac{\alpha_h(V_1)}{\alpha_m(V_1) + \alpha_h(V_1) + 2\beta_m(V_1)} \tag{8.12}$$

and from m_2h_1 into m_1h_1 it is

$$p_5 = \frac{2\beta_m(V_1)}{\alpha_m(V_1) + \alpha_h(V_1) + 2\beta_m(V_1)} . \tag{8.13}$$

Partitioning the unit line into three intervals, r_3 is generated to decide into which state the channel has switched: if $0 < r_3 < p_3$, the open state m_3h_1 is chosen; if $p_3 < r_3 < p_3+p_4$, it will be m_2h_0, and if $p_3 + p_4 < r_3 \leq 1$, the state m_1h_1 is chosen. In this manner, the life history of one particular sodium channel is simulated as the membrane is voltage-clamped (Fig. 8.7). A patch of membrane with more than one channel, where the channels are assumed to act independenty—no substantial evidence for coupling among neighboring channels has emerged—can be simulated with a more efficient procedure that keeps track of what fraction of the population is in what state (Skaugen and Walloe, 1979; Chow and White, 1996).

As witnessed by the bottom trace in Fig. 8.7, the changing conductance associated with a handful of opening and closing channels can be quite different from that of the macroscopic conductance expected by the continuous Hodgkin–Huxley equation. Yet for 600 channels, the behavior approximates that of the continuous deterministic $m^3(t)h(t)$ formulation. This is not surprising, since the central limit theorem tells us that the standard deviation of the mean of a population of n independent variables increases like $\sqrt{n}$ as $n \rightarrow \infty$. Thus, the relative deviation from the mean decreases as $1/\sqrt{n}$ for large values of n. For 600 sodium channels, corresponding to a few square micrometer of squid membrane, the total conductance differs little from that obtained by the continuous m^3h formulation. Suitably normalized, this is also the conductance obtained when experimentally averaging over hundreds of single-channel openings, as shown empirically in Fig. 8.6B.

A similar Monte-Carlo procedure can be used to calculate the behavior of a piece of squid axon membrane in the presence of a constant stimulus current (Strassberg and DeFelice, 1993). Here, the potassium channels are modeled using the simplest possible linear and noncooperative kinetic scheme with five states (four closed and one open) with the appropriate values of $\alpha_n(V_m)$ and $\beta_n(V_m)$ (with all rate constants as specified in Chap. 6). This patch of membrane is assumed to be studded with a constant density of channels (18

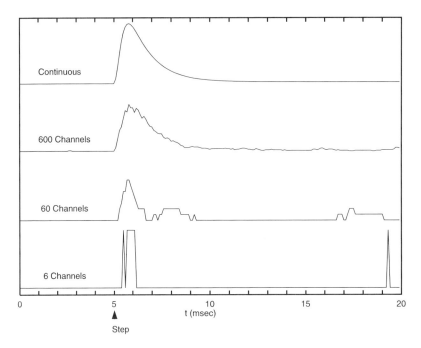

Fig. 8.7 SIMULATED LIFE HISTORY OF INDIVIDUAL SODIUM CHANNELS The membrane potential in a simulated membrane patch containing a variable number of Na$^+$ channels was stepped from $V_0 = 0$ to $V_1 = 50$ mV at 5 msec (arrow). The normalized conductance associated with the eight-state Markov model shown in Fig. 8.5 was evaluated numerically for several trial runs (see Strassberg and DeFelice, 1993). As the number of channels is increased from 6 to 600, the graded and deterministic nature of the (normalized) sodium conductance emerges from the binary and stochastic single-channel behavior. The top trace shows the conductance computed using the continuous time-course (approximating $(1 - e^{-t/\tau_m})^3 e^{-t/\tau_h}$) formalism of Hodgkin and Huxley (1952). This figure should be compared against the experimentally recorded sodium current through a few channels in Fig. 8.6B. Reprinted in modified form by permission from Strassberg and DeFelice (1993).

potassium and 60 sodium channels per square micrometer) each characterized by its own state. As the size of the patch is increased from 1 to 10^4 μm^2 (always assuming that all channels see the same membrane potential) in the presence of a constant injected current density (to normalize for the membrane area) a steady progression from random openings to a highly regular train of action potentials can be observed (Strassberg and DeFelice, 1993). For patches containing on the order of 1000 binary channels, repetitive spikes with approximately the same firing frequency (90 ± 10 Hz) as in the continuous Hodgkin–Huxley model occur (top trace in Fig. 8.8).

We saw in Fig. 6.10 how adding noise to the input current linearized the abrupt discharge curve associated with the space-clamped Hodgkin–Huxley system. A similar linear behavior has been obtained in computer simulations of the complete f–I curve in the presence of 1000 and fewer discrete Na$^+$ channels (Skaugen and Walloe, 1979). It is only when increasing the number of Na$^+$ channels to 10,000 (while keeping $\overline{G}_{Na}$ constant) that the f–I curve with the sudden onset of the firing characteristic of the deterministic Hodgkin–Huxley equations is recovered.

The simulations discussed here were carried out at V_{rest}. At more depolarized potentials the amplitudes of the channel fluctuations are larger. Indeed, it is plausible to imagine a scenario where an extended and deterministic input hovering just around firing threshold

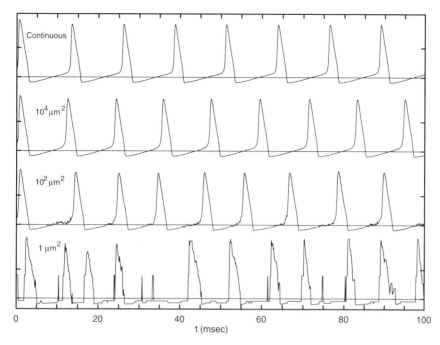

Fig. 8.8 ACTION POTENTIALS AND SINGLE CHANNELS Computed membrane potential (relative to V_{rest} indicated by horizontal lines) in different size patches of squid axon membrane populated by a constant density of Na^+ and K^+ channels. The space-clamped membrane is responding to a current injection of 100 pA/μm^2. The transitions of each all-or-none channel are described by its own probabilistic Markov model (the eight-state model in Fig. 8.5 for the Na^+ channel and the simplest possible five-state linear model for the K^+ channel). For patches containing less than 100 channels (bottom trace), firing can be quite irregular. For patches containing dozens or fewer channels it becomes impossible to define action potentials unambiguously, since the opening of one or two channels can rapidly depolarize the membrane (not shown). As the membrane potential acts on 1000 or more binary and stochastic channels, the response becomes quite predictable, and merges into the behavior expected by numerical integration of the Hodgkin–Huxley equations for continuous and deterministic currents (top trace). The density is set to 60 Na^+ channels and 18 K^+ channels per square micrometer, each with a single channel conductance γ of 20 pS. All other values are as specified in the standard Hodgkin–Huxley model. Reprinted in modified form by permission from Strassberg and DeFelice (1993).

causes stochastic spiking by virtue of the fact that even in a spatially extended system, fluctuations in a small number of channels are enough to exceed V_{th} and initiate firing. Indeed, pyramidal cells responding to just suprathreshold, maintained current injections are known to show much more jitter in their exact timing compared to a temporally modulated current input (Mainen and Sejnowski, 1995). As sustained inputs are unlikely to occur under ecological conditions, it remains unclear to what extent the stochastic nature of ionic channels influences computations performed on physiological input.

For simulated squid axon patches of less than 1 μm^2, the opening of any one channel has an appreciable effect on the membrane potential. The all-or-none behavior of the action potential is less and less evident in these cases, since the simultaneous opening of a few channels depolarizes the membrane significantly. Given this variability in the amplitude of the membrane response, it is difficult to define a firing frequency for a very small number of channels.

We should emphasize here the obvious, namely, that individual channels do not show any threshold behavior. They do not change their configuration abruptly at some fixed voltage level. Rather, the probability of opening increases continuously with voltage. The macroscopically observed voltage (or current) threshold occurs when the sum of all inward currents through the Na^+ channels just about exceeds the sum of the outward currents generated by the potassium and leak conductances and by the local circuit current.

8.3.2 Spontaneous Action Potentials

Much more tantalizing behavior can be seen in the absence of any direct stimulus. Figure 8.9 illustrates the flat voltage trajectory associated with the continuous Hodgkin–Huxley equations when no external current is supplied. This stationary behavior also occurs for

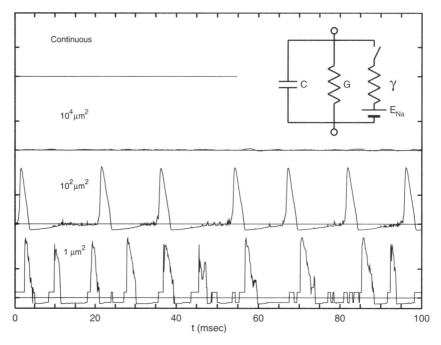

Fig. 8.9 SPONTANEOUS ACTION POTENTIALS The same model as is Fig. 8.8 is rerun, but now in the absence of *any* current stimulus. The bottom three traces show Monte-Carlo sample trials for axonal membrane patches of different surface areas A populated with a constant density of Na^+ and K^+ channels (60 and 18 channels per square micrometer, respectively). Because the opening of a single sodium channel (of conductance γ) can give rise to a very large depolarization, the spontaneous openings of two sodium channels is enough to trigger the opening of the remaining sodium channels and the membrane generates a spike. These spikes are Poisson distributed (once a refractory period has been accounted for; Chow and White, 1996). The top trace shows the solution of the continuous and deterministic Hodgkin–Huxley equation: V in the absence of a current input remains flat. The circuit inset provides the basic intuition why a single channel, in the presence of a small leak conductance G, can cause a large membrane depolarization. These simulations predict that in high-impedance systems with an active membrane, spontaneous spiking can be observed under certain conditions in the absence of any synaptic input. The membrane leak conductance G for these three conditions corresponds to 3 pS, 300 pS, and 30 nS, as compared to a single-channel conductance of $\gamma = 20$ pS for both channel types. For more details see legend to Fig. 8.8. Reprinted in modified form by permission from Strassberg and DeFelice (1993).

simulations with several hundred thousand discrete probabilistic ionic channels (third trace from the bottom in Fig. 8.9). As the number of channels is reduced—by reducing the size of the space-clamped patch—a much more interesting behavior emerges: the axonal membrane "spontaneously" generates a train of remarkably regular action potentials.

What causes microscopic events, such as the openings of a few channels, to be so drastically amplified that they can lead to observable macroscopic phenomena? Upon inspection of the $V(t)$ trace for the smallest 1-μm^2 patch at the bottom of Fig. 8.9, we see many binary transitions from a value close to E_K to a value a few millivolts above V_{rest}. These events are caused by the abrupt opening of single sodium channels. That one ionic channel can have this large effect can be understood in terms of a very simple circuit (inset in Fig. 8.9). We assume the membrane to be at rest, with a steady-state membrane leak conductance $G = AG_m$, where A is the surface area of the patch and $G_m = 3$ pS/μm^2. Each channel opening contributes $\gamma = 20$ pS toward the total conductance, an amount independent of the surface area. The voltage trajectory following the opening of a single channel is given by

$$V(t) = \Delta V(1 - e^{-t/\tau}) \tag{8.14}$$

with $\Delta V = E_{Na}/(1 + AG_m/\gamma)$ and $\tau = AC_m/(AG_m + \gamma)$. For this simple model, ΔV in response to a single sodium channel opening is 100 mV. However, $V(t)$ in the lower trace in Fig. 8.9 hovers just above E_K, due to a few open potassium channels. They add enough background conductance for ΔV to be on the order of 20 mV, with the membrane potential reaching its equilibrium within a fraction of a millisecond. When two sodium channels (out of the 60 present in the 1-μm^2-large patch) open coincidentally, they rapidly depolarize the membrane beyond V_{th} associated with the deterministic equations, thereby initiating an action potential.

The fact that the random opening of channels can induce "spontaneous" spikes is ultimately caused by the fact that while the leak conductance (and the membrane capacitance) is assumed to scale with the membrane area A, the channel conductance is independent of A. The probability that two out of 60 independently acting sodium channels are open is so high, even at E_K, that the membrane is either always refractory or spiking, leading to an average firing frequency of around 90 Hz. And all of this in the absence of any synaptic or current input. For larger values of A, the membrane depolarization produced by the simultaneous opening of a small number of channels becomes so small that the membrane potential effectively remains always around zero (third voltage trace in Fig. 8.9).

In an elegant study demonstrating how concepts from physics can be applied to neurobiological problems, Chow and White (1996) derive an analytical expression for the probability of spontaneous spiking in analogy to the classical problem of determining the probability that a particle in a double-well potential makes the transition from one well to the other. Under the assumption that sodium activation occurs at a much faster time scale than changes in V, n, or h, they incorporate fluctuations in m into a stochastic conductance term that is added to the conventional deterministic g_{Na} term. Chow and White find that the probability of spontaneous firing decays exponentially as the area of the membrane patch increases. Furthermore, the interspike intervals are exponentially distributed once the refractory period is accounted for (Chap. 15). In other words, the spontaneous spikes are Poisson distributed.

Under what circumstances could the random opening of one or a few ionic channels cause such spontaneous action potentials to occur? From an electrical point of view, this requires a small input conductance (on the order of the single channel conductance or less) as well as a low membrane capacitance. Assuming a relatively conservative value of the membrane resistance of $R_m = 30,000 \ \Omega \cdot cm^2$, implies that the entire leak conductance of a spherical 10-μm-diameter cell is about five times the open-channel conductance of a single sodium channel.

The occurrence of action potentials following the opening of a single ionic channel in cells with very high input resistances (tens of gigaohms) has been experimentally observed in chromaffin cells of the adrenal medulla (Fenwick, Marty, and Neher, 1982), rat olfactory receptor neurons (Lynch and Barry, 1989), and in hippocampal cultured neurons (Johansson and Arhem, 1994). Whether or not the opening of these channels is truly spontaneous, that is, due to thermal fluctuation in the configuration of the channel protein, or triggered by some factor in the extra- or intracellular milieu is very difficult to establish experimentally. Similar phenomena might well occur in thin spines, distal dendrites, or even the cell body of small interneurons, in particular if shielded by membrane invaginations or the cellular nucleus. Does the nervous system exploit such local sources of microscopic noise for its own, functional purposes?

8.4 Recapitulation

Underlying the entire gamut of electrochemical events in the nervous system are proteins inserted into the bilipid membrane, so-called ionic channels, which allow specific ions passage across the bilipid membrane. Given their small electrical conductance (between 5 and 200 pS), it required the development of the giga-seal technique by Neher and Sakmann to readily observe their behavior. From a functional point of view, the key properties of channels are that they possess one or a very small number of open, conductive states and that the transitions among closed, open, and inactivated states are governed in a probabilistic manner by the amplitude of the applied membrane potential (for voltage-dependent channels) or the presence of various agonists (for ligand-gated channels).

A very large effort in the field is directed toward identifying and characterizing the molecular sequence of these channels and in relating specific structural features of the channel protein to its voltage and ionic selectivity and, ultimately, to its function. Such a detailed molecular understanding goes hand in hand with the construction of ever more complex kinetic models, describing the transitions of a single channel among a large number of internal states in terms of probabilistic Markov models.

Numerical studies have related the microscopic, stochastic chatter of individual ionic channels opening and closing to the observance of macroscopic currents changing in a highly deterministic and graded manner. A membrane at rest and studded with a few thousand discrete channels behaves in general little different from the deterministic Hodgkin–Huxley equations. Deviations are only expected to occur when the membrane potential is close to threshold or when the channels are embedded into a small patch of neuronal membrane with a low leak conductance. In the latter case, the single-channel conductance can have the same magnitude as the passive leak conductance, and the opening of one or a few channels can depolarize the membrane beyond V_{th}. The resultant spikes are Poisson distributed. In other words, the microscopic behavior of individual molecules is amplified and causes a macroscopic event, an action potential. Such a mechanism, known to occur in certain experimental preparations, could be of functional relevance in very small cells or in electrically decoupled, distal parts of the dendritic tree as a "random event" generator.

In the remainder of this book we will typically deal with large enough membrane areas—and therefore channel numbers—that we are usually justified in treating electrical events in terms of deterministic, continuous currents, rather than in terms of probabilistic, all-or-none ionic channels.

9

BEYOND HODGKIN AND HUXLEY:
CALCIUM AND CALCIUM-
DEPENDENT POTASSIUM
CURRENTS

The cornerstone of modern biophysics is the comprehensive analysis by Hodgkin and Huxley (1952a,b,c,d) of the generation and propagation of action potentials in the squid giant axon. The basis of their model is a fast sodium current I_{Na} and a delayed potassium current I_K (which here we also refer to as I_{DR}). The last 40 years of research have shown that impulse conduction along axons can be successfully analyzed in terms of one or both of these currents (see Sec. 6.6). Nonetheless, their equations do not capture—nor were they intended to capture—a number of important biophysical phenomena, such as adaptation of the firing frequency to long-lasting stimuli or bursting, that is, the generation of two to five spikes within 5–20 msec. Moreover, the transmission of electrical signals within and between neurons involves more than the mere circulation of stereotyped pulses. These impulses must be set up and generated by subthreshold processes.

The differences between the firing behavior of most neurons and the squid giant axon reflect the roles of other voltage-dependent ionic conductances than the two described by Hodgkin and Huxley. Over the last two decades, more than several dozen membrane conductances have been characterized (Hagiwara, 1983; Llinás, 1988; Hille, 1992). They differ in principal carrier, voltage, and time dependence, dependence on the presence of intracellular calcium and on their susceptibility to modulation by synaptic inputs and second messengers. Our knowledge of these conductances and the role they play in impulse formation has accelerated rapidly in recent years as a result of various technical innovations such as single-cell isolation, patch clamping, and molecular techniques. We will here describe the most important of these conductances and briefly characterize each one. In order to understand more completely the functional role of these conductances in determining the response of the cell to input, empirical equations that approximate their behavior under physiological conditions must be developed and compared against the physiological preparations. In a remarkable testimony to the power and the generality of the Hodgkin–Huxley approach, the majority of such phenomenological models has used their methodology of describing individual ionic conductances in terms of activating and inactivating particles with first-order kinetics (see Chap. 6). Theirs is also our method of choice for quantifying the membrane conductance that we use throughout this book.

Because of the large number of ionic conductances that have been described throughout the nervous systems of different animals, we would certainly exhaust the patience of the reader by listing and describing them all (as Llinás, 1988, has done). Instead, we focus on a few key conductances, in particular those that appear to be common to many neurons in most animals, from slugs and flies to mammals.

9.1 Calcium Currents

In Chaps. 6 and 8, we characterized two currents, an inward one carried by sodium ions and an outward one mediated by a potassium current. The missing member in the triumvirate of ions crucial for understanding the dynamics of the membrane potential is calcium. Ionized calcium is the most common signal transduction element in all of biology. Ca^{2+} is required for the survival of cells; yet an excess of calcium ions can lead to cell death (Johnson, Koike, and Franklin, 1992). Ca^{2+} plays a crucial role in triggering synaptic release upon the invasion of an action potential at a presynaptic site, it is the key determinant for axonal growth and muscle contraction, and it constitutes the crucial signal that initiates synaptic plasticity. Indeed, Ca^{2+} can be thought of as coupling the membrane potential to cellular output such as secretion, contraction, growth, and plasticity; that is, linking computation— as expressed by means of rapid and highly localized changes in $V_m(t)$—to action. Whenever a neuron decides to do something, Ca^{2+} is the key.

Scores of mechanisms exist to precisely adjust the concentration of free intracellular calcium $[Ca^{2+}]_i$. As we will discuss in more detail in Chap. 11, the influx of Ca^{2+} into the intracellular cytoplasm is primarily regulated by voltage-dependent calcium channels and by the NMDA synaptic receptor complex (see Sec. 4.6) in the membrane as well as by release from intracellular storage sites in the endoplasmic reticulum. These organelles are spread like a vast three-dimensional spider web throughout neurons and are thought to warehouse Ca^{2+} ions. Because the release from these intracellular storage sites usually occurs on a slow time scale, we will not discuss it here, referring the reader to Clapham (1995) for further references.

What are some of the important functional properties shared by all calcium conductances? Most importantly, the associated calcium current is always activated by depolarization, is always inward, and has some degree of inactivation that occurs on a much slower time scale than does inactivation of the sodium current. What is strikingly different among the different calcium currents is their sensitivity to depolarization. Some currents activate in response to a depolarization of a few millivolts, while others require 20 mV or even more. Some show rapid and some very slow voltage-dependent inactivation. At least three distinct flavors of voltage-dependent calcium currents have been reasonably well characterized, although as the molecular structure of calcium channels becomes better understood, we will probably be able to distinguish many more subtypes (Yang et al., 1993; McCleskey, 1994; Catterall, 1995, 1996). Finally, calcium currents are no aberrations found only in a few exceptional neurons. All nerve cells show evidence of calcium currents, many simultaneously expressing two or more types (Tsien et al., 1988). While calcium currents appear to be largely absent in axons, they can be found throughout the dendritic tree, the cell body, and at presynaptic sites.

9.1.1 Goldman–Hodgkin–Katz Current Equation

Up to now, we have used Ohm's law to describe the relationship between sustained current flow and applied membrane potential. However, given the extreme imbalance between the typical extracellular calcium concentration, in the low millimolar range, and the intracellular concentration, ranging between 10 and 100 nM, it can be difficult to measure a reversal potential. Indeed, one would hardly expect calcium channels to generate much *outward* current beyond the calcium equilibrium potential. Even though the effective driving potential might be very large at very depolarizing values, only a few calcium ions are avilable to move into the extracellular space, making the outward current very close to zero.

The standard theoretical account for the current flowing under these conditions, based on the electrodiffusion model, was introduced into the literature by Goldman (1943) and Hodgkin and Katz (1949) (see also Secs. 4.5 and 11.3). It makes no assumption about pores, rather hypothesizing that ions move independently through the membrane along a constant electrical field (that is, a linear drop in membrane potential). Using this phenomenological model, the ionic current flowing across the membrane can be derived (see Jack, Noble, and Tsien, 1975 or Hille, 1992) to be

$$I_{Ca} = P_{Ca} 2vF \frac{[Ca^{2+}]_i - [Ca^{2+}]_o e^{-v}}{1 - e^{-v}} \tag{9.1}$$

where $v = 2V_m F/RT$ is the normalized membrane potential, with $F = 96,480$ C/mol being Faraday's constant, R the gas constant, and T the absolute temperature. P_{Ca} is the *permeability* of the membrane to calcium and is defined as the constant of proportionality between the molar flux density of calcium across a unit area of membrane S and the concentration difference across the membrane

$$S = -P_{Ca} \Delta [Ca^{2+}] \tag{9.2}$$

where the membrane permeability has units of cm/sec. (This is essentially Fick's first law of diffusion; see Eq. 11.14.) Equation 9.1 is known as the *Goldman-Hodgkin-Katz* (GHK) current equation.

If the ionic milieu on both sides of the membrane is the same, the GHK current equation is linear in the membrane potential, with a reversal potential of zero. As the external concentration increases relative to the intracellular one, the $I-V$ curve becomes shallower and shallower for positive potentials, since it becomes increasingly more difficult for the Ca^{2+} ions to move against their concentration gradient.

For large, positive values of v, the current is linear in v, as for very negative values (e.g., $V_m < -15$ mV in the case of Fig. 9.1). The ratio of the slopes in the two domains is simply the ratio of the concentration values across the membrane.

Matching Eq. 9.1 against a real calcium current requires further considerations since, for instance, calcium channels are mildly permeable to potassium. (The ratio of the two permeabilities is 1/1000; see Lee and Tsien, 1982, and Fig. 9.1.)

Whether or not the calcium current should be described by Ohm's law or by the GHK current equation has yet to be carefully evaluated from an experimental point of view. The various numerical models that match simulations against experimental records

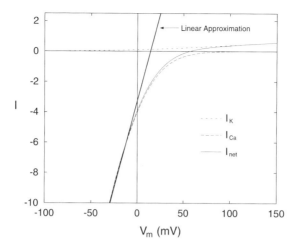

Fig. 9.1 THEORETICAL $I-V$ RELATIONSHIP FOR A CALCIUM CURRENT   Due to the 40,000-fold concentration difference between extracellular Ca^{2+} (around 2 mM) and intracellular Ca^{2+} (around 50 nM), the phenomenological constant field equation of Goldman, Hodgkin, and Katz (Eq. 9.1) predicts a strongly rectifying $I-V$ relationship for calcium. If one takes account of the fact that calcium channels are also weakly selective for potassium ions, the reversal potential shifts to somewhat lower values and the curve becomes less steep, although still rectifying. A linear relationship, based upon approximating the slope of the $I-V$ curve for values $V_m < -20$ mV is superimposed. Under physiological conditions, the membrane potential almost always resides in this range. Reprinted in modified form by permission from Hille (1992).

have used either GHK or variants thereof (Llinás, Steinberg and Walton, 1981a,b; Lytton and Sejnowski, 1991; Huguenard and McCormick, 1992; DeSchutter and Bower, 1994a; Borg-Graham, 1997) or Ohm's law (Hudspeth and Lewis, 1988b; Traub and Miles, 1991; McCormick, Huguenard, and Strowbridge, 1992; Yamada, Koch, and Adams, 1998). In defense of the linear model it should be noted that under physiological conditions the cell spends almost all of its time at membrane potentials < -15 mV, a region in which the $I-V$ relationship can be approximated very well by a linear one (Fig. 9.1). Cells only briefly transgress to more positive potentials.

   With this background, let us now review three of the most important calcium currents. These are not the only types of calcium currents that have been dissected out of the physiological responses of neurons. (Other currents include the P and the R calcium currents; Mintz, Adams, and Bean, 1992; Randall and Tsien, 1995.). Indeed, it remains unclear how many distinct types of ionic channels permeable to calcium ions exist in the nervous systems. This number could, potentially, be much larger than we assume today.

9.1.2 High-Threshold Calcium Current

This current, labeled L for long lasting, was for many years the only known calcium current. Different from the fast sodium current of the squid axon, $I_{Ca(L)}$ is only activated at the very depolarized levels expected during action potentials. Its halfway activation point lies around 10–0 mV (Fig. 9.2), implying that around V_{rest}, $I_{Ca(L)}$ is totally deactivated. The L current does show inactivation that is not, however, dependent on the membrane potential but depends, instead, on the intracellular calcium concentration just below the membrane. It is halfway inactivated at quite low values of $[Ca^{2+}]_i$ (tens of micromolars). Because activation of $I_{Ca(L)}$ does not follow the onset of depolarization instantaneously but appears with a delay, the activation particle m is usually taken to the second, third, or even higher power (Hagiwara and Byerly, 1981; Johnston and Wu, 1995). Recalling the GHK equation 9.1, we can express the current as

$$I_{Ca(L)} = m(V_m)^k h([Ca^{2+}]_i) I_{Ca} \tag{9.3}$$

with k an integer ≥ 2, and I_{Ca} the GHK expression for the current as a function of the voltage (see also Kay and Wong, 1987; Hille, 1992).

Because of the existence of specific pharmacological L channel blockers, it is known that such L channels live in proximal dendrites and at the cell body of cortical pyramidal and thalamic cells (Galvan et al., 1986; Llinás, 1988; Fisher, Gray and Johnston, 1990; Westenbroek, Ahlijanian and Catterall, 1990).

9.1.3 Low-Threshold Transient Calcium Current

The low-threshold transient or T calcium current is activated at lower voltages than the L current. Typically, its half-activation point is at -40 mV. At normal resting levels, this current is turned off. It does inactivate slowly but strongly as a function of voltage, as evident in Fig. 9.3A. Under physiological conditions, the T current can be triggered by hyperpolarizing the membrane potential—thereby completely removing inactivation—and subsequently allowing synaptic input (or a depolarizing current injection) to activate it. $I_{Ca(T)}$ can generate an all-or-nothing electrical event (usually called an *LT spike*; see below) similar to a classical action potential, but much broader, on top of which two or more sodium spikes are riding (e.g., Fig. 6.1H).

In thalamic relay cells, this current has been modeled (McCormick, Huguenard, and Strowbridge, 1992; McCormick and Huguenard, 1992; Huguenard and McCormick, 1992) as

$$I_{Ca(T)} = m(V_m)^2 h(V_m) I_{Ca} . \tag{9.4}$$

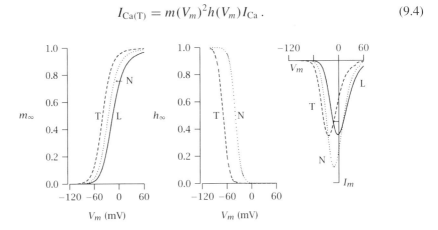

Fig. 9.2 ACTIVATION AND INACTIVATION RANGES FOR CALCIUM CURRENTS Steady-state activation (leftmost plot) and inactivation (middle plot) variables as a function of the membrane potential for the three calcium currents discussed in the text. Inactivation of the L calcium current does not depend on voltage but on the intracellular calcium concentration. The rightmost plot shows their *I–V* relationships. It is likely that intermediate forms of calcium current exist without specific pharmacological blockers to uniquely identify them. Reprinted by permission from Johnston and Wu (1995).

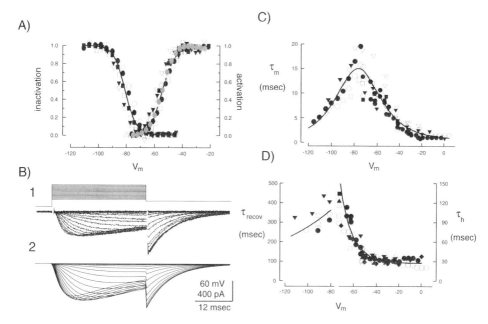

Fig. 9.3 **ACTIVATION AND INACTIVATION OF THE TRANSIENT CALCIUM CURRENT** Properties of the transient low-threshold calcium current found in rat thalamic relay cells as modeled by Huguenard and McCormick (1992) using a standard Hodgkin–Huxley-like formalism (Eq. 9.4). Data for six thalamic cells are shown. **(A)** Steady-state activation and inactivation as a function of V_m. **(B)** I_T is activated by stepping the potential for 30 msec to various potentials, followed by repolarization to -80 mV (1—experimental data; 2—model). **(C)** Time constant of activation and **(D)** of inactivation and recovery. Reprinted by permission from Huguenard and McCormick (1992).

The steady-state activation m_∞ and inactivation h_∞ as well as their associated time constants τ_m and τ_h are illustrated in Fig. 9.3 as a function of the membrane potential for the thalamus current. (For an alternate formulation see Wang, Rinzel, and Rogawski, 1991.) Different from I_{Na} in the squid axon, $I_{Ca(T)}$ does not possess a *window current*. The window current refers to current flowing at steady-state, determined by the amount of overlap between activation and inactivation at any particular membrane potential. For the T current, the window current $\overline{P}_{Ca(T)} m_\infty^2 h_\infty I_{Ca}$, is always zero or close to zero for any sustained voltages; hence its name (Fig. 9.3A). Physiological data point toward the T current as playing a prominent role in distal dendrites (Christie et al., 1995).

A significant difference between low- and high-threshold calcium currents is that the latter require depolarization beyond 0 mV, something only expected during action potentials, while the former are activated at much more modest levels of depolarization.

9.1.4 Low-Threshold Spike in Thalamic Neurons

The role that the T current might play in information transmission was first elucidated in two classical papers by Jahnsen and Llinás (1984a,b). Thalamic relay cells occupy a unique position in the central nervous system. With the exception of olfaction, all sensory systems access cortex by projecting through the thalamus (for a review see Sherman and Koch, 1986, 1998). Furthermore, about one-half of all synaptic contacts onto thalamic cells originate in the lower layers of the cortex. The thalamus is also the recipient of diffusely projecting input from the brainstem. The various systems have more than ample opportunity to control what is transmitted through the thalamus (Crick, 1984a).

Intracellular recordings from thalamic cells in slice as well as in the intact animal (Jahnsen and Llinás, 1984a,b; Coulter, Huguenard, and Prince, 1989; Hernández-Cruz and Pape, 1989; Guido et al., 1995; Guido and Weyand, 1995) have revealed two functional, quite distinct modes of operation which are characterized by whether or not the transient calcium current is activated (Fig. 9.4).

If the membrane potential at the soma is around −75 mV or below—held there by modulatory input from the brain stem that is known to hyperpolarize thalamic cells tonically—the inactivation of $I_{Ca(T)}$ is removed and any sufficiently large depolarizing synaptic input will activate the low-threshold calcium current (Fig. 9.4A). Similar to the fast Hodgkin–Huxley-like action potential, this further depolarizes the membrane, activating additional $I_{Ca(T)}$, and so on. The result is an all-or-none action potential, the LT spike, which is relatively broad, around 50 msec. This is due to a two order of magnitude difference between fast activation and slow inactivation (Fig. 9.3D). As can be seen in Fig. 9.4A, this broad depolarization triggers a series of very fast conventional sodium spikes riding on top of the LT spike. The cell is said to *burst*; consequently, this manner of operating is known as the *burst mode*. Subsequent to the LT spike, the membrane is profoundly refractory for a period of 100 msec or longer before it can generate the next calcium spike. During the intervening period, any synaptic input will have little effect on the cell. If, conversely, the membrane is at rest or slightly depolarized, $I_{Ca(T)}$ is inactivated and any input will cause either only a subthreshold depolarization (Fig. 9.4B) or a series of conventional action potentials (Fig. 9.4C). Stronger depolarization is encoded in a higher firing frequency: the cell operates in its *relay mode*.

Steriade and his colleagues (Steriade and McCarley, 1990; Steriade, McCormick, and Sejnowski, 1993) have provided much evidence that the relatively minor dc shifts in V_m

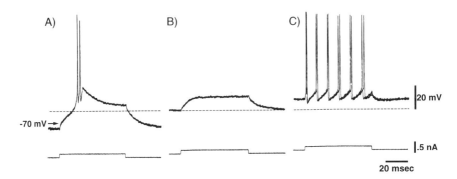

Fig. 9.4 Bursting in Thalamic Neurons The majority of thalamic cells can be in one of two functionally quite distinct states. **(A)** If the membrane potential is tonically hyperpolarized to −75 mV or beyond—due to modulatory activity from the brainstem or inhibitory synaptic input—a transient low-threshold calcium current is de-inactivated. A depolarizing current step (indicated in the lower trace) triggers a slow all-or-none electrical event due to inward calcium current, the so-called LT spike. Riding on top of the LT action potential are a number of conventional, fast sodium spikes. The LT spike concludes with a profound hyperpolarization that prevents excitatory synaptic input during this time from generating output spikes. **(B)** In an intermediate state, the cell responds to the same input in a linear fashion, with a slowly rising and decaying depolarization. **(C)** If the membrane is tonically depolarized to −60 mV or above, the same input causes the cell to fire a series of conventional fast sodium action potentials. This constitutes the so-called *relay mode*. Reprinted in modified form by permission from Jahnsen and Llinás (1984a).

which bring about the switch from one mode into the other are controlled by modulatory input from a small number of discrete sites in the brainstem. It is known that thalamic cells in anesthetized as well as in awake behaving cats can respond to visual stimuli using burst as well as isolated spikes, preferentially generating bursts during the early phase of fixation (Guido and Weyand, 1995; Guido et al., 1995). We will return in Sec. 16.3 to the possible significance of bursting for information processing.

9.1.5 N-Type Calcium Current

This is a current that has properties intermediate between the L and the T calcium currents, hence its name N for neither L nor T. It is activated at potentials intermediate between the low- and high-threshold currents (Fig. 9.2). It does show slower, voltage-dependent inactivation dynamics than the T current and has a specific pharmacological channel blocker. The N current is modeled as

$$I_{Ca(N)} = m(V_m)^3 h(V_m) I_{Ca} \tag{9.5}$$

(see Sutor and Zieglgänsberger, 1987; Lytton and Sejnowski, 1991).

Westenbroek and colleagues (1992) stained the entire rat brain with antibodies to a part of the N channel. As expected, most gray matter through the central nervous system was stained, in particular proximal and distal dendrites and presynaptic terminals. Thus, N channels appear to complement the locations of L channels, which are restricted to the cell bodies and the proximal dendrites of pyramidal cells (Westenbroek, Ahlijanian, and Catterall, 1990).

9.1.6 Calcium as a Measure of the Spiking Activity of the Neuron

No matter through which type of calcium channel Ca^{2+} ions enter the neuron, they will now diffuse throughout the available space and interact with enzymes and proteins in the membrane and the cytoplasm. We defer until Chap. 11 an in-depth account of the mathematics of diffusion and binding. Nonetheless, let us here briefly allude to a computationally interesting role of intracellular calcium.

Around the intracellular mouth of calcium channels, Ca^{2+} exit the 5 Å pore at rates exceeding one million per second. These ions diffuse throughout the neuron, binding at available sites in the cytoplasm and the intracellular organelles (this binding can induce release of Ca^{2+} ions from these organelles). While the majority of Ca^{2+} ions is bound, a small amount remains unbound and slowly accumulates inside the cytoplasm. Each action potential admits further calcium ions which further increases $[Ca^{2+}]_i$. This buildup is illustrated in numerical simulations of the spherical cell body of bullfrog sympathetic ganglion cells (Fig. 9.5; see Gamble and Koch, 1987, and Sec. 9.4 below). The Ca^{2+} ions rushing in through calcium channels for the few milliseconds that these channels are open during an action potential, increase $[Ca^{2+}]_i$ in the core of the cell by between 3 and 4 nM. To a first order, this increase is relatively independent of the spiking frequency (here varying between 10 and 50 Hz). While 3 nM might appear as a tiny increase, it is averaged over the entire cell body and will be much larger in local regions. The increase needs to be judged against the low resting level of $[Ca^{2+}]_i$, between 10 and 50 nM.

Experimental calcium measurements in cell bodies of the sea snail *Aplysia* (Gorman and Thomas, 1980), of bullfrog sympathetic ganglion cells (Smith, MacDermott, and Weight, 1983; Hernández-Cruz, Sala, and Adams, 1990) and of cortical pyramidal cells (Fig. 9.6; Helmchen, Imoto, and Sakmann, 1996) have confirmed this hypothetical buildup, which

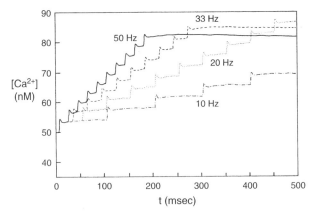

Fig. 9.5 INTRACELLULAR CALCIUM ACCUMULATION FOLLOWING SPIKING ACTIVITY Increase in intracellular free calcium in simulated bullfrog sympathetic ganglion cells in response to repetitive fast synaptic input at different input frequencies. For a more detailed description of this model see Sec. 9.4. The increase of $[Ca^{2+}]_i$ in the central 19-μm-radius core of the 20-μm-radius cell body is relatively independent of the spiking frequency (each synaptic input triggers one spike). In other words, the calcium concentration provides an index for the spiking activity of the cell in recent times. (The time constant of this integrated measure depends on the cellular geometry; the larger the volume, the longer the effective time constant; see Sec. 11.7.2.) This "activity" measure can be read out by any calcium-dependent enzyme or protein. Reprinted by permission from Gamble and Koch (1987).

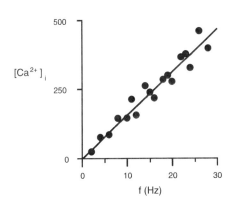

Fig. 9.6 CALCIUM BUILDUP PROPORTIONAL TO NEURONAL ACTIVITY Following each action potential in a layer 5 pyramidal neuron, calcium rushes into the cell and accumulates, here recorded using a calcium-dependent fluorescent dye in the proximal apical dendrite (Helmchen, Imoto, and Sakmann, 1996). This increase in calcium reaches an equilibrium with a time constant of 200 msec. The final level (in nM) is linearly related to the firing frequency of the cell (evoked by current injections), with a slope of 15 nM/Hz (solid line). Reprinted in modified form by permission from Helmchen, Imoto, and Sakmann (1996).

reaches a plateau between 0.1 and 1 sec, depending on the geometry of the cell. It gradually subsides due to extrusion processes that remove the excess calcium from the cytoplasm.

The integrated measure of past neuronal activity can be *read out* by any enzyme or protein that is sensitive to calcium, such as a calcium-dependent potassium channel (see below), or any of the myriad cellular processes that are so exquisitely sensitive to calcium concentration (see Chaps. 11 and 20). This allows the neuron or a local circuit to up or downregulate its own activity (a sort of *gain control* operation). Indeed, experimental evidence has accumulated in favor of this *calcium set-point* hypothesis, in which the concentration of free, intracellular calcium determines the amplitude of various ionic currents (Johnson, Koike, and Franklin, 1992; Turrigiano, Abbott, and Marder, 1994). An elegant modeling study has shown us how a simple neuronal system, here modeled by the Morris-Lecar equations (Sec. 7.2), can be modified to adjust its maximal conductances $\overline{G}$ as a function of integrated calcium activity until the system expresses stable spiking patterns (LeMasson, Marder, and Abbott, 1993).

9.2 Potassium Currents

A variety at least as great as among the calcium channels, if not greater, is displayed by ionic channels permeable to potassium ions. Under physiological conditions, current flow through these channels is outward, that is, upward in the normal convention. Given the physiological reversal potential of K^+ in the neighborhood of -100 mV, potassium currents can be thought of as stabilizing the membrane potential, since activating a potassium current pulls the membrane potential to hyperpolarizing levels. As we saw already in the case of I_{DR}, the delayed rectified potassium current in the squid axon (Sec. 6.2.1), potassium currents serve to keep the action potential brief. In the absence of such a potassium current, which rapidly repolarizes the membrane, the action potential would be much longer and would follow the time course of inactivation of the sodium channel.[1] As we will see, potassium currents serve to delay the generation of an action potential following synaptic or current input as well as to reduce the firing frequency in the face of a constant input.

Like the majority of ionic currents, potassium currents—with the noticeable exception of the anomalous or inward rectifier current (Hagiwara, 1983; Doupnik, Davidson, and Lester, 1995)—require depolarization to activate. Around V_{rest}, they—as well as all sodium and calcium currents—are very small, minimizing the possibly wasteful antagonistic inward and outward current flow and, thereby, metabolic cost. In a sense, biology might have discovered a principle that designers of portable computers only implemented within the last few years: when not in use, reduce the clock rate of the microprocessors as much as possible in order to conserve precious battery power.

9.2.1 Transient Potassium Currents and Delays

Next to I_{DR}, arguably the most important potassium current is a transient, inactivating potassium current, known as I_A. First characterized by Connor and Stevens (1971b), it lives in a rather narrow voltage range between -70 and -40 mV, activating within 5–10 msec. Different from the delayed rectifier, I_A inactivates on a slower time scale, typically on the order of 100 msec. Based on their voltage-clamp data, Connor and Stevens (1971c) described I_A on the basis of Ohm's law,

$$I_A = \overline{G}_{K(A)} m(V_m)^4 h(V_m)(V_m - E_K).$$ (9.6)

The fourth power yields the observed sigmoidal activation curves, while the single inactivation particle reflects the exponential decay of the current. The time constants of activation and inactivation depend only weakly on the membrane potential.

The functional relevance of I_A (Connor and Stevens, 1971c) can be understood by reference to Fig. 9.7. As we learned in Chaps. 6 and 7, the squid axonal membrane shows an abrupt onset of spiking: in the standard model, the minimum spiking frequency is 53 Hz. We can understand why either by looking at the exact equations or by considering phase space analysis. In terms of biophysics, the Hodgkin–Huxley equations are unable to

---

1. An exception to this are the nodes of Ranvier of myelinated fibers, in which a vigorous leak current serves to partially repolarize the membrane; Sec. 6.6.

generate spikes at very low frequencies, since too slow a rate of depolarization would lead to inactivation of the sodium current. In Chap. 7 we used phase space analysis to argue that onset of oscillations with an infinitely long period requires a saddle-node bifurcation that can be obtained with the help of an inactivation variable that depends steeply on the membrane potential (witness the $\dot{w} = 0$ nullcline in Fig. 7.9).

I_A exactly fulfills this condition: it is active in a narrow subthreshold range. In cortical pyramidal cells—such as the one modeled in Fig 9.7—this current will try to hyperpolarize the membrane potential subsequent to a current step (hence the bump in Fig. 9.7A; Segal and Barker, 1984; Schwindt et al., 1988; Zona et al., 1988). In this voltage range, I_A slowly inactivates, thereby contributing less and less outward current: V_m slowly drifts upward until the spiking threshold is reached and a spike is triggered. A similar mechanism generates delays of up to 3–4 sec in molluscan neurons (Getting, 1983). It has also been hypothetized that a class of cells in the cat lateral geniculate nucleus, *lagged cells*, which show a much delayed response to a visual stimulus compared to nonlagged geniculate cells, implement this delay via a transient A current (Saul and Humphrey, 1990; McCormick, 1991).

I_A can exert another profound effect on the firing behavior of cells. As is apparent upon inspection of the $f-I$ curve of a patch of squid axon membrane (Fig. 6.10A), the bandwidth

A)

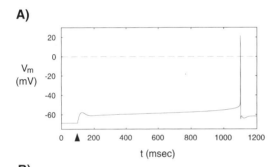

B)

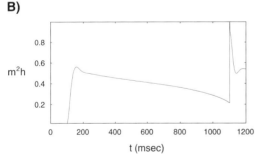

C)

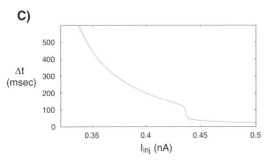

Fig. 9.7 DELAYED SPIKING DUE TO TRANSIENT POTASSIUM CURRENT The A current implements a delay element. **(A)** Somatic membrane potential in the standard pyramidal cell model in response to a 0.32-nA current step injected at 100 msec. V_m depolarizes and activates the transient potassium I_A current. At these potentials I_A slowly inactivates, as evident in **(B)**, showing the normalized A conductance. (An m^2h instead of the m^4h formalism of Eq. 9.6 was used.) This gradual inactivation of a potassium current allows the membrane potential to creep upward until, 1 sec after the current injection, an action potential is generated. Without I_A the membrane potential would either remain subthreshold or generate a spike within a few tens of milliseconds. **(C)** Relationship between the injected current and the delay between current onset and spike initiation.

of firing is very limited because of the high onset firing frequency. The remedy for this is the introduction of I_A, allowing the membrane to generate action potentials across a much larger firing frequency range than before.

The A current is just one member of a larger class of transient inactivating potassium currents that share similar pharmacological and functional properties (Rudy, 1988). Other members of this club include I_D (D for delay) described in hippocampal cells (Storm, 1988) and I_{As}, found in thalamic neurons (McCormick, 1991). Both currents have very long activation times, causing a delay in the onset of spiking in response to a current injection of up to 10 sec.

We conclude that I_A and its siblings subserve at least two functions: (1) to allow cells to fire at very low frequencies, thereby increasing the bandwidth of the firing frequency code significantly, and (2) to implement a delay element that does not require ultraslow intrinsic time constants.

9.2.2 Calcium-Dependent Potassium Currents

The amplitude of several potassium currents is influenced by $[Ca^{2+}]_i$, the intracellular concentration of Ca^{2+} ions. These currents are collectively referred to as $I_{K(Ca)}$ to emphasize that they are modulated by calcium but that the current itself is carried by K^+ ions. The function of these currents is to shape the membrane potential following individual action potentials (Fig. 9.8), to read out the previous spiking history of the cell, shaping adaptation accordingly, and to instantiate, in conjunction with a calcium current, a resonance-like behavior (Hudspeth and Lewis, 1988b; Sec. 10.4.3).

We here discuss two potassium currents: I_C and I_{AHP} (Latorre et al., 1989).

I_C is a fairly large, outward current whose activity depends on $[Ca^{2+}]_i$ as well as on V_m. I_C quickly activates upon the entry of calcium through calcium channels during an action potential. Similar to the delayed rectifier, it rapidly repolarizes the membrane potential, shutting off in the process in readiness for the next spike. It is not easy to study the dynamics

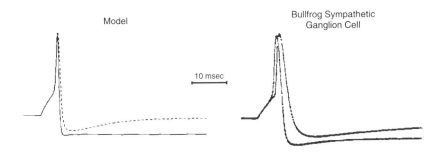

Fig. 9.8 CALCIUM AND THE ACTION POTENTIAL Effect of blocking calcium channels on computer-simulated and observed action potentials in type B bullfrog sympathetic ganglion cells (Yamada, Koch, and Adams, 1998). In both cases, the standard action potential (induced by a brief current pulse) repolarizes more quickly and is followed by a longer afterhyperpolarization than the action potential (dotted lines) in the absence of calcium currents (achieved in the experiment by adding cadmium to the bathing solution and in the model by reducing $\overline{G}_{Ca}$ to 5% of its normal value). The effect of blocking Ca^{2+} on the peak amplitude of the spike is minimal: the spike itself is mediated by Na^+ and not by Ca^{2+} ions. Reprinted in modified form by permission from Yamada, Koch, and Adams (1998).

of this, or any other calcium-dependent potassium current, since its time course is highly sensitive to the calcium concentration in the immediate neighborhood of the underlying channel. Because Ca^{2+} will decay very differently in thin dendrites than in a large cell body, I_C will also have a different time course, even if the dynamics of the membrane potential are the same in both cases.

The microscopic gating of the ionic channels underlying this current has been studied in some detail, giving rise to complex kinetic gating schemes with two or more calcium ions necessary for the channel to open (McManus and Magleby, 1988). A much simpler gating scheme for the underlying conductance g_C that reproduces much of the time and voltage dependency involves calcium ions binding to a single activation particle, which effects the transition from closed to open channel in a voltage-dependent manner (Adams et al., 1986; Yamada, Koch, and Adams, 1998):

$$I_C = \overline{G}_C m(V_m, [Ca^{2+}]_i)(V_m - E_K). \tag{9.7}$$

If I_{DR} and I_C were the only currents activated by a single action potential, the spike afterhyperpolarization (AHP) would promptly return to rest with a time course dictated by the membrane time constant. Yet neurons throughout the nervous system (Connors, Gutnick, and Prince, 1982; Pennefather et al., 1985; Lancaster and Adams, 1986) show a second component of the AHP, reflecting a much smaller and slower calcium-dependent potassium conductance. The associated current, termed I_{AHP}, is small and depends only on the concentration of intracellular calcium just below the membrane. The secret to its success is that it is sustained: when calcium rushes in during an action potential, I_{AHP} turns on and effectively subtracts from the depolarizing current from an electrode or from synaptic input. In hippocampal pyramidal cells, the amplitude of the hyperpolarization following spiking activity increases with increasing number of action potentials, showing that I_{AHP} increments with each spike (Madison and Nicoll, 1984). Blocking this current prevents firing frequency adaptation from occurring. The current is described as

$$I_{AHP} = \overline{G}_{AHP} m([Ca^{2+}]_i)(V_m - E_K). \tag{9.8}$$

Experimentally, I_{AHP} can be distinguished from I_C by the use of naturally occurring toxins that are added to the solution bathing the neurons: while I_C is blocked in many cells by charybdotoxin, a peptide from scorpion venom, I_{AHP} can be blocked by injecting a toxin from bee venom, apamin. This reveals the tight coevolution of predators and the nervous systems of the animals they hunt.

9.3 Firing Frequency Adaptation

Throughout neocortex, hippocampus, and thalamus, cells receive a clearly defined input from a single site in the pontine brainstem, the *locus coeruleus*. Each of the small number of coeruleus neurons (about 2×1600 in rat and $2 \times 13,000$ in humans) projects in a very diffuse manner throughout the brain. A single neuron might enervate frontal cortex, thalamus as well as occipital cortex (Foote and Morrison, 1987). Excitation in these cells has been related to increased vigilance and arousal in both rats and monkeys (Foote, Aston-Jones and Bloom, 1980).

Upon stimulation, the terminals of these neurons release the neurotransmitter *noradrenaline* (also termed norepinephrine) onto their postsynaptic targets (see Table 4.1). This leads to a quite pronounced and long-lasting change in the excitability of these cells, an effect

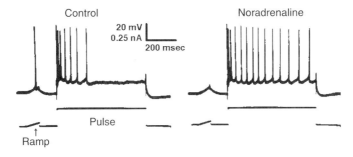

Fig. 9.9 SLOW MODULATION OF FIRING FREQUENCY   Change in firing frequency adaptation in CA1 pyramidal cells from a hippocampal slice during application of noradrenaline. A brief just-threshold current ramp is applied first, followed by a 650-msec-long, suprathreshold current pulse. The principal action of noradrenaline, mediated by a second messenger, is to reduce firing frequency adaptation by blocking the slow calcium-dependent potassium conductance I_{AHP}. A side effect is a slight increase in the firing threshold of the cell (that is, the brief current ramp does not trigger an action potential in the presence of noradrenaline). This modulation in the firing properties can persist for seconds or even minutes. Hippocampal pyramidal cells receive a well-defined noradrenergic input from brainstem neurons. Reprinted by permission from Madison and Nicoll (1986).

that has been studied in isolation in pyramidal cells in the neocortical and the hippocampal slice preparation (Madison and Nicoll, 1986; Nicoll, 1988; Schwindt et al., 1988).

Application of noradrenaline to pyramidal cells causes two changes. One is a small (about 2–3 mV) hyperpolarization, leading to a measurable *increase* in the spiking threshold (Fig. 9.9). A much more dramatic effect can be observed when the cell is sufficiently excited to fire trains of spikes. Normally, these cells adapt, that is, their firing rate in response to a constant current step slows down (as in "Control" in Fig. 9.9). In the presence of noradrenaline, spike discharge accommodation is almost completely blocked. Its action is specific in that neither the shape of the action potential itself nor the fast afterhyperpolarization phase is affected. Indeed, experiments have shown (Madison and Nicoll, 1986) that noradrenaline specifically blocks I_{AHP} for many tens of seconds. This effect is mediated by a metabotropic receptor (Sec. 4.4 and Table 4.2). In this case, noradrenaline binds to β_1 adrenergic receptors, releasing a second messenger inside the cell which directly or indirectly phosphorylates the channel proteins underlying I_{AHP}. An added complication is that the small hyperpolarizing action of noradrenaline—caused by an *increase* in a potassium conductance— is mediated by α adrenergic receptors. Any neuroactive substance in the brain will often initiate a series of changes in the electrical behavior of its postsynaptic targets via multiple actuators (here receptors).

Noradrenaline is not alone in being able to control spike rate adaptation in cortical cells. Acetylcholine, mediated by a dense cholinergic input to the hippocampus from cells in the medial septal nucleus, has a similar effect, although its action is even more complex than that of noradrenaline (it appears to modify at least four distinct potassium currents; Cole and Nicoll, 1984; McCormick and Prince, 1986; Nicoll, 1988; Schwindt et al., 1988). The pyramidal cells that receive such a cholinergic input can have their spike rate accommodation switched off for tens of seconds in the absence of any direct change of the membrane potential.

The lesson is that multiple, diffusely projecting systems in the brain can independently control the excitability of neurons via slow modulation of membrane conductances. Their mode of action is somewhat similar to a *broadcast* that is sent out to all processors in a massively parallel computer. In other words, all processors will receive the same signal. Because a single neuron in the locus coeruleus projects widely throughout the brain, this blockage of spike rate adaptation is not very precise in terms of space nor in terms of time (since it typically requires a fraction of a second to set in and may last for many, many seconds). From a computational point of view, these diffuse systems that target specific conductances should be thought of as modifying the basic electrical properties of the processing circuits on a relatively global scale, possibly for controlling the level of arousal or vigilance and for providing a single slow global feedback signal during learning.

9.4 Other Currents

Besides the handful of voltage-dependent membrane currents discussed so far, there exist a plethora of other currents that are responsible, in isolation or in concert with others, for shaping various phases of cellular excitability (Llinás, 1988). They include I_M, a slowly changing voltage-dependent potassium current found in pyramidal cells, which contributes to the accommodation of action potential discharge (Halliwell and Adams, 1982; Brown, 1988), as well as an entire family of inward rectifier potassium currents, which are activated upon hyperpolarization of the membrane potential (Hagiwara, 1983; Spain, Schwindt, and Crill, 1987).

A smaller and less well understood family of channels is primarily permeable to chloride ions, by far the most abundant anions in biological systems (Hille, 1992). In general, the reversal potential for Cl^- is within 10–15 mV of the resting potential of neurons, and activation of a chloride current will repolarize the cell and prevent it from firing. The voltage and time dependencies of chloride conductances are only minor and slow. Therefore, they probably contribute to the leak conductance that is usually assumed to be independent of the applied membrane potential. There also exists a calcium-activated chloride conductance with only a weak voltage dependency, which could play a similar role to calcium-dependent potassium conductances (Mayer, 1985).

A possible important current is the *sustained* or *persistent sodium current* $I_{Na,p}$. It has been clearly documented in cortical pyramidal cells (Stafstrom et al., 1985; French et al., 1990) due to the sensitivity to TTX that it shares with its fast sibling, I_{Na}. This current turns on within a few milliseconds upon depolarization, but either does not inactivate at all or only very slowly, thereby providing a steady inward current. Because $I_{Na,p}$ is activated near the resting potential, it tends to amplify small EPSPs. The firing frequency of the cell will be quite sensitive to the total amount of this current, since too much $I_{Na,p}$ will prevent repolarization following action potentials.

9.5 An Integrated View

For the most part, this chapter has described individual currents and their possible functional roles. Before we go on to other matters, we would like to give the reader a holistic view of how these different currents interrelate and how they function within an entire system. As example, we will discuss the simulacrum of type B bullfrog sympathetic ganglion cells of Adams et al., (1986) and Yamada, Koch, and Adams (1998). These neurons make only a lowly contribution to the overall scheme of things, enabling the animal to control its glands for slime production. From an electrophysiological point of view, though, these cells are

A)

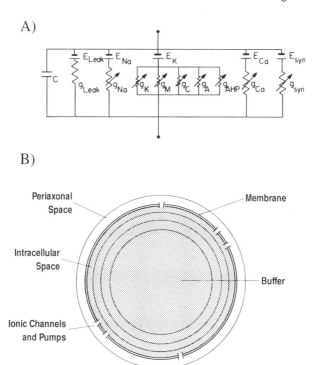

B)

Fig. 9.10 Modeling a Bullfrog Sympathetic Ganglion Cell (A) Electrical circuit used in the Yamada, Koch, and Adams (1998) simulacrum of type B bullfrog sympathetic ganglion cells. Given their spherical shape and the complete absence of any dendrites, a single lumped compartment that includes seven time-, voltage-, and calcium-dependent membrane conductances is entirely sufficient to model their electrical behavior. (B) Underlying assumptions for modeling the extracellular accumulation of potassium (in a single extracellular compartment) and the series of shells surrounding a well-mixed central core to account for Ca^{2+} diffusion inside the cell. A calcium buffer is distributed throughout the cell. Not drawn to scale. Reprinted by permission from Yamada, Koch, and Adams (1998).

ideal since they are easy to cultivate, big (with a mean diameter of 35 μm), and devoid of any dendritic processes.

The electrical structure of the model, shown schematically in Fig. 9.10, includes the seven time-, voltage-, and calcium-dependent conductances found in these cells, in addition to a passive RC contribution: the fast sodium I_{Na} and the delayed rectifier potassium I_{DR} currents underlying spike production, an L type calcium current I_{Ca}, the two forms of calcium-dependent potassium currents discussed above, I_C and I_{AHP}, the transient potassium current I_A, and a fifth potassium current, the small and slowly activating voltage-dependent M current. The description of these currents is derived from voltage-clamp data and is based on the conventional Hodgkin–Huxley rate formalism and the use of Ohm's law. (for the numerical values of all parameters, consult the appendix in Yamada, Koch, and Adams, 1998). Because of the spherical geometry of these cells, they are electrically compact and can be modeled using a single compartment.

Sympathetic ganglion cells are *not* compact from the point of view of the spatio-temporal distribution of the Ca^{2+} ions entering the cell via calcium channels. As treated in detail in

Chap. 11, the diffusion equation is solved in spherical coordinates, using a series of thin shells surrounding a well-stirred, central core. Ca^{2+} ions can bind to a stationary buffer distributed throughout the cytoplasm using a second-order reaction (see Sec. 11.4.1). In the outermost compartment just below the membrane, calcium ions are extruded from the cell via a first-order calcium pump. Finally, in order to account for the accumulation of potassium ions in the pericellular space just outside the neuronal membrane—leading to a significant reduction of the potassium battery E_K following spike activity—the cell is assumed to be surrounded by a single compartment, whose concentration of K^+ ions is determined by a simple uptake mechanism. We leave the mathematics of solving two coupled sets of nonlinear equations for $V_m(t)$ and $[Ca^{2+}](x, t)$ to Appendix C.

Figures 9.11 and 9.12 illustrate the basic performance of the model in response to a short current pulse that triggers an action potential. The spike itself (Fig. 9.8) is similar to the squid axon spike (Fig. 6.6) with the noticable exception of the second, long-lasting phase of the afterhyperpolarization, mediated by I_{AHP}. Moreover, in these cells I_C, rather than I_{DR}, is largely responsible for repolarizing the potential following the excursion of V_m to 0 mV and beyond (as can be ascertained by comparing the peak amplitudes of I_C and I_{DR}).

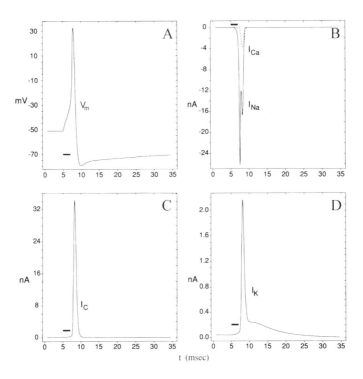

Fig. 9.11 RESPONSE OF THE SYMPATHETIC GANGLION CELL MODEL: MAJOR CURRENTS Response of the bullfrog sympathetic ganglion cell model (see Fig. 9.10) at rest to a 2-msec-long current pulse of 1.75-nA amplitude (indicated by black bars). This panorama depicts the major currents involved in spike initiation and repolarization seen in (**A**) (identical to the spike in Fig. 9.8). Notice the fast and slow components of the afterhyperpolarization. The two inward currents I_{Na} and I_{Ca} are plotted in (**B**). Only 9% of the incoming charge is carried by Ca^{2+} ions, emphasizing the fact that the primary function of calcium ions is to shape neuronal excitability, rather than to be directly involved in spike production. (**C**) Fast calcium-dependent potassium current I_C that contributes substantially more to spike repolarization than the delayed rectifier current I_{DR}, shown in (**D**). Reprinted by permission from Yamada, Koch, and Adams (1998).

The amplitude of I_{AHP} is minute compared to I_C; yet, because it is maintained, it has a crucial role to play in determining spike accommodation (Adams et al., 1986). Changes in the driving potential can be quite significant, as seen in the transient 50 mV reduction in E_{Ca} (Fig. 9.12B) caused by the very brief elevation of $[Ca^{2+}]_i$ just below the membrane to micromolar values.

The effect of blocking either I_M or I_{AHP} or both on the average firing frequency is the subject of Fig. 9.13. Blocking I_{AHP} leaves the current threshold for spike initiation unchanged but does increase the gain of the $f-I$ curve by about a factor of 2. Exactly the same result is observed in a quite different system, vestibular neurons in the brainstem (du Lac, 1996; Fig. 21.3) where pharmacological elimination of the slow calcium-dependent potassium current increases the gain of the $f-I$ curve by a factor varying between 1.38 and 2.28.

Blocking but I_M in the bullfrog leaves the slope of the discharge curve relatively unchanged while reducing I_{th} to 5% of its original value. Removing both conductances

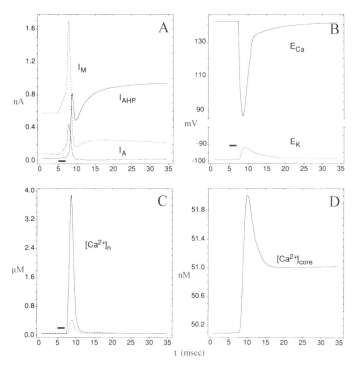

Fig. 9.12 RESPONSE OF THE SYMPATHETIC GANGLION CELL MODEL: MINOR CURRENTS AND CALCIUM CONCENTRATION Response of the bullfrog sympathetic ganglion cell to the same stimulus as in Fig. 9.11. **(A)** Three smaller potassium currents I_M, I_{AHP}, and I_A. The latter does not appear to play any significant role in these cells. I_M is active at rest. The primary effect of blockage of I_M is a reduction in current threshold I_{th}, while blockage of I_{AHP} leads to a partial loss of spike firing adaptation. Blockage of both currents occurs *in vivo* during stimulation of cholinergic input to the ganglia. The changes in the Nernst potentials for E_{Ca} and E_K are shown in **(B)**. The primary rapid and the secondary slow components in the change in E_{Ca} reflect the rapid increase in free calcium concentration in the outermost shell just below the membrane plotted in **(C)**. Activation of I_{Ca} leads to a rapid influx of Ca^{2+} ions that quickly bind to the buffer. The slow decay results from calcium loss via the pump as well as diffusion toward the core of the cell. $[Ca^{2+}]_i$ for the tenth shell below the membrane (dotted line) is also illustrated in **(C)**. The calcium concentration in the central core is shown in **(D)**. Reprinted by permission from Yamada, Koch, and Adams (1998).

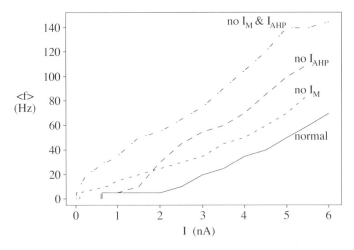

Fig. 9.13 CHANGING THE GAIN OF THE DISCHARGE CURVE Effect of blocking either I_M or I_{AHP} or both on the f–I curve of the model bullfrog sympathetic ganglion cell. Mean firing rate, averaged over 200 msec, following current steps of variable amplitude. Removing I_M primarily affects I_{th} but leaves the slope intact, while eliminating I_{AHP} increases the gain of the discharge curve by a factor of 2. Blocking the two simultaneously leads to the summation of both effects. The control of the gain of a neuron via control of the amount of the slow calcium-dependent potassium current present might be a crucial biophysical operation underlying different computations. Reprinted by permission from Yamada, Koch, and Adams (1998).

lowers the current threshold but increases the slope, leading to the dramatic removal of spike adaptation observed experimentally (Adams et al., 1986).

If I_{AHP} can be controlled in individual neurons, rather than by a globally acting neuro-transmitter such as noradrenaline, as introduced above, it offers the possibility of selectively up or down regulating the gain of a neuron without affecting the spike generation mechanism itself. Given the importance of receptive field models that incorporate a multiplicative effect on the gain of individual neurons (leading to so-called *gain fields*; Zipser and Andersen, 1988; but see Pouget and Sejnowski, 1997), this might be a crucial canonical biophysical operation.

9.6 Recapitulation

In this chapter, we summarized some of what is known about ionic currents populating the nervous system other than I_{Na} and I_{DR} of squid axon fame. In particular, we introduced calcium currents. Ca^{2+} ions are crucial for the life and death of neurons; they link rapid changes in the membrane potential, used for computation, with action, such as neurotrans-mitter release at a synapse, muscle contraction, or biochemical changes underlying cellular plasticity. Three broad classes of calcium currents are known: L, T, and N currents, although it is unclear whether these are just three samples along a continuum of such currents. Because of the substantial imbalance between the concentration of Ca^{2+} inside and outside the cell, electrical rectification needs to be taken account of. Typically, this requires the use of the Goldman-Hodgkin-Katz equation to describe calcium currents, rather than the linear Ohm's law. The accumulation of calcium ions following action potential activity has the useful function that the intracellular calcium concentration in the cell represents an index of the recent spiking history of the cell. Various proteins and enzymes, such as calcium-dependent potassium channels, might be able to read out this variable and use it to normalize excitability or for gain control.

While calcium ions themselves play almost no role in generating conventional action potentials, calcium currents can support a slower all-or-none electrical event, the LT spike, leading to a burst of fast sodium spikes. It may be possible that these and other forms of bursts represent events of special significance in the nervous system.

We introduced some of the many potassium currents that have been identified, focusing on the transient inactivating potassium current I_A. It serves to linearize the discharge curve around threshold and implements a delay element between the onset of depolarizing input and spike initiation. It does so via a voltage-dependent current whose time constant is considerably faster than the duration of the delay.

Also important are the calcium-dependent potassium currents, collectively referred to as $I_{K(Ca)}$. Calcium ions that move into the cell during the depolarizing phase of the spike turn on these hyperpolarizing outward currents, rendering subsequent spiking that much more difficult. That this might have profound significance for the nervous system can be observed in hippocampus, neocortex, and elsewhere. The release of noradrenaline by brainstem fibers abolishes spike frequency accommodation, greatly increasing the excitability of these cells. Interestingly, the effect of blocking only I_{AHP} appears to be specific to increasing the slope of the cell's $f-I$ curve, that is, increasing its gain.

We keep on drawing analogies between the operations the nervous system carries out to process information and digital computers. There are, of course, very deep, conceptual differences between the two. One is the amazing diversity seen in biology. An I_A current in a neocortical pyramidal cell is not the same as the I_A current in hippocampal pyramid cells or in thalamic relay cells. Each has its own unique activation and inactivation range, kinetics, and pharmacology, optimized by evolution for its particular role in the survival of the organism. A case in point appears to be calcium channels. So far, molecular biological techniques have revealed about a dozen different genes coding for two of the five subunits of the calcium channel, a number that is likely to increase over time. This provides the nervous system with the wherewithal to generate a very large number of calcium channels with different functional properties (Hofmann, Biel, and Flockerzi, 1994). This tremendous diversity could be necessary to allow each organism to optimize its complement of ionic currents for its particular operations depending on its particular and unique developmental history. This is, indeed, very far removed from the way we design and build digital integrated circuits using a very small library of canonical circuit elements.

10

LINEARIZING
VOLTAGE-DEPENDENT
CURRENTS

We hinted several times at the fact that a small excitatory synaptic input in the presence of voltage-dependent channels will lead to a local depolarization, followed by a hyperpolarization (Fig. 10.1). Those of us who built our own radios will recognize such an overshooting response as being indicative of so-called RLC circuits which include resistances, capacitances as well as *inductances*. As a reminder, a linear inductance is defined as a circuit element whose instantaneous $I–V$ relationship is,

$$V(t) = -L\frac{dI(t)}{dt} \tag{10.1}$$

where L is the inductance measured in units of henry (abbreviated as H). Although neurobiology does not possess any coils or coillike elements whose voltage is proportional to the current change, membranes with certain types of voltage- and time-dependent conductances can behave as *if* they contained inductances. We talk of a *phenomenological inductance*, a phenomenon first described by Cole (1941) and Cole and Baker (1941) in the squid axon (see Cole, 1972). Under certain circumstances, discussed further below, such damped oscillations can become quite prominent.

This behavior can be obtained in an entirely linear system, as can be observed when reducing (in numerical simulations) the amplitude of the synaptic input (or step current): even though the voltage excursion around steady-state becomes smaller and smaller, the overshoot persists (Fig. 10.1). It is not due to any amplification inherent in such a membrane but is caused by its time- and voltage-dependent nature. Such a linear membrane, whose constitutive elements do not depend on either voltage or time, and which behaves like a bandpass element, has been called *quasi-active* (Koch, 1984) to distinguish it from a truly *active*, that is, nonlinear membrane.

In this chapter, we will explain in considerable detail how an inductance-like behavior can arise from these membranes by linearizing the Hodgkin–Huxley equations. Experimentally, this can be done by considering the *small-signal* response of the squid giant axon and

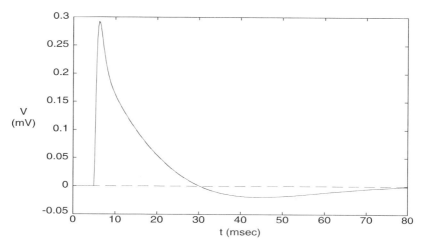

Fig. 10.1 A Quasi-Active Membrane A small ($g_{\text{peak}} = 1$ nS) excitatory synaptic input close to the soma of a simulated pyramidal cell (Fig. 18.1B) gives rise to an EPSP followed by a small hyperpolarization due to activation of somatic potassium currents. This overshoot is not due to any nonlinearity in the system since it can be observed for arbitrarily small inputs. (If the synaptic input is reduced by a factor of 100 to $g_{\text{peak}} = 0.01$ nS and the associated EPSP rescaled by 100 and superimposed onto the larger amplitude response, the curves cannot be distinguished.) The system behaves like a linear electrical circuit containing an *inductance*-like element and the associated membrane is called *quasi-active*.

comparing it against the theoretical predicted value, a further test of the veracity of the Hodgkin–Huxley equations, which they passed with flying colors. For more details on the linearization procedure we will be using, consult the original references (Chandler, FitzHugh, and Cole, 1962; Sabah and Leibovic, 1969; and the very clear Mauro et al., 1970). Although such a linearization has been applied to membranes other than the squid axon (Clapham and DeFelice, 1976, 1982) we will primarily discuss this well-explored model system.

10.1 Linearization of the Potassium Current

In order to demonstrate the principle behind this linearization, we will describe the procedure in detail for the potassium current of the Hodgkin–Huxley equation (Eq. 6.4 and Fig. 10.2A):

$$I_{\text{K}} = \bar{g}_{\text{K}} n^4 (V - E_{\text{K}}) . \tag{10.2}$$

Let us consider small variations of this current around some fixed potential V_r. We can express such a variation as

$$\delta I_{\text{K}} = \left(\frac{\partial I_{\text{K}}}{\partial V}\right)_r \delta V + \left(\frac{\partial I_{\text{K}}}{\partial n}\right)_r \delta n , \tag{10.3}$$

where both derivatives are evaluated at the potential around which the system is linearized, hence the subscript r. Retaining only the first-order terms and neglecting all higher order ones (such as $(\delta n)^2$ or $\delta n \delta V$) is at the heart of the linearization procedure. We have

$$\left.\frac{\partial I_{\text{K}}}{\partial V}\right|_r = \bar{g}_{\text{K}} n_\infty^4 (V_r) = G_{\text{K}} \tag{10.4}$$

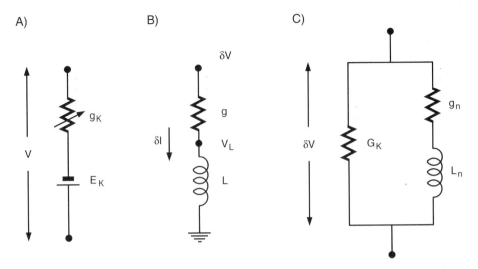

Fig. 10.2 **PHENOMENOLOGICAL INDUCTANCE ASSOCIATED WITH THE LINEARIZED POTASSIUM CURRENT** **(A)** Equivalent electrical circuit for a noninactivating potassium current. The conductance g_K is given by $\bar{g}_K n^4$, with n the voltage- and time-dependent activation constant. **(B)** Conductance g in series with an inductance L. The voltage across both components is δV and the current through them δI. V_L is the voltage at the intermediate location. **(C)** Equivalent circuit when the potassium current (in panel A) is linearized around some fixed membrane potential. The response to this circuit mimics the small-signal response of the potassium current I_K around some potential V_r. It includes a pure conductance, $G_K = \bar{g}_K n^4$, in parallel with a conductance g_n in series with an inductance L_n. For $V_m > E_K$, all three components are positive.

(see Fig. 10.2C) and

$$\frac{\partial I_K}{\partial n} = 4\bar{g}_K n^3 (V - E_K) . \tag{10.5}$$

We eliminate the variation in the rate constant δn by recalling Eq. 6.6 governing its dynamics,

$$\frac{dn}{dt} = \alpha_n(1 - n) - \beta_n n . \tag{10.6}$$

For small variations we can expand this to

$$\frac{d\delta n}{dt} = \delta\alpha_n - (\delta\alpha_n + \delta\beta_n)n - (\alpha_n + \beta_n)\delta n . \tag{10.7}$$

Since α_n and β_n only depend on the membrane potential, we can express their variations as

$$\delta\alpha_n = \left(\frac{d\alpha_n}{dV}\right)\delta V \tag{10.8}$$

and

$$\delta\beta_n = \left(\frac{d\beta_n}{dV}\right)\delta V . \tag{10.9}$$

Replacing these two expression into Eq. 10.7 leads to

$$\frac{d\delta n}{dt} = \left(\frac{d\alpha_n}{dV}\right)\delta V - (\alpha_n + \beta_n)\delta n - n\left(\frac{d(\alpha_n + \beta_n)}{dV}\right)\delta V . \tag{10.10}$$

This can be formally expressed as

$$\left(\frac{d}{dt} + \alpha_n + \beta_n\right)\delta n = \left[\left(\frac{d\alpha_n}{dV}\right) - n\left(\frac{d(\alpha_n + \beta_n)}{dV}\right)\right]\delta V. \tag{10.11}$$

When replacing this result in Eq. 10.3 and exploiting Eqs. 10.4 and 10.5, we arrive at the following equation for the first-order variation of the potassium current

$$\delta I_K = \left[G_K + \frac{a}{d/dt + b}\right]\delta V, \tag{10.12}$$

with two parameters that need to be evaluated at the membrane potential at which the current is linearized,

$$a = 4\bar{g}_K n^3(V - E_K)\left[\left(\frac{d\alpha_n}{dV}\right) - n\left(\frac{d(\alpha_n + \beta_n)}{dV}\right)\right] \tag{10.13}$$

and

$$b = \alpha_n + \beta_n. \tag{10.14}$$

In order to understand Eq. 10.12 let us apply Kirchhoff's voltage law to the one branch circuit shown in Fig. 10.2B. With δI the current through both components we can write

$$(\delta V - V_L) + (V_L - 0) = \delta V. \tag{10.15}$$

Eliminating the voltage V_L, we are left with

$$\left(g + L\frac{d}{dt}\right)\delta I = \delta V. \tag{10.16}$$

Comparing this result with the second component on the right-hand side of Eq. 10.12 we can see that they are very similar. Indeed, the voltage δV across the electrical circuit with two branches drawn in Fig. 10.2C is the same as δV in Eq. 10.12 if

$$g_n = \frac{a}{b} = \frac{4\bar{g}_K n^3(V - E_K)\left[(d\alpha_n/dV)_r - n\,(d(\alpha_n + \beta_n)/dV)_r\right]}{\alpha_n + \beta_n} \tag{10.17}$$

and if we set the inductance to

$$L_n = \frac{1}{a} = \frac{1}{g_n(\alpha_n + \beta_n)}. \tag{10.18}$$

For a small perturbation δV around V_r, the potassium current responds as though the conductance G_K (in units of S·cm²) is shunted by the conductance g_n (also in S·cm²) in series with the inductance L_n (in units of sec·cm²/S, that is, in H·cm²). By examining the slopes of the on and off rates α_n and β_n (Eqs. 6.10 and 6.11), we see that for the entire range of relevant voltages, $V \geq E_K$, all three electrical components are always positive.

Mauro et al., (1970) captured the key point of the previous analysis in a graphical manner; see Fig. 10.3. As before, we assume that we are only considering the small-signal response (this is, after all, what any linearization procedure amounts to) of the potassium current around some fixed potential V_r (positive to E_K). Injecting a current step $I_{inj} > 0$ at this potential will turn on the potassium current I_K (Fig. 10.3A). Because the underlying potassium conductance requires some time to increase to its final value, the initial change in membrane potential δV is given by the injected current divided by the conductance G_K. In the absence of any capacitance, this causes the membrane to depolarize instantaneously by I_{inj}/G_K, moving in the direction of the dotted slope (the vector labeled $\mathbf{t}_0$ in Fig. 10.3A). However, this depolarization will eventually led to an increase in the potassium conductance, which causes the voltage change δV to *decrease* with time, as indicated by the trajectory in Fig. 10.3C, to finally approach the new steady-state value of the membrane potential $V_{r'}$ (around $\mathbf{t}_2$). In other words, as expected of an inductance—inducing a voltage proportional to the change in current—δV rises instantaneously and then falls off gradually. This analysis also makes obvious the fact that this inductive behavior is caused by both the time and the voltage dependency of the membrane conductance.

When the current I_{inj} terminates, the inverse occurs (Fig. 10.3B). δV first undershoots V_r, relaxing with a time lag back to its earlier steady-state value V_r. The response is inductive for both on and off steps (provided that $V_r > E_K$). If the capacitive nature of the neuronal membrane is included, the phenomenological inductance combines with the capacitance to give a damped oscillatory response (Fig. 10.3C).

10.2 Linearization of the Sodium Current

In a similar manner, we can compute the small-signal response of the sodium current,

$$I_{Na} = \bar{g}_{Na}m^3h(V - E_{Na}) . \tag{10.19}$$

By taking variations and retaining only the first-order terms we arrive at

$$\delta I_{Na} = \left(\frac{\partial I_{Na}}{\partial V}\right)_r \delta V + \left(\frac{\partial I_{Na}}{\partial m}\right)_r \delta m + \left(\frac{\partial I_{Na}}{\partial h}\right)_r \delta h \tag{10.20}$$

that is,

$$\delta I_{Na} = \bar{g}_{Na}m^3h\delta V + 3\bar{g}_{Na}m^2h(V - E_{Na})\delta m + \bar{g}_{Na}m^3(V - E_{Na})\delta h . \tag{10.21}$$

Following the same procedure to eliminate the variational variables δm and δh as for the potassium current leaves us with an equation for the small-signal change in the sodium current:

$$\delta I_{Na} = G_{Na}\delta V + \frac{g_m(\alpha_m + \beta_m)}{d/dt + \alpha_m + \beta_m}\delta V + \frac{g_h(\alpha_h + \beta_h)}{d/dt + \alpha_h + \beta_h}\delta V \tag{10.22}$$

with

$$G_{Na} = \bar{g}_{Na}m^3h \tag{10.23}$$

$$g_m = \frac{3\bar{g}_{Na}m^2h(V - E_{Na})\left[(d\alpha_m/dV)_r - m\,(d(\alpha_m + \beta_m)/dV)_r\right]}{\alpha_m + \beta_m} \tag{10.24}$$

and

$$g_h = \frac{\bar{g}_{Na}m^3(V - E_{Na})\left[(d\alpha_h/dV)_r - h(d(\alpha_h + \beta_h)/dV)_r\right]}{\alpha_h + \beta_h} \tag{10.25}$$

all evaluated at $V = V_r$. As illustrated in Fig. 10.4A, this describes the current flow in a three-branch electrical circuit with inductances

$$L_m = \frac{1}{g_m(\alpha_m + \beta_m)} \tag{10.26}$$

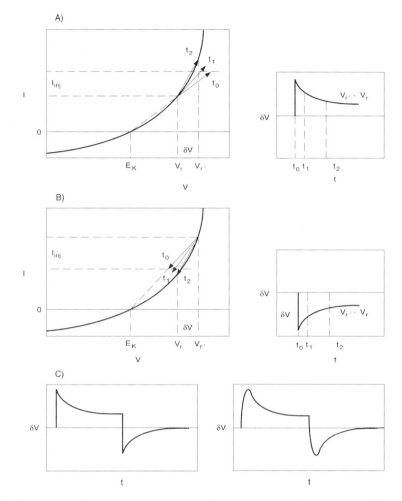

Fig. 10.3 GRAPHICAL INTERPRETATION OF THE LINEARIZATION PROCEDURE Following the lead of Mauro et al., (1970), we here illustrate the linearization procedure for the potassium current I_K. **(A)** Steady-state $I–V$ relationship of a potassium current is indicated schematically and we are considering the small-signal response around the potential V_r. Not shown is the time dependency of the underlying potassium conductance. Injecting a positive current I_{inj} leads to an instantaneous jump of the membrane potential to a new value (which lies along the dotted line going through the reversal potential E_K) given by I_{inj}/G_K (the vector labeled t_0). As the potassium conductance gradually adapts to its new value and increases, the voltage change δV becomes smaller, until it converges to its new value $V_{r'}$ (from t_0 to t_1 to t_2; left panel). This inductive behavior would not be evident in a stationary nonlinear membrane. **(B)** Upon termination of the current step I_{inj}, the voltage change δV first overshoots before settling to its old value. **(C)** Complete voltage response to the current injection in the absence of any membrane capacitance (left panel). When the capacitance is added to the circuit, a damped oscillatory response results (right panel).

and

$$L_h = \frac{1}{g_h(\alpha_m + \beta_m)} \, . \tag{10.27}$$

The two inductive branches reflect the separate contributions from the activating and the inactivating components of the sodium conductance.

A)

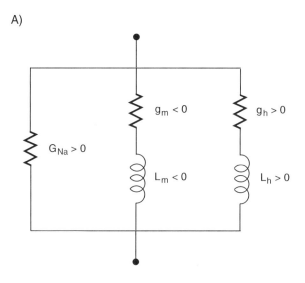

B)

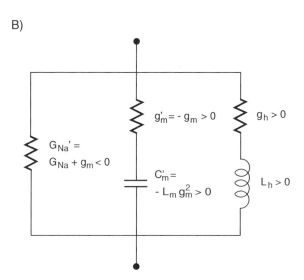

Fig. 10.4 RLC Circuit for the Linearized Sodium Current (A) Electrical RLC circuit obtained by linearizing the Hodgkin-Huxley sodium current at some potential below E_{Na}. Both elements associated with the activation variable are negative. (B) These negative elements can be transformed into a positive conductance g'_m in series with a positive phenomenological capacitance C'_m, with a shunt conductance G'_{Na} that is negative.

At this point, a problem arises. For the voltage range of interest, that is, $E_K \leq V \leq E_{Na}$, the resistive and inductive components of the inactivating branch of the circuit, that is, g_h and L_h, are positive, but the corresponding circuit elements of the m branch are negative: $g_m < 0$ and, therefore, $L_m < 0$. The negative conductance arises as a consequence of the negative slope of the I_{Na}–V curve.

We can fix this problem by transforming the negative components into an equivalent RC system (Fig. 10.4B) with

$$G'_{Na} = G_{Na} + g_m \tag{10.28}$$

$$g'_m = -g_m \tag{10.29}$$

and a positive capacitance, instead of a negative inductance,

$$C'_m = -L_m \cdot g_m^2 . \tag{10.30}$$

This capacitance is also a phenomenological one, in the sense that activating the sodium current leads to similar behavior (for small perturbation) as would the existence of an additional capacitive branch.

For $V < E_{Na}$, the conductance g'_m and the capacitance C'_m are positive, but the shunting conductance G'_{Na} is negative. This is to be expected if we only treat I_{Na} in isolation, with no counteracting current. Indeed, in order for the circuit to be stable, the total steady-state conductance must be positive (Koch, 1984). This can be assured if the linearized potassium current as well as the leak current are all jointly considered, as we will do now.

10.3 Linearized Membrane Impedance of a Patch of Squid Axon

The final RLC circuit, mimicking the small-signal response of a patch of squid axon membrane, includes five branches (Fig. 10.5A): a pure capacitance due to the physical capacitance of the bilipid membrane, a single positive conductance lumping the steady-state contributions of the leak, sodium, and potassium currents, the two phenomenological inductive branches and the one phenomenological capacitive branch. The values of these electrical elements vary with V_r, as discussed in Mauro et al., (1970) and in the legend to Fig. 10.5.

We know from electrical circuit theory that RLC circuits can show oscillatory behavior with the presence of one or more resonances. This is best investigated by computing the complex impedance $\tilde{K}(f)$ of the circuit in Fig. 10.5A. Writing down an expression for the total current (see Eqs. 10.12 and 10.22) directly in Fourier space, we have

$$\delta\tilde{V}(f) = \tilde{K}(f) \cdot \left(\delta\tilde{I}_K(f) + \delta\tilde{I}_{Na}(f) + \delta\tilde{I}_{leak}(f)\right) . \tag{10.31}$$

With $d\delta I(t)/dt$ transforming into $if \cdot \delta\tilde{I}(f)$, we obtain an expression for the impedance as the ratio of $\delta\tilde{V}(f)$ and $\delta\tilde{I}(f)$:

$$\tilde{K}(f) = \frac{\beta_3(if)^3 + \beta_2(if)^2 + \beta_1(if) + \beta_0}{\alpha_4(if)^4 + \alpha_3(if)^3 + \alpha_2(if)^2 + \alpha_1(if) + \alpha_0} . \tag{10.32}$$

The α_i's and β_i's are constants, which depend on the values of the electrical components (Koch, 1984). Figure 10.5B shows the calculated amplitude of the impedance if the membrane is linearized around the resting potential, and Fig. 10.5C shows the experimentally recorded amplitude. Both sets of curves have the same characteristic bandpass response. For the low-temperature curves, sinusoidal inputs in the neighborhood of 50–70 Hz are more attenuated than slower or more rapidly varying current inputs. The linearized squid axon membrane behaves similar to a passive membrane for sinusoidal inputs with 200 Hz or faster components.

A)

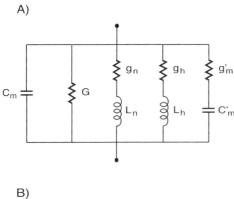

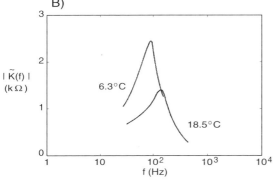

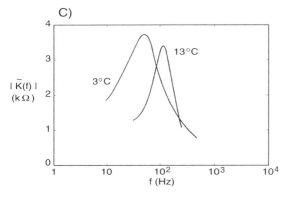

Fig. 10.5 Linearized Squid Axon Membrane (**A**) Electrical RLC circuit mimicking the small-signal response of a patch of squid axon obtained by adding δI_K, δI_{Na}, as well as the contribution from the leak current δI_{leak}. Note that G is positive. If the Hodgkin–Huxley equations with standard parameters (at 6.3° C; see Chap. 6) are linearized around rest, the following numerical values are obtained (for details, see the appendix in Koch, 1984): $C_m = 1$, $G = 0.246$, $g_n = 0.849$, $g_h = 0.072$, $g'_m = 0.432$, $L_n = 6.43$, $L_h = 119.0$ and $C'_m = 0.102$ (units are $\mu F/cm^2$ for the capacitances, $H \cdot cm^2$ for the inductances, and mS/cm^2 for the conductances). The combined steady-state conductance of the system at rest is 1.167 mS/cm^2. (**B**) Amplitude of the membrane impedance of this circuit, corresponding to the impedance of the small-signal response of the squid axon at two different temperatures. As expected of a *quasi-active* membrane, the transversal membrane impedance displays a resonance around 67 Hz. Reprinted in modified form by permission from Mauro et al., (1970). (**C**) Amplitude of the small-signal impedance measured in the space-clamped squid axon at two different temperatures. Reprinted in modified form by permission from Mauro et al., (1970).

The range of validity of the linearized membrane is a few millivolts around the resting potential (for more details, see Sabah and Leibovic, 1969; Koch, 1984). This is evident in the experimental record from the squid axon displayed in Fig. 10.6A. Here, a current step of increasing magnitude is injected into the space-clamped squid giant axon (see Fig. 6.2A for the experimental arrangement). As is apparent, the voltage has a characteristic ringing response as well as displaying undershooting at the offset of the current step. This ringing becomes more pronounced for larger, but still subthreshold, current stimuli, revealing nonlinear components. A similar experiment is repeated in Fig. 10.6B, where an action potential is elicited at the top of the oscillation.

It makes sense to associate a *resonant frequency* with the membrane impedance, defined

A)

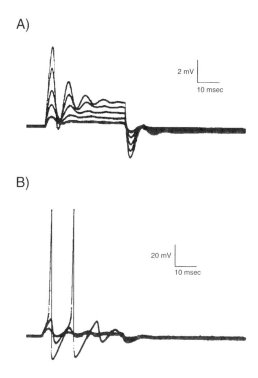

B)

2 mV

10 msec

20 mV

10 msec

Fig. 10.6 SMALL-SIGNAL VOLTAGE RE-SPONSE IN SQUID AXON Membrane potential in a space-clamped giant axonal fiber (of 0.4-mm diameter). (**A**) Membrane potential in response to depolarizing current steps of different amplitudes. Notice the damped oscillations during and at the offset of the current step. The maximum current strength applied is barely subthreshold in the last case. Reprinted in modified form by permission from Mauro et al., (1970). (**B**) A suprathreshold current of sufficient amplitude is applied to evoke a second action potential that arises from the second oscillation. At higher current amplitudes, a train of spikes can be generated with a moderate increase in repetition rate relative to the subthreshold oscillation frequency. Reprinted in modified form by permission from Mauro et al., (1970).

here as the frequency of the peak response of $\tilde{K}(f)$. If the standard Hodgkin–Huxley equations at 6.3° C are linearized around rest, $f_{max} = 67$ Hz (Fig. 10.6B). This frequency is relatively stable for larger current steps, as is evident in Fig. 10.5B, varying no more than 15% and roughly coinciding with the frequency of the oscillatory components seen in the experimental records. Furthermore, f_{max} is only marginally higher than the natural spiking frequency of an "infinite" train of action potentials (53 Hz; see Fig. 6.9B).

So far, we only considered a patch of squid membrane (or, equivalently, a space-clamped axon). On the basis of the propagation factor $\gamma(f)$ defined in Sec. 2.3.1, the input and transfer impedances as well as the frequency-dependent space constant of the linearized squid axon can be computed. It is possible to prove (Koch, 1984) that if the membrane impedance is a bandpass, then so will these three functions for an infinite cable. For instance, a 1 μm thin infinite cable with the quasi-active membrane shown in Fig. 10.5A has a dc space constant of 175 μm. In any passive cable structure, the space constant at higher frequency will always be less than the dc space constant (Fig. 2.8), due to the unavoidable charge leakage via the membrane capacitance. While this capacitance is also present, the inductances "present" in the linearized squid axon can boost $\lambda(f)$ for an intermediate frequency range (for the definition of the frequency-dependent space constant in an infinite cable see Eq. 2.36). Thus, $\lambda(f)$ has a pronounced peak (345 μm) at 74 Hz. The sharpness of the tuning curve (as expressed by its half width at half height) as well as the resonant frequency f_{max} increase monotonically with the channel density of both potassium and sodium channels as well as with the distance between the injection and the recording sites (Koch, 1984).

Figure 10.7A shows the computed amplitude of the somatic input impedance of our standard model of the layer 5 pyramidal cell as a function of the frequency of the applied sinusoidal current injection. While its dendritic tree is passive, the soma contains seven

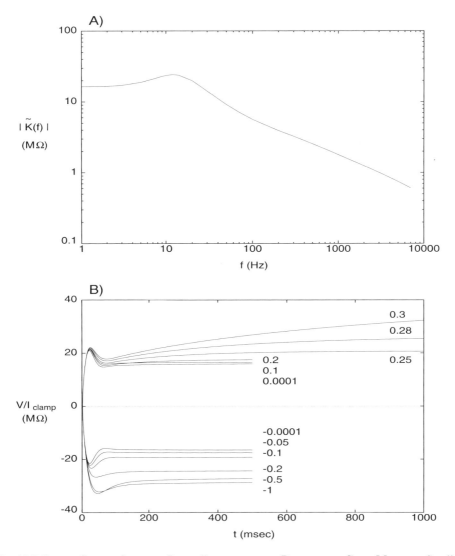

Fig. 10.7 SMALL-SIGNAL SOMATIC INPUT IMPEDANCE OF PYRAMIDAL CELL MODEL Small-signal response of the voltage-dependent model of the layer 5 pyramidal cell. **(A)** Amplitude of the somatic input impedance $|\tilde{K}_{ss}(f)|$ defined as the peak somatic voltage change relative to the resting potential (at -65 mV) in response to a small sinusoidal current (peak amplitude of 10 pA) of frequency f injected into the soma. The peak occurs around 12 Hz. The roll-off at high frequency is due to the distributed capacitance. **(B)** The range of this linear approximation is illustrated by plotting the normalized membrane change in response to current steps I_{clamp}. For currents less then 0.1 nA, the responses superimpose and the depolarizing response is a mirror image of the hyperpolarizing response. Larger currents increasingly deviate from this linear behavior. For current steps of 0.30 nA, the membrane potential slowly drifts upward, eventually initiating an action potential.

voltage-, time-, and calcium-dependent membrane conductances. $\tilde{K}$ is computed using a small-signal response. The amplitude of $\tilde{K}$ is 16.5 MΩ at dc and peaks at 24.1 MΩ around 12 Hz. At these frequencies, the transient potassium current I_A is relatively inactivated (the inactivation rate constant has a time constant of 100 msec). The decay evident at high frequencies is due to the membrane capacitance and has a dependency close to $1/\sqrt{f}$, as expected in an infinite cable. (At these high frequencies, the neuron can be considered to have a practical infinite extent relative to $\lambda(f)$; Fig. 2.9.) The range of validity of this small-signal approximation to the neuronal response can be directly read off Fig. 10.7B. For current steps up to 0.1 nA, this relatively large cell is in its linear range. Larger currents, in particular if they are depolarizing, will trigger a sufficient amount of sodium current to bring the cell out of its linear range. For a step of 0.295 nA amplitude, the potential will slowly creep upward until an action potential is initiated (not shown; see Sec. 17.3).

10.4 Functional Implications of Quasi-Active Membranes

What are some of the computational implications for neurons with quasi-active membranes? The fact that the input and transfer impedances and the frequency-dependent space constant have peaks at nonzero frequencies has interesting consequences for information processing. Let us allude to three examples.

10.4.1 Spatio-Temporal Filtering

In a dendritic tree endowed with a purely passive membrane, $\tilde{K}_{ij}$ and λ are monotonically decreasing functions of the frequency of the applied current. As illustrated with the help of the dynamic morphoelectrotonic transforms in Fig. 3.8, voltage attenuation is substantially more pronounced at higher than at lower frequencies. For a neuron in the visual pathway, say in the retina, this effect will partly determine its receptive field, since a visual stimulus with a high frequency content can only influence the cell's firing if it is closer to the cell body than more slowly varying synaptic input of the same amplitude. This effect is caricaturized in Fig. 10.8A. In a sort of precursor to METs, Koch (1984) determined the fraction of the dendritic tree of a cat retinal ganglion cell whose voltage attenuation to the soma, A_{is}, for two different frequencies is less than 4. This amounts to a functional definition of the *receptive field* of the cell (assuming that the synaptic inervation density is constant across its dendrites). If the dendritic tree is entirely passive, the "receptive field" of this cell is large for dc synaptic input, shrinking to a much smaller size for 100 Hz input.

If, to the contrary, the cell is endowed with the quasi-active membrane shown in Fig. 10.5A, the opposite behavior occurs. Because the membrane impedance is higher for high-frequency inputs, the fraction of the dendritic tree for which $A_{is} \leq 4$ is large for 100 Hz input, and small for either dc or 200 Hz input (Fig. 10.8B). Assuming that the processing between the photoreceptor and the ganglion cell is linear for low-contrast visual stimuli, this neuron would be tuned to transient input and would respond less to sustained stimuli, a form of spatio-temporal tuning (Derrington and Lennie, 1982; Enroth-Cugell et al., 1983).

10.4.2 Temporal Differentiation

Computing the temporal derivative of some function $V(t)$ is equivalent to multiplying its Fourier transform $\tilde{V}(f)$ by if. Such a differentiation operation can be implemented by filtering $V(t)$ with a function whose Fourier transform is proportional to f.

While the transfer impedance of any neuronal membrane will ultimately decay with increasing frequency (due to the presence of the capacitance), the transfer impedance in a cable endowed with a quasi-active membrane shows a high-pass behavior over a limited

A)   B)

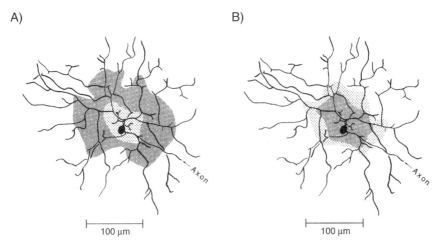

Fig. 10.8 Spatio-Temporal Filtering in Retinal Ganglion Cells In a cell endowed with a quasi-active membrane, high-frequency synaptic input can propagate further than sustained input. This effect is shown here graphically by determining the fraction of the dendritic tree whose voltage attenuation to the soma is less than some threshold, here $A_{is} \leq 4$ in a cat retinal ganglion cell reconstructed from Boycott and Wässle (1974). This can be thought of as the receptive field of the cell. **(A)** The membrane is entirely passive. The large, heavily dotted area indicates the region within which the voltage attenuation for stationary inputs to the soma is less than 4. The smaller, lightly dotted area illustrates the same concept for a sinusoidal input of 100 Hz. **(B)** If the quasi-active membrane obtained from the linearized Hodgkin–Huxley equations (Fig. 10.5A) is added to the passive component in panel A, the previous situation is reversed. Now a sinusoidal current input at 100 Hz (lightly dotted region) attenuates much less than sustained input (heavily dotted region). For 200-Hz input the region of small attenuation will be less than for the region for sustained input. Such a cell would be more responsive to transient than to dc signals. Reprinted by permission from Koch (1984).

frequency range. If the signal is restricted to this part of the spectrum, then such a membrane can implement a temporal derivative operation.

   Koch (1984) simulated this behavior in an extended, unbranched cable with the quasi-active membrane of Fig. 10.5A. Injecting a current that peaks at 5 msec and decays within 20 msec (whose spectrum is monotonically decreasing and has decayed to 50% of its dc value at 70 Hz, the peak frequency of the transfer) the voltage recorded 350 μm away is qualitatively very similar to the temporal derivative of the current injected into the cable. The approximation is worse for faster synaptic input (since a larger fraction of its spectrum will lie beyond the peak at f_{max}) and better for slowly changing input. What is appealing about this linear operation is that it can be implemented in a very compact manner in a single cable.

10.4.3 Electrical Tuning in Hair Cells

Receptor cells in auditory and electroreceptive systems of many species show *electrical resonance*: upon appropriate sensory stimulation, their membrane potential oscillates. This has been explored in detail for hair cells in the amphibian auditory system. *Hair cells*, whose function is to translate the acoustic stimuli into electrical ones (for an incisive review of their biophysical properties, see Hudspeth, 1985), are organized tonotopically along the basilar membrane in the cochlea according to their characteristic frequency (the frequency of the acoustic stimuli the cell optimally responds to). Upon stimulation with a current step, the membrane potential responds with exponentially damped oscillations (Crawford and Fettiplace, 1981; Lewis and Hudspeth, 1983; see Fig. 10.9A). The oscillation frequency is a nonlinear function of the membrane potential. The frequency around the resting potential indicates the cell's natural frequency, that is, the frequency of sound or vibration to which the cell is most sensitive. In the bullfrog, this ranges from 80 to 160 Hz and in the turtle from 100 to about 700 Hz.

Hudspeth and Lewis (1988a,b) used the patch-clamp technique to identify the biophysical mechanism responsible for this resonance. They characterized two key currents, a noninactivating calcium current I_{Ca}, and a calcium- and voltage-dependent potassium current $I_{K(Ca)}$ (which they modeled using a linear, five-state kinetic scheme that requires two intracellular Ca^{2+} ions for the channel to open, and a third Ca^{2+} ion that can prolong the channel opening; Hudspeth and Lewis, 1988a). Using a Hodgkin–Huxley-like model, which also incorporates the capacitive and leak currents, they showed that electrical tuning could be explained in its majority by the interdependence of these currents (Hudspeth and Lewis, 1988b). Injection of a step current depolarizes the membrane, activating I_{Ca}. Calcium ions rush in and rapidly turn on $I_{K(Ca)}$. This outward current then brings the membrane potential down again, turning itself off in the process. If the current step persists, the cycle can now begin anew. Varying the kinetics and the density of the calcium-dependent calcium channels controls the frequency of resonance.

The input to hair cells is provided by activation of the mechanically sensitive cilia (hairs) that modulate a conductance in their membrane. If this conductance is included in the model and stimulated using sinusoidally varying acoustical input, the resulting tuning curve (Fig. 10.9B) closely mimics the electrical resonance frequency around the resting potential. We conclude that in the amphibian cochlea, hair cells behave like small electrical resonant elements—bandpass filters— whose tuning properties are due to the interplay of an inward and an outward current. Note that Hudspeth and Lewis (1988b) used a nonlinear model to explain the electrical resonance. Previous efforts to model hair cells using RLC circuits as described above (Crawford and Fettiplace, 1981; Art and Fettiplace, 1987) did, however, capture most of the qualitative aspects of these oscillations.

Have such resonating neuronal elements been found anywhere else? Llinás, Grace, and Yarom (1991), recording in layer 4 neurons in the cortical slice preparation, found that a constant current step evoked subthreshold oscillations in the 10–50 Hz frequency range. Because these cells are small, with smooth and aspiny dendrites, Llinás and coworkers argue that they are inhibitory interneurons. In contradistinction, other neurons in the cortex, termed *chattering cells*, show an oscillatory firing pattern in the 20–70 Hz band in response to visual stimuli or suprathreshold current injection, but no such oscillatory changes in the membrane potential in response to subthreshold current injections (Gray and McCormick, 1996). The function of these oscillations is at present not known, although much has been speculated on this topic (for an overview, see Koch, 1993).

10.5 Recapitulation

Although it can be argued that a linear analysis of a nonlinear phenomenon does not do justice to it, it will certainly help us to understand certain aspects of the mechanism underlying the phenomenon. This is true when considering certain resonant or oscillatory behaviors evident in nerve cells.

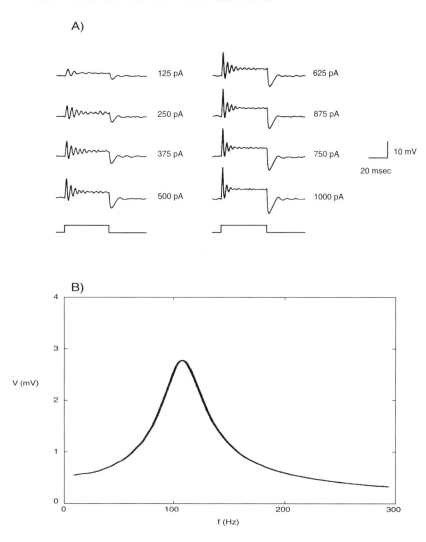

Fig. 10.9 ELECTRICAL RESONANCE IN HAIR CELLS Hair cells in the cochlea of amphibians behave like electrical band-pass elements, preferentially responding to an acoustic stimulus of a particular frequency. (**A**) Membrane potential recorded in a solitary hair cell of the bullfrog (Hudspeth and Lewis, 1988b). Upon injection of a 50-msec-long current step, the membrane responds by generating exponentially damped oscillations whose frequency varies with the membrane potential. The frequency (93 Hz) around the resting potential (-49 mV) determines the frequency of sound or vibration the cell is most sensitive to. As the average membrane potential increases to -41 mV (by using larger current steps, whose amplitude is indicated to the right of each trace), the frequency of the damped oscillations increases to 240 Hz. Reprinted by permission from Hudspeth and Lewis (1988b). (**B**) The acoustic stimulus couples to a mechanically triggered membrane conductance. If this conductance is added to a model that includes a calcium and a calcium-dependent potassium current, it is sufficient to explain the resonant property of isolated hair cells. Here the peak-to-peak amplitude of the receptor potential in response to a sinusoidal ±20-nm displacement of the hair bundle is plotted as a function of the stimulus frequency. Thus, each individual cell acts like a bandpass element, here optimally sensitive to input around 112 Hz. Reprinted by permission from Hudspeth and Lewis (1988b).

As Detwiler, Hodgkin and McNaughton (1980) pointed out, a time- and voltage-dependent potassium current that is activated by depolarization or an inward current that is activated by hyperpolarization (as the h part of the fast sodium current) does behave, under certain restricted conditions, like an inductance in series with a conductance would. That is, one can mimic the small-signal behavior of such a current by an electrical circuit that includes such an inductance. This explains why EPSPs or the depolarization in response to even very small current injections will be followed by a hyperpolarizing overshoot (provided the experimental setup is sensitive enough to record such small voltages superimposed onto the background noise). The membrane impedance of a membrane that includes resistances, capacitances, as well as phenomenological inductances (RLC circuit) peaks at some nonzero resonant frequency f_{max}. Such a quasi-active membrane with a bandpass response will cause the membrane potential to show damped oscillations at or close to the resonant frequency f_{max} in response to a current step. This has been confirmed empirically when recording the response of the squid giant axon to small, subthreshold current steps.

One consequence of a bandpass-like membrane impedance is that the transfer impedance and the frequency-dependent space constant associated with an infinite cable covered by such a quasi-active membrane can also show resonant behavior. This implies that input signals with frequencies around f_{max} are treated preferentially in terms of a smaller voltage attenuation to the soma (or other measures of synaptic efficiency) than inputs at faster or slower frequencies: the cable is spatio-temporally tuned to some frequency range. If the transfer impedance of a cable with a quasi-active membrane can be approximated by a bandpass, injecting a current whose dominant signal content lies in the spectrum between dc and f_{max} induces a voltage change that approximates the continuous temporal derivative of the injected current. In other words, a small piece of quasi-active dendritic cable can implement a temporal derivative operation.

Hudspeth and Lewis (1988b) used a nonlinear, squid axonlike membrane description of a calcium and a calcium-dependent potassium current (in series with leak and capacitive currents) to model the electrical behavior of bullfrog hair cells in the cochlea. Confirming earlier linear RLC analyses, each hair cell by itself acts as an electrical resonant element best tuned to respond to sounds in the f_{max} frequency band. The kinetics and density of $I_{K(Ca)}$ control f_{max} as well as the tuning of the bandpass. The morale is that individual neurons can implement a variety of different linear and nonlinear computational operations on the basis of the cornucopia of known membrane currents.

11

DIFFUSION, BUFFERING,
AND BINDING

In Chap. 9 we introduced calcium ions and alluded to their crucial role in regulating the day-to-day life of neurons. The dynamics of the free intracellular calcium is controlled by a number of physical and chemical processes, foremost among them *diffusion* and *binding* to a host of different proteins, which serve as calcium *buffers* and as calcium *sensors* or *triggers*. Whereas buffers simply bind Ca^{2+} above some critical concentration, releasing it back into the cytoplasm when $[Ca^{2+}]_i$ has been reduced below this level, certain proteins— such as calmodulin—change their conformation when they bind with Ca^{2+} ions, thereby activating or modulating enzymes, ionic channels, or other proteins.

The calcium concentration inside the cell not only determines the degree of activation of calcium-dependent potassium currents but—much more importantly—is relevant for determining the changes in structure expressed in synaptic plasticity. As discussed in Chap. 13, it is these changes that are thought to underlie learning. Given the relevance of second messenger molecules, such as Ca^{2+}, IP_3, cyclic AMP and others, for the processes underlying growth, sensory adaptation, and the establishment and maintenance of synaptic plasticity, it is crucial that we have some understanding of the role that diffusion and chemical kinetics play in governing the behavior of these substances.

Today, we have unprecedented access to the spatio-temporal dynamics of intracellular calcium in individual neurons using *fluorescent calcium dyes*, such as fura-2 or fluo-3, in combination with *confocal* or *two-photon microscopy* in the visible or in the infrared spectrum (Tsien, 1988; Tank et al., 1988; Hernández-Cruz, Sala, and Adams, 1990; Ghosh and Greenberg, 1995). The data gleaned from these methods of directly visualizing calcium in the dendritic tree, cell body, and presynaptic terminals are quantitative enough to allow detailed comparison with numerical models, commencing with the work of Hodgkin and his colleagues in the squid axon (Blaustein and Hodgkin, 1969; Baker, Hodgkin, and Ridgway, 1971) and continuing with the study of calcium transients in the presynaptic terminal (Llinás, Steinberg and Walton, 1981a,b; Zucker and Stockbridge, 1983; Stockbridge and Moore, 1984; Simon and Llinás, 1985; Zucker and Fogelson, 1986; Parnas, Hovav, and Parnas, 1989; Yamada and Zucker, 1992) as well as in dendrites, cell bodies and spines (Connor and Nikolakopoulou, 1982; Gamble and Koch, 1987; Sala and Hernández-Cruz, 1990; Holmes and Levy, 1990; Carnevale and Rosenthal, 1992; Zador and Koch, 1994; DeSchutter and Bower, 1994a,b; DeSchutter and Smolen, 1998; Borg-Graham, 1998; Yamada, Koch, and Adams, 1998).

In this chapter, we study the physics of diffusion and buffering, dwelling at some length on the fundamental limitations imposed by physics on how fast substances can diffuse. We will develop certain analogies between the reaction-diffusion equation and the cable equation and discuss the potential relevance of these processes for information processing and computation. While these results apply to any substance diffusing inside cells and binding to other molecules, we will focus on Ca^{2+} ions, given their crucial role in the day-to-day life of neurons and, in fact, in all cells. The control and regulation of calcium in neuronal structures is a vast subject, which we can only briefly hint at. For more details on calcium signaling in neurons the reader is urged to consult Hille's monograph (1992), the articles by Ghosh and Greenberg (1995) and by Clapham (1995), or the insightful Meyer and Stryer (1991) review. Finally, for the computational methods, consult DeSchutter and Smolen (1998). We end by discussing some general aspects of information processing using calcium.

Let us commence by deriving the diffusion equation that underlies the distribution of ions inside and outside cells.

11.1 Diffusion Equation

Historically, the diffusion equation was first derived from a macroscopic, deterministic point of view by Fick (1855), who postulated a force arising between molecules proportional to their concentration difference. Although such a phenomenological approach does lead to the correct equation, we now know that no such force exists. Instead, diffusion is due to the constant agitation of molecules arising from their thermal energy. The physically correct probabilistic interpretation of this *Brownian motion* was advanced by Einstein (1905), who used the theory of random walk. Because of the simplicity and elegance of his arguments, we will develop here a heuristic derivation of the diffusion equation using this stochastic perspective.

11.1.1 Random Walk Model of Diffusion

The starting point of the molecular theory of heat is that molecules and small particles at the absolute temperature T have, on average, a kinetic energy $kT/2$ associated with movement along each dimension. The resulting velocities of these particles in solution do not allow them to travel very far since they collide within picoseconds with other molecules, changing their direction of travel as a result. This incessant and random agitation of molecules due to thermal energy is the basis of diffusion.

Imagine a particle moving along an one-dimensional line. Both space and time are discretized, such that the particle can only be at locations $\pm m\Delta x$, where m is an integer. Each time step Δt, the particle can move either with 50% probability to the right or with 50% probability to the left, in both cases by the amount Δx. The particle cannot remain in place or move by more than Δx. Finally, we assume that the particle has no memory of any locations it has previously visited.

These rules allow us to define a probability distribution $p(x, t)$ that the particle will be at position x at time t. The only way the particle could have arrived at position x is if it was at either one of the two adjacent positions in the previous time step. That is,

$$p(x, t) = \frac{1}{2} \left(p(x - \Delta x, t - \Delta t) + p(x + \Delta x, t - \Delta t) \right). \tag{11.1}$$

Subtracting $p(x, t - \Delta t)$ from both sides as well as dividing by Δt yields

$$\frac{p(x, t) - p(x, t - \Delta t)}{\Delta t} =$$

$$\frac{(\Delta x)^2}{2 \Delta t} \frac{p(x - \Delta x, t - \Delta t) - 2p(x, t - \Delta t) + p(x + \Delta x, t - \Delta t)}{(\Delta x)^2} \tag{11.2}$$

The left-hand side of this equation is nothing but the first-order approximation of the first temporal derivative of p, while the term on the right-hand side corresponds to the second-order approximation of the second spatial derivative of $p(x, t - \Delta t)$. In the limit as Δx and Δt approach zero in such a way that

$$\frac{(\Delta x)^2}{2 \Delta t} = D \tag{11.3}$$

converges toward a finite constant (with units of cm^2/sec), we end up with the partial differential *diffusion equation*

$$\frac{\partial p(x, t)}{\partial t} = D \frac{\partial^2 p(x, t)}{\partial x^2}. \tag{11.4}$$

It describes how the probability function $p(x, t)$ evolves over time and is also known under the generic name of *Fokker-Planck equation*. We will reencounter this class of equations in Chap. 15.

The simplest interpretation of diffusion can be given if the single particle is replaced by a cloud of N independently moving particles that all start off at the origin, $x = 0$. What is the expected mean position of this cloud of particles and the variance around this mean position? According to our simple model, $x_i(t)$ for particle i at time $t = n \Delta t$ can only differ from the previous position by $\pm \Delta x$,

$$x_i(n \Delta t) = x_i((n - 1) \Delta t) \pm \Delta x. \tag{11.5}$$

The average position of the cloud, $\langle x(n \Delta t) \rangle$, at time $n \Delta t$ is given by summing over all the individual particles,

$$\langle x(n \Delta t) \rangle = \frac{1}{N} \sum_{i=1}^{N} x_i(n \Delta t) = \frac{1}{N} \sum_{i=1}^{N} [x_i((n - 1) \Delta t) \pm \Delta x]. \tag{11.6}$$

Since we assumed earlier that each particle has an equal chance of moving to the left or to the right (and that no boundaries impede the motion of the particles), the last term in brackets averages zero for large numbers of particles. This equation then tells us that the mean position of the cloud at time $n \Delta t$ does not change from its previous position, $\langle x(n \Delta t) \rangle = \langle x((n - 1) \Delta t) \rangle$. If the cloud starts off at the origin, we conclude that the average location of the cloud will never move,

$$\langle x(n \Delta t) \rangle = 0. \tag{11.7}$$

If we incorporate a systematic drift into the motion of the particles, for instance, by applying an electric field that imposes a preferred direction of motion (assuming that the moving particle is charged), this result will have to be modified. However, in the absence of such a bias, the cloud will remain centered at the origin.

How much does the cloud spread out over time? As anybody can observe when placing a drop of ink into a water glass, at first a small part of the water is stained an intense blue. Later on, the spot of blue becomes less intense and larger, dispersing eventually throughout the entire glass. One measure of this spreading is the variance of the mean position of the cloud. This is defined as

$$\langle x(n\Delta t)^2 \rangle = \frac{1}{N} \sum_{i=1}^{N} x_i (n\Delta t)^2$$

$$= \frac{1}{N} \sum_{i=1}^{N} [x_i((n-1)\Delta t)^2 \pm 2\Delta x \cdot x_i((n-1)\Delta x) + (\Delta x)^2] \tag{11.8}$$

(we here used Eq. 11.5). Because the random walk is unbiased, the second term in this equation will be zero on average, resulting in

$$\langle x(n\Delta t)^2 \rangle = \langle x((n-1)\Delta t)^2 \rangle + (\Delta x)^2 . \tag{11.9}$$

We can, of course, recursively apply the same decomposition for $\langle x((n-1)\Delta t)^2 \rangle$, ending up in n steps with $\langle x(0)^2 \rangle$. Since the entire population started out at the origin, this last term in zero. Setting $t = n\Delta t$ and remembering that $(\Delta x)^2/(2\Delta t) = D$, we conclude

$$\langle x(t)^2 \rangle = 2Dt . \tag{11.10}$$

Thus, while the mean position of the cloud does not move, the variance or width of the cloud increases linearly with time and its standard deviation as the square root of time. Interpreted in terms of single particles, Eq. 11.7 implies that the mean position of a single particle does not change while the probability of finding it at larger and larger distances from the origin increases as the square root of time (Eq. 11.10).

11.1.2 Diffusion in Two or Three Dimensions

The same concept can be applied to the motion of a particle in more than one dimension. As long as the motion along the x direction is independent of the motion in the y direction, the variance of motion in the plane is the sum of the variances of motion in the two independent directions,

$$\langle r(t)^2 \rangle = \langle x(t)^2 \rangle + \langle y(t)^2 \rangle = 4Dt . \tag{11.11}$$

A computer simulation of such a random walk is illustrated in Fig. 11.1. Given the square-root relationship between time and distance traversed, it is possible for the particle to explore short distances much more thoroughly than large distances. Thus, the typical pattern evident in Fig. 11.1: the particle tends to return to the same region many times before eventually wandering away. When it does move away, it blindly chooses another region to explore, with no regard for whether or not it had previously visited that area. This reflects the fact that the particle does not move down any gradient in response to a "diffusive" force as postulated by Fick (1855), but acts in a probabilistic manner to spread over the available space; its tracks do not uniformly fill up the available space. For a particle diffusing in three dimensions, exactly the same principle applies, except that now

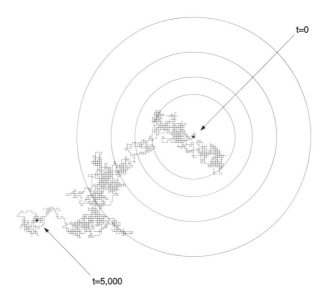

Fig. 11.1 **DIFFUSION OF A PARTICLE IN THE PLANE** Computer simulation of a single particle executing a random walk on a rectangular grid. At each point in time, it has a probability of 0.25 of moving in any one of four directions. In the limit of an infinitesimally fine grid, this approximates diffusion in a plane. The particle started at the center and after 5000 iterations ended up at the lower left corner. The radii of the four circles correspond to the variance in the locations of the particle after 500, 1000, 2000, and 4000 iterations. The expected time-averaged location of the particle is always at the center of the bull's eye.

$$\langle r(t)^2 \rangle = \langle x(t)^2 \rangle + \langle y(t)^2 \rangle + \langle z(t)^2 \rangle = 6Dt \,. \tag{11.12}$$

The *random walk* approach to diffusion has been very fruitful in mathematical physics. It is straightforward to derive equations corresponding to particles moving down a gradient caused, for instance, by an applied electric field, or if the space the particle moves in has absorbing or reflecting boundaries such as a membrane. For more details on the mathematics of random walk see the idiosyncratic, but immensely informative, monograph by Mandelbrot (1977); for a delightful text on the use of random walk techniques to describe the motion of bacteria and other small organisms see Berg (1983).

11.1.3 Diffusion Coefficient

In the above, we assumed—see Eq. 11.3—that D is purely dictated by the temporal and spatial discretization steps. However, for real particles in solution—where gravity can be neglected—D must clearly depend on certain physical attributes of the particles themselves as well as on properties of the solution. What precisely determines D and how large is it for real ions, such as Ca^{2+}, and big molecules, such as calmodulin which binds free Ca^{2+} in neurons?

Einstein (1905), arguing that diffusion is thermal agitation of particles opposed by friction, derived a general relationship between diffusion and friction. For particles that can be approximated as spheres much larger than the size of individual water molecules (which are on the order of 1 Å), D is determined by the *Einstein-Stokes relation* as

$$D = \frac{kT}{6\pi \eta r_S} \tag{11.13}$$

where r_S is the radius of the diffusing particle and η the viscosity of water. The proportionality of D to the absolute temperature T simply reflects the thermal origin of diffusion. For room temperature (20° C), this equation reduces to $D = \alpha/r_S$, with $\alpha = 2.15 \ \mu m^2/msec$ and the radius r_S specified in units of angstrom (Å). (For a very thorough discussion of this, see Hille, 1992.) The diffusion coefficient is usually specified in cm^2/sec; however, we will adopt units more convenient to neuronal structures and measure D in units of $\mu m^2/msec$.

The Einstein-Stokes relationship—based on classical hydrodynamics— is accurate for large particles, but works less well for small ions. For instance, for the 0.99 Å radius calcium ion, D should be 2.14 $\mu m^2/msec$ but is, in fact, 0.6 $\mu m^2/msec$ (Blaustein and Hodgkin, 1969). This mismatch is partially caused by the formation of a hydration shell around the ion, that is, water molecules that lie in direct contact with each ion in solution. For larger and heavier molecules, this is less of a consideration and their radius can be thought of as being approximately proportional to the cubic root of their molecular weight. Table 11.1 lists the diffusion coefficients of a number of ions and molecules for diffusion in an aqueous medium characteristic of the intracellular cytoplasm.

11.2 Solutions to the Diffusion Equation

Before we discuss the impulse response or Green's function of the diffusion equation, let us sketch out its macroscopic phenomenological derivation. The mathematical theory of diffusion in an isotropic milieu is based on the simplest phenomenological expression possible—also called Fick's first law of diffusion (Fick, 1855)—that the rate of transfer $S(x, t)$ of a diffusing substance across a surface of unit area (also called the flux; Eq. 9.2) is proportional to the concentration gradient measured normal to the surface,

$$S(x, t) = -D\frac{\partial C(x, t)}{\partial x} \tag{11.14}$$

TABLE 11.1
Diffusion Coefficients

Ion or molecule	Diffusion coefficient
H^+	9.3[1]
NO	3.8[2]
Na^+	1.33[1]
K^+	1.96[1]
Ca^{2+}	0.6[3]
IP_3	0.2[4]
Calmodulin	0.13
CaM kinase II	0.034

Diffusion coefficients in an aqueous environment for different ions and second messenger molecules in units of $10^{-5} \ cm^2/sec$, that is $\mu m^2/msec$.
[1] Table 10.1 in Hille (1992).
[2] In rat cortex; Meulemans (1994); see also Wise and Houghton (1968).
[3] Blaustein and Hodgkin (1969).
[4] Allbritton, Meyer, and Stryer (1992).

where D is the constant of proportionality and the minus sign arises because diffusion occurs against a concentration increase. Since the amount of substance diffusing across the boundary is proportional to the area of the boundary, the diffusion coefficient has dimensions of $\mu m^2/msec$.

Because this amount corresponds to a particular number of molecules diffusing, we refer to it using the colloquial, but entirely appropriate, term *stuff* for lack of a more specific term and express it in units of gram molecules or moles (mol) while the concentration C is given in molars (M), that is moles of molecules per liter.[1] The extracellular concentration of Ca^{2+} ions is about 2 mM, while its intracellular concentration at rest is around 10 to 20 nM, a difference of five orders of magnitude.

Consider a cylindrical cable where the concentration C only varies along the x dimension, reducing the problem of solving the diffusion equation from three dimensions to one. The justification of this is identical to the one we made for one-dimensional cable theory in Chap. 2, namely, that the radius of the cylinder is much shorter than its length and that the equilibrium time for radial diffusion is short compared to longitudinal diffusion.

If the concentration inside a compartment centered at x and with boundaries at $x+\Delta x$ and $x - \Delta x$ (Fig. 11.2) varies by $\partial C/\partial t$, the change in the amount of stuff in this compartment is given by

$$\frac{\Delta x \pi d^2}{2} \frac{\partial C(x,t)}{\partial t}.\tag{11.15}$$

This change should be identical to the net rate of transfer across one boundary minus the rate of transfer across the other,

$$\frac{\pi d^2}{4}(S(x - \Delta x, t) - S(x + \Delta x, t)).\tag{11.16}$$

Setting these two expressions equal leads to

$$\frac{\partial C(x,t)}{\partial t} = \frac{S(x - \Delta x, t) - S(x + \Delta x, t)}{2\Delta x}.\tag{11.17}$$

In the limit of an infinitesimally small interval ($\Delta x \to 0$), the expression $S(x + \Delta x, t) - S(x - \Delta x, t)$ converges to $2\Delta x\, \partial S(x,t)/\partial x$. Applying Eq. 11.14 we arrive once again at the *diffusion equation*

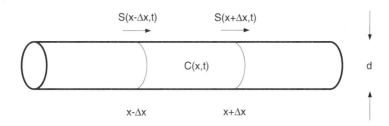

Fig. 11.2 **DIFFUSION IN A CYLINDER** Diffusion of some substance into and out of a cylindrical compartment with boundaries at $x - \Delta x$ and $x + \Delta x$. $S(x, t)$ corresponds to the rate of transfer across the cross section of area $\pi d^2/4$, and $C(x, t)$ corresponds to the concentration. As in one dimensional cable theory, under certain conditions the three-dimensional diffusion equation can be reduced to a one-dimensional one.

---

1. To remind the reader, one mole of water, that is, 18 grams molecules, corresponds to 6.023×10^{23} H_2O molecules. Its concentration is 1000 grams per liter divided by 18 grams per mole, that is 55 M.

$$\frac{\partial C(x, t)}{\partial t} = D\frac{\partial^2 C(x, t)}{\partial x^2}. \tag{11.18}$$

This is a linear partial differential equation of the parabolic type, with a unique solution if both an initial and a boundary condition are specified. We wish to point out already here, although we will not exploit this similarity until Sec. 11.7, that Eq. 11.18 is isomorphic to the linear cable equation (Eq. 2.7) if $r_m \to \infty$ (Fig. 11.3).

11.2.1 Steady-State Solution for an Infinite Cable

What is the steady-state behavior of the diffusion equation? If, in analogy with the cable equation, we "clamp" the concentration at the origin $C(x = 0, t)$ for all times to a fixed value C_0, Eq. 11.18 reduces to an ordinary differential equation,

$$\frac{d^2 C(x)}{dx^2} = 0. \tag{11.19}$$

Its solution for an infinite cable is $C(x) = C_0$ for all values of x. After enough time has passed, the concentration in the entire cable rises to the concentration at the origin. This behavior is in marked contrast to the exponential decay of the potential in response to a current step (Eq. 2.12) and is a consequence of the fact that none of the diffusing substance "leaks" out across the walls of the cylinder.

11.2.2 Time-Dependent Solution for an Infinite Cable

The simplest time-dependent solution is the one for the concentration change along an infinite one-dimensional cable if an amout S_0 of calcium ions is injected instantaneously

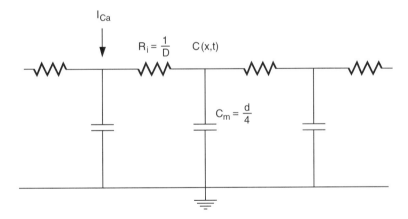

Fig. 11.3 EQUIVALENT ELECTRICAL CIRCUIT ASSOCIATED WITH THE DIFFUSION EQUATION Lumped electrical circuit representation associated with the diffusion of a substance in an elongated fiber. The input here is defined as $2I_{Ca}/Fd$, where d is the diameter of the process, F is Faraday's constant, and I_{Ca} is the calcium current flowing across the membrane. With the membrane capacity c_m set to 1 and $r_a = 1/D$, this circuit can be mapped onto the lumped electrical circuit approximating the cable equation (Fig. 2.3) in the absence of any membrane conductance (that is, $r_m \to \infty$). As we will show in Sec. 11.7, this analogy to the cable equation can be extended to the presence of a fast buffer and ionic pumps in the membrane. In the limit of infinitely small mesh size, the solution to this circuit approximates the continuous diffusion equation (Eq. 11.18).

into the cylinder at $x = 0$. We can show by simple differentiation that the resulting spatio-temporal evolution of C is given by

$$C_\delta(x, t) = \frac{S_0}{\sqrt{2\pi}} \frac{1}{(2Dt)^{1/2}} e^{-\frac{x^2}{4Dt}} .$$ (11.20)

The associated concentration profile is plotted in Fig. 11.4 for different values of t and x. In comparing this function with the impulse response function of the infinite, passive cable (Eq. 2.31 and Fig. 2.6), we notice the absence of an e^{-t} term (caused by the lack of a finite, membrane leak conductance for the diffusion equation). As a consequence, the total amount of substance is conserved, $\int C_\delta(x, t)dx = S_0$, at all times t. The absence of this exponential causes the Green's function for the diffusion equation to decay less rapidly than the corresponding Green's function for the cable equation. In particular, and not surprisingly, given our earlier result concerning the variance in the mean position of a cloud of randomly moving particles, the variance of the impulse response function increases linearly with time. We expand upon this point in the following section.

11.2.3 Square-Root Relationship of Diffusion

For any fixed time t, Eq. 11.20 can be expressed as a Gaussian, as can be seen upon inspection of Fig. 11.4A,

$$C_\delta(x, \sigma) = \frac{S_0}{\sigma\sqrt{2\pi}} e^{-\frac{x^2}{2\sigma^2}}$$ (11.21)

with

$$\sigma = \sqrt{2Dt} .$$ (11.22)

The variance of the Gaussian increases linearly with t, or the standard deviation increases with the square root of time. This has important consequences.

Let us consider the concentration in a semi-infinite cylinder, where the concentration at one end is held fixed: $C(x = 0, t) = C_0$ for all times. In the case of the cable equation, this would be equivalent to clamping the voltage at one end of the cable. We can solve for $C(x, t)$ directly by using either Fourier or Laplace transforms (Crank, 1975),

$$C(x, \sigma) = C_0 \, \text{erfc} \left(\frac{x}{\sigma\sqrt{2}} \right)$$ (11.23)

where $\text{erfc}(x)$ is the complementary error function defined as

$$\text{erfc}(z) = \frac{2}{\sqrt{\pi}} \int_{z}^{+\infty} e^{-y^2} dy$$ (11.24)

with $\text{erfc}(0) = 1$. The propagation of this concentration increase along the cylinder—illustrated in Fig. 11.5—involves only the dimensionless parameter $x/(\sigma\sqrt{2})$. Due to the square-root relationship between σ and t, it follows that the time required for any location to reach a given concentration is proportional to the square of the distance. If we ask at what time $t_{1/2}(x)$ the concentration at x reaches half of the source concentration, that is, $C_0/2$, we solve for t in Eq. 11.23, with its left-hand side set to $C_0/2$. By consulting tables for the error function (such as in Crank, 1975), we see that $\text{erfc}(0.5) \approx 1/2$ within a few percent. In other words, $x/(2\sqrt{Dt_{1/2}}) \approx 1/2$, or

$$t_{1/2} \approx \frac{x^2}{D} .$$ (11.25)

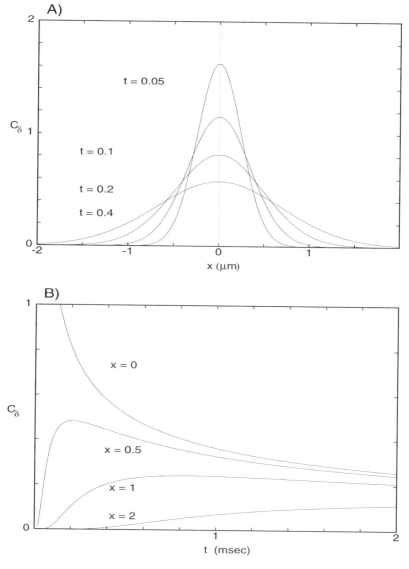

Fig. 11.4 IMPULSE RESPONSE OF THE DIFFUSION EQUATION IN AN INFINITE CABLE Concentration $C_\delta(x, t)$ in an infinite cylinder in response to an instantaneous injection of substance at $t = 0$ at the origin $x = 0$ as a function of space (**A**) or time (**B**; see Eq. 11.20). This Green's function decays more slowly than the Green's function of the linear cable equation (Eq. 2.31). As is clear from panel **A**, for any fixed time t the impulse response function can be described as a Gaussian, whose variance increases linearly with time.

Note that the approximation only involves the numerical factor in front of the square. We can reformulate this equation by stating that for a given duration t from the onset of the concentration step, the distance $x_{1/2}$ at which the concentration has reached half of its peak value is given by

$$x_{1/2} \approx \sqrt{Dt} \,.$$

(11.26)

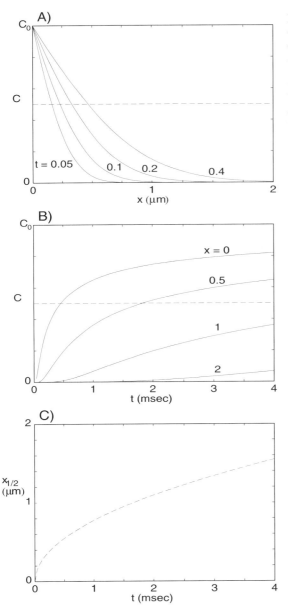

Fig. 11.5 THE SQUARE-ROOT LAW OF DIFFUSION At $t = 0$, the calcium concentration at the origin of a semi-infinite cable is clamped to C_0. **(A)** Evolving concentration profile along the cable. **(B)** As time goes by, the concentration throughout the cable slowly rises to C_0 (different from the solution of the cable equation). At any one instance, we can ask at what location $x_{1/2}$ along the cable the concentration first reaches $C_0/2$ (dashed line). This relationship, shown in **(C)**, is a square-root one. It takes four times as long to diffuse twice the distance (Eq. 11.26). This imposes a fundamental physical constraint on how fast any substance can diffuse in one or more spatial dimensions.

In three-dimensional space, the right-hand side needs to be multiplied by $\sqrt{3}$ (as in Eq. 11.12). If considering diffusion of ions within the porous and highly restricted extracellular space, further adjustments are necessary (see Sec. 20.2).

As an example, let us assume that the concentration of calcium at one end of a one-dimensional cylinder is clamped to $1\,\mu M$. Assuming that no buffer impedes the diffusion of the calcium ions along the cable (and that $D_{Ca} = 0.6\,\mu m^2/msec$), a calcium front—defined here as the time at which the concentration at any one location first reaches $0.5\,\mu M$—moving down the cylinder takes about 1 msec to cover the first 0.77 μm, 10 msec to propagate

2.5 μm away from the origin and has traveled only 7.7 μm after 100 msec (Fig. 11.5). In neuronal tissues, these times are considerably further reduced due to the binding of calcium to intracellular buffers (Sec. 11.4).

The linear diffusion equation, as a member of the family of parabolic differential equations which also includes the linear cable equation (Sec. 2.3), does not admit to any solution $C(x, t) = C(x - vt)$ that propagates with fixed velocity v along the cable. This behavior is in stark contrast to the linear relationship between distance and time for action potentials propagating down the axon or for the wave of calcium release occurring in eggs following fertilization, making the cells impervious to the entry of additional sperms.

The square-root behavior imposes a fundamental limitation on the time required for concentration changes of calcium or any other intracellular messenger to affect distant sites (in the absence of active transport systems) and is apparent everywhere. For instance, if we keep the concentration of calcium in the shell just below the membrane of a spherical cell constant, the center of the sphere will have reached one half the shell concentration after about $0.04d^2/D$ msec, where d is the diameter of the sphere (in micrometers). For additional solutions to the diffusion equation under various boundary conditions consult Crank (1975) and the monograph by Cussler (1984).

11.3 Electrodiffusion and the Nernst-Planck Equation

In the derivation of the cable equation in the second chapter, we had assumed that the concentration of ions does not vary along the longitudinal direction of the cable. Thus, ions are only propelled along the cable by the voltage gradient, giving rise to the $\partial^2 V/\partial x^2$ term. In the previous section, we discussed the effective motion of ions due to diffusion, down the concentration gradient.

In general, of course, we need to include the movement of ions caused by concentration differences as well as by drift along the electric field. Even if a dendrite is initially at equilibrium with respect to the spatial distribution of sodium, potassium, calcium, and chloride ions, which are most important for fast signaling, the influx and efflux of ions across the membrane disturb this equilibrium. This is particularly true for very small volumes, such as thin dendrites or spines where the influx of even a moderate amount of calcium significantly increases $[Ca^{2+}]_i$, creating concentration gradients that propel calcium ions down this gradient (in addition to any existing voltage gradient).

In order to account for these effects, we need to combine Fick's law (Eq. 11.14) with Ohm's law, something done by Nernst (1888, 1889) as well as Planck (1890). Assuming that the longitudinal current and ionic concentrations are uniform across the cross section of the cylindrical dendrite or axon and that the random, diffusional motion superimposes linearly onto the electrotonic motions of ions driven by a true "force," we can write the one-dimensional *Nernst-Planck electrodiffusion equation* (Hille, 1992),

$$I_{i,k}(x, t) = -z_k F D_k \frac{\partial C_k(x, t)}{\partial x} - \frac{z_k^2 F^2 D_k C_k(x, t)}{RT} \frac{\partial V_m(x, t)}{\partial x} \tag{11.27}$$

where $I_{i,k}(x, t)$ is the axial current for the ionic species being considered (here labeled k), z_k is its valence, D_k its diffusion coefficient, R the gas constant, and F Faraday's constant.

The $z_k F D_k$ term in front of the concentration gradient converts "stuff" into a current, while the constants in front of the potential gradient express implicitly how the electrical conductance relates to the concentration and the mobility of the ions (see also Eq. 11.13).

Thus, ions drift down the potential gradient, while simultaneously spreading due to diffusion. Hodgkin has remarked that while diffusion is like a hopping flea, electrodiffusion is like a flee that is hopping in a breeze.

As an aside, let us note that application of this kinetic equation in the direction perpendicular to the membrane directly yields Eq. 4.3 for the synaptic reversal or Nernst potential of an ionic current at equilibrium. Since the net current across the membrane must be zero, we can rearrange Eq. 11.27,

$$\frac{dV_m(r)}{dr} = -\frac{RT}{z_k F}\frac{1}{C_k(r)}\frac{dC_k(r)}{dr} = -\frac{RT}{z_k F}\frac{d}{dr}\log C_k(r) \tag{11.28}$$

with r the spatial variable across the membrane. Integrating this across the membrane yields the expression for the Nernst potential associated with each ionic species k.

Let us return to our main argument. Due to reasons of conservation, the change in concentration of the ions in an infinitesimally small cylindrical segment of length δx and diameter d must be balanced by the sum of the transmembrane current $i_{m,k}(x, t)$ per unit length (suitably weighted by the surface-to-volume ratio $4/d$; Fig. 11.2) and the difference between the ingoing and outgoing axial currents, or

$$\frac{4}{d}i_{m,k}(x, t) + \frac{\partial I_{i,k}(x, t)}{\partial x} + z_k F\frac{\partial C_k(x, t)}{\partial t} = 0. \tag{11.29}$$

Taking the spatial derivative of both sides of the electrodiffusion equation and extracting the expression for the change in concentration from Eq. 11.29 leads to

$$\frac{\partial C_k(x, t)}{\partial t} = -\frac{4}{z_k F d}i_{m,k}(x, t) + D_k\frac{\partial^2 C_k(x, t)}{\partial x^2} +$$

$$\frac{z_k F D_k}{RT}\frac{\partial}{\partial x}\left(C_k(x, t)\frac{\partial V_m(x, t)}{\partial x}\right). \tag{11.30}$$

The terms on the right-hand side of this nonlinear coupled partial equation correspond to the transmembrane current (given either by Ohm's law with the appropriate reversal potential or by the GHK current equation), the familiar diffusional term, and a voltage gradient term. In general, this equation must be solved for each species of ions that is present at relevant concentrations (assuming that different ionic species move independent of each other, something that may not always be true).

Since Eq. 11.30 is a single equation in two unknowns $C_k(x, t)$ and $V_m(x, t)$, it needs to be supplemented by an equation specifying the membrane potential, here expressing the fact that $V_m(t)$ is determined by the change in the total charge—added over all ionic species k weighted by their valence—divided by the membrane capacitance C_m in addition to an offset term, or

$$V_m(x, t) = V_{\text{rest}} + \frac{Fd}{4C_m}\sum_k z_k\left(C_k(x, t) - C_{k,\text{rest}}\right) \tag{11.31}$$

where $C_{k,\text{rest}}$ is the resting concentration of the kth ionic species.

In principle we need to solve Eqs. 11.30 and 11.31 to describe the dynamics of the membrane potential in extended cable structures properly (supplemented by an additional constraint at branching points; Qian and Sejnowski, 1989). However, this does not come cheaply. When linearly coupling two differential equations, the temporal discretization step

Δt required for an accurate evolution of the system is, in general, the smaller of the two time steps associated with the individual equations. In other words, the discretization necessary to solve Eq. 11.27 has to be much finer in both time and space than the discretization required for solving the cable equation with constant concentrations, resulting in much longer running times for the numerical algorithm.

11.3.1 Relationship between the Electrodiffusion Equation and the Cable Equation

In "large" neuronal processes, the intracellular concentration of various ions changes by relatively small amounts, implying that the diffusional contributions are negligible and that the longitudinal currents are purely resistive. It is straightforward to obtain the familiar cable equation as a special case of Eq. 11.30 by assuming that the axial ionic concentration gradients can be neglected, $\partial C_k(x, t)/\partial x \approx 0$. In combination with Eq. 11.29, Eq. 11.30 reduces to

$$C_m \frac{\partial V_m}{\partial t} + \sum_k i_{m,k}(x, t) = \frac{d}{4} \frac{F^2}{RT} \sum_k D_k z_k^2 C_k \frac{\partial^2 V_m}{\partial x^2} .\qquad(11.32)$$

This corresponds to the cable equation if we identify the intracellular resistance R_i with

$$R_i = \frac{RT}{F^2} \frac{1}{\sum_k D_k z_k^2 C_k} .\qquad(11.33)$$

Qian and Sejnowski (1989) compare numerical simulations of the propagating action potential in the squid giant axon (with a 0.476 mm diameter) using the electrodiffusion equation against the cable model solution of Cooley and Dodge (1966). With the diffusion coefficients for sodium and potassium reported in Table 11.1, they calculate R_i from Eq. 11.33, obtaining 33.4 $\Omega \cdot$cm for the potassium resistance and 267 $\Omega \cdot$cm for the sodium resistance, for a total value of $R_i = 29.7$ $\Omega \cdot$cm. This last value is reasonably close to the internal resistance of 35.4 $\Omega \cdot$cm used by Hodgkin and Huxley (1952d). The final solutions are indistinguishable from those of Cooley and Dodge (1966). Reducing the diameter several hundredfold to 1 μm only leads to a maximum relative concentration change of 1.4%, too small to have any significant effect. Thus, as in a metal wire, charged carriers[2] move under the influence of Ohm's law and diffusion does not play any (significant) role.

The concentration change due to synaptic input scales as $1/r$, implying that for submicrometer dimensions, diffusional transport of ions will contribute substantially to the total current. Chapter 12 deals with one such case, dendritic spines, where a single excitatory or inhibitory synaptic input can change the concentration of calcium or chloride significantly on a very rapid time scale. Here the predictions of the electrodiffusion model can deviate substantially from those of the cable model.

11.3.2 An Approximation to the Electrodiffusion Equation

From a physical point of view, the electrodiffusion equation constitutes a better description of current flow in a neuronal process than the cable equation. Thus, it behooves us to routinely use the former, rather than the latter, for simulating events in dendrites. Yet, because of the additional computational load imposed by solving Eq. 11.30 using a finer spatio-temporal discretization grid than the one required for solving the cable equation, almost

---

2. Electrons in one case and Na^+, K^+, and other ions in the other.

nobody has done this. Qian and Sejnowski (1989) offer a remedy for this by describing a fast approximation for solving the electrodiffusion equation in extended cable structures. It involves replacing the constant intracellular resistance R_i by batteries and resistances for each ionic species being considered (Fig. 11.6). At each time step, this algorithm

1. Calculates the intracellular concentration of each ionic species in each compartment by integrating over both the transmembrane and the intraaxial currents between compartments.

2. Computes the new value of the Nernst reversal potential for each ionic species for each compartment across the membrane (E_{Na} and E_K in Fig. 11.6).

3. Replaces the single intracellular resistance R_i of the cable equation with individual longitudinal resistances,

$$R_{i,k} = \frac{RT}{F^2} \frac{1}{D_k z_k^2 C_k} \tag{11.34}$$

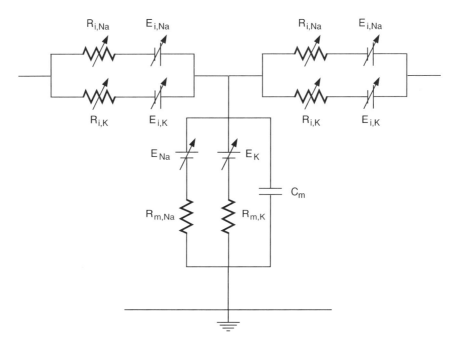

Fig. 11.6 APPROXIMATING THE ELECTRODIFFUSION EQUATION Solving the electrodiffusion equation, a better description of the transport of charged ions than the cable equation, is computationally more demanding than solving the cable equation (due to the higher spatial and temporal discretization required). Qian and Sejnowski (1989) advocate the usage of an approximation to Eq. 11.30, illustrated here. The crucial step is to replace (at each time step and for each compartment) the single axial resistance R_i of the cable equation (Fig. 2.3) with a resistance $R_{i,k}$ and battery $E_{i,k}$ for each ionic species being considered. The transmembrane reversal batteries E_{Na} and E_K are given by the Nernst equation (Eq. 4.3). This procedure is illustrated for the thick squid giant axon using a symmetrical discretization procedure with $R_{i,Na} = 267\ \Omega\cdot cm$ and $R_{i,K} = 33.4\ \Omega\cdot cm$. Given the tiny changes in intracellular sodium and potassium concentration, $E_{i,Na} \approx E_{i,K} \approx 0$. Even for a much thinner axon of $1\ \mu m$-diameter, the maximal value of these batteries is 0.4 mV. Only for very thin distal dendrites or spines will qualitative differences to the cable model appear. Reprinted in modified form by permission from Qian and Sejnowski (1989).

and batteries between compartment j and $j + 1$,

$$E_{i,k} = \frac{RT}{F z_k} \log \frac{C_k(j)}{C_k(j + 1)} \tag{11.35}$$

for each ionic species k.

For most situations that Qian and Sejnowski considered, this algorithm agrees well with the solutions to the full electrodiffusion equation, only doubling execution times in comparison to solving the cable equation. When in doubt whether or not diffusion of ions will affect the solution of the cable equation, we would recommend this algorithm to the reader (for more details, see Qian and Sejnowski, 1989).

11.4 Buffering of Calcium

In a classical experiment, Hodgkin and Keynes (1957) used radioactive ^{45}Ca to track the diffusion of calcium in squid axon fibers. From the observed broadening of the radioactive patches the effective diffusion constant was estimated to be about one-tenth of the diffusion coefficient D_{Ca} in aqueous solution. This is in contrast to the behavior of potassium ions under similar circumstances. This showed that once calcium enters the intracellular cytoplasm it is not free to diffuse. Indeed, 95% and more of the entering calcium is quickly bound by a host of systems, ranging from a set of different protein buffers to cellular organelles, such as mitochondria and the smooth endoplasmic reticulum. Mitochondria have a considerable ability to accumulate (and release) Ca^{2+}. The endoplasmic reticulum in neurons—related to the sarcoplasmic reticulum in skeletal muscle cells, which is responsible for the release and subsequent reuptake of Ca^{2+} during muscle contractions—also accumulates calcium, yet at a relatively slow rate. Since the calcium uptake of these organelles takes place on a time scale of seconds and longer (Rasgado-Flores and Blaustein, 1987), we focus our discussion on the dynamics of calcium binding to protein buffers. (For an overview of neuronal calcium homeostasis, see Carafoli, 1987; McBurney and Neering, 1987; Blaustein, 1988; Clapham, 1995.)

A large number of Ca^{2+}-binding proteins, such as calmodulin, calbindin, and parvalbumin, are present at high concentrations in nerve cells. The most important neuronal calcium buffer, *calmodulin*, a 15,000 dalton regulatory protein, is present in brain tissue at a concentration of 30–50 μM and acts as an internal *calcium sensor* (Manalan and Klee, 1984; see Fig. 11.7). It is located in the cell bodies, dendrites, and postsynaptic densities of most neurons in the central nervous system, but not in axons. Each calmodulin (CaM) molecule has four Ca^{2+}-binding sites. At resting levels of calcium concentration, none or only one of these sites is occupied. As the concentration of free calcium rises to micromolar levels, the four binding sites are occupied successively. The fully bound calcium-calmodulin complex in turn can bind to a large number of regulatory proteins to alter their function, such as calmodulin-dependent protein kinases (CaM kinases), protein phosphatases, and adenylate cyclases.

These enzymes, some of which have been implicated in the induction of long-term potentiation (Miller and Kennedy, 1986; Kennedy, 1989, 1992; Ghosh and Greenberg, 1995), can trigger the modification of other synaptic proteins locally at the synapse or can mediate more general cellular responses by activating molecules involved in the regulation of gene expression. As the concentration of calcium in the cytoplasm drops, the calcium ions are progressively released from the buffer and are free to wander about. The resulting

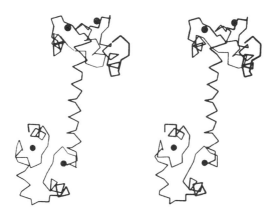

Fig. 11.7 STRUCTURE OF CALMODULIN Calmodulin is the most important calcium receptor protein. It is ubiquitous in eukaryotic cells and is found in the brain at concentrations of 30–50 μM. The 15,000-dalton protein can bind four Ca^{2+} ions in a cooperative manner with dissociation constants in the μM range. In the absence of calcium, the protein has some structure. Binding of Ca^{2+} to all four sites induces a conformational change, leading to a dumbbell-shaped molecule. Two heads, each of which contains the two bound calcium ions, are interconnected by a long α chain. In this form, the calcium-calmodulin complex can interact with a large number of other proteins, giving rise to a complex regulatory network. Reprinted by permission from Babu et al., (1985).

changes, which were triggered by the initial binding of Ca^{2+} ions to calmodulin, can outlast, by far, the calcium transient (Sec. 20.1). Before we come to this, let us first quantify the dynamics of various types of binding.

11.4.1 Second-Order Buffering

In one of the simplest chemical reactions possible, a single calcium ion binds to a single buffer molecule B, resulting in a bound buffer-calcium complex B·Ca. If the *forward binding rate* with which this reaction proceeds is f (in units of bindings per second per molar) and the *backward rate* is b (in units of bindings per second), we have

$$\text{B} + \text{Ca}^{2+} \quad \rightleftharpoons \quad \text{B} \cdot \text{Ca} . \tag{11.36}$$

In terms of associated kinetic equations, we can write

$$\frac{d[\text{Ca}^{2+}]}{dt} = b[\text{B} \cdot \text{Ca}] - f[\text{B}][\text{Ca}^{2+}]$$

$$\frac{d[\text{B}]}{dt} = b[\text{B} \cdot \text{Ca}] - f[\text{B}][\text{Ca}^{2+}] \tag{11.37}$$

$$T_{\text{B}} = [\text{B} \cdot \text{Ca}] + [\text{B}]$$

where [B] and [B·Ca] denote the concentrations of the buffer and the buffer-calcium complex. The last equation expresses the fact that only a fixed amount T_{B} of buffer molecules exists, where T_{B} corresponds to the total amount of the buffer. f and b tell us something about the speed with which the buffering reaction occurs. The above type of reaction is known as a *second-order reaction*, since two substances participate, each with 1 mol, resulting in a quadratic term in the associated kinetic Eq. (11.37) (Pauling and Pauling, 1975).

Setting the temporal derivative to zero yields the steady-state distribution of the buffer-calcium complex,

$$[B \cdot Ca] = \frac{T_B[Ca^{2+}]}{K_d + [Ca^{2+}]} \tag{11.38}$$

where $K_d = b/f$ is the *dissociation constant* of the buffer expressed in molars. If the calcium concentration has reached K_d, exactly half the available buffer is bound to calcium. The lower K_d, the lower the calcium concentration at which the buffer begins to bind Ca^{2+} ions. K_d therefore tells us something about the *affinity* of the buffer. The K_d of most calcium binding proteins is in the low to mid micromolar range.

If the calcium concentration is much less than K_d, Eq. 11.38 can be approximated by

$$[B \cdot Ca] \approx [Ca^{2+}]\frac{T_B}{K_d} . \tag{11.39}$$

At these low calcium concentrations $[B \cdot Ca]$ is proportional to $[Ca^{2+}]$. We will explore the consequences of this in Sec. 11.7.

For $[Ca^{2+}] < K_d$, the Ca^{2+} *binding ratio*, that is the ratio of bound calcium to free calcium, is given by

$$\beta = \frac{[B \cdot Ca]}{[Ca^{2+}]} = \frac{T_B}{K_d} . \tag{11.40}$$

This number is usually large, upward of 20 (Allbritton, Meyer, and Stryer, 1992; Neher, 1995; Gabso, Neher, and Spira, 1997), implying that 95% or more of all calcium is bound to intracellular buffers. Clearly the cell cares a great deal about regulating the amount of free calcium ions sloshing around the neuron.

11.4.2 Higher Order Buffering

As mentioned above, the ubiquitous calcium buffer protein calmodulin has four calcium binding sites on each molecule. A general four-site system is defined by more than a dozen rate constants. In particular, binding calcium to one site can induce a conformational change of the molecule, affecting the binding of calcium to the remaining sites. Calmodulin appears to display positive *cooperativity*, such that binding of calcium to the first two binding sites increases the affinity for the remaining two sites. Furthermore, the binding constants for the four sites differ significantly from each other. For the sake of simplicity, we will assume that all four sites are independent of each other and have identical binding constants. The K_d of calmodulin is on the order of 10 μM (Manalan and Klee, 1984; Klee, 1988), such that at 40–50 μM levels of Ca^{2+}, most of the calcium-binding sites will be occupied. For n binding sites, we can write down a series of n second-order equations of the from

$$B \cdot Ca_{m-1} + Ca^{2+} \rightleftharpoons B \cdot Ca_m \tag{11.41}$$

with $m = 1, 2, \cdots, n$. By repetitively applying the associated kinetic equations, we can express the steady-state concentration of all intermediate species and of the fully bound buffer-calcium complex as

$$[B \cdot Ca_n] = T_B \frac{([Ca^{2+}]/K_d)^n}{1 + \sum_{m=1}^{n}([Ca^{2+}]/K_d)^m} \tag{11.42}$$

where T_B is the total concentration of the buffer in its various guises.

In the case of calmodulin, the forward rate constant f is about 50 per second and per micromolar, the backward rate b is about 500 per second, implying a dissociation constant

$K_d = 10\,\mu M$. Therefore, as long as $[Ca^{2+}]$ is less than a few micromolar, the denominator in Eq. 11.42 is close to 1 and the steady-state concentration of the fully bound calmodulin complex will be proportional to the fourth power of calcium. Each additional binding step to another protein or enzyme will potentiate or sharpen the polynomial relationship between $[Ca^{2+}]$ and the buffer, since n increases by 1 in the above equation.

Computationally, higher order binding processes can be thought of as a squaring (for $n = 2$) or an nth-order polynomial operation (Koch and Poggio, 1992). Gamble and Koch (1987) showed numerically that a fast burst of action potentials (10 spikes at 333 Hz) to a synapse located on a spine elevates intracellular calcium in the spine head by about a factor of 5 in comparison to the peak calcium evoked after 10 spikes at 50 Hz (see also Chap. 12). This difference in levels of free calcium in the low- versus high-frequency input situation is amplified 1000-fold if the concentration of fully bound calmodulin is considered; in the first case 1000 times more $[CaM \cdot Ca_4]$ is evoked than in the latter case (roughly about a factor of 5^4). If a certain level of concentration of some critical substance is required to initiate some reaction, multiple binding steps will tend to lead to an all-or-none threshold behavior.

11.5 Reaction-Diffusion Equations

Let us now combine the diffusion equation for calcium with the kinetic equations expressing the binding of calcium to a single buffer via a second-order reaction. We assume that the buffer itself is stationary and does not diffuse and denote the intracellular calcium concentration as $[Ca^{2+}](x, t)$. The equations governing the evolution of the calcium and the bound buffer concentration, frequently referred to as *reaction-diffusion* equations, are of the form

$$\frac{\partial [Ca^{2+}](x, t)}{\partial t} = D \frac{\partial^2 [Ca^{2+}](x, t)}{\partial x^2} - \frac{\partial [B \cdot Ca](x, t)}{\partial t}$$

$$\frac{\partial [B \cdot Ca](x, t)}{\partial t} = f[B](x, t)[Ca^{2+}](x, t) - b[B \cdot Ca](x, t) \qquad (11.43)$$

$$T_B = [B \cdot Ca] + [B]$$

As the reader can see, these coupled partial differential equations are nonlinear, due to the fact that we assumed cooperative binding between calcium and buffer (expressed in the $[B] \cdot [Ca^{2+}] = (T_B - [B \cdot Ca]) \times [Ca^{2+}]$ term), making their analytical solution difficult (Chap. 14 in Crank, 1975, catalogues the major known solutions). If additional buffers are included or calcium binds to calmodulin or other proteins with multiple binding sites, further equations are required.

Coupled reaction-diffusion equations are well known in developmental biology. They were first invoked by the computer science pioneer Turing (1952) to explain the regular structures in biological systems and have been used to explain the stripes of the zebra and of certain fish (Kondo and Asai, 1995) and aspects of embryonic development in frogs and fruit flies (Kauffman, 1993; Meinhardt, 1994).

A particularly simple and insightful solution to Eq. 11.43 can be obtained if the concentration of the buffer-calcium complex is proportional to the concentration of calcium. This occurs if the binding of calcium to the buffer is very fast compared to the time scale of diffusion *and* if the calcium concentration is much less than the K_d of the buffer (*instantaneous buffer model*).

The first condition implies that the buffer is always at equilibrium relative to the diffusional time scale, while the second condition implies that $[B \cdot Ca] = \beta[Ca^{2+}]$ (Eq. 11.39, where β is defined in Eq. 11.40). Both assumptions allow us to reduce the nonlinear reaction-diffusions Eqs. 11.43 to a single linear equation with

$$\frac{\partial[Ca^{2+}](x,t)}{\partial t} = D\frac{\partial^2[Ca^{2+}](x,t)}{\partial x^2} - \beta\frac{\partial[Ca^{2+}](x,t)}{\partial t} \qquad (11.44)$$

and, therefore,

$$\frac{\partial[Ca^{2+}](x,t)}{\partial t} = \frac{D}{1+\beta}\frac{\partial^2[Ca^{2+}](x,t)}{\partial x^2}. \qquad (11.45)$$

In other words, in the presence of fast buffering dynamics and relatively small amounts of calcium, the spatio-temporal dynamics of calcium ions are governed by the canonical diffusion equation, except that the original diffusion coefficient is replaced by a smaller one,

$$D_{\text{eff}} = \frac{D}{1+\beta} = \frac{D}{1 + T_B/K_d}. \qquad (11.46)$$

Diffusion is slowed down, since the Ca^{2+} ions bind to the buffer and are therefore not available to diffuse. In the case of intracellular cytoplasm containing calmodulin at $T_B = 100\ \mu M$ (with K_d in the 5–10 μM range), the effective diffusion coefficient D_{eff} for calcium is at least 10 times slower than the coefficient measured in aqueous solution, in agreement with the Hodgkin and Keynes (1957) experiment mentioned. Due to the square-root relationship between time and distance, this translates into a substantially reduced ability of calcium to act as a fast intracellular messenger if distances larger than a few micrometers are involved.

We conclude that a nondiffusible buffer will always slow down the diffusive spread of calcium.

11.5.1 Experimental Visualization of Calcium Transients in Diffusion-Buffered Systems

One system where the observed calcium dynamics have been compared with numerical calculations are bullfrog sympathetic ganglion cells. These relatively large, spherical cells devoid of dendrites (described in Sec. 9.5) are an ideal test bed for applying confocal laser-scanned microscopy to record the dynamics of intracellular calcium using the fluorescent Ca^{2+} indicator fluo-3. In a pioneering application of this technology, Hernández-Cruz, Sala, and Adams (1990) measured $[Ca^{2+}]_i$ in these cells after application of a voltage-clamp pulse to briefly activate voltage-dependent calcium channels. Because of tradeoffs between temporal and spatial resolution of this method, they recorded the free calcium concentration across a narrow one-dimensional slot that extends across the width of the cell, thereby achieving a 5 msec temporal resolution (Fig. 11.8).

The influx of calcium across the membrane leads to a wave of calcium (Fig. 11.8A and E). Although much of this calcium becomes bound to the resident buffers as well as to the calcium indicator dye, enough remains for the calcium wave to reach the center of the cell after about 300 msec. $[Ca^{2+}]_i$ then equilibrates across the cell (Fig. 11.8F, G, and H), slowly returning to its resting state after 6–8 sec. These basic features were qualitatively reproduced by a radial reaction-diffusion model (Sala and Hernández-Cruz, 1990). It is similar to the model of Yamada, Koch, and Adams (Sec. 9.5) except that it makes no attempt to simulate the voltage dynamics and that it includes *mobile buffers*, that is, buffers that—with or

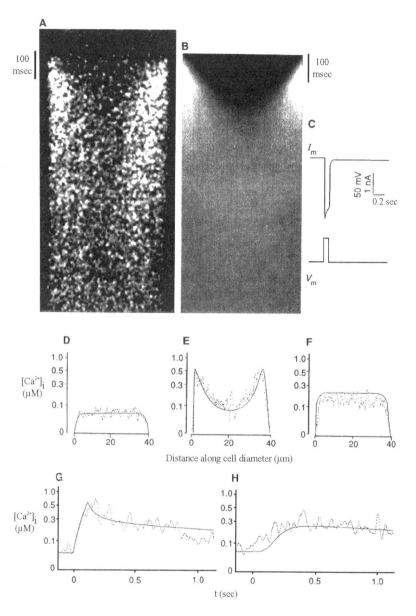

Fig. 11.8 COMPARING EXPERIMENTAL CALCIUM TRANSIENTS AGAINST A MODEL Experimental measurements of intracellular calcium dynamics in bullfrog sympathetic ganglion cells (Hernández-Cruz, Sala, and Adams, 1990) compared to numerical solutions of the associated one-dimensional radial reaction-diffusion equations (Sala and Hernández-Cruz, 1990). The data were obtained by monitoring the calcium dye fluo-3 along a thin strip across the diameter of these spherical cells with the aid of confocal microscopy. Measured (**A**) and simulated (**B**) radial spread of Ca^{2+}. Time extends downward and space across. (**C**) A 100-msec-long voltage-clamp step to +10 mV is used to trigger an influx of Ca^{2+} via voltage-dependent calcium channels (I_m; see also the vertical bars in **A** and **B**). Calcium concentration profiles as a function of space and time extracted from the data in **A** (dotted traces) or the model in **B** (solid traces). (**D**) Resting calcium concentration. $[Ca^{2+}]_i$ is reduced below the membrane due to the presence of calcium pumps in the model. Spatial profile 100 msec (**E**) and 1 sec (**F**) after the stimulus onset. Temporal dynamics of free calcium 2.5 μm below the membrane (**G**) and at the center (**H**) of the 40-μm-diameter cell. Reprinted by permission from Hernández-Cruz, Sala, and Adams, (1990).

without bound calcium—can diffuse inside the cell (with diffusion coefficients of 0.25 and 0.1 μm^2/msec). This causes the decay of calcium below the shell to occur at two different time scales (Fig. 11.8G): a fast one, over within 300 msec, which is due to the diffusional redistribution of Ca^{2+} throughout the cell, and a much slower phase, lasting for seconds, which reflects the slow buffering, the diffusion of the buffer, and the extrusion of calcium across the membrane by the calcium pumps (see the following section).

The qualitative match between observation and model evident in Fig. 11.8 underscores that the evolution of calcium in these simple structures can be understood in terms of diffusion and buffering.

11.6 Ionic Pumps

The crucial component of the system controlling the homeostasis of sodium, potassium, chloride, and calcium are specialized membrane-bound molecules that act as *ionic pumps*. Also known as ion *transporters*, they maintain the ionic gradients across the membrane that enable neurons to signal and to generate and propagate action potentials. In conjunction with buffers and other uptake systems, they also provide exquisite regulation of the intracellular concentration of free Ca^{2+}.

The single most important ion transporter is probably the Na^+–K^+ *pump* (Hille, 1992). It is driven by the energy derived from hydrolysis of ATP. Three Na^+ ions are pumped out of the cell for every two K^+ ions moved into the cell. This results in a net accumulation of charge, that is, in a small but measurable current. The pump is therefore known as *electrogenic* and is ubiquitous in the membranes of all cells, neuronal or not. Given the fact that most cells have stable resting potentials that have to be continuously maintained in the face of EPSPs and IPSPs, action potentials, and so on, the Na^+–K^+ pump consumes a lot of power. Roughly half the metabolic energy used in the retina of a rabbit (Ames et al., 1992) and in the mammalian brain in general (Ames, 1997) has been attributed to it.

Two major transport systems are responsible for the net outward movement of Ca^{2+} across the neuronal membrane against the large concentration gradient (Dipolo and Beauge, 1983; Blaustein, 1988; McBurney and Neering, 1987). One system, the Na^+–Ca^{2+} *exchanger*, exploits the energy gained when moving three Na^+ ions inward by moving one calcium ion out of the cell, thereby generating one excess charge for each Ca^{2+} ion that is removed. This pump has a maximal rate of Ca^{2+} removal of 2–3 nmol per square centimeter of membrane area per second and a K_d in the low micromolar range (that is, the pump is operating at half its maximum if $[Ca^{2+}]_i = K_d$). DiFrancesco and Noble (1985) have developed a model of the cardiac Na^+–Ca^{2+} exchanger, which has also been applied to neurons (Gabbiani, Midtgaard, and Knöpfel, 1994).

A second pump system requires energy in the form of one ATP molecule for each calcium ion pumped out. This ATP-driven calcium pump can be considered to be a pump with a higher affinity ($K_d = 0.2\ \mu M$) yet a lower capacity compared to the Na^+–Ca^{2+} exchanger, with a maximal rate of removal of about 0.2 nmol/cm^2/sec. Its Ca^{2+} dependence is frequently approximated by a Michaelis-Menten equation (see below; Garrahan and Rega, 1990). While both systems operate continuously, the ATP-driven pump can quickly turn on following Ca^{2+} influx subsequent to an action potential, while the lower affinity but much higher capacity system is primarily responsible for maintaining resting levels of Ca^{2+} over longer times.

From the point of view of charge entering or leaving the cell, both ionic pumps have to be treated as ionic currents, the ATP-driven pump acting as an outward current (calcium is removed from the cytoplasm) and the sodium-calcium exchanger acting as an inward current (since three positive charges are moved inside for every two charges being removed). In general, their contributions will be small (but see DeSchutter and Smolen, 1998).

One simple way to model the dynamics of these pumps is via saturable first-order Michaelis-Menten kinetics,

$$\frac{\partial [Ca^{2+}]_{pump}}{\partial t} = \frac{4 P_m}{d} \frac{[Ca^{2+}]}{1 + [Ca^{2+}]/K_{d-pump}} \tag{11.47}$$

where P_m is given by the number of calcium ions that can be pumped out per square micrometer of neuronal membrane divided by K_{d-pump}. The $4/d$ factor takes account of the fact that the molecules acting as the pump are inserted into the membrane area (length $\times \pi d$) that encloses the volume (length $\times \pi d^2/4$) containing the calcium ions (surface-to-volume ratio). The decrease in calcium concentration due to the action of the pump is directly proportional to the calcium concentration if $[Ca^{2+}] \ll K_d$.

11.7 Analogy between the Cable Equation and the Reaction-Diffusion Equation

As witnessed in previous chapters, a great deal of knowledge and intuition has accumulated about the behavior of the membrane potential in one-dimensional cables and dendritic trees. Can we transfer any of this to the solutions of the reaction-diffusion equations? In particular, can we define appropriate space and time constants to characterize the spatio-temporal dynamics of calcium—or any other substance—in response to synaptic input? Drawing upon the study by Zador and Koch (1994), we show how the techniques developed for one-dimensional cable theory can be applied to reaction-diffusion equations.

Our starting point is the distribution of calcium ions in a cylinder following the influx of a calcium current $I_{Ca}(x, t)$ across the membrane. This current can flow through voltage- or ligand-activated channels. As in one-dimensional cable theory, we neglect the radial components of diffusion, assuming that their associated time constants are much, much faster than the longitudional ones (e.g., Rall, 1969b).

The inflowing calcium ions diffuse to neighboring locations, bind to various buffers, and can be pumped back out of the cable. The buffer itself can also diffuse with a diffusion coefficient D_B. The expressions governing the resulting change in the concentration of calcium $[Ca^{2+}](x, t)$ and bound calcium-buffer $[B \cdot Ca](x, t)$ is

$$\frac{\partial [Ca^{2+}](x, t)}{\partial t} = D \frac{\partial^2 [Ca^{2+}](x, t)}{\partial x^2} - f[Ca^{2+}](x, t)[B](x, t) + b[B \cdot Ca](x, t)$$

$$- \frac{\partial [Ca^{2+}](x, t)_{pump}}{\partial t} - \frac{2 I_{Ca}(x, t)}{F d}$$

$$\frac{\partial [B \cdot Ca](x, t)}{\partial t} = D_B \frac{\partial^2 [B \cdot Ca](x, t)}{\partial x^2} + f[B](x, t)[Ca^{2+}](x, t)$$

$$- b[B \cdot Ca](x, t)$$

$$T_B = [B \cdot Ca] + [B] \tag{11.48}$$

In the first equation, the first term on the right-hand side corresponds to the diffusive contribution to the change in calcium concentration, the second and third terms are caused by the removal of calcium due to its binding with the buffer, the fourth corresponds to the reduction in $[Ca^{2+}]_i$ due to the action of the calcium pump, and the last term converts the inward (that is, negative) calcium current (carrying $2e$ charge per ion) into a calcium concentration (the $4/d$ factor accounts for the area-to-volume ratio). The second equation specifies the change in bound calcium-buffer concentration as a function of diffusion and buffer binding and unbinding. The last equation stipulates that in the absence of any buffer sources and sinks and assuming that the buffer diffuses at the same pace as the bound calcium-buffer complex, the total buffer concentration T_B is constant.

Note the nonlinear coupling between the two variables (the $[Ca^{2+}] \times [B]$ term), rendering the solution to these partial differential equations difficult.

11.7.1 Linearization

As we will show now, under certain limiting conditions, Eqs. 11.48 can be reduced to a *single, linear* partial differential equation, formally equivalent to the cable equation. This reduction is based on the *instantaneous buffer* assumption; that is, the kinetics of buffering are much faster than diffusion. Since the former occurs on a microsecond to millisecond time scale (Falke et al., 1994) and the latter requires 10–100 msec, this is a very valid assumption and implies that

$$f[Ca^{2+}][B] = b[B \cdot Ca] \tag{11.49}$$

holds everywhere.

From a mathematical point of view, Eqs. 11.48 constitute a *singularly perturbed system*, in which one variable evolves much faster than the others. Other instances of such systems are the Hodgkin–Huxley and the FitzHugh–Nagumo equations (Keener, 1988; Wagner and Keizer, 1994). The concentration of the bound buffer at any instant can be approximated by its steady-state distribution (Eq. 11.38),

$$[B \cdot Ca](x, t) = \frac{T_B[Ca^{2+}](x, t)}{K_d + [Ca^{2+}](x, t)}. \tag{11.50}$$

Assuming furthermore that the calcium concentration is less than the K_d of the binding process (see Eq. 11.39) and that the pump is not saturated (that is, $[Ca^{2+}](x, t) < K_{d-pump}$), Eqs. 11.48 can be reduced (Zador and Koch, 1994; Wagner and Keizer, 1994) to

$$\frac{d(1 + \beta)}{4} \frac{\partial[Ca^{2+}](x, t)}{\partial t} = \frac{d(D + \beta D_B)}{4} \frac{\partial^2[Ca^{2+}](x, t)}{\partial x^2}$$

$$- P_m[Ca^{2+}](x, t) - \frac{I_{Ca}(x, t)}{2F} \tag{11.51}$$

Equation 11.51 should be very familiar to us, since it is the cable equation in disguise. To recall, the cable equation in an infinite cylinder in response to an injected current density $I_{inj}(x, t)$ is (see Eq. 2.7),

$$C_m \frac{\partial V(x, t)}{\partial t} = \frac{d}{4R_i} \frac{\partial^2 V(x, t)}{\partial x^2} - \frac{V(x, t)}{R_m} + I_{inj}(x, t). \tag{11.52}$$

If the following identifications are made, these two linear equations are identical:

$$R_m^{-1} \longleftrightarrow P_m$$

$$R_i^{-1} \longleftrightarrow D + \beta D_B$$

$$C_m \longleftrightarrow \frac{d(1 + \beta)}{4}$$

$$I_{\text{inj}} \longleftrightarrow -\frac{I_{Ca}}{2F} \tag{11.53}$$

That the pump acts like a membrane conductance R_m^{-1} is straightforward enough to understand: the more pump molecules are present, the more calcium will "leak" out of the cell. Equation 11.51 was derived under a low calcium constraint. For higher concentrations, the pump saturates (Eq. 11.47) and acts like a constant hyperpolarizing current, removing calcium ions at a constant rate.

Just as the axial resistance determines the spread of the voltage along the longitudinal axis, so does the diffusion constant determine the rate of calcium flux along the longitudinal axis (see also Fig. 11.3). The effect of a diffusible buffer is to increase the effective diffusion constant by an additive term $\beta D_B = T_B D_B / K_d$. There are now two sources of calcium mobility: direct diffusion of Ca^{2+} ions and diffusion of bound calcium riding "piggyback" along with the buffer. The net effect of buffering and diffusion of the bound buffer can be expressed by a revised *effective diffusion coefficient*

$$D_{\text{eff}} = \frac{D + \beta D_B}{1 + \beta}. \tag{11.54}$$

(For a generalization, see Wagner and Keizer, 1994.) For the observed large binding ratio β, diffusion is dominated by the diffusion of the calcium-buffer complex.

Because calcium is measured using calcium-dependent fluorescent dyes, which themselves act as buffers for calcium ions, the perturbation of the Ca^{2+} signals by the measurement act must be taken into account (Neher, 1995).

As illustrated by Fig. 11.3, the basic diffusion equation includes an effective membrane capacitance of unity. The effect of a fast buffer is to boost this capacitance by an additive term given by the binding ratio β. Similar to a capacitance, the buffer acts to slow down changes in the calcium concentration. Note that the buffer does not affect the steady-state distribution of calcium in response to a sustained calcium current injection I_{Ca}.

11.7.2 Chemical Dynamics and Space and Time Constants of the Diffusion Equation

Further exploiting the analogy between the two equations (Zador and Koch, 1994; see also Kasai and Petersen, 1994), we know that the response of the reaction-diffusion equation to a stationary calcium current I_{Ca} in an infinite cable will be a decaying exponential, allowing us to define a space constant,

$$\lambda_{r-d} = \sqrt{\frac{d(D + \beta D_B)}{4 P_m}}. \tag{11.55}$$

Here r–d stands for "reaction-diffusion." We can also define a time constant associated with the linearized reaction-diffusion equation (Eq. 11.51) as

$$\tau_{r-d} = \frac{d(1 + \beta)}{4P_m} \qquad (11.56)$$

(see Table 11.2), with

$$D_{\text{eff}} = \frac{\lambda_{r-d}^2}{\tau_{r-d}}. \qquad (11.57)$$

This similarity allows us to apply the results we derived in Chaps. 2 and 3 directly to write down equivalent expressions for the linearized reaction-diffusion equation. In particular, we can introduce the transfer "resistance" $\tilde{K}_{ij}$, defined as the ratio of the sustained change in calcium concentration at location j in response to the sustained calcium current I_{Ca} injected at location i. If locations i and j are a distance x_{ij} apart in an infinite cylinder, we have (see Eq. 3.23),

$$\tilde{K}_{ij} = \tilde{K}_{ii} e^{-x_{ij}/\lambda_{r-d}} \qquad (11.58)$$

with the steady-state input resistance defined as (Eq. 3.24; see also Carnevale and Rosenthal, 1992),

$$\tilde{K}_{ii} = \frac{1}{2Fd^{3/2}\pi \sqrt{(D + \beta D_B)P_m}}. \qquad (11.59)$$

The unit of the chemical input resistance is M/A (injecting so many amperes of current increases the concentration by so many molar). From all of this we can infer a number of interesting facts.

1. As pointed out above, the buffering scheme will only affect the transient behavior, not the sustained response, acting like a capacitance. The more buffer that is present, the larger T_B and therefore β and thus the longer τ_{r-d}.

2. The scaling behavior of the sustained response in a cable of diameter d is identical to that of λ and $\tilde{K}_{ii}$ for the cable equation (Fig. 11.9). Because the constant in front of the pump term scales with the ratio of surface area to volume, that is, as $4/d$, and D, D_B and β are independent of d, λ_{r-d} scales as $\sqrt{d}$ and $\tilde{K}_{ii}$ as $d^{-3/2}$. Thus, injecting a calcium current into a small volume will give rise to a much larger change in calcium concentration than injecting the identical current into a larger volume.

$$\lambda_{r-d} \ll \lambda. \qquad (11.60)$$

3. What does scale differently is the time constant. While τ_m is independent of the radius of the neuronal process, τ_{r-d} increases linearly with d (Eq. 11.56 and Fig. 11.9). Calcium

TABLE 11.2
Space and Time Constants and Input Resistance for the Linearized Reaction-Diffusion and Cable Equations

	Space constant λ	Time constant τ	Input resistance $\tilde{K}_{ii}$
Cable eq.	$\sqrt{dR_m/4R_i}$	$R_m C_m$	$d^{-3/2}\sqrt{R_m R_i/\pi}$
Reaction-diffusion eq.	$\sqrt{d(D + \beta D_B)/4P_m}$	$d(1 + \beta)/4P_m$	$d^{-3/2}/2F\pi\sqrt{(D + \beta D_B)P_m}$

Definition of space and time constants as well as the steady-state input resistance for the cable and the linearized reaction-diffusion equation for an infinite cylinder. $\beta = T_B/K_d$ characterizes the buffer and P_m the ionic membrane pump. See Zador and Koch (1994).

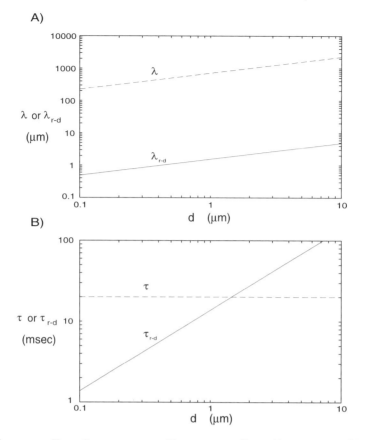

Fig. 11.9 SPACE AND TIME CONSTANTS AS A FUNCTION OF CABLE DIAMETER   (**A**) Space and (**B**) time constants for the linearized reaction-diffusion equation (Eq. 11.51) (solid line) and the cable equation (Eq. 2.7) (dashed line) as a function of the diameter of the infinite cable. Notice that the electrical space constant is much larger than the chemical space constant, with important functional consequences. While this is also true for the time constants for thin fibers, the time constant of the reaction-diffusion equation scales with the diameter and can therefore exceed τ_m in large structures. The parameters are chosen to mimic the binding of calcium to calmodulin. Numerical values are defined in Table 11.3. Reprinted by permission from Zador and Koch (1994).

dynamics will be slower in thicker cables than in thinner ones. (We are not, of course, accounting for radial diffusion within the cylinder, since we are only considering the one-dimensional diffusion equation.)

4. Table 11.3 lists some typical values for the space and time constants and the input resistance in an infinite cable for both Eqs. 2.7 and 11.51. What is immediately apparent is the substantial difference between the space constant of the cable equation and that of the reaction-diffusion equation,

In other words, while voltage can act over substantial distances, the effect of the concentration of calcium (or other second messengers) is much more local. Reducing the density of calcium pumps by one, two, or even three orders of magnitude does not affect this difference dramatically. Ultimately this is due to the fact that the membrane conductances dominating the resting levels of the membrane potential (mainly K^+

TABLE 11.3
Numerical Values of λ, τ, and $\tilde{K}_{ii}$ for the Linearized
Reaction-Diffusion and Cable Equations

Diameter	λ_{r-d}	τ_{r-d}	$\tilde{K}_{ii}(r-d)$	λ	τ_m	$\tilde{K}_{ii}$
0.1	0.49	1.38	84.6	224	20	14,200
1.0	1.54	13.8	26.8	707	20	450
10.0	4.88	137.5	0.085	2236	20	14.2

Space and time constants and the sustained input resistance of the reaction-diffusion and the cable equations in an infinite cable of indicated diameter (in μm) for calcium (with $D = 0.6\ \mu m^2$/msec) binding to diffusible calmodulin (with $T_B = 100\ \mu M$, $K_d = 10\ \mu M$, $\beta = 10$, and $D_B = 0.13\ \mu m^2$/msec) and in the presence of a high-affinity calcium pump ($P_m = 0.2\ \mu m$/msec). This is compared against a standard dendritic cable with $R_i = 100\ \Omega\cdot$cm, $R_m = 20{,}000\ \Omega\cdot cm^2$ and $C_m = 1\ \mu F/cm^2$. Notice the dramatic difference between λ_{r-d} and λ and the different scaling behavior of τ_{r-d} and τ_m. The space constants are in units of μm and the time constants in msec. The chemical input resistance is in units of nM/fA and the electrical input resistance in MΩ.

conductances) are small relative to the interaxial resistance, while relatively more Ca^{2+} ions pass through the membrane via the pumps than flow longitudinally.

5. The dynamics of the two processes are comparable for small cylinders, but can differ greatly—due to the dependency of τ_{r-d} on the geometry—for thicker dendrites (Table 11.3 and Fig.11.9). Due to the smaller surface-to-volume ratio, the time constant of the reaction-diffusion equation can be quite slow, in the hundreds of milliseconds.

How well does the linearized reaction-diffusion equation hold up in practice? This was evaluated by observing the spread of calcium along the axon of an *Aplysia* neuron (Gabso, Neher, and Spira, 1997). Following the injection of calcium from a micropipette into the axon, calcium was tracked with the help of another fluorescent dye, fura-2, for tens of seconds as the calcium ions diffused away from the injection site for a few hundreds of micrometers.

As expressed by Eq. 2.31 and illustrated in Fig. 2.7A, the impulse response function of the cable equation in an infinite cylinder for any fixed point in time is a Gaussian. This also holds true for the linearized reaction-diffusion system we are considering,

$$[Ca^{2+}]_\delta(x, t) = \frac{A_0}{\sqrt{t}} e^{-x^2/(D_{\text{eff}} t)} e^{-t/\tau_{r-d}} \tag{11.61}$$

where the effective diffusion coefficient D_{eff} is defined in Eq. 11.57. Gabso, Neher, and Spira (1997) fit a Gaussian through the spatial profile of the calcium signal at different times (Fig. 11.10A). If the spatio-temporal dynamics of [Ca^{2+}] follow Eq. 11.51, then the square of the standard deviation of the Gaussian should increase linearly in time, which it does (Fig. 11.10B). The slope of the curve gives the effective diffusion coefficient of calcium and the calcium bound to any buffers intrinsic to the axon as well as to the fura-2. Note that in this study, the calcium signal was purposely kept below 0.5 μM.

While the analogy between the cable equation and the reaction-diffusion equation breaks down for large values of calcium concentration and for more complex buffering schemes, the behavior of [Ca^{2+}] will not deviate qualitatively from that described by the linear Eq. 11.51.

This is demonstrated in Fig. 11.11, which plots the normalized calcium concentration at two locations in an infinite cable for a series of calcium injections. For small currents, the system operates in the low calcium limit, and the dynamics of [Ca^{2+}], obtained by solving the nonlinear coupled partial differential Eqs. 11.48, are fitted well by the linear

A)

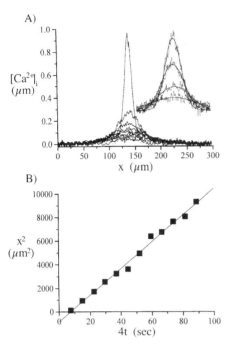

B)

Fig. 11.10 CALCIUM SPREAD ALONG AN AXON
Experimental determination of the calcium signal (recorded using the calcium-dependent fluorescent dye fura-2) following a brief intracellular injection of calcium at one point into the axon of a cultured metacerebral *Aplysia* neuron by Gabso, Neher, and Spira (1997). **(A)** Spatial profile of the $[Ca^{2+}]_i$ signal for different times. A baseline value was subtracted from each curve so that they all approach zero for large distances. Images were acquired every 1.8 sec, with the second, third, fifth, and seventh measurements displayed in the enlarged inset. A Gaussian was fitted through each curve (smooth lines in the inset). **(B)** The square of the standard deviation of the Gaussian was plotted as a function of time. If the spatio-temporal calcium dynamics follows the linearized reaction-diffusion equation (Eq. 11.51) these points should fall on a straight line whose slope is the effective diffusion coefficient D_{eff} of free and bound calcium, here equal to 0.112 $\mu m^2/msec$. Reprinted by permission from Gabso, Neher, and Spira (1997).

approximation of Eq. 11.51. As the current is made larger, the buffer saturates and becomes ineffective, since it no longer absorbs any of the inflowing calcium ions. Indeed, following Eqs. 11.37 and 11.50, at high calcium concentrations the entire buffer concentration is taken up by the bound calcium-buffer complex. Formally, this corresponds to $\beta = 0$ in Eq. 11.51, which explains the greatly spedup chemical dynamics (τ_{r-d} decreases by about one order of magnitude). However, the important point to note is that the behavior of the solutions to the full equations is bracketed by the solutions to the linear equation with $\beta = 0$ and 10, without deviating in any significant way from them (for instance, they are all monotonic, saturating functions).

The morale is that without further significant nonlinearites, our conclusions regarding the space and time constants associated with the reaction-diffusion equation do not change dramatically.

11.8 Calcium Nonlinearities

It is known that a number of different cell types exhibit all-or-none calcium events that occur over and over again (Berridge, 1990; for an excellent review of this topic see Meyer and Stryer, 1991). Such oscillations in the calcium concentration, whose duration lasts on the order of seconds or longer, can be observed in response to hormones. They have very sharp onsets and vary in frequency in a monotonic manner with the concentration of the hormone. Although these calcium spikes are on the order of three to four orders of magnitude slower than voltage spikes, they do share a number of features with their faster cousins, in particular being caused by a highly nonlinear positive feedback mechanism. Indeed, equations similar to the Hodgkin–Huxley equations have been used to replicate these experimental findings in a computer model (Meyer and Stryer, 1991; Ogden, 1996).

A)   B)

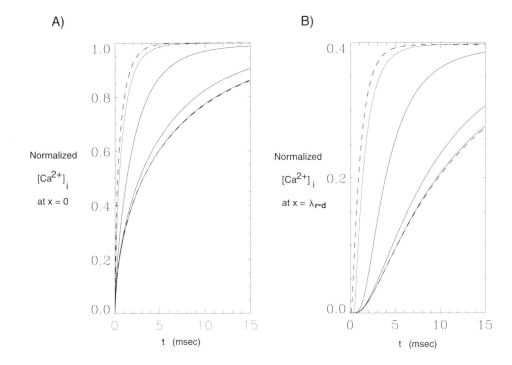

Fig. 11.11 **EFFECT OF BUFFER SATURATION ON CALCIUM DIFFUSION**   Effect of saturating the buffer in an infinite cable. Plotted is the normalized calcium concentration (**A**) at the origin of the current input as well as (**B**) one space constant λ_{r-d} away. The dashed curves represent solutions to the linearized reaction-diffusion equation (Eq. 11.51) for $\beta = 10$ (lower dashed curve) and for the fully saturated buffer (corresponding to $\beta = 0$ in Eq. 11.51). The solid curves correspond to the solution of the nonlinear coupled Eqs. 11.48 for a peak calcium current of 10^{-4}, 10^{-3}, 10^{-2}, and 10^{-1} nA amplitude (from bottom to top). The behavior of the full system is bracketed by the two linear solutions; no qualitatively new behavior appears. Reprinted by permission from Zador and Koch (1994).

In some model systems, waves of increased $[Ca^{2+}]_i$ have been observed (Cornell-Bell et al., 1990; Dupont and Goldbeter, 1992; Sanderson, 1996). Applying the excitatory neurotransmitter glutamate to hippocampal *astrocytes*—non-neuronal supporting cells, which occur in great numbers throughout the brain—triggers such calcium waves, propagating at constant speeds of about 20 μm/sec throughout the cytoplasm of the astrocyte. Frequently, these waves propagate across adjacent astrocytes. The possible significance of such waves for neuronal signaling is not known.

While such nonlinear calcium events are quite intriguing, their long time scale makes it unlikely that they play a significant role in the rapid computations that we are primarily interested in.

11.9 Recapitulation

Diffusion is a fundamental fact of life for molecules in the intracellular or extracellular cytoplasm. Through its random action, it acts to move substances throughout the cell.

From a computational point of view, the most important fact about diffusion is that it places strong constraints on how rapid calcium or other second messenger molecules can affect things far away. The distance over which some concentration increase diffuses is proportional to the square root of the time that has passed. In the absence of any calcium nonlinearities and active transport processes, this square-root law fundamentally limits the ability of the calcium signal to implement the fast type of information processing operations required for many perceptual, cognitive, or motor tasks. Recognizing a friend's face, shifting visual attention from one location to a neighboring one, or raising one's hand to catch a ball can all be accomplished within a few hundred milliseconds.

The calcium that rushes into the cell via ionic channels is tightly regulated. The vast majority is bound to a host of intracellular buffers, such that only one out of 20 Ca^{2+} ions is free to interact with other molecules, severely limiting the effective diffusion coefficient of calcium.

While calcium ions diffuse along some process, their concentration rapidly decreases. This is especially true for a substance that diffuses in the three-dimensional extracellular tissue; its spatial concentration profile decreases sharply with distance from the source (as e^{-r^2}). A relevant case are certain unconventional neuroactive substances, such as *nitric oxide*, that can diffuse across the membrane cytoskeleton (Secs. 20.2 and 20.3).

Of course, these constraints do not argue against the use of the local intracellular calcium concentration for computing and for short-term memory storage (Sobel and Tank, 1994). When calcium and its protein targets are in close spatial proximity, the rate at which calcium can bind to this protein limits the speed of the computational operation being implemented (Sec. 20.1). This allows chemical switching to proceed in the submillisecond domain. Using concentration changes for implementing rapid operations does impose stringent conditions on a fast local input and a fast local read-out mechanism.

In general, movements of ions due to the inhomogeneous distribution of the various relevant ions (Ca^{2+}, Na^+, K^+ and Cl^-) must be incorporated into the cable equation, leading to the Nernst-Planck electrodiffusion equation. However, as long as the diameter of the neuronal process is above a fraction of a micrometer, this equation is well approximated by the cable equation. Only when studying very small processes, such as dendritic spines or very thin dendrites, does the longitudinal diffusion of the carriers need to be taken into account.

If the buffering reaction is substantially faster than diffusion and if the calcium concentration is small (technically, if $[Ca^{2+}]_i < K_d$ of the pump and of the buffer), the coupled system of reaction-diffusion equations can be reduced to a single linear partial differential equation, which is formally equivalent to the cable equation. This allows us to define space and time constants and input resistances in analogy to these parameters in passive dendrites. One important insight is that $\lambda_{r-d} \ll \lambda$, implying that from the point of view of spatial compartmentalization, the presence of reasonable amounts of calcium pumps and buffers in the dendritic tree will fractionate the tree into a series of small and relatively independent compartments. In each of these subunits, independent calcium-initiated chemical computations could be carried out. This is in contrast to the relatively smaller attenuation experienced by the membrane potential in a dendritic tree. It may well be possible that the architecture and morphology of the dendritic tree reflects less the need for electrical computations but more its role in isolating and amplifying chemical signals. We will study a beautiful instance of this in the following chapter on dendritic spines.

Because of the mathematical equivalence between electrical, chemical, and even biochemical networks (Busse and Hess, 1973; Eigen, 1974; Hjelmfelt and Ross, 1992; Barkai and Leibler, 1997) that derives from their common underlying mathematical structure, appropriate sets of reaction-diffusion systems can be devised that emulate specific electrical circuits. In principle, computations can be carried out using either membrane potential as the crucial variable—controlled by the cable equation—or concentration of calcium or some other substances—controlled by reaction-diffusion equations (for examples of this, see Poggio and Koch, 1985). The principal differences are the relevant spatial and temporal scales, dictated by the different physical parameters, as well as the dynamical range of the two sets of variables. Given neuronal noise levels, the membrane potential can be considered to vary by a factor of hundred or less, while the concentration of calcium or other substances can vary by three or more orders of magnitude during physiological events.

12

DENDRITIC SPINES

Dendritic spines, sometimes also called dendritic thorns, are tiny, specialized protoplasmic protuberances that cover the surface of many neurons. First described by Ramón y Cajal (1909; 1991) in light-microscopic studies of Golgi stained tissue, they are among the most striking subneuronal features of many neurons. Indeed, the presence of a high density of dendritic spines allows the unambiguous classification of neuronal types into *spiny* and *aspiny*, *sparsely spiny*, or *smooth* neurons. Over 90% of all excitatory synapses that occur in the cortex are located on dendritic spines. Spines can be found in all vertebrates as well as in invertebrates (e.g., the dendrites of Kenyon cells in the mushroom bodies in the olfactory system of the insect brain). The intimate association of spines with synaptic traffic suggests some crucial role in synaptic transmission and plasticity.

Because of their submicrometer size (see below), physiological hypotheses as to the function of dendritic spines have only very recently become accessible to the experimentalist. For the previous two decades, spine properties have been investigated through analytical and computational studies based on morphological data, providing a very fertile ground for the crosspollination of theory and experiment. (For a very readable historical account of this see Segev et al., 1995.) The recent technical advances in the direct visualization of calcium dynamics in dendrites and spines are now permitting direct tests of some of these theoretical inferences (Guthrie, Segal, and Kater, 1991; Müller and Connor, 1991; Yuste and Denk, 1995; Denk, Sugimori, and Llinás, 1995; Svoboda, Tank, and Denk, 1996). As discussed in this chapter and, more extensively, in Chap. 19, the theoretical models that have endowed spines with active properties giving rise to all-or-none behavior (Perkel and Perkel, 1985; Shepherd et al., 1985; Segev and Rall, 1988; Baer and Rinzel, 1991) have, in general, been confirmed experimentally.

Historically, the possibility of implementing synaptic memory by modulating the electroanatomy of spines was recognized early on (Chang, 1952) and was subsequently analyzed in depth by Rall (1970, 1974, 1978) and many others. Because small changes in the spine morphology can lead to large changes in the amplitude of the EPSP induced by the excitatory synapse on the spine, spines have been considered to contribute to the modulation of synaptic "weight" during long-term potentiation (see Chap. 13). In a radical extension of this original idea, Crick (1982) advanced the "twitching spine" hypothesis, the notion that spines, using actin-based contractile machinery in the spine neck, could subserve very fast changes (less than a second) in synaptic efficacy, and that this might provide a mechanism for short-term memory. Remarkably, it has now been shown in cultured hippocampal neurons that spines can change their shape within seconds (Fischer et al., 1998).

Within the last years the focus of research has shifted away from the electrical properties of spines toward their ability to provide a local and isolated biochemical compartment for calcium and other second messengers underlying the induction and expression of synaptic plasticity (for the first clear statement of this, see Shepherd, 1972, 1978; for reviews, see Koch and Zador, 1993; Harris and Kater, 1994; Shepherd, 1996). Furthermore, spines have also been treated as devices subserving highly localized "pseudological" computations.

We begin this chapter by reviewing the natural history of spines before we analyze their electrical properties and show how dendrites studded with spines can be reduced to smooth dendrites in terms of the solution of the associated linear cable equation. We defer a thorough discussion of the threshold type of logical computations that can be carried out in spines with HH-like membranes to Sec. 19.3.2. Finally, we apply the concepts introduced in the previous chapter to understand why spines can serve as a microenvironment for highly local calcium changes that serve to isolate synapses on the spine from the dendrite and vice versa.

12.1 Natural History of Spines

In the following, we review some of the pertinent facts concerning the structure, size, and distribution of dendritic spines.

12.1.1 Distribution of Spines

Neurons in different parts of the brain can be classified into two groups according to whether or not their dendrites are studded with spines. *Spiny* neurons include pyramidal and stellate cells and account for about three-quarters of the neurons in the neocortex, while *smooth* cells, whose dendrites carry few or no spines, make up the remainder. Smooth cells include basket cells, chandelier or axo-axonic cells, and double-bouquet cells, and stain for the inhibitory neurotransmitter GABA (Douglas and Martin, 1998). It is at present a mystery why one class of cortical cells, inhibitory interneurons, should have no spines, while excitatory cells have so many.

It is important to realize that in many parts of the brain outside of the cortex, inhibitory cells can be covered with spines. Two of the best known examples are the Purkinje cells in the cerebellum (Llinás and Walton, 1998) and the principal cells in the neostriatum (Kawaguchi, Wilson, and Emson, 1990; Wilson, 1998). Both are covered by spines, both use an inhibitory neurotransmitter, and both constitute the only output of their respective systems (Fig. 12.1).

Indeed, the proper generalization appears to be that spiny cells are the principal output cell class of their particular brain area (frequently also referred to as *projection cells*; Shepherd, 1998).

Spines are numerous. A large layer 5 pyramidal cell in the visual cortex may have as many as 15,000 spines, averaging about two spines per micrometer of dendrite (Larkman, 1991), while the density for CA1 pyramidal cells varies from about one to five spines per micrometer of dendrite, depending on the staining method used (Harris and Stevens, 1989; Amaral, Ishizuka, and Claiborne, 1990). The inhibitory projection cells in the neostriatum have a peak density of four to six spines per micrometer of dendrite (Wilson et al., 1983). The record, though, is held by cells in the human cerebellum, where individual Purkinje

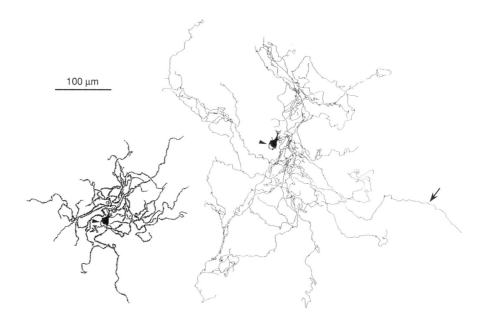

Fig. 12.1 Spiny Inhibitory Neuron Dendritic (left) and axonal (right) arborization of a projection cell in the rat neostriatum, part of the basal ganglia, that was injected with biocytin and reconstructed. The dendrites are covered by up to five spines per micrometer. In the human, three-quarters of the approximately 110 million neostriatum neurons are of this type. They make both local connections, as illustrated here, and leave the neostriatum (elongated arrow) to terminate in the globus pallidus. These axons are the only output of the neostriatum and have an *inhibitory* effect on their postsynaptic target. Spiny cells in neocortex and hippocampus, in contrast, are excitatory. Reprinted by permission from Kawaguchi, Wilson, and Emson (1990).

cells are studded by up to 200,000 spines, each spine carrying a single excitatory synapse from a parallel fiber (Braitenberg and Atwood, 1958; Harris and Stevens, 1988b).

12.1.2 Microanatomy of Spines

Spines are found in a wide variety of shapes (Jones and Powell, 1969; Peters and Kaiserman-Abramof, 1970; Woolf, Shepherd, and Greer, 1991a,b; Figs. 12.2–12.4), ranging from short and stubby, through the archetypal "mushroom shaped," to long and thin ones. Many spines branch. In the case of spines on CA3 hippocampal pyramidal cells receiving mossy fiber inputs, up to 16 branches have been observed to emerge from a single dendritic origin (Chicurel and Harris, 1992).

Given the diversity of spine shapes (Fig. 12.4), it is not easy to define the "average" spine. However, most spines show a clearly differentiated "neck" and a "head." Their exact morphology can best be appreciated via three-dimensional reconstructions of serial electron micrographs of the entire cell (a heroic task) as carried out by White and Rock (1980) in the mouse somatosensory cortex, by Wilson et al., (1983) in the neostriatum, and by Harris and Stevens (1988a,b, 1989) in the cerebellum and the cortex. For CA1 pyramidal cells in the rat hippocampus, Harris and Stevens (1989) found wide variability in the dimensions of spines, with spine necks ranging in length from 0.08 to 1.58 μm and in diameter from 0.04 to 0.46 μm. The total volume ranges from 0.04 to 0.56 μm^3 (Fig. 12.4; see also Segev

Fig. 12.2 STEREO VIEW OF A SPINY DENDRITE   Stereoscopic view of a portion of a spiny dendrite (of the type shown in Fig. 12.1) in the rat neostriatum. The large number of spines springing from the dendrite can be viewed in depth by focusing the eyes at a point at infinity (that is, by looking through the image). The reconstruction has been performed using high-voltage electron-microscopic axial tomography from a single 3-μm-thick section. The cube has a dimension of 0.5 μm on each side. Reprinted by permission from Wilson et al., (1992).

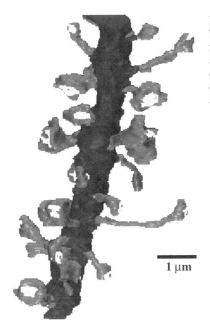

1 μm

Fig. 12.3 DENDRITIC SPINES IN THE HIPPOCAMPUS Three-dimensional reconstruction of a piece of dendrite from a CA1 pyramidal cell of the rat hippocampus. There are between one and three spines per micrometer of dendrite. The diversity in the morphologies and dimensions of spines is striking. Reprinted by permission from Harris and Stevens (1989).

et al., 1995). Larger spine heads are associated with larger synapses—as measured by the size of the associated postsynaptic density—and more vesicles in the presynaptic axonal varicosity.

These dimensions indicate how spines bridge the gap between molecular and cellular scales: at a resting concentration of 80 nM and for the average spine head volume of 0.05 μm^3, only about two unbuffered Ca^{2+} ions are expected to be found in a spine head.

Dendritic spines appear to be filled with complex cellular machinery (for an overview, see Harris and Kater, 1994), most notably a specialized form of smooth endoplasmic reticulum termed the *spine apparatus*. The membranes making up the spine apparatus are closely apposed to the plasma membrane of the spine neck and appear to sequester calcium (Fifková, Markham, and Delay, 1983; Burgoyne, Gray, and Barron, 1983).

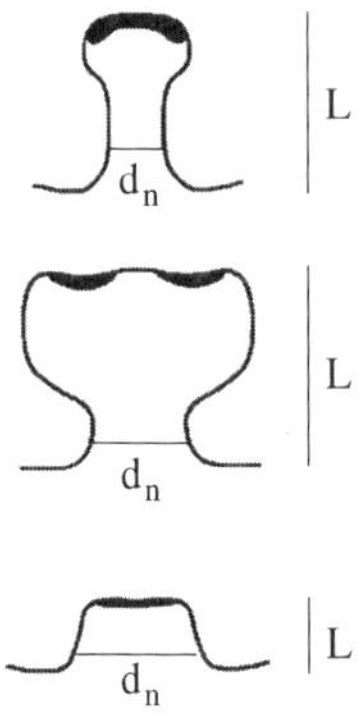

	Type of Spine	d_n (μm)	L (μm)	Area (μm^2)	Volume (μm^3)
L	Thin	0.10 ± 0.03	0.98 ± 0.42	0.59 ± 0.29	0.04 ± 0.02
L	Mushroom	0.20 ± 0.07	1.50 ± 0.25	2.70 ± 0.93	0.29 ± 0.13
L	Stubby	0.32 ± 0.13	0.44 ± 0.15	0.45 ± 0.14	0.03 ± 0.01

Fig. 12.4 DIVERSITY OF SPINE SHAPES Three most frequent types of spine shapes in the cortex with some of their dimensions: spine neck diameter d_n, total length L, surface area, and volume. The length of the spine neck l_n is 0.51 ± 0.34 μm for thin spines and 0.43 ± 0.21 μm for mushroom-type spines. All values are averages and standard deviations. Black areas indicate the postsynaptic densities (PSD) of the asymmetric excitatory synapses on the spines. Frequently more than one synapse is located on a spine. Stubby spines are also known as "sessile." The data are measured from electron-microscopic reconstructions of spines on adult rat CA1 hippocampal pyramidal cells. Reprinted in modified form by permission from Harris, Jensen, and Tsao (1992).

Other cellular organs, such as mitochondria, microtubules, or ribosomes, are usually absent. An exception are the spines on granule cell dendrites, which make reciprocal dendro-dendritic synapses with projection neurons in the mammalian olfactory bulb (Rall et al., 1966; Cameron, Kaliszewski, and Greer, 1991; Sec. 5.3). Here the mitochondria are most likely needed to provide energy to subserve presynaptic functions. Although spines lack neurofilaments, spine heads contain a dense network of actin filaments (Fischer et al., 1998). In the neck, actin filaments are oriented lengthwise along the spine apparatus (Fifková, 1985). A number of proteins known to be involved in actin-mediated activities, such as neuronal myosin (Drenckhahn and Kaiser, 1983), fodrin (Carlin, Bartelt, and Siekevitz, 1983) and calmodulin (Caceres et al., 1983), have been found in dendritic spines.

The close association of spines with the terminal boutons of axons prompted early speculation that spines might conduct impulses between neurons (Ramón y Cajal, 1909). Electron microscopic studies have since confirmed that spines are indeed the major postsynaptic target of excitatory (asymmetric, type I) synaptic input (Gray, 1959). For instance, in cat visual cortex, 93% of the afferent geniculate terminals in layer 4 is located on spines (LeVay, 1986), while a similarly high percentage is observed in the Schaffer projection from CA3 cells onto CA1 pyramidal cells in the hippocampus. Note, however, that not all excitatory pathways terminate on spines. For instance, about 70% of the excitatory synapses from layer 6 pyramidal cells terminates directly on the dendrites of layer 4 spiny stellate cells, and only a bit less than one-third of layer 6 synapses chooses spines as their termination zones (Ahmed et al., 1994). As mentioned before, excitatory projections onto inhibitory cells always make a synapse directly onto the dendrites since these cells lack spines.

Interestingly, the association between spines and synapses is limited to excitatory traffic. Inhibitory (symmetric, type II) profiles are only observed at a small fraction (between 5 and 20%) of spines in the cortex, but usually in conjunction with excitatory synapses and never by themselves (Jones and Powell, 1969; Dehay et al., 1991; Fifková, Eason, and Schaner, 1992).

12.1.3 Induced Changes in Spine Morphology

Both the absolute number and the shape of spines can change—sometimes quite drastic-ally—in the young and in the mature animal, and also as a function of external stimulation.

During development, absolute spine density can more than double relative to densities in the adult. Yet this increase does not occur homogeneously across all spines. Rather, thin, branched, and certain types of mushroom spines increase fourfold in density (between 2 weeks postnatal and adulthood in the rat hippocampus), while stubby spines decrease by more than half (Harris and Stevens, 1989; Harris, Jensen, and Tsao, 1992; see also Schüz, 1986; Papa et al., 1995; and Fig. 12.4). In adult female rats, the dendritic spine density on CA1 hippocampal pyramidal cells varies by 30% or more over the five-day estrus cycle, while pyramidal cells in the CA3 region show little statistically significant variation (Woolley et al., 1990). It is not known whether the number of synapses also varies by this fraction, or whether it is just the spines that wax and wane, but this certainly makes for a very dynamic environment.

Other studies have shown that the shape of spines—in particular the length and diameter of the neck—can change in response to behavioral or environmental cues such as light, social interaction, or one-shot learning or exploratory motor activity (Purpura, 1974; Coss and Globus, 1978; Bradley and Horn, 1979; Brandon and Coss, 1982; Rausch and Scheich, 1982; Lowndes and Stewart, 1994; Moser, Trommald, and Andersen, 1994).

As discussed in the following chapter, long-term potentiation (LTP), the best studied form of synaptic plasticity, is induced by brief high-frequency electrical stimulation. In the hippocampus, stimulating cells in this manner causes alterations in the spine structure (Van Harrefeld and Fifková, 1975; Lee et al., 1980; Greenough and Chang, 1985; Desmond and Levy, 1990; Calverly and Jones, 1990; Harris, Jensen and Tsao, 1992). Some of the reported changes include larger spine heads, distortions in the shape of the spine stem, an increased incidence of concave spine heads, and an increase in the number of shaft synapses. It is not known what role—if any—the changes in spine shape play in the increase in synaptic efficacy (see also Desmond and Levy, 1988). Remarkably, some of these changes in spine shape occur within seconds, mediated by actin filaments (Fischer et al., 1998).

12.2 Spines only Connect

Most of the early hypotheses considered the establishment of physical contact with presy-naptic terminals as the main function of spines. (For a modern view of this, see Swindale, 1981.) It was argued that because dendritic space is scarce, spines provide additional membrane areas for synapses to make contact with. This idea has been largely dismissed, however, because electron microscopic views of spiny dendrites show that the dendritic membrane between spines often lacks synapses.

On the basis of their three-dimensional electron-microscopic reconstructions, Harris and Stevens (1988a) estimate that 29–45% of the dendritic membrane area of Purkinje cells would have been covered by synapses if all spines had been deleted and the associated synapses moved onto the dendrites. For CA1 pyramidal cell dendrites, only 5–9% of the total dendritic surface area would have been covered by the spine synapses (Harris and

Stevens, 1988a), arguing against the hypothesis that spines are necessary to supplement the dendritic membrane area available for synaptic contacts. This is corroborated by the observation (Schüz and Dortenmann, 1987) that the density of excitatory synapses on nonspiny dendrites in the cortex exceeds the spine density of pyramidal cell dendrites.

These results do not, however, address whether spines play a role in streamlining the layout of axonal and dendritic processes in the three-dimensional neuropil. It is certainly true that for a fixed dendritic radius, spiny dendrites sample a larger brain volume than dendrites devoid of spines (indeed, the latter tend to be thicker than the former; Harris and Kater, 1994).

In order to understand the functional theories that have been proposed to explain the existence of spines, we need to study their electrical properties and their ability to compartmentalize various important biological molecules.

12.3 Passive Electrical Properties of Single Spines

Rall (1974, 1978) was the first researcher to analyze the properties of passive spines, followed by many others (Wilson, 1984; Turner, 1984; Koch and Poggio, 1983a,b; Brown et al., 1988). We here follow the derivation of Koch and Poggio (1983b).

12.3.1 Current Injection into a Spine

The principal idea can be understood in terms of the simplified electrical circuit of a passive spine attached to a dendrite (Fig. 12.5). Any current injected into the spine head must flow either across the spine head resistance R_h and capacitance C_h or down through the spine neck impedance and into the dendrite. Neglecting capacitive and cable properties of the neck—due to its tiny size—we model the spine neck resistance as that of a cylinder of length l_n, diameter d_n, and resistivity R_i,

$$R_n = \frac{4 R_i l_n}{\pi d_n^2}.$$ (12.1)

The input impedance of the spine head membrane can be described by the complex function

$$\tilde{K}_h(f) = \frac{R_h}{1 + i f \tau_m}$$ (12.2)

where $\tau_m = R_h C_h = R_m C_m$. Because R_h is inversely proportional to the spine head area, for average spine dimensions, $R_h > 1000 \text{ G}\Omega$.

The input impedance at the spine head $\tilde{K}_{sp,sp}(f)$ is determined by current flowing either through the spine head or through the spine neck. Remembering the way parallel resistances add, we have

$$\frac{1}{\tilde{K}_{sp,sp}(f)} = \frac{1}{\tilde{K}_{dd}(f) + R_n} + \frac{1}{\tilde{K}_h(f)}$$ (12.3)

where $\tilde{K}_{dd}(f)$ is the dendritic input impedance. Because for the entire relevant frequency range (up to hundreds of kilohertz) the spine head impedance is so much bigger than the sum of the spine neck resistance and the dendritic input impedance, we obtain the simple additive relationship between spine and dendritic input impedance,

$$\tilde{K}_{sp,sp}(f) \approx \tilde{K}_{dd}(f) + R_n.$$ (12.4)

A)

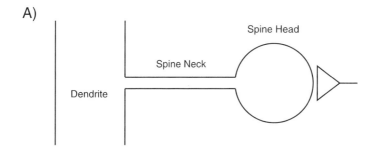

B)

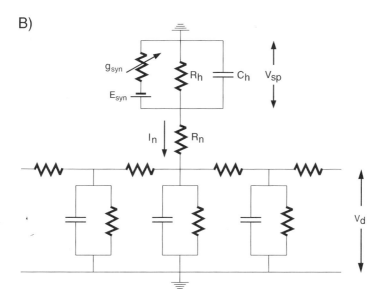

Fig. 12.5 ELECTRICAL MODEL OF A PASSIVE SPINE (A) Schema of a spine with a single synaptic input on a passive dendrite. Drawn to scale, the dendrite is 0.63 μm thick, the spine head has a diameter of 0.7 μm, and the spine neck has dimensions of $d_n = 0.1$ μm and $l_n = 1$ μm. With $R_i = 200$ Ω·cm, the spine neck resistance is 254 MΩ (Eq. 12.1). With a spine neck half as thick, R_n quadruples to just beyond 1 GΩ. (B) Lumped electrical model of such a spine, described by an RC "head" compartment attached through an ohmic neck resistance R_n to the parent dendrite. Due to its small size, the cable and capacitive properties of the neck can be neglected.

Solving directly for the transfer impedance from the spine to the parent dendrite on the basis of Fig. 12.5 while assuming that no current loss occurs across the spine head membrane up to very high frequencies, leads to

$$\tilde{K}_{sp,d}(f) \approx \tilde{K}_{dd}(f).$$ (12.5)

Numerically, the two last equations hold to within a small fraction of a percent (Koch and Poggio, 1983b). By exploiting transitivity (Eq. 3.29), we can write an expression for the transfer impedance from the spine to any point i in the dendritic tree as

$$\tilde{K}_{sp,i}(f) = \frac{\tilde{K}_{sp,d}(f)\tilde{K}_{di}(f)}{\tilde{K}_{dd}(f)} \approx \tilde{K}_{di}(f).$$ (12.6)

In other words, the depolarization at any point i in a passive dendritic tree due to a spine input is identical to the EPSP caused by the same current injected into the dendrite at the base of the spine. Or, put more succinctly, electrically speaking, "spines don't matter in the linear case" (Koch and Poggio, 1985c). The reason why any current injected into the spine will reach the dendrite is that practically no current is lost across the membrane of the spine head and neck.

However, it should be noted that the membrane potential in the spine neck might need to exceed some critical threshold value to be able to initiate some biochemical event (such as those leading to long-term potentiation, as described in Chap. 13). If a synapse were made directly on the dendrite, it would require a far larger conductance change to achieve the same EPSP amplitude than a synapse on a spine. Thus, spines might matter for essentially "local" reasons (see below and Segev et al., 1995).

12.3.2 Excitatory Synaptic Input to a Spine

As discussed above, spines always appear to carry at least one excitatory synapse (Fig. 12.5). We can express the current flowing across this synapse as

$$I_{\text{syn}}(t) = g_{\text{syn}}(t)(E_{\text{syn}} - V_{sp}(t)) \tag{12.7}$$

where V_{sp} is the EPSP at the spine head. Applying Ohm's law, we arrive at

$$V_{sp}(t) = K_{sp,sp}(t) * \left(g_{\text{syn}}(t)(E_{\text{syn}} - V_{sp}(t))\right) . \tag{12.8}$$

Dealing first with stationary (or slowly varying) synaptic input, we can express the steady-state amplitude of the spine EPSP as

$$V_{sp} = \frac{g_{\text{syn}} \tilde{K}_{sp,sp} E_{\text{syn}}}{1 + g_{\text{syn}} \tilde{K}_{sp,sp}} \tag{12.9}$$

where $\tilde{K}_{sp,sp}$ corresponds to the amplitude of the steady-state spine input resistance. By the judicious use of Ohm's and Kirchhoff's laws, we can derive a similar expression for V_d, the EPSP in the dendrite just below the spine,

$$V_d = \frac{g_{\text{syn}} \tilde{K}_{dd} E_{\text{syn}}}{1 + g_{\text{syn}} \tilde{K}_{sp,sp}} = \frac{g_{\text{syn}} \tilde{K}_{dd} E_{\text{syn}}}{1 + g_{\text{syn}}(\tilde{K}_{dd} + R_n)} . \tag{12.10}$$

Let us consider these equations in the case of very small and very large synaptic input.

If the product of the synaptic conductance change g_{syn} and the spine input resistance $\tilde{K}_{sp,sp}$ is much less than 1, the numerator in Eqs. 12.9 and 12.10 can be approximated by 1 and the spine and dendritic EPSPs can be expressed as

$$V_{sp} \approx g_{\text{syn}} \tilde{K}_{sp,sp} E_{\text{syn}} \tag{12.11}$$

and

$$V_d \approx g_{\text{syn}} \tilde{K}_{dd} E_{\text{syn}} . \tag{12.12}$$

In other words, if the synaptic-induced conductance change g_{syn} is small relative to the spine input conductance $1/\tilde{K}_{sp,sp}$, the action of the synapse can be approximated by a constant current source of amplitude $g_{\text{syn}} E_{\text{syn}}$. As we saw above, spines do not matter in this limit. More specifically, changes in spine morphology have no effect on the dendritic potential just below the spine. Note, however, that this does *not* imply that the spine EPSP is identical to the dendritic EPSP; they are not. Indeed, the dendritic EPSP will be attenuated with respect to the spine EPSP by a factor of $V_d/V_{sp} = \tilde{K}_{dd}/(\tilde{K}_{dd} + R_n)$ (Kawato and Tsukahara, 1983; Turner, 1984; Koch and Poggio, 1985c).

At the other extreme, if g_{syn} is large (that is, if $g_{syn}\tilde{K}_{sp,sp} \gg 1$), the voltage at the spine head saturates and the spine potential converges toward the synaptic reversal potential

$$V_{sp} \to E_{syn} \tag{12.13}$$

while the dendritic EPSP converges to

$$V_d \to \frac{\tilde{K}_{dd}E_{syn}}{\tilde{K}_{sp,sp}} = \frac{\tilde{K}_{dd}E_{syn}}{\tilde{K}_{dd} + R_n}. \tag{12.14}$$

Because in this regime the synapse-spine complex acts as a voltage—rather than a current—source, that is, like a battery, changes in the geometry of the spine neck will—via changes in R_n—affect the amount of synaptic current entering the spine head. Under these conditions, changes in spine morphology will change the effective "weight" of the synapse.

Figure 12.6A indicates graphically how stretching or squishing the spine neck, keeping its membrane area constant, affects the somatic EPSP for a sustained synaptic input of $g_{peak} = 1$ nS. Spine geometry only plays a role for very thin and elongated spines, when $g_{peak} > 1/R_n$ (the curve bends around $d_n = 0.05$ μm, where $g_{peak} \approx 1/R_n$). For transient inputs, very similar considerations apply (Fig. 12.6A and B). Owing to the very small membrane area of the spine head, the total spine capacity is exceedingly small, leading to very large impedance values even at high frequencies. Thus, relative independence of the dendritic membrane potential for small conductance inputs and saturation for large inputs also hold for very rapid inputs (Fig. 12.6). Spine neck geometry can affect the weight of the synapse significantly if the spine starts out very thin and elongated and becomes very short and chubby. (However, a factor of 5 change of the somatic EPSP requires an order of magnitude change in both l_n and d_n.)

Depending on the product of the synaptic conductance change and the spine input resistance, $g_{syn}\tilde{K}_{sp,sp} = g_{syn}(\tilde{K}_{dd} + R_n)$, the spine-synapse complex will thus tend to act either more as a current or as a voltage source, constraining the extent to which the morphology of the spine can alter the synaptic efficiency of the synapse. An additional requirement for the spine shape to influence the synaptic weight is that R_n needs to be large compared to the dendritic input resistance. Otherwise, modulating R_n will have little or no effect on $\tilde{K}_{sp,sp}$ (Eq. 12.10).

We can now pose the critical question: are spines elongated and thin enough to be able to effectively modulate synaptic weight? Technical advances in the last decade have provided sufficient data to estimate the spine operating regime. Probably the most complete data are available at the Schaffer collateral input to region CA1 pyramidal cells in the hippocampus. Here, the experimental estimates of g_{syn} are 0.05–0.2 nS for the fast voltage-independent AMPA synaptic component (Bekkers, Richerson, and Stevens, 1990; Malinow and Tsien, 1990) and less than 0.5 nS for the NMDA component (Bashir et al., 1991).

Based on their reconstruction of spines in the same region, Harris and Stevens (1988b, 1989) estimate the spine neck conductance $G_n = 1/R_n$ to lie mainly between 18 and 138 nS, making the critical ratio g_{syn}/G_n very small. Even taking account of the *spine apparatus*, the smooth endoplasmic reticulum that partly occludes the spine neck, does not change this conclusion appreciably. Svoboda, Tank, and Denk (1996), on the basis of high-resolution two-photon microscopy, directly estimated the diffusive and resistive coupling between the spine head and its parent dendrite, concluding with a lower bound on G_n of 7 nS.

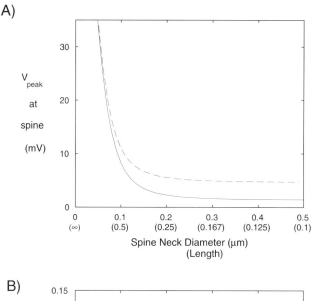

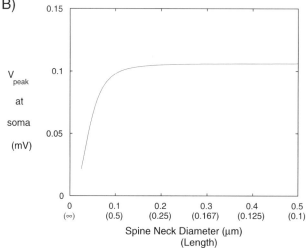

Fig. 12.6 How Do Changes in Spine Neck Geometry Affect Synaptic Weight? Neck length l_n of a single spine is varied in the presence of a single fast ($t_{peak} = 0.5$ msec) voltage-independent excitatory synapse ($g_{peak} = 1$ nS) on a spine along the upper portion of the apical tree in the layer 5 pyramidal cell (around location $a1$ in Fig. 3.7). We here assume that the total surface area of the spine neck remains constant (at 0.157 μm^2) as d_n (upper part of label on x axis) is changed, implying that the spine neck length l_n (lower part of label on x axis) changes inversely. It follows from Eq. 12.1 that the spine neck resistance scales as d_n^{-3}. The resultant EPSP is computed at the spine head itself (in **A**) as well as at the soma (in **B**). Experimental estimates of the spine neck resistance R_n are below 150 MΩ, corresponding here to values of $d_n > 0.15$ μm, falling on the portion of the curve that is relatively flat. Given the large spine neck conductance relative to the peak synaptic conductance change (estimated to be about 0.2 nS for the AMPA component), changes in spine geometry are likely to have only a minor effect on the size of the associated EPSP. In other words, postsynaptic changes in spine geometry are unlikely to implement synaptic "weight" changes. The dotted curve in **A** corresponds to the steady-state change in membrane potential in response to a sustained conductance input of the same amplitude as the transient input. When the spine is shielded from any dendritic capacitance, that is, for $d_n \to 0$, the capacitance of the spine itself can be neglected.

In agreement with earlier estimates of spine dimensions (Wilson, 1984; Turner, 1984; Brown et al., 1988), we conclude that the hippocampal projection onto spines on CA1 pyramidal cells acts as a current source and that changing the morphology of these spines most likely does not affect the associated dendritic EPSP. A similar conclusion has also been reached for a reconstructed spiny stellate cell in somatosensory cortex (Segev et al., 1995).

A different way to understand this is to ask: What effective synaptic conductance does an excitatory synapse directly on the dendrite need to possess in order to inject the same current as an excitatory synapse of amplitude g_{syn} on a spine? Neglecting all capacitive effects and any other synaptic input, the single spine will inject the current $I_n = g_{syn}E_{syn}/(1+R_n g_{syn})$ into the dendrite. Thus, a synapse on a spine can be mimicked by a dendritic synapse of amplitude

$$g'_{syn} = \frac{g_{syn}}{1 + R_n g_{syn}} . \tag{12.15}$$

We see that unless the product of the spine neck resistance with the synaptic conductance change is on the order of unity (or larger), the effect of the spine can be neglected.

The observation that spine necks probably do not contribute toward modulating synaptic weight agrees with experimental evidence regarding the mechanisms underlying the expression of LTP in the hippocampus. While the specific pathways and sites of action remain controversial, there is little to suggest that a postsynaptic change in the electrical impedance of the spine is involved (see the next chapter). This conclusion is also likely to hold in the neocortex, unless synaptic conductance changes are substantially larger and spines thinner and longer than in hippocampus.

12.3.3 Joint Excitatory and Inhibitory Input to a Spine

As mentioned above, it has been a consistent observation that between 5 and 15% of all spines in the cortex carry symmetrical, GABAergic synaptic terminals. Almost always, these same spines are also contacted by excitatory synapses. Such a pair of excitatory and inhibitory synapses localized on a single spine would be the ultimate in highly specific microcircuitry.

Koch and Poggio (1983a,b; see also Diamond, Gray, and Yasargil, 1970) proposed that such a *dual* synaptic arrangement can implement a temporally and spatially very specific veto operation on the basis of the nonlinear interaction between excitation and inhibition of the silent or shunting type (see Sec. 5.1). Indeed, they found that fast $GABA_A$ inhibition can reduce the EPSP in the spine to very small levels provided that it arrived within a very tight time window around the onset of excitation (Fig. 12.7A). According to these numerical solutions to the linear cable equation, such a microcircuit implements an AND-NOT gate with temporal discrimination at the 0.1 msec level.

The degree of temporal specificity, caused by the exceedingly small capacitance in the spine head, can be quantified by the use of the *input delay* in the spine $D_{sp,sp}$ (Secs. 2.4 and 3.6.2). Following Agmon-Snir and Segev (1993), this corresponds to the delay between the centroid of the synaptic current and the centroid of the local EPSP. On the basis of the electrical model of the spine illustrated in Fig. 12.5 and neglecting the tiny current loss across the spine head membrane, we have

$$D_{sp,sp} \approx \frac{\tilde{K}_{dd}}{\tilde{K}_{dd} + R_n} D_{dd} \tag{12.16}$$

A)

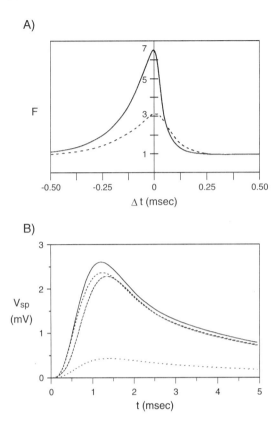

B)

Fig. 12.7 AND-NOT GATING AT THE LEVEL OF A SINGLE SPINE   Nonlinear interaction of the AND-NOT type between a fast excitatory synaptic input (of the AMPA type) and inhibition at a single spine. **(A)** The effectiveness of silent GABA$_A$ inhibition in reducing excitation when both are colocalized either on the spine (solid curve) or on the dendrite at the base of the spine (dotted curve). The curve shows the F factor (Eq. 5.14), that is, the ratio of EPSP without inhibition to the mixed excitatory-inhibitory potential, as a function of the onset of inhibition relative to excitation. The half-widths of the curves are 0.12 and 0.22 msec for the spine and dendrite case, respectively. The linear cable equation for a spine located on a pyramidal cell was solved with $t_{peak} = 0.25$ msec for both excitation and inhibition, $E_i = 0$ and $E_e = 80$ mV. Reprinted in modified form by permission from Koch and Poggio (1983b). **(B)** Time course of the spine membrane potential (relative to rest) for the nonlinear Nernst-Planck electrodiffusion model. Here a small EPSP (solid line; $t_{peak} = 1$ msec, $G_{Na} = 0.1$ nS) can be effectively vetoed by a hyperpolarizing potassium conductance increase (due to a GABA$_B$ inhibition with $t_{peak} = 1$ msec, $G_K = 1$ nS; lower dotted curve), while the same or a hundred times stronger conductance change associated with silent inhibition of the GABA$_A$ type (intermediate dashed curves) had almost no effect. Reprinted in modified form by permission from Qian and Sejnowski (1990).

where D_{dd} is the local dendritic delay at the base of the dendrite. If the spine neck is sufficiently thin and elongated, the time window during which inhibition can effectively veto excitation can be a fraction of the corresponding dendritic time window. This effect is particularly strong for spines close to the soma—providing an isolated environment for excitation and inhibition to interact on a very small time scale—while the local spine delay for spines in the distal dendritic tree will be close to the local dendritic delay $(\tilde{K}_{dd}/R_n \to \infty)$.

However, as described in Sec. 11.3, for very thin processes it is important to account for the change in the relevant ionic concentrations. In the case of a spine, even small synaptic inputs can significantly alter the reversal potentials for chloride and potassium, both across the membrane and between the spine head, neck, and the dendrite. This requires solving the nonlinear Nernst-Planck equation as advocated by Qian and Sejnowski (1989, 1990). They studied the interaction between excitation and inhibition on a spine and found that for reasonable inhibitory conductance changes associated with chloride ions (such as would be caused by activation of $GABA_A$ receptors), the intracellular concentration of chloride rapidly increases, bringing the synaptic reversal potential of inhibition substantially above the resting potential, thereby rendering it ineffective (Fig. 12.7B). Indeed, massive activation of $GABA_A$ receptors in thin processes will lead to a depolarization as the chloride concentration gradient decreases, shifting E_{Cl} to progressively more depolarized values. Such "paradoxical" EPSPs have indeed been observed in distal dendrites of pyramidal cells following intense $GABA_A$ activation (Staley, Soldo, and Proctor, 1995).

While shifts in the potassium reversal battery will also occur, rendering $GABA_B$ inhibition equally ineffective for large conductance changes, the situation is quite different for small inputs. Here E_K moves toward the resting potential of the spine, causing a profound reduction in the amplitude of the EPSP without leading to a hyperpolarizing response (Fig. 12.7B). This leads Qian and Sejnowski (1990) to argue that should inhibition play a role in selectively vetoing excitation on a spine or on thin dendrites, that inhibition must be of the $GABA_B$ type (for large dendrites, the advantage reverts to silent $GABA_A$ inhibition). We find no fault in their arguments.

Must dual-input cortical spines necessarily be functional? Dehay et al., (1991) throw the entire idea into doubt by serially reconstructing selected spines that receive geniculate input in cat primary visual cortex. Here, as elsewhere, about 8% of all spines in the input layer (layer 4) carry both an excitatory as well as an inhibitory synaptic input. Because Emerson et al., (1985) as well as Koch and Poggio (1985b) hypothesized that direction and orientation selectivity in visual cortex neurons is mediated by very specific interactions among excitation and shunting inhibition possibly localized on spines, Dehay and colleagues investigated whether geniculo-cortical synapses on spines are a preferred target for inhibition. Using a laborious electron-microscopic reconstruction technique, they showed that these synapses do not have a higher likelihood of being paired with an inhibitory synapse than the majority of synapses originated among cortical cells. Furthermore, in certain cases Dehay et al., (1991) observed an axon from an inhibitory cell making both a synapse on a spine and, a few micrometers away, another synapse on the parent dendrite, revealing a lack of spatial specificity.

At least in cat visual cortex, it appears that dual-input spines could simply be the logical consequence of an imprecise developmental rule that specifies that inhibitory synapses should primarily innervate dendrites and that a small fraction fails to do so.

12.3.4 Geniculate Spine Triad

In other parts of the brain, however, the association between dual synaptic input and spines is much stronger. We encountered already one such *microcircuit* (Rall et al., 1966; Shepherd, 1978) in the mammalian olfactory bulb (Sec. 5.3), which mediates self and lateral inhibition. We will here discuss another instance of a synaptic microcircuit, involving dual input to a spinelike structure.

The mammalian lateral geniculate nucleus (LGN), the midway station for visual input between the retina and the visual cortex, is the site of a very complex synaptic structure known as *glomerulus* (Fig. 12.8). At the heart of each glomerulus lies a terminal from a retinal axon. It is surrounded by the complex intertwining of very thin dendritic processes of a geniculate interneuron with those of a geniculate relay cell (Hamori et al., 1974; Hamos et al., 1985). This circuit is also termed a *triad* because of the intimate association among three different synaptic terminals: the retinal input makes a synapse both on the spine of a geniculate relay cell and on the interneuron. The GABAergic interneuron, in turn, makes a synapse on the geniculate spine.

Retinal output is organized in multiple parallel channels of information. One of these, the X cells, which are particularly sensitive to high spatial frequencies, tends to make the majority of synaptic contacts onto geniculate relay cells and interneurons in association with these spine triads (Wilson, Friedlander, and Sherman, 1984). At least in cat, the Y cells, the other major retinal output pathway, tend to make their synapses directly onto the dendrites of their geniculate target relay cells, bypassing the inhibitory interneurons (Sherman and Guillery, 1996).

Passive cable modeling of geniculate relay cells with the spine-triad local circuit using excitation and shunting inhibition (Koch, 1985) shows that if the size of the inhibitory

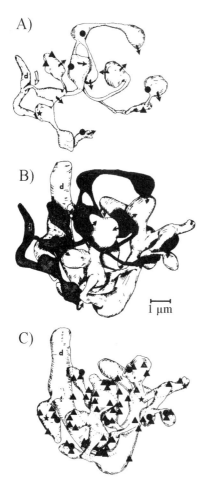

A)

B)

1 μm

C)

Fig. 12.8 Synaptic Spine Triads in the LGN
These drawings, based on laborious electron-microscopic reconstructions of cells in the cat LGN (Hamos et al., 1985), illustrate the very intricate and baroque synaptic microcircuits that can be found in the nervous system. The arrangement shown, known as a *glomerulus*, includes **(A)** a dendrite (marked d) of a local inhibitory interneuron from which a dozen very thin processes spring forth. The full circles indicate the location of excitatory synapses from a retinal axon. The interneuron in turn inhibits the geniculate relay cell at the nine sites indicated with full arrows. **(B)** This interneuron (in black) is shown grappling with five spinelike appendages on a dendrite of a geniculate relay cell (in white), which are shown more fully in **(C)**. The full circles indicate nine retinal synapses, and the filled triangles 40 inhibitory synapses from other interneurons. The *triad* is a synaptic arrangement in which an excitatory input (here, an optic nerve axon) excites both an interneuron as well as the geniculate relay cell. The interneuron, in turn, directly inhibits this site. Reprinted by permission from Hamos et al., (1985).

conductance change due to the $GABA_A$ receptors is neither too small nor too large, it can effectively and specifically veto the retinally induced EPSP without inhibiting EPSPs from neighboring synapses. Thus, although the inhibition is postsynaptic, from a functional point of view it acts similar to *presynaptic inhibition*.

Because of the expected 1 or 2 msec delay between retinal input triggering an EPSP in the interneuron and the opening of postsynaptic $GABA_A$ receptors on the spine in response to this input, inhibition is likely to be not very effective at low firing rates (caused, for instance, by a low contrast stimulus). Inhibition would be strongly activated for sustained, high-contrast input. Because the spine-triad circuit is essentially limited to X cells, Y cells should not show any comparable localized and specific type of inhibition. Experimentally, it is known that geniculate cells are more phasic than their retinal counterparts (Cleland, Dubin, and Levick, 1971) and that inhibition is independent of stimulus contrasts for Y cells but increases with increasing stimulus contrast in X cells (Berardi and Morrone, 1984). Finally, it is clear that the level of activity in the interneuron itself will be crucial in determining whether or not it can inhibit the retinal input (Bloomfield and Sherman, 1989). Such context-dependent inhibition could be important, for instance, during saccadic suppression.

We conclude that the spine-triad microcircuit on geniculate X cells appears to implement a specific type of activity-dependent inhibition (Hamori et al., 1974; Koch, 1985).

12.4 Active Electrical Properties of Single Spines

In the mid-1980s, several independent groups worked on the idea of synaptic amplification in dendritic spines endowed with *active, regenerative* electrical properties (Perkel and Perkel, 1985; Miller, Rall and Rinzel, 1985; Shepherd et al., 1985; Pongracz, 1985; Segev and Rall, 1988). At first viewed with skepticism by experimentalists, recent calcium-imaging experiments have shown that spines on hippocampal pyramidal as well as cerebellar Purkinje cells can show all-or-nothing responses (Yuste and Denk, 1995; Denk, Sugimore and Llinás, 1995; see Sec. 12.6.2). It remains to be seen whether these spikes are actually restricted to the spines or whether they invade the neighboring dendrites.

The basic idea is simple and will be explained with reference to the painstakingly executed Segev and Rall (1988; see also Rall and Segev, 1987) study. They endowed the head of a single spine—on an otherwise passive dendrite—with the fast sodium and delayed-rectifier potassium conductances found in the squid axon (adjusted for 20° C and with a tenfold increase in peak conductances; see Fig. 12.9). With a single-channel conductance of about 15 pS (Table 8.1) this corresponds to 1050 sodium channels per spine. Using a very fast synaptic input ($t_{peak} = 0.035$ msec), Segev and Rall found that the amplitude of the synaptic input g_{peak} had to exceed a very sharp threshold in order to generate an all-or-nothing action potential in the spine (Fig. 12.10). Such a tongue-twisting "spine spike" will be initiated once V_{sp} exceeds a critical voltage threshold V_{th} (see the discussion in Secs. 6.3.1 and 19.2). At V_{th}, the sum of the synaptic and active spine currents exceeds the current I_n flowing through the spine neck and onward into the dendrite.

In the suprathreshold domain, the amplification that can be obtained with a small number of strategically placed active channels on a spine is evident in the lower graph of Fig. 12.10. The 90 mV spine action potential is attenuated across the 1000 MΩ neck but still leads to a 5 mV EPSP at the base of the spine. Contrariwise, a passive spine would have depolarized to 27 mV, of which only about 1 mV would have survived the attenuation through the spine neck (dotted lines in Fig. 12.10). Thus, a tiny 1 μm^2 patch of active channels leads to a fivefold voltage amplification.

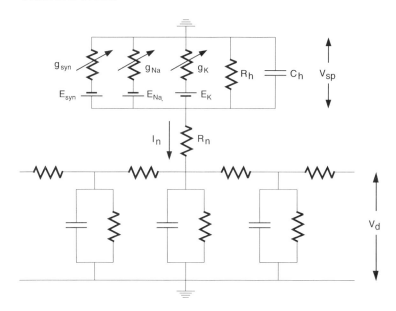

Fig. 12.9 MODEL OF AN ACTIVE SPINE Lumped electrical model of an active spine, as used in the Segev and Rall (1988) study. The spine head is endowed with squid axon membrane (at 20° C, but with a tenfold increase in the maximal conductances $\overline{G}_{Na}$ and $\overline{G}_K$). With $C_m = 1\ \mu F/cm^2$, the passive time constant in the spine at rest is 1.4 msec. The synaptic input is very fast, with $t_{peak} = 0.035$ msec and $E_{syn} = 100$ mV. A tenfold slower synaptic input is expected to yield a similar behavior in the presence of a tenfold slower membrane time constant. The input resistance at the base spine is 262 MΩ.

The amplitude (as well as the integrated area) of the dendritic EPSP is independent of variations in g_{peak} (as long as g_{peak} is above threshold; Fig. 12.10), since the maximal value of V_{sp} is determined by the height of the action potential, which is rather fixed. Given the large variability in the postsynaptic amplitude of individual synaptic inputs at cortical synapses (see, Fig. 4.4), active spines can thereby eliminate a source of uncertainty associated with passive spines (as well as amplify their responses).

Figure 12.11 illustrates the dependency of V_{sp} and V_d on the spine neck resistance. For small values of R_n (here below 620 MΩ), the spine input impedance $\tilde{K}_{sp,sp}$ is not sufficiently high in order for the local EPSP to exceed V_{th}: the peak value of both V_{sp} and V_d in the presence of sodium and potassium channels is little different from the peak value of the potential in the passive situation (right panels in Fig. 12.11). For $R_n = 630$ MΩ, a spike is generated after a short delay, while for the high resistance neck (curve b), an action potential is established in a very secure manner and with a larger amplitude. Seemingly paradoxically, however, the amplitudes of the resulting EPSPs at the dendrite are reversed. While a larger spine neck resistance increases the spine input impedance (witness Eq. 12.4), thereby facilitating spike generation, it also causes a larger voltage attenuation across the spine neck. In the limit of an infinitely thin spine neck, the smallest input will cause V_d to "lock up" at E_{syn}; yet none of that will be visible at the base of the spine. (For all the subtleties of this dependency, see Segev and Rall, 1988.) As the lower right panel in Fig. 12.11 best summarizes, amplification over the passive spine is obtained within a narrow range of R_n. Given the inverse quadratic dependency of R_n on d_n (Eq. 12.1), this translates into a very strong dependency of V_d on the spine neck geometry.

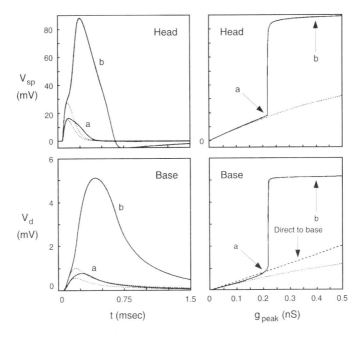

Fig. 12.10 ACTIVE SPINES AS A FUNCTION OF THE INPUT AMPLITUDE   Time course and peak depolarization in the spine and dendrite shown in Fig. 12.9 as a function of the amplitude of the fast (t_{peak} = 0.035 msec) voltage-independent synaptic conductance change g_{peak}. The two left panels illustrate $V_{sp}(t)$ and $V_d(t)$ for two values of g_{peak} (0.2 nS for a and 0.4 nS for b) while the right panels plot the peak values of V_{sp} and V_d as a function of the input amplitude. R_n is always set to 1 GΩ. Dotted curves are for a passive spine, whereas the solid curves are for an active spine. Close inspection of these plots reveals some minimal evidence of active channels for the smaller inputs of the two. If the input is doubled, a very robust all-or-none spike is initiated. Further increases of g_{peak} have no further effect on the peak of either V_{sp} or V_d. The dashed line in the lower right panel shows the peak of V_d when the same synaptic input is applied directly to the passive dendrite, rather than to the spine. Active spines can be used to amplify synaptic input. Reprinted by permission from Segev and Rall (1988).

At present, we do not know whether sodium channels are present on dendritic spines. However, phenomena similar to those discussed here could be instantiated with slower calcium-mediated all-or-none spine spikes. Two-photon microscopy experiments using calcium dyes have provided tantalizing preliminary evidence for all-or-none calcium events following synaptic input onto spines of hippocampal and Purkinje cells (see Sec. 12.6.2).

What Segev and Rall (1988; see also Perkel and Perkel, 1985; Miller, Rall, and Rinzel, 1985; Shepherd et al., 1985) demonstrate is that active spines can significantly amplify (two to tenfold) synaptic input at minimal metabolic costs (only requiring the insertion of voltage-dependent channels into a very small membrane patch the size of a spine head). We conclude that active spines could provide the biophysical substrate for *information storage* as well as *information processing*.

The strong dependency of V_d on parameters associated with spine neck geometry provides the cell with a sensitive mechanism to modulate the synaptic weight of individual inputs. Only within a narrow range of R_n will the synaptic input be high; outside of this range, the dendritic EPSP will be much reduced. At the same time, this sensitivity also represents the Achilles heel of the hypothesis: as evidenced by the large degree of variability in the

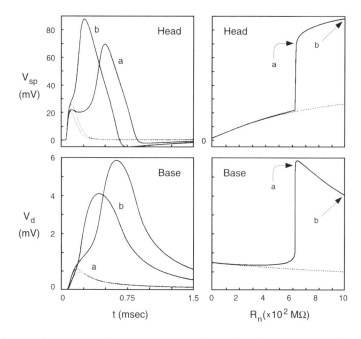

Fig. 12.11 ACTIVE SPINES AS A FUNCTION OF THE SPINE NECK RESISTANCE   Time course and peak depolarization in the spine and dendrite shown in Fig. 12.9 as a function of the spine neck resistance (for $g_{peak} = 0.37$ nS and $t_{peak} = 0.035$ msec). The two left panels illustrate $V_{sp}(t)$ and $V_d(t)$ for $R_n = 630$ MΩ (curve a) and $R_n = 1000$ MΩ (curve b) while the right panels plot the peak values of V_{sp} and V_d as a function of R_n. Dotted curves are for a passive spine, whereas the solid curves are for an active spine. Large amplification of the dendritic EPSP with respect to a passive spine only occurs in a relatively narrow range of R_n values: if the neck is too short or too thick, the electrical decoupling of excitable channels at the spine head from the rest of the cell is insufficient to initiate an action potential, while for very thin and long spine necks (implying large values of R_n), the attenuation across the spine neck is excessive. This dependency on spine neck geometry could be exploited in synaptic plasticity. Reprinted by permission from Segev and Rall (1988).

spine neck geometry (Fig. 12.4), the brain might not be able to control spine neck geometry with sufficient accuracy to exploit it in a consistent manner.

What remains unclear from an experimental point of view is whether action potentials are restricted to the spines themselves or whether they spread to the parent dendrite. As witnessed by Fig. 12.11, a true spiking spine requires a very thin or elongated neck to achieve the necessary isolation from the parent dendrite.

More interesting from the point of view of neuronal computation are the combinatorial possibilities for nonlinear interactions among active spines on distal dendrites. Under certain circumstances, one or a small number of simultaneous inputs to active spines can trigger action potentials in neighboring spines. We will pick up the threads of this story in Sec. 19.3.2.

12.5 Effect of Spines on Cables

Given the large spine density (up to 10 spines per micrometer of dendrites), 50% or more of the total neuronal membrane area resides in spines. It is therefore important to account for this additional area to properly analyze the propagation of electrical signals in the dendritic tree. Although these effects provide no insight into the function of spines, they may be ironically the only ones that we can infer with confidence to be important.

Accounting for the effect that spines have on cable propagation is based on the assumption that for current flow from the dendrite back into the spine the membrane area of the spine can be incorporated into the membrane of the parent dendrite. As we expect from basic cable theory and as discussed more fully below, there exists almost no voltage attenuation from the dendrite back into the spine (contrary to the voltage attenuation from the spine into the dendrite). This isopotentiality is the reason that the spine area can be lumped into the overall area of the dendrite. Several different methods to deal with this problem have been proposed, based on the intuition that in the presence of a large number of spines, the total capacitance will increase while the membrane resistance will decrease.

One method (Stratford et al., 1989) transforms the spiny dendrite (with diameter d and length l) into a single "equivalent" (and smooth) cylinder, whose diameter d' and length l' are larger than those of the parent dendrite by some fraction $\mathcal{F}$, with

$$\mathcal{F} = \frac{A_{\text{dend}} + A_{\text{spine}}}{A_{\text{dend}}} \tag{12.17}$$

where A_{dend} is the total membrane area of the "stem" dendrite without any spines, and A_{spine} is the total membrane area of all spines (spine necks and spine heads) on the dendrite. The dimensions of this equivalent cable are

$$l' = l \mathcal{F}^{2/3} \tag{12.18}$$

and

$$d' = d \mathcal{F}^{1/3} . \tag{12.19}$$

Solving for the membrane potential in a cable studded by n spines requires the solution of $2n + 1$ coupled equations (assuming that each spine is modeled using two compartments, one for the neck and one for the head). The transformation described here reduces this to solving the cable equation in a single, unbranched cylinder. The effective area of the cable is scaled by $\mathcal{F}$, its infinite input resistance by $\mathcal{F}^{-1/2}$, and its electrotonic length of the equivalent cable by $\mathcal{F}^{1/2}$. The time constant remains unchanged. The transformed segment behaves like the spiny original under most circumstances (Stratford et al., 1989).

An alternative transformation method (Shelton, 1985; Holmes, 1989; Segev et al., 1992) preserves the dimensions of the dendrite but scales C_m and R_m appropriately by

$$R'_m = \frac{R_m}{\mathcal{F}} \tag{12.20}$$

and

$$C'_m = C_m \mathcal{F} . \tag{12.21}$$

Both methods are mathematically equivalent.

A novel analytical method to incorporate spines into the continuous cable equation was developed by Baer and Rinzel (1991). It is powerful, since it can deal with passive as well as active spines. Baer and Rinzel replace the cable equation for a single passive cable by two equations, one for the voltage along the dendrite $V_d(x, t)$ and one for the voltage in the spines $V_{sp}(x, t)$. The density of spines per unit electrotonic length of cable is $s(x)$. A single spine receives synaptic input current $I_{\text{syn}}(x, t)$ (which can vary with the position of the spine along the cable), generating a local current across the spine head I_{ionic} that can include Hodgkin–Huxley or other voltage-dependent currents. Spines are independent of each other, only interacting with the dendrites via the neck current I_n that flows through the spine neck. The associated cable equation can be expressed as

$$\tau \frac{\partial V_d}{\partial t} = \lambda^2 \frac{\partial^2 V_d}{\partial x^2} - V_d + s(x) R_\infty I_n \tag{12.22}$$

$$C_h \frac{\partial V_{sp}}{\partial t} = -I_{ionic} - I_n - I_{syn}(x, t) \tag{12.23}$$

with an additional equation for the neck current,

$$I_n = \frac{V_{sp} - V_d}{R_n}. \tag{12.24}$$

Here τ, λ and R_∞ are the time and space constants and the input resistance of a semi-infinite cable in the absence of any spines, and C_h is the tiny spine head capacitance (Fig. 12.5).

Baer and Rinzel (1991) go on to solve these equations for active as well as passive spines. In the former case, wavelike propagation can occur along the passive cable, supported by the amplifying spines. In the latter case, both equations are linear and $I_{ionic} = V_{sp}/R_h$. Assuming a constant density of spines, $s(x) = s$, and steady-state conditions, they compute the effective electrotonic length of a spiny dendrite as

$$L' = L\sqrt{1 + s R_\infty/(R_n + R_h)} \tag{12.25}$$

and the infinite input resistance as

$$R'_\infty = \frac{R_\infty}{\sqrt{1 + s R_\infty/(R_n + R_h)}}. \tag{12.26}$$

Since under almost all circumstances $R_h \gg R_n$, these reduce to

$$L' = L\mathcal{F}_{1/2} \tag{12.27}$$

and

$$R'_\infty = R_\infty \mathcal{F}^{-1/2} \tag{12.28}$$

as in the more heuristic methods discussed at the beginning of this section.

In summary, all of the spine-replacement methods discussed here conclude that the collective effect of dendritic spines is to increase the effective membrane capacitance and decrease the membrane resistance, leading to a lower input resistance and an increased electrotonic length compared to a smooth dendrite of equal diameter and physical length (Wilson, 1988). Indeed, Jaslove (1992) argues that these population effects—influencing the degree of spatio-temporal integration occurring in the dendrites—are the primary functional reason for why some cells are studded with spines while others are devoid of them.

12.6 Diffusion in Dendritic Spines

So far, we have focused exclusively on the electrical properties of spines. Yet, over the last decade, an alternative view of spines has emerged which emphasizes their effect on chemical, rather than electrical, signaling. As discussed in Chap. 13, chemical dynamics of intracellular calcium and other second messengers in the spine are of particular importance in the induction of long-term potentiation.

12.6.1 Solutions of the Reaction-Diffusion Equation for Spines

Compartmental modeling of calcium diffusion and binding following synaptic input to the spine (Robinson and Koch, 1984; Coss and Perkel, 1985; Gamble and Koch, 1987; Wickens, 1988) played a trail-blazing role here, followed a few years later by the experimental study of calcium dynamics in single spines (Müller and Connor, 1991; Guthrie, Segal and Kater, 1991; Jaffe, Fisher, and Brown, 1994; Eilers, Augustine, and Konnerth, 1995; Yuste and Denk, 1995; Denk, Sugimore, and Llinas, 1995; Svoboda, Tank, and Denk, 1996). More recent modeling studies have, of course, become much more sophisticated and use a very fine grained spatial resolution to keep track of substances in the spine head and neck (Holmes and Levy, 1990; Zador, Koch, and Brown, 1990; Koch, Zador, and Brown, 1992; DeSchutter and Bower, 1993; Gold and Bear, 1994; Zador and Koch, 1994; Woolf and Greer, 1994; Jaffe, Fisher and Brown, 1994).

The principal ideas are really very simple, in particular in light of the linearized reaction-diffusion equation discussed in Sec. 11.7. Using the geometry of Fig. 12.12A, and in the presence of low calcium concentration, a linear nonsaturable calcium pump in the spine neck and head membrane, and a fast second-order nondiffusible buffer, the reaction-diffusion equations (Eqs. 11.48) in the presence of calcium pumps can be reduced to the linear partial differential equation (Eq. 11.51), formally identical to the cable equation.

This allows us to define the "chemical input resistance" (Carnevale and Rosenthal, 1992; Zador and Koch, 1994), in analogy to the standard electrical input resistance, as the change in calcium concentration in response to a current of calcium ions (Table 11.2). Both the chemical and the electrical spine input resistances are much larger than the associated dendritic input resistances. At the spine head, the input resistance is 4.2×10^{-2} μM/fA while the dendritic input resistance is 4.8×10^{-3} μM/fA. In other words, a sustained calcium current injected into the spine head leads to a tenfold larger increase in calcium concentration than the same current applied at the dendritic shaft.

A key difference between the electrical and the chemical properties of spines arises from the effect of the spine neck. In our analysis, as well as in almost all published studies of the electrical properties of spines, the electrical cable properties associated with the spine neck are neglected, since no significant current will cross the membrane of the 1-μm long "cable," a direct consequence of the large value of λ. In order to estimate the loss of calcium current through the spine neck membrane, we compute the space constant of Eq. 11.55 associated with the linearized reaction-diffusion equation. Reading off from Fig. 11.9, λ_{r-d} for a 0.1 μm thin cable is 0.46 μm (0.27 μm without diffusible buffer). Because the dendritic shaft is more than $2\lambda_{r-d}$ away from the spine head, the calcium concentration at the shaft is expected to be tenfold lower than at the head. This is in dramatic contrast to the almost complete lack of electrical current attenuation experienced between the spine head and the base (λ is close to three orders of magnitude larger than λ_{r-d}; Fig. 11.9).

These principles are illustrated in simulations of the fully nonlinear calcium dynamics thought to underly the induction of LTP (Zador, Koch, and Brown, 1990; Brown et al., 1991b). In the model, synaptic input activates a fast non-NMDA conductance (Eq. 4.5) as well as an NMDA voltage-gated conductance (Eq. 4.6; Fig. 12.12A). 2% of the NMDA current is assumed to be carried by Ca^{2+} ions. Subsequent to entry into the intracellular cytoplasm, these ions can bind to one of the four calmodulin binding sites (Fig. 11.7),

A)

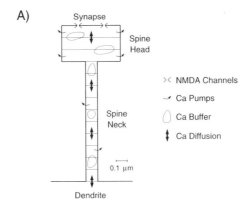

B)

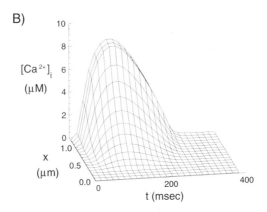

Fig. 12.12 Sᴘᴀᴛɪᴏ-Tᴇᴍᴘᴏʀᴀʟ Dʏɴᴀᴍɪᴄs ᴏғ Cᴀʟᴄɪᴜᴍ ɪɴ ᴀ Sᴘɪɴᴇ   (**A**) Compartmental representation for solving the reaction-diffusion equation associated with the system controlling intracellular
calcium in response to a mixed excitatory voltage-independent non-NMDA and a voltage-dependent
NMDA synaptic input. Opening of the NMDA channels causes an influx of calcium ions. These
are removed by the action of membrane-bound pumps in the spine head and neck, by binding to
cytoplasmic calcium buffers, as well as by diffusion into the dendrite. The drawing is to scale, with
$d_n = 0.1\ \mu$m. The individual compartments are shown by thin lines. (**B**) Spatio-temporal dynamics
of Ca^{2+} in response to a train of three presynaptic stimuli (at 100 Hz) while the membrane potential
in the spine was simultaneously clamped to -40 mV (mimicking a somatic voltage-clamp protocol).
Changes in $[Ca^{2+}]_i$—induced by the calcium influx through the NMDA channel—are restricted
mainly to the spine head. x indicates the distance from the dendritic shaft (with 0 corresponding to
the shaft and $x = 1.3\ \mu$m to the subsynaptic cleft). Reprinted by permission from Zador, Koch, and
Brown (1990).

they can be pumped out of the cell via two different pumps, and they can diffuse along
the spine neck and out into the dendrite (assumed to be an infinite capacity sink with a
resting calcium level of 50 nM). Simulated calcium dynamics following a train of synaptic
stimuli are shown in Fig. 12.12B. Due to the tiny volume of the spine, the small Ca^{2+}
influx following synaptic stimulation leads to a large transient increase in the spine calcium
concentration (here, $[Ca^{2+}]_{sp} = 10\ \mu M$ corresponds to about 350 Ca^{2+} ions). We conclude
that spines can amplify the small incoming calcium signal dramatically. Yet due to the large
mismatch in volumes and the associated small chemical input impedance at the dendrite, the

large peak changes in $[Ca^{2+}]_i$ at the spine cause the dendritic level of $[Ca^{2+}]_i$ to fluctuate only by several tens of nanomolars.

Due to the difference in input impedances between the spine and the dendrite, both calcium and voltage will be severely attenuated when going from the spine to the dendrite, even though no current is lost across the spine neck membrane in the case of the cable equation. This is illustrated in a different manner in Fig. 12.13, showing both voltage and calcium attenuation along the spine, that is, the ratio of the voltage (or calcium concentration) at one location to the voltage (or calcium concentration) at another location.

Quite a dramatic difference emerges when considering antidromic attenuation. When voltage clamping the dendrite to any particular value, the voltage at the spine head will be attenuated by the factor (Eq. 3.34)

$$A_{d,sp} = \frac{\tilde{K}_{dd}}{\tilde{K}_{sp,d}} . \tag{12.29}$$

Since $\tilde{K}_{sp,d} \approx \tilde{K}_{dd}$ (Eq. 12.5) to a pretty good approximation, no voltage attenuation is expected into the spine, as confirmed by the upper dashed line in Fig. 12.13. However, due to the sustained loss of calcium ions as they are being pumped through the spine neck membrane into the extracellular cytoplasm, the effective value of $\tilde{K}_{sp,d}$ for the reaction-

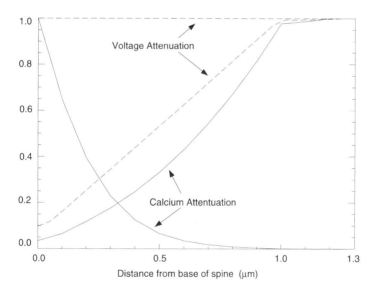

Fig. 12.13 VOLTAGE AND CALCIUM ATTENUATION IN AND OUT OF A SPINE Voltage and calcium attenuation for two different simulations as a function of the distance from the dendrite (at 0; the spine neck ends at 1 μm and the spine head extends from 1 to 1.3 μm). The upward diagonal dashed and solid curves illustrate voltage and calcium attenuation from the spine into the dendrite. These curves are taken from the simulation shown in Fig. 12.12B and indicate peak calcium and voltages following three presynaptic stimuli to the spine synapse. Both variables attenuate about tenfold. In a different simulation illustrating antidromic *steady-state* behavior, either the calcium (downward going solid curve) or the voltage (constant dashed line) is clamped to a fixed value in the dendrite. Due to the presence of calcium pumps in the spine neck, calcium rapidly attenuates and reaches baseline levels at the spine head. The membrane potential, however, does not attenuate across the spine neck due to the negligible current loss across the spine neck and head membranes. Reprinted by permission from Koch and Zador (1993).

diffusion system is very low. As a consequence, when the calcium concentration in the dendrite is "clamped" to 1 μM, the calcium attenuation into the spine head is very large: it remains protected from the high dendritic calcium values by the presence of calcium pumps in the membrane of the spine neck (Zador, Koch, and Brown, 1990; Koch, Zador, and Brown, 1992), providing a graphic illustration of the tiny size of the associated space constant. Given the square-root dependency of λ_{r-d} on the density of pump molecules in the membrane (Table 11.3), this density would have to change substantially before any significant effect of dendrite calcium on the spine head would be seen. Or, short and sweet, "spines are electrically coupled but chemically isolated from changes occurring in the dendrite." Of course, the presence of the spine apparatus or other organelles would further impede the flow of calcium ions into the spine head. What applies to calcium will, of course, also apply to other second messenger systems (see also Woolf and Greer, 1994).

The value of many of the relevant biophysical and biochemical parameters can only be specified within a factor of 2, 5, or even 10. For instance, relatively small changes in the spine neck geometry or in the buffer concentration can cause the peak calcium concentration in the spine head to vary quite a bit. Whether this sensitivity of peak $[Ca^{2+}]_{sp}$ subserves a function, as contended by Gold and Bear (1994) (Fig. 13.7) is not known.

One obvious effect that should be emphasized is that the electrical as well as the chemical input impedance increases with increasingly longer and thinner spines. If the establishment of LTP were to depend on exceeding either a voltage threshold or a critical concentration of $[Ca^{2+}]_i$, the thin and elongated spines would be in a better position than the short, stubby ones to achieve this. Conversely, since the synaptic weight depends little on the spine neck—for the range of geometries reported in hippocampus and for passive spines—stubby spines could be considered to be permanently modified, while elongated or thin spines would constitute a reservoir that can be recruited for long-term synaptic potentiation as well as for long-term depression. Indeed, several reports have emphasized that high-frequency synaptic stimulation that causes LTP leads to an excess of short or thick spines (Lee et al., 1980; Chang and Greenough, 1984).

Although we mainly use arguments from linear cable theory, the simulations in Figs. 12.12 and 12.13, based as they are on the peak transient calcium concentration changes in the full nonlinear model, confirm these expectations.

12.6.2 Imaging Calcium Dynamics in Single Dendritic Spines

While it remains technically impossible to record spine EPSPs directly, a number of groups are "pushing the technological envelope" by imaging calcium activity in dendrites and in individual spines via calcium-sensitive fluorescent dyes (Müller and Connor, 1991; Guthrie, Segal, and Kater, 1991; Regehr and Tank, 1992; Jaffe, Fisher, and Brown, 1994; Yuste and Denk, 1995; Segal, 1995a; Eilers, Augustine, and Konnerth, 1995; Denk, Sugimore, and Llinás, 1995; for a review see Denk et al., 1996). The more recent studies make use of confocal laser scanning or two-photon fluorescence microscopy. They confirm that in the absence of calcium entry through voltage-dependent calcium channels, spines do isolate in both directions: a high calcium concentration in the dendrite is frequently not paralleled by an equally high concentration change in the spine (Fig. 12.14C). Control experiments with injected cobalt indicate that the lag is not due to a physical diffusion barrier between the dendrite and the spine (Guthrie, Segal, and Kater, 1991), supporting the idea that calcium-dependent processes, such as calcium pumps or other uptake systems, were responsible for isolating the spine head. Conversely, synaptic stimulation leads to restricted calcium

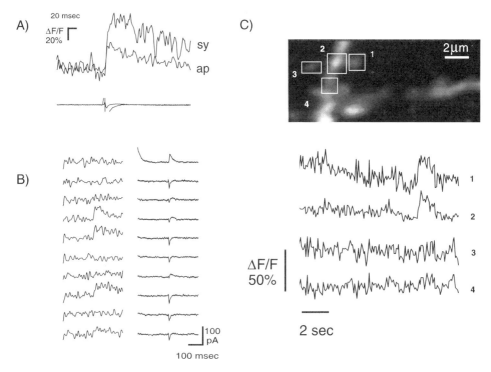

Fig. 12.14 IMAGING CALCIUM IN A SINGLE SPINE   Fluorescence signal from a single spine on a CA1 pyramidal cell in a rat hippocampus slice. (**A**) Using two-photon microscopy, Yuste and Denk (1995) record the increase in fluorescence of a calcium indicator, due to an increase in free intracellular calcium, evoked either by spontaneous synaptic activity (sy) or by the antidromic invasion of a postsynaptic action potential (ap). Following synaptic input, calcium rises within 2 msec, the temporal resolution of the measurement. The influx of calcium in response to a postsynaptic spike is mediated by voltage-dependent calcium currents in the dendrites and spines. (**B**) Stochastic failure of synaptic transmission at the same spine can also be visualized. Left column: fluorescence measurements in response to subthreshold synaptic stimulation. Right column: simultaneous current recording at the soma. In only three cases is a clear synaptic response evident at the single spine. Only in about 20 to 40% of the cases will a presynaptic spike cause release of a synaptic vesicle and a postsynaptic response. (**C**) Spontaneous synaptic activity causes an increase in $[Ca^{2+}]_i$ at a single spine (1) and the adjacent dendritic shaft (2), while a neighboring spine (3) or dendrite (4) shows no such increase. Later during the same recording session, a synaptic-evoked calcium accumulation was restricted to the spine (3). Reprinted by permission from Yuste and Denk (1995).

increase in individual spines but not in the dendrite, confirming the basic aspects of the models discussed above. Indeed, the resolution of these methods is such that stochastic failure of synaptic transmission can be observed in the calcium response at a single spine (Fig. 12.14B).

   In real life, things become more interesting, but also more complicated, by the presence of high-threshold noninactivating voltage-dependent calcium channels in dendrites and spines: both somatic depolarization via an intracellular electrode and antidromic spike invasion lead to calcium increases in dendrites and in spines throughout the cell (Fig. 12.14A; Markram, Helm, and Sakmann, 1995; Sec. 19.1). With the exception of Robinson and Koch (1984) and Gamble and Koch (1987), previous compartmental models have assumed that all calcium entry occurs via the activated NMDA synaptic receptors at the stimulated spine. Since antidromic spike invasion does occur under physiological conditions and may indeed be facilitated by the presence of sodium channels on the apical dendrites (Stuart and Sakmann, 1994; Secs. 19.1 and 19.2.2), it is unclear what this implies about the specificity of LTP, which is generally believed to be restricted to the stimulated synapse (see the following chapter). It might well be that under physiological conditions the critical calcium concentration required to initiate the biochemical cascade that leads to the enhancement of the synapse cannot be achieved unless calcium accumulation from synaptic (calcium entry via NMDA and non-NMDA receptors; Schneggenburger et al., 1993) as well as nonsynaptic (voltage-dependent calcium channels and calcium-triggered calcium release; Regehr and Tank, 1992) sources cooperate.

12.7 Recapitulation

The small extent of dendritic spines, somewhat smaller than the average-sized *Escherichia coli* colon bacterium, precluded until very recently direct experimental access, making the function and properties of dendritic spines a favorite subject among modelers. Hypotheses concerning their function can be divided into two major categories: spines as devices for the induction and/or the expression of synaptic plasticity and spines as devices subserving specific computations. Other proposals have been advanced, such as that spines primarily serve to connect axons with dendrites (Swindale, 1981), that spines serve to increase the effective membrane capacity of a dendrite (Jaslove, 1992), or that spines serve to protect the dendrite from high, and therefore possibly toxic, overdoses of calcium during synaptic activation (Segal, 1995b). While each of these ideas may contain some grain of truth, we here emphasize those properties of spines of direct relevance to information processing and storage.

That changes in the electrical resistance of the spine neck can modulate the EPSP amplitude of a synapse on this spine has been a popular idea since it was discussed in detail by Rall (1970, 1974). It now appears, at least in the case of hippocampal CA1 pyramidal cells (but most likely also for neocortical spines), that the spine neck conductance is too large relative to the synaptic-induced conductance change to be able to effectively modulate the dendritic EPSP for *passive* spines. Furthermore, as covered in the following chapter, the weight of the evidence leans in favor of a pre- rather than a postsynaptic site as the locus where the long-term changes in synaptic weight are affected. It is most likely in the induction phase of Hebbian long-term changes in synaptic plasticity that spines play a critical role.

Both theoretical and experimental evidence is accumulating that spines may create an isolated biochemical microenvironment for inducing changes in synaptic strength. In the induction of associative (or Hebbian) LTP, for example, the spine could restrict changes in postsynaptic calcium concentration to precisely those synapses that met the criteria for potentiation. Furthermore, changes in spine shape could control the peak calcium concentration induced by synaptic input: all other factors being equal, long and skinny spines would have higher peak calcium levels than short and stubby spines. This could provide an alternative explanation for some of the environmental effects on spine morphology. The biochemical compartmentalization provided by dendritic spines could, of course, be equally crucial for a number of other diffusible second messengers, such as IP_3, calmodulin, cyclic AMP, and others. We conclude that while spines may not play any role in the expression of synaptic plasticity, they may be crucial in the induction phase by offering a protected microenvironment for calcium and other messenger molecules.

Spines can provide a very rich substrate for computation if they are endowed with regenerative, all-or-none electrical properties as experimental data are now suggesting. The elevated spine input resistance, as compared to the input resistance of the parent dendrite, and the partial electrical isolation of the spine from the rest of the cell, could make spines a favorable site for the initiation of action potentials. The proposal that spines contain voltage-dependent membrane conductances, first advanced by theoreticians and now experimentally supported by evidence for calcium spikes, implies that a dendritic spine is a basic computational gate with two states, on or off. Modeling studies, treated in Sec. 19.3.2, demonstrate that a population of strategically placed active spines can instantiate various classes of quasi-Boolean logic. The combination of excitatory and inhibitory synapses on a single spine—whether passive or active—observed in certain brain areas can only enhance this computational power. With up to 200,000 spines on certain cell types, it remains to be seen whether the nervous system makes use of this vast computational capacity.

13

SYNAPTIC PLASTICITY

Animals live in an ever-changing environment to which they must continuously adapt. Adaptation in the nervous system occurs at every level, from ion channels and synapses to single neurons and whole networks. It operates in many different forms and on many time scales. Retinal adaptation, for example, permits us to adjust within minutes to changes of over eight orders of magnitude of brightness, from the dark of a moonless night to high noon. High-level memory—the storage and recognition of a person's face, for example—can also be seen as a specialized form of adaptation (see Squire, 1987).

The ubiquity of adaptation in the nervous system is a radical but often underappreciated difference between brains and computers. With few exceptions, all modern computers are patterned according to the architecture laid out by von Neumann (1956). Here the adaptive elements—the random access memory (RAM)—are both physically and conceptually distinct from the processing elements, the central processing unit (CPU). Even proposals to incorporate massive amounts of so-called *intelligent RAM* (IRAM) directly onto any future processor chip fall well short of the degree of intermixing present in nervous systems (Kozyrakis et al., 1997). It is only within the last few years that a few pioneers have begun to demonstrate the advantages of incorporating adaptive elements at all stages of the computation into electronic circuits (Mead, 1990; Koch and Mathur, 1996; Diorio et al., 1996).

For over a century (Tanzi, 1893; Ramón y Cajal, 1909, 1991), the leading hypothesis among both theoreticians and experimentalists has been that *synaptic* plasticity underlies most long-term behavioral plasticity. It has nevertheless been extremely difficult to establish a direct link between behavioral plasticity and its biophysical substrate, in part because most biophysical research is conducted with *in vitro* preparations in which a slice of the brain is removed from the organism, while behavior is best studied in the intact animal. In mammalian systems the problem is particularly acute, but combined pharmacological, behavioral, and genetic approaches are yielding promising if as yet incomplete results (Saucier and Cain, 1995; Cain, 1997; Davis, Butcher, and Morris, 1992; Tonegawa, 1995; McHugh et al., 1996; Rogan, Stäubli, LeDoux, 1997). Even in "simple" invertebrate systems, such as the sea slug *Aplysia* (for instance, Hawkins, Kandel, and Siegelbaum, 1993), it has been difficult to trace behavioral changes to their underlying physiological and molecular mechanisms. Thus, the notion that synaptic plasticity is the primary substrate of long-term learning and memory must at present be viewed as our most plausible hypothesis (Milner, Squire, and Kandel, 1998). The role of nonsynaptic plasticity in behavior has been much less investigated (but see Sec. 13.6).

What do we mean by *synaptic strength*? Recall from Chap. 4 the quantal model of synaptic transmission. There, the coupling strength between two neurons was described in terms of three variables: the *number of release sites*, n, the *probability of release* of a vesicle following a presynaptic action potential p, and some measure of the *postsynaptic response to a single vesicle q* (Sec. 4.2.2). Depending on the circumstances, q can be a current, a voltage, or a conductance change—or even some indirect measure, such as the change in the fluorescence of some calcium-sensitive dye. In this chapter, we will refer to p and n as presynaptic variables, and to q as postsynaptic, recognizing that in some cases they may have other interpretations. Taken together, these three measures define the *average synaptic efficiency*.

Under many experimental conditions, synaptic efficacy is not stationary but changes with activity. Figure 13.1 illustrates how the response in a hippocampal pyramidal neuron *in vitro* depends on the history of synaptic usage. Here the stimulus was a spike train recorded *in vivo* with an extracellular electrode from the hippocampus of an awake behaving rat, and "played back" *in vitro*. The synaptic responses vary twofold or more, in a reliable and reproducible manner. The observed variability results from the interaction of a number of separate forms of rapid use-dependent changes in synaptic efficacy. Similar forms of plasticity have been observed at synapses throughout the peripheral and central nervous systems of essentially all organisms studied, from crustaceans to mammals.

The responses shown in Fig. 13.1 represent the complex interactions of many use-dependent forms of synaptic plasticity, some of which are listed in Table 13.1. Some involve an increase in synaptic efficacy, while others involve a decrease. They differ most strikingly in duration: some (e.g., facilitation) decay on the order of about 10–100 msec, while others (e.g., long-term potentiation, or LTP) persist for hours, days, or longer. The spectrum of time constants is in fact so broad that it covers essentially every time scale, from the fastest (that of synaptic transmission itself) to the slowest (developmental).

These forms of plasticity differ not only in time scale, but also in the conditions required for their induction.[1] Some—particularly the shorter lasting forms—depend only on the history of presynaptic stimulation, independent of the postsynaptic response. Thus facilitation, augmentation, and posttetanic potentiation (PTP) occur after rapid presynaptic stimulation, with more vigorous stimulation leading to more persistent potentiation. Other forms of plasticity depend on some conjunction of pre- and postsynaptic activity. The most famous example is LTP (see Sec. 13.3.1), which obeys Hebb's rule in that its induction requires simultaneous pre- and postsynaptic activation.

Changes in efficacy can be understood in terms of the quantal model of synaptic transmission. As we shall see, shorter lasting forms of plasticity—those that depend exclusively on the history of presynaptic activity—typically involve a change in one of the presynaptic parameters, the probability p of release. The mechanisms underlying long-lasting forms of plasticity such as LTP remain controversial, but changes in p may also be involved.

In this chapter we limit our discussion to a select subset of the vast literature on synaptic plasticity, focusing particularly on the mammalian central nervous system. We begin with a brief review of quantal synaptic transmission. Next we discuss potentiation and depression, starting with the most rapid forms (PPF and PPD) and continuing to the longer lasting forms

---

1. *Induction* refers to the conditions that trigger a change in synaptic efficacy, while *expression* refers to the manifestation of the change within the synaptic machinery.

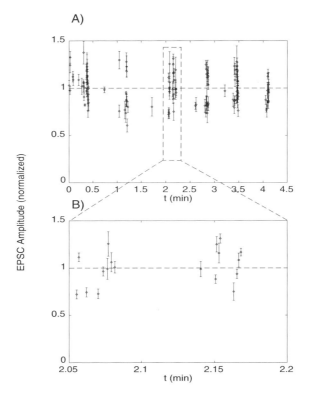

Fig. 13.1 SYNAPTIC RESPONSE DEPENDS ON THE HISTORY OF PRIOR USAGE   Excitatory postsynaptic currents (EPSCs) recorded from a CA1 pyramidal neuron in a hippocampal slice in response to stimulation of the Schaffer collateral input. The stimulus is a spike train recorded *in vivo* from the hippocampus of an awake behaving rat, and "played-back" at a reduced speed *in vitro*. The presynaptic spikes have an average interspike interval of 1950 msec, which varies from a low at 35 msec to a maximum at 35 sec. The normalized strength of the EPSC varies in a deterministic manner depending on the prior usage of the synapse. For a constant synaptic weight, the normalized amplitudes should all fall on the dashed line. **(A)** EPSC as a function of time. The mean and the standard deviation (4 repetitions) are shown. Note that the response amplitude varies rapidly by more than twofold. **(B)** Excerpt at a high temporal resolution. Unpublished data from L. E. Dobrunz and C. F. Stevens, printed with permission.

(LTP and LTD). Subsequently, we treat the computational implications of these various forms of synaptic plasticity. Finally, we close with a brief digression about nonsynaptic plasticity. We will not discuss developmental plasticity, or plasticity in subcortical areas or in nonmammalian preparations. For more information on these topics, or more in depth reviews of topics considered in this chapter, the following reviews provide good starting points (Madison, Malenka, and Nicoll, 1991; Ito, 1991; Hawkins, Kandel, and Siegelbaum, 1993; Zola-Morgan and Squire, 1993; Bliss and Collingridge, 1993; Miller, 1994; Schuman and Madison, 1994a; Malenka, 1994; Bear and Malenka, 1994; Linden and Connor, 1995; Carew, 1996).

13.1 Quantal Release

Recall the simple form of the quantal hypothesis described in Sec. 4.2.2. The time-averaged response size R is given by

TABLE 13.1
Different Forms of Synaptic Plasticity

Phenomenon	Duration	Locus of induction
Short-term enhancement		
Paired-pulse facilitation (PPF)	100 msec	Pre
Augmentation	10 sec	Pre
Posttetanic potentiation (PTP)	1 min	Pre
Long-term enhancement		
Short-term potentiation (STP)	15 min	Post
Long-term potentiation (LTP)	>30 min	Pre and post
Depression		
Paired-pulse depression (PPD)	100 msec	Pre
Depletion	10 sec	Pre
Long-term depression (LTD)	>30 min	Pre and post

Synaptic plasticity occurs across many time scales. This table lists some of the better studied forms of plasticity together with a very approximate estimate of their associated decay constants, and whether the conditions required for induction depend on pre- or postsynaptic activity, or both. This distinction is crucial from a computational point of view, since Hebbian learning rules require a postsynaptic locus for the induction of plasticity. Note that for LTP and LTD, we are referring specifically to the form found at the Schaffer collateral input to neurons in the CA1 region of the rodent hippocampus; other forms have different requirements.

$$R = npq \, . \tag{13.1}$$

This model, with modifications, has been applied with remarkable success to chemical synapses throughout the nervous system.

The number n of release sites, and their anatomical correlate, are different at different synapses. At one extreme, an axon can make a single anatomical synapse with one independent site of vesicular release onto the postsynaptic target (corresponding to $n = 1$; Fig. 13.2A). Single synapses are common in the hippocampus: an axon from a CA3 hippocampus pyramidal cell usually only forms a single synapse with a CA1 neuron (Sorra and Harris, 1993). Frequently, the same axon makes several, independent anatomical synapses with the dendrites of the same cell (Markram, Helm, and Sakmann, 1995; see the $n = 3$ case in Fig. 13.2B). From the point of view of network connectivity, all three synapses correspond to a single functional connection, since stimulation of the presynaptic axon excites all three synaptic terminals equally. Some pathways implement a fail-safe strategy by having a very large number of release sites n. Well-known examples of such "supercharged" (and presumably highly reliable) connections include retinal axons synapsing onto geniculate relay cells (Sherman and Koch, 1998) and the climbing fiber to Purkinje cell synapse (Llinás and Walton, 1998). In the latter case, a single climbing fiber innervates the soma and main dendrite of a cerebellar Purkinje cell by making up to 200 synaptic contacts. Finally, in a number of pathways a single axon terminal forms multiple independent release sites within a single synaptic contact (Fig. 13.2C). The best example of such a synapse is the *neuromuscular junction*, where the terminal axon field of a motor axon forms on the order of 1000 release sites with the muscle (Katz, 1966).

As emphasized repeatedly in this book, the number of functional contacts made by cortical and hippocampal neurons can be quite small (from one to about a dozen). What is the significance of small n? Recall that release from each site is probabilistic. At a synapse with n sites of release probability p, the probability that the synapse fails to release transmitters following a stimulus is

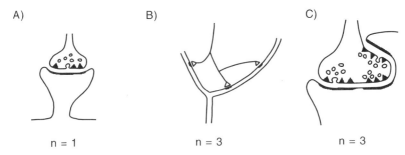

A)   B)   C)

n = 1   n = 3   n = 3

Fig. 13.2 QUANTAL MODEL OF SYNAPTIC RELEASE Basic unit of synaptic physiology, the *synaptic release site*. In response to a presynaptic spike, the vesicle fuses with the presynaptic membrane and releases its content, the neurotransmitter molecules, into the synaptic cleft. At each such release site, only a *single* vesicle is released (or fails to release) in response to a presynaptic spike. Synaptic transmission is both quantal and probabilistic, since the probability of release p is typically low (30% or less). If a vesicle is released, it induces a mean postsynaptic response q. **(A)** A common form of excitatory synaptic connections between a pair of central neurons: one anatomical synapse makes a single release site ($n = 1$) on its postsynaptic target. **(B)** Frequently, a single axon makes a small number of independent anatomical synapses on the dendrite of another cell (here $n = 3$). **(C)** At some synapses, such as the neuromuscular junction, thousands of release zones act independent of each other. Other examples include the calyx synapse between the cochlear nerve and one of the brainstem auditory nuclei. Reprinted in modified form by permission from Korn and Faber (1991).

$$P_{\text{failure}} = (1 - p)^n . \tag{13.2}$$

That is, the failure probability falls exponentially as the number of release sites increases. Moreover, fluctuations in the response size R are inversely proportional to $\sqrt{n}$.

For synapses such as the neuromuscular junction, with thousands of release sites, the probability of failure under physiological conditions is very low, and fluctuations are small. At central synapses, by contrast, failure is a very real possibility. For a synapse with a single release site and $p = 0.3$, failures occur 70% of the time. Hence transmission becomes unreliable (as illustrated in Fig. 4.3).

13.2 Short-Term Synaptic Enhancement

As suggested by Table 13.1, any distinction between short- and long-term forms of enhancement is somewhat arbitrary. It is nevertheless useful to distinguish forms that operate on a time scale from about a millisecond to about a minute, that is, facilitation and PTP, from longer lasting forms like LTP.

13.2.1 Facilitation Is an Increase in Release Probability

Synaptic *facilitation* was first described at the frog neuromuscular junction (Feng, 1941; Katz, 1966; Mallart and Martin, 1968; Magleby, 1987), but has subsequently been observed at nearly all synapses studied. Figure 13.3 illustrates one form of short-term facilitation in the mammalian central nervous system known as *paired-pulse facilitation* (PPF), since it is observed after a single pair of stimuli is delivered to a synapse. In the experiment shown here, an extracellular stimulus activated a relatively large (but undetermined) number of fibers, and the response was recorded in a single region CA1 pyramidal neuron. Because a

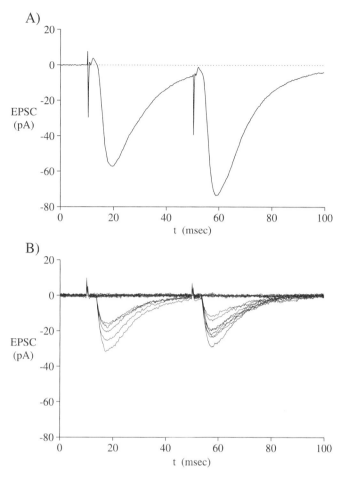

Fig. 13.3 **PAIRED-PULSE FACILITATION** Paired-pulse facilitation (PPF) at the CA3 to CA1 synapse (made by the Schaffer collaterals; see Fig 13.4) in the hippocampus slice. **(A)** Demonstration of PPF at the population level. Excitatory synaptic current recorded in a hippocampal region CA1 pyramidal cell following the response to a pair of extracellular stimuli (40 msec apart) delivered to the Schaffer collateral inputs. With this technique, many synapses are simultaneously activated. Note that the second response is larger than the first. This increase in synaptic response decays away with a time constant of about 100 msec. **(B)** PPF is demonstrated at an individual synapse via 21 consecutive trials under conditions of minimal stimulation. In each trial, two stimuli were delivered, separated by 40 msec. Five out of 21 trials lead to a signal for the first stimulus, but eight out of 21 on the second. Even though the variability in the amplitude of the postsynaptic response is high, the mean amplitudes of the responses in the first and second pulses are the same. The notches around 10 and 50 msec are stimulus artifacts. Unpublished data from L. E. Dobrunz, printed with permission.

typical CA3 *Schaffer collateral* makes only about one or two synapses onto its CA1 target cell, this electrical stimulus will activate a comparable number of release sites. (For an overview of the hippocampal circuitry, see Brown and Zador, 1990 and also Fig. 13.4.)

Consider a hypothetical example in which there are n release sites, each of which has the same initial probability of release p_0. q corresponds to the average excitatory postsynaptic current (EPSC) under voltage clamp. Under these conditions the mean response size to the extracellular stimulus is

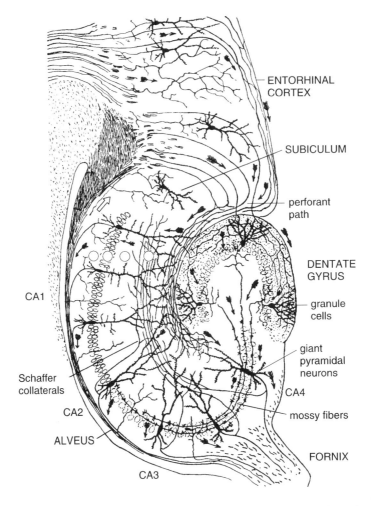

Fig. 13.4 HIPPOCAMPAL CIRCUITRY Neuronal elements of the hippocampal formation in rodents as drawn by Ramòn y Cajal at the turn of the century (when it was called Ammon's horn). This cortical structure is implicated in the transfer from short- to long-term memory. Granule cells in the dentate gyrus send their output axons, so-called *mossy fibers*, to pyramidal cells in the CA3 region. These pyramidal cells in turn project, with so-called *Schaffer collaterals*, onto pyramidal cells in the CA1 region. The majority of LTP and LTD research has been carried out at either the mossy fiber to CA3 synapse or the Schaffer collateral to CA1 synapse. Reprinted by permission from Brown and Zador (1990).

$$R = np_0 q \qquad (13.3)$$

(Fig. 13.3A). If a brief high-frequency (*tetanic*) stimulus is now delivered, the probability of release jumps to p_1, and the mean response to the same test stimulus jumps to $np_1 q$. Subsequently, the probability of release—and therefore the response size—slowly decays to baseline with a characteristic time constant on the order of hundreds of milliseconds, much slower than the onset time constant.

That the higher response magnitude is due to an increase in the release p and not to an increase in n or q is illustrated clearly by Fig. 13.3B. Here a *minimal stimulation* protocol

(Raastad, Storm, and Andersen, 1992; Allen and Stevens, 1994; Dobrunz and Stevens, 1997) is employed, so that only a single release site is activated. As seen in Fig. 13.3B, the first response at low stimulus rates frequently does not lead to a postsynaptic response; it fails to cause the release of even a single vesicle, and is therefore termed a *failure*. The failure rate f depends on p; if we assume one release site per synapse, $f = 1 - p$. In this particular experiment 15 out of 21 trials produced no response ($f = 0.71$) on the first stimulus, so on the assumption of a single release site, $p = 1 - 0.71 = 0.29$, that is, only three out of ten presynaptic spikes will cause the release of a synaptic vesicle. Note that even if release does occur, the amplitude of the average postsynaptic response R is itself quite variable (see Fig. 4.4), because of the underlying variability in q (Bekkers, Richerson, and Stevens, 1990).

If a pair of stimuli are now delivered in rapid succession (here 40 msec), the probability of release on the second trial p_1 is higher than that on the first (in Fig. 13.3B, p increased from 0.29 to 0.38). This increase when a second pulse follows soon after a first, is the single-synapse correlate of the paired-pulse facilitation shown at the population level in Fig. 13.3A. The release probability decays back to the initial probability with the same time constant as does the population amplitude.

The onset of facilitation is nearly as rapid as can possibly be resolved: it appears after a single stimulus (but see Dobrunz, Huang, and Stevens, 1997). The decay of PPF is slower, on the order of hundreds of milliseconds, and can be described by a simple exponential,

$$p(t) = p_0 + (p_f - p_0)e^{-t/\tau_f} \tag{13.4}$$

where p_0 and p_f are the probabilities of release before and after facilitation, respectively, and τ_f is the characteristic decay time.

The implications of facilitation for neuronal computation remain unclear. One might speculate that paired-pulse facilitation acts as a kind of "burst-filter": the probability of at least one successful release will be much higher during a burst of action potentials, that is, a handful of spikes within 10–30 msec, than if the same number of action potentials were uniformly distributed. In other words, a burst of spikes can be interpreted as a high-fidelity signal (see also Sec. 16.2).

13.2.2 Augmentation and Posttetanic Potentiation

While some facilitation is induced after a single stimulus, the degree of facilitation increases with the number of stimuli. As the number and frequency of stimuli increase, another form of potentiation, *augmentation*, is induced. Further stimulation brings into play a third form, termed *posttetanic potentiation* (PTP). These different forms of short-term potentiation have been best characterized at the neuromuscular junction (Magleby, 1987; see also Hirst, Redman, and Wong, 1981; Langdon, Johnson, and Barrionuevo, 1995). They differ from one another most notably by their characteristic time constant. For instance, the time course of augmentation can be expressed in the same form as the dynamics of facilitation above,

$$p(t) = p_0 + (p_a - p_0)e^{-t/\tau_a} \tag{13.5}$$

where p_a is the probability of release after augmentation and τ_a the characteristic decay time (on the order of seconds). The formulation for PTP is identical, except that the time constant is on the order of a minute or so.

Although facilitation, augmentation, and PTP all operate by increasing the probability of release, they do so by distinct mechanisms distinguishable not only by their kinetics, but also by their pharmacology (see, e.g., Zucker, 1996). Nevertheless, these processes cannot be completely independent, if only because p cannot exceed unity. How these processes interact remains unclear even at the neuromuscular junction where they have been best studied.

The equations describing these forms of potentiation can be interpreted in terms of the kinetics of some factor governing release probability. It is tempting to identify that factor as intracellular calcium concentration in the presynaptic terminal. As we shall see in the next section, the situation is rather more complex.

13.2.3 Synaptic Release and Presynaptic Calcium

As briefly alluded to in Chap. 4, synaptic release is caused by a rapid increase in the concentration of intracellular calcium that follows the invasion of the presynaptic terminal by an action potential. This increase in calcium concentration triggers fusion of a vesicle inside the presynaptic terminal with the presynaptic membrane and the subsequent release of neurotransmitters into the cleft (Fig. 4.2). Sometimes release of a single quantum occurs *spontaneously*, that is, independently of any presynaptic stimulus. The rate of spontaneous release at any one particular synapse is very low, less than one per minute. The calcium that rushes into the presynaptic terminal following the presynaptic spike raises the probability of release dramatically—by perhaps five orders of magnitude—over the very low spontaneous rate, but only for a very brief period (several hundred microseconds).

The molecular machinery coupling calcium in the presynaptic terminal with vesicle fusion and release is only beginning to be understood (see Zucker, 1996; Bauerfeind, Galli, and De Camilli, 1996; Rothman and Wieland, 1996; Sudhof, 1995; Matthews, 1996). The first indications that calcium was involved in synaptic transmission came from experiments in which the concentration of extracellular calcium was manipulated. Evoked release can be abolished by eliminating extracellular calcium, and reduced by lowering it. A component of spontaneous release persists even in the absence of extracellular calcium. More recently, technical advances have made it possible to measure presynaptic calcium directly (Delaney and Tank, 1994; Regehr, Delaney, and Tank, 1994; Zucker, 1996; Sabatini and Regehr, 1996; Wu and Saggau, 1995; Helmchen, Borst, and Sakmann, 1997).

Dodge and Rahamimoff (1967) fit the dependence of the amplitude of the observed postsynaptic response on the extracellular calcium concentration $[Ca^{2+}]_o$ and magnesium concentration $[Mg^{2+}]_o$ (which antagonizes calcium) with the following equation:

$$R(t) \propto \left(\frac{[Ca^{2+}]_o / K_c}{[Ca^{2+}]_o / K_c + [Mg^{2+}]_o / K_m + 1} \right)^z \tag{13.6}$$

where K_c and K_m are constants, and z is a parameter fit to the data. The best fits were obtained for z between 3 and 4. This equation can be derived by assuming that in order for release to occur, calcium must bind to z independent sites, and that K_c and K_m are the equilibrium constants for calcium and magnesium.

The Dodge-Rahamimoff relation relates release probability to the concentration of extracellular calcium. The dependence of release probability on the concentration of *internal* calcium is a thornier issue. For example, Eq. 13.6 was derived under equilibrium assumptions; but the rapidity with which an action potential triggers release (on the order of 0.15 msec; Llinás, Steinberg, and Walton, 1981b; Sabatini and Regehr, 1996) indicates that the source of calcium influx must be just tens of nanometers from the synaptic vesicles. Calcium concentration cannot equilibrate this quickly. This has led to the notion of Ca^{2+}

microdomains (Llinás, Steinberg, and Walton, 1981b; Llinás, Sugimori, and Silver, 1995)—
neighborhoods of high calcium concentration near the presynaptic calcium channels. Thus,
it appears that calcium flux, rather than equilibrium bulk calcium in the presynaptic terminal,
triggers fast vesicular fusion.

The dependence of the release probability on presynaptic calcium concentration suggests
that *residual calcium* might also underlie various components of short-term plasticity (Katz
and Miledi, 1968). According to this hypothesis, the enhanced release probability following
a train of action potentials results from an increased level of calcium in the presynaptic
terminal which, by itself, is insufficient to sustain release, but which adds to calcium from
subsequent releases and thereby results in a higher release probability. In the original
form of this hypothesis, the residual calcium simply added to the flux from an action
potential. More recent evidence indicates that the increase in resting calcium concentration
associated with short-term facilitation is too slight (1 μM; Delaney, Zucker, and Tank,
1989; Delaney and Tank, 1994), compared with the hundreds of micromolars during the
action potential (Lando and Zucker, 1994) for any additive effect to be relevant. The target
of the residual calcium therefore appears to be low affinity targets that modulate release
probability.

13.3 Long-Term Synaptic Enhancement

As noted above, the distinction between short and long-term enhancement is somewhat
arbitrary. In what follows we will consider those forms that last up to a few minutes
(facilitation, augmentation, and posttetanic potentiation) "short," while those that last longer
"long." There is, however, a more fundamental basis upon which to distinguish them: all
the short lasting forms of enhancement appear to depend only on the state of the presynaptic
terminal for induction, while the longer lasting forms often require some involvement from
the postsynaptic side (but see, e.g., Williams and Johnston, 1989; Nicoll and Malenka, 1995).

13.3.1 Long-Term Potentiation

Long-term potentiation is a rapid and sustained increase in synaptic efficacy following a
brief but potent stimulus. First described in the mammalian hippocampus at the perforant
path input to the dentate gyrus (Bliss and Lomo, 1973), it has since been observed at
diverse synaptic pathways, in the hippocampus, the neocortex, and elsewhere. LTP can last
for hours, days, weeks, or longer.

LTP research is very popular: between 1990 and 1997, over 2000 papers on LTP were
published. The excitement stems in large part from the hope that LTP is a model for learning
and memory, offering the most direct link from the molecular to the computational and
behavioral levels of analysis. The field of LTP is also very controversial, so that there is
only a surprisingly small number of completely accepted findings. Good reviews can be
found in the literature (Madison, Malenka, and Nicoll, 1991; Tsumoto, 1992; Johnston et
al., 1992; Bliss and Collingridge, 1993; Malenka, 1995).

Although LTP has been found in many neuronal structures, it has been best studied in
the hippocampus, at the synapses made from region CA3 pyramidal cells via the Schaffer
collateral pathways onto region CA1 pyramidal cells (Fig. 13.4), so we will focus on this
synapse. Care must be exercised when comparing LTP obtained in different systems, since
different forms of LTP coexist even in the hippocampus, with some independent of NMDA

receptor activation (Harris and Cotman, 1986; Johnston et al., 1992; Nicoll and Malenka, 1995).

As emphasized above, it is important to distinguish the rules and mechanisms underlying the *induction* of LTP (or any other form of synaptic plasticity) from the those governing its *expression*. Three basic facts about the induction of LTP at the CA3 to CA1 synapse are clear and essentially undisputed.

1. Under physiological conditions, induction typically requires (nearly) simultaneous presynaptic neurotransmitter release and postsynaptic depolarization. Because of this interesting fact, the mechanism of LTP has been interpreted as Hebbian (Sec. 13.5.1, Kelso, Ganong, and Brown, 1986; Malinow and Miller, 1986; Wigstrom et al., 1986). Figure 13.5 illustrates the dependence of LTP on postsynaptic depolarization. Stimulation at a low constant rate produces a baseline EPSP whose magnitude does not change following a postsynaptic depolarization. A rapid presynaptic stimulus by itself during simultaneous postsynaptic hyperpolarization also fails to induce LTP, although

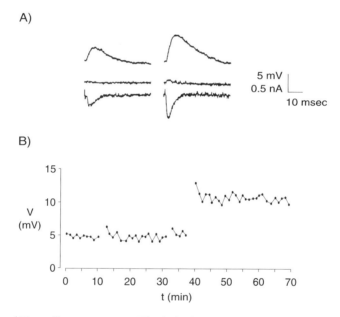

Fig. 13.5 LONG-TERM POTENTIATION The induction of LTP requires simultaneous pre- and postsynaptic activities, as demonstrated in a seminal study (Kelso, Ganong, and Brown, 1986) at the Schaffer collateral synapse onto CA1 pyramidal cells in a hippocampal slice (Fig. 13.4). **(A)** Averaged synaptic responses recorded under current clamp (upper traces) or voltage clamp (middle traces show membrane potential control and lower traces show the inward EPSC). The responses in the control period are plotted on the left, while the enhanced responses, 20 min after vigorous synaptic input is paired with a large depolarizing current to the soma of the cell, forcing it to spike, are indicated on the right. **(B)** Peak value of this EPSP as a function of time for different experimental manipulations. Neither postsynaptic depolarization (achieved by injecting a large depolarizing current into the soma) by itself (first trace), nor presynaptic stimulation paired with clamping the soma to −80 mV for two different synaptic inputs (second and third traces) is sufficient to induce LTP, as seen by the constant response amplitudes. Only when presynaptic input is paired with the postsynaptic depolarization is a long-term enhancement of the synaptic weight observed. Reprinted by permission from Kelso, Ganong, and Brown (1986).

it does lead to PTP. Only when vigorous synaptic input is paired with postsynaptic depolarization (here via an intracellular electrode) is LTP induced: the synaptic response to the same baseline stimulation doubles in size and remains elevated for the duration of the experiment, over an hour. Similar *in vivo* experiments suggest that the change can persist for weeks or longer. The depolarization may arise from the stimulus itself (if it activates a sufficiently large number of fibers), from activation of some other synaptic input, or from somatic depolarization via an intracellular electrode. The depolarization is necessary to relieve the Mg^{2+} block of NMDA receptors (see below).

2. The second widely accepted fact about the induction of LTP at the CA3 to CA1 synapse is that it requires activation of NMDA receptors (Collingridge, Kehl, and McLennan, 1983). NMDA receptors are unique among synaptic receptors in that they are directly gated by both voltage and neurotransmitter, so that they pass current only when the membrane is depolarized sufficiently to relieve a block by magnesium ions. The induction of LTP is inhibited by agents that block the NMDA receptor, such as AP5. NMDA receptor activation is not required for either normal synaptic transmission in the hippocampus or the maintenance of LTP: once LTP has been induced, blockage of NMDA-mediated synaptic transmission by AP5 does not inhibit the expression LTP. This is also true in neocortex: activation of NMDA receptors is required for LTP induction (Artola and Singer, 1987; Malenka, 1995).

3. The third fact about LTP induction is its dependency on a localized increase in the postsynaptic concentration of calcium (Lynch et al., 1983; Malenka et al., 1988). When calcium buffers that bind any excess calcium are injected into the postsynaptic cell, the induction of LTP is blocked (Barrionuevo and Brown, 1983).

A simple model for the induction of LTP that accounts for these observations is illustrated in Fig. 12.12. In this model, excitatory NMDA and non-NMDA receptors are colocalized (Bekker and Stevens, 1989) at single synapses made by the Schaffer fibers onto spines of CA1 pyramidal cells. Release of neurotransmitters always activate the current through the AMPA receptor-gated channel, but activates the current through the NMDA receptor-gated channel only when there is sufficient postsynaptic depolarization to remove the Mg^{2+} blocking the channel. (Release of a single quantum of neurotransmitter is not believed to result in sufficient depolarization at the spine head to relieve the Mg^{2+} block.) Influx of calcium through open NMDA channels causes a large, localized and transient increase in the concentration of postsynaptic calcium (see Sec. 12.6.2; Holmes and Levy, 1990; Zador, Koch, and Brown, 1990; Guthrie, Segal, and Kater, 1991; Yuste and Denk, 1995; Svoboda, Tank, and Denk, 1996). This accounts for the "input specificity" of LTP, that is, the fact that LTP is expressed only at those synapses where the conditions for induction are satisfied, since only here is $[Ca^{2+}]_i$ large enough to trigger the biochemical cascade that finally leads to the establishment of LTP. More recent research, alluded to in Sec. 20.3, shows that under some conditions synapse specificity may break down (Bonhoeffer, Staiger, and Aertsen, 1989; Schuman, 1997; Engert and Bonhoeffer, 1997). Thus the induction of LTP occurs at the postsynaptic site and requires the *conjunction* of pre- and postsynaptic activity.

The site of expression of LTP remains controversial. Many hypotheses have been proposed (Lynch and Baudry, 1987; Kauer, Malenka, and Nicoll, 1988; Muller, Joly, and Lynch, 1988; Edwards, 1991, 1995; Stevens and Wang, 1994; Liao, Hessler, and Malinow, 1995; Kullmann and Siegelbaum, 1995), which can be summarized as involving a change in either the presynaptic element, the postsynaptic element, or both. Conjectures involving a presynaptic locus include an increase in (1) the number of release sites, (2) the probability of release, (3) the amount of neurotransmitter loaded into each vesicle; hypotheses involving the postsynaptic terminal include (4) an increase in receptor affinity or density, and the (5) recruitment of previously "silent" synapses. As discussed in the previous chapter (Sec. 12.3.2), the earlier hypothesis that changes in the geometry of the postsynaptic spine affect its synaptic weight does not appear to be correct, at least in the case of passive spines on hippocampal pyramidal cells.

If the induction of LTP occurs postsynaptically but its expression presynaptically, something needs to signal this information back across the synaptic terminal. Several possibilities exist. The one that has attracted most attention is *retrograde messenger* molecules, that is, substances whose production is triggered postsynaptically when LTP is induced and that diffuse backward across the synapse. Proposed retrograde messengers include *arachidonic acid*, and diffusible second messengers *nitric oxide* and *carbon monoxide* (Williams et al., 1989; Gally et al., 1990; Schuman and Madison, 1994a; Schuman, 1995; see also Sec. 20.3).

13.3.2 Short-Term Potentiation

For completeness we also mention *short-term potentiation* (STP), sometimes referred to as *decremental LTP*. Much less is known about short term potentiation than about its long-term counterpart (Davies et al., 1989; Colino, Huang, and Malenka, 1992; Kullmann et al., 1992). First detected in experiments investigating different protocols for inducing LTP, it was treated largely as an irritant, a potential source of artifact in experiments on LTP. It is defined operationally in terms of its time constant: it is the potentiation that persists longer than the minute or two of PTP, but not as long as LTP; typically it decays with a time constant of about 15 min. Both its induction and its expression appear to be largely postsynaptic.

13.4 Synaptic Depression

We treat short- and long-term synaptic depression together here, not because they are less interesting or important than enhancement, but because they are not as well understood.

The most rapid forms of *synaptic depression* involve a decrease in the probability of transmitter release. Immediately following a release of a vesicle (but not a failure) at a single site, there is often a 5–10 msec effective "dead time" during which release cannot occur (Stevens and Wang, 1995; Dobrunz, Huang, and Stevens, 1997). Unlike synaptic enhancement, the amount of depression depends not on the number of presynaptic action potentials, but rather on the number of vesicles released. A longer lasting form of depression occurs (on the order of seconds) following depletion of the available pool of vesicles (Stevens and Tsujimoto, 1995; Dobrunz and Stevens, 1997).

While it has long been recognized that what goes up must come down, for years the counterpart to long-term potentiation, termed *long-term depression* (LTD), could not be reliably induced in the CA1 region of the hippocampus (although it had been described in the cerebellum and the dentate gyrus; see Levy and Steward, 1983; Ito, 1991). With the advent of a reliable protocol for the induction of long-term depression (Dudek and Bear, 1992) in both hippocampus and neocortex, LTD can be easily obtained, reducing the synaptic weight to about half of its pre-LTD value (Fig. 13.6). LTD is now receiving the

A)

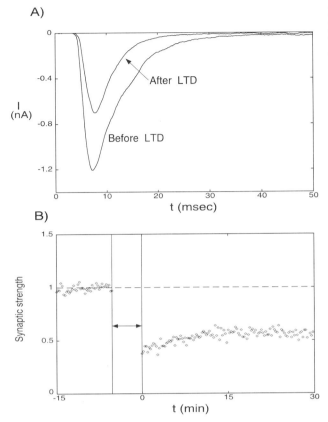

Fig. 13.6 Long-Term De-pression Induction of long-term depression in a pair of cultured hippocampal neurons. LTD is induced by a moderate amount of synaptic stimulation via a stimulus electrode while holding the postsynaptic membrane at −50 mV (double arrowhead in **B**). (**A**) Sample trace of an EPSC before and after the induction of LTD. (**B**) Time course of synaptic strength (expressed as the ratio of peak EPSP after induction to peak EPSP before induction) as a function of time. Averaged over nine such pairs, LTD reduced the normalized synaptic strength to 0.44 ± 0.06. Reprinted by permission from Goda and Stevens (1996).

same scrutiny as LTP (for surveys, see Tsumoto, 1992; Dudek and Bear, 1992; Artola and Singer, 1993; Malenka, 1994; Linden and Connor, 1995).

Just as with long-term potentiation, the requisite conditions for induction must be met at the postsynaptic terminal; in fact, the conditions required for the induction of LTD are remarkably similar to those for LTP. Some forms of LTD require NMDA receptor activation; the essential requirement always appears to be sufficient postsynaptic depolarization in order to elevate $[Ca^{2+}]_i$ in the postsynaptic terminal above resting levels, but below that required for the induction of LTP (Lisman, 1989; Artola and Singer, 1993; Malenka and Nicoll, 1993; Linden and Connor, 1995; Cummings et al., 1996; but see Neveu and Zucker, 1996). In other words, there exists a critical threshold of free intracellular calcium concentration that governs the increase or decrease of the synaptic weight in a highly specific manner.

The mechanism underlying LTD in the hippocampus is a *decrease* in the probability of synaptic release (Stevens and Wang, 1994, 1995; Bolshakov and Siegelbaum, 1994, 1995). If, as suggested above, the mechanism underlying LTP turns out to be an increase in p, LTP and LTD would display an elegant symmetry, each exerting differential control on the probability knob.

13.5 Synaptic Algorithms

Conjectures going back to the turn of the century, Tanzi (1893; reviewed in Brown, Kairiss, and Keenan, 1991b) implicate synapses as the locus for physiochemical changes underlying learning. Wood-Jones and Porteus (1928) even speculate about the mechanism underlying these changes. But early work remained silent about the conditions under which the changes in efficacy would occur.

13.5.1 Hebbian Learning

The modern approach to synaptic algorithms can be traced to Hebb's very influential monograph (Hebb, 1949), in which he prescribed

> When an axon of cell A is near enough to excite cell B or repeatedly or consistently takes part in firing it, some growth process or metabolic change takes place in one or both cells such that A's efficiency, as one of the cells firing B, is increased.

Hebb was therein the first to propose explicitly the conditions under which the change in efficacy would occur. It is because his proposal provides an activity-based rule for increasing efficacy that it is called an "algorithm." In mathematical terms, it is usually expressed as

$$\Delta w_{ij} \propto V_i V_j \tag{13.7}$$

where Δw_{ij} is the change in the strength of the synaptic coupling w_{ij} between the presynaptic neuron i and the postsynaptic cell j; V_i and V_j represent the activities of these neurons. Such a pure *Hebbian* rule has been used extensively in associative, or content-addressable, memory networks (Steinbuch, 1961; Hopfield, 1982, 1984).

In the simplest mapping of these symbols onto biophysics, $w_{ij} = npq$ (Eq. 13.1), V_i corresponds to the presynaptic spiking frequency, and V_j to the postsynaptic membrane potential at the spine or in the dendrite just below. Yet it is far from clear what the exact relationship between the algorithmic variables and their biophysical counterpart is. For instance, does V_j really correspond to the instantaneous dendritic membrane potential or is it more akin to some low-pass filtered version of V_m?

Hebb's original proposal is incomplete, since it offers no prescription for determining under what conditions synaptic efficacy decreases (the weights can only increase) and because it is unclear what types of memories can and cannot be recovered using this rule. Subsequent theoretical work (e.g., Stent, 1973; Sejnowski, 1977a,b; Palm, 1982; Linsker, 1988; for a survey, see Hertz, Krogh, and Palmer, 1991) elaborated on the conditions under which the synaptic weight should change.

One popular modern variant of Hebb's original rule is known as the *covariance rule* (Sejnowski, 1977a), so named because it is formally identical to a statistical covariance. It can be written simply as

$$\Delta w_{ij} \propto (V_i - \langle V_i \rangle)(V_j - \langle V_j \rangle) \tag{13.8}$$

where $\langle V_i \rangle$ corresponds to the average activity of the presynaptic neuron over some suitable time interval (similarly for $\langle V_j \rangle$). If the actual presynaptic activity is less than its recent average while the postsynaptic activity is elevated, or vice versa, the synaptic weight will decrease.

Note that we did not specify over what duration the averages should be taken. If the average is short—on the order of seconds to minutes—this formulation can be used to describe LTP and LTD, while if it is longer it can be used to describe developmental effects (see e.g., Miller, 1994). The covariance rule says the coupling between neurons i and j should increase if their activities are correlated; but that otherwise it should decrease. This rule has been used by Hopfield (1984) and many others as the basis for autoassociative memories.

One appealing feature of formulating Hebb's rule according to Eq. 13.8 is that it can be expanded to yield terms that have plausible biophysical interpretations. Multiplying the right-hand side of Eq. 13.8, and assigning separate positive constants $k_1 \cdots k_4$ to the terms, we have

$$\Delta w_{ij} = k_1 V_i V_j - k_2 V_i \langle V_j \rangle - k_3 V_j \langle V_i \rangle - k_4 \langle V_i \rangle \langle V_j \rangle . \qquad (13.9)$$

The first term is the usual Hebbian one, corresponding to an increase following simultaneous pre- and postsynaptic activity. The second and third terms correspond to homo- and heterosynaptic LTD.[2] The last term corresponds to a tonic increase. While this formulation is clearly an oversimplification, the correspondence to interpretable biophysical phenomena— coupled with its widespread use in the field of neural networks—makes it a very useful starting point.

Recent evidence (in particular Engert and Bonhoeffer, 1997) suggests that synaptic specificity in LTP may break down at short distances. That is, excitatory synapses within 50–70 μm of the potentiated synapse on the same neuron also have their synaptic weight increased—despite a lack of presynaptic activity. If this is also true in the intact animal, the above learning rules have to be modified to account for less specific synaptic storage schemes (as in Eq. 20.6).

13.5.2 Temporally Asymmetric Hebbian Learning Rules

Mapping V_i onto the presynaptic spiking frequency (an averaged quantity) as above implies that the exact temporal relationship between the arrival times of the presynaptic spike and the postsynaptic depolarization does not matter. Experimental evidence indicates that it does, with powerful functional consequences.

Evidence from both hippocampal and neocortical pyramidal cells indicates that in order for the synaptic weight to increase, presynaptic activity has to precede the postsynaptic one (Levy and Steward, 1983; Gustafsson et al., 1987; Debanne, Gähwiler, and Thompson, 1994). Particularly compelling evidence comes from a recent experiment by Markram et al., (1997) in which they systematically varied the relationship between the presynaptic spike arriving at the synapse and the timing of the postsynaptic action potential that propagates back into the dendritic tree to the postsynaptic site (for more details see Chap. 19). If the presynaptic spike precedes the postsynaptic one, as should occur if the presynaptic input participates in triggering a spike in the postsynaptic cell, long-term potentiation occurs, that is, the synaptic weight increases. If the order of temporal arrival times is reversed (e.g., from +10 to −10 msec), the synaptic weight decreases. This enables the system to assign credit to those synapses that were actually responsible for generating the postsynaptic spike.

What this teaches us is that the "static" Hebb learning rule (e.g., in Eq. 13.8) needs to be replaced by a dynamic version that is asymmetric in time, that is, a positive delay between the arrival times of the presynaptic and postsynaptic spikes does not have the same effect on the synaptic weight changes as the reverse situation.

The use of temporally asymmetric Hebbian learning rules can induce associations over time, and not just between simultaneous events. As a result, networks of neurons endowed

---

2. In homosynaptic LTP, the activated input by itself depolarizes the postsynaptic site sufficiently to induce LTP, while in heterosynaptic LTP a second simultaneously active synaptic input needs to provide the requisite depolarization.

with such synapses can learn sequences (Minai and Levy, 1993), enabling them to predict the future state of the postsynaptic neuron based on past experience (Abbott and Blum, 1996). Asymmetric Hebbian rules have been applied to a host of biological learning paradigms such as bee foraging and reinforcement learning in the basal ganglia (Montague and Sejnowski, 1994; Montague et al., 1995; Montague, Dayan, and Sejnowski, 1996) and the learning of fine temporal discriminations at the single-neuron level (Gerstner et al., 1996).

13.5.3 Sliding Threshold Rule

A quite distinct theoretical framework for synaptic plasticity has its origin in a model for developmental plasticity in the visual cortex. It is known as the *sliding threshold* or the *BCM* theory after the initials of the authors who proposed this synaptic rule (Bienenstock, Cooper, and Munro, 1982; see also Bear, Cooper, and Ebner, 1987).

As in a standard learning rule, the synaptic modification is Hebbian, that is, the weight change is proportional to the product of the pre- and postsynaptic activities. The exact form of Δw_{ij} is given by the product of the presynaptic activity V_i and a function ϕ of the postsynaptic response V_j and a variable threshold θ_m,

$$\Delta w_{ij} = V_i \phi(V_j, \theta_m). \tag{13.10}$$

Figure 13.7A illustrates ϕ as a function of the postsynaptic activity V_j; its key feature is that ϕ is zero at zero, becomes negative, and changes sign at a critical threshold θ_m. Thus, BCM predicts that synaptic input activity that is too weak (that is, that lies below θ_m) will cause LTD, while strong synaptic input leads to LTP. Some of the evidence discussed in Sec. 13.3.1 is in agreement with this. Dudek and Bear (1992) provided further support by varying the frequency of presynaptic stimulation (roughly proportional to the V_i term in Eq. 13.8) over two orders of magnitude and observing LTD at low frequencies, but LTP at higher ones (Fig. 13.7B).

The threshold θ is required to be a supralinear function of the time averaged postsynaptic activity (that is, it must grow more than a linear function; typically, $\theta_m = \langle V_i^2 \rangle$ is chosen). This has the important consequence that the value of the threshold must be the same at all synapses onto a particular neuron; yet the associated w_{ij}'s can still change at different rates depending on the level of presynaptic activity (Eq. 13.8).

The sliding threshold results in a stable activity level for the cell. If the activity is too low, θ_m decreases until the appropriate w_{ij}'s have increased to bring the postsynaptic activity up and vice versa. Thus, the threshold modification serves to stabilize the synaptic population, a crucial property of any developmental rule. To what extent the threshold separating LTD from LTP induction changes as a function of the activity of the cell is not known experimentally. A number of possible molecular mechanisms, based on the influx of calcium into the spine, exist that can instantiate such a threshold (see Bear, 1995, for a discussion).

13.5.4 Short-Term Plasticity

The computational implications of rapid forms of plasticity have not received nearly as much attention as those of long-term forms, perhaps because of the paucity of network models that make use of real dynamics. Nevertheless, the rapid forms exert large effects on the magnitude of the synaptic response (Dobrunz and Stevens, 1996; Fig. 13.1). Changes of this magnitude—as large or larger than those typically induced by LTP—suggest an important functional role for these rapid forms.

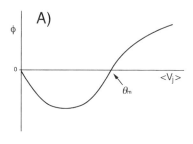
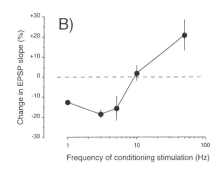

Fig. 13.7 "Sliding Threshold" Theory of Synaptic Learning The Bienenstock, Cooper, and Munro (1982) learning rule, proposed within the context of developmental learning, is a Hebbian rule with a twist. The synaptic weight change is proportional to the product of the presynaptic activity and a function ϕ (Eq. 13.10). This function depends on the postsynaptic activity in the manner illustrated in (A). A critical feature is the modifiable threshold θ_m: postsynaptic activity less than this threshold causes LTD, while higher activity leads to an increased synaptic weight (LTP). The sliding threshold changes as a supralinear function of the time-averaged postsynaptic activity. (B) Direct experimental evidence bolstering the arguments for the existence of ϕ with a similar shape as used in the BCM model (Dudek and Bear, 1992). Here, 900 spikes are generated in the Schaffer collaterals to the CA1 pyramidal cells in a hippocampal slice via electrical stimulation, varying at frequencies between 0.5 and 50 Hz. Presynaptic stimulation frequencies below 10 Hz always lead to a reduction in the slope of the EPSP (relative to baseline), that is, in LTD, which persists without any sign of recovery for at least one hour. Higher stimulation frequencies lead to LTP. These effects are dependent on NMDA receptor activation. Reprinted by permission from Dudek and Bear (1992).

A novel way of thinking about short-term changes in synaptic weight has come about by work among teams of experimentalists and theoreticians (Markram and Tsodyks, 1996; Tsodyks and Markram, 1997; Abbott et al., 1997; for a summary see Zador and Dobrunz, 1997). Short-term depression (see Table 13.1 and Fig. 13.1) affects the postsynaptic response to a regular train of spikes at a fixed frequency f. While the response to the first spike is large, subsequent responses will be diminished until they reach a steady-state (Fig. 13.8A). For firing rates above 10 Hz, the asymptotic relative synaptic amplitude per impulse $A(f)$ (with $A \leq 1$) is approximately inversely proportional to the stimulus rate,

$$A(f) \approx \frac{C}{f} \tag{13.11}$$

with C some constant. This depression generally recovers within a second or so. The postsynaptic response per unit time is therefore

$$f \times A(f) \approx f \times \frac{C}{f} \approx C . \tag{13.12}$$

That is, the steady-state synaptic response is independent of the stimulus rate, rendering the synapse very sensitive to *changes* in the stimulus rate. The instantaneous response to a rapid increase Δf in its presynaptic firing rate is given by

$$\Delta f \times A(f) \approx C \frac{\Delta f}{f} . \tag{13.13}$$

As a consequence, the transient change in the postsynaptic response will be proportional

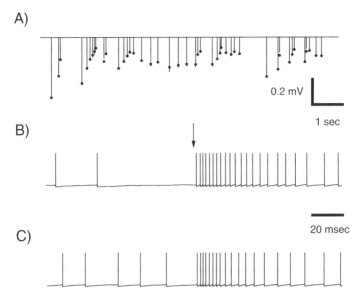

Fig. 13.8 Short-Term Depression and Adapting Synapses Short-term synaptic depression can implement a Weber-Fechner-like law at the synaptic level, whereby the postsynaptic response is proportional to the fractional change in firing rate (Abbott et al., 1997). **(A)** Field potentials, a measure of the postsynaptic response, recorded in layer 2/3 of rat visual cortex slices evoked by a Poisson train of extracellular stimulation in layer 4. The lines show the data and the dots the fit of a mathematical model. Onset and recovery from depression are clearly seen. **(B)**, **(C)** Responses of a simulated integrate-and-fire neuron with depressing synapses to a sudden step increase in the afferent firing rate at the time indicated by the arrow. In **(B)** the firing rate of the afferents increased from 25 to 100 Hz and in **(C)** from 50 to 200 Hz. Because in both cases $\Delta f/f = 3$, the postsynaptic response is nearly equal, only signaling relative changes. The time scale applies to both **B** and **C**. Unpublished data from L.F. Abbott, printed with permission.

to the relative change in firing frequency: increasing the firing rate fourfold, from 25 to 100 Hz, has as much effect as going from 50 to 200 Hz (Fig. 13.8B and C). As Abbott and his colleagues (1997) point out, this behavior is reminiscent of the Weber-Fechner law of psychophysics, stating that humans are sensitive to relative and not absolute changes in signal intensity (e.g., in irradiance or sound amplitude).

This is rather elegant. The very hardware used to carry out computations (synapses) continuously adapts to their input, only signaling relative changes, enabling the system to respond in a very sensitive manner in the face of a constantly and widely varying external and internal environment. In contrast, digital computers are carefully designed to avoid adaptation and other usage-dependent effects from occurring. Interestingly, single-transistor learning synapses—based on the floating-gate concept underlying erasable programmable ROM digital memory—have now been built in a standard CMOS process (Diorio et al., 1996; Koch and Mathur, 1996). Similar to synapses, they can change their effective weights in a continuous manner while they carry out computations. Whether they will find widespread applications in electronic circuits remains to be seen.

A particularly intriguing finding comes from Markram and Tsodyks (1996), who studied the interaction of LTP and short-term plasticity. They found that LTP has no effect on the steady-state response to a train of stimuli, but does affect the transient component. They suggested that the effect of LTP is to "redistribute" the synaptic response in time, increasing the response to the first few impulses in a stimulus train at the expense of the next few, but leaving the asymptote unaffected. These experiments suggest that the effects of long-term plasticity might be mediated by modifications of short-term plasticity. Attempts to construct a general framework for computing with dynamic synapses are underway (Maass and Zador, 1998).

13.5.5 Unreliable Synapses: Bug or Feature?

We have seen that single synapses in the mammalian cortex appear to be unreliable: release at single sites can occur as infrequently as one out of every 10 times (or even less) that an action potential invades the presynaptic terminal (Fig. 4.3). This should be contrasted with the reliability of a transistor in today's highly integrated silicon circuits, which is very, very close to 1. (The probability of failure of a digital CMOS inverter can be estimated to be less than 10^{-14} per switching event.)

It is natural to wonder whether synaptic unreliability is an unfortunate but necessary property that the brain must accept due to biophysical constraints (in particular, the problem of packing on the order of one billion synapses, each firing at least several times each second, into one cubic millimeter of cortical gray matter). It might be possible that synapses this small simply cannot be made reliable. Alternatively, might there be some computational advantage to this unreliability? Or, formulated as a *bon mot*, is the lack of synaptic reliability a "bug" or a "feature"?

If there are indeed constraints on the potential reliability of cortical synapses, they do not involve the fidelity with which a single presynaptic action potential can be converted into vesicular release. We know this because the probability of release at unreliable synapses can sometimes increase nearly to unity, following synaptic enhancement. However, it may be that the limit is not on the fidelity of transduction, but on the total number of vesicles that can be released in some interval. Thus it is possible that the release cannot be sustained during periods of high presynaptic activity if, for example, there is a limit to the uptake rate at which released vesicles can be recycled. This issue may be resolved experimentally if synapses that are as compact as hippocampal synapses are found for which release is reliable even during periods of sustained activity (Smetters and Zador, 1996).

An alternative view is that there is some computational advantage to having unreliable synapses. The theme running through this chapter—that change in the probability of release is the mechanism underlying many forms of plasticity—suggests one possible advantage. It appears that release probability is a parameter that can be modified conveniently and dynamically on a short time scale. In this view the lack of reliability is required to give a synapse its large dynamic range, since varying either the number of release sites n, or the postsynaptic response q, over an equally large range is much more demanding. Only if most synapses have relatively low release probabilities can modulation of p implement changes in efficacy. The tradeoff is between reliability and the bandwidth of modulation of the postsynaptic response.

13.6 Nonsynaptic Plasticity

It should not come as a surprise to us that neuronal plasticity is not restricted to synapses. The most obvious—and best understood—examples of nonsynaptic plasticity are those governing short-term changes in neuronal firing. Indeed, the adaptation in firing frequency in response to a constant current input seen in most pyramidal cells can be considered to be a form of nonsynaptic plasticity (Fig. 13.9). Like synaptic adaptation, firing adaptation occurs on a spectrum of time scales, from milliseconds to seconds. As discussed in Sec. 9.3, adaptation is mediated by changes in a slow and calcium-dependent potassium current. Changes in firing patterns over days or longer, corresponding to developmental time scales, have also been observed (Spitzer, 1994; Turrigiano, Abbott, and Marder, 1994).

There also is evidence for changes in specific ionic currents during associative conditioning. Alkon and his colleagues have established a direct link in the living animal—in his case the sea snail *Hermissenda*—between a classical conditioning task and changes in two potassium currents (for a review, see Alkon, 1987). Rotation of this mollusk, as would occur naturally during turbulence in the ocean, elicits a clinging response of its "foot." During associative conditioning, a gradient of light is paired with rotation and the animal learns to "associate" the light with the rotation. Following training, the light stimulus by itself triggers the foot clinging response. A large component of this response can be traced back to an enhanced photoresponse in the type B photoreceptor (in *Hermissenda*, these photoreceptors generate action potentials and receive direct and indirect synaptic input from hair cells sensitive to the rotation of the animal). Specifically, after training, a transient and inactivating A-like potassium current and a calcium-dependent potassium current in the cell body of the photoreceptor are reduced by 30–40% (Alkon, Lederhendler. and Shoukimas, 1982; Alkon et al., 1985). These changes occur postsynaptic to inputs from other photoreceptors and from the hair cells, last for days, and are evident after blockage of all synaptic input, underlining the fact that learning affects not only synapses but also membrane currents at the soma and elsewhere.

Theoretical work in this area, exploiting both supervised and unsupervised learning rules to modify voltage-dependent membrane conductances in the soma and dendrites to achieve

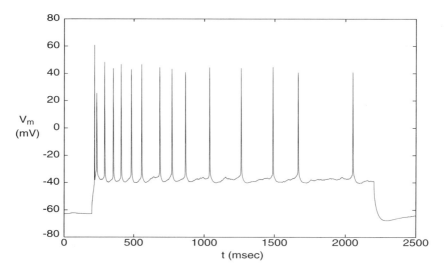

Fig. 13.9 Firing Frequency Adaptation Is a Form of Nonsynaptic Plasticity Many cortical neurons show *spike frequency adaptation* in response to a prolonged current step. Conceptually, this can be thought of as a form of learning, in this case learning to adapt out the sustained, and therefore predictable, component of the current. This figure portrays the spike train of a region CA1 pyramidal neuron in response to a 2-sec, 10-pA current step. Note that the spike rate begins very high and declines steadily. Unpublished data from A. Zador, printed with permission.

a particular behavior, is still in its infancy (Zador, Claiborne, and Brown, 1992; Bell, 1992; Koch et al., 1996).

13.7 Recapitulation

Behavioral plasticity or adaptation is critical to an organism's survival. Adaptation occurs throughout the nervous system and on many different time scales, from milliseconds to days and even longer. Changes in synaptic strength are widely postulated to be the primary biophysical substrate for many forms of behavioral plasticity, including learning and memory, although our understanding of the link remains far from complete. Use-dependent forms of synaptic plasticity have been characterized in many preparations.

Synaptic strength or weight can be characterized by the triplet (n, p, q), where n is the number of release sites, p the probability of synaptic release, and q the amplitude of the postsynaptic response following the docking of a single vesicle. The induction of most short-term forms of plasticity depends only upon the history of activity in the presynaptic terminal, while longer term forms require that appropriate conditions be met at both the pre- and the postsynaptic sites. It is therefore only these longer term forms of plasticity that can implement Hebbian type of learning. Biophysical modulation of p underlies most short-term forms, and appears to account for at least a component of some long-term forms as well.

By far the best studied biophysical model of long-term synaptic change is LTP. The induction of LTP requires a conjunction of presynaptic neurotransmitter release, combined with postsynaptic depolarization of the postsynaptic site. There appears to be quite a requirement for the presynaptic input to precede firing activity in the postsynaptic cell. These experimentally observed forms of plasticity can be described at a more abstract level in terms of synaptic algorithms, in particular by temporally asymmetric Hebbian learning rules. Such formulations are useful because a great deal is known from the literature on artificial neural networks about the computational possibilities of Hebbian synapses.

It is important to realize the prevalence of usage-dependent forms of synaptic plasticity. While digital transistors have been designed to be as constant as possible at switching speeds of hundreds of megahertz over the lifetime of the processor, a single synapse will vary its weight considerably in response to two or more consecutive spikes. These short-term changes can take many forms, including depression and facilitation. In at least one case, the three different types of synaptic input to layer 4 spiny cells show the entire gamut of short-term plasticity: none, short-term depression and short-term enhancement (Stratford et al., 1996). In summary, synaptic properties show a complex dependency on their previous history of usage and on the postsynaptic activity. We are only beginning to understand the computational significance of such dynamic switching elements.

14

SIMPLIFIED MODELS
OF INDIVIDUAL NEURONS

In the previous thirteen chapters, we met and described, sometimes in excruciating detail, the constitutive elements making up the neuronal hardware: dendrites, synapses, voltage-dependent conductances, axons, spines and calcium. We saw how, different from electronic circuits in which only very few levels of organization exist, the nervous systems has many tightly interlocking levels of organization that codepend on each other in crucial ways. It is now time to put some of these elements together into a functioning whole, a single nerve cell. With such a single nerve cell model in hand, we can ask functional questions, such as: at what time scale does it operate, what sort of operations can it carry out, and how good is it at encoding information.

We begin this Herculean task by (1) completely neglecting the dendritic tree and (2) replacing the conductance-based description of the spiking process (e.g., the Hodgkin-Huxley equations) by one of two canonical descriptions. These two steps dramatically reduce the complexity of the problem of characterizing the electrical behavior of neurons. Instead of having to solve coupled, nonlinear partial differential equations, we are left with a single ordinary differential equation. Such simplifications allow us to formally treat networks of large numbers of interconnected neurons, as exemplified in the neural network literature, and to simulate their dynamics. Understanding any complex system always entails choosing a level of description that retains key properties of the system while removing those nonessential for the purpose at hand. The study of brains is no exception to this.

Numerous simplified single-cell models have been proposed over the years, yet most of them can be reduced to just one of two forms. These can be distinguished by the form of their output: *spike* or *pulse* models generate discrete, all-or-none impulses. Their output over time can be treated as a series of delta functions $\sum_i \delta(t - t_i)$. Implicitly this assumes that no information is contained in the spike height or width. The original model of a neuron in McCulloch and Pitts (1943) as well as the venerable *integrate-and-fire* unit are instances of pulse models. In *firing rate* neurons, the output is a continuous firing rate, assumed to be a positive, bounded, and stationary function of the input. Examples of these are the units at the heart of Hopfield's (1984) associative memory network.

Yet before we can delve into more detail we need to introduce the deep issue of the proper output representation of a spiking cell, which relates directly to the question of the neuronal code used to transmit information among neurons.

14.1 Rate Codes, Temporal Coding, and All of That

How the neuronal output is represented, as a series of discrete pulses or as a continuous firing rate, relates to the *code* used by the nervous system to transmit information between cells. So let us briefly digress and talk about neuronal codes.

In a typical physiological experiment, the same stimulus is presented multiple times to a neuron and its response is recorded (Fig. 14.1). One immediately notices that the detailed response of the cell changes from trial to trial. Characterizing and analyzing the stochastic components of the neuronal response is important enough that we dedicate the following chapter to it.

Given the pulselike nature of spike trains, the standard procedure to quantify the neuronal response is to count how many spikes arrived within some sampling window Δt and to divide this number by the number of presentations. This yields the conditional probability that a spike occurred between t and $t + \Delta t$ given some particular stimulus.

In the limit of very small sampling windows—such that the probability for more than one spike occurring within Δt is vanishingly small—and infinitely many trials, the probability of spiking is given by $f(t)\delta t$, where $f(t)$ is the *instantaneous firing rate* of the neuron (in units of spikes per time). Plotting $f(t)$ as a function of the time after onset of a stimulus gives rise to the *poststimulus time histogram* (PSTH; Fig. 14.1).

It is important to understand the artificial nature of this construct $f(t)$. The nervous system has to make a decision based on a single spike train and not on the average of tens or more spike trains. A visual neuron in the fly does not have the opportunity to see the hand that is about to swat it approach ten times before it makes the decision to initiate an escape response! A neuron can only observe $\sum_i \delta(t - t_i)$ from its presynaptic partners, and not $f(t)$.

Under certain conditions this might be different. If a cell, say in the cortex, has access to the spiking output of many cells with the same receptive field properties, the temporal average of the single presynaptic neuron can be replaced by an ensemble average over a population of neurons, thereby approximating $f(t)$. In many cases *population rate coding* cannot occur for lack of a sufficient large cell population to average over. In the insect, for instance, a very small number of clearly identifiable neurons code for particular features of the sensory input and no ensemble averaging occurs.

Given the stochastic nature of spike trains, a common assumption is that the averaged firing rate of a neuron constitutes the primary variable relating neuronal response to sensory experience (Adrian and Zotterman, 1926; Adrian, 1932; Lettvin et al., 1959; Barlow, 1972). This belief is supported by the existence of a quantitative relationship between the averaged firing rate of single cortical neurons and psychophysical judgments made by a monkey. That is, the animal's behavior in a visual discrimination task can be statistically predicted by counting spikes over a long interval (typically 1 sec or more) in a single neuron in visual cortex (Werner and Mountcastle, 1963; Barlow et al., 1987; Newsome, Britten, and Movshon, 1989; Vogels and Orban, 1990; Zohary, Hillman, and Hochstein, 1990; Britten et al., 1992).

In these experiments, the rate is estimated by averaging over a window that is large compared to the time in which the sensory stimulus itself changes,

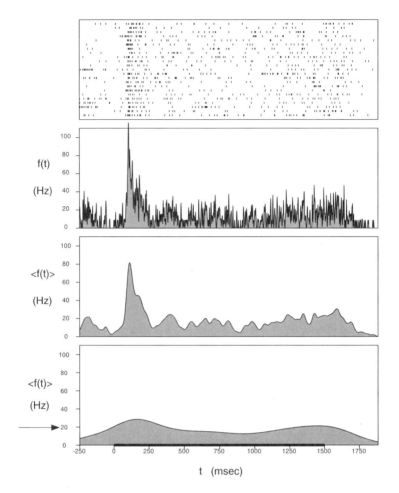

Fig. 14.1 What Is the Firing Rate Definition of the *firing rate*. The starting point is numerous trials in which the same stimulus is repeatedly presented to the animal and the spikes generated by some cell are recorded. These are shown in the *raster diagram* at the top, taken from a cell in cortical area V4 in the awake monkey. The stimulus—a grating—is flashed on at 0 and lasts until 1500 msec. Twenty-three of these trials are averaged, smoothed with a Gaussian of 2-msec standard deviation σ and normalized. This averaging window is so small that it effectively defines the instantaneous firing rate $f(t)$. These plots are known as *poststimulus time histograms* (PSTHs). The two lower plots illustrate an *average firing rate* $\langle f(t) \rangle$ obtained from the raster diagrams using Gaussian smoothing with σ set to 20 and 200 msec. In many experiments, only the average number of spikes triggered during each trial, corresponding to a very large value of σ (see arrow at 19.5 Hz), is used to relate the cellular response to the behavior of the animal. It is important to realize that a single neuron only sees spike trains and not a smoothly varying firing rate. Unpublished data from D. Leopold and N. Logothetis, printed with permission.

$$\langle f(t) \rangle = \frac{1}{n} \sum_{j=1}^{n} \frac{1}{\Delta T} \int_{t}^{t+\Delta T} \sum_{i=1}^{n_j} \delta(t' - t_{ij}) dt' \qquad (14.1)$$

where the first sum j is executed over the n identical trials and the second sum over all n_j spikes at time t_{ij} during the jth trial. Notice that the δ terms have the dimension of spikes per δt, so that the average $\langle f(t) \rangle^1$ has the correct units associated with a rate. Instead of a rectangular window, a frequent alternative is smoothing the spike train using a Gaussian convolution kernel with standard deviation σ,

$$\langle f(t) \rangle = \frac{1}{n} \sum_{j=1}^{n} \frac{1}{\sigma \sqrt{2\pi}} \int_{-\infty}^{+\infty} e^{-\frac{(t-t')^2}{2\sigma^2}} \sum_{i=1}^{n_j} \delta(t' - t_{ij}) dt' \qquad (14.2)$$

(Fig. 14.1). Because of the success in linking $\langle f(t) \rangle$ with behavior, it has been assumed by some that only the mean rate, averaged over a significant fraction of a second or longer, is the relevant code in the nervous system and that the detailed time course of spikes is not.

The past decade has witnessed a revival in the question to what extent an average rate coding neglects information. (For an expose of these ideas, see the superb textbook by Rieke et al., 1996.) On the basis of signal-processing and information-theoretical approaches, we know that individual motion-selective cells in the fly (Bialek et al., 1991), single afferent axons in the auditory system in the bullfrog (Rieke, Warland, and Bialek, 1993) and single neurons in the electrosensory system in weakly electric fish (Wessel, Koch, and Gabbiani, 1996) can carry between 1 and 3 bits of sensory information per spike, amounting to rates of up to 300 bits per second. This information is encoded using the instantaneous rate with a resolution of 5 msec or less. And the elegant experiments of Markram and his colleagues (1997), demonstrating the effect a short delay between a presynaptic and a postsynaptic spike arriving at a synapse can have on its weight (Sec. 13.5.4), provide a biophysical rationale for why timing at the 10 msec level is crucial for synaptic plasticity.

We summarize this vast body of work (see Rieke et al., 1996) by concluding that in many instances $\langle f(t) \rangle$—averaged over a 5–10 msec time frame—appears to be the relevant code used to transmit information:

More complex neuronal codes do exist and are frequently referred to under the catchall term of *temporal coding*. (For an exhaustive listing of possible codes, see Perkel and Bullock, 1968.) However, because of the implication that rate codes do not preserve detailed timing information, we prefer the term coined by Larry Abbott (personal communication), *correlation coding*.

In an instantaneous firing rate code, the generation of each spike is independent of other spikes in the trains (neglecting the refractory period and bursting), only a single number, the rate matters. In a correlation code this assumption is abandoned in favor of coupling among pairs, triplets, or higher order groupings of spikes.

To give one example of a correlation code, let $f(t)$ in response to some stimulus be a maintained response. We assume that this cell is very noisy with distinct spike patterns on individual trials. Averaging over them all leads to a flat response of amplitude f_c. In a rate code, f_c is the only information available to a postsynaptic neuron. Closer inspection of the microstructure of spiking reveals that the intervals between consecutive spikes are not independent of each other, but that two short spike intervals are always followed by a long one. Any code that fails to exploit these higher order correlations among four consecutive spikes would miss something. For experimental evidence of such codes see the references (Segundo et al., 1963; Chung, Raymond, and Lettvin, 1970; Optican and Richmond, 1987; Eskandar, Richmond, and Optican, 1992; Richmond and Optican, 1992).

A generic problem with the assumption of correlation codes is the question of decoding. It is unclear what sort of biophysical mechanisms are required to exploit the information

---

1. Sometimes also written as $\langle f(t) \rangle_T$ to express its dependency on the size of the averaging window.

hidden in such correlations among spikes. They might be prohibitively complicated to implement at the membrane level.

So far we have said little about *population coding*. Once again, one can distinguish two broad types of codes, correlated ones and noncorrelated ones (Abbott, 1994). The latter are straightforward: here the information from numerous neurons is combined into a population code but without taking account of any correlations among neurons (Knight, 1972a,b). There is plenty of good evidence for such codes in a great variety of different sensory and motor systems, ranging from the four cricket cercal interneurons encoding the direction the wind is blowing from (Theunissen and Miller, 1991) to the larger ensembles encoding the direction of sound in the barn owl (Knudsen, du Lac, and Esterly, 1987; Konishi, 1992) and eye movements in the mammalian superior colliculus (Lee, Rohrer, and Sparks, 1988) to the posterior parietal cortex in the monkey encoding our representation of space (Pouget and Sejnowski, 1997).

Correlation population codes exploit the exact temporal relationships among streams of action potentials. One way to discover such codes is to record from two or more neurons simultaneously and to measure their cross-correlation function. For instance, it may be that two presynaptic neurons always generate spikes within 1 or 2 msec of each other. Much technological advance has occurred in this area in recent years, so that multi-unit recordings have now become routine.

In a variety of sensory systems in both invertebrates and vertebrates, physiological evidence indicates that cross correlations among groups of cells appear to encode various stimulus features (Freeman, 1975; Abeles, 1982a, 1990; Strehler and Lestienne, 1986; Eckhorn et al., 1988; Gray et al., 1989; Bialek et al., 1991; Eskandar, Richmond, and Optican, 1992; Konishi, 1992; Singer and Gray, 1995; Decharms and Merzenich, 1996; Wehr and Laurent, 1996). The best evidence to date linking neuronal synchronization directly to behavior comes from the bee's olfactory system (Stopfer et al., 1997). When pharmacological agents were used to block cellular synchronization—without interrupting neuronal activity *per se*—fine olfactory discrimination was disrupted.

Much theoretical work using pulse-coded neural networks has focused on the idea that spike coincidence across neurons encodes information (Sejnowski, 1977a; Abeles, 1982a, 1990; Amit and Tsodyks, 1991; Koch and Schuster, 1992; Griniasty, Tsodyks, and Amit, 1993; Abbott and van Vreeswijk, 1993; Zipser et al., 1993; Softky, 1995; Hopfield, 1995; Maass, 1996; van Vreeswijk and Sompolinsky, 1996). Indeed, it has been proposed that the precise temporal correlation among groups of neurons is a crucial signal for a number of perceptual processes, including figure-ground segregation and the binding of different attributes of an object into a single coherent percept (Milner, 1974; von der Malsburg, 1981; von der Malsburg and Schneider, 1986), selective visual attention (Niebur and Koch, 1994), and even the neuronal basis of awareness (Crick and Koch, 1990; for a review see Koch, 1993).

From our point of view as biophysicists, a correlation population code based on coincidence detection has the significant advantage that it is straightforward to implement at the membrane level (witness Fig. 21.2).

The question of rate versus correlation coding remains with us. Yet this stark either-or dichotomy is not very useful. Clearly, the timing of spikes, at least at the 10 msec level, is important. And in some systems, detailed information across groups of cells will also prove to be of relevance, although for which properties and at what time scale is an open question.

It therefore behooves us to study how accurately and reliably neurons can generate individual action potentials and how robust these are to noise. We here lay the groundwork by introducing pulse neurons as well as firing rate models and describing their basic properties. We believe that such single-cell models represent the most reasonable tradeoff between simplicity and faithfulness to key neuronal attributes. The following chapter will deepen our discussion of stochastic aspects of neuronal firing. We will also discuss firing rate models.

14.2 Integrate-and-Fire Models

We turn to a very simple, but quite powerful model of a spiking cell with a long and distinguished history, first investigated by Lapicque (1907, 1926) before anything specific was known about the mechanisms underlying impulse generation. In its vanilla flavored version, it is known as the *integrate-and-fire* model (Stein, 1967a,b; Knight, 1972a; Jack, Noble, and Tsien, 1975; Tuckwell, 1988b; frequently also referred to as *voltage threshold* or *Lapicque's* model in the older literature). Its simplicity rivals physics' linear oscillator model, yet it does encapture the two key aspects of neuronal excitability: a passive, integrating subthreshold domain and the generation of stereotypical impulses once a threshold has been exceeded.

In the world of high speed electronics, integrate-and-fire models have their counterpart in the class of one-bit analog-digital converters known as *oversampled Delta-Sigma modulators* (Wong and Gray, 1990; Aziz, Sorensen, and van der Spiegel, 1996).[2]

The *nonleaky* or *perfect integrate-and-fire* unit consists but of a single capacitance for integrating the charge delivered by synaptic input in addition to a fixed and stationary voltage threshold V_{th} for spike initiation (Fig. 14.2). The *leaky* or *forgetful integrate-and-fire* model includes a resistance, accounting for leakage currents through the membrane. While integrate-and-fire models do not incorporate the detailed time course of the action potential, the effect of adaptation can be included. Indeed, the current-frequency relationship of such an integrate-and-fire cell with a handful of parameters can be very close to that of a much more complex, conductance-based cell model.

14.2.1 Perfect or Nonleaky Integrate-and-Fire Unit

We will be considering a number of variants of integrate-and-fire "units." All are characterized by a subthreshold domain of operation and a voltage threshold V_{th} for spike generation. The perfect integrate-and-fire unit deals with subthreshold integration via a single capacitance C. While unphysiological, it is mathematically tractable, which is why it is frequently invoked for pedagogical purposes. For the sake of mathematical convenience we assume the input to be a current $I(t)$, arising either from synaptic input or from an intracellular electrode. The generalization to a conductance-based input is straightforward.

The voltage trajectory of the perfect integrator is governed by the first-order differential equation

$$C\frac{dV(t)}{dt} = I(t).$$ (14.3)

Together with an initial condition Eq. 14.3 specifies the subthreshold time course of the membrane potential.

Once the potential reaches V_{th}, a pulse is triggered and the charge that has accumulated on the capacitance is shunted to zero (through the open switch in Fig. 14.2A). This would normally be accomplished by the various conductances underlying spiking. Sweeping the

---

2. A significant body of mathematics has sprung up around these $\Delta\Sigma$ modulators that should be explored for its relevance to neuroscience; see, for example Norsworthy, Schreier, and Temes (1996).

A)

Perfect Integrate-and-Fire Unit

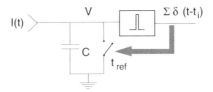

B)

Leaky Integrate-and-Fire Unit

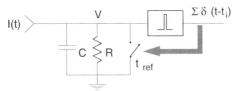

C)

Adapting Integrate-and-Fire Unit

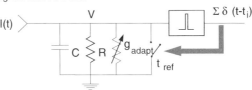

Fig. 14.2 INTEGRATE-AND-FIRE MODELS Three basic variants of integrate-and-fire units. Common to all are passive integration within a single compartment for the subthreshold domain and spike generation accomplished with a voltage threshold V_{th}. Whenever the membrane potential $V(t)$ reaches V_{th}, a pulse is generated and the unit is short-circuited. For a duration t_{ref} following spike generation, any input $I(t)$ is shunted to ground (corresponding to an absolute refractory period). **(A)** The *perfect* or *nonleaky* integrate-and-fire model contains but a capacitance. **(B)** The *leaky* or *forgetful* integrate-and-fire unit accounts for the decay of the membrane potential by an additional component, a leak resistance R. **(C)** The *adapting* integrate-and-fire unit with six free parameters (Eqs. 14.12 and 14.13) shows firing rate adaptation via the introduction of g_{adapt}, corresponding to a calcium-dependent potassium conductance (in addition to the absolute refractory period). Each spike increments the amplitude of the conductance; its value decays exponentially to zero between spikes.

charge to "ground" has the effect of instantaneously resetting $V(t)$ to zero. Because the model has no pretense of mimicking the currents involved in shaping the action potential, spike generation itself is not part of the model. Formally, we model the action potential by assuming that at the instant t' at which $V(t') = V_{th}$ (or the first time $V(t)$ exceeds V_{th} for models with instantaneously rising EPSPs) an output pulse, described by the delta function $\delta(t - t')$, is generated. The successive times t_i of spike occurrence are determined recursively from the equation

$$\int_{t_i}^{t_{i+1}} I(t)dt = CV_{th}.$$ (14.4)

As discussed in section 6.4, a canonical way in which experimentalists characterize a cell's behavior is by determining its *discharge* or f–I curve, the relationship between the amplitude of an injected current step and the average firing frequency (defined over an interval longer than the interspike interval; as in Eq. 14.1, $\langle f \rangle$ is computed as the inverse of the interspike interval).

In response to a sustained current, the membrane potential will charge up the capacitance until V_{th} is reached and V is reset to zero. The larger the current, the smaller the intervals between spikes and the higher the firing rate, according to

$$\langle f \rangle = \frac{I}{C V_{th}} . \tag{14.5}$$

Three features are worthwhile here. (1) The firing rate is linearly related to the input current (Fig. 14.3B). (2) Arbitrarily small input currents will eventually lead to a spike, since no input is forgotten. (3) The output spike train is perfectly regular. Of course, real neurons rarely, if ever, respond to a sustained current injection with a regular discharge of spikes but instead show substantial variability in the exact timing of the spikes. This is particularly true of neurons recorded *in vivo* (Holt et al., 1996). The following chapter will deal with this situation.

A)

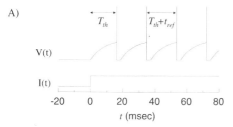

B)

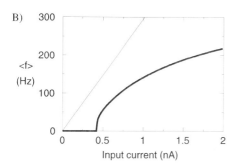

C)

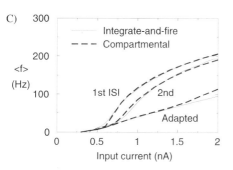

Fig. 14.3 SPIKING IN A LEAKY INTEGRATE-AND-FIRE MODEL Average firing frequency, determined as the inverse of the interspike interval, as a function of the amplitude of a maintained current input, for a leaky integrate-and-fire unit (Fig. 14.2B). (**A**) Exemplar trace of such a unit receiving a current step input with $I = 0.5$ nA. Before the membrane potential has time to reach equilibrium, the unit spikes. $V_{th} = 16.4$ mV, $C = 0.207$ nF, $R = 38.3$ MΩ, and $t_{ref} = 2.68$ msec. (**B**) f–I or *discharge* curve for the same leaky unit with refractory period (Eq. 14.11). The slope is infinite at threshold ($I_{th} = V_{th}/R$). The firing rate saturates at $1/t_{ref}$. For comparison, the f–I curve of the nonleaky unit without refractory period with constant slope $1/(V_{th}C)$ is superimposed. (**C**) An adapting conductance (with $G_{inc} = 20.4$ nS and $\tau_{adapt} = 52.3$ msec) is added to the leaky integrate-and-fire unit (see Fig. 14.2C) and the resulting f–I curve is compared against the discharge curve of the biophysical detailed compartmental model of the layer 5 pyramidal cell (Fig. 17.10). The degree of matching between simple and very complex models is quite remarkable and supports our contention that suitably modified integrate-and-fire models do mimic numerous aspects of the behavior of neurons. Adaptation is already evident when considering the first interspike interval (between the first and second spikes). Adaptation linearizes the very steep f–I curve around I_{th} (compare with B).

The dynamic firing range of nerve cells is limited by the fact that the sodium current responsible for spiking must recover from inactivation. Potassium currents furthermore limit the peak firing range. The effect of the absolute refractory period is mimicked by postulating that following spike generation, the membrane potential is set to zero for a fixed duration t_{ref}; any current arriving within this window is shunted away. This introduces a nonlinear saturation into the f–I curve of the perfect integrator,

$$\langle f \rangle = \frac{I}{CV_{th} + t_{ref}I} . \tag{14.6}$$

The output of such an integrator to an arbitrary input current consists of a series of impulses, $\sum_i \delta(t - t_i)$, all of which are spaced at least t_{ref} apart.

14.2.2 Forgetful or Leaky Integrate-and-Fire Unit

The model considered so far will sum linearly multiple subthreshold inputs irrespective of their temporal relationship because no account is made of a leak. A more realistic behavior is obtained by incorporating a leak resistance into the subthreshold domain (Fig. 14.2B),

$$C\frac{dV(t)}{dt} + \frac{V(t)}{R} = I(t) . \tag{14.7}$$

Having first met this equation in Chap. 1 (Eq. 1.5), we know that the evolution of the subthreshold voltage is completely characterized by convolving $I(t)$ with the associated Green's function, $e^{-t/\tau}$ (with $\tau = RC$). The time course of the membrane potential of the leaky integrate-and-fire unit to a step of constant current I, switched on at $t = 0$ and remaining on, can be obtained by solving Eq. 14.7,

$$V(t) = IR(1 - e^{-t/\tau}) + V(t = 0)e^{-t/\tau} . \tag{14.8}$$

The membrane charges exponentially up to its stationary value $V = IR$.

The integrator model will only follow this equation as long as the voltage remains below V_{th}, since upon reaching the threshold a spike is initiated and the voltage is reset to zero (Fig. 14.3A). The minimal sustained current necessary to trigger an action potential, that is, the threshold current, is

$$I_{th} = \frac{V_{th}}{R} . \tag{14.9}$$

For any current I larger than I_{th}, an output impulse will be generated at time T_{th}, such that $IR(1 - e^{-T_{th}/\tau}) = V_{th}$ holds. Inverting this relationship yields the time to spike as

$$T_{th} = -\tau \ln\left(1 - \frac{V_{th}}{IR}\right) . \tag{14.10}$$

Solving this equation for the minimal duration needed for a sustained current of a fixed amplitude to generate a spike generates what is known as the *strength-duration curve* (Noble and Stein, 1966; Jack, Noble, and Tsien, 1975). Since the voltage is reset following an impulse and if we assume that the input current persists, the membrane will again charge up to the threshold, triggering the next spike $T_{th} + t_{ref}$ later (Fig. 14.3A).

If we take proper account of the refractory period by assuming that for t_{ref} following each spike all input current is simply lost (due to the shunting effect of the conductances underlying the afterhyperpolarization), the continuous firing rate as a function of the injected current will be (Fig. 14.3B)

$$\langle f \rangle = \frac{1}{T_{th} + t_{ref}} = \frac{1}{t_{ref} - \tau \ln (1 - V_{th}/IR)}. \qquad (14.11)$$

For currents below I_{th} no spike is triggered and at $I = I_{th}$, the slope of the f–I curve is infinite. For large currents, the firing rate saturates at the inverse of the refractory period (Fig. 14.3B). In the absence of a refractory period, the slope of the f–I curve levels off to a constant value of $1/(V_{th}C)$, identical to the slope of the nonleaky unit.[3] Its steepness can be increased by reducing the threshold voltage or by decreasing the membrane capacitance. Due to the refractory period, the f–I curve gently bends over to level off at $f_{max} = 1/t_{ref}$ for (unphysiologically) high current levels.

14.2.3 Other Variants

Besides the generic version of the integrate-and-fire model discussed above, a number of variants are in use.

1. In order to better account for the 50–100 msec time course of adaptation, Wehmeier and colleagues (1989) introduced a purely time-dependent shunting conductance g_{adapt} (with a reversal potential equal to the resting potential, here assumed zero). Each spike increases this conductance by a fixed amount G_{inc}; between spikes, g_{adapt} decreases exponentially with a time constant τ_{adapt}. Such an effective calcium-dependent potassium conductance imitates both the absolute and the relative refractory period following spike initiation. We will refer to such a unit as an *adapting* integrate-and-fire model (Fig. 14.2C). Note that a refractory period t_{ref} is still necessary in order to mimic the very short-term aspect of adaptation. In the subthreshold domain, this unit is described by

$$C\frac{dV}{dt} = -\frac{V(1 + Rg_{adapt})}{R} + I \qquad (14.12)$$

$$\tau_{adapt}\frac{dg_{adapt}}{dt} = -g_{adapt}. \qquad (14.13)$$

If V reaches V_{th} at time t', a spike is generated at this point in time and $g_{adapt}(t')$ is incremented by G_{inc}. This model is completely characterized by six parameters: $V_{th}, C, R, t_{ref}, G_{inc}$ and τ_{adapt}.

2. An alternative to this output-dependent membrane conductance is to increase the voltage threshold following each spike in a deterministic manner, for instance, using the rule

$$V_{th}(t) = V_{th,0}(1 + \alpha e^{-(t-t')/\tau_{adapt}}), \qquad (14.14)$$

where $t - t'$ is the time from the last impulse, $V_{th,0}$ the threshold in the absence of any adaptation, and α the maximal normalized voltage threshold (Calvin and Stevens, 1968; Holden, 1976).

3. In order to account for those neurons that do not show any profound afterhyperpolarization following spiking, the membrane potential can be reset to a value closer to V_{th} (e.g., 20% of V_{th}) instead of zero. This is equivalent to resetting the potential to zero but adding a constant current and can have a considerable effect on the jitter in pulse timing (Troyer and Miller, 1997).

---

3. This can be seen upon developing the ℓn term in Eq. 14.11 into a Taylor series, with $\ell n(1 + x) \approx x - x^2/2$ for $|x| \ll 1$.

4. In a strategy to imitate the seemingly random nature of spike times, some authors resort to drawing the voltage threshold from some probability density distribution (Holden, 1976; Gestri, Masterbroek, and Zaagman, 1980). Yet in real neurons, the spiking mechanism itself appears to be quite reliable (Calvin and Stevens, 1968; Mainen and Sejnowski, 1995). In the case of the perfect integrator, a random threshold and a constant input can be shown to be equivalent to a random input and a constant threshold. Frequently, the former situation is both mathematically and computationally easier to deal with than the latter (for more details, see Gabbiani and Koch, 1998).

What all of these models share is a mechanism for passive integration of synaptic inputs, a voltage threshold for spike initiation, and a lack of specific spiking currents.

In Chap. 17 we will learn that an action potential in a full-blown model of a pyramidal cell (with eight voltage-dependent conductances) is, indeed, generated whenever the somatic membrane potential exceeds -49 mV. This is because the synaptic current flowing into the soma—caused by rapid EPSPs on the time scale of milliseconds—primarily charges up the capacitance. Relative to this rapid charging current, the ionic currents in the subthreshold regime change on a much slower time scale.

How well does this $f-I$ curve compare against curves obtained from much more detailed and sophisticated models? Presaging Sec. 17.5, we plot the $f-I$ curve of the layer 5 pyramidal cell, including the effect of firing rate adaptation, in Fig. 14.3C. Notice the very low slope of the $f-I$ curve around threshold, in contrast to the infinite slope of the $f-I$ curve of the leaky integrate-and-fire unit. If an adapting conductance is incorporated into the leaky integrator (Eqs. 14.12 and 14.13 and Fig. 14.2C), it is surprising how well this single-cell model (with just six degrees of freedom) resembles the much more detailed compartmental model based on membrane conductances.

Due to the presence of the leak term, integrate-and-fire models have been difficult to fully characterize analytically but have also been surprisingly successful in describing neuronal excitability. They have been applied to model the firing behavior of numerous cell types: neurons in the limulus eye (Knight, 1972b), α motoneurons (Calvin and Stevens, 1968), neurons in the visual system of the housefly (Gestri, Masterbroek and Zaagman, 1980), cortical cells (Softky and Koch, 1993; Troyer and Miller, 1997), and others.

While singing the praise of integrator models, it must be pointed out that many cells do not behave like integrate-and-fire units. For instance, cerebellar Purkinje cells (Jaeger, DeSchutter, and Bower, 1997) or the many types of oscillating neurons that constitute the central pattern generators found throughout the animal kingdom (Marder and Calabrese, 1996) have such strong inherent nonlinearities, generated by powerful intrinsic currents, that any attempts to directly map their behavior onto this class of models would fail miserably. Approximations are possible, though. For instance, bursting (Chap. 16) could be treated by letting the rapid Na$^+$ spikes be handled by the integrate-and-fire threshold mechanism. The slow dynamics governing at what instant the burst is triggered are generated by incorporating voltage-dependent conductances into the unit.

14.2.4 Response Time of Integrate-and-Fire Units

When a spiking, nonadapting membrane (such as the squid axon) receives a sustained suprathreshold current input, its membrane potential never reaches an equilibrium but moves along a limit cycle. That is, it undergoes periodic changes in its state variables (Chap. 7). When the integrate-and-fire neuron spikes, and the state variable is reset, it loses memory of the previous input current and begins to respond to the new current by charging toward the threshold. It follows from these considerations that if there is a step change in current, the integrate-and-fire neuron must converge to its limit cycle by the end of the *first* interspike interval after the change, because everything during this first interval is exactly the same as during subsequent intervals.

Figure 14.4 compares the step response of an integrate-and-fire unit, a compartmental model of a cortical pyramidal neuron, an experimental record derived from a neuron in cat visual cortex, and a mean-rate neuron. The first interspike interval already reflects the new firing rate—the convergence occurs on as short a time interval as can be defined (that is, the interspike interval). The inclusion of adaptation currents (Fig. 14.3B) does not substantially affect this analysis. Cortical cells (Fig. 14.4C) also reach their maximum firing rate by the first interspike interval. Thereafter, the firing rate decreases slowly due to the temporal dynamics of adaptation.

Just because the subthreshold dynamics are governed by τ does not imply that the neuron must respond to a suprathreshold input with the same dynamics. Returning to expression 14.10 for the time to spike T_{th}, we notice that the larger the injected current, the sooner the cell spikes (Fig. 14.5A). Furthermore, T_{th} actually decreases as the input resistance, and therefore τ increases (Fig. 14.5B). This can easily be explained by recalling that T_{th} is the time it takes for the membrane potential to reach the fixed threshold V_{th}. Increasing the input resistance will shorten this time, even if overall it would have taken the membrane longer to reach its ultimate steady-state value RI (which is never reached since a spike is triggered and the membrane potential reset once V hits V_{th}).

14.3 Firing Rate Models

The potential in a continuous firing rate unit, such as those at the heart of most neural networks, has the same dynamics as that in the leaky integrate-and-fire unit,

$$C\frac{dV(t)}{dt} = -\frac{V(t)}{R} + I(t).$$
(14.15)

A subtle but far-reaching difference is that the instantaneous output of this unit $f(t)$ is a continuous but nonlinear function of $V(t)$:

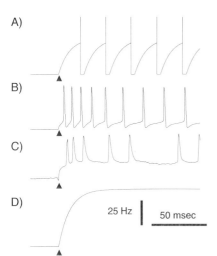

A)

B)

C)

D)

25 Hz | 50 msec

Fig. 14.4 SPIKING CELLS Sample spike rasters in response to a step current injection. **(A)** Leaky integrate-and-fire unit with refractory period spiking in response to a current step of 1.6 nA (for parameters, see legend to Fig. 14.3A). The arrows indicate the time at which the current injection commenced. **(B)** Somatic membrane potential in the layer 5 pyramidal cell model in response to a 1.5-nA current input. **(C)** Response of a cell in the primary visual cortex of the anesthetized adult cat to a 0.6-nA current injection (from Ahmed et al., 1993). The firing rate does not increase gradually; the effect of the change in current is fully visible in the first interspike interval. **(D)** Output of a nonadapting firing rate model with $\tau = 20$ msec. In the linear regime of the cell's $f–I$ curve, the firing rate can be considered to be a low-pass-filtered version of the step input.

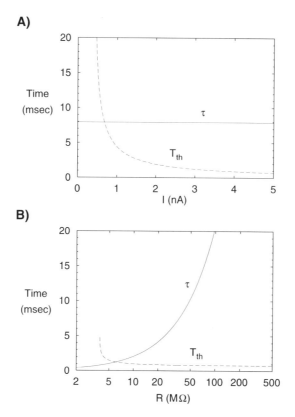

Fig. 14.5 **INTEGRATE-AND-FIRE UNITS CAN RESPOND MUCH FASTER THAN** τ Membrane time constant $\tau = RC$ of a leaky integrate-and-fire unit and the time T_{th} it takes such a unit, starting at $V = 0$, to reach V_{th} and spike (Eq. 14.10). **(A)** As the amplitude of the injected current increases, the unit can spike very rapidly. The true time to spike is $\leq T_{th}$, since the unit's initial state is usually $V > 0$. The parameters are as in Fig. 14.3A. **(B)** Input resistance R is varied over two orders of magnitude. As τ diverges, T_{th} converges to CV_{th}/I, the time it takes for the voltage across a capacitance to reach V_{th}. While this may be counterintuitive, it follows from the fact that T_{th} is the time it takes to reach a fixed threshold value, while τ dictates the approach toward a steady-state value beyond V_{th}. The injected current $I = 4.3$ nA. All else as in the upper panel. For small values of R, the current fails to reach I_{th} and T_{th} diverges.

$$f = g(V). \tag{14.16}$$

Most commonly g is a *sigmoidal* function, that is, a monotonically increasing, positive, and saturating function. A popular choice is

$$f = \frac{1}{1 + e^{-2\beta V}}, \tag{14.17}$$

where β is some positive constant (Fig. 14.6). Other functions have also been used (e.g., $\tanh V$ or V^2). V is sometimes identified with the so-called *generator potential*, the somatic membrane potential when the spikes are disabled (e.g., by blocking the fast sodium conductance). Under certain conditions, the generator potential correlates well with the firing rate when spikes are blocked (Katz, 1950; Granit, Kernell, and Shortess, 1963; Stein, 1967a; Dodge, Knight, and Toyoda, 1968). If f is thought of as the firing rate of the unit, then the function g can be identified with the $f-I$ curve of the cell under investigation and can be directly fitted against experimental data. Given the continuous and smooth nature of g, no true current threshold exists.

A circuit implementation of a firing rate neuron is shown in Fig. 14.6. Here the ideal operational amplifier is assumed to draw no input current (that is, it has an infinite input impedance), converting the difference between V and ground into the potential $f = g(V)$.

In this class of models the firing rate changes smoothly in response to a rapid change in I. For small steps in input current $g(V)$ is approximately linear. In the linear regime, the firing rate is given by convolving the input current with a first-order low-pass filter with time constant $\tau = RC$, and is therefore a smoothed version of the input.

14.3.1 Comparing the Dynamics of a Spiking Cell with a Firing Rate Cell

As we mentioned above, the voltage in an integrate-and-fire unit in response to a sustained suprathreshold input moves along a limit cycle. This is not true for the potential in a firing rate model. V reaches its equilibrium value in a time dictated by τ; the firing rate follows V without further delay.

If τ is increased in either model by increasing the neuron input resistance R, the rate of change of the subthreshold voltage also increases. This allows the spiking neuron to reach threshold more quickly (Fig. 14.7). In contrast, it will take the firing rate unit longer to reach its steady-state voltage because the equilibrium voltage is also increased, implying that the dynamics in the firing rate model slow down as τ increases (Fig. 14.7). In the extreme case, when $\tau \to \infty$ and the leaky integrate-and-fire model converges to a perfect integrator model, the firing rate model does not even asymptotically approach an equilibrium, while the spiking unit settles into its steady-state limit cycle faster than it does for a finite τ. Similarly, decreasing τ will not make a spiking neuron respond faster.

A spiking mechanism can therefore speed up a neuron's response to a step change in the input. In a population of spiking cells with uniformly distributed initial conditions, there will be cells whose T_{th} will be arbitrarily close to zero. Knight (1972a) made this point, proving rigorously that an infinite population of either perfect or leaky integrate-and-fire units uniformly distributed in phase will respond instantaneously to any suprathreshold stimulus. This is true for increases or decreases in input. In theory, how fast a neuron that receives input from such a population could detect a change in its inputs is limited only

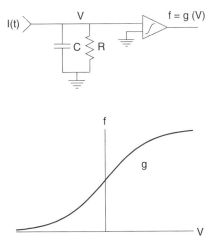

Fig. 14.6 Continuous Firing Rate Single-Cell Model As in the integrate-and-fire models (Fig. 14.2), the input current in a firing rate neuron charges up an RC element. The instantaneous firing rate is obtained by passing the membrane potential V through a smooth, stationary nonlinearity g. In the circuit idiom used here, this nonlinearity is implemented by an ideal operational amplifier. The associated adapted $f-I$ curve, specified by Eq. 14.17, $g(V) = g(RI)$, is a continuous, positive, and saturating function with no threshold.

A)

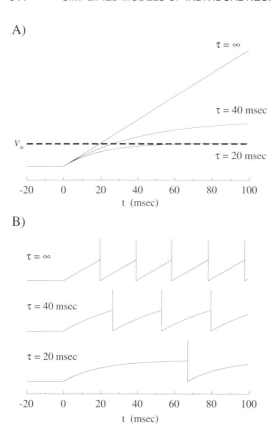

B)

Fig. 14.7 RESPONSE TIME IN SPIKING AND FIRING RATE MODELS Effect of changing the membrane time constant on (**A**) a nonspiking (or firing rate) neuron and (**B**) an integrate-and-fire neuron. The input current to the cell changes from zero to 0.85 nA at time 0. Subthreshold parameters were the same in both panels ($C = 1$ nF; R was varied from 20 MΩ to infinity; no refractory period). The subthreshold voltage of the integrate-and-fire model in **B** is exactly the same as for the nonspiking model in **A** until the threshold V_{th} (dashed line in **A**) is crossed. Increasing R increases the rate of change of voltage, but also increases the equilibrium voltage. Nonspiking neurons therefore converge more slowly as the time constant increases. As $\tau \to \infty$ (when the leaky integrate-and-fire unit turns into a perfect one) the integrate-and-fire model responds earlier.

by the number of statistically independent inputs (Panzeri et al., 1996). This is unlike the response of one or even a population of firing-rate neurons. Because of the low-pass filtering stage, their firing rate cannot change instantaneously.

It has been argued that the temporal dynamics of neurons embedded within a network are primarily determined by the time course of the synaptic currents. This has as the consequence that the subthreshold RC time constant should more properly be replaced by one or more synaptic time constants (e.g., Amit and Tsodyks, 1991; Amit and Brunel, 1993; Burkitt, 1994; Suarez, Koch, and Douglas, 1995; Brunel, 1996). A very simple, yet physiologically correct way to express the firing rate is

$$ f = h_{ss}(I(t)), \qquad (14.18) $$

where h_{ss} is the steady-state f–I discharge curve of the cell, and $I(t)$ is the total current flowing into the soma. This equation dispenses with the subthreshold voltage since it only plays a very limited role for suprathreshold currents. The subthreshold domain will come into play for inputs hovering just around threshold (e.g., the lower trace in Fig. 14.7B).

In order to turn this into a complete single-cell model, two additional ingredients are needed (Abbott, 1994).

1. How to obtain the current $I(t)$? The simplest manner is to use the standard neural network "linear sum over all inputs" formulation (see Eq. 14.20). We can be more accurate and incorporate the nonlinear effects occurring in the dendritic tree (synaptic interaction,

synaptic saturation, and so on). This is discussed at length in Sec. 18.4. Almost any degree of biophysical complexity can be accounted for. What cannot readily be included are back-propagating action potentials and the like (Chap. 19) since we assume that the coupling between dendrites and the spike-initiation zone is one-way only.

2. What is the relationship between the instantaneous firing rate $f(t)$ and the steady-state rate? The net effects of synaptic time constants and adaptation currents can all be lumped into a single, low-pass temporal filter, which from a phenomenological point of view, can be described by a first-order differential equation,

$$\tau_{\text{eff}} \frac{df(t)}{dt} = h_{ss}(I(t)) - f(t) \tag{14.19}$$

where τ_{eff} is the above-mentioned effective time constant (in the 20–30 msec range, in particular if one wants to account for both NMDA and adaptation currents), not related to the passive membrane time constant τ_m. This constitutes an alternative procedure to Eqs. 14.15 and 14.16 to define a firing rate model (Abbott, 1994). It gives rise to the same phenomenological equation but with a different physiological interpretation.

We conclude that because the discharge rate is often quite linear over the relevant range of firing rates (Granit, Kernell, and Shortess, 1963; Ahmed et al., 1993), a *linear* threshold unit with no threshold may often be a satisfactory approximation for the firing rate of a real neuron.

14.4 Neural Networks

Neural networks consist of a large number of neurons connected using a scalar *synaptic weight* w_{ij}. Almost always, the single-cell model used is a mean rate one, from the earliest publications of Steinbuch (1961) to Wilson and Cowan (1972), Hopfield (1984), and more recent work (summarized in Arbib, 1995).

14.4.1 Linear Synaptic Interactions Are Common to Almost All Neural Networks

The evolution of the network, often termed *neurodynamics*, is governed by a coupled system of single-cell equations. For any one cell i (out of n such cells) it takes the form

$$C \frac{dV_i(t)}{dt} = -\frac{V_i(t)}{R} + I_i + \sum_{j=1}^{n} w_{ij} f_j(t) \tag{14.20}$$

where f_j is the firing rate of the jth neuron and I_i corresponds to the external current injected into the cell. f_i is related to V_i via Eq. 14.16, that is, via a stationary nonlinearity

$$f_i(t) = g(V_i(t)). \tag{14.21}$$

In our usual circuit idiom (Fig. 14.8) in which both f_i and V_i are voltages, the synaptic coupling between the two neurons, that is, w_{ij}, has the dimension of a conductance. Negative, hyperpolarizing synapses are implemented by inverting the output of the amplifier. The evolution of the circuit in Fig. 14.8 can be expressed by an equation of the form shown in Eq. 14.20.

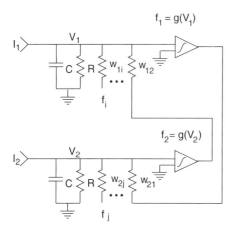

$f_1 = g(V_1)$

Fig. 14.8 TWO INTERACTING FIRING CELLS IN A NEURAL NETWORK Circuit model of two interacting mean rate neurons. Such continuous output units constitute the standard working horse of neural networks. Common to all is that the coupling among neurons is characterized by a scalar w_{ij} that can take on any real value, depending on whether the synapse is inhibitory or excitatory. The interaction among synaptic inputs is strictly linear. Local learning rules of the type discussed in Sec. 13.5 are used for determining the amplitude of the w_{ij}'s. A qualitatively very similar model of linear synaptic interactions has been used in the neural network community for studying the computational power of networks of spiking units.

Different from the biophysics of synaptic conductance inputs (Sec. 4.5), a change δf_j in the firing activity of the jth presynaptic neuron leads to a change $w_{ij}\delta f_j$ in the *current* delivered to the operational amplifier. In effect, the synaptic inputs act as current sources, and no nonlinear interaction among synaptic inputs exists.

Firing rate models incorporating a low-pass filter to capture the passive properties of the underlying membrane have been applied widely in abstract neural network analysis (Wilson and Cowan, 1972; Cohen and Grossberg, 1983; Hopfield, 1984; Arbib, 1995; an excellent textbook covering this area in a relatively intuitive manner is the one by Hertz, Krogh, and Palmer, 1991) and in the analysis of the dynamics of real neurobiological networks (Knight, Toyoda, and Dodge, 1970; Abbott, 1991; Traub and Miles, 1991; Amit and Tsodyks, 1991; Carandini and Heeger, 1994).

The study of the evolution of networks of spiking neurons—usually either of the integrate-and-fire or of the Hodgkin-Huxley variety—has only began of late because the discontinuous nature of their output and the attendant mathematical difficulties (Hansel and Mato, 1993; van Vreeswijk and Abbott, 1993; Usher et al., 1994; van Vreeswisk and Sompolinsky, 1996; Maass, 1996). Given the revival of interest in the role of spike timing in computation, more progress is likely to be just around the corner. The equations of motion of such a network remain governed by linear synaptic interactions that usually take the form

$$C\frac{dV_i(t)}{dt} = -\frac{V_i(t)}{R} + I_i + \sum_{j=1}^{n} w_{ij} \sum_{k} \delta(t - t_{jk}), \qquad (14.22)$$

with an auxiliary equation for the voltage to deal with the refractory period (or adaptation) following spike generation (Usher et al., 1994). The second sum on the right-hand side includes all times t_{jk} at which the jth neuron generated a spike. Each time this happens all of the postsynaptic targets of j receive a current pulse of amplitude w_{ij}. (Propagation delays can easily be incorporated into this notation.)

14.4.2 Multiplicative Interactions and Neural Networks

The bottom line of the previous section is that the vast majority of neural networks have been built on the assumption of linearity of synaptic interactions. The nonlinearity that is necessary for any true information processing to occur resides solely in the firing mechanism at the output of the cell (V_{th} for spiking models and g for rate neurons).

In view of the many different types of nonlinear interactions that can occur in the dendritic tree, including AND-NOT interactions (Sec. 5.1), NMDA synapses (Sec. 5.2), and the voltage-dependent calcium and sodium membrane conductances found in the dendritic tree as discussed as length in Chap. 19, it appears wise to cast around for some canonical single-cell model that captures some of these nonlinearities yet is simple enough to be still amenable to analysis.

From a computational point of view, the simplest nonlinearity is *multiplication*, as in

$$\mathcal{M}(x_1, x_2) = \alpha x_1 \cdot x_2 \qquad (14.23)$$

(with $\alpha \neq 0$). A substantial body of evidence supports the presence of multiplicative-like operations in the nervous system (Koch and Poggio, 1992). Physiological and behavioral data strongly suggest that the optomotor response of insects to moving stimuli is mediated by a correlation-like operation. Psychophysical work in humans from a number of independent groups strongly supports models of the correlation type, albeit with spatio-temporal filters different from those in insects. Simple cells recorded in the primary visual cortex of the cat (Emerson, Bergen, and Adelson, 1992) appear to encode some of the stages in the most popular of these algorithms, the *spatio-temporal energy* model of motion perception (Adelson and Bergen, 1985). From a mathematical point of view, all these motion algorithms can be implemented by the multiplication of two variables (Buchner, 1984; Hildreth and Koch, 1987; Poggio, Yang, and Torre 1989; Koch and Poggio, 1992).

Another instance of a multiplication-like operation in the nervous system is the modulation of the receptive field location of neurons in the posterior parietal cortex by the eye and head positions of the monkey (Andersen, Essick, and Siegel, 1985; Zipser and Andersen, 1988; Van Opstal and Hepp, 1995; Brotchie et al., 1995). This operation serves to transform the image from a retinal coordinate system into one that takes eye and body positions into account. A final example is the output of an identified neuron—the *descending contralateral movement detector*—in the visual pathway of the locust that signals rapidly approaching objects. Its firing rate can be accurately described as the product of the angular image velocity and a term that depends exponentially on the angular size of the approaching object on the animal's retina (Hatsopoulos, Gabbiani, and Laurent, 1995).

The generalization of multiplicative algorithms are *polynomial* ones. The output of a polynomial cell consists of the sum of contributions from a set of products,

$$\mathcal{P}(x) = a_1 + b_1 x_1 + b_2 x_2 + c_1 x_1^2 + c_2 x_1 x_2 + \cdots + d_1 x_1^2 x_2 + \cdots, \qquad (14.24)$$

where $x = (x_1, x_2, \ldots)$ represents the synaptic input and the scalar $\mathcal{P}(x)$ the output of the unit. Single-cell models that implement such a function are known as *sigma-pi* or *higher order*[4] units (Feldman and Ballard, 1982; Volper and Hampson, 1987; Mel, 1992). If the highest order product in Eq. 14.24 is quadratic, that is, only terms such as $x_i x_j$ and x_i^2 are represented and d_1 and all higher coefficients are all identical to zero, the neuron is known as a *second-order* or *multiplicative* unit.

The evolution of the membrane potential is identical to the one used in the linear threshold ones. For a second-order multiplicative unit it is

$$C \frac{dV_i(t)}{dt} = -\frac{V_i(t)}{R} + I_i + \sum_{j=1}^{n} w_{ij} f_j(t) + \sum_{j=1}^{n} \sum_{k=1}^{n} w'_{ijk} f_j(t) f_k(t). \qquad (14.25)$$

---

4. *Order* refers to the maximal number of variables multiplied in each term (3 in the case of Eq. 14.24).

Associated with this single-cell model are second-order "synapses" w'_{ijk}. Such a synapse only contributes to the postsynaptic potential if both inputs j and k are simultaneously active. In principle, up to n^2 synapses exist per neuron. For real neurons with dendritic trees, the connectivity matrices w_{ij} and w'_{ijk} are not independent of each other but are functions of the cable properties and of the specific synaptic architecture used. In general, they are also functions of time.

As an illustrative example, let us reinterpret the interaction occurring between an excitatory and a silent inhibitory synapse in a passive dendritic tree of a direction-selective cell—assuming that the conductance inputs g_e and g_i are small—in terms of a multiplicative unit (Sec. 5.1). We assume that the excitatory postsynaptic conductance change g_e is proportional to the presynaptic firing frequency f_e of the neuron e and that the amplitude of the silent inhibition g_i is proportional to the firing frequency f_i of an inhibitory unit synapsing onto the cell. With $E_i = 0$, we can reexpress the steady-state potential of the direction-selective cell j in Eq. 5.13 as

$$V_j = w_{je} f_e - w_{jee} f_e^2 - w_{jei} f_e f_i . \qquad (14.26)$$

The synaptic strengths are specified in terms of the input and transfer resistances and batteries as $w_{je} = E_e \tilde{K}_{es}$, $w_{jee} = E_e \tilde{K}_{es} \tilde{K}_{ee}$ and $w_{jei} = E_e \tilde{K}_{is} \tilde{K}_{ie}$. Note the unconventional nature of the dimensions of f and w. Equation 14.26 directly illustrates the connection between the nonlinear interactions among synapses and multiplicative neural network units.

Mel (1992, 1993, 1994) argues persuasively that NMDA synaptic input in combination with voltage-dependent sodium and calcium conductances distributed throughout the dendritic tree implements something akin to Eq. 14.24 (Sec. 5.2). Synaptic input distributed in spatial clusters throughout the dendritic tree approximates a polynomial, in the sense that simultaneous excitation of m neighboring synapses (where m can be large) causes a larger somatic response than activation of a similar number of synapses distributed in a diffuse fashion throughout the tree (Figs. 5.7–5.9). Different from Eq. 14.26, this operation is both more robust and less specific, since the absence of any one particular input will have little effect on the overall output (as indicated by the broad peak in Fig. 5.8B).

It is easy to see the power of polynomial units in the case of binary Boolean functions and circuits. For instance, a single such unit can implement an exclusive-or function, something that a single-layer neural network of linear threshold units is unable to do. (For a rigorous investigation of this, see Bruck, 1990; Bruck and Smolensky, 1992; for background material see Koch and Poggio, 1992). Of course, neurons always have the sigmoidal f–I nonlinearity to fall back upon (g in Eq. 14.16 or h_{ss} in Eq. 14.18).

14.5 Recapitulation

We started off this chapter by defining the instantaneous firing frequency $f(t)$. It is a fictive variable that can be obtained by averaging the spiking response of a single neuron to multiple presentations of the same stimulus. Due to the stochastic nature of the neuronal response the exact microstructures of spike trains are rarely reproducible from trial to trial. This is why the temporal average $\langle f(t) \rangle$ of the firing rate is the most common variable measured during neurophysiological experiments. In a handful of experiments $\langle f \rangle$, evaluated over fractions of a second or longer, has been directly related to the behavior of the animal.

There is no question that firing rate codes that preserve temporal information at the 5–10 msec level are used in the nervous system. To what extent more complex correlation codes—exploiting information encoded among n-tuples of spikes from one neuron or spiking information across multiple neurons—exist remains a subject of considerable debate.

What is the simplest model of a single neuron that captures some of its key operations? Two families of models are in common use today: integrate-and-fire and continuous firing rate models. While the former retains the timing information of individual action potentials, the latter assumes that it is only the average or mean firing rate of a neuron that matters to its postsynaptic targets. Both neglect the dendritic tree and both eliminate the complex time course of the sodium and potassium membrane conductances underlying spiking.

The key insight behind the various guises of the integrate-and-fire model is that from a phenomenological point of view the neuron possesses two domains of operation, a subthreshold and a suprathreshold one. In the subthreshold domain, synaptic inputs are integrated and decay away; their temporal evolution is governed by the time constant τ. Once the voltage threshold is reached, a pulse is generated and the membrane potential is reset. Different versions of integrate-and-fire models, incorporating various mechanisms to account for adaptation, can be well fitted to the discharge curves of cortical and other cells. It will be argued in Sec. 17.3 that firing in response to fast synaptic input in complex, conductance-based single-cell models is, indeed, initiated whenever a voltage threshold is exceeded.

In response to a suprathreshold stimulus, these units, in accordance with their biological counterparts, can spike in a time $T_{th} \ll \tau$. A network of integrate-and-fire units can respond almost instantaneously to a stimulus. The take home lesson is that the dynamics of the subthreshold domain do not carry over into the suprathreshold domain.

In firing rate neurons the continuous output variable is an instantaneous function of the voltage. Since the evolution of the voltage is dictated by a time constant, the firing rate will always be low-pass filtered with respect to the input current, distinct from the response of real neurons, and different from integrate-and-fire units. If one would like to retain the continuous nature of the firing rate model, a more physiologically correct way to achieve this would be to make the steady-state firing rate a function of the total current (synaptic, dendritic, or otherwise) at the cell body. The output of such a neuron can be interpreted as the firing rate associated with a population of spiking cells.

At the heart of the vast majority of neural networks lies the assumption that synaptic inputs interact in a linear manner. The nonlinearity that is necessary for computation is relegated to the firing mechanism at the output. A biophysically more faithful and more complex model that incorporates multiplicative interactions among synaptic inputs is the polynomial or sigma-pi unit.

Multiplication is a key operation underlying many neuronal operations. Chapter 5 treated the evidence in favor of the view that a dendritic tree endowed with NMDA synapses and voltage-dependent membrane conductances (see Chap. 19 as well) can implement a robust version of such a polynomial unit. The nonlinear operations underlying the polynomial interactions do *not* depend on the threshold occurring at the cell body but precede it. The computational power of such neurons is considerably beyond that of their feeble-minded linear threshold counterparts.

15

STOCHASTIC MODELS
OF SINGLE CELLS

The majority of experiments in neurophysiology are based upon *spike trains* recorded from individual or multiple nerve cells. If all the action potentials are taken to be identical and only the times at which they are generated are considered, the experimentalist obtains a discrete series of time events $\{t_1, \cdots, t_n\}$, where t_i corresponds to the occurrence of the ith spike, characterizing the spike train. This spike train is transmitted down the axon to all of the target cells of the neuron, and it is this spike train that contains all of the relevant information that the cell is representing (assuming no dendro-dendritic connections).[1]

As alluded to in the preceding chapter, there are two opposing views of neuronal coding, with many interim shades. One view holds that it is the firing *rate*, averaged over a suitable temporal window (Eqs. 14.1 or 14.2), that is relevant for information processing. The dissenting view, *correlation coding*, argues that the interactions among spikes, at the single cell as well as between multiple cells, encodes information.

A key property of spike trains is their seemingly *stochastic* or random nature, quite in contrast to switching in digital computers.[2] This randomness is apparent in the highly irregular discharge pattern of a central neuron to a sensory stimulus whose details are rarely reproducible from one trial to the next (Figs. 14.1 and 15.1). The apparent lack of reproducible spike patterns (but see Fig. 15.11) has been one of the principal arguments in favor of the hypothesis that neurons only care about the firing frequency averaged over very long time windows. Such a *mean rate code* is very robust to "sloppy" hardware but is also relatively inefficient in terms of transmitting the maximal amount of information per spike. Encoding information in the intervals between spikes is obviously much more efficient, in particular if correlated across multiple neurons. Such a scheme does place a premium on postsynaptic neurons that can somehow decode this information.

Because little or no information can be encoded into a stream of regularly spaced action potentials, this raises the question of how variable neuronal firing really is. How

---

1. Slower processes, such as extracellular potassium accumulation or local depletion of calcium, could also play a role in information transmission among cells. Since very little is known about these, we neglect them for now; see, however, the penultimate chapter.

2. We use the term "seemingly" on purpose. It may well be that what we think of as noise is the signal we are seeking!

350

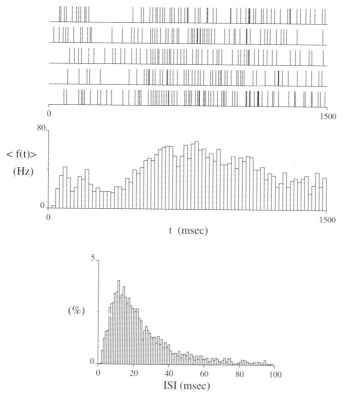

Fig. 15.1 VARIABILITY OF NEURONAL SPIKING A high-contrast bar is swept repeatedly over the receptive field of a cortical cell in the awake macaque monkey. Much variability in the microstructure of spiking is evident from trial to trial. The poststimulus time histogram in the middle corresponds to the averaged firing rate $\langle f(t) \rangle$ (using 20 msec bins) taken over 40 trials. The lower plot illustrates the associated interspike interval (ISI) histogram. It shows a lack of very short intervals, indicative of a refractory period, and an exponentially decreasing likelihood of finding very large gaps between spikes. The lack of reproducibility of the detailed spike pattern is the primary reason arguing for the idea of a mean rate code (Eq. 14.1). Yet neurons deep within the cortex can faithfully reproduce the microstructure of spiking over several hours (Fig. 15.11). From W. Newsome, K. Britten, personal communication.

can the observed randomness be explained on the basis of the cell's biophysics and synaptic input (Calvin and Stevens, 1968; Softky and Koch, 1993; Shadlen and Newsome, 1994; Softky, 1995; König, Engel, and Singer, 1996)? The mathematical theory of stochastic point processes and the field of statistical signal processing offer a number of tools adequate for analyzing the properties of spike trains. We will study these here and will relate them to simple models of biophysics. This will enable us to infer something about the integrative mechanisms underlying neuronal firing activity.

In the previous chapter we exclusively dealt with deterministic input $I(t)$ to spiking neurons. Given the highly stochastic nature of neurons evident in Fig. 15.1, it is imperative that we now begin to deal with *random variables*, that is, observables that take on discrete (such as the number of spikes) or continuous (such as the membrane potential) values with certain probabilities. For an introduction into the field, we refer the reader to Papoulis (1984); for monographs on stochastic neural activity, see Holden (1976), Tuckwell (1988c), and Smith

(1992). Gabbiani and Koch (1998) should be read as a complementary text to this chapter, as they discuss in more detail signal processing approaches toward spike train analysis and their numerical implementation into a suite of freely available MATLAB routines.

15.1 Random Processes and Neural Activity

For starters, let us assume that we are dealing with discrete random variables. We consider a *stochastic* or *random process*, that is, a family of such random variables parameterized by time t. In other words, at every instant t the random process has the discrete value $N(t)$, while the entire process is specified by $\{N(t), t \geq 0\}$. In the following, we are frequently faced by questions of the form: "What is the probability that N is as large as some specific n_{th}?" In order to answer these, we need to specify a probability distribution from which the random variable N is drawn. Before we do so, let us remind our readers of several important concepts.

One is the idea of a *point process*, which is simply a process that can be mapped onto a set of random points on the time axis. Two examples of point processes are the occurrence of action potentials and the times at which a synaptic vesicle is relased at some synapse. A point process is described by a series of delta pulses, $\sum \delta(t - t_i)$.

We can introduce a new random variable, defined as the interval between consecutive events, $T_i = t_{i+1} - t_i$, to help us introduce the notion of a *renewal process*. This is a point process in which the random variables T_i are independent and identically distributed. The notion of a renewal process comes from industrial practice: the probability that some machine will fail within some interval is identical for all machines and will not vary from one interval to the next. Applied to a spike train, it would imply that the chance of finding some particular interspike interval (Fig. 15.1) is independent of whether a short or a long interspike interval preceded it. This does tend to be true to a first approximation for many cortical cells. On the other hand, it is certainly not true for bursting cells, where the probability for finding a long interspike interval after two or three very short ones is strongly enhanced (see the following chapter). Renewal processes are much simpler to describe from a mathematical point of view than nonrenewal processes. Lastly, a *stationary* stochastic process (or a random process) is one whose statistical properties do not depend on the time of observation but remain constant in time.

Since true stationarity is very difficult to verify in practice, it is common to use the less stringent concept of a *wide-sense* or *weakly stationary* process. This is a random process whose mean is constant and whose autocorrelation function only depends on $t - t'$, that is, a process whose first- and second-order statistical properties are time invariant.

Let us now focus on the most common discrete renewal process, the *Poisson process*.

15.1.1 Poisson Process

A Poisson process is the simplest possible random process with no memory and is characterized by a single parameter, the *rate* or *mean frequency* μ. It is of great relevance to neurobiology, since a number of discrete biophysical events appear to follow a Poisson distribution closely. The best studied example of a Poisson process is the spontaneous release of individual packets of the neurotransmitter acetylcholine at the frog neuromuscular junction (Fatt and Katz, 1952; for an overview see Stevens, 1993, and Chap. 4.2). At the postsynaptic terminal this release gives rise to so-called *miniature endplate potentials*, whose arrival times can be accurately modeled by a Poisson process. As we will see below, the distribution of action potentials in cortical cells can be approximated to a certain degree by a modified Poisson process.

A number of different, but equivalent, definitions of a Poisson process exists. We define $\{N(t), t \geq 0\}$ to be a simple *Poisson process* with mean rate μ if:

1. Given any $t_0 < t_1 < t_2 < \cdots < t_{n-1} < t_n$, the random variables $N(t_k) - N(t_{k-1})$, $k = 1, 2, \cdots, n$, are mutually independent.

2. For any $0 \leq t_1 < t_2$ the average number of events occurring between t_1 and t_2 is $\mu(t_2 - t_1)$.

The first condition specifies that the number of events occurring in one interval is independent of the number of events occurring in any other interval, provided they do not overlap. The second property tells us that the expected number of events is proportional to the rate times the duration of the interval.

As no point in time is singled out in this definition of the Poisson process, it is a *stationary* process.

It follows from these conditions (Feller, 1966) that the actual number of events, $N(t_2) - N(t_1)$, is a random variable with the Poisson probability distribution

$$\Pr\{N(t_2) - N(t_1) = k\} = \frac{(\mu(t_2 - t_1))^k e^{-\mu(t_2 - t_1)}}{k!} \tag{15.1}$$

with $k = 0, 1, 2, \cdots$.

The parameter μ specifies the average number of events per unit time. With $t_1 = t$ and $t_2 = t + \Delta t$, we have for the probability that exactly k events occur in the interval Δt,

$$\Pr\{N(t + \Delta t) - N(t) = k\} = \frac{(\mu \Delta t)^k e^{-\mu \Delta t}}{k!}. \tag{15.2}$$

If $\mu \Delta t \ll 1$, that is, if much less than one event is expected to occur in the interval Δt, the e^{-x} term in Eq. 15.2 can be expanded into a Taylor series $1 - x + x^2/2 - x^3/3 + \cdots$. We then have for the probability that none or a single event occurs in the interval Δt,

$$\Pr\{N(t + \Delta t) - N(t) = 0\} = e^{-\mu \Delta t} \approx 1 - \mu \Delta t \tag{15.3}$$

$$\Pr\{N(t + \Delta t) - N(t) = 1\} = \mu \Delta t e^{-\mu \Delta t} \approx \mu \Delta t. \tag{15.4}$$

These estimations are accurate within $(\Delta t)^2$, that is, the error in these estimates goes as a polynomial of order $(\Delta t)^2$ to zero as $\Delta t \to 0$. In other words, as the product of the average firing frequency and the interval is made smaller and smaller the approximation becomes better and better. Figure 15.2 illustrates how the events generated by a Poisson process are distributed compared to the spike distribution taken from a neuron. Both processes have the same overall rate. On account of the refractory period, we modified the Poisson process to prevent two spikes from directly following each other.

A commonly used procedure for numerically generating Poisson distributed action potentials with mean rate μ is based on this approximation. A random number r, uniformly distributed between 0 and 1, is generated for every interval Δt; if $r \leq \mu \Delta t$, a spike is presumed to have occurred between t and $t + \Delta t$; otherwise, no spike is generated. (This assumes that the probability of generating a spike in the interval considered, that is, $\mu \Delta t$ is much less than one.)

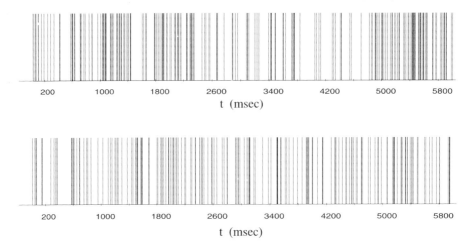

Fig. 15.2 SPIKE TRAINS AND THE RANDOM POISSON PROCESS Spike train from an extracellularly recorded cell in the parietal cortex of a behaving monkey (upper trace; data kindly provided by J. Fuster) and from a synthetic Poisson process with a refractory period of 1 msec (lower trace). The mean firing rate of both processes is 30 Hz. Notice the relatively rare occurrences of large gaps between spikes. In a Poisson process, the probability of occurrence of gaps of duration T decreases as $e^{-T\mu}$.

15.1.2 Power Spectrum Analysis of Point Processes

A number of signal processing techniques rely on evaluating average spike train properties in response to random stimulus ensembles. This works best if the system at hand can be described as a linear, time-invariant one. Under these conditions, the *power spectrum* $S(f)$ of the associated spike train reveals details about the filter function that is used by the system (e.g., does the neuron act as a low-pass or as a band-pass filter?).

A basic result of the theory of stochastic processes (Papoulis, 1984) is that the power spectrum of a Poisson process is flat at all frequencies except for a delta peak at the origin. To be more precise, the spectrum associated with an infinite train of delta impulses distributed according to a Poisson process of rate μ is

$$S(f) = \mu + 2\pi\mu^2\delta(f).\tag{15.5}$$

This is in accordance with our intuition that all spectral components should be equally represented in a completely random spike train; no particular frequency (such as $1/\mu$) is singled out. The *autocorrelation function*, defined as the inverse Fourier transform of the power spectrum, has a similar shape, taking on a constant value of μ^2 everywhere except at the origin, where it is a $\delta(t)$ function.

Neurons, however, do not fire totally without memory, due to the presence of a refractory period. A mathematically convenient manner in which this can be accounted for (Perkel, Gerstein, and Moore, 1967; Bair et al., 1994) is by introducing the *renewal density function*, which denotes the probability for a spike to occur between $t_1 + t$ and $t_1 + t + dt$, given that a spike was just generated at t_1. For a pure Poisson process, the renewal density function is a constant equal to the rate. In our case, however, the renewal density has a dip around zero, indicating a reduced probability of spiking due to the refractory period. This can easily be measured experimentally. Assuming that the dip due to the refractory period can be accounted for by a Gaussian of variance σ_{ref}^2, the associated power spectrum is,

$$S(f) = \mu(1 - \sqrt{2\pi}\mu\sigma_{ref}e^{-2(\pi f\sigma_{ref})^2}) + 2\pi\mu^2\delta(f).\tag{15.6}$$

To ensure that the power is always positive, the maximum firing rate must be limited: $\mu \leq 1/(\sqrt{2\pi}\sigma_{\text{ref}})$. This spectrum, parameterized by the mean rate μ and the width of the refractory period σ_{ref} is constant for large values of f but dips toward its minimum at $f = 0$ (Fig. 15.3). A longer refractory period (that is, a larger value of σ_{ref}) causes a deeper trough at low frequencies. Note that this result appears at odds with intuition, since a refractory period seems to demand a dip in the neighborhood of the inverse of the smallest interspike interval, that is, at high temporal frequencies.

A mathematically more general manner to account for the refractory period is by postulating that the interspike intervals are independently and identically distributed random variables, which themselves are the sum of a random refractory period and a statistically independent interval due to a known stationary process (Franklin and Bair, 1995).

Spike train analysis of cortical cells (Bair et al., 1994) showed that the spectrum of 40% of these neurons can be fitted using Eq. 15.6 (left column in Fig. 15.3). The good match between the analytical model, experimental data, and synthetic spike train supports the hypothesis that once the refractory period is accounted for, a Poisson hypothesis constitutes a first-order description of cortical spike trains.

15.2 Stochastic Activity in Integrate-and-Fire Models

Let us now take an integrate-and-fire model, as described in the previous chapter, and bombard it with stochastic, Poisson distributed, synaptic inputs. In what is to come, we will neglect the membrane leak, as well as adaptation and the refractory period, to get at the principle underlying stochastic activity in these models. Using the nonleaky

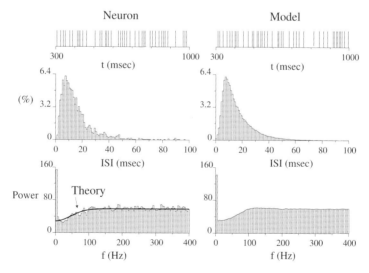

Fig. 15.3 POWER SPECTRUM OF CORTICAL CELLS Representative spike train, interspike interval histogram, and power spectrum for a neuron recorded from cortical area MT in a behaving monkey (left column) compared against a modeled point process (right column). The synthetic data assume that spikes are Poisson distributed with an absolute refractory period drawn from a Gaussian distribution (of 5 msec mean and 2 msec standard deviation). Equation 15.6 has been fitted to the neuron's spectrum (curve labeled "Theory") with $\mu = 58$ Hz, and $\sigma_{\text{ref}} = 3.5$ msec). The spectrum of many nonbursting cells can be fitted against this simple model, which satisfactorily captures many aspects of their random behavior. Reprinted in modified form by permission from Bair et al., (1994).

integrate-and-fire model allows us to derive closed-form expressions for many variables of interest. Conceptually, $R = 0$ corresponds to a cell for which the time between spikes is much shorter than the membrane time constant, so that the decay of the membrane potential can be neglected. We later investigate $R \neq 0$ using numerical simulations.

We assume that the integrator cell receives excitatory synaptic input from a Poisson distributed random process $N_e(t)$ with mean rate μ_e and synaptic weight a_e. Conceptually, each synaptic input can be thought of as dumping a_e amount of charge onto the capacitance C, raising the membrane potential by a_e/C toward threshold. We assume here that C has been folded into the synaptic weight a_e; it will not be included in the equations to come (equivalently, $C = 1$). The random process $V(t)$, that is, the membrane potential, shows random jumps of amplitude a_e. In the absence of a threshold, the membrane potential can be described as

$$V(t) = a_e N_e(t) \,. \tag{15.7}$$

15.2.1 Interspike Interval Histogram

Let us proceed methodically by first starting with trivial cases, working up to more complex situations. In this spirit, we assume that $a_e \geq V_{\mathrm{th}}$, that is, each synaptic input is, by itself, sufficient to cause a spike. (Think of a single retinal ganglion cell making a few hundred synaptic contacts on a single geniculate relay cell.) Trivially, the probability distribution of the output spike will be identical to that of the synaptic input.

If the perfect integrator generated a pulse at $t = 0$ and was reset, what is the waiting time T_1 for the next output pulse to occur? In the case of $a_e \geq V_{\mathrm{th}}$, this is equivalent to asking what is the waiting time for the next synaptic input to occur. This function can be computed if we know the probability for no event to occur between $(0, t)$, which is

$$\Pr\{T_1 > t\} = \Pr\{N(t) - N(0) = 0\} = e^{-\mu_e t} \,. \tag{15.8}$$

The distribution function for the next spike to occur is one minus this result, that is, $1 - e^{-\mu_e T}$; the longer one waits, the more likely an input is to occur. The probability density function $p_1(t)$ for t is the temporal derivative of the distribution function,

$$p_1(t) = \mu_e e^{-\mu_e t} \,. \tag{15.9}$$

The choice of our initial observation point, here $t = 0$, does not affect this result since the Poisson process is ahistoric. This is an important point to note; the density function does not depend on whether or not an event just occurred. (Of course, given the existence of a refractory period, this will not be true for real neurons at very small time scales.) Therefore, $p_1(t)$ corresponds to the density of time intervals between adjacent pulses.

The average duration between events is

$$\langle t \rangle = \int_0^\infty t p_1(t) dt = \frac{1}{\mu_e} \tag{15.10}$$

justifying our interpretation of μ_e as the mean rate.

What happens if we assume that n_{th} inputs are needed to trigger a pulse (with $n_{\mathrm{th}} = \lfloor V_{\mathrm{th}}/a_e + 1 \rfloor$, where $\lfloor x \rfloor$ is the largest integer less than or equal to x)? This is similar to a Geiger counter that is rigged to click every time it detects n_{th} radioactive decays. Answering this question requires us to compute the time T_{th} in which we expect a Poisson process with rate μ_e to generate exactly n_{th} events. This is given by the probability for $n_{\mathrm{th}} - 1$ events to occur between now and t, multiplied by the probability for a single event to occur in the interval $[t, t + \delta t)$,

$$\Pr\{T_{\text{th}} > t + \delta t\} = \frac{(\mu_e t)^{n_{\text{th}}-1} \cdot e^{-\mu_e t}}{(n_{\text{th}} - 1)!} (\mu_e \delta t) . \tag{15.11}$$

The probability density function of T_{th} is the limit of this expression divided by δt as $\delta t \to 0$,

$$p_{n_{\text{th}}}(t) = \frac{\mu_e \cdot (\mu_e t)^{n_{\text{th}}-1} \cdot e^{-\mu_e t}}{(n_{\text{th}} - 1)!} . \tag{15.12}$$

This expression for the waiting time is known as the n_{th}th-order *gamma density*.

The density function $p_{n_{\text{th}}}(t)$ is obtained experimentally by binning consecutive interspike intervals from spike trains, as in Figs. 15.1 and 15.4. In this guise it is called the *interspike interval* (ISI) histogram. The mean interspike interval is

$$\langle T_{\text{th}} \rangle = \frac{n_{\text{th}}}{\mu_e} \tag{15.13}$$

A)

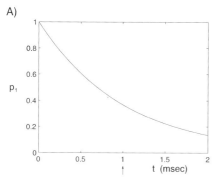

B)

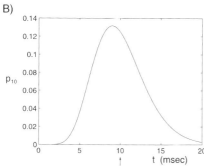

C)

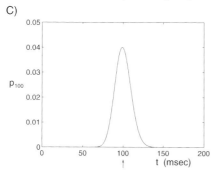

Fig. 15.4 INTERSPIKE INTERVAL DISTRIBUTION
In the interspike interval distribution the occurrences of the intervals between adjacent spikes is histogrammed. This is done here analytically for pulses generated by a perfect integrate-and-fire unit that receives Poisson distributed input with a constant rate of $\mu = 1000$ (per second). The threshold of the unit varies, with **(A)** $n_{\text{th}} = 1$ (each input generates an output pulse), **(B)** $n_{\text{th}} = 10$, and **(C)** $n_{\text{th}} = 100$. The ISIs are gamma distributed (Eq. 15.12). As expected, the mean output rate decreases as the number of synaptic inputs needed to reach threshold n_{th} increases (the mean interspike interval is at 1, 10, and 100 msec for the three panels; see arrows). The normalized standard deviation, called the *coefficient of variation* C_V (Eq. 15.15) scales as $1/\sqrt{n_{\text{th}}}$. In other words, the relative jitter in the timing of output pulses becomes smaller as n_{th} becomes larger. If the cell is integrating over a significant number of small inputs, the ISI distribution approaches the normal distribution (as in **C**).

and the variance is

$$\text{Var}[T_{\text{th}}] = \frac{n_{\text{th}}}{\mu_e^2}. \tag{15.14}$$

Figure 15.4 shows waiting time densities for a nonleaky integrate-and-fire neuron receiving excitatory input at a constant rate μ_e in which the effective voltage threshold is increased by a decade, corresponding to $n_{\text{th}} = 1, 10,$ and 100. As n_{th} of the gamma distribution increases, the probability density of the interspike intervals rapidly tends toward a Gaussian distribution.

15.2.2 Coefficient of Variation

The most common way to quantify interspike variability is via the normalized standard deviation, that is, the square root of the variance of the ISI histogram divided by its mean,

$$C_V = \frac{(\text{Var}[T_{\text{th}}])^{1/2}}{\langle T_{\text{th}} \rangle}. \tag{15.15}$$

This measure is known as the *coefficient of variation*. For the perfect integrator model considered here, we have

$$C_V = \frac{1}{n_{\text{th}}^{1/2}}. \tag{15.16}$$

For $n_{\text{th}} = 1$, associated with an exponential ISI (Fig. 15.4A), $C_V = 1$. As the number of random inputs over which the unit integrates (or averages) becomes larger, the normalized ISI becomes narrower and narrower and C_V smaller and smaller (Fig. 15.4C). The relationship between spike train variability and C_V is illustrated graphically in Fig. 15.5.

This result can be explained qualitatively by invoking the *central limit theorem*, which states that as the number n of independent random variables x_i goes to infinity, the random variable defined by the mean of all x_i's, $\langle x \rangle = (1/n) \sum_{i=1}^{n} x_i$, has an asymptotically Gaussian distribution, with a mean identical to the mean of the population and with a standard deviation scaling as $1/\sqrt{n}$ of the standard deviation of the individual x_i's. If we were to compute the average height of all students in one class, there would be a great deal of variability around the mean. Yet the C_V associated with the average height of all men in the United States is minute. In other words, if a neuron can only be brought to fire action potentials by summing over dozens or more of independent synaptic inputs, it should fire very regularly. And as we noted above, a neuron with little variability in its interspike interval cannot readily exploit temporal coding (since little information can be encoded in a regular interspike interval).

Incorporating an absolute refractory period t_{ref} into the integrator shifts the interspike interval probability distribution by the same amount to the right. It gives the perfect integrator cell a characteristic time scale, so we cannot expect it to have identical statistics at all firing rates. The new C_V is

$$C_V = \frac{1}{n_{\text{th}}^{1/2}} \frac{\langle T_{\text{th}} \rangle - t_{\text{ref}}}{\langle T_{\text{th}} \rangle}. \tag{15.17}$$

The effect of the refractory period is to regularize the spike train, lowering its variability. This is particularly true for high firing rates, as $\langle T_{\text{th}} \rangle$ approaches t_{ref}.

If we were to measure the interspike interval variability generated by a sustained current injection into an integrate-and-fire cell (Fig. 15.5E) or into a squid giant axon

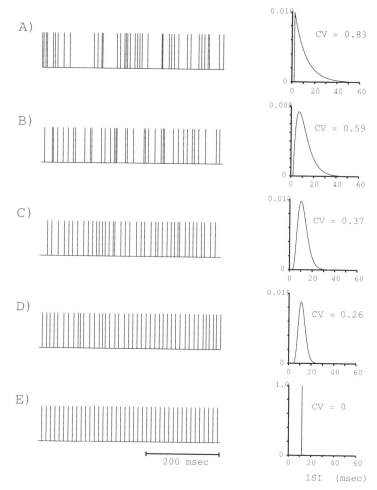

Fig. 15.5 SPIKE TRAIN INTERSPIKE VARIABILITY Sample spike trains and interspike interval histograms for a perfect integrator model with an absolute refractory period t_{ref} of 2 msec for Poisson distributed synaptic input. The mean interspike interval is in all cases identical to 12 msec. **(A)** Each synaptic input gives rise to a spike. The ISI is a shifted exponential; the deviation of the associated C_V from unity reflects the regularizing effect of the refractory period. **(B)**–**(D)** n_{th} is increased to 2, 5, and 10, respectively. To retain the same average firing frequency, the input firing frequency was also increased by the same amount. The ISI can be described by a gamma density (Eq. 15.12). **(E)** Response to a sustained current injection.

(see Fig. 6.9), it would tend to zero. Biological pacemaking systems like the heart or the rhythm underlying breathing have C_V's in the low percent range. In some systems the C_V can be a miniscule fraction of one percent. The neurons in the weakly electric fish *Eigenmannia* that are responsible for triggering the clocklike electric organ discharge in the 0.1–1 kilohertz frequency range have a spike jitter of less than 1 μsec (Kawasaki, Rose, and Heiligenberg, 1988).

In all of these examples, $C_V \leq 1$, with the upper bound given by a pure Poisson process. Yet in many instances interspike interval distributions of real neurons have C_V's greater than 1 (Wilbur and Rinzel, 1983; see also Sec. 16.1). This can be achieved in a so-called *doubly stochastic Poisson* process (Saleh, 1978), where one stochastic process is used to generate the rate of a second one. Such a process was postulated to characterize—successfully—a number of properties of retinal ganglion cell spike trains at low light levels (Saleh and Teich, 1985). In this limit, the distribution of photons absorbed at the retina is expected to be Poisson. If each photon results in a slowly decaying input current to a ganglion cell that generates on average two spikes per incoming photon, the interspike interval distribution of the model will have a C_V greater than 1. The cause of this additional variability lies in the random number of spikes generated for each incoming Poisson pulse (Gabbiani and Koch, 1998).

Special care must be taken when analyzing the variability of bursting cells (see the following chapter). If a bursty cell switches between bursting and nonbursting, the variability of its interspike intervals can be larger than that of a Poisson process, that is, $C_V > 1$ (Wilbur and Rinzel, 1983). Numerical techniques have been developed to account for the highly correlated adjacent interspike intervals that occur during bursting (a measure termed C_{V2}; Holt et al., 1996).

15.2.3 *Spike Count and Fano Factor*

So far, we have restricted ourselves to a consideration of the temporal jitter. However, neurons show considerable variation in the number of spikes triggered in response to a particular stimulus. This can be quantified by the ratio of the variance of the number of spikes generated within some observational window T to the mean number of spikes within the same time period, the so-called *index of dispersion* or *Fano factor* (Fano, 1947),

$$F(T) = \frac{\text{Var}[N(T)]}{\langle N(T) \rangle}. \tag{15.18}$$

Counting Poisson distributed spikes of mean rate μ over an observational interval of duration T, leads to a mean number of spikes $T\mu$ with variance $T\mu$. In other words, for an ideal Poisson process, no matter for what duration it is observed, $F(T) = 1$.

Experimentally, $F(T)$ should be estimated by computing the average number of spikes and their variance for a fixed counting time T using many repetitions of the same stimulus. The counting time should now be systematically varied over the range of interest.[3] The Fano factor is usally plotted in log-log coordinates, and the slope of the best fit through these points is reported. For a true Poisson process, the slope is 1. Slopes bigger than 1 imply that the variance in the number of events grows faster than the mean, indicating long-term correlations in the data.[4]

As illustrated in Fig. 15.6, the $F(T)$ factor in different neural systems is frequently close to unity for observational times on the order of 100 msec, compatible with the Poisson hypothesis (modified by a refractory period that lowers $F(T)$ to a value slightly less than unity; Teich, 1992). Yet for large windows, say in the second to minute domain, long-term trends in the data are very apparent (revealed by the very large $F(T)$ values). These are most likely due to state changes in the underlying neural networks. Obviously, modeling data over these time spans forces abandonment of the Poisson hypothesis. In some sensory systems $F(T)$ can be considerably less than 1 (van Steveninck et al., 1997; Berry, Warland, and Meister, 1997).

---

3. A frequently used approximation is to estimate $(\langle N \rangle, \text{Var}[N])$ for a fixed trial, say of 1 sec duration. This procedure is repeated under circumstances (that is, by varying the stimulus intensity) that give rise to different values of $\langle N \rangle$.

4. Indeed, the Fano factor can be related to the autocorrelation function of the underlying point process (Gabbiani and Koch, 1998).

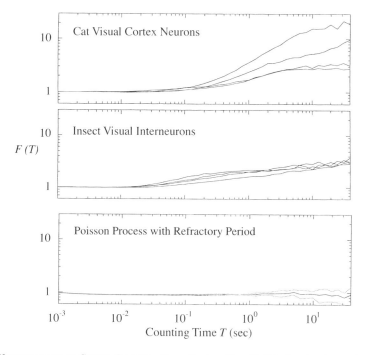

Fig. 15.6 VARIABILITY IN SPIKE COUNT   Fano Factor (Eq. 15.18), that is, the variance of the number of spikes normalized by their mean, within an observational counting window of duration T in different spiking systems. The top panel shows $F(T)$ for the spontaneous activity of four cells taken from the visual cortex of the anesthetized cat (Teich et al., 1997). For $T < 0.1$ sec, the variability is close to unity, as expected from a Poisson process. However, for larger observational windows, $F(T)$ strongly increases, indicative of long-term correlations in the firing behavior, possibly due to slow state changes in the cat. The qualitatively same behavior for spontaneous activity can be observed in the middle panel, plotting $F(T)$ for four visual interneurons in the cricket (Turcott, Barker, and Teich, 1995), although $F(T)$ is not as large as in the cortex. The lower plot illustrates $F(T)$ (along with its standard deviation) associated with 10 simulated runs of a Poisson process with a refractory period. Since a refractory period imposes some degree of order, $F(T)$ is less than that of a pure Poisson process but still close to unity. Reprinted by permission from Teich, Turcott, and Siegel (1996).

In general, the jitter in the timing of events need not be related to the jitter in the number of events. Under conditions when the spiking can be described as a *renewal process*, that is, if the interspike intervals are independently and identically distributed, the distribution of spike counts for long observation times will be approximately Gaussian distributed. Since the same is true for the interspike interval distribution (e.g., Fig. 15.4C), these two measures are related via

$$F = C_V{}^2 \tag{15.19}$$

in the limit as the observational period $T \rightarrow \infty$ (Cox, 1962). This is a property of spike trains that is not widely appreciated.

Many studies have measured interspike intervals as well as spike count variability, although usually not simultaneously, of spontaneous or stimulus-induced activity. One of the aims is to infer something about the dynamics of the spiking threshold or about the different states the neuron can be in (Werner and Mountcastle, 1963; Smith and Smith, 1965; Calvin and Stevens, 1967, 1968; Noda and Adey, 1970; Burns and Webb, 1976;

Vogels, Spileers, and Orban, 1989; Snowden, Treue, and Andersen, 1992; Softky and Koch, 1993; Usher et al., 1994; Turcott, Barker, and Teich, 1995; Teich, Turcott, and Siegel, 1996; Teich et al., 1997). It was only recently that attempts have been made to reconcile the observed high variability of neurons with the known biophysical properties of single cells and network properties. In order to discuss some of these, let us now throw inhibition into the brew.

15.2.4 Random Walk Model of Stochastic Activity

We do this in the form of adding an inhibitory synaptic process of rate μ_i and weight $a_i > 0$. Now the random process $V(t)$ undergoes random "up" jumps (of amplitude a_e) and "down" jumps (of amplitude a_i; Fig. 15.7). In the absence of a threshold, the potential can be modeled by

$$V(t) = a_e N_e(t) - a_i N_i(t), \tag{15.20}$$

with $V(0) = 0$ and $V < V_{th}$. The expected value of this process follows from the fact that Eq. 15.20 is linear, as

$$\langle V(t) \rangle = a_e \langle N_e(t) \rangle - a_i \langle N_i(t) \rangle \tag{15.21}$$

$$= a_e \mu_e t - a_i \mu_i t = \mu t, \tag{15.22}$$

where $\mu = a_e \mu_e - a_i \mu_i$ is termed *drift*. The variance of the voltage over time is

$$\text{Var}[V(t)] = a_e^2 \text{Var}[N_e(t)] + a_i^2 \text{Var}[N_i(t)] \tag{15.23}$$

$$= a_e^2 \mu_e t + a_i^2 \mu_i t = \sigma^2 t, \tag{15.24}$$

where $\sigma^2 = a_e^2 \mu_e + a_i^2 \mu_i$ is called the *variance parameter*. The drift corresponds to the net input current into the unit. As long as it is different from zero, the average membrane potential as well as the jitter will diverge (in the absence of either a leak or a threshold). Note that the standard deviation around $\langle V \rangle$ increases as the square root of time, reminiscent of the jitter in the motion of a particle diffusing in one or more dimensions (Sec. 11.1). This is not surprising, since both represent instances of *random walks*. (For an introduction into the theory of the random walk, see Feller, 1966, or Berg, 1983.)

The probability distribution of $V(t)$ in the absence of any threshold and membrane leak can be computed analytically. We spare the reader its derivation, referring him or her instead to Eq. 9.42 in the monograph by Tuckwell (1988b). Figure 15.7A illustrates one particular realization of the random process $V(t)$, together with $\langle V(t) \rangle$ and its associated jitter.

Will the membrane potential ever reach threshold V_{th} and generate a pulse? And how long is this event expected to take? This question, known as the "time of first passage to threshold" problem, has a nontrivial but well characterized probability density function $f_{th}(t)$ associated with it (see Eq. 9.54 in Tuckwell, 1988b and Fig. 15.7B). If T_{th} is the random time taken for V to go from its initial state at $V = 0$ to $V = V_{th}$, one can estimate that

$$\Pr\{T_{th} < \infty\} = \int_0^\infty f_{th}(t)dt \tag{15.25}$$

or

$$\Pr\{T_{th} < \infty\} = \begin{cases} 1 & \text{if } \mu \geq 0 \\ (a_e \mu_e / a_i \mu_i)^{V_{th}} & \text{if } \mu < 0. \end{cases} \tag{15.26}$$

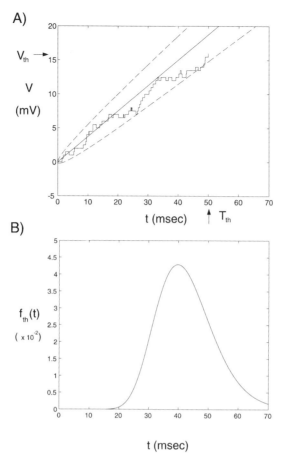

Fig. 15.7 RANDOM WALK OF THE MEMBRANE POTENTIAL   Illustration of the random walk model of excitatory and inhibitory synaptic input into a nonleaky integrate-and-fire unit as pioneered in this form by Stein (1965). The cell receives Poisson distributed excitatory input (with rate $\mu_e = 1000$ Hz), each increasing the voltage by 0.5 mV, and Poisson distributed inhibitory input (at a rate of 250 Hz), each input decreasing the potential by 0.5 mV. Threshold is reached at 16 mV. **(A)** One instantiation of such a random walk, together with the expected mean potential and its standard deviation. The unit generates a spike at around $T_{th} = 50$ msec. **(B)** Probability density $f_{th}(t)$ for the first passage to threshold, that is, for the time it takes before the voltage threshold is reached for the first time.

If the drift is positive, the unit will eventually spike with probability 1. If the net input is negative, then one must wait for a fluctuation in the input to elevate V to threshold. This is less and less likely as the effective inhibition exceeds excitation with a finite chance that threshold is never reached.

In the former case, that is, for $a_e \mu_e > a_i \mu_i$, we can easily compute the first and second moments associated with T_{th}. Specifically, the mean time to fire is

$$\langle T_{th} \rangle = \frac{V_{th}}{\mu} \tag{15.27}$$

and the variance of this mean is

$$\mathrm{Var}[T_{th}] = V_{th} \frac{a_e \mu_e + a_i \mu_i}{\mu^3} . \tag{15.28}$$

Because we assume that the unit had started out at $V = 0$ (implying that it had just spiked and was reset), T_{th} corresponds to the average interval between adjacent spikes. While we are unable to predict the exact trajectory of the membrane potential, we can perfectly well predict certain time-averaged aspects of spike trains.

The normalized "width" of the ISI, expressed by the coefficient of variation, is

$$C_V = \frac{(\text{Var}[T_{th}])^{1/2}}{\langle T_{th} \rangle} = \left(\frac{1}{V_{th}} \frac{a_e \mu_e + a_i \mu_i}{a_e \mu_e - a_i \mu_i} \right)^{1/2}. \tag{15.29}$$

Setting $\mu_i = 0$ in these expressions gives us back our earlier equations. Adding in inhibition leads to the expected result that the mean membrane potential decreases and that $\langle T_{th} \rangle$ increases. A bit less expected is that the variance of $\langle V \rangle$ as well as the jitter associated with $\langle T_{th} \rangle$ both increase. Intuitively, this can be explained by noting that we are adding a new source of unpredictability, even if it is inhibitory. Indeed, by adding enough inhibition one can always achieve a situation in which the drift is close to zero but the jitter becomes very large. This is termed *balanced inhibition* or *balanced excitation and inhibition* and was first investigated by Gerstein and Mandelbrot, 1964 (for more recent accounts, see Shadlen and Newsome, 1994; Tsodyks and Sejnowski, 1995 and, in particular, van Vreeswijk and Sompolinsky, 1996). Since the amount of inhibition increases in parallel with the amount of excitation, threshold is not reached by integrating a large number of small inputs over time (as with $\mu > 0$), but by a large enough "spontaneous" positive fluctuation in excitation, complemented by a large enough downward fluctuation in inhibition. This is of course much less likely to occur than if the drift is large (and positive).

A straightforward generalization of these results involves more than two synaptic processes. Conceptually, one process that is Poisson distributed with rate μ and amplitude a can always be replaced by n processes, all independent of each other, firing at rate μ/n and amplitude a, or by n independent processes of rate μ but of amplitude a/n. Indeed, all we need to assume is that the synaptic inputs impinging onto the cell derive from the superposition of arbitrary point processes that are independent of each other. If these processes are stationary (see above), the superposition of such processes will converge to a stationary Poisson process (Cox and Isham, 1980).

Finally, we do not want to close this section without briefly mentioning *Wiener processes* or *Brownian motion*. What makes the mathematical study of the random walk models discussed so far tortuous is that each sample path is discontinuous, since upon arrival of a synaptic input, $V(t)$ changes abruptly by $\pm a$. The associated differential-difference equations are less well understood than standard differential equations. Introduced into the neurobiological community by Gerstein and Mandelbrot (1964), Wiener processes are thoroughly studied and many of the relevant mathematical problems have been solved. A Wiener process can be obtained as a limiting case from our standard random walk model by letting the amplitude a of each synapse become smaller and smaller, while the rate at which the input arrives becomes faster and faster. In the appropriate limit one obtains a continuous sample path, albeit with derivatives that can diverge (indeed the "derivative" of the path is *white noise*). The crucial difference as compared to the discrete Poisson process is that for any two times t_1 and t_2, $V(t_2) - V(t_1)$ is a continuous Gaussian random variable with zero mean and variance $(t_2 - t_1)$ (whereas for the random walk models, $V(t_2) - V(t_1)$ takes on discrete values). Because the basic intuition concerning the behavior of a simple nonleaky integrate-and-fire model can be obtained with the random walk model discussed above, complemented by computer simulations that are to follow, we shall not treat Wiener processes here, referring the reader instead to Tuckwell (1988b).

15.2.5 Random Walk in the Presence of a Leak

So far, we neglected the presence of the membrane leak resistance R, which will induce an exponential voltage decay between synaptic inputs, leading to "forgetting" inputs that arrived in the past compared to more recent arrivals. It was Stein (1965) who first incorporated a decay term into the standard random walk model. Heuristically, we need to replace Eq. 15.19 by

$$C\frac{dV}{dt} = -\frac{V}{R} + a_e\frac{dN_e}{dt} - a_i\frac{dN_i}{dt}\,. \tag{15.30}$$

Because the trajectories of both N_e and N_i are discontinuous, jumping by ±1 each time a synaptic input arrives, dN/dt can be thought of as a series of delta functions, increasing V abruptly by a_e (or decreasing by a_i). In the absence of any threshold, the mean membrane potential is

$$\langle V(t)\rangle = R(a_e\mu_e - a_i\mu_i) = R\mu \tag{15.31}$$

where the drift $\mu = a_e\mu_e - a_i\mu_i$ has the dimension of a current.[5] Equation 15.31 was derived under the assumption that the initial value of the membrane potential has decayed away as $e^{-t/\tau}$, with $\tau = RC$.

Different from the nonleaky case, where the membrane potential diverges (as $\langle V(t)\rangle = \mu t/C$ in physical units; see Eq. 15.21), the membrane leak stabilizes the membrane potential at a level proportional to the drift, that is, the mean input current. The variance will also remain finite

$$\mathrm{Var}[V(t)] = \frac{1}{2}\frac{R}{C}(a_e{}^2\mu_e + a_i{}^2\mu_i) = \frac{1}{2}\frac{R}{C}\sigma^2\,. \tag{15.32}$$

It is straightforward to generalize this to the case of n inputs.

What about the mean time to spike? Unfortunately, even after several decades of effort, no general expressions for the probability density of T_{th} and related quantities, such as the ISI distribution, are available and we have to resort to numerical solutions. Qualitatively, the leak term has little effect on the ISI for high rates, because there is not sufficient time for any significant fraction of the charge to leak away before V_{th} is reached. In other words, we can neglect the effect of a leak if the interspike interval is much less than τ.

At low firing rates, the membrane potential "forgets" when the last firing occurred, so that the subsequent firing time is virtually independent of the previous time. In this mode, the leaky integrator operates as a "coincidence" detection device for occasional bursts of EPSPs. More interspike intervals will occur at large values of t than expected from the model without a leak and the associated C_V value is larger than for the nonleaky integrator.

A number of attempts have been made to include stochastic synaptic activity into distributed cable models, including solving the cable equation in the presence of white noise current (Tuckwell and Wan, 1980; see also Chap. 9 in Tuckwell, 1988b). Since the complexity of these models usually precludes the development of useful intuition concerning the computational properties of nerve cells, we will not discuss them any further and refer the reader to Tuckwell (1988b,c).

---

5. In the previous pages we assumed—implicitly, for the most part—that each synaptic input increases the membrane potential by a_e/C although we did not explicitly write out the dependency on C.

15.3 What Do Cortical Cells Do?

So, how random are cortical cells? And is their observed degree of variability compatible with their known biophysics?

Before we discuss this, a word of caution. The majority of experimental studies assume that the statistics of spikes do not vary significantly during the period of observation. Since this degree of stationarity is very difficult to verify empirically, a more reasonable concept is that of a weakly stationary process, for which only the mean and the covariance need to be invariant under time translation. For a spike train generated in response to a physiological stimulus (such as a flashed bar of light) nonstationarity is practically guaranteed since both single-cell and network effects conspire to lead to firing frequency adaptation. This is reflected in the poststimulus time histogram, which usually is not flat (e.g., Fig. 15.1). Several techniques exist to deal with such nonstationary data (for instance, by only using the adapted part of the spike train or by normalizing for the local rate $\mu(t)$; Perkel, Gerstein, and Moore, 1967; Softky and Koch, 1993).

Investigating the response of cat motoneurons in the spinal cord to intracellular current injections, Calvin and Stevens (1967, 1968) concluded that in two out of five cells they considered in great detail, the low degree of observed jitter in spike timing ($C_V \approx 0.05$) could be explained on the basis of noisy synaptic input charging up the somatic potential until it reaches V_{th} and triggers a spike (Fig. 15.8A). This is expected if the cell behaved like an ideal integrate-and-fire unit under the random walk assumptions (Eq. 15.21).

15.3.1 Cortical Cells Fire Randomly

What about cells in the cerebral cortex? How randomly do they fire? And what does their surprisingly high degree of randomness imply about the neural code used for information transmission? Anecdotal evidence (Fig. 15.8B) illustrates that under certain conditions, the somatic membrane potential of cortical neurons does behave as expected from the random walk model discussed above, integrating a large number of small inputs until threshold is reached. But is this really compatible with their variability? Although the firing variability of thalamic and cortical cells has been studied experimentally for quite some time, this was caried out predominantly during spontaneous firing, when rates are very low (Poggio and Viernstein, 1964; Noda and Adey, 1970; Burns and Webb, 1976; Teich, Turcott, and Siegel, 1996). Softky and Koch (1992, 1993) measured the firing variability of cells from cortical areas V1 and MT in the awake monkey responding to moving bars and random dots with maintained firing rates of up to 200 Hz. Applying appropriate normalization procedures and excluding bursting cells, Softky and Koch found that C_V was close to unity—as expected from a Poisson process (Fig. 15.9).

This poses somewhat of a paradox. At high enough firing rates, when the decay of the membrane potential can be neglected, C_V is inversely proportional to the square root of n_{th}, the number of excitatory inputs necessary to trigger threshold (Eq. 15.16). If n_{th} is large, as commonly held (that is, 20, 50, or 100),[6] the cell should fire in a very regular manner, which cortical cells patently do not do. On the basis of detailed compartmental simulations of a cortical cell with passive dendrites, Softky and Koch (1993) showed that

---

6. Based on evidence from spike-triggered averaging, in which spikes recorded in one neuron are correlated with monosynaptic EPSPs recorded in another pyramidal cell; the amplitudes usually fall below 0.5 mV (Mason, Nicoll, and Stratford, 1991; for a review see Fetz, Toyama, and Smith, 1991). This implies that on the order of 100 simultaneous excitatory inputs are required to bring a cortical cell above threshold.

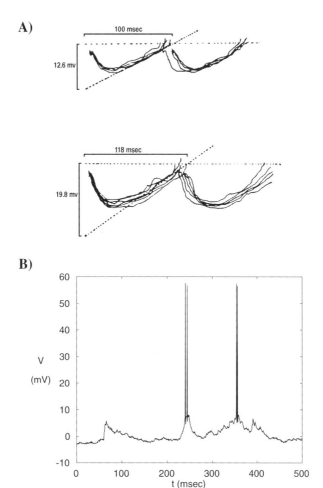

Fig. 15.8 DOES THE MEMBRANE POTENTIAL DRIFT UP TO V_{th}? Intracellular membrane potential in two different cell types. **(A)** Somatic potential in a motoneuron in the spinal cord of an anesthetized cat in response to a constant intracellular current injection (Calvin and Stevens, 1968). Portions of the membrane potential between adjacent spikes are superimposed, matched against a theoretical model in which the depolarization to threshold is expected to be linear in time (Eq. 15.22) and evaluated against a fixed V_{th} (dotted horizontal line). This model explains the jitter in spike timing in two out of five motoneurons investigated entirely on the basis of synaptic input noise. Reprinted in modified form by permission from Calvin and Stevens (1968). **(B)** Intracellular recording from a complex cell in cat visual cortex (Ahmed et al., 1993), responding to a bar moving in the preferred direction. The depolarizing event around 80 msec was not quite sufficient to generate a spike. This cell does appear to integrate over many tens of milliseconds, as witnessed by the slow rise and fall of the membrane potential. It remains a matter of intense debate whether spiking in cortical cells can be explained by treating the cell as integrating over a large number of small excitatory and inhibitory inputs or whether cortical cells detect coincidences in synaptic input at a millisecond or finer time scale. Unpublished data from B. Ahmed and K. Martin, printed with permission.

neither (1) distributing synaptic inputs throughout the cell, (2) including Hodgkin-Huxley-like currents at the soma, (3) incorporating a "reasonable" amount of synaptic inhibition, nor (4) using weakly cross-correlated synaptic inputs changed this basic conclusion appreciably.

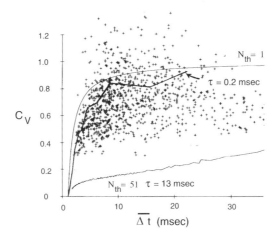

Fig. 15.9 FIRING VARIABILITY OF CELLS IN MONKEY CORTEX Comparison of the variability of spike timing, quantified using the coefficient of variation C_V, as a function of the mean interspike interval for cortical cells and modeled data. The scattered crosses are from V1 and MT cells recorded in the behaving monkey and are normalized for their mean firing rates. Their C_V is high, close to unity. Superimposed are the C_V curves computed for a random walk leaky integrate-and-fire model using conventional assumptions (lower trace; $n_{th} = 51, \tau = 13$ msec, and $t_{ref} = 1$ msec) and less conventional ones (middle bold trace; $n_{th} = 51$ and $\tau = 0.2$ msec). The thin, top curve corresponds to the limiting case of a pure Poisson spike train ($n_{th} = 1$) in combination with $t_{ref} = 1$ msec. The effect of this absolute refractory period is to render the spike timing very regular at very high firing rates. Reprinted in modified form by permission from Softky and Koch (1993)

Cortical cells cannot integrate over a large number of small synaptic inputs and still fire as irregularly as they do. Instead, either the time constant of the cell has to be very small, in the 1 msec range (in which case synaptic input that arrived more than 2 or 3 msec ago has been forgotten), or n_{th} must be very small, in the neighborhood of 2–5 (Fig. 15.9).

Measuring variability in the spike count confirmed this high degree of randomness. Log-log plots of the mean versus the variance in the number of spikes in the cortex of the monkey consistently give rise to slopes of around 5/4, implying large values for $F(T)$ (Tolhurst, Movshon, and Dean, 1983; Zohary, Hillman, and Hochstein, 1990; Snowden, Treue, and Andersen, 1992; Softky and Koch, 1993; Teich, Turcott, and Siegel, 1996). As before, the standard integrator model of pyramidal cells should lead to much smaller values of the Fano factor $F(T)$.

15.3.2 Pyramidal Cells: Integrator or Coincidence Detector

Softky and Koch (1992, 1993) argue for a *coincidence detection* model in which distal dendrites generate fast sodium action potentials which are triggered when two or more excitatory synaptic inputs arrive within a millisecond of each other. A handful of these active synaptic events, if coincident at the soma, will trigger a spike. (This effectively reduces n_{th} to the required small values.) Indeed, Softky (1994, 1995) claims that the experimental evidence for active dendritic conductances is compatible with submillisecond coincidence detection occurring in the dendrites of pyramidal cells (see Sec. 19.3.3).

This has rekindled a debate over the extent to which pyramidal cells either integrate over many small inputs on a time scale of tens of milliseconds or detect coincidences at the millisecond or faster time scale (Abeles, 1982b; Shadlen and Newsome, 1994; König, Engel, and Singer, 1996). In the former case, the code would be the traditional noisy rate code that is very robust to changes in the underlying hardware. If neurons do use precise spike times—placing great demands on the spatio-temporal precision with which neurons need to be wired up—the resultant spike interval code can be one or two orders of magnitude more efficient than the rate code (Stein, 1967a,b; Rieke et al., 1996).

As illustrated in Fig. 15.10A, the Geiger counter model leads to very regular spiking, since the cell integrates over a large number of inputs (here $n_{th} = 300$). High irregularity of the spike discharge can be achieved in several ways. If the neuron has a very short time constant, it will only fire if a sufficient number of spikes arrive within a very small time. Such a coincidence detector model is illustrated in Fig. 15.10B and is simulated by considering V to be the sum of all EPSPs arriving within the last millisecond. If at least 35 EPSPs are coincident within this time, an output spike is generated. This model is, of course, sensitive to disturbances in the exact timing of the synaptic input. It is precisely this sensitivity that can be used to convey information.

Shadlen and Newsome (1994), Tsodyks and Sejnowski (1995), and van Vreeswijk and Sompolinsky (1996) seek an explanation of the high variability in a *balanced inhibition* random walk, in which the drift is zero (as in Gerstein and Mandelbrot, 1964), that is, the average net current is zero, excitation balancing inhibition. The unit spikes whenever a fluctuation leads to a momentary drop in synaptic inhibition in combination with a simultaneous excess of excitation. (This is achieved in Fig. 15.10C by having the 150 inhibitory cells fire at the same rate as the 300 excitatory cells, but with twice their postsynaptic weight.)

It can be shown analytically (van Vreeswijk and Sompolinsky, 1996) that large networks of simple neurons, randomly and sparsely interconnected with many relatively powerful excitatory and inhibitory synapses whose activity is approximately "balanced," display highly irregular patterns of spiking activity characteristic of *deterministic chaos*. Such networks can respond very rapidly to external input, much faster than the time constant of the individual neurons. van Vreeswijk and Sompolinsky (1996) argue that irregular spiking is a robust emergent network property not depending on intricate cellular properties. Yet this ability to respond rapidly comes at a price of continual activity, requiring a constant expenditure of metabolic energy. Experimentally, it remains an open question to what extent individual cells receive balaced input, that is to what extent sustained changes in excitatory input are opposed by equally powerful sustained changes in inhibition.

A related solution to the dilemma of obtaining high variability while retaining the integrating aspect of cortical cells is correlated synaptic inputs. If excitatory synaptic inputs have a tendency to arrive simultaneously, less temporal dispersion will occur and inputs will be more effective. Voltage-dependent sodium conductances in the dendritic tree will be particularly sensitive to such simultaneous inputs, effectively lowering the number of inputs needed to reach threshold.

Including more complex spike generation dynamics, such as bursting (Wilbur and Rinzel, 1983; but see Holt et al., 1996), or incorporating short-term adaptation into the synapses (Abbott et al., 1997; Sec. 13.5.4) will increase C_V and $F(T)$. The somewhat arbitrary choice of resetting the membrane potential following spike generation to the unit's resting potential offers yet another way to increase variability. Intracellular data from cortical cells frequently reveal the lack of any hyperpolarization following an action potential. Bell et al., (1995) and Troyer and Miller (1997) argue that the membrane potential should be reset to a much more positive value, thereby lowering the effective n_{th}.

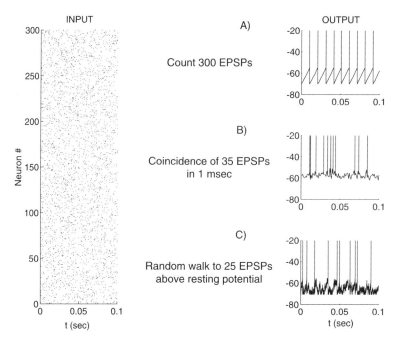

Fig. 15.10 THREE MODELS OF SYNAPTIC INTEGRATION The input to the simulated neuron is shown in the left panel: 300 excitatory Poisson distributed synaptic inputs, firing at a mean rate of $\mu_e = 100$ Hz. The output units on the right also fire at this average rate. (**A**) Ideal integrate-and-fire model. Each synaptic input increases V by 0.05 mV, corresponding to $n_{th} = 300$. As expected, the output pulses have a clocklike regularity. (**B**) In this cartoon of a coincidence detector model, only inputs within the previous millisecond contribute toward V (corresponding to an effective submillisecond time constant). If $n_{th} = 35$ EPSPs arrive "simultaneously" (here, within 1 msec), the unit spikes. Notice the elevated mean membrane potential. In this model, the timing of spikes can code information. (**C**) Standard random walk model with a balance of excitation and inhibition. In addition to the 300 excitatory inputs, 150 inhibitory inputs have been added, firing at the same rate; $a_e = 0.6$ mV (corresponding to an effective $n_{th} = 25$) and $a_i = 1.2$ mV. This model achieves a high variability but at a cost of substantial inhibition. Since no information is encoded in the time of arrival of spikes, it is very robust. Adding a leak term to the first and third models does not affect these conclusions substantially. Reprinted by permission from Shadlen and Newsome (1994)

What is clear is that the simple Geiger counter model of synaptic integration needs to be revised. A plethora of mechanisms—such as synaptic inhibition, short-term synaptic depression, correlated synaptic input, a depolarizing reset, active dendritic processing on a fast time scale—will interact synergistically to achieve the observed random behavior of cortical cells.

Lest the reader forget the cautionary note posted earlier, the preceding paragraphs apply specifically to cortical cells. Given the great diversity apparent in the nervous system, the question of the variability of neuronal firing and its implication for the code must be investigated anew in each specific cell type. In cells where synaptic input but modulates the strongly nonlinear dynamics of firing, as in oscillatory or bursting cells (Jaeger, DeSchutter, and Bower, 1997; Marder and Calabrese, 1996), the source of variability could be quite distinct from those discussed here.

15.3.3 Temporal Precision of Cortical Cells

Yet under some conditions, cortical spike trains can look remarkably consistent from trial to trial. This is exemplified in a quite dramatic manner by Fig. 15.11, from an experiment carried out in a monkey that had to respond to various patterns of random dot motion (Newsome, Britten, and Movshon, 1989). If one averages—as is frequently done in practice—over many such presentations of different clouds of randomly moving dots, the spiking activity of the cell looks uneventful (beyond the initial transient; left column). Yet if one pulls out all of the responses to exactly the same random dot movie, a highly repeatable pattern of spiking becomes visible. Such reliable responses could be observed in the majority of MT cells examined (Bair and Koch, 1996). These stimulus-induced temporal modulations disappear for coherent motion, that is, when the entire cloud of dots moves as one across the receptive field, explaining why they are not visible in Fig. 15.1 (Bair, 1995).

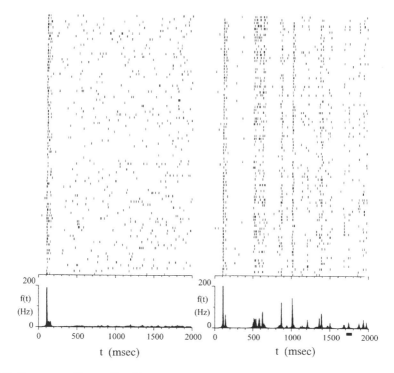

Fig. 15.11 CORTICAL CELLS CAN FIRE IN A VERY PRECISE MANNER   Responses of an MT cell in a behaving monkey to patterns of randomly and independently moving dots. The left column shows the cortical response to different random dot patterns. Averaged over 90 trials, the cell's response is well described by a random point process with $\mu = 3.4$ Hz (excluding the initial "flash" response; see the bottom panel). When only trials over the 2-hour-long experiment that were stimulated by the identical random dot movie are considered (right column), strikingly repeating patterns can be observed. (Viewing this figure obliquely best reveals the precision of spiking.) Notice the sharp peak following the 1-sec mark or that nearly all spikes in the final 400 msec cluster into six vertical streaks. Despite this precision, observed in the majority of MT cells, spike count variability $F(T)$ is still very high. The poststimulus time histogram, using an adaptive square window, is shown at the bottom. Reprinted by permission from Bair and Koch (1996).

When contemplating Fig. 15.11, it should be kept in mind that at the beginning of the experiment, the animal is thirsty and eager to perform, while at the end it is satiated and not very motivated. Furthermore, the animal continuously makes small eye movements, termed microsaccades. Yet despite this, a cell removed at least six synapses from the visual input fires in a very repeatable pattern throughout this time.

Bair and Koch (1996) quantify the degree of precision by computing the temporal jitter, that is, the standard deviation in time of the onset of periods of elevated firing (such as the "event" indicated by a black bar near 1740 msec in Fig. 15.11; see also Sestokas and Lehmkuhle, 1986). For this event, the jitter is 3.3 msec. For 80% of all cells, the most precise response has a jitter of less than 10 msec, while in some cases it is as low as 2–4 msec.

Although somewhat counterintuitive, even these spike trains can be generated by a Poisson process, but in this case with a time-varying rate $\mu(t)$ (called *inhomogeneous* Poisson process). As in a homogeneous Poisson process, the number of spikes in any one time interval is independent of the number of spikes in any other nonoverlapping interval. The time intervals between adjacent spikes also remain independent of each other but are not identically distributed anymore (Tuckwell, 1988b). In the case of the data shown in Fig. 15.11, the Fano factor $F(T)$ is close to unity when evaluated for the duration of the stimulus (Britten et al., 1993).

We conclude that individual cortical cells are able to reliably follow the time course of events in the external world with a resolution in the 5–10 msec range (see also Mainen and Sejnowski, 1995).

Let us hasten to add that much evidence has accumulated (as cited in Sec. 14.1) indicating that the precision of spike generation in one cortical cell with respect to another cell's firing can be even higher (e.g., Lestienne and Strehler, 1987). Best known is the demonstration of repeating patterns among simultaneously recorded cells by Abeles et al., (1993). Pairs of cells may fire action potentials at predictable intervals that can be as long as 200 msec, two orders of magnitude longer than the delay due to a direct synaptic connection, with a precision of about 1 msec. This is the key observation at the heart of Abeles's *synfire chain* model of cortical processing (Abeles, 1990).

Any rash judgment on whether or not the detailed firing pattern is noise or signal should be tempered by Barlow's (1972) statement about the firing of individual nerve cells that "their apparently erratic behavior was caused by our ignorance, not the neuron's incompetence."

15.4 Recapitulation

In this chapter we tried to address and quantify the stochastic, seemingly random nature of neuronal firing. The degree of randomness speaks to the nature of the neural code used to transmit information between cells. Very regularly firing cells are obviously not very good at encoding information in their timing patterns; yet this will not prevent information from being encoded in their mean firing rates, albeit at a lower rate (in terms of bits per spike) but in a robust manner.

The spiking behavior of cells has traditionally been described as a random point process, in particular as a renewal process with independent and identically distributed interspike intervals. Cortical cells firing at high rates are at least as variable as expected from a simple Poisson process. Variability is usually quantified using two measures: C_V to assess the interspike interval variability and $F(T)$ for spike count variability, with both taking on unity for a Poisson process. If spike trains are generated by a renewal process, $F = C_V^2$ in the limit of large observational intervals.

The power spectrum of cortical spike trains is flat with a dip around the origin, as expected from a Poisson process modified by a refractory period. (This refractory period only comes into play at very high firing rates, serving to regularize them.) The rate of the spiking process is usually not constant in time, but is up or down regulated at the 5–10 msec level. The principal deviation from Poisson statistics is the fact that adjacent interspike intervals are not independent of each other (even when neglecting bursting cells), that is, spike trains cannot really be described by a renewal process.

As always in science, this conclusion gives birth to intertwining considerations. What type of models of synaptic integration give rise to the high degree of randomness apparent in neuronal firing and how do these constrain the nature of the neuronal code? The standard Geiger counter model predicts that when an integrate-and-fire unit needs to integrate over a large number of small synaptic inputs, it should fire very regularly. Since this is patently not true in the cortex, it needs to be abandoned. Out of the many alternatives proposed, two divergent views crystallize. One school retains the idea that neurons integrate over large number of excitatory inputs with little regard for their exact timing by invoking a large degree of inhibitory inputs (following the random walk model advocated by Gerstein and Mandelbrot, 1964), a depolarizing reset, or correlated synaptic input. The other school sees cortical cells as coincidence detectors, firing if small numbers of excitatory events arrive simultaneously at the millisecond (or even submillisecond) scale (via powerful and fast dendritic nonlinearities). Under these circumstances, detailed timing information can be used to transmit information in a manner much more efficient, yet also more demanding, than in a mean rate code. Only additional experimental evidence can resolve this issue.

Of course, at small enough time scales or for a handful of spikes, the debate loses its significance, since it becomes meaningless to define a rate for a 20 msec long segment of a spike train with just two action potentials.

What this dispute shows is that integrate-and-fire units serve as gold standard against which models of variability are evaluated. Given their relative simplicity—compared against the much more complex conductance-based models of firing—this is quite re-markable.

The highly variable character of cortical firing allows neurons to potentially pack one or more bits of information per action potential into spike trains (as done in sensory neurons closer to the periphery). Information theory as applied to a band-limited communication channel has taught us that the optimal code—optimal in terms of using the entire bandwidth available—looks completely random, since every redundancy has been removed to increase the efficiency. One could infer from this that neurons make optimal use of the limited bandwidth of axons using a sophisticated multiplexed interspike interval code from which all redundancies have been removed, and that neurons, properly decoded, maximize the existing channel bandwidth. To what extent they actually do for physiologically relevant stimuli remains an open issue.

16

BURSTING CELLS

Some neurons throughout the animal kingdom respond to an intracellular current injection or to an appropriate sensory stimulus with a stereotypical sequence of two to five fast spikes riding upon a slow depolarizing envelope. The entire event, termed a *burst*, is over within 10–40 msec and is usually terminated by a profound afterhyperpolarization (AHP). Such *bursting cells* are not a random feature of a certain fraction of all cells but can be identified with specific neuronal subpopulations. What are the mechanisms generating this *intrinsic firing pattern* and what is its meaning?

Bursting cells can easily be distinguished from a cell firing at a high maintained frequency by the fact that bursts will persist even at a low firing frequency. As illustrated by the thalamic relay cell of Fig. 9.4, some cells can switch between a mode in which they predominantly respond to stimuli via single, isolated spikes and one in which bursts are common. Because we believe that bursting constitutes a special manner of signaling important information, we devote a single, albeit small chapter to this topic. In the following, we describe a unique class of cells that frequently signal with bursts, and we touch upon the possible biophysical mechanisms that give rise to bursting. We finish this excursion by focussing on a functional study of bursting cells in the electric fish and speculate about the functional relevance of burst firing.

16.1 Intrinsically Bursting Cells

Neocortical cells are frequently classified according to their response to sustained current injections. While these distinctions are not all or none, there is broad agreement for three classes: *regular spiking*, *fast spiking*, and *intrinsically bursting* neurons (Connors, Gutnick, and Prince, 1982; McCormick et al., 1985; Connors and Gutnick, 1990; Agmon and Connors, 1992; Baranyi, Szente, and Woody, 1993; Nuñez, Amzica, and Steriade, 1993; Gutnick and Crill, 1995; Gray and McCormick, 1996). Additional cell classes have been identified (e.g., the *chattering cells* that fire bursts of spikes with interburst intervals ranging from 15 to 50 msec; Gray and McCormick, 1996), but whether or not they occur widely has not yet been settled. The cells of interest to us are the intrinsically bursting cells.

In response to a just-threshold current stimulus, an intrinsically bursting (IB) cell fires a single burst. If the current amplitude increases, some IB cells burst repeatedly at a constant frequency of between 5 and 12 Hz (as in Fig. 16.1C), while others respond by repetitive single spikes. It is not uncommon for an IB cell to flip back and forth between these two firing modes. Even more complex behavior, such as the initiation of a long train of bursts in response to a brief hyperpolarizing current stimulus, has been reported (Silva, Amitai, and Connors, 1991). The spikes themselves are conventional sodium carried ones. Their amplitude during the burst frequently decreases—presumably because the sustained depolarization partially inactivates I_{Na}—while their duration increases.

Intrinsic bursting cells show a unique laminar distribution and morphology. In studies using intracellular staining, IB cells correspond to large pyramidal neurons restricted—with few exceptions—to layer 5, with stereotyped dendritic morphologies and axonal projection patterns. IB cells have an extended apical bush in layers 1 and 2, while nonbursting cells in layer 5 lack this feature (Fig. 16.1D; Larkman and Mason, 1990; Gray and McCormick, 1996; Yang, Seamans, and Gorelova, 1996).

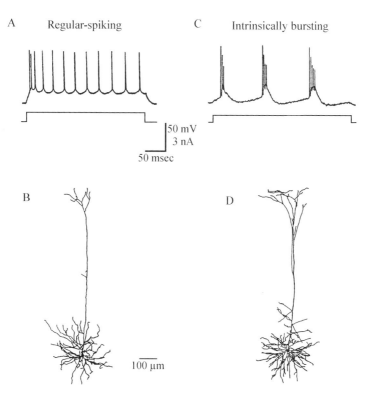

Fig. 16.1 Intrinsically Bursting and Regular Spiking Pyramidal Cells Structure-function relationship among pyramidal cells from rodent cortex. Typical firing patterns seen in response to an intracellularly injected current step in (**A**) *regular spiking*, and (**C**) *intrinsically bursting* layer 5 pyramidal cells from mouse cortex (Agmon and Connors, 1992). Drawings of biocytin-filled layer 5 (**B**) regular spiking and (**D**) intrinsically bursting (IB) pyramidal cells from the rat. IB cells tend to have an extensive apical dendrite (reaching all the way up to layer 1) while regular spiking cells show a smaller degree of dendritic arborization. The output targets of the two cell types are quite distinct, with bursting cells projecting outside the cortex proper. Unpublished data from L. J. Cauller, I. Bülthoff, and B. W. Connors, printed with permission.

Neurons whose cell bodies are located in layer 5 constitute the primary output of the neocortex. Whenever the cortex decides to do something, the message has to be sent outside the cortical system via layer 5 cells. This includes the one-million-axon (in humans) pyramidal tract that leaves the motor cortex and adjacent areas and projects into the spinal cord (Porter and Lemon, 1993).

Some evidence suggests that the cells projecting outside the cortex are of the bursting type. IB cells project to the pons and to the superior colliculus, while nonbursting layer 5 pyramidal cells project to the contralateral cortex via the corpus callosum (Wang and McCormick, 1993; Kasper et al., 1994). It is known that many of the identified layer 5 cells in the rat motor cortex that project to the spinal cord have the firing characteristics of IB cells (Tseng and Prince, 1993). It is obviously important to know whether all cells projecting to subcortical (and nonthalamic) targets are of the bursting type.

16.2 Mechanisms for Bursting

Bursting cells have been analyzed using biophysically based conductance models as pioneered by Traub and his colleagues (Traub and Llinás, 1979; Traub, 1979, 1982; Traub et al., 1991; summarized in the monograph by Traub and Miles, 1991) and from the perspective of dynamical systems theory (Chap. 7; Rinzel, 1987; Wang and Rinzel, 1996; Rinzel and Ermentrout, 1997).

There is likely to be a plurality of mechanisms that cause bursts. In Sec. 9.1.4 we discussed the critical role of a somatic low-threshold calcium current responsible for mediating bursts in thalamic cells (Fig. 9.4).

Central to the genesis of bursting in cortical cells are fast, sodium-driven action potentials at the cell body, which not only propagate forward along the axon, but also invade the dendritic tree (Rhodes and Gray, 1994). Backpropagation is aided by a low density of dendritic sodium channels that boost the signal. (For a more detailed discussion of this, see Secs. 19.1 and 19.2.) Dendritic calcium conductances build up local depolarization. In the meantime, somatic repolarization establishes a pronounced potential difference between the cell body and the dendritic tree, giving rise to ohmic currents. These will depolarize the soma and the spike initiating zone in the form of an *afterdepolarizing potential* (ADP). It is this ADP that causes the rapid sequence of fast sodium spikes. The entire chain of events is finished off by a hyperpolarization induced by a combination of voltage- and calcium-dependent potassium currents.

In this view of bursting, dendritic calcium currents are at the root of the prolonged depolarization that triggers the fast, somatic sodium spikes. However, it appears that in some systems one or more sodium currents, in combination with the "right" dendritic tree morphology, can by themselves initiate bursting (Turner et al., 1994; Franceschetti et al., 1995; Azouz, Jensen, and Yaari, 1996).

Mainen and Sejnowski (1996) derive what this right morphology should be. They simulate the dynamics of spiking in a number of reconstructed cortical cells with different dendritic anatomies and endowed with the same distribution and density of voltage-dependent sodium, potassium, and calcium currents (Fig. 16.2). Changing only the dendritic tree geometry, but not the mix of voltage-dependent conductances, gives rise to the entire spectrum of intrinsic responses observed in the cortex: from nonadapting and adapting cells to bursting ones. The large dendritic tree associated with layer 5 pyramidal cells prolongs the sodium spike propagating back into the dendritic arbor sufficiently for bursts to occur,

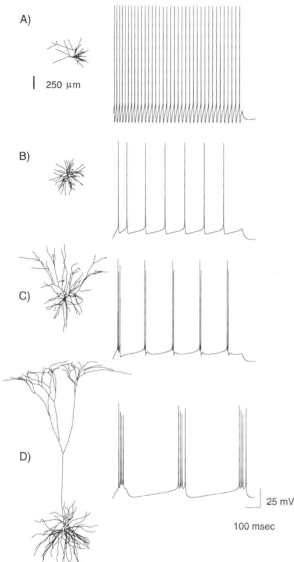

Fig. 16.2 CELLULAR MORPHOL-OGY AND INTRINSIC FIRING PAT-TERNS Different reconstructed neurons are endowed with the same densities of voltage-dependent conductances in soma and dendritic tree in a computer model: a fast sodium current, three potassium currents, and a high-threshold calcium current (Mainen and Sejnowski, 1996). The distinct firing patterns, evoked in response to sustained current injection (of 50, 70, 100, and 200 pA for **A** to **D**, respectively), run the gamut of adapting, nonadapting, and bursting behavior seen in the cortex. The four cells are (**A**) layer 3 aspiny stellate cell from rat cortex, (**B**) layer 4 spiny stellate, (**C**) layer 3 pyramid, and (**D**) layer 5 pyramid from cat visual cortex. Reprinted by permission from Mainen and Sejnowski (1996).

while no such behavior is observed in the much more compact population of stellate or pyramidal cells in superficial layers.

These results go hand in hand with the observed correlation between cellular morphology and intrinsic firing pattern reported above and emphasizes the often subtle and complex interplay between membrane-bound conductances and cable structure.

16.3 What Is the Significance of Bursting?

Anytime two spatially overlapping but morphologically distinct cell classes project to different locations (as the layer 5 bursting and nonbursting cells), one should assume that they encode and transmit different features (since otherwise the job could be done with the help of a single cell class whose axons branch to innervate both areas). Viewed in this light, the existence of a cell type that responds to sustained current steps in a bursty manner and projects outside the cortical system demands an explanation in functional terms.

One broad hypothesis, first entertained by Crick (1984a), is that bursting constitutes a specialized signal, a particular code symbol, distinct from isolated spikes, conveying a particular type of information.

The relevance of bursts for conveying information reliably has been ascertained for the electrosensory system of the electric fish (Gabbiani et al., 1996). Although of a quite distinct evolutionary lineage than the cerebral cortex, it is profitable to review this study briefly. Given the conservative nature of biology, reinventing and reusing the same mechanism over and over again, it would not be surprising if its lesson also applied to the cortex (Heiligenberg, 1991).

Gabbiani et al., (1996) subject the fish *Eigenmannia* to random amplitude modulations of an electric field in the water whose carrier frequency is equal to the fish's own electrical organ discharge frequency that has been pharmacologically disabled (Fig. 16.3A), while recording from pyramidal cells in the electrosensory lateral line lobe (ELL). Signal detection theory can be used to determine which particular features of the electric field are represented by the firing of these cells. Assuming that the ELL pyramidal cells implement a linear summation and threshold operation, it can be inferred that the cells respond best to up or down strokes of the electric field.

Slightly more than half of the spikes generated by the pyramidal cells occur in short bursts (on average three spikes with a mean spike separation of 9 msec; Fig. 16.3A). The signal-to-noise ratio of this signal detection operation—based only on those spikes that make up a burst—is substantially higher than the signal-to-noise ratio of the signal based on isolated spikes (Fig. 16.3B). In other words, spikes taken from a burst signal the presence of a particular stimulus feature, here an "abrupt" change in the electric field, more reliably than isolated spikes. Intracellular evidence implicates the apical dendritic tree of the ELL pyramidal cells as the locus where the burst computation is carried out (Turner et al., 1994).

It is possible that bursty neurons communicate their messages to their target cells with the aid of two output channels multiplexed onto a single axon, one employing isolated spikes and the other bursts. Depending on the type and amount of short-term synaptic plasticity present at the output synapses of bursting cells, they could transmit one or the other channel (Sec. 13.5.4; Abbott et al., 1997; Lisman, 1997).

This would work in the following fashion. Initially the output synapses of bursting cells have a very low probability of synaptic release p_0. A single spike is unlikely to cause a release of a vesicle but will, very briefly, enhance the probability of release for the next spike. Such transient but powerful facilitation would lead to isolated spikes being effectively disregarded while bursts would cause a synaptic release with probability close to 1. The price the nervous system would pay for the usage of bursts is their limited temporal resolution as compared to individual spikes (in the Gabbiani et al., 1996, study, the interval between the first and the last spike is 18 ± 9 msec).

Conversely, a synapse with high initial probability of release and a powerful but short acting short-term depression will disregard the additional spikes contained within a burst. This hypothesis predicts that synapses made by the axon of a bursting cell have low p_0 and show a very short lasting but powerful facilitation not present at the synapses of nonbursting cells.

Koch and Crick (1994) speculate that bursting could be one of the biophysical correlates of short-term memory (via elevated, presynaptic calcium concentration during such a burst). The stimulus feature represented by a population of bursting cells would correspond to the feature "stored" in short-term memory. Because of the close association between short-

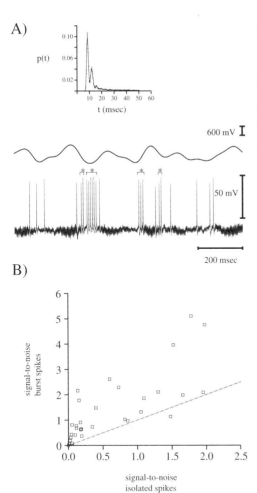

A)

B)

Fig. 16.3 RELIABLE ENCODING OF SENSORY INFORMATION IN BURSTS (A) In experiments carried out in the electric fish *Eigenmannia*, Gabbiani and colleagues (1996) stimulate its electrosensory system using zero-mean random amplitude modulations (thick smooth line) superimposed onto a fixed carrier frequency of an electric field while recording from pyramidal cells in the electrosensory lateral line lobe (ELL). A large fraction of all spikes belongs to bursts (∗) that show up as a prominent peak in the interspike intervals (here bursts are defined as spikes separated by <14 msec; see the inset). (B) On the basis of signal detection theory individual pyramidal cells can be shown to detect either up or down strokes of the electric field amplitude with up to 85% accuracy. The signal-to-noise ratio of this pattern discrimination operation, which assumes that ELL pyramidal cells perform a linear threshold computation, is enhanced (by more than a factor of 2) when considering only spikes from bursts compared to isolated spikes. In other words, burst spikes encode information in a more reliable manner than isolated action potentials. Reprinted by permission from Gabbiani et al., (1996).

term memory and awareness, it is possible that bursting cells are preferentially involved in mediating the neuronal correlate of awareness, in particular considering their strategic location in the cortical layer projecting to the tectum, pulvinar, and other extracortical targets (Crick and Koch, 1995).

Quite a different explanation for the utility of a particular subclass of spike patterns has been advanced in Traub et al., (1996). In some cells, *spike doublets*, consisting of two fast spikes separated by a stereotyped 4–5 msec interspike interval, are frequently observed. Spike doublets are characteristic of GABAergic, axo-axonic cortical interneurons (Buhl et al., 1994). Computer modeling of the underlying cortical networks show that such doublets mediate the frequently observed temporal synchronization of distant neuronal populations. Their oscillatory activities are synchronized with a near-zero phase lag, even though they are separated by propagation delays of 5 msec or more.

16.4 Recapitulation

Bursts, that is, two to five closely spaced, sodium-dependent fast action potentials riding on top of a much slower depolarization, are a dominant feature of a number of cell classes, not only in the cortex but also among thalamic relay neurons and elsewhere. In the cortex, intrinsically bursting cells have a unique morphology and are confined to layer 5, where they constitute the dominant output cell class.

The biophysical mechanisms underlying bursting are diverse. A low-threshold calcium current at the soma is the principal agent for bursting in thalamic cells (Sec. 9.1.4). Bursts in pyramidal cells can originate when sodium spikes propagate back into the dendritic tree, causing parts of the arbor to be depolarized. Under the right circumstances, for instance, when amplified by dendritic sodium and/or calcium currents, this signals returns to the soma in the form of an afterdepolarization that triggers several fast sodium spikes.

It has been argued that bursting represents a special code, quite distinct from the firing of an isolated action potential. This hypothesis has some experimental support from the electrosensory system of the electric fish. Spikes taken from bursts signal the presence of particular features in the input in a much more reliable manner than isolated spikes.

This implies that spiking cells are not restricted to using an asynchronous binary code but might employ a two-channel multiplexing strategy, with the decoding being carried out by rapidly adapting synapses.

A non-exclusive alternative is that bursts serve to synchronize distant neuronal populations.

17

INPUT RESISTANCE,
TIME CONSTANTS,
AND SPIKE INITIATION

This chapter represents somewhat of a technical interlude. Having introduced the reader to both simplified and more complex compartmental single neuron models, we need to revisit terrain with which we are already somewhat familiar.

In the following pages we reevaluate two important concepts we defined in the first few chapters: the somatic *input resistance* and the neuronal *time constant*. For passive systems, both are simple enough variables: R_{in} is the change in somatic membrane potential in response to a small sustained current injection divided by the amplitude of the current injection, while τ_m is the slowest time constant associated with the exponential charging or discharging of the neuronal membrane in response to a current pulse or step. However, because neurons express nonstationary and nonlinear membrane conductances, the measurement and interpretation of these two variables in active structures is not as straightforward as before. Having obtained a more sophisticated understanding of these issues, we will turn toward the question of the existence of a current, voltage, or charge threshold at which a biophysical faithful model of a cell triggers action potentials. We conclude with recent work that suggests how concepts from the subthreshold domain, like the input resistance or the average membrane potential, could be extended to the case in which the cell is discharging a stream of action potentials.

This chapter is mainly for the cognoscendi or for those of us that need to make sense of experimental data by comparing them to theoretical models that usually fail to reflect reality adequately.

17.1 Measuring Input Resistances

In Sec. 3.4, we defined $\tilde{K}_{ii}(f)$ for passive cable structures as the voltage change at location i in response to a sinusoidal current injection of frequency f at the same location. Its dc component is also referred to as input resistance or R_{in}. Three difficulties render this definition of input resistance problematic in real cells: (1) most membranes, in particular at the soma, show voltage-dependent nonlinearities, (2) the associated ionic membrane conductances are time dependent and (3) instrumental aspects, such as the effect of the impedance of the recording electrode on R_{in}, add uncertainty to the measuring process. While we will not deal with the last problem (see Smith et al., 1985 for further information). we do have to address the first two issues.

In order to gain a better understanding of the input resistance, we need to focus on the steady-state current-voltage relationship illustrated in Fig. 17.1. $I_\infty^{static}(V_m)$, or sometimes simply $I_\infty(V_m)$, is obtained by clamping the membrane potential at the soma to V_m, letting the system adapt and measuring the clamp current flowing across the somatic membrane.

Using the standard Hodgkin-Huxley formalism, we can express the clamp current as

$$I_\infty(V_m) = \sum_i g_i(V_m)(V_m - E_{rev,i}) = \sum_i \overline{g_i} m_{\infty,i}^{a_i} h_{\infty,i}^{b_i}(V_m - E_i) \qquad (17.1)$$

where the summation index i includes all conductances, whether or not they are voltage-dependent (for more details, see Chap. 6). Equation 17.1 does not include an additive term corresponding to the current flowing into or out of the dendrites. Not surprisingly, and

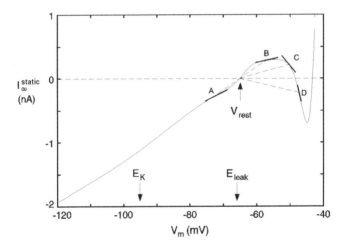

Fig. 17.1 STATIONARY CURRENT-VOLTAGE RELATIONSHIP Stationary $I-V$ relationship for the layer 5 pyramidal cell model of Fig. 3.7. Operationally, $I_\infty^{static}(V_m)$ is measured by clamping the somatic membrane potential to V_m, letting the system adapt and measuring the clamp current flowing across the somatic membrane. The currents to the right of their reversal potentials (indicated by arrows) are outward going, that is, positive. Thus, for $V_m \geq E_K$, all potassium currents are positive while for the entire voltage range shown here, I_{Na} is negative. Typical of excitable cells, the $I-V$ relationship is highly nonlinear, in particular for depolarizing membrane potentials. It intersects the zero current axis at three points. The resting potential of this system, specified by the leftmost zero crossing is $V_{rest} = -65$ mV. The *chord conductance* is defined as the ratio of the current at any holding potential V_m to the voltage difference between V_m and the resting potential V_{rest}. For points A, B, and C it is positive, and for D, negative. The *slope conductance* is defined as the local tangent to the $I-V$ curve, that is as the slope of the curve. The slope conductance at A and B is positive and at C and D negative. For the linear portion of the $I-V$ curve, the chord conductance is almost identical to the slope conductance (e.g., point A). The inverse of the *slope conductance* is usually what is meant by "input resistance."

different from a linear ohmic resistance, the total somatic membrane current is a nonlinear function of the membrane potential, in particular for depolarizing values. The resting state of the cell is given by the leftward (Fig. 17.1) stable zero crossing at -65 mV.

We can better understand $I_\infty(V_m)$ in Fig. 17.1 by recalling that for any current that can be expressed as the product of a driving potential and a conductance, values of V_m to the left of the associated reversal potential correspond to an inward, that is, negative current. Conversely, holding the potential to a more depolarized value than the reversal potential causes an outward (positive) current to flow.

For example, the long slope to the left of V_{rest} is due to the passive leak conductance and the anomalous rectifier current I_{AR}, the hump between -65 and -48 mV is caused by activation of the delayed-rectifier K^+ current, and the sharp negative peak around -45 mV in Fig. 17.1 by I_{Na}. The large outward current at more positive values of V_m is almost exclusively carried by the noninactivating I_{DR}, overshadowing the inactivating I_{Na}. This I–V curve gives us the opportunity to define two quite different conductances.

17.1.1 Membrane Chord Conductance

The steady-state membrane *chord conductance* is defined as the ratio of the total current flowing at any particular potential V_m to the applied membrane potential relative to the resting potential,

$$G_{\text{in}}^{\text{chord}} = \frac{I_\infty(V_m) - I_\infty(V_{\text{rest}})}{V_m - V_{\text{rest}}} = \frac{I_\infty(V_m)}{V_m - V_{\text{rest}}}. \tag{17.2}$$

The chord conductance derives its name from its construction: conceptually, it can be obtained by pulling a chord or string between the points, $(V_{\text{rest}}, 0)$ and $(V_m, I_\infty(V_m))$. Physically, the chord conductance tells us about the sign and the absolute value of the total membrane current flowing at any one potential. If the I–V relationship has several stable resting points, the chord conductance can be defined around each resting point.

17.1.2 Membrane Slope Conductance

The steady-state membrane *slope conductance* is defined as the local slope of the I–V relationship at the operating point V_m,

$$G_{\text{in}}^{\text{static}} = \frac{d I_\infty^{\text{static}}(V_m)}{d V_m} = \sum_i \left[g_i(V_m) + \frac{dg_i(V_m)}{dV_m}(V_m - E_i) \right] \tag{17.3}$$

(for a single compartmental model). Analytically, the slope conductance corresponds to the linear term in the Taylor series approximation of I_∞^{static} around V_m. The slope conductance of the entire membrane consists of the sum of the individual membrane conductances $g_i(V_m)$ plus a derivative term to account for the membrane nonlinearities.

When using a current pulse to measure $G_{\text{in}}^{\text{static}}$ (Fig. 17.2), one must be careful to wait long enough until all the slow membrane currents have reached equilibrium.

The Inverse of the Slope Conductance Is the Input Resistance

Because the slope conductance characterizes the response of the membrane current to small changes in the membrane potential relative to V_m, its inverse is usually what is meant by *input resistance* R_{in} (which is the convention we will follow). Figure 17.2 illustrates this procedure graphically. Injecting a small current into the soma of the pyramidal cell model causes the membrane potential to peak before it settles down to its final value. Dividing

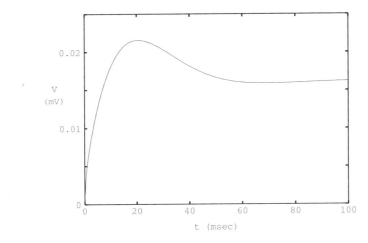

Fig. 17.2 MEASURING THE INPUT RESISTANCE   Somatic membrane potential in response to a current step of 1-pA amplitude in the pyramidal cell model. At $V_{\text{rest}} = -65\,mV$, the steady-state slope conductance $G_{\text{in}}^{\text{static}} = 1/0.0165 = 60.6$ nS, corresponding to an input resistance R_{in} of 16.5 MΩ. The initial overshoot is due to activation and subsequent inactivation of the transient A current. Blocking I_A causes this hump to disappear and increases R_{in} to 38 MΩ. Reprinted by permission from Bernander (1993).

this steady-state change in potential by the injected current gives the input resistance as 16.5 MΩ. The hump disappears if the transient A current is blocked. Now the potential rises smoothly to its final value.

For a linear I–V relationship the second term in Eq. 17.3 disappears and the slope conductance is equal to the chord conductance and corresponds to $1/\bar{K}_{ii}$ (e.g., point A in Fig. 17.1). In other words, for linear systems these difficulties with the definition of the input resistance do not arise.

In principle, $G_{\text{in}}^{\text{static}}$ can be measured under either voltage clamp or current clamp. For instance, to compute the slope conductance at rest, we can inject a small current δI into the cell and measure the steady-state voltage change δV (as in Fig. 17.2) or, conversely, we can clamp the potential from V_{rest} to a slightly displaced value $V_{\text{rest}} \pm \delta V$ and measure the change in clamp current δI. For small enough values of δI and δV, both methods converge to the same result.

Negative Slope Conductance

Because the magnitude of the second term in Eq. 17.3 can often be much larger than the first term but of opposite sign, the slope conductance can be negative, zero, or positive. To understand the significance of this, let us look at the example in Fig. 17.3. At rest, the slope conductance at the soma is 57 nS. In the presence of excitatory NMDA synaptic input to the dendritic tree, the soma depolarizes by about 3 mV. Injecting a hyperpolarizing current to bring the membrane potential back to its resting potential now yields a reduced slope conductance of 50.3 nS. Quadrupling the amount of NMDA input drops the slope conductance to 20 nS.

This result appears paradoxical: adding membrane conductance, that is, opening ionic channels, causes a decrease in the local input conductance. It can be explained in a qualitative manner by appealing to Eq. 17.3. (We here gloss over the fact that the NMDA synapses in Fig. 17.3 are distributed throughout the cell, while Eq. 17.3 applies to a single compartment.)

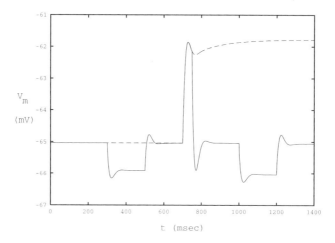

t (msec)

Fig. 17.3 NEGATIVE SLOPE CONDUCTANCE   Activating NMDA synapses can lead to a paradoxical *decrease* in the slope conductance $dI_\infty(V_m)/dV_m$. At rest, the slope conductance is 57 nS (as determined by the amplitude of the current step divided by the amplitude of the hyperpolarizing potential after it has converged at around 500 msec). At $t = 700$ msec, synaptic input of the NMDA type that is distributed throughout the dendritic tree is activated, depolarizing the soma by 3 mV (dashed line). Because this shift in the potential activates or inactivates some of the voltage-dependent somatic conductances, confounding the measurement of $dI_\infty(V_m)/dV_m$, the potential is brought back with the help of a hyperpolarizing clamp current (solid line). Repeating the current pulse experiment reveals a slope conductance of 50.3 nS, even though the total membrane conductance has increased with activation of the NMDA synapses.

The first term corresponds to the total amount of NMDA conductance added. This positive term is overwhelmed by the $[dg_i(V_m)/dV_m](V_m - E_i)$ term. Due to the negative slope region of the NMDA receptor (see Fig. 4.8A) and the reversal potential $E_i = 0$, this term is negative and dominates the total conductance in this voltage regime. We shall promptly discuss the relevance of the slope conductance to the spiking threshold.

In our discussion of input conductance, we neglected any temporal aspects of the I–V relationship. Both chord as well as slope conductances can also be defined for the instantaneous I–V relationship or for any other point in time. (For more details see the discussion in Jack, Noble, and Tsien, 1975.)

17.2 Time Constants for Active Systems

The membrane time constant τ_m is a widely used measure of the dynamics of neurons. For a single RC compartment or an infinite, homogeneous cable, τ_m is simply RC. But—as discussed at length in Sec. 3.6.1—complexities arise when one considers extended passive structures. In general, for a passive compartmental model of a cell, a finite number of distinct equalizing time constants τ_i can be defined (Eq. 3.49).

These can be evaluated directly, without solving the cable equation, by efficient matrix inversion techniques (see Appendix C; Perkel, Mulloney, and Budelli, 1981). Usually, it is difficult to infer more than the first two or three (at best) of these time constants from the voltage record (e.g., Fig. 3.12). The amplitude and distribution of these time constants are related in interesting ways to the neuronal geometry and its possible symmetries. (For a detailed discussion see Rall and Segev, 1985; Rall et al., 1992; Holmes, Segev, and Rall, 1992; Major, Evans, and Jack, 1993a,b; Major, 1993.)

A more problematic issue is the fact that real neurons show a passive response only under very limited conditions. Usually, one or more voltage-dependent membrane conductances are activated around V_{rest}. For instance, a mixed Na^+/K^+ current I_{AR} becomes activated upon *hyperpolarization*, counter to the action of most other conductances which activate upon depolarization. The action of this *anomalous* or *inward rectifier* (Chap. 9) is to reduce R_{in} and τ_m as the membrane is hyperpolarized (e.g., Spruston and Johnston, 1992). Figure 17.4 illustrates this in an exemplary way in an intracellular recording from a superficial pyramidal cell (Major, 1992). Hyperpolarizing the membrane potential from -62 to -112 mV reduces the slowest decay time constants by a factor of 3 (from 21.5 to 6.9 msec) and the input resistance by a factor of 2 (from 88.7 to 37.1 MΩ).

One possibility for obtaining the "true" value of the voltage-independent components would be to block all contaminating ionic currents by the judicious use of pharmacological channel blockers. Yet this hardly corresponds to the natural situation of a neuron receiving synaptic input.

The problem is, of course, that one cannot associate a single time constant with an active voltage-dependent membrane. Most researchers neglect this observation and fit an exponential to portions of the voltage tail of the cell's response to small current pulses (Fig. 17.4; Iansek and Redman, 1973; Durand et al., 1983; Rall et al., 1992). This yields

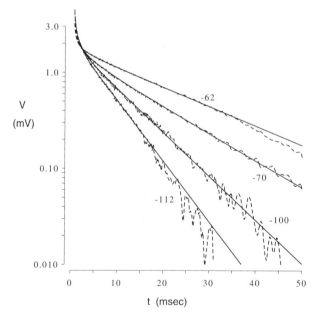

Fig. 17.4 VARIABLE MEMBRANE TIME CONSTANT Experimental record from a layer 2 pyramidal cell in the visual cortex of a rat brain slice taken by a sharp intracellular electrode. Shown is the membrane potential in response to a very short current pulse on a semilogarithmic scale at four different holding potentials. The solid lines correspond to the voltages reconstructed by "peeling" the experimental waveforms (Sec. 3.6.1). As the membrane is depolarized from -62 to -112 mV, the slowest decay—corresponding to τ_m—decreases from 21.5 to 6.9 msec, a reduction by a factor of 3. This fact makes the exact determination of a passive time constant difficult. Reprinted by permission from Major (1992).

estimates of τ_m of between 14 and 16 msec for the pyramidal cell model at rest, matching the observed time constant of 23 msec for this cell recorded in the anesthetized cat reasonably well. In the following chapter we will study how up or down regulating the synaptic input can change τ_m by an order of magnitude or more.

17.3 Action Potential Generation and the Question of Threshold

Throughout the previous chapters, we always assumed that the criterion for spike initiation is that the membrane potential exceeds a particular threshold value. To what extent is this simple assumption warranted? After all, it could well be possible that the threshold needs to be breached within a certain time window or that even further conditions, such as a minimal current, need to be fulfilled. We will now investigate this question in more detail for a membrane patch model. The more involved question of threshold generation in active cable structures is picked up in Chap. 19.

But first a preamble dealing with the instantaneous current-voltage relationship and related matters. Much of this section is based on material from Noble and Stein (1966) and Koch, Bernander, and Douglas (1995).

17.3.1 Current-Voltage Relationship
In order to understand under what conditions the cell can be thought of as having a voltage or a current threshold, we need to reconsider the steady-state current-voltage relationship $I_\infty^{\text{static}}(V_m)$ (Fig. 17.1) as well as the momentary or *instantaneous* I–V curve, denoted by $I_0(V_m)$ (Jack, Noble, and Tsien, 1975; Fig. 17.5). Numerically, this is obtained by instantaneously moving the somatic membrane potential from its resting value to a new value and assuming that all active conductances remain unchanged with the notable exception of the fast sodium activation particle m. Given the very fast time constant of sodium activation, m is practically always at equilibrium with respect to the voltage, while all other conductances have not had time to change from their values at V_{rest}.

In the subthreshold domain, both the steady-state and the instantaneous curves have very similar shapes. For the pyramidal cell model of Fig. 3.7, both currents possess the same intersection with the zero current axis at V_{rest}, are outward (positive) for more depolarized potentials, and become negative around -48 mV. However, the steady-state current quickly becomes positive again, attaining very large values. This large outward current is almost exclusively carried by the noninactivating delayed rectifying potassium current, I_{DR}, which overshadows the inactivating I_{Na}. Conversely, the instantaneous current remains negative until it reverses close to the sodium reversal potential of 50 mV. This current is dominated by I_{Na}, though with increasing depolarization the driving potential of the potassium currents and the anomalous rectifier current increase while the driving potential of I_{Na} decreases.

17.3.2 Stability of the Membrane Voltage
If the transient activation and inactivation of currents throughout the dendritic tree are neglected, the firing behavior of the cell can be defined by the nonlinear and stationary I–V curve, $I = I_\infty^{\text{static}}(V_m)$ (Fig. 17.5) in parallel with a membrane capacity. If a current I_{clamp} is injected into this circuit,

$$C\frac{dV_m}{dt} = I_{\text{clamp}} - I_\infty^{\text{static}}(V_m) \qquad (17.4)$$

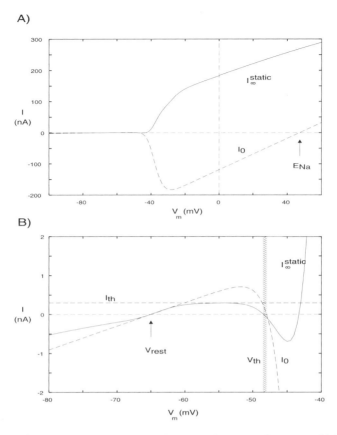

Fig. 17.5 STEADY-STATE AND INSTANTANEOUS CURRENT-VOLTAGE CURVES (A) Somatic membrane potential V_m was voltage clamped and the steady-state current $I_\infty^{static}(V_m)$ (same as in Fig. 17.1) at the soma of the pyramidal cell was recorded. The instantaneous $I–V$ curve $I_0(V_m)$ is obtained by instantaneously displacing the membrane potential from V_{rest} to V_m and measuring the initial current. All somatic membrane conductances retain the values they had at V_{rest}, with the sole exception of the fast sodium activation process—due to its very fast time constant (50 μsec) we assume that it reaches its steady-state value at V_m. Note the very large amplitudes of I_0 (due to I_{Na} activation) and I_∞^{static} (due to I_{DR} activation). I_0 crosses over close to the reversal potential for I_{Na}. (B) Detail of A in the vicinity of the resting potential and spike threshold. Both curves reverse at V_{rest}. The slope of I_∞^{static} corresponds to the input (slope) conductance rest. Both curves cross zero at around -48 mV. The amplitude of $I_\infty(V_m)$ at the local peak around -54 mV represents the current threshold I_{th} for spike initiation, while the location of the middle zero crossing of I_0 corresponds to the voltage threshold V_{th} for spike initiation (indicated by the thin stippled area). Reprinted by permission from Koch, Bernander, and Douglas (1995).

determines the dynamics of the somatic membrane potential. Different from the phase space models treated in Chap. 7, this system has a single independent variable V_m and the associated phase space consists of a line. Its equilibrium points are defined by $I_{clamp} = I_\infty^{static}(\overline{V}_m)$. In the absence of an externally injected current, these are given by the zero crossings of Fig. 17.1. However, not all of these points are stable. In a manner analogous to Sec. 7.1.2, we analyze their stability by considering small variations of the membrane potential δV around these singular points $\overline{V}_m$. This leads to

$$C\frac{d\delta V}{dt} = -\frac{dI_\infty^{static}}{dV}\delta V \,. \tag{17.5}$$

In other words, in regions where the *slope conductance* $dI^{\text{static}}_{\infty}/dV$ is positive, the dynamics follow an equation of the type $\delta\dot{V} = -\alpha\delta V$ (with $\alpha > 0$) and the system will settle down to the equilibrium point (that is, $\delta V \to 0$). Under steady-state conditions, $I^{\text{static}}_{\infty}$ shows three zero crossings, one at V_{rest}, one around -48.5 mV, and one at -43 mV. The slope at the leftmost zero crossing is positive, implying that V_m at V_{rest} is stable to small perturbations. Displacing the system by $\delta V > 0$ away from V_{rest} causes a positive, outward ionic current to flow, a current that will drive the membrane potential back to V_{rest}. Conversely, a hyperpolarizing deflection will lead to an inward current that will tend to repolarize the soma.

By the same logic, the second equilibrium point at -48.5 mV is not stable. If the membrane is displaced by $\delta V_m < 0$ away from this point, a positive, outward current will flow, causing the membrane potential to decrease and to move further from this zero crossing until the system comes to rest at V_{rest}. If $\delta V_m > 0$, an inward current will flow, which further depolarizes the system. Since the smallest voltage perturbation will carry V_m away from this zero crossing, this point is unstable.

The rightmost zero crossing at -43 mV is never attained under physiological conditions, since for these values of V_m, $I^{\text{static}}_{\infty}$ lies in a realm of phase space not accessible to the system under normal conditions.

17.3.3 Voltage Threshold

To understand the origin of the voltage threshold, we make use of very simple considerations discussed more fully in Noble and Stein (1966) and Jack, Noble, and Tsien (1975). We assume that the somatic membrane receives a rapid and powerful current input caused, for instance, by numerous highly synchronized excitatory synaptic inputs (as in Fig. 17.6A) and we neglect the extended cable structure of the neuron.

This has the effect of charging up the membrane capacitance very rapidly, thereby displacing the voltage to a new value without, at first, affecting any other parameters of the system. In the limit of an infinitely fast input, the system moves along the instantaneous I–V curve I_0. If the input is small enough and displaces the potential to between V_{rest} and -48 mV, I_0 will be outward, driving the membrane potential toward less depolarized values and ultimately back to V_{rest}. This voltage trajectory corresponds to a subthreshold EPSP.

If, however, the input pulse takes V_m instantaneously beyond the zero crossing at -48 mV, I_0 becomes negative, driving the membrane to more depolarized values. In a positive feedback loop, this will further increase the inward current. The net effect is that the membrane depolarizes very rapidly: the membrane generates an action potential (Fig. 17.6B). Because during this time the other membrane currents will have started to change, I_0 can no longer be used to determine the subsequent fate of the potential. Nonetheless, the zero crossing of the instantaneous I–V curve determines the voltage value at which the membrane is able to generate sufficient current to drive the action potential without further applied current. We therefore identify this zero crossing of the instantaneous I–V curve with the voltage threshold for spike initiation V_{th} (about -48 mV).

Our argument predicts that any somatic current—whether delivered via an electrode or via synaptic input—that is sufficiently fast and powerful to depolarize the somatic membrane beyond V_{th} will initiate an action potential. We confirmed this by plotting in Fig. 17.6A the somatic potential $V_m(t)$ in response to rapid, synchronized synaptic input to the dendritic tree that is sufficient to fire the cell at a brisk rate of 21 Hz. Figure 17.6B shows a single spike in response to a brief but strong current pulse injected directly into

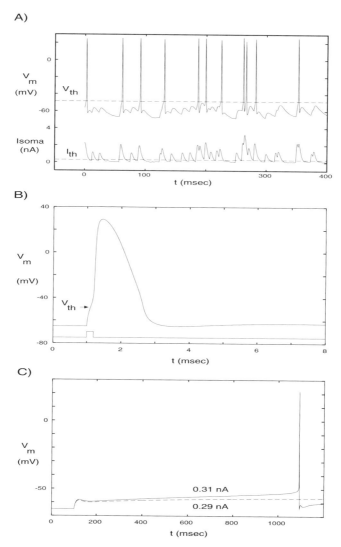

Fig. 17.6 DIFFERENT PATHS TO SPIKE INITIATION Generation of a somatic spike in the pyramidal cell model can occur in different ways, depending on the dynamics of the input. **(A)** Somatic voltage trace in response to random synaptic input to the dendritic tree. 100 fast (t_{peak} = 0.5 msec) AMPA synapses fired synchronously according to a Poisson process at a rate of 75 Hz. The current I_{soma} delivered by this synchronized synaptic input to the cell body is plotted at the bottom. This current was measured by clamping the soma to V_{rest} and measuring the clamp current. Even though I_{th} is frequently transiently exceeded by a large amount, the somatic membrane potential is usually not sufficiently depolarized to exceed V_{th} and initiate a spike. The variability in the amplitude of the EPSCs derives from the fact that following each synchronized synaptic event, a new set of 100 excitatory synapses is randomly chosen. A single action potential in response to **(B)** a 0.2-msec 10-nA current pulse (indicated in the lower trace) and **(C)** a sustained current step. For I_{clamp} = 0.29 nA the potential levels out at −56.5 mV, while for 0.31 nA a spike occurs with a delay of 1 sec. The first two plots demonstrate conditions under which action potentials are generated if a voltage threshold V_{th} is exceeded, while the last plot corresponds to a paradigm where spike initiation occurs if a current threshold I_{th} is exceeded. Reprinted by permission from Koch, Bernander, and Douglas (1995).

the soma. Careful inspection of both voltage trajectories indicates that each and every time $V_m(t)$ reaches $V_{th} = -48$ mV, an action potential is initiated (see dotted lines and the arrow in Fig. 17.6A, B).

Figure 17.6A reveals that many somatic EPSPs fail to trigger spikes. This fact can be exploited to operationalize the voltage threshold by histogramming the peak values of subthreshold somatic EPSPs. If spike initiation truly occurs at V_{th}, such a histogram should show peak EPSP amplitudes up to V_{th}, followed by an abrupt absence of any local maxima in V_m until, at high values, the peak voltages of the action potential itself appear. This is exactly the distribution seen in Fig. 17.7, with a complete lacuna of any local peaks in the somatic membrane potential between -48.5 and $+20$ mV.

This histogram was obtained using rapid synaptic input. Slow[1] synaptic input will not lead to such a crisp threshold. Experimentally, using rapid and powerful synaptic input that hovers around threshold should optimize the conditions for detecting a voltage threshold using this technique.

17.3.4 Current Threshold

Figure 17.6C illustrates a different path along which spikes can be initiated. If a small, sustained current I_{clamp} of 0.29 nA current is injected into the soma, the somatic membrane potential responds with a slight overshoot, due to the activation and subsequent inactivation of the transient A current, before settling to a steady-state somatic potential (at -56.5 mV). Increasing I_{clamp} by 20 pA is sufficient to initiate an action potential 1 sec after the onset of the current input. The extended delay between input and spike is primarily caused by I_A, as shown by Connor and Stevens (1971a,b,c). As a consequence, the cell can spike at very low frequencies (Fig. 9.7).

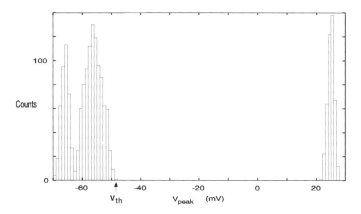

Fig. 17.7 OPERATIONALIZED DEFINITION OF A VOLTAGE THRESHOLD   The pyramidal cell model was bombarded for 20 sec with fast excitatory synaptic input (of the type shown in Fig. 17.6A, with the cell firing at 21 Hz) and the amplitude of the local maxima of the somatic membrane potential histogrammed. The three peaks correspond to the noisy synaptic background around $V_{rest} = -65$ mV (leftmost peak), synchronized EPSPs that failed to elicit action potentials (central peak), and action potentials (rightmost peak). The maximum value of the central hump is -48.6 mV, providing a convenient way in which the voltage threshold V_{th} can be measured. Reprinted by permission from Koch, Bernander, and Douglas (1995).

---

1. Fast and slow is always relative to the dynamics of the large sodium current determining spike initiation.

From an operational point of view, if the current delivered to the somatic membrane varies very slowly, the membrane can be thought of as having a current threshold $I_{th} = 0.295$ nA. To explain this, we turn to the steady-state I–V plot in Fig. 17.5. Injecting a sustained and depolarizing current step of amplitude I_{clamp} into the soma corresponds to shifting the I_∞^{static} curve downward by an equal amount. If I_{clamp} is small, the principal effect of this is to move the location of the zero crossings of I_∞^{static}. In particular, the leftmost zero crossing— and therefore the new somatic resting potential— shifts to the right and the middle zero crossing shifts to the left. At a critical value of I_{clamp}, labeled I_{th}, the two zero crossings merge. At this point the system still has an equilibrium point. If the current is increased any further, this portion of the I_∞^{static} curve drops below zero and loses its equilibrium. The inward current generated by the system forces the membrane to depolarize, driving V_m toward E_{Na} and initiating an action potential. In fact, for sustained current injections the membrane has no stable equilibrium point, displaying instead a stable limit cycle: as long as the suprathreshold current persists, the membrane generates a series of spikes. From these considerations we predict that the current threshold corresponds to the local maximum of the I_∞ curve, here equal to 0.295 nA, very close to the value obtained by integrating the appropriate equations (Fig. 17.6C).

Note that the steady-state somatic membrane potential at which threshold current is reached (about -54.2 mV) is *different* from V_{th} because of the nonlinear and nonstationary somatic membrane. In a linear and stationary system, such as in integrate-and-fire models (Eq. 14.9) a linear relationship holds between the two,

$$V_{th} = R_{in}I_{th} .$$ (17.6)

17.3.5 Charge Threshold

Hodgkin and Rushton (1946; see also Noble and Stein, 1966) proposed that the threshold condition for the excitation of a cable is that a constant amount of charge Q_{th} be applied. Is this true for our cell?

When a significant amount of current—either from synaptic input or from a micro-electrode—is very rapidly delivered to the cell body, the voltage change will primarily be determined by the capacitive current, since the membrane currents take time to change. Under these conditions, the cell should also show a *charge threshold*. In order to understand this, let us follow a train of thought laid out by Noble and Stein (1966). Because we are considering rapid events, the steady-state I–V in Eq. 17.4 must be replaced by the instantaneous I–V curve:

$$C\frac{dV_m}{dt} = I_{clamp} - I_0(V_m)$$ (17.7)

(as before, we here neglect the dendritic tree). By rearranging and integrating with respect to voltage, we obtain an expression for how long the current step has to be applied (from $V_{rest} = 0$) until the voltage threshold has been reached,

$$T_{pulse} = C \int_0^{V_{th}} \frac{dV_m}{I_{clamp} - I_0(V_m)} .$$ (17.8)

The integration can be carried out for any I–V relationship that does not depend on time, such as I_0 or I_∞ (Fig. 17.5). If for short times, the injected current I_{clamp} is much bigger than the subthreshold ionic currents I_0 (on the order of 1 nA or larger; see Figs. 17.5A and 17.6), I_0 can be neglected. The total charge delivered by a step current of amplitude I_{clamp} and duration T_{th} is

$$Q = I_{\text{clamp}} T_{\text{pulse}} \approx C V_{\text{th}} = Q_{\text{th}} . \qquad (17.9)$$

Put differently, for short, intense stimuli, a constant charge Q_{th} must be supplied to the capacitance in order to reach threshold (Noble and Stein, 1966).

We numerically evaluate this for a reduced one-compartment model of the pyramidal cell (Koch, Bernander, and Douglas, 1995) as well as for the full model by injecting I_{clamp} current into the soma and recording the duration of the pulse T_{pulse} necessary to generate an action potential. This threshold charge is plotted in Fig. 17.8 as a function of the amplitude of the injected current step. For the one-compartment model Q starts out high but decreases rapidly and saturates at $Q_{\text{th}} = 7.65$ pC.

If a 7.65 pC charge is placed onto a 0.508 nF capacitance (its value in the one-compartment model), the potential changes by 15.1 mV, very close to V_{th}. We conclude that in a single-compartment active model, a charge and a voltage threshold both imply the same thing: for rapid input the cell spikes whenever the input current dumps Q_{th} charge onto the capacitance, bringing the membrane potential to V_{th}. The existence of a charge threshold immediately implies an inverse relationship between the amplitude of the applied current and T_{th}. It is known that both the Hodgkin–Huxley equations for a clamped axon and the actual space-clamped giant squid axon membrane give rise to such a *strength-duration* curve (Hagiwara and Oomura, 1958; Cooley and Dodge, 1966; Noble and Stein, 1966).

This simple story becomes more complicated when we repeat the same experiment in the 400 compartment model (Fig. 17.8). The amount of charge delivered by the electrode that is necessary for the cell to spike decreases monotonically with increasing current. It only flattens out for unphysiologically large amplitudes. Under these conditions, no single

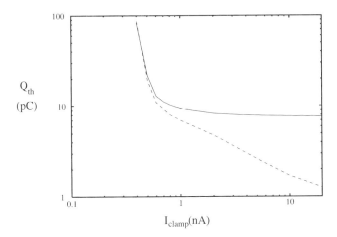

Q_{th} (pC)

I_{clamp}(nA)

Fig. 17.8 Is There a Charge Threshold for Spiking? Relationship between the charge Q_{th} injected by a current pulse of constant amplitude I_{clamp} and duration T_{pulse} necessary for spiking to occur. For a single-compartment model of the pyramidal cell (where the effect of the entire passive dendritic tree is mimicked by adjusting the capacitance and the leak resistance at the soma), Q_{th} rapidly reaches an asymptote (solid line). That is, for moderately strong current inputs a spike is initiated whenever the limiting charge Q_{th} is delivered to the capacitance; this charge obeys $Q_{\text{th}} = C V_{\text{th}}$. In other words, for a single compartment the existence of the voltage threshold for spiking is equivalent to a charge threshold. For weaker inputs more charge is required. No such constant threshold condition is observed for the full pyramidal cell model (dashed line). Due to the presence of the distributed capacitances in the dendritic tree, action potential initiation does not obey a charge threshold condition.

equation governing the evolution of the somatic membrane potential can be written because of the presence of the many equalization time constants between the soma and the dendritic compartments. Only if the current injection is very large, with the associated very short durations <0.05 msec, can one neglect the dendritic capacitance and observe a fixed Q_{th}. We conclude that in any cell with a substantial dendritic tree, the occurrence of an action potential is not caused by the charge at the spike initiating zone exceeding some fixed threshold value.

17.3.6 Voltage versus Current Threshold

An *advocatus diaboli* might argue that the cell always possesses a voltage threshold, with the value of this threshold increasing from -48 mV for very fast inputs to -54.2 mV for slow ones. Yet in our eyes a single, but variable threshold is conceptually less elegant than two fixed but different thresholds (V_{th} and I_{th}). Furthermore, given the very broad, local peak of I_∞^{static} in Fig. 17.5B, V_{th} is ill defined, in particular in the face of ever existing noise.

As we will see in the following chapter, the number of simultaneously activated somatic EPSPs needed to trigger a spike is within 10% of the number predicted based on V_{th}. As long as the synaptic current delivered to the soma I_{soma} has a fast rise time compared to the activation and inactivation time constants of all the somatic currents with the exception of sodium activation (that is, on the order of 1 msec) and is powerful enough to displace V_m past V_{th}, the cell will spike. To what extent synaptic input approximates such behavior or can be better described by a slowly varying current input under *in vivo* physiological conditions depends on many factors beyond the scope of this chapter.

We are not arguing that every neuron must have I–V curves that are qualitatively similar to the ones in our model of a pyramidal cell. For instance, the steady-state I–V curve for the Hodgkin–Huxley membrane patch model is a monotonically increasing function of V_m with no local maxima (Fig. 6.9A). In this case $I_\infty^{static}(V_m)$ cannot be used to predict I_{th}. The instantaneous I–V curve I_0 of a patch of squid axon around rest (Fig. 6.6) does, on the other hand, look qualitatively similar to the $I_0(V_m)$ curve of the pyramidal cell and so can be used to estimate in a qualitative manner V_{th} (see Sec. 6.3.1).

17.4 Action Potential

We would be remiss if, after this extensive discussion of the conditions under which spikes can be initiated, we would not describe the action potential itself. Its form is highly stereotypical, with little variation from one spike to the next (Fig. 17.6). The spike amplitude from threshold at -49 mV to the peak at 30 mV is about 80 mV, and the full width at half height is 0.9 msec. These numbers compare favorably with experimentally recorded spikes from slice and *in vivo* (Bindman and Prince, 1983; Spain Schwindt, and Crill, 1990, 1991; Hirsch and Gilbert, 1991; Pockberger, 1991). The rapid upstroke during the initial part of the action potential, on the order of several hundred millivolts per millisecond, is caused by the sodium-mediated inward current charging up the membrane capacitance. The smaller the capacitance (as, for instance, in small cells) or the larger the density of fast, sodium channels, the faster the upstroke and, ultimately, the steeper the slope of the discharge curve. Ultimately, it is the amplitude of the upstroke that limits the speed with which action potentials can be generated.

Two ionic currents are the primary culprits in generating and shaping the action potential. The largest is the current supporting the action potential itself I_{Na} peaking at 70 nA. Due to adaptation, peak I_{Na} for consecutive spikes is reduced to 50 nA. The second largest current is the delayed-rectifier potassium current, I_{DR} coming in at 23 nA. The gargantuan size of these two ionic currents, compared to anything the synapses in the dendritic tree can deliver to the soma, explains the almost complete independence of the shape of the action potential to the way in which the spike was triggered. The total calcium current flowing is miniscule compared to I_{Na}. Its primary function is not to drive the action potential itself but to activate calcium-dependent potassium currents controlling adaptation.

17.5 Repetitive Spiking

Until now, we have only considered the events leading to a single action potential. Yet, cells usually spike repeatedly. The response of the full pyramidal cell model to a sustained current injection is shown in Fig. 17.9. Similar to the modified Morris-Lecar equations (Eqs. 7.16 and 7.17; see also Figs. 7.10 and 7.11) and different from the squid giant axon, the pyramidal cell model can fire at very low frequencies: injecting 0.31 nA of sustained current causes the cell to spike about once every second. The current responsible for these very long interspike intervals is the inactivating potassium current I_A (Connor and Stevens, 1971a,b,c; Getting, 1989; for a detailed discussion see Fig. 9.7 and related text). Due to its hyperpolarizing reversal potential, this potassium current reduces the evoked depolarization (see the bump in Fig. 17.2). Because I_A inactivates with increasing depolarization, different from the large, delayed rectifier potassium current, it gradually decreases in size, enabling V_m to slowly creep upward until threshold is reached.

17.5.1 Discharge Curve
If the input is powerful enough to trigger two or more spikes, the firing frequency adapts within 50 msec to between one-half and one-third of its initial value. Adaptation can be

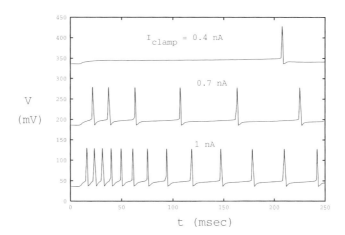

Fig. 17.9 REPETITIVE FIRING Generation of several action potentials in the pyramidal cell model in response to sustained current injections at the cell body. The minimum current step that evokes a spike, rheobase, is $I_{th} = 0.295$ nA. Injecting 0.4 nA evokes repetitive spikes with very long interspike intervals, caused by the presence of the transient I_A current. At higher spiking frequencies, the firing adapts due to the calcium-dependent potassium currents. The voltage curves are shifted for viewing purposes. No noise has been simulated, approximating the situation one would encounter in cultured cells. Reprinted in modified form by permission from Bernander (1993).

visualized as reducing the gain of the $f-I$ curve (Fig. 17.10A; see also Fig. 14.2C). As discussed in Sec. 9.2.2, it is caused by one or more calcium-dependent potassium currents, acting as a sustained negative input that offsets the positive current delivered via the electrode. At a very low spiking frequency, no adaptation occurs since $[Ca^{2+}]_i$ decays to its resting levels prior to the next occurrence of a spike.

The curve for the first ISI shows a biphasic shape. Initially, the slope is low from I_{th} to around 0.6 nA (*primary slope*). For larger input currents the slope is much larger (*secondary slope*). This portion of the $f-I$ curve can be fitted rather well by a logarithmic relationship between f and I (Figs. 6.10 and 17.10A), as first observed for the uniformly polarized Hodgkin-Huxley equations (Agin, 1964). The firing frequency associated with the second ISI displays a very similar pattern, except that the primary slope spans a larger current range.

Though the shape of the adapted $f-I$ curve in Fig. 17.10A appears deceptively linear, the slope varies by more than a factor of 2. Figure 17.10A shows the linear fit:

$$f = 49.5 \cdot (I - 0.22) \tag{17.10}$$

with the constant of proportionality corresponding to about 50 spikes per second per nanoampere of injected current, in good agreement with experimental *in vitro* (Mason and Larkman, 1990) and *in vivo* (Jagadeesh, Gray, and Ferster, 1992) observations in cat cortex. The figure also shows a logarithmic fit,

$$f = 75 \cdot (\ln I + 0.59) \,. \tag{17.11}$$

This fit is somewhat more accurate, except at very low input currents. While the difference between these two fits may seem negligible, the logarithmic fit will be more useful in the following.

Note that the discharge curve does *not* show any strong evidence of saturation in the firing frequency, a saturation that is a key property of almost all nonlinear neural network models. It is in any case unclear from where, under physiological conditions, synaptic currents greater than 2 nA would originate.

17.5.2 Membrane Potential during Spiking Activity

How can the somatic membrane potential be characterized while the system is spiking repeatedly, that is, moving along its stable attractor? A very useful heuristic measure is the time-averaged somatic membrane potential

$$\langle V_m \rangle = \frac{1}{T} \int_0^T V_m(t) dt \tag{17.12}$$

where T includes several interspike intervals in response to a sustained current I. In order to avoid the confounding effects of adaptation, $\langle V_m \rangle$ should be evaluated only after adaptation is complete. Plotting $\langle V_m \rangle$ as a function of I (solid curve in Fig. 17.10B) results in a curve with a linear relationship close to the resting potential, whose inverse slope is identical to the cell's input conductance.

Conceptually—as well as by visual inspection—this curve naturally falls into two sections. Below the point at which the cell ceases to spike and the curve shows a cusp, that is, for $I < I_{th}$, it becomes identical to the inverse of the steady-state $I-V$ curve (Fig. 17.5B). The portion of the $\langle V_m \rangle$ versus I curve that lies beyond I_{th} shows two remarkable features. Firstly, the dynamic range of $\langle V_m \rangle$ over the physiological range of input currents is small.

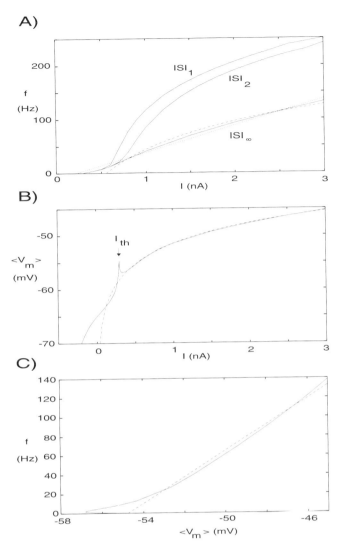

Fig. 17.10 DISCHARGE CURVE AND AVERAGED MEMBRANE POTENTIAL Characterization of the firing behavior of the pyramidal cell model. (**A**) Discharge curve computed as the inverse of the first, second, and fully adapted interspike intervals (solid) as a function of the amplitude of the current step (see also Fig. 14.2C). Linear (dotted) and logarithmic (dashed) fits are shown for the steady-state curve (Eqs. 17.10 and 17.11). (**B**) Introducing the averaged somatic membrane potential $\langle V_m \rangle = (1/T) \int_0^T V_m(t)dt$ enables the definition of a *dynamic I–V* relationship between a sustained input current and $\langle V_m \rangle$ (solid line). For $I < I_{th}$, this curve coincides with the conventionally defined inverse of the $I_\infty^{static}(V_m)$ curve (Fig. 17.5). The dashed curve is a logarithmic fit, indicating a diodelike relationship between the average membrane voltage and the current. (**C**) Relationship between $\langle V_m \rangle$ and the output spike frequency. Remarkably, the total dynamic range of $\langle V_m \rangle$ is only about 12 mV for physiological firing frequencies. The dashed line indicates the best linear fit (Eq. 17.15). Reprinted by permission from Koch, Bernander, and Douglas (1995).

As the adapted output spike frequency f ranges from 5 to 150 Hz, $\langle V_m \rangle$ is confined to a fairly limited range of about 12 mV. Secondly, the upper portion of the curve can be fitted very well by a logarithmic relationship, except for the "hook" around I_{th} (dashed curve in Fig. 17.10B).

This relationship can also be viewed with $\langle V_m \rangle$ as the independent variable. From this perspective, the current I can be thought of as a *cumulative spike current*, and the I–$\langle V_m \rangle$ curve as a *dynamic I–V relationship* $I_\infty^{dynamic}(\langle V_m \rangle)$ (dynamic because the somatic membrane potential travels on a stable limit cycle in response to the sustained input) that reverses at V_{rest}. $I_\infty^{dynamic}(\langle V_m \rangle)$ is quite distinct from $I_\infty^{static}(V_m)$. In the latter, the membrane potential is clamped to V_m and the current flowing at this voltage is evaluated, while in the dynamic case, a sustained current is injected while the membrane is left free to spike. The dynamic $I–V$ curve is relatively straightforward to measure experimentally.

The logarithmic relationship implies a *diodelike* behavior, with the current depending exponentially on the average membrane potential,

$$I_{spike} = e^{(\langle V_m \rangle + 51.5)/5.50} \tag{17.13}$$

(where V_m is expressed in units of millivolts; dashed line in Fig. 17.10B). In analogy to the steady-state slope conductance G_{in}^{static} (Eq. 17.3), we can define the *dynamic input conductance* as

$$G_{in}^{dynamic} = \frac{1}{R_{in}^{dynamic}} = \frac{dI_\infty^{dynamic}}{d\langle V_m \rangle} = \frac{dI_{spike}}{d\langle V_m \rangle} \tag{17.14}$$

(Fig. 17.11B). The cumulative effect of all currents active during spiking resembles a voltage-gated hyperpolarizing conductance that shunts large input currents, thereby stabilizing the membrane potential. In the absence of this "spike" current the membrane would depolarize to much higher values. For instance, with an input conductance of $G_{in}^{static} = 60.6 \, nS$ at V_{rest}, a 2 nA injected current step should depolarize the membrane to -34 mV in the absence of any active currents, while the actual potential $\langle V_m \rangle$ is approximately -48 mV. We conclude that the dynamic conductance $G_{in}^{dynamic}$ acts similarly to an imperfect voltage clamp, providing a stable sink for currents originating in the dendrites, a rather unusual way to look at the spike generation mechanism (Koch, Bernander, and Douglas, 1995).

Finally, we can combine the logarithmic relationship between injected current and firing frequency (Eq. 17.11) with the exponential dependence of I_{spike} on $\langle V_m \rangle$ (Eq. 17.13) to arrive at a linear relationship between the averaged membrane potential and the firing frequency,

$$f(\langle V_m \rangle) = 13.6 \cdot (\langle V_m \rangle + 54.8). \tag{17.15}$$

This relationship is useful for analog electronic circuit implementation of neurons (Mahowald and Douglas, 1991; Douglas, Mahowald, and Mead, 1995), since it allows one to interpret the membrane potential across some transistor as an effective firing frequency without adding additional circuitry to implement some output nonlinearity. As can be seen in Fig. 17.10C, the true spike frequency curve (as ascertained from the simulations) lies above Eq. 17.15 for small firing frequencies. Before concluding, several remarks pertinent to the definitions of these dynamics variables are in order.

1. The low-pass nature of the dendritic tree, caused by the distributed capacitance, will filter out the high-frequency components associated with the rapid up and down swing

A)

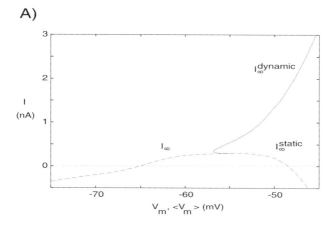

B)

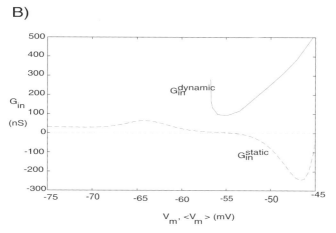

Fig. 17.11 Dynamic Discharge Curve Time-averaged membrane potential $\langle V_m \rangle$ permits the introduction of a dynamic I–V curve as the relationship between the average somatic membrane potential in the full pyramidal cell model and the total current flowing across the somatic membrane. **(A)** The static I–V curve is computed under *voltage* clamp by measuring the clamp current and its dynamic counterpart under *current* clamp (it is the inverse of the curve in Fig. 17.10B). In the subthreshold regime $I < I_{th}$ the two curves coincide (labeled I_∞). **(B)** The sustained and dynamic somatic input conductances as a function of V_m or $\langle V_m \rangle$ defined as the slopes of the I–V curves in **A**. For $V < V_{th}$, $G_{in}^{dynamic} = G_{in}^{static}$. The dynamic input conductance allows the introduction of a time constant that characterizes how rapidly the adapted firing rate responds to new input. Reprinted by permission from Koch, Bernander, and Douglas (1995).

of the action potentials as they propagate from the soma into the apical tree. Thus, $\langle V_m \rangle$ is qualitatively closer to the membrane potential seen by dendritic sites than $V_m(t)$. Also, slowly activating or inactivating conductances do not care about rapid fluctuations in V_m, but are responsive to a low-pass filtered or average potential.

2. $I_\infty^{dynamic}(\langle V_m \rangle)$, in combination with the cell's discharge curve, leads to a linear relationship between the average membrane potential $\langle V_m \rangle$ and the firing rate in the fully adapted case (Eq. 17.15).

3. The time-averaging technique discussed here represents one way in which the dynamics of a complex neuron can be reduced to the mean-field dynamics of the standard neural network unit (Sec. 14.4). It is not advisable to use $\langle V_m \rangle$ when considering fast events that depend on the exact timing (e.g., fast dendritic spikes discussed in Sec. 19.3).

4. The ratio of the membrane capacitance and $G_{in}^{dynamic}$ can be thought of as a *dynamic time constant* $\tau_{dynamic}$. It is a measure of how rapidly the averaged membrane potential, and therefore the firing rate, can vary in response to a change in I_{soma}. As can be seen in Fig. 17.11B, if sufficient current is injected for the cell to fire in a sustained manner at 65 Hz, $\langle V_m \rangle = -50$ mV, increasing $G_{in}^{dynamic}$ by about a factor of 4 relative to rest. This implies that the effective time constant of the cell has decreased to about 4 msec. Due to the presence of the many membrane conductances generating and modulating action potentials, the firing rate can respond much faster to a rapid increase in I_{soma} than suggested by the conventional definition of τ_m. For higher firing rates, $\tau_{dynamic}$ can be close to 2 msec, implying that neurons can respond at the millisecond or faster scale (Softky, 1994). Whether cells make use of this very high bandwidth is another matter.

Our enthusiasm for such a new measure must be tempered by two considerations. First, the $I_\infty^{dynamic}(\langle V_m \rangle)$ curve is defined for the adapted firing frequency, and adaptation takes between 20 and 100 msec to occur. A second difficulty is that $\tau_{dynamic}$ is defined for a fictive, continuous firing rate and for a constant current input. This is a meaningless quantity on a time scale less than the interspike interval. This sampling problem renders the exact definition of a time course problematic at very low firing frequencies.

17.6 Recapitulation

This chapter treated the characterization of a nonlinear and nonstationary neuron from the point of view of the soma on the basis of two measures: the relationship between the current flowing across the somatic membrane and the voltage and the local slope of such an *I–V* curve. The inverse of this last variable is the slope or input resistance. The traditional *I–V* curve $I_\infty^{static}(V_m)$ is defined under voltage clamp conditions: the voltage is fixed and the current flowing at this voltage is recorded. The local peak around V_{rest} defines the minimal current I_{th} necessary to initiate spiking. Under the assumption that the activation and inactivation variables of the voltage-dependent currents change much more slowly than the activation $m(t)$ of the sodium current, it is possible to define an instantaneous current-voltage curve $I_0(V_m)$. (Conceptually, this can be measured by yanking the membrane potential from V_{rest} to V_m and instantaneously recording the resulting membrane current.) Stability considerations of the sort explored in Chap. 7 dictate that the zero crossing of I_0, if the slope is negative, corresponds to the voltage V_{th} that needs to be exceeded for a spike to occur.

We conclude that *fast* and *powerful* versus *sustained* but *peri-threshold* stimuli define two different threshold conditions: in the first case V_m must exceed V_{th} while in the latter case I_{soma} must exceed I_{th}. The larger the current delivered to the soma by a synaptic input, the faster V_{th} can be reached and the sooner the cell spikes. This process is not limited by τ but by the amplitude of the upstroke of the action potential (that is, by the amout of sodium current and the somatic capacitance). For cells with no or little distributed capacitances, that is, with small dendritic trees, a voltage threshold is equivalent to a charge threshold, in which a fixed amount of charge needs to be placed onto the somatic capacitance in order to reach V_{th}.

Introducing $\langle V_m \rangle$ as the temporal average of the somatic membrane potential (including spikes) allows us to extend most of the above concepts into the spiking regime. Biophysically, $\langle V_m \rangle$ approximates the average membrane potential experienced by a distal dendritic site, where the high-frequency components of the back-propagating somatic action potential have been filtered out by the capacitances distributed throughout the cable.

The dynamic current-voltage curve $I_\infty^{\text{dynamic}}(\langle V_m \rangle)$ is defined as the inverse of the relationship between the sustained current injected via a microelectrode and the average membrane potential (Koch, Bernander, and Douglas, 1995). It can be thought of as a single, effective *spike* current that serves to stabilize the membrane potential in the neighborhood of V_{th} and that reverses at V_{rest}. This current can be described rather well by an exponential relationship: increasing $\langle V_m \rangle$ by 3.7 mV roughly doubles I. The fact that this current behaves similarly to a diode in the forward direction makes such a relationship particularly easy to implement using electronic circuits. A similar exponential relationship also exists for a patch of squid axon described by the Hodgkin–Huxley equations (not shown).

One caveat. These dynamical measures are based on time-averaged quantities; they cannot be used under conditions where the detailed time course of spiking is thought to be relevant for postsynaptic processing.

18

SYNAPTIC INPUT TO
A PASSIVE TREE

Now that we have quantified the behavior of the cell in response to current pulses and current steps as delivered by the physiologist's microelectrode, let us study the behavior of the cell responding to a more physiological input. For instance, a visual stimulus in the environment will activate cells in the retina and its target, neurons in the lateral geniculate nucleus. These, in turn, make on the order of 50 excitatory synapses onto the apical tree of a layer 5 pyramidal cell in primary visual cortex such as the one we use throughout the book, and about 100–150 synapses onto a layer 4 spiny stellate cell (Peters and Payne, 1993; Ahmed et al., 1994, 1996; Peters, Payne, and Rudd, 1994). All of these synapses will be triggered within a fraction of a millisecond (Alonso, Usrey, and Reid, 1996). Thus, any sensory input to a neuron is likely to activate on the order of 10^2 synapses, rather than one or two very specific synapses as envisioned in Chap. 5 in the discussion of synaptic AND-NOT logic.

This chapter will reexamine the effect of synaptic input to a realistic dendritic tree. We will commence by considering a single synaptic input as a sort of baseline condition. This represents a rather artificial condition; but because the excitatory postsynaptic potential and current at the soma are frequently experimentally recorded and provide important insights into the situation prevailing in the presence of massive synaptic input, we will discuss them in detail. Next we will treat the case of many temporally dispersed synaptic inputs to a leaky integrate-and-fire model and to the passive dendritic tree of the pyramidal cell. In particular, we are interested in uncovering the exact relationship between the temporal input jitter and the output jitter.

The bulk of this chapter deals with the effect of massive synaptic input onto the firing behavior of the cell, by making use of the convenient fiction that the detailed temporal arrangement of action potentials is irrelevant for neuronal information processing. This allows us to derive an analytical expression relating the synaptic input to the somatic current and ultimately to the output frequency of the cell. Under these conditions, continuous input variables—the firing rates of the presynaptic neurons—are mapped onto a continuous output frequency.

18.1 Action of a Single Synaptic Input

In general, it is very difficult to quantify the effect that an individual synapse has on the potential at the soma or on the firing frequency of the cell using a single measure. The reason is, of course, that synaptic input needs to be treated as a time-varying conductance change somewhere in the dendritic tree acting upon a nonlinear and nonstationary system. In neural network models, this is quite a different manner, since the complexity of synapses is reduced to a scalar weight w_{ij}, all synaptic input arrives at a single compartment, and the spiking nonlinearity is a stationary one (Chap. 14).

18.1.1 Unitary Excitatory Postsynaptic Potentials and Currents

Figure 18.1 illustrates the local and somatic EPSPs in response to a single excitatory, voltage-independent synapse at one of three different locations in the tree (apical and basal dendrite and the soma; see Fig. 18.6). The resultant EPSP is sometimes also called *unitary* EPSP. We include the situation of an excitatory synapse at the soma—even though no or only very few excitatory synapses are located at the cell body of pyramidal neurons—because it represents a sort of baseline against which the other cases can be judged. Furthermore, a significant fraction of excitatory synapses on smooth, that is, inhibitory, interneurons is made onto or close to the soma.

The local EPSPs peak within 1–1.2 msec, gradually returning to the resting potential with a time course dictated by τ_m. The somatic EPSPs undershoot V_{rest} before returning to base levels. This small hyperpolarizing response is caused by activation of two potassium

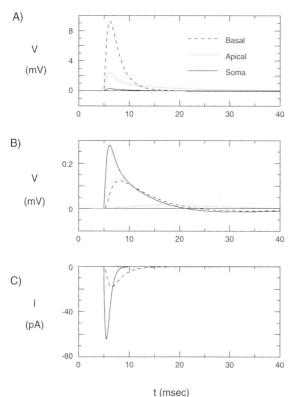

Fig. 18.1 Voltage-Independent Unitary Synaptic Input Excitatory postsynaptic potential (EPSP) at the site of the synapse (**A**) and at the soma (**B**). A single fast AMPA synapse ($g_{\text{peak}} = 1$ nS; $t_{\text{peak}} = 0.5$ msec) was either located in the apical tree (arrow in Fig. 18.6; dotted line), toward the end of a basal dendrite (arrow in Fig. 18.6; dashed line), or at the soma (solid line). The undershoot in the somatic EPSPs is caused by activation of potassium currents. (**C**) Excitatory postsynaptic current (EPSC) at the soma for the same three synaptic inputs while the membrane potential at the soma is clamped to V_{rest}. The resulting current flows inward, plotted negative by convention, and peaks prior to the peak in the associated somatic EPSP.

currents I_A and I_M. Phenomenologically, the undershoot arises from the linear but inductive nature of the potassium currents for small voltage excursions around rest (Sec. 10.1 and Fig. 10.5).

Figure 18.1 also depicts I_{clamp} in response to the three synaptic inputs when the somatic potential is clamped to $V_{rest} = -65$ mV. The peak current varies fiftyfold, from -1.2 pA for the distal apical input to -64 pA for the somatic one. The somatic EPSCs peak earlier than the associated somatic EPSPs, compatible with the notion that the current leads the voltage expected of an RLC circuit dominated by the resistive and capacitive components.

Note that the EPSC does *not* correspond to the current flowing during the somatic EPSP, since in the former case the somatic potential is clamped (explaining the lack of a positive, outward current during the late phases of the EPSC). The actual current flowing across the membrane during the EPSP will show such an outward component due to the hyperpolarizing swing of the membrane potential. Yet, because the membrane potential during the EPSP does not deviate much from V_{rest} under the conditions extant in Fig. 18.1, the EPSC is qualitatively similar to the current flowing during the EPSP.

Activating a single synapse whose conductance is halfway split between voltage independent AMPA receptors and voltage-dependent NMDA receptors perturbs this picture but little. Its primary effect is to prolong the duration of the EPSP and the EPSC. The reason for the discrepancy between the small effect of voltage-dependent input in the situation considered here and the much larger effect apparent in the experiment of Stern, Edwards, and Sakmann (1992) pictured in Fig. 4.9 is that in the former the local EPSP barely reaches 8 mV, an insufficient depolarization to remove the magnesium blockage of the NMDA receptor channels. Thus, blocking the NMDA component (as in Fig. 4.9B) has little effect. In the experiment, focal stimulation recruited a much larger number of synapses which locally depolarized the membrane sufficiently to activate NMDA input.

How do these simulated EPSPs and EPSCs stack up against experimental data from layer 5 pyramidal cells in the rat visual cortex? Figure 18.2 histograms the time course and the peak amplitudes of about 1000 spontaneous EPSCs (clamped around the resting potential) and close to 500 spontaneous EPSPs (from the monumental efforts of Smetters, 1995).

The time courses of experimentally recorded somatic EPSPs and EPSCs are comparable to those of Fig. 18.1, compatible with a very fast ($t_{peak} = 0.5$ msec), AMPA-mediated conductance change and a slow membrane time constant. The amplitude histograms of the EPSCs encompass the three simulated values from Fig. 18.1C. (The locations of the synapses that gave rise to the data in Fig. 18.2 are not known, but are assumed to be distributed throughout the dendritic tree.) The somatic EPSPs are, however, much larger than the simulated ones, due to a mismatch in input impedances in the two cases. This is expected, since the simulated pyramidal cell receives 2 Hz synaptic background activity, which substantially lowers the input impedance everywhere (see the following section).

Close inspection of Fig. 18.2 reveals a systematic dependency of the rise time of the EPSPs with their width. Rall (1967; Rall et al., 1967) first derived the relationship expected between the rise time of a unitary EPSP and its halfwidth; for more distant input, the rise time of the somatic EPSP will be longer and it will be more spread out. Plotting these two measures of dispersion against each other, known as a *shape index plot* (Rall et al., 1967), can reveal something about the distance between the synapse and the soma. As applied to spinal motoneurons, these techniques played an important role in establishing the importance of dendritic location (Redman and Walmsley, 1983).

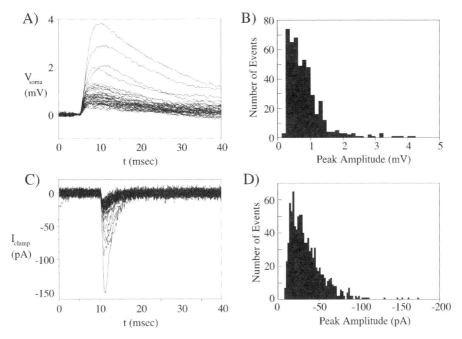

Fig. 18.2 **EXPERIMENTALLY RECORDED UNITARY EPSPs AND EPSCs** Spontaneous (**A**) EPSPs and (**C**) EPSCs recorded from layer 5 pyramidal cells in slices of rat visual cortex in the presence of bicuculline to block synaptic inhibition. The histograms of (**B**) the peak amplitude of the EPSPs (n = 479) and (**D**) the EPSCs (n = 953) reveal a broad distribution, due to the fact that the synaptic origins of the observed events are distributed throughout the dendritic tree. Reprinted by permission from Smetters (1995).

18.1.2 Utility of Measures of Synaptic Efficacy

As discussed extensively in Sec. 3.2, Rall (1962, 1964) showed that under certain conditions a dendritic tree can be reduced to a single finite cable, the *equivalent cylinder*. In particular, at each branch point the $d^{3/2}$ law, governing the relationship between parent and daughter branches, must hold and all dendrites must terminate at the same electrical distance from the soma. Unfortunately though, dendritic branch points of pyramidal cells, according to Larkman et al., (1992; see also Hillman, 1979; Brown, Fricke, and Perkel, 1981; Johnston, 1981) do not obey the $d^{3/2}$ law. Furthermore, the apical tree is electrically much longer than the basal dendrites, violating another condition needed to reduce the cell to a single cylinder.

In Sec. 3.5, we introduced several measures of synaptic "efficiency," such as the electrotonic distance between a site i and the soma L_{is} (a cumulative measure of the normalized electrotonic distance between i and the soma) and the logarithm of the voltage attenuation L_{is}^v (Eq. 3.38). Recall that in an infinite cable, $L = L^v$. In Table 18.1 these factors are evaluated for the apical and basal synaptic input of Fig. 18.1, where $L_{is}^{v,t}$ is the logarithm of the actual voltage attenuation experienced by the synaptic input. (L_{is}^v is computed from the dc input and transfer resistances as in Eq. 3.38.) Because the dendritic tree cannot be approximated by an equivalent cable but has pronounced asymmetries—current flows much more easily toward than away from the soma—the electrotonic distance L is pretty much useless as a quantitative indicator for attenuation. The log attenuation predicted by the stationary input and transfer resistances does a much better job, coming within 14% of the true log attenuation. This difference is, of course, due to the transient nature of the synaptic input.

The discrepancies between the actual attenuation of the EPSPs and the attenuation predicted by the electrotonic length apparent in Table 18.1 provide a convincing numerical argument *against* using the electrotonic distance L—derived from solving the linear cable equation in finite cables—as a measure of the coupling of a synapse to the soma.

18.1.3 What Do Unitary EPSPs and EPSCs Tell Us about the Threshold?

As we learned in the previous chapter, the pyramidal cell model possesses a voltage threshold V_{th} (around -49 mV) for rapid synaptic input. We should therefore be able to estimate the number of simultaneously excitatory synaptic inputs needed to reach threshold n_{th} as the effective threshold voltage, $V_{th} - V_{rest} = 16$ mV, divided by the peak unitary somatic EPSP amplitude. For an excitatory input at the soma, $V_{peak} = 0.28$ mV (solid trace in Fig. 18.1), leading to $n_{th} = 57$. From numerical simulations, we know that the actual number is $n_{th} = 64$. This estimate does not take the changing driving potential into account. (At -65 mV, the driving potential for the synaptic input is 16 mV larger than at threshold.)

The effect of synaptic saturation (Fig. 4.11) at high-impedance sites, such as the distal, apical, or the basal tree, complicates this simple analysis. Many distal sites are prevented from generating sufficiently large somatic EPSPs to ever reach threshold. At the soma, the input (slope) conductance of 60.6 nS is large enough compared to the peak conductance change of a single synapse (0.5 nS) for saturation not to play a role. Furthermore all voltage-dependent currents but sodium activation can be considered to be stationary, thereby approximating the conditions under which V_{th} was estimated. We conclude that the unitary EPSP amplitude can be used to predict the voltage threshold remarkably well (see also Koch, Bernander, and Douglas, 1995).

Applying the same logic to estimate I_{th} predicts that nine such somatic inputs, each delivering 33 pA (Fig. 18.1C), suffice to reach I_{th}, underestimating the actual number by a factor of 7. But, as we pointed out in Sec. 17.3.4, the concept of a current threshold is only applicable for sustained inputs. The highly transient EPSCs cannot be used to estimate the threshold. (This was illustrated in Fig. 17.5A, in which the total synaptic current flowing onto the spike initiating membrane I_{soma} exceeds I_{th} numerous times without initiating a spike.)

We conclude that when considering rapid synaptic input, the somatic potential due to synaptic activity is a more relevant variable than the current obtained under voltage clamp for predicting when the cell will trigger an action potential.

TABLE 18.1
Voltage Attenuation of Synaptic Input

Position	$\tilde{K}_{ii}$	$\tilde{K}_{is}$	$\tilde{K}_{ss}$	L	L_{is}^v	$L_{is}^{v,t}$
Basal	387.79	18.47	16.48	0.24	3.32	4.32
Apical	341.68	10.47	16.48	0.56	3.48	5.07

Actual and expected voltage attenuation between the two synaptic sites used in Fig. 18.1 and the soma as computed from the stationary input ($\tilde{K}_{ii}$ in MΩ) and transfer resistance ($\tilde{K}_{is}$ in MΩ). $e^{L_{is}}$ is the attenuation expected from the electrotonic length measurement (Eq. 3.33); L_{is}^v is defined via Eq. 3.38 as the logarithm of the voltage attenuation. In an infinite cable $L = L^v$. Due to the pronounced asymmetries in the cable structure, these two measures here differ markedly. The natural logarithm of the actual voltage attenuation $L_{is}^{v,t}$, as defined by the peak EPSP values in Fig. 18.1, is larger than L_{is}^v due to the distributed capacitances that preferentially remove high temporal frequencies (in particular for the distal, apical input).

18.2 Massive Synaptic Input

Now that we have seen the effect of a single synapse, we are ready to deal with the much more physiological situation in which a hundred or more synapses are simultaneously activated. How will this affect the somatic potential and current and, ultimately, the firing frequency of the cell? Because of the nonlinear relationship between synaptically induced conductance changes and the membrane potential, this question is not easy to answer in general. Let us start by studying the situation when the neuron is bombarded with numerous individual excitatory and inhibitory inputs.

18.2.1 Relationship between Synaptic Input and Spike Output Jitter

When a cortical neuron is presented with the appropriate stimulus, it rapidly and reliably increases its probability of firing. Individual cortical neurons respond to a dynamic random dot motion stimulus with a highly reproducible temporal modulation of their firing rate, precise to a few milliseconds (Fig. 15.11). The time that it takes for such a cell to significantly modulate its firing rate—as determined by averaging over many presentations of an identical stimulus—is almost always less than 10 msec and occurs in neurons that are at least six synapses removed from the periphery. This precision is surprising because the propagation of an input through a multilayer network with continuous mean rates causes rise times to become increasingly shallow (due to the low-pass filtering effect treated in Sec. 14.2.4).

Yet, as the sensory triggered "wave" of activity propagates through many layers of the cortex, the signal does not appear to become appreciably rounded off, but is only delayed between consecutive stages by about 10 msec. This has been assessed directly by comparing the latency and rise time of neurons in different cortical areas to the same stimulus. In the case of an awake monkey, neurons in the primary visual cortex responded to a visual stimulus with an average 10–90% rise time of 8 msec while neurons in a subsequent processing stage, cortical area V4, show a virtually identical rise time of 7 msec but with an additional 26 msec latency (Marsalek, Koch, and Maunsell, 1997).

Let us address this important problem in the following manner. Suppose an instantaneous sensory event in the world triggers a volley of activity in n excitatory synaptic inputs. We will assume that the arrival time of the input is centered around $t = 0$ and that its standard deviation in time, here called input jitter, is σ_{in}. If we further suppose that these n synapses are sufficient to trigger one (or more) spikes (that is, $n > n_{th}$) we can compute the standard deviation in time, termed the output jitter σ_{out}, of the spike triggered in response to this input. This will help us understand to what extent temporal jitter in the input is preserved, amplified, or reduced by multiple processing stages.

We first deal with a simplified problem by assuming that the probability density of the arriving synaptic inputs $p_{in}(t)$ is the uniform density on the interval [0,1]. ($p_{in}(t) = 1$ for $t \in [0, 1]$, and 0 otherwise.) Different from before, we assume that each input only triggers a single synaptic event of postsynaptic strength a_e. These inputs are integrated onto the capacitance of a nonleaky integrate-and-fire unit (of voltage threshold $V_{th} = a_e n_{th}$). Together with the uniform density, this assures us that at t = 1 the voltage will be exactly na_e (in the absence of a threshold).

The probability that the voltage at time t has attained the value $n_{th}a_e$ is given by the *beta density* (Papoulis, 1984)

$$p_{out}(t) = \frac{(n+1)!}{n_{th}!\,(n - n_{th})!} t^{n_{th}} (1 - t)^{(n - n_{th})} . \tag{18.1}$$

The standard deviation of this distribution is

$$\sigma = \sqrt{\frac{(n_{th} + 1)(n + 1 - n_{th})}{(n + 2)^2(n + 3)}}. \tag{18.2}$$

In the case of the uniform probability density, $\sigma_{in} = 1/2\sqrt{3}$, allowing us to rescale Eq. 18.2 in terms of the output jitter,

$$\sigma_{out} = \sigma_{in} 2\sqrt{3} \sqrt{\frac{(n_{th} + 1)(n + 1 - n_{th})}{(n + 2)^2(n + 3)}}. \tag{18.3}$$

What happens if the number of synaptic inputs n greatly exceeds the number of inputs required to bring the cell to fire n_{th}? Making the approximations that $n + 3 \approx n + 2 \approx n$, that $n + 1 - n_{th} \approx n$, and $n_{th} + 1 \approx n_{th}$, we arrive at

$$\sigma_{out} \approx \sigma_{in} 2\sqrt{3} \frac{1}{n} \sqrt{\frac{V_{th}}{a_e}}. \tag{18.4}$$

Or, the output jitter is inversely proportional to the number of excitatory synaptic inputs. This makes sense, since the time it takes to reach threshold will be proportional to the drift that is dictated by n. The jitter in the time to threshold is inversely related to the drift. Of course, our derivation only holds under conditions when the membrane leak can be neglected (that is, when most of the synaptic input arrives within a fraction of τ).

Detailed simulations using a leaky integrate-and-fire model as well as our customary layer 5 pyramidal cell compartmental model bear this out (Fig. 18.3A; for the details, consult Marsalek, Koch, and Maunsell, 1997). In the presence of massive excitatory input ($n = 250$, with $n_{th} = 66$; see previous section), whose arrival time has a Gaussian distribution in time, the relationship between σ_{in} and σ_{out} is a linear one, with a slope of around 0.116, as expected from the approximation in Eq. 18.4. Adding inhibitory synaptic input increases this slope, but not substantially (Fig. 18.3B). Any more inhibition and the cell fails to reach threshold in a growing fraction of all trials.

In the case of massive synaptic input, the temporal output jitter will be smaller than the input jitter. It follows that for a cascade of spiking cells, the output jitter converges to zero (Marsalek, Koch, and Maunsell, 1997; see also Abeles et al., 1994; Hermann, Hertz, and Prügel-Bennett, 1995). In real networks the jitter is unlikely to become vanishingly small since we neglect several additional sources of timing variability. The two most dominant sources are likely to be:

1. The inhomogeneous spike propagation times between consecutive layers of neurons due to variations in the diameter and length of the associated axons and axonal terminations (wiring jitter). We considered these axonal delays in Fig. 6.17 and found that they are small (on the order of ±0.5 msec).

2. Jitter in the delay between presynaptic spikes and the opening of the postsynaptic synaptic channels (synaptic jitter).

We conclude that the output jitter in a network with many layers of spiking neurons (such as in Abeles's (1990) synfire-chain model) converges to a fixed but small number, bounded by the "jitter" in the anatomical connections and in the synaptic transduction process.

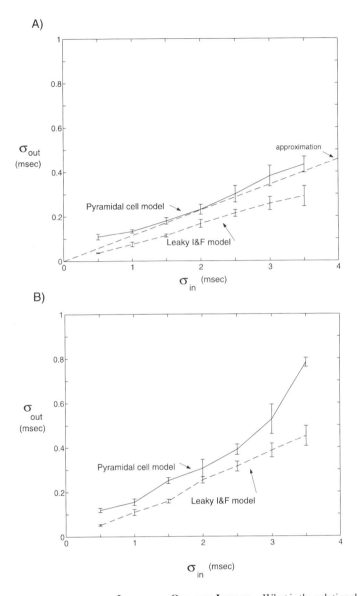

Fig. 18.3 RELATIONSHIP BETWEEN INPUT AND OUTPUT JITTER   What is the relationship between the standard deviation in time of synaptic input σ_{in} and the standard deviation in time of the resultant output spike σ_{out}? Marsalek, Koch, and Maunsell (1997) treated this question using a leaky integrate-and-fire model and the more complex layer 5 pyramidal cell compartmental model. (**A**) 250 excitatory fast voltage-independent synaptic inputs are triggered once. The probability density function $p_{in}(t)$ is Gaussian with standard deviation σ_{in}. For the integrate-and-fire model, synapses are assumed to be linear current sources, with $a_e = 0.23$ mV; $\tau = 10$ msec. For the compartmental model, synapses are of the fast AMPA type, distributed throughout the dendritic tree. In both cases, the effective voltage threshold $V_{th} = 16$ mV and $n_{th} = 66$. Equation 18.4 predicts a linear relationship between σ_{in} and σ_{out} with slope 0.116 (line labeled "approximation"). Error bars correspond to the standard deviation from five runs with 50 spikes each. (**B**) A qualitatively similar result is obtained if 62 inhibitory synapses are added (hyperpolarizing current synapses with $a_i = 0.23$ mV for the integrator and GABA$_A$ synapses for the more detailed model). Any more inhibition, and the cell frequently fails to trigger a spike. Thus, a spiking cell with a passive dendritic tree can reduce input jitter provided that n is much bigger than n_{th}. Reprinted by permission from Marsalek, Koch, and Maunsell (1997).

18.2.2 Cable Theory for Massive Synaptic Input

Standard one-dimensional passive cable theory (Jack, Noble, and Tsien, 1975; Rall, 1989) assumes that the electrotonic structure of the cell does not change in response to synaptic input. As long as synaptic input is treated as a de- or hyperpolarizing current this is certainly true. The independence of the electrical structure from the input allows us to characterize the spatio-temporal integrative properties of the cable or the cell in terms of the transfer and input impedances as well as space and time constants.

Yet as we emphasized from the beginning of this book, even the activity of a single synapse has an impact on the input impedance $K_{ii}(t)$ (e.g., Eq. 4.24). While the actual size of this effect is negligible for one or a few isolated synaptic inputs, it can dominate the behavior of the system for large inputs. This is particularly true given that patch-clamp recordings of pyramidal and Purkinje cells find very high values for the passive membrane resistance (on the order of 100,000 $\Omega \cdot$cm^2; see Appendix A).

Let us exemplify our argument with some numbers. Assume a small, spherical cell with a radius equal to $r = 15\ \mu$m. Its input conductance $G_{in} = 4\pi r^2 / R_m$ amounts to only 0.28 nS and its time constant to 100 msec for $R_m = 100,000\ \Omega \cdot$cm^2. Activation of a single synapse with $g_{peak} = 0.5$ nS will immediately, albeit briefly, triple the effective input conductance and reduce the time constant to a third of its original value. A second synapse will not find the same electrotonic structure present as the first one did. It is immaterial for this argument whether the synapse considered is excitatory or inhibitory, as long as it increases the local membrane conductance. The fact that synaptic inputs can significantly affect the electrotonic structure of the cell depends on the fact that R_m is high relative to the inverse of the synaptic-induced conductance change. For low values of R_m, say in the neighborhood of 2,000 $\Omega \cdot$cm^2, many more synaptic inputs would have to be activated in order for the synaptic conductance to dominate the conductance contributed by the passive membrane of the cell.

As expressed in a perfunctory manner in Sec. 4.9.1, Eq. 4.24 describes the dependency of $K_{ii}(t)$ on the exact timing of the synaptic input $g_{syn}(t)$. Clearly, for two or more synapses, the dependency can be considerably more complex, since the input impedance and other cable properties also depend on the relative timing between the inputs and whether or not they are correlated.

In order to study the effect of massive synaptic input on the cell, we will disregard such temporal effects by assuming that the inputs arriving at different synapses are independent of each other and replace the series of presynaptic spikes by an instantaneous rate f_i. We also disregard any detailed dynamics of the spiking by a "suitable" temporal averaging procedure that reduces the total synaptic input from n independent synapses to any one particular compartment to a single number,

$$\overline{g} = \sum_{i=0}^{n} \overline{g_i} \langle f_i \rangle \tag{18.5}$$

where $\overline{g_i}$ represents the integrated conductance for a single input,

$$\overline{g_i} = \int_{0}^{+\infty} g_i(t)dt . \tag{18.6}$$

When $g_i(t)$ has the form of an α function (Eq. 4.5), $\overline{g_i} = e g_{peak} t_{peak}$. If the synapse is of the NMDA type, this average conductance will be voltage dependent, rendering the following equations slightly more complex. Obviously, this averaging method does not incorporate any short-term synaptic changes observed in brain slices (Sec. 13.5.4), such as short-term depression and facilitation.

With this averaging procedure in place, we can replace the different types of synapses in each compartment (AMPA, NMDA, GABA$_A$, and so on) as well as the leak conductance by a single "effective" conductance,

$$G_{\text{eff}} = \sum_{i=1}^{n} \overline{g_i} \langle f_i \rangle + G_m A \qquad (18.7)$$

in series with a single effective battery,

$$E_{\text{eff}} = \frac{\sum_{i=1}^{n} E_i \overline{g_i} \langle f_i \rangle + E_{\text{leak}} G_m A}{G_{\text{eff}}} \qquad (18.8)$$

where A is the membrane area of the compartment needed to convert the leak conductance per unit area, G_m, into an absolute conductance. Equation 18.7 tells us that as the presynaptic firing frequency $\langle f_i \rangle$ increases, so does the postsynaptic membrane conductance, regardless of whether the synapses are excitatory or inhibitory. Only the conglomerate reversal potential E_{eff} is influenced by the signs and amplitudes E_i of the synaptic batteries: if inhibitory cells fire more strongly than excitatory ones, the effective synaptic reversal potential will be pulled toward more hyperpolarizing values and *vice versa*. We conclude that the effective membrane resistance should not be thought of as a fixed parameter, but as a dynamic variable that can change within a fraction of a second, in contrast to the intracellular resistivity and the membrane capacity, which appear much more difficult to modulate rapidly.

For a single infinite cylinder with the appropriate boundary condition, we can write down a slightly modified cable equation,

$$\frac{d}{4 R_i G_{\text{eff}}} \frac{\partial^2 V_m(x,t)}{\partial x^2} = \frac{C_m}{G_{\text{eff}}} \frac{\partial V_m(x,t)}{\partial t} + (V_m(x,t) - E_{\text{eff}}) \qquad (18.9)$$

which contains a single term accounting for all synaptic conductances. Defining the length constant of the cable (Eq. 2.13) as

$$\lambda = \sqrt{\frac{d}{4 R_i G_{\text{eff}}}} \qquad (18.10)$$

and the time constant as

$$\tau_m = \frac{C_m}{G_{\text{eff}}} \qquad (18.11)$$

we arrive back at the standard normalized cable equation (Eq. 2.7),

$$\lambda^2 \frac{\partial^2 V_m(x,t)}{\partial x^2} = \tau \frac{\partial V_m(x,t)}{\partial t} + (V_m(x,t) - E_{\text{eff}}). \qquad (18.12)$$

In the remainder of this chapter, we will deal with three distinct scenarios of time-averaged massive synaptic input. We start off by analyzing the effect of diffuse synaptic activity that occurs throughout the cell before we turn toward an analysis of the effect of massive excitatory and shunting inhibitory synaptic activity on the cell.

18.3 Effect of Synaptic Background Activity

Neurons, just like people, do not exist in isolation but are embedded within a tightly interwoven network of other nerve cells. A typical neocortical pyramidal cell is the recipient of anywhere between 5000 and 20,000 synapses from other neurons, while Purkinje cells in the cerebellum may receive up to 200,000 synapses. In the lightly anesthetized animal as well as in the awake behaving animal, these cells are "spontaneously" active, which means in practice that they generate action potentials that cannot readily be accounted for by the presence of any simple sensory stimulus (such as a bright bar or a loud tone). The origin of this spontaneous activity is presently unknown but could be due to several sources: (1) the remnants of spontaneous activity at the sensory periphery that has percolated to the more central stages (e.g., photon shot noise in the photoreceptors), (2) the spontaneous spiking due to channel fluctuations (Sec. 8.3.2), or (3) the spontaneous release of synaptic vesicles in the absence of significant presynaptic activity.

Experimentally recorded values of the spontaneous firing in cortex range, from low values of 0.25–2.5 spikes per second in the anesthetized cat visual cortex (Gilbert, 1977; Leventhal and Hirsch, 1978) to larger values of 5–10 Hz in extrastriate areas, such as the cat motor cortex (Woody, Gruen, and McCarley, 1984), the rat sensory-motor cortex (Bindman and Prince, 1983), the auditory cortex in rhesus monkeys and baboon (Abeles, 1990), and the inferior temporal cortex of the behaving monkey (R. Desimone, personal communication).

These nonzero rates of background activity are in sharp contrast to the situation prevailing for brain slices or for cultured neurons. The removal of afferent fibers (by cutting to extract the slice from the brain), combined with the absence of the right combination of neuromodulators and growth factors, makes neurons rather reticent about firing in the absence of a direct electrical or pharmacological stimulus.

What consequence does this background activity, expressed via the time-averaged rate $\langle f_b \rangle$, have on any one neuron? Barrett (1975) first brought up the possibility that synaptic background activity might affect the electrical properties of α motoneurons (see also Rall, 1974). The subject remained dormant until a number of groups took up the issue via detailed compartmental simulations (Holmes and Woody, 1989; Bernander et al., 1991; Rapp, Yarom, and Segev, 1992) and analytical investigations (Abbott, 1991).

We will summarize the results of these studies by focusing on the changes in R_{in}, τ_m, L, and V_{rest} wrought by varying the background firing rate $\langle f_b \rangle$.

18.3.1 Input Resistance

The input resistance of the finite cable (looking toward the sealed end boundary at the other terminal) is specified by Eq. 2.21 as

$$R_{in} = \frac{2 R_i^{1/2}}{\pi d^{3/2}} \frac{\coth(L)}{G_{eff}} , \tag{18.13}$$

where $L = \ell/\lambda$ is the electrotonic length of the cable. In the limit of an infinite high background activity, that is, for $\langle f_b \rangle \to \infty$, R_{in} decreases as $\langle f_b \rangle^{-1/2}$. Basic intuition tells us that additional synaptic input, independent of whether excitatory or inhibitory, increases the membrane conductance and thereby drives down the effective membrane resistance R_{eff} (Fig. 18.4A).

The same plot also shows R_{in} for the compartmental model of the pyramidal cell with the standard complement of voltage-dependent somatic membrane conductances ("active" neuron) and for the same model but in the absence of these active conductances ("passive" model). (The geometry of the single cylinder was adjusted to coincide with the passive pyramidal cell model at $\langle f_b \rangle = 0$.) All three models share the same features: as $\langle f_b \rangle$ is turned up from 0 to 2 Hz, the input resistance drops by a factor of 10, from 193 to 20 MΩ, in the passive case, and by a factor of 4, from 50 to 12 MΩ, in the active case. The deviation

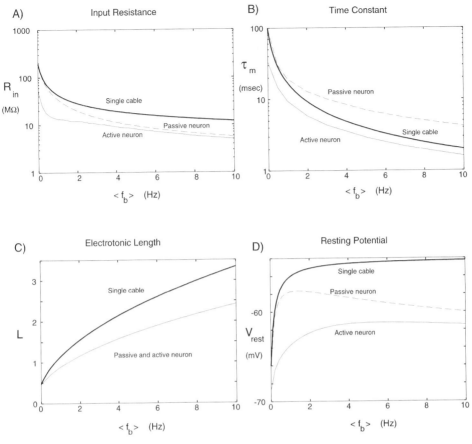

Fig. 18.4 EFFECT OF SYNAPTIC BACKGROUND ACTIVITY ON R_{in}, τ_m, L, AND V_{rest} Illustration of the effect of varying the synaptic background activity $\langle f_b \rangle$ on four variables characterizing spatio-temporal integration in three systems; a finite, passive cylinder (3060 μm long and 5.8 μm thick; with sealed-end boundary condition; bold line labeled "single cable"), the complete model of the layer 5 pyramidal cell with active conductances ("active" neuron) and the same model but with all active somatic conductances removed ("passive" neuron). All cases show the same effect: as $\langle f_b \rangle$ increases, the membrane becomes more and more leaky, and R_{in} as well as τ_m decrease and the electrotonic length of the passive cable or the electrotonic distance between a point in the layer 1 portion of the distal apical tree and the soma in the pyramidal cell becomes ever larger. $\langle f_b \rangle$ is very close to zero in brain slices, while $\langle f_b \rangle > 1$ Hz corresponds to the situation in the living brain. The behavior of the cable is given by the equations in the text. R_{in}, τ_m, and V_{rest} are all measured at the soma. Inhibitory and excitatory synapses are assumed to have identical rates $\langle f_b \rangle$. Note the logarithmic scale on the upper two panels. Reprinted by permission from Bernander (1993).

of the compartmental model from the cylinder at high values of $\langle f_b \rangle$ is largely explained by the inhomogeneous synaptic distribution along the pyramidal cell (inhibitory synapses cluster in a neighborhood of the soma while excitation is more distal). Further differences result from the voltage-dependent components.

The dependency of R_{in} and other electrical parameters on $\langle f_b \rangle$ becomes more complex in the presence of voltage-dependent NMDA input. As discussed in Sec. 17.1.2 and Fig. 17.3, incorporating NMDA conductances into the membrane can have the paradoxical effect of reducing the slope conductance. For the situation discussed here, this implies that under certain conditions the input resistance can actually increase as the background firing rate is increased (Bernander, 1993).

18.3.2 Time Constant

Equation 18.11 characterizes the dependency of the passive time constant on $\langle f_b \rangle$. We expect that as $\langle f_b \rangle \to \infty$, τ_m goes to zero as $1/\langle f_b \rangle$. The fractional decrease in the somatic time constant is similarly large in the compartmental model: from 100 to 12.8 msec for the passive case and from 33.7 to 5.9 msec in the active case (Fig. 18.4B). This decrease in τ_m as $\langle f_b \rangle$ increases implies that the pyramidal cell becomes much more sensitive to the degree of temporal dispersion in any synaptic input. To demonstrate this, Bernander and colleagues (1991) computed n_{th}, the average number of excitatory non-NMDA synapses, distributed throughout the dendritic tree, necessary to trigger a spike. These synapses were either all activated simultaneously or spread out over a 25 msec extended time window (with each synapse activated only once).

It is apparent from Fig. 18.5 that for the desynchronized case n_{th} increases significantly as $\langle f_b \rangle$ increases. Under slice conditions, 115 synchronized synapses are necessary to bring the cell above threshold, whereas 145 are needed if the input is desynchronized. This small difference between synchronized inputs and inputs arriving smeared out in time is due to the long integration period of the cell ($\tau_m = 100$ msec). For a 1 Hz background activity, 113 synchronized or 202 desynchronized inputs are needed to fire the cell. At $\langle f_b \rangle = 7$ Hz, when 35,000 synaptic events are bombarding the cell every second, 3.5 times as many

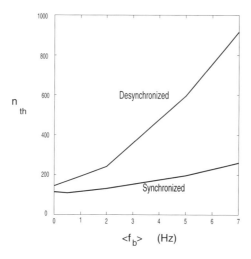

Fig. 18.5 TEMPORAL INTEGRATION The ability of the standard layer 5 pyramidal cell to distinguish coincident arriving synaptic input from a desynchronized one increases as the spontaneous background activity $\langle f_b \rangle$ increases and τ_m decreases (Fig. 18.4B). The minimal number of excitatory voltage-independent synaptic inputs n_{th} (with $g_{peak} = 0.5$ nS) needed to trigger a spike at the soma as a function of $\langle f_b \rangle$. The synapses are spread over the dendritic tree in accordance with the known distributions and are either all activated simultaneously or activated once over a 25-msec-wide window, independently of each other. For $\langle f_b \rangle \approx 0$ (a situation prevalent for most *in vitro* studies), the difference between synchronized and desynchronized inputs is minor but becomes substantial when the overall network activity is high. The average background activity can therefore be seen as the signal controlling the temporal tuning properties of the cell. Each simulation is run numerous times using a different distribution of synapses. Reprinted by permission from Bernander et al., (1991).

desynchronized inputs are needed to trigger the cell than if the inputs arrive together. As $\langle f_b \rangle$ increases, the cell becomes more and more selective to differences in the arrival times of synaptic inputs.

18.3.3 Electroanatomy

According to Eq. 18.10, the electrotonic length of a finite cable should increase as $\langle f_b \rangle^{1/2}$. As the membrane becomes more and more leaky, distant points are, electrically speaking, less and less coupled to the soma. Because the dendritic tree contains no active conductances, the electrotonic distance from the soma to a point in the apical bush (about 1 mm away from the soma) is the same in the active and the passive model. Given the square root dependency of L on $\langle f_b \rangle$, this distance, computed as the sum of all branch segments on the path between the location and the soma increases by a factor of 2.5 as $\langle f_b \rangle$ increases from 0 to 2 Hz.

A graphic demonstration of the effect of varying $\langle f_b \rangle$ on the electroanatomy of the cell is given in Fig. 18.6 using the graphical morphoelectronic transform (MET) (see Sec. 3.5.4) of Zador (1993) and Zador, Agmon-Snir, and Segev (1995). Each dendritic compartment is "stretched" or "shrunk" such that its length in the graph is proportional to its electrotonic

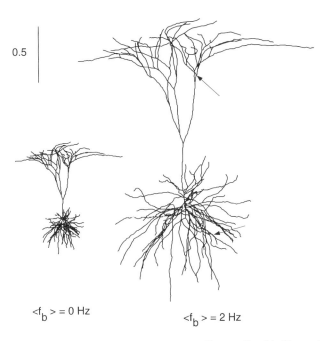

0.5

$<f_b> = 0$ Hz

$<f_b> = 2$ Hz

Fig. 18.6 VARIABLE ELECTROANATOMY OF PYRAMIDAL CELL   Graphic illustration of the change of the electrotonic dimensions of the pyramidal cell with synaptic background activity with the aid of a morphoelectronic transform. The extent of the individual dendrites is proportional to their electrotonic length; the distance of each compartment from the soma is proportional to the electrotonic distance of that point from the soma. In the absence of any background activity—approximating the conditions prevalent in cultured or in slice neurons—the pyramidal cell is very compact. As $\langle f_b \rangle$ is increased to 2 Hz, distal sites on long and thin dendrites are more and more decoupled from the soma. The scale bar corresponds to the distance over which the sustained voltage decays by $e^{0.5}$ in an infinite cable. The arrows point to the location of the apical and basal synapses used in Fig. 18.1. Reprinted by permission from Bernander et al., (1991).

length L_i. L increases from 1.2 to 2.6 for the most distal compartment in layer 1 and from 0.2 to 0.7 for a synapse at the tip of a basal dendrite. That basal dendrites stretch more than apical ones is due to their higher synaptic innervation and the corresponding lower effective values of R_m.

Yet another way to visualize the change in electrotonic properties is to plot the somatic EPSPs evoked by a fast non-NMDA excitatory synapse located either in the distal apical tree, along a basal dendrite, or at the soma (Fig. 18.7). While the amplitude of the somatic and basal EPSPs decrease by a factor of 2 or 3 as $\langle f_b \rangle$ changes from 0 to 5 Hz, the effect on the distal apical synaptic input is much more dramatic. For $\langle f_b \rangle = 5$ Hz, almost no deviation from the somatic resting potential is seen.

18.3.4 Resting Potential

Of the four variables considered here, only E_{eff} depends on the balance of excitatory and inhibitory input (Eq. 18.8). For large values of $\langle f_b \rangle$, V_{soma} depolarizes by close to 10 mV (Fig. 18.4D). This is a consequence of the fact that 80% of the synapses are excitatory and the assumption that the synaptic background activity is the same for excitatory as well as for inhibitory afferents. The somatic potential does not increase indefinitely with $\langle f_b \rangle$, but saturates due to the interaction between the net excitatory and inhibitory synaptic currents and the voltage-dependent potassium currents I_A and I_M, driving the potential to -95 mV (for more details, see Bernander, 1993).

It is known that the resting potential differs significantly between neurons recorded *in vivo* and *in vitro*. Numerous slice studies of pyramidal cells report resting potentials in the range of -84 to -67 mV (with a mean of -74 mV; Thomson, Girdlestone, and West,

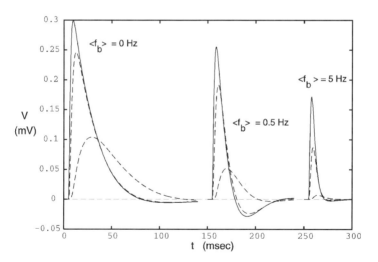

Fig. 18.7 SYNAPTIC INPUT IN THE PRESENCE OF BACKGROUND ACTIVITY Somatic EPSP in response to activation of a single non-NMDA synapse at one of three locations in the pyramidal cell (from top to bottom: soma, basal, and apical tree; see Fig. 18.6) at three different levels of synaptic background activity. The largest EPSPs are obtained under slice conditions ($\langle f_b \rangle = 0$). For 25,000 synaptic events per second ($\langle f_b \rangle = 5$ Hz), the peak somatic, basal, and apical EPSPs are reduced by factors of 1.7, 2.8, and 14.9 relative to their peak values at $\langle f_b \rangle = 0$. The middle set of EPSPs corresponds to the panel of Fig. 18.1B. Notice the much shorter rise and decay times at higher background frequencies. Reprinted in modified form by permission from Bernander (1993).

1988; Mason and Larkman, 1990; Spain, Schwindt, and Crill, 1990; Hirsch and Gilbert, 1991; Mason, Nicoll, and Stratford, 1991), while *in vivo* V_{rest} is usually in the range of -81 to -57 mV (with a mean of -64.5 mV; Bindman and Prince, 1983; Bindman, Meyer, and Prince, 1988; Holmes and Woody, 1989; Pockberger, 1991). Whether this difference is caused primarily by the difference in synaptic background activity remains open.

18.3.5 Functional Implications

The principal effects of modulating the synaptic background activity on the spatio-temporal properties of individual neurons are clear-cut: as $\langle f_b \rangle$ increases from no activity to physiological values in the 5–10 Hz range, input resistance, and space and time constants decrease. Functionally, the cell becomes more and more extended and sensitive to temporal synchrony. This conclusion is not dependent on the numerical details and occurs in simple, analytical cable models (Abbott, 1991; Bernander, 1993) as well as in detailed compartment models of Purkinje (Rapp, Yarom, and Segev, 1992) and pyramidal cells (Bernander et al., 1991). This phenomenon is contingent upon two key assumptions: (1) that the effective membrane resistance in the absence of synaptic background activity is high and (2) that on the order of several thousand (or more) spontaneous synaptic events occur every second. If these conditions are not met, background activity will have either no or only a negligible effect on the spatio-temporal properties of the cell. Note well that these effects would not be observed if synaptic inputs were treated as current inputs.

While it is undoubtedly important to understand the effect of $\langle f_b \rangle$ at the level of an individual cell, a more functionally relevant question is to what extent the spontaneous firing activity is under the control of other brain systems. In particular, to what extent and how fast can neuromodulators up or down regulate $\langle f_b \rangle$ for entire groups of neurons? If they can, the phenomena discussed here could have a number of interesting implications for neuronal information processing strategies.

For instance, in the dark few cells in the visual cortex will be active and the membrane time constant τ_m will be long, enabling the cell to integrate the input over a long time. Conversely, for bright, high contrast visual stimuli, the overall network activity may be much higher, leading to a small value of τ_m and short integration times. This *adaptive gain-control* mechanism is somewhat akin to that used by the network of coupled rods in the retina (Detwiler, Hodgkin, and McNaughton, 1980). As exemplified in Fig. 18.6, larger values of $\langle f_b \rangle$ decouple distal sites from the soma. Because the apical tufts in layers 1 and 2 constitute the dominant target zones for cortico-cortical feedback connections (Friedman, 1983; Rockland, 1994; Salin and Bullier, 1995), a large degree of afferent activity would decouple this feedback by increasing its distance from the soma, thereby making the cell more responsive to direct sensory input arriving on the proximal parts of the apical trunk.

It is obvious that the properties of individual neurons influence the behavior of the embedding neuronal network, such as whether or not it converges to a fixpoint and its convergence time. We have here an instance of the converse, where a collective property of the network—its average activity—modulates the properties of individual members of the network. Particularly intriguing is the fact that the temporal resolution of neurons increases as the mean network activity increases. This illustrates the strong two-way interdependency of the many structural levels of neurobiological hardware, in stark contrast to computer hardware. An intriguing but open question is the issue of stability: under what conditions can a network that modulates the electrical properties of its elements arrive at a stable equilibrium point, that is, how can homeostasis be achieved?

18.4 Relating Synaptic Input to Output Spiking

So far, we have always evaluated the effect of synaptic input to the somatic membrane potential, since it is here—or in the axonal hillock close to the cell body—that the output of the cell—spikes—is generated. But, ultimately, we wish to understand the relationship between synaptic input and the firing activity of the cell.

This is a difficult problem due to the nonstationary nonlinearities at the soma. If EPSPs arrive while the somatic potential hovers just below the spiking threshold, it might be enough to push the potential above threshold, while the arrival of EPSP following an action potential might have a negligible effect on the output frequency. Furthermore, it also makes a difference whether the synapse is activated only once or repeatedly. We will sidestep these problems by adopting the stance we did in the previous sections, that is, we assume that the synaptic inputs are independent of each other and that each individual rate can be replaced by a suitable average (Eq. 18.5) which we assume for simplicity's sake to be directly proportional to the presynaptic firing frequency.

As opposed to conventional cable theory, we will not compute the somatic potential evoked by synaptic input but the current $I_{syn,s}$, flowing longitudinally down the dendrite into the soma. It is this current that will be seen by the spiking mechanism at the cell body and axon hillock and that will ultimately cause the cell to generate axon potentials. Knowledge of $I_{syn,s}$, in conjunction with the f–I curve, allows us to go directly from synaptic input to firing output. We here follow the exposition of Bernander, Douglas, and Koch (1994).

18.4.1 Somatic Current from Distal Synaptic Input

We follow Abbott (1991) in computing the current flowing out of one terminal of a single cable when one or more synapses are activated at the other terminal at $X = L$. (The cable has the electrotonic length $L = \ell/\lambda$ and diameter d.) We can think of the $X = 0$ terminal as being connected to an RC compartment representing the soma. Activating the synapses causes ionic current to flow into the cable and along the transversal intracellular resistivity toward the low-impedance cell body.

Because we are averaging over a fraction of a second or longer we can neglect the capacitive term in our modified cable equation (Eq. 18.12), and—following Eq. 2.2— express the current $I_{syn,s}$ flowing into the soma by the derivative of the voltage along the cable divided by the longitudinal resistance,

$$I_{syn,s} = -\frac{1}{r_a} \frac{dV}{dx}\Big|_{X=0}. \tag{18.14}$$

The excitatory synapses increase the membrane conductance at $X = L$ by $\overline{g_e} \langle f_e \rangle$ (with $\langle f_e \rangle$ the presynaptic firing rate of the excitatory synapses and $\overline{g_e}$ the time-averaged conductance per area; Eq. 18.6) in series with the synaptic reversal potential E_e. In order to compute the current, we need to know the voltage. As in Eq. 2.18, the voltage in this finite cable can be expressed as

$$V(X) = \alpha \cosh(L - X) + \beta \sinh(L - X). \tag{18.15}$$

Solving this requires two boundary conditions. The voltage at the synaptic terminal is given by the usual synaptic equation (Eq. 4.12) as

$$V(L) = \frac{R_{in}\overline{g_e} \langle f_e \rangle E}{1 + R_{in}\overline{g_e} \langle f_e \rangle}. \tag{18.16}$$

As argued in the preceding chapter, the spike mechanism at the soma acts as a sort of voltage clamp, preventing the membrane potential from making long-term excursions beyond V_{th}. For the sake of convenience, Abbott (1991) clamps the potential at the $X = 0$ terminal to 0. This also specifies the input resistance R_{in} at the synaptic terminal as the resistance associated with a killed-end terminal (Eq. 2.24),

$$R_{in} = R_\infty \tanh(L) = r_a \lambda \tanh(L) \,. \tag{18.17}$$

Remembering that the derivative of $\sinh(L - X)$ is $\cosh(L - X)/\lambda$ and that $\sinh^2(x) - \cosh^2(x) = -1$, we arrive at the following expression for the current flowing into the voltage-clamped terminal at $X = 0$,

$$I_{syn,s} = -\frac{\overline{g_e} \langle f_e \rangle E_e}{\cosh(L) + \lambda r_a \overline{g_e} \langle f_e \rangle \sinh(L)} \,. \tag{18.18}$$

If the input is small relative to the local input resistance, the current is linearly related to the input frequency. As $\langle f_e \rangle$ increases, the local postsynaptic potential starts to saturate, limiting the amount of current that enters through the synapse. Finally, the driving potential $E_e - V(L)$ is close to zero and no amount of presynaptic firing activity will push any additional current through the synapse and toward the other terminal (Fig. 18.8). The limiting current (in the sense that it cannot be exceeded without amplification) at the $X = 0$ terminal becomes

$$I_{syn,s}^{sat} = -\frac{\pi d^2 E_e}{4\lambda R_i} \frac{1}{\sinh(L)} \,. \tag{18.19}$$

Two important remarks. The saturation of the net current delivered by the synaptic input to the soma has very little to do with current losses along the cable. If no current leaks across the membrane by postulating $G_{eff} \to 0$, $I_{syn,s}^{sat}$ only increases by a factor of $\sinh(L)/L$.

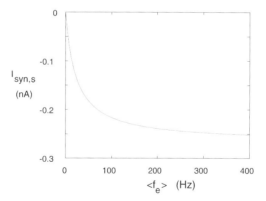

Fig. 18.8 LONGITUDINAL CURRENT IN A SINGLE CABLE Excitatory synaptic input is applied to one end of the same finite cable (of electrotonic length $L = 1.138$ used in Fig. 18.4), and the other terminal, conceptually corresponding to the soma, is clamped to the resting potential. The longitudinal current $I_{syn,s}$ flowing from the synapses into the $X = 0$ terminal clamped to 0 (Eq. 18.18), is computed as a function of the averaged presynaptic firing rate of the excitatory input $\langle f_e \rangle$. The saturating characteristic has little to do with current leaking out across the membrane. Indeed, preventing any current from leaking across the membrane only increases the maximal current $I_{syn,s}^{sat}$ by 23%. The culprit is local saturation of the membrane potential at the synaptic terminal. No amount of synaptic input can generate more current $I_{syn,s}$ (unless the membrane contains amplifying currents, as discussed in the following chapter).

Also, $I_{syn,s}$ does *not* correspond to the current that is computed with the aid of the transfer resistances (see the closing comments to Sec. 3.5.3).

18.4.2 Relating f_{out} to f_{in}

The key idea is to assume that the spike generation process at the cell body and axonal hillock can be treated as a stationary nonlinearity whose exact form is given by the discharge curve, that is, by the plot of the output frequency f_{out} against the amplitude of the injected current step (see Figs. 17.10A and 18.10A). It does not matter to this spike triggering mechanism whether the current it sees is injected from a microelectrode or is delivered from a synapse via the dendritic tree. All that is relevant is that a particular current I can be associated to a particular output spiking frequency along the $f–I$ curve. We incorporate the spike frequency adaptation into our consideration by directly making use of the fully adapted $f–I$ curve.

Knowing the relationships between the steady-state input frequency $\langle f_{in} \rangle$ and the somatic current $I_{syn,s}$ (as in Eq. 18.18) on the one hand, and the adapted $f–I$ curve on the other allows us to derive an unique function relating $\langle f_{in} \rangle$ to f_{out},

$$f_{out}(\langle f_{in} \rangle) = f_{out}(I_{syn,s}(\langle f_{in} \rangle))$$ (18.20)

(see Fig. 18.9).

Bernander Douglas, and Koch (1994) use this procedure to derive the input-output relationship for the layer 5 pyramidal cell. $I_{syn,s}$ is estimated on the basis of two different

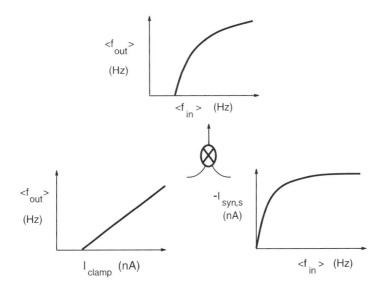

Fig. 18.9 COMPLETE MEASURE OF SYNAPTIC EFFICIENCY   Schematic illustrating the procedure linking the time-averaged presynaptic input frequency $\langle f_{in} \rangle$ to the adapted output frequency f_{out} of the cell via Eq. 18.20. It requires knowledge of the relationship between the presynaptic firing frequency $\langle f_{in} \rangle$ and the current flowing from the synapse into the soma $I_{syn,s}$ (lower right) and the discharge curve $f–I$ (lower left; here a linear function, although this is not necessary). The former function accounts for properties of the synapses and the dendritic tree, the latter for the somatic spike-generating properties. The resulting function $f_{out}(\langle f_{in} \rangle)$ accounts for the synaptic conductance input, synaptic saturation, and active, voltage-dependent membrane conductances. The intermediate variable is current, and not—as in the standard cable equation—voltage. Reprinted by permission from Bernander (1993).

methods. Knowledge of $I_{syn,s}$ requires that the somatic membrane potential $V_m(t)$ be known. In the derivation of Eq. 18.18 it was assumed that $V(0)$ was clamped to 0. When the cell spikes, this is patently wrong. Since the considerations of this section are predicated on using a time-averaged input frequency, why not use the time-averaged membrane potential $\langle V_m \rangle$ of Eq. 17.12? As we discussed in Sec. 17.5.2, this quantity changes only little for large swings in spiking activity. Indeed, the cumulative action of the voltage-dependent somatic currents is to clamp, albeit imperfectly, $\langle V_m \rangle$ to around -50 mV, corresponding to moderate values of f_{out} in the 50-Hz range. Bernander and colleagues (1994) compute the clamp current required to hold the somatic potential to -50 mV for a fixed synaptic input and subtract the clamp current needed to hold V_{soma} to -50 mV in the absence of any synaptic input. This method was also used experimentally to measure $I_{syn,s}$ in response to a constant synaptic input in α motoneurons (Powers, Tobinson, and Konodi, 1992). In the second method, $I_{syn,s}$ is computed directly (without voltage clamping the soma) as the current flowing between the first segment of the apical tree and the soma. Both methods give identical results.

As can be seen in Fig. 18.10B and C, when 500 excitatory non-NMDA synapses are either placed directly onto the cell body or spread throughout the basal dendrites (almost exclusively restricted to layer 5), $I_{syn,s}$ is linear in $\langle f_{in} \rangle$ over the range of interest. Saturation only becomes evident as the same 500 synapses are moved to layer 4 or to the more distal parts of the apical tree in superficial layers. When all inputs are confined to the superficial layers 1, 2, and 3, $I_{syn,s}$ is at most 0.65 nA. The saturation effect is even more extreme when synaptic input is restricted to the top two layers 1 and 2. Even if all 500 synapses are activated at an unphysiological 500 Hz, at most 0.25 nA of current reaches the soma (and this across an apical dendrite that has an unusually thick trunk of about 4.4 μm). This strong saturation is due to the fact that the distal input is associated with high input impedances, driving the local potential quickly to the synaptic reversal potential. No amount of further synaptic input can increase the synaptic current. Saturation would *not* occur if synaptic input were treated as a pure current. As in the case of the single cable, the minuscule current contributed from distal sites is not due to leakage through the membrane as the current spreads to the soma. Indeed, eliminating leakage by setting $R_m \to \infty$ causes the maximal current to increase by only 2% above base level.

In the last panel in Fig. 18.10, these somatic currents are translated on the basis of the adapted discharge curve (Fig. 18.10A) into an output firing rate. Because of the current threshold, synaptic input that delivers less than $I_{th} = 0.29$ nA current to the soma will be cut off. For instance, by itself, input to layers 1 and 2 will not lead to any maintained spike discharge, no matter how big the synapses and how vigorous their presynaptic activity, while the addition of synapses to layer 3 leads to a very weak discharge. And yet, these dendrites make up 26% of the surface area of the tree. Above threshold, the adapted $f-I$ curve is reasonably linear (Eq. 17.10), with a slope of 50 spikes per second per nanoampere of current, leading to linear input-output relationships for synaptic input to the soma or the basal dendrites (contributing 62% of the total membrane area of the neuron). As a sanity check, Bernander and colleagues explicitly computed the adapted firing rate as a function of the input and superimposed this onto Fig. 18.10D, with little difference.

18.4.3 Functional Considerations

The standard measures of the synaptic efficiency of cable theory are voltage or charge attenuation (see Chap. 3). The advantage of using the total current $I_{syn,s}$ flowing from one or more synapses into the soma (Powers, Tobinson, and Konodi, 1992; Bernander, Douglas,

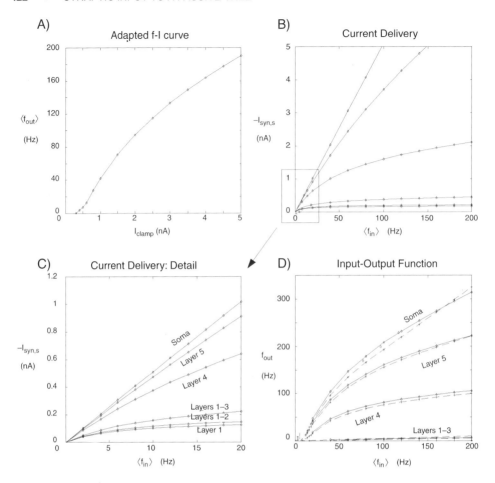

Fig. 18.10 RELATING SYNAPTIC INPUT TO THE FIRING FREQUENCY   Characterizing the input-output behavior of the layer 5 pyramidal cell using the procedure illustrated in Fig. 18.9. **(A)** For ease of comparison, its adapted f–I discharge curve. **(B)**, and **(C)** The input, 500 excitatory non-NMDA synapses, is placed onto the soma or the portion of the dendrites running through a specific cortical layer. The postsynaptic conductance increase is related linearly to the presynaptic firing rate $\langle f_{in}\rangle$. These synapses cause the current $I_{syn,s}$ to flow into the soma. The different curves correspond to synaptic placement (from top to bottom) onto the soma, layer 5, layers 1 to 3, layers 1 and 2, and layer 1 only. Similar to Eq. 18.18, dendritic saturation causes the distal input to saturate. **(D)** $I_{syn,s}(\langle f_{in}\rangle)$ is passed through the cell's discharge curve to predict the relationship between $\langle f_{in}\rangle$ and f_{out} (dashed curves). Computing this function by direct simulation (solid curves) leads to very similar results. Because $I_{syn,s}$ for synaptic input from layers 1 and 2 is less than I_{th} (about 0.295 nA), these synapses—assuming a passive dendritic membrane—cannot contribute by themselves to the maintained discharge. Experimentally, it is known that synaptic input to the apical tuft *can* trigger spikes in the cell body far away (Cauller and Connors, 1992, 1994). This argues for the existence of dendritic voltage-dependent currents that counteract saturation and amplify $I_{syn,s}$ (see the following chapter).

and Koch, 1994) is that this measure includes the effects of synaptic nonlinearities and interactions among different inputs. Furthermore, as we will show in the following chapter, $I_{syn,s}$ can also be derived in the presence of voltage-dependent dendritic conductances, when all other measures break down. If two synaptic sites are well separated such that the voltage in one does not readily influence the voltage at the other location (for instance, input into the basal dendrites coupled with input to the apical tuft), the total current at the soma will be the sum of the individual $I_{\text{syn},s}$ components.

Computing the current flowing into the soma is relatively straightforward as long as V_{soma} is fixed (see the derivation of Eq. 18.18). However, this is not the case if $I_{\text{syn},s}$ is above I_{th}, leading to spike generation. Under these conditions, $V_{\text{soma}}(t)$ moves on a stable limit cycle, making it difficult to characterize synaptic input using a single voltage measure, such as the peak somatic EPSP. Yet the method advocated here works despite these complexities (as witnessed by the close match in Fig. 18.10D between the predicted input-output curves and the actual ones) by making use of the time-averaged somatic potential $\langle V_m \rangle$ (Eq. 17.12). Its key assumption is that the system is in equilibrium, that is, the input can be characterized by an averaged input rate and the postsynaptic firing rate has adapted (typically within 100 msec). In a more sophisticated model, adaptation could be incorporated by making the f–I curve time dependent. Likewise, short-term synaptic facilitation and depression could also be dealt with.

As applied to the pyramidal cell, the main conclusion is that in the absence of voltage-dependent dendritic conductances distal synaptic input by itself can only deliver a very limited amount of current to the soma. In conjunction with other synaptic input, such as to the basal dendrites, distal input could increase the maintained rate by about 12 spikes per second.

In the neocortex, much of the synaptic input into layer 1 originates in higher cortical processing stages (Friedman, 1983; Rockland, 1994; Salin and Bullier, 1995). In the absence of amplifying dendritic conductances, cortico-cortical feedback could only weakly modulate the firing of the pyramidal cell. Yet Cauller and Connors (1992, 1994) have provided experimental evidence that synaptic input onto the layer 1 portion of layer 5 pyramidal cells in the somatosensory cortex can evoke action potentials. This, together with the evidence reviewed in the following chapter, argues that voltage-dependent conductances in the dendritic tree serve to amplify distal inputs.

18.5 Shunting Inhibition Acts Linearly

Inhibitory synaptic input with a reversal potential close or equal to the local resting potential, *shunting inhibition*, first mentioned toward the end of Chap. 1 and treated in considerable detail in Sec. 5.1.4, acts nonlinearly, akin to a division (Blomfield, 1974). This type of nonlinear operation has been proposed to be the crucial biophysical mechanism underlying retinal direction selectivity (Torre and Poggio, 1978; Koch, Poggio, and Torre, 1982). More recently, both Carandini and Heeger (1994) and Nelson (1994) have made shunting inhibition, activated by a recurrent feedback loop, the centerpiece of a gain normalization circuit in the visual cortex (Fig. 1.11) and in electric fish. However, none of these studies considered the effect that the spiking mechanism has on the divisive action of shunting inhibition. When this is done, a surprising conclusion emerges (Holt and Koch, 1997).

For the sake of clarity, let us first consider shunting inhibition to the leaky integrate-and-fire model of Eq. 14.7. Using our standard synaptic averaging technique, shunting inhibition can be mimicked by decreasing the leak resistance appropriately, with the new value given by

$$g_{\text{leak}} = g_i + \frac{1}{R} \tag{18.21}$$

where R is the value of the resistance in the absence of any input. The result can be inspected in Fig. 18.11A. Increasing g_i shifts the curve toward the right, with the slopes remaining constant, contrary to what one would expect based on models that only treat the subthreshold domain (as in Fig. 1.10 and throughout Chap. 5). Adding a refractory period and adaptation does not change this conclusion. The same is also true if one considers the much more realistic scenario of GABA$_A$ inhibitory receptors, with a reversal potential of -70 mV, which are distributed at and around the soma. As in the previous section, the total amount of inhibition is postulated to be proportional to the presynaptic firing frequency (Eq. 18.5). Holt and Koch (1997) compute the average f_{out}–f_{in} curve by varying the input frequency of the excitatory voltage-independent synaptic input distributed throughout the tree. Changing the amount of inhibition leads to a clear shift in the curves with little effect on the slopes (Fig. 18.11B).

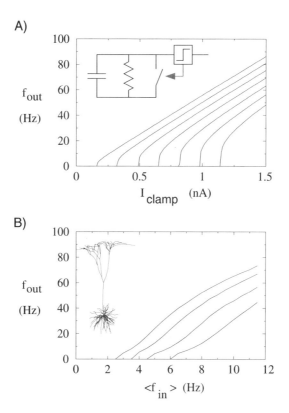

Fig. 18.11 SHUNTING INHIBITION AND SPIKING Shunting inhibition has a subtractive rather than a divisive effect on firing rates. This is demonstrated in two different single-cell models. **(A)** Discharge curves for a leaky integrate-and-fire unit with different values for the leak conductance $g_{leak} = g_i + 1/R$ (for $R = 62.5$MΩ). g_i is the amplitude of the inhibitory conductance change whose reversal potential is equal to the unit's resting potential (here zero). Varying g_{leak} in steps of 10 nS from 10 to 70 nS (from left to right) shifts the curve, rather than changing the slope of the discharge curve. **(B)** The same observation is made in the pyramidal cell model, with GABA$_A$ inhibition around the soma and excitatory voltage-independent input distributed throughout the cell. The fully adapted postsynaptic firing rate is plotted as a function of the average input frequency to the excitatory synapses for four different settings of presynaptic firing rates to the GABA$_A$ synapses (0.5, 2, 4, and 6 Hz, left to right). Reprinted by permission from Holt and Koch (1997).

This effect is most easily understood in the integrate-and-fire model. In the absence of any spiking threshold, the membrane potential in response to a synaptic input current of the shunting type

$$I_{\text{syn}} = g_i (V - E_i) = g_i V \tag{18.22}$$

rises until

$$V = \frac{I_{\text{syn}}}{g_{\text{leak}}} \tag{18.23}$$

(Fig. 18.12). Under these conditions, the steady-state leak current is proportional to the input current and the divisive effect is observed (Eq. 1.35). However, in the integrator model, V never rises above the spiking threshold V_{th}. No matter how large the input current, the leak current can never be larger than $V_{\text{th}} g_{\text{leak}}$. The effect of the leak conductance can be replaced by a current whose amplitude is equal to the time-average value of the current through the leak conductance: $\langle I_{\text{leak}} \rangle = g_{\text{leak}} \langle V \rangle$. This trick reduces the leaky integrate-and-fire unit to a perfect integrator. The cell will still fire at exactly the same rate because the same charge $\int (I_{\text{syn}} - I_{\text{leak}}) dt$ is deposited on the capacitor in the same time interval, although for the leaky integrator the deposition rate is not constant.

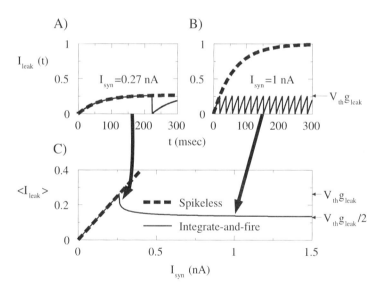

Fig. 18.12 WHY SHUNTING INHIBITION HAS A SUBTRACTIVE EFFECT (A) Time-dependent current across the leak conductance I_{leak} (in nA and equal to $V(t) g_{\text{leak}}$) in response to a constant 0.5-nA current injected into a leaky integrate-and-fire unit with (solid line) and without (dashed line) a voltage threshold V_{th}. The sharp drop in I_{leak} occurs when the cell fires, since the voltage is reset. **(B)** Same for 1-nA current. Note that I_{leak} in the presence of a voltage threshold has a maximum value well below I_{leak} in the absence of a voltage threshold. **(C)** Time-averaged leak current $\langle I_{\text{leak}} \rangle$ (in nanoamperes) as a function of input current, computed from Eqs. 18.24 and 18.25. Below threshold, the spikeless model and the integrate-and-fire model have the same $\langle I_{\text{leak}} \rangle$, but above threshold $\langle I_{\text{leak}} \rangle$ is reduced considerably. For I_{syn} just greater than threshold, the cell spends most of its time with $V \approx V_{\text{th}}$, so $\langle I_{\text{leak}} \rangle$ is high. For high I_{syn}, the voltage increases approximately linearly with time and V has a sawtooth waveform, as shown in **B**. This means that $\langle I_{\text{leak}} \rangle = (\max I_{\text{leak}})/2 = V_{\text{th}} g_{\text{leak}}/2$. Reprinted by permission from Holt and Koch (1997).

For a constant input that is just above threshold, $\langle V \rangle$ will be close to V_{th} and $\langle I_{leak} \rangle$ will be large. For larger synaptic input currents, the time-averaged membrane potential becomes smaller and smaller (since V has to charge up from the reset point) and, therefore, the time-averaged leak current *decreases* for increasing inputs (compare Fig. 18.12A and B). It can be shown that

$$\langle I_{leak} \rangle = I_{syn} \tag{18.24}$$

if $I_{syn} < V_{th} g_{leak}$. Otherwise

$$\langle I_{leak} \rangle = V_{th} g_{leak} \left(\frac{I_{syn}}{V_{th} g_{leak}} + \frac{1}{\ln \left(1 - V_{th} g_{leak} / I_{syn} \right)} \right). \tag{18.25}$$

For even quite moderate levels of I_{syn} just above $V_{th} g_{leak}$, the lower expression is approximately equal to $g_{leak} V_{th}/2$, independent of I_{syn} (Fig. 18.12C), and the leak conductance can be replaced—to a good first approximation—by a constant offset current. The discharge curve for the resulting perfect integrate-and-fire neuron is

$$f_{out} = \frac{I_{syn} - \langle I_{leak} \rangle}{C V_{th}} = \frac{I_{syn}}{C V_{th}} - \frac{g_{leak}}{2C}. \tag{18.26}$$

The same mechanism explains the result for the compartmental models. Here, the voltage does rise above the firing threshold but because the spike repolarization currents are large, any current from synapses flowing into the soma is ignored during the brief spike. The time-averaged somatic membrane potential (Fig. 17.10B) always remains close to V_{th}.

Two caveats. These results hold because inhibition is at or very close to the soma. For more distal inhibition, as for instance in retinal ganglion cells, the local potential at the inhibitory synapse it not limited by V_{th} and shunting inhibition will work in a more divisive mode. Yet for such distal sites, its effect on the spike triggering zone far away will be small. The results discussed here assume that the synaptic input changes slowly. It is at present unclear what the effect of very high, but transient input frequencies to the inhibitory synapses will be.

18.6 Recapitulation

The past few pages treated synaptic input arriving at the passive dendritic tree of a spiking cell. Only the cell body is assumed to contain voltage-dependent membrane conductances. We first discussed unitary EPSPs and EPSCs, concluding that the former can be used to predict quite well the voltage threshold V_{th} of the cell. We then investigated the exact relationship between the temporal jitter in a normally distributed set of synapses σ_{in} and the jitter in the output spike σ_{out}. Surprisingly, we discovered that $\sigma_{out} \propto \sigma_{in}$, with the constant of proportionality being much less than unity. Once again (e.g., Chap. 14) for strong synaptic input (that is, in the above threshold domain), the passive membrane time constant plays only a minor role in determining the temporal behavior of the cell. This has important implications, in particular for the faithful conservation of timing relationships across many layers of neurons (Abeles, 1990; Marsalek, Koch, and Maunsell, 1997).

For the remainder of the chapter we assume that synaptic input is not correlated at a fine time scale and that, indeed, we can define an average input frequency $\langle f \rangle$. This allows us to treat massive synaptic input in three separate cases.

Neurons receive between 10^3 and 10^5 synapses onto their dendritic tree from other neurons that fire spontaneously between 1 to 10 or more spikes per second in the behaving animal. This massive synaptic background activity has several effects on the spatio-temporal behavior of the cell. As this background firing rate goes up, the electrotonic dimensions of the cell increase (that is, distal synaptic input becomes more decoupled) and the input resistance and time constant decrease by one or more orders of magnitude. This implies, among other things, that the cell becomes more sensitive to temporal coincidences as the overall network activity increases. Given the rapid changes in firing activity in large parts of the brain, as assessed by EEG and other macroscopic techniques recording large-scale brain activity, this implies that the collective behavior of the network can directly influence the spatio-temporal integrative properties of individual neurons.

We also described a method that characterizes the efficiency of massive synaptic input, involving the current $I_{syn,s}$ that is delivered by synaptic input to the soma. Deriving $I_{syn,s}$ for synaptic input to a particular spatial region in the tree firing at a frequency $\langle f_{in} \rangle$ has the major advantage that the resultant function can be combined with the $f-I$ curve measured at the soma to yield a complete relationship between the sustained presynaptic input firing frequency $\langle f_{in} \rangle$ and the adapted output firing rate f_{out}.

When this method is applied to study synaptic input into the distal part of the apical tree of a pyramidal cell, it is seen that the current from these sites (1) saturates already at very small presynaptic firing rates and (2) influences the maintained firing rate only marginally. This saturation is only to a very minor extent caused by current leaking across the membrane between the synaptic site and the soma. As we will see in the next chapter, introducing voltage-dependent currents into the dendrite can both prevent saturation and amplify these small currents, effectively turning the apical tree into a better "wire."

Finally, we reexamined the effect of shunting inhibition. While it acts in a divisive manner in subthreshold cable models, it behaves in a linear, subtractive manner when spiking is accounted for. This is due to the fact that the average somatic membrane potential does not exceed the voltage threshold, limiting the maximal inhibitory current. Thus, at least for models where inhibition acts in a maintained manner at or close to the cell body, shunting inhibition does not implement a divisive normalization operation as postulated by many.

However, at the next level of organization, the small recurrently connected network, linear inhibition can act again in a divisive manner (Douglas et al., 1995). We conclude with the observation that a biophysical mechanism that implements one type of operation in the subthreshold domain might implement a quite different one in the suprathreshold domain. The morale is that it is dangerous in neurobiology to study any one mechanism at only a single, isolated level of complexity. Phenomena at multiple levels, such as ionic channel, synapse, dendrite, neuron, small network, and so on, interact in highly nonlinear and nonintuitive ways. This is, of course, a characteristic of any evolved systems and makes them so interesting.

19

VOLTAGE-DEPENDENT EVENTS
IN THE DENDRITIC TREE

So far, we worked under the convenient fiction that active, voltage-dependent membrane conductances are confined to the spike initiation zone at or close to the cell body and that the dendritic tree is essentially passive. Under the influence of one-dimensional passive cable theory, as refined by Rall and his school (Chaps. 2 and 3), the passive model of dendritic integration of synaptic inputs has become dominant and is taught in all the textbooks. Paradoxically, from the earliest days of intracellular recordings from the fat dendrites of spinal cord motoneurons with the aid of glass microelectrodes, active dendritic responses had been witnessed (Brock, Coombs, and Eccles, 1952; Eccles, Libet, and Young, 1958). Today, there exists overwhelming evidence for a host of voltage-dependent sodium and calcium-conductances in the dendritic tree. In the following section we summarize the experimental evidence and discuss current biophysical modeling efforts focusing on the question of the existence and genesis of fast all-or-none electrical events in the dendrites. We then turn toward possible functional roles of active dendritic processing.

One word of advice. It has been argued that linear cable theory as applied to dendrites and taught in the first chapters of this book is irrelevant in the face of all this evidence for active processing and can be relegated to the dustbin. However, this would be a mistake. Under many physiological conditions these nonlinearities will not be relevant. Even if they are, the resistive and capacitive cable properties of the dendrites profoundly influence the initiation and propagation of dendritic action potentials and other active phenomena. Thus, for a complete understanding of the events in active dendritic trees we need to be thoroughly versed in cable theory.

19.1 Experimental Evidence for Voltage-Dependent Dendritic Membrane Conductances

The issue of dendritic all-or-none electrical events must be seen as separate from the broader question of the existence and nature of active, that is, voltage-dependent, membrane conductances in the dendritic tree.

The last decade has seen a proliferation of studies of active dendrites, propelled by a number of simultaneous technical advances: (1) patch-clamp recordings in brain slices that allow for very high-quality records of somatic and dendritic electrical signals (Edwards et al., 1989; Stuart and Spruston, 1995), (2) calcium-dependent fluorescent dyes that enable the dynamics of intracellular free calcium in dendrites and even spines (e.g., Fig. 12.14) to be accurately recorded (Tsien, 1988, 1989), and (3) new forms of visible and infrared microscopy for high-resolution imaging of dendrites in brain slices as well as in the living animal (Fine et al., 1988; Denk, Strickler, and Webb, 1990; Dodt and Zieglgänsberger, 1994).

We cannot do justice to this gargantuan literature. Instead, we follow the excellent survey article by Mainen and Sejnowski (1998) and provide this brief synopsis. We will expand upon this, as necessary, further below.

1. Ionic channels from all major families of sodium, calcium, and potassium channels have been identified in dendrites. This includes both a very rapid and a sustained sodium current, low- and high-voltage activated calcium currents as well as sustained and transient potassium currents (Johnston et al., 1996; Hoffman et al., 1997).

2. Some types of channels are confined to specific compartments, such as the soma, initial segment, dendritic shaft, or even spines (Westenbroek, Merrick, and Catterall, 1989; Westenbroek, Ahlijanian, and Catterall, 1990; Westenbroek et al., 1992; Turner et al., 1994; Denk, Sugimori, and Llinás, 1995).

3. Voltage-gated channel densities estimated by patch-clamp recording in dendrites generally are less than 10 channels per square micrometer of membrane. This should be contrasted with the 1000–2000 fast sodium channels per square micrometer found at the axonal node of Ranvier (Ritchie, 1995). A possible exception appear to be A-like transient potassium channels in the distal dendritic tree of pyramidal cells (Hoffman et al., 1997).

19.1.1 Fast Dendritic Spikes

The earliest evidence for rapid all-or-none events in the dendrites comes from hippocampal pyramidal and cerebellar Purkinje cells (Spencer and Kandel, 1961; Llinás ct al., 1968; Llinás and Nicholson, 1971). The later observations were particularly convincing and conjured up a view of the Purkinje cell where the somatic membrane produces conventional fast sodium spikes, while the extended dendritic tree can generate fast sodium spikes, slower calcium spikes (see Fig. 6.1G), as well as so-called *plateau potentials*, long-lasting and graded depolarizations (Llinaás and Sugimori, 1980, 1992; Regehr, Konnerth, and Armstrong, 1992).

Of great interest has been the question of whether conventional action potentials, triggered at or near the cell body, propagate not only forward along the axon but also backward, into the dendritic tree. It was Ramón y Cajal (1909, 1991) who postulated the law of *dynamic polarization*, which stipulates that dendrites and cell bodies are the receptive areas for the synaptic input, and that the resulting output pulses are transmitted unidirectionally along the axon to its targets.

The classical description of spiking in a neuron derives from motoneurons (Coombs, Curtis, and Eccles, 1957; Fatt, 1957; Fuortes, Frank, and Becker, 1957). Here, action potential initiation normally occurs in a proximal segment of the axon, in a specialized region between the soma and the axon proper, called *axon hillock* or *initial segment*, that is also the target of a particular type of inhibitory synapse (Farinas and DeFelipe, 1991). Yet when the motoneuron is strongly stimulated, spikes can be initiated in the dendrites.

Both *orthodromic*, that is, from the dendrites into the soma and axon, and *antidromic spike propagation*, from the cell body back into the dendrite has been inferred in pyramidal cells (e.g., Turner et al., 1991). Unambiguous observation of antidromic spike invasion had to await dual somatic-dendritic recordings in neocortical and hippocampal pyramidal cells from mammalian brain slices (Stuart and Häusser, 1994; Stuart and Sakmann, 1994; Spruston et al., 1995; reviewed in Stuart et al., 1997; Fig. 19.1 and discussed further below).

These findings clearly violate Ramón y Cajal's law of dynamic polarization. Spike initiation in the dendrite has been observed in several *in vitro* preparations under frequently unphysiological conditions (e.g., clamping V_{soma} to between -90 and -100 mV, blocking all inhibition, and so on; Regehr et al., 1993; Turner, Meyers, and Barker, 1993). Whether or not anti- or orthodromic dendritic spikes can be observed under more natural conditions, in particular in response to sensory or motor stimuli in an intact animal remains an open issue (Svoboda et al., 1997).

19.2 Action Potential Initiation in Cable Structures

A full grasp of the interplay between synaptic and voltage-gated currents on the one hand and the neuronal geometry on the other requires a detailed understanding of the nature of spike initiation. What determines the conditions under which an action potential can be triggered in excitable cable structures? The intuitive answer is to look to the properties

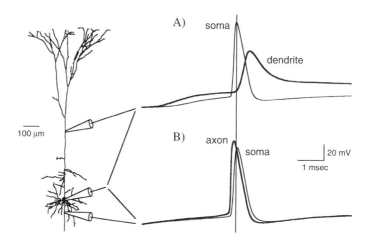

Fig. 19.1 ACTION POTENTIAL INITIATION IN A PYRAMIDAL CELL   Drawing of a layer 5 neocortical pyramidal cell from a 4-week-old rat slice preparation. The approximative locations of the sites of the dendritic, somatic, and axonal recordings are indicated. **(A)**, **(B)** Simultaneous patch-pipette recordings of action potentials elicited by threshold synaptic stimulations (Stuart et al., 1997). Dendritic and axonal recordings (heavy lines) are 270 μm and 17 μm, respectively, from the soma and are from different cells. V_{th} at the soma is 13 mV **(A)** and 15 mV **(B)** relative to the resting potential. The vertical line corresponds to the occurrence of the peak of the somatic membrane potential. Here, the action potential is always triggered in the axon, subsequently propagating back through the soma into the dendritic tree. This is termed *antidromic spike invasion*. Reprinted by permission from Stuart et al., (1997).

(density, kinetics, and type) of the excitable channels at the site of spike initiation. In this view, the membrane voltage needs to exceed a critical value V_{th} at which point the net ionic current becomes inward (Sec. 17.3.3). This inward current will further depolarize the membrane, thereby starting the regenerative process that underlies the initiation of the action potential, as covered in considerable detail in Chap. 6 (recall, in particular, Fig. 6.6).

However, these considerations hold only for isopotential structures where the ionic current can only flow across the membrane. What effect do cable properties have upon V_{th} (Rushton, 1937; Noble and Stein, 1966; Jack, Noble, and Tsien, 1975; Regehr and Armstrong, 1994)? In the following, we will first dwell on the theoretical issue of how V_{th} for rapid Hodgkin–Huxley-like spikes depends on the local cable geometry before we treat detailed biophysical modeling studies that reproduce antidromic spike invasion.

19.2.1 Effect of Dendritic Geometry on Spike Initiation

A significant portion of the excitable current in a cable that is generated at the site of current injection flows longitudinally away from this site, rather than depolarizing the local membrane. To compensate for this current loss, the voltage threshold in the cable must exceed $V_{th}^{uniform}$, the voltage threshold in the corresponding isopotential structure. (For an early discussion of this, see Hodgkin and Rushton, 1946; Noble and Stein, 1966.) Indeed, the geometry of the neuron strongly affects the condition for spike initiation and, as we shall show below, there are cases where excitation can never be realized in cable structures while spikes could be obtained in the corresponding uniform case (see also Chap. 9 in Jack, Noble, and Tsien, 1975).

Figure 19.2A illustrates that V_{th} does, indeed, strongly depend on cable properties, as well as on the site of input application. Using the standard Hodgkin–Huxley equations at $20°$ C, $V_{th}^{uniform}$ is approximately 8 mV. In the other extreme of an infinitely long cylinder, the voltage threshold $V_{th}^{infinite}$ is about 17.5 mV (independent of the diameter d), more than twice that of the isopotential case. For cylinders of finite length L, V_{th} lies between these two extremes. As expected, V_{th} is closer to the uniform isopotential case when L is small and it approaches $V_{th}^{infinite}$ in cylinders with large L values.

In any finite cable structure, V_{th} will always depend on the site of polarization X, and sometimes quite strongly. For short cylinders (e.g., $L = 0.5$ or 1), the lowest V_{th}, closest to $V_{th}^{uniform}$, is at the center of the cylinder. In other words, in these cylinders the midpoint is, effectively, the most isopotential site. For exactly the same reason, the local delay D_{ii} at the center of these cylinders is the largest (Agmon-Snir and Segev, 1993; see also Sec. 2.4). When one moves toward the ends of the cylinder, V_{th} increases monotonically to reach a maximum at the sealed-end boundaries. The situation changes for electrically longer cylinders (e.g., $L = 2$ or 4) where the lowest V_{th} is near (but not at) the terminals rather than at the center, as was the case for shorter cylinders. The effect of the sealed-end terminals causes current to be discharged more slowly near the boundaries compared to other sites, lowering V_{th}. The same result holds also for the site where the local delay is maximal; in long cylinders D_{ii} is maximal near both ends rather than at the center of the cylinder (Agmon-Snir and Segev, 1993).

Observing V_{th} at the midpoint of a cable with $L = 4$ (or at the ends of the cable with $L = 2$), one notices that the voltage threshold is essentially equal to $V_{th}^{infinite}$. This implies that for spike initiation in a Hodgkin–Huxley cable, being two space constants away from the sealed-end boundary approximates an infinitely long cylinder. One may conclude that, for neurons with cable lengths of approximately $L = 1$, V_{th} is the lowest at (or very near) the

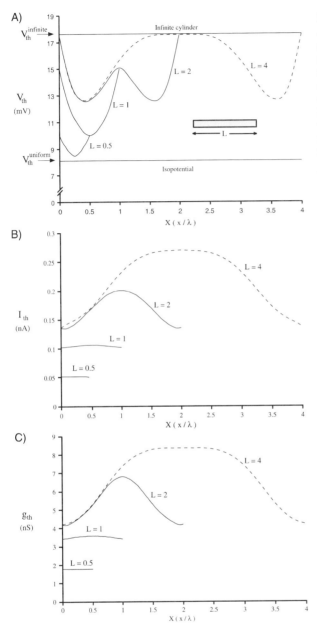

Fig. 19.2 Spike Initiation in a Single Cylinder (A) Voltage, (B) current, and (C) conductance thresholds for spike initiation in four different uniform cables of electrotonic length L endowed with Hodgkin–Huxley-like membrane as a function of the normalized location X of the input. (A) The input is a current stimulus of 0.5-msec duration (one-third of the passive membrane time constant) that was applied to different sites along the cable. The minimal voltage above rest at which a spike is initiated is defined to be V_{th}. The voltage threshold in any one cable can vary by up to 33% with X. (B) I_{th} is defined as the minimal amplitude of a sustained current able to initiate a spike. The dependency of the current threshold on the location of the stimulus is quite distinct from the dependency of V_{th} on X. For cables around $L = 1$, I_{th} only depends weakly on X. (C) The peak conductance of a single fast excitatory synaptic input was varied at different locations until a spike was initiated. This corresponds to a more physiological condition than the two previous panels. Qualitatively, g_{th} behaves as I_{th} ($t_{peak} = 0.3$ msec; $E_e = 0$ mV). All parameters of the Hodgkin–Huxley equations are as in Chap. 6, except that the temperature was set to 20° C. The cylinders have sealed end boundaries and a 2 μm diameter. Reprinted by permission from Rapp (1997).

midpoint of the structure. For neurons with a radial geometry, such as alpha motoneurons, cortical pyramidal neurons, spiny stellate cells, and so on (where the soma is situated near the "center of gravity" of the neuron), the geometry *per se* implies that the voltage threshold is lowest near the soma.

We argued in Sec. 17.3.4 that for slowly varying inputs, spike initiation occurs whenever a minimal current amplitude I_{th} (rather than a voltage threshold) is exceeded. How does this quantity vary with the length of the cylinder and the location of current injection? Based on Ohm's law and the fact that the input impedance in a finite cylinder is always lowest at the center, $I_{th}(X)$ is expected to behave differently from $V_{th}(X)$. Figure 19.2B shows that this is indeed the case. Unlike V_{th}, which is always maximal at the ends of the cylinder, I_{th} is always minimal at these sites and peaks in the middle of the cable. And unlike V_{th}, which does not depend on the diameter of the fiber, but only on X, L, and the nature of the membrane itself, I_{th} is proportional to the total membrane area and thus scales with the diameter of the fiber. We conclude that in a finite-length segment of squid axon, the site for minimal threshold current for spike initiation is at the terminals.

As an aside, note that certain symmetry conditions are evident in Fig. 19.2. While V_{th} at the midpoint of cable of length L is always equal to V_{th} at the ends of the cable of length $L/2$, I_{th} at the center of the cable of length L is exactly twice that at the end of the cable of length $L/2$. This is the direct result of the symmetry of current flow when injected to the center of the cable (see also Rall, 1977).

Finally, Fig. 19.2C covers the more physiological situation of determining the minimal amplitude of a fast excitatory synaptic input sufficient to trigger a spike. The dependency of g_{th} on X and L parallels that of the current threshold.

Figure 19.3 treats spike initiation in a weakly excitable branched cable, corresponding to the situation prevalent in the apical dendrites of *in vitro* pyramidal cells (Stuart and Sakmann, 1994; Regehr and Armstrong, 1994; Magee and Johnston, 1995a,b; Spruston et al., 1995; Mainen et al., 1995; Rapp, Yarom, and Segev, 1996; Stuart et al., 1997). In particular, a low but uniform density of rapidly inactivating voltage-dependent Na^+ channels was reported along the apical dendrite up to at least 200 μm away from the cell body, channels that were no different from those found at the soma. The density is equivalent to about 4 mS/cm^2, or 1/30 its value in the squid giant axon (hence the attribute *weakly excitable*).

To illustrate the general principle involved, we chose an excitable cable whose total length is $L = 1$ and which branches at its midpoint (at $X = 0.5$) into two cylinders of unequal diameter (Fig. 19.3) such that Rall's (1959) $d^{3/2}$ rule holds.[1] Since the boundary condition is identical for both dendrites (sealed end) and their electrotonic length is also the same (0.5), the entire three-part cable can be reduced to a single equivalent cylinder (Sec. 3.2). This explains why V_{th} as a function of X should be symmetrical around the midpoint of the cable (lower curve in Fig. 19.3). Qualitatively, this curve looks similar to those seen in Fig. 19.2A except that, because of the much lower channel density, V_{th} is much higher than for a Hodgkin–Huxley cable and the points of lowest voltage threshold are at $X = 0.3$ and 0.7 (when delivering input simultaneously to both daughter branches).

Things change from the case of a strongly excitable cable when input is only delivered to a single daughter, say, the thick one (of 1.7 μm diameter). Now V_{th} diverges to infinity near the terminal of the cable. No input to the distal portion of the thick daughter branch can trigger an action potential (see the striped portion of the schematic in Fig. 19.3) because most of the—already small—excitable current generated locally flows within the cable and not across the membrane encasing the cable, and the regenerative process of the spike fails to start. This failure is much more dramatic in the thin ($d = 0.72$ μm) daughter branch where action potentials can only be triggered near the branch point. The large impedance load imposed by the rest of the dendritic tree onto thin branches implies that such arbors are unfavorable sites for spike initiation. In contrast, thicker (more proximal) branches in the same tree are more likely to be excited above threshold by local polarization.

---

1. One complication that we will not treat here is that the exact definition of what is meant by an action potential is rather involved for weakly excitable cables, given the smooth and continuous transition between a subthreshold and a suprathreshold response (see Rapp, 1997).

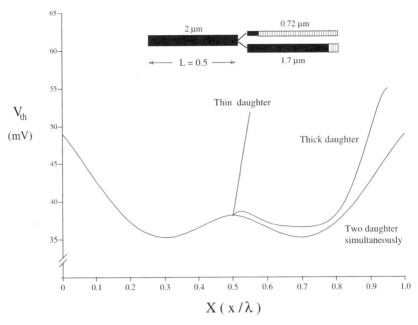

Fig. 19.3 SPIKE INITIATION IN A WEAKLY EXCITABLE BRANCHED DENDRITE Threshold conditions for spike initiation in a weakly excitable branched cable, mimicking the conditions believed to occur in the apical dendrite of neocortical pyramidal cells (Rapp, Yarom, and Segev, 1996). The parent dendrite splits into two 0.5λ-long cylinders of different diameters (see inset). The entire membrane is covered with sodium (and potassium) channels whose density is 1/15 of those of the standard Hodgkin–Huxley squid axon density. The voltage threshold (relative to V_{rest}) required to initiate a spike, is plotted as a function of the relative location of the stimulating electrode. The stimulus is a 15.3-msec-long current pulse delivered into the parent dendrite (left portion of the curve), into one or the other daughter branches, or into both branches simultaneously (right portion of both curves). No spikes can be initiated in the dashed portion of the two daughter branches in the inset. The membrane time constant is 46 msec. $\overline{G}_{Na} = 8$ mS/cm^2, $\overline{G}_K = 2.4$ mS/cm^2. The kinetics of the conductances were adapted to mimic neocortical pyramidal cells. Reprinted by permission from Rapp (1997).

As pointed out above, this situation changes when the stimulus is delivered simultaneously to both daughter branches. The lower right limb in Fig. 19.3 represents the case where the input current is applied simultaneously to the two daughter branches at sites with equal X values. The total current I_{stim} injected into each cable is split in proportion to their relative input conductances, that is, the current injected into the branch of diameter d_1 is equal to $I_{stim} d_1^{3/2} / (d_1^{3/2} + d_2^{3/2})$.

The result is striking (although directly expected from cable theory) in that a spike is now initiated in both branches, even at the distal sites of the thin branch. The reason for this improved efficiency of spike initiation is that the cable structure becomes effectively more isopotential, that is, less current flows longitudinally and more via the membrane. As we saw in Fig. 19.2, the excitation threshold is reduced for isopotential conditions. Finally, when receiving a distributed input as in Fig. 19.3, the threshold conditions in the branched cable are identical to those of the associated equivalent cylinder.

19.2.2 Biophysical Modeling of Antidromic Spike Invasion

To better understand these principles and to develop our intuition concerning anti- and orthodromic spike propagation, let us spend some time with Rapp, Yarom, and Segev (1996; see also Mainen et al., 1995, who come to much the same conclusions).

This study aims to reproduce the electrical events occurring in large layer 5 pyramidal cells recovered from the young rat cortex on the basis of a compartmental model. The apical tree extends by about 1 mm from the soma. Importantly, a large fraction of the associated axon is incorporated into the model: near the soma the axon's diameter is 2.8 μm, thinning to 0.5 μm in its fine collaterals.

What active currents need to be incorporated into this model? Pharmacological blockage of all calcium currents shows that they affect fast action potential propagation only minimally (Markram, Helm, and Sakmann, 1995), as does the sustained sodium current. The principal actor is a very rapid sodium current that peaks and inactivates within 1 msec following a voltage step (Stuart and Sakmann, 1994). Its kinetics, as well as that of a delayed rectifying potassium current, is modeled by an appropriately modified Hodgkin–Huxley scheme.

In a first step, Rapp, Yarom, and Segev (1996) homogeneously spread the Na^+ current throughout the dendrites, soma, and axon. At the low densities reported from the dendrites (3–4 channels per square micrometer, corresponding to 4.5–6 mS/cm^2; see Table 8.1) no full-blown action potentials can be generated at either dendritic or somatic sites. Increasing this density to an effective value of 12 mS/cm^2 enables a strong somatic input to generate a spike. This spike propagates back into the dendrites without any decrement, contrary to the experimental observation. Furthermore, dendritic stimulation leads to a local regenerative response that strongly attenuates on its way to the soma, failing to trigger a somatic spike.

This asymmetric behavior can be explained by recalling the asymmetry inherent in the voltage attenuation in a dendritic tree (Sec. 3.5.2 and Fig. 3.7): the voltage attenuates much more strongly when moving from a high- to a low-impedance site (that is, from the dendritic tree to the soma) than the other way around.

From the early motoneuron work cited above, it was known that action potentials can be trigged at lower voltages in the axon than in the soma. Dual intracellular recordings in hippocampal cells suggest that the normal site of initiation of fast action potentials is neither the soma nor the initial segment but the axon itself, possibly at the first node of Ranvier well away from the cell body (Colbert and Johnston, 1996). Indeed, the number of sodium channels measured in this study at the initial segment and soma is similar to that reported in the apical dendrites, around 3 to 4 per square micrometer. This not only suggests that the densities of sodium channels are substantially higher in the axon than at the soma (up to 2000 channels per square micrometer have been reported at the nodes of myelinated fibers; see Sec. 6.6) but also that their activation curve is shifted to more negative membrane potentials. Accordingly, Rapp, Yarom, and Segev (1996) use a low value for the sodium channels in the dendrites and a high value for the axon as well as shifting the sodium activation curve for the axonal sodium channels m_∞ by 10 mV to the left (see legend to Fig. 19.4). An alternative to the latter is to employ the same activation curves for dendrite and axon but a much, much higher axonal density (Mainen et al., 1995).

In order for a dendritic input to trigger a spike at the soma, the difference between V_{th} in the dendrite (which can be substantial; recall Fig. 19.3) and V_{th} in the axon needs to exceed the voltage attenuation between these two sites. This will allow dendritic input to trigger an axonal spike while remaining subthreshold within the dendritic tree, as demonstrated in Fig. 19.4. Panel A illustrates that the spike is initiated in the axon, subsequently propagating back through the soma into the apical dendrite at a speed of 150 μm/msec. In agreement with the data, the back-propagating spike is attenuated by a factor of 2 in amplitude as well as becoming broader in time. The former is due to the low density of sodium channels while the latter is caused by the distributed dendritic capacitance. Blocking the dendritic sodium channels does reveal, however, their boosting action on the back-propagating signal (Fig. 19.4C).

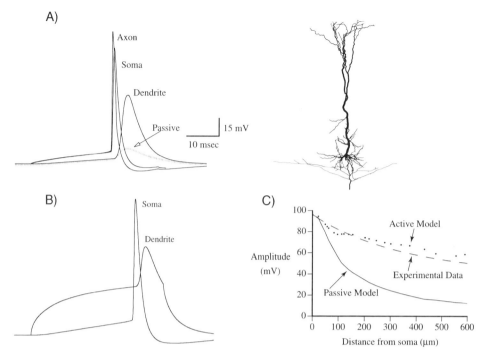

Fig. 19.4 SIMULATING ANTIDROMIC SPIKE INVASION Compartmental model of a layer 5 pyramidal cell from rat somatosensory cortex (shown at right; the vertical extent from the soma to the dendrite at the top is 1 mm) by Rapp, Yarom, and Segev (1996) with weakly excitable dendrites and cell body but a highly excitable axon. **(A)** Application of a 150-pA, 40-msec current pulse to the soma triggers an action potential at the axon (axon trace recorded 50 μm away from the soma). Within 0.5 msec it reaches the cell body and is propagating back into the dendrites (recorded 550 μm from the soma). Blocking the sodium current along the apical dendrite (dashed trace labeled "Passive") reveals the active nature of the back-propagating response. **(B)** Delivering a 300-pA, 40-msec current pulse to the apical dendrite causes a spike to be triggered at the axon and not locally, in the dendrite. This is due to (1) the passive load imposed by the dendritic arbor, (2) the lowered spiking threshold in the axon (m_∞ is shifted by 10 mV to more negative values compared to the dendritic sodium activation curve), and (3) the high sodium channel density at the axon ($\overline{G}_{Na} = 140$ mS/cm^2). **(C)** Peak amplitude of the spike propagating back along the apical dendrite as a function of the distance from the soma. The dashed line corresponds to the experimental data of Stuart and Sakmann (1994) and is superimposed onto the output of the model (dots). When dendritic sodium channels are blocked ("Passive Model"), the active nature of the back-propagating response is revealed. $\overline{G}_{Na}$ ranges from 8 mS/cm^2 in the soma to 4 mS/cm^2 in the distal dendrites. Reprinted by permission from Rapp, Yarom, and Segev (1996).

The passive electrical load imposed by the dendrites and cell body, in combination with a weakly excitable dendritic tree and highly excitable axon, explains why spikes are always triggered in the axon and propagate back into the tree. How much the spike propagates back into the tree is controlled by a number of factors, such as the presence of transient potassium currents in the dendrites (Hoffman et al., 1997), inactivation of the sodium current by previous spikes (Migliore, 1996), and synaptic inhibition (Buzsaki et al., 1996; Tsubokawa and Ross, 1996).

Before we proceed further, it needs to be pointed out that most of the data upon which the modeling discussed in this section was based was obtained from pyramidal cells in relatively young rat slices. Developmental changes in channel densities and distribution (Huguenard,

Hamill, and Prince, 1988), local "hot spots" of sodium or calcium channels, the presence of neuromodulators, and other factors could well allow synaptic input to locally trigger action potentials in the dendritic tree.

19.3 Synaptic Input into Active Dendrites: Functional Considerations

From a computational point of view, passive dendrites can only instantiate a few elementary operations: (1) low-pass filtering, (2) saturation, and (3) multiplicative-like interactions among synaptic inputs. A much richer repertoire of nonlinear and nonstationary operations is possible if voltage-dependent membrane conductances are present in the dendritic tree.

In the remainder of this chapter, we discuss a number of specific operations that have been postulated to occur in excitable dendrites. The first one very briefly deals with the possible functional role of back-propagating spikes. The next two exploit precise coincidence detection using rapid and locally generated action potentials in spines and dendrites. The last two, involving graded amplification of distal synaptic input and multiplication with active conductances, consider dendritic all-or-none events as epiphenomena, focusing instead on the underlying voltage-dependent currents.

19.3.1 Back-Propagating Spike as Acknowledgment Signal

The role of back-propagating action potentials is puzzling if considered within the context of information processing, but makes more sense as an "acknowledgment signal" for synaptic plasticity and learning. We reviewed in Sec. 13.5.2 evidence by Markram et al., (1997) and earlier studies in favor of this hypothesis. By controlling the spike generation times in a pair of excitatory coupled neurons, the amplitude and time course of synaptic coupling between the neurons can be up or down regulated. If the presynaptic spike precedes the postsynaptic one, as should occur if the presynaptic one participates in triggering a spike in the postsynaptic cell, long-term potentiation occurs, that is, the synaptic weight increases. If the temporal order of arrival times is reversed (e.g., from +10 to -10 msec), long-term depression results. This provides the biophysical substrate for a novel class of powerful dynamic Hebbian synaptic learning rules for learning associations over time (e.g., Montague et al., 1995; Abbott and Blum, 1996; Gerstner et al., 1996).

19.3.2 Implementing Logic Computations with Spikes in Spines

In the chapter on dendritic spines (Sec. 12.4) we treated the idea of "spiking spines," that is dendritic spines that are endowed with a sufficient density of either sodium or calcium channels to be able to trigger all-or-none electrical events in the spine in response to a fast synaptic input (Segev and Rall, 1988). As summarized in Fig. 12.11, there exists a range of spine neck resistance values R_n, for which the induced EPSP in the passive dendrite just below the spine V_d is greatly amplified over the EPSP that would have been obtained if the spine had been passive. If R_n is too small, the high load of the dendritic tree increases V_{th} beyond the point at which synaptic input can trigger a spike. If R_n is too high, a spike is triggered but is highly attenuated by the time it reaches the dendrite.

Section 12.6.2 dealt with the experimental evidence using fluorescent dyes for all-or-none calcium events in individual spines (reviewed in Denk et al., 1996). Given the great difficulty in recording the membrane potential in spines and very thin dendrites, it will be very demanding to verify directly the existence of fast sodium-mediated spikes in dendritic spines.

In Chap. 12 we restricted ourselves to treating single spines. More interesting from the point of view of neuronal computation are the combinatorial possibilities for nonlinear interactions among multiple active spines. As demonstrated by compartmental modeling (Shepherd et al., 1985; Shepherd and Brayton, 1987; Rall and Segev, 1987; Shepherd, Woolf, and Carnevale, 1989; see also Baer and Rinzel, 1991), one or a small number of simultaneous inputs to active spines can trigger action potentials in neighboring spines, providing a rich substrate for computations.

Zador, Claiborne, and Brown (1992) and Fromherz and Gaede (1993) have explored similar ideas relating to the implementation of the exclusive OR (XOR) operation using voltage-dependent currents in dendrites rather than in spines.

Active spines can interact with each other as long as the dendritic membrane potential V_d in response to synaptic input to some active spines is sufficient to trigger an action potential in those spines that did not receive any input. Recall that the voltage attenuation from the dendrite into the spine is basically nonexisting (Fig. 12.13). Thus V_d only needs to be large enough to exceed V_{th} in the spine (which decreases monotonically as R_n increases; Segev and Rall, 1988). A large value of V_d can be more easily obtained in the distal dendritic tree with its associated high input impedance.

Given the very high density of spines (up to 10 spines per micrometer; Sec. 12.1.1), what type of interactions might occur on distal dendrites? Rall and Segev (1987) explore this issue on the basis of an idealized neuron modeled by an equivalent cable with a trunk and five levels of symmetrical branches. The two most distal levels of dendrites are each endowed with 45 passive and 5 active spines. The parameters are chosen in such a way that a fast ($t_{peak} = 0.2$ msec) AMPA input to the spine triggers a local action potential, giving rise to a dendritic EPSP of 8.5 mV and a somatic EPSP of 24 μV (Fig. 19.5A; the somatic EPSP for an identical input to a passive spine is half of this amplitude). The local V_d is not sufficient to bring the other four active spines on this branch to threshold. However, two simultaneous inputs to two active spines will do the job, firing off the other three active spines that did not receive synaptic input (Fig. 19.5B). With an appropriately set threshold, this branch implements a logic OR, such that any two inputs to active spines will generate the same amplified output response (the peak somatic response in B is about 84 μV).

Triggering all five active spines simultaneously depolarizes the neighboring sibling branch sufficiently to trigger all of its active spines (Fig. 19.5C). It typically takes a millisecond or less for the spines that did not receive any direct synaptic input to trigger action potentials. In Fig. 19.5D, any two inputs to active spines on one branch and input to any three passive spines are sufficient to trigger all active spines on the two distal spines. Because the firing is more synchronous than in Fig. 19.5C, it sets off—with a delay—the active spines in the parent branch. Finally, Fig. 19.5E illustrates what could occur with inhibition targeting specific active spines, preventing a local action potential from being generated.

In principle, very specific higher order logical interactions could be implemented by active spine clusters. In practice, though, it is unclear whether the brain possesses specific enough developmental and learning rules that would guide an axon to form a synapse on one specific spine on one specific branch. The demanding requirements of which spine needs to be activated at what point in time in order to obtain a specific interaction renders them possibly not robust enough to be implemented in the nervous system. Further below we will deal with a much more robust mechanism, which exploits voltage-dependent inward

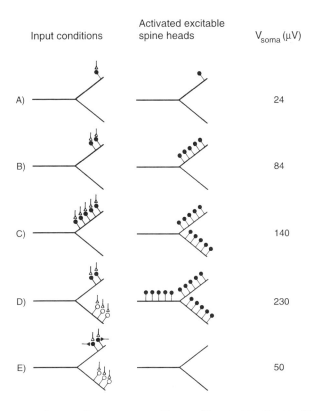

Fig. 19.5 **INTERACTING ACTIVE SPINES IN THE DISTAL DENDRITIC TREE**   Clusters of highly excitable dendritic spines can implement very specific interactions. Rall and Segev (1987) simulate an idealized dendritic tree with five levels of symmetrical branching, of which only the last two are shown. Each of the three branches carries five spines that include sufficient sodium and potassium channels to give rise to a rapid action potential and 45 passive spines. **(A)** Rapid synaptic input to a single active spine is sufficient to trigger a local action potential (see Fig. 12.10) that depolarizes the soma by 24 μV, while input to a passive spine would only depolarize the soma by 12 μV. **(B)** Two simultaneous inputs to two active spines raise the dendritic membrane potential sufficiently to trigger spikes in the remaining three active spines in this branch, but not in its sibling. **(C)** This can be accomplished by exciting all five active spines. **(D)** Two synaptic inputs to active spines on one branch and three to passive spines on the other will give rise to a wave of depolarization that is synchronous enough (compared to **C**) to trigger the active spines in its parent branch. **(E)** If fast inhibition is paired with fast excitation, no spikes are triggered, demonstrating the potentially large number of very specific interactions that can be implemented with active groups of spines. The open question is whether this mechanism is robust enough to be implemented within the central nervous system. The excitable membrane uses standard Hodgkin–Huxley-type Na$^+$ and K$^+$ conductances at five times their densities and adjusted for 20 C°. Reprinted by permission from Rall and Segev (1987).

currents—but not dendritic spiking—to implement a form of multiplication in spirit related to the computations presented here.

19.3.3 Coincidence Detection with Dendritic Spikes

One of the guiding themes of Chap. 15 has been the question to what extent individual nerve cells can be treated as "integrators," adding up many small synaptic inputs over tens of milliseconds, or as "coincidence detectors," firing in response to the exact temporal relationship at the millisecond scale (see, in particular, Sec. 15.3.2). This issue is intimately connected to how much information (in terms of bits per spike or the reconstruction error; Gabbiani and Koch, 1998) single neurons can convey about a stimulus in the world. For stimuli varying on a time scale of a few milliseconds, submillisecond resolution at the single cell level should be superior in terms of efficiency of information encoding compared to representing the same information in the firing rate of a population of units that integrate over 20, 50, or more msec (Knight, 1972a; Softky, 1995). Of course, high-fidelity neurons place substantial demands on developmental mechanisms to arrange for the required precise interneuronal wiring. The advantage of a simple code carried by a population of integrator unit lies in their robustness to noise and in their simpler self-assembly requirements.

As investigated already by Segev and Rall (1988; see the preceding section), dendritic spikes provide a convenient biophysical mechanism for high-fidelity coincidence detection (Softky and Koch, 1993; Softky, 1994). We here discuss a recent model of a pyramidal cell discriminating spikes with a resolution in the 0.5 msec range (see also Sec. 5.2.3).

Softky (1995) endows the layer 5 pyramidal cell's dendrites with weakly excitable Hodgkin–Huxley conductances that are barely sufficient to sustain local regenerative events in the basal dendrites. This amounts to choosing $\overline{G}_{\text{Na}}$ to be 6 mS/cm^2, a twentyfold reduction of the maximal conductance in the squid axon and about one standard deviation above the mean value of 4 mS/cm^2 reported in rat neonatal apical dendrites (Stuart and Sakmann, 1994). Potassium conductances (with $\overline{G}_K = 1.8$ mS/cm^2) and time constants are chosen such that a single excitatory AMPA synaptic input (with $t_{\text{peak}} = 0.24$ msec) to the center of each thin terminal basal dendritic branch (Fig. 19.6A) yields a subthreshold response, while two simultaneous inputs are sufficient to trigger a very rapid all-or-none dendritic event of 40 mV or more amplitude. This signal is attenuated fiftyfold at the soma (Fig. 19.6B). If the same two EPSPs are separated by a mere 2 msec, the peak somatic EPSP is three times smaller. That the local potential in the basal dendrites can be exquisitely sensitive to the arrival times of synaptic input was already evident in the small value of the local delay $D_{ii} = 1$ msec (Fig. 3.13), which is much less than $\tau_m = 20$ msec.

This small time window for significant interaction among synaptic input is put to use by Softky by adding a third, inhibitory synaptic input (of the GABA$_A$ type with $t_{\text{peak}} = 0.4$ msec) to the same basal branch. Conjoint activation of all three inputs more or less nulls out the membrane potential, while a mere 1 msec delay between the simultaneous onset of the pair of excitatory inputs and inhibition fails to prevent the dendritic spike (Fig. 19.6C).

Softky (1995) simulates a triplet of synaptic inputs, with both fast excitatory inputs arriving simultaneously, followed 1 msec later by fast inhibition, placed at the midpoint of each of the terminals of the basal tree. Each triplet occurs independently of the others and is activated at a rate of 25 Hz. The result is that the pyramidal cell fires vigorously at about 55 Hz (Fig. 19.7A). What happens if the arrival times of all three elements is jittered in time (using a Gaussian distribution of standard deviation σ and assuming that on average the two excitatory inputs coincided and are followed 1 msec later by inhibition)? One such trial run with $\sigma = 0.5$ msec jitter is illustrated in Fig. 19.7B. The cell's firing is greatly reduced since dendritic spikes are absent. Figure 19.7C reports the average firing rate of the cell to repeated trials for a range of values of σ. The cell's response drops steeply with increasing σ, with a half-width at half-height of about 0.27 msec.

There is little indication in the somatic voltage record (Fig. 19.7A and B) of the very strong and transient dendritic signals that give rise to this temporal sensitivity.

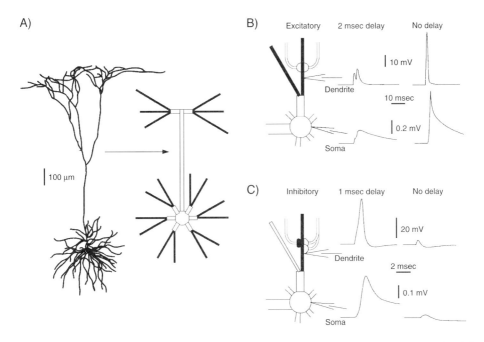

Fig. 19.6 COINCIDENCE DETECTION WITH DENDRITIC SPIKES Dendritic spikes initiated in response to local synaptic input can implement submillisecond temporal discrimination (Softky, 1994, 1995; see also Segev and Rall, 1988). (**A**) The standard layer 5 pyramidal cell is outfitted with weakly excitable dendrites and a strongly excitable cell body. Two fast excitatory synapses and one fast inhibitory synapse are placed along the center of each distal basal dendritic compartment. (**B**) The conductances are adjusted such that activation of a fast excitatory input to a basal dendrite, followed 2 msec later by a second such input, gives rise to a small local EPSP, while simultaneous activation of both triggers a very rapid all-or-none event, three times stronger than if the same synaptic events were separated by 2 msec. (**C**) If inhibition follows the two excitatory inputs by a mere 1 msec, it has little effect on the peak EPSP. Yet when all three synaptic elements are activated simultaneously, V_m remains close to rest. The voltage-dependent membrane conductances are of the Hodgkin–Huxley type with $\overline{G}_{Na} = 6$ mS/cm^2 in the dendrites. Reprinted by permission from Softky (1995).

The reason we discuss this study is that it demonstrates how, on the basis of a bio-physically plausible parameter range, temporal coincidence detection with submillisecond resolution might occur. This mechanism could be exploited by the nervous system to implement a temporal code in which the exact (to within 0.5 msec or less) time of spiking relative to other neurons carries all of the relevant information (Softky, 1994). Such a temporal encoding strategy appears to be used in the insect olfactory system (Laurent, 1996; Wehr and Laurent, 1996). Given the combinatorial possibilities for expressing messages by shifting spike times around, it is considerably more powerful than a firing rate code that averages spiking from a population of cells over 50 msec or more.

As pointed out above, this does place a premium on the nervous system having learned how to compensate for axonal propagation and other delays in the system. The study by Markram et al., (1997) provides hope for the existence of such a mechanism at the level of an individual synapse, opening the door to learning rules that explicitly deal with delays.

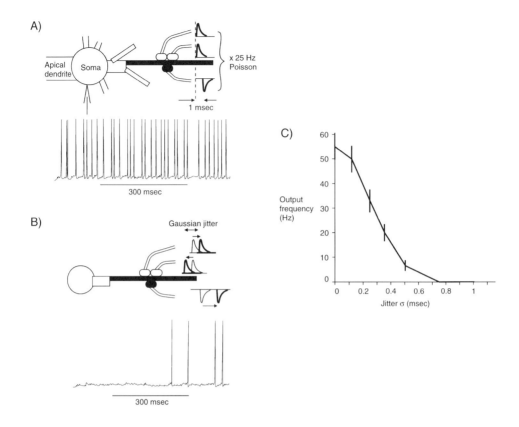

Fig. 19.7 SUBMILLISECOND COINCIDENCE DETECTION The model of Softky (1995) responds sensitively to jitter in the arrival times of synaptic input. This mechanism could be exploited to encode information in the relative spiking times of a group of neurons. **(A)** If all basal dendrites are endowed with weak Hodgkin–Huxley conductances and with the three synaptic elements illustrated in Fig. 19.6 that are activated independently of each other at 25 Hz, the somatic EPSPs due to the dendritic spikes summate, causing the cell to fire vigorously at 55 Hz. Note that one "synaptic event" here implies two simultaneously activated excitatory inputs followed 1 msec later by inhibition. **(B)** Adding jitter to the activation times of all synapses (the jitter is drawn from a Gaussian distribution centered at zero and with a standard deviation of σ) eliminates dendritic spikes, causing the cell to fire only weakly (illustrated here for $\sigma = 0.5$ msec). **(C)** Output firing rate as a function of σ; an average jitter of 0.27 msec in all three synaptic activation times causes the output to drop twofold. Reprinted by permission from Softky (1995).

19.3.4 Nonlinear Spatial Synaptic Interactions Using Active Currents

While the previous section dealt with temporal interaction among synaptic inputs in an active dendritic tree, let us now turn toward spatial interactions.

Chapter 5 treated nonlinear interaction among synaptic inputs. Due to the dependency of the NMDA membrane conductance on the postsynaptic membrane potential (Eq. 4.6), the EPSP in response to the simultaneous (or near-simultaneous) activation of two or more NMDA synapses can be much larger than the sum of the EPSPs due to the activation of each input by itself. The closer these synapses are spaced to each other, the larger the EPSP (Fig. 19.8B). Cooperativity among NMDA synapses needs to be contrasted to the suppressive interactions among neighboring voltage-independent synapses in a passive dendritic tree. Due to the reduction in the driving potential and the absence of any mitigating increases in the synaptic-induced conductance change, the resultant EPSP will always be less than the sum of the individual synapses (Fig. 19.8A). Mel (1992, 1993) shows how this sensitivity to synaptic placement can be exploited to implement a robust multiplicative operation in the dendritic tree.

How does the addition of voltage-dependent dendritic calcium and sodium conductances impact the neuron's sensitivity to spatial clustering of synapses? Mel (1993) investigates this problem by randomly distributing a total of 100 synapses in clusters of k synapses each over the dendritic tree, where k ranges from 1 to 15 (see also Sec. 5.2.1). Each synapse is activated independently by a 100 Hz Poisson point process using one of several different postsynaptic scenarios. In one, all synaptic input is of the AMPA type, which is voltage independent, but the dendritic tree of the layer 5 pyramidal cell (Fig. 5.6) is endowed with

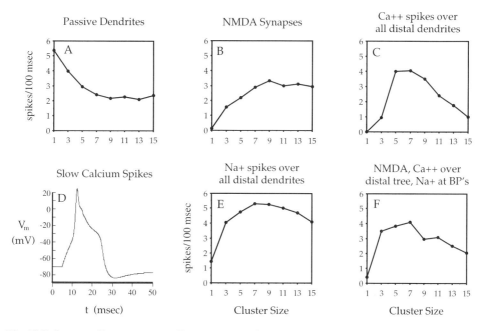

Fig. 19.8 SPATIAL CLUSTERING OF SYNAPSES IN AN ACTIVE DENDRITIC ARBOR   Effect of spatial proximity of synaptic input on the cellular response in the presence of voltage-dependent dendritic membrane conductances. Mel (1993) distributes 100 synapses in 100/k clusters of k synapses each throughout the layer 5 pyramidal cell (Fig. 5.6) and activates each of them independently with a 100-Hz presynaptic train of spikes. The cellular response, expressed as the number of somatic action potentials within 100 msec, is averaged over 100 different random synaptic distributions. **(A)** For a passive dendritic tree and voltage-independent synaptic input, synaptic saturation dictates that spreading the synapses as far away from each other will maximize the cell's response. **(B)** Shifting 90% of the synaptic conductance to NMDA receptors favors activating spatially adjacent synapses. **(C)** In the presence of g_{Ca} and g_{AHP} in the distal portion of the tree, a similar cluster sensitivity results (here for AMPA input only). **(D)** This synaptic input frequently triggers a slow all-or-none dendritic event as portrayed here. **(E)** Replacing the slow dendritic g_{Ca} with fast Hodgkin–Huxley-like sodium spikes makes little difference as does the scenario summarized in **(F)**, a combination of calcium conductances in the distal dendritic tree with sodium spikes at dendritic branch points in the presence of strong synaptic input of the NMDA type. Reprinted in modified form by permission from Mel (1993).

a L-type high-threshold, noninactivating calcium conductance g_{Ca} in conjunction with a calcium- and voltage-dependent potassium conductance g_{AHP} (as well as radial diffusion and calcium buffering). These conductances are evenly distributed throughout the tree with the exception of the most proximal 60 μm stretch of each basal dendrite and the first 450 μm long section of the apical shaft, which are devoid of any such conductances. Under these circumstances, a sufficiently strong synaptic input can trigger a slow all-or-none dendritic event (Fig. 19.8D).

This gives rise to a very pronounced cluster sensitivity (Fig. 19.8C). Arranging the synaptic input into 20 clusters of five synapses each quadruples the cell's spiking response at the soma compared to clustering the same synapses in 33 groups of three synapses each. Replacing the dendritic g_{Ca} by a rapidly activating sodium conductance in conjunction with a delayed-rectifier potassium conductance yields fast dendritic spikes that lead to an inverted U-form cluster dependency (Fig. 19.8E). Finally, a last block of trials consists of the slow and distal calcium-conductance scenario, coupled with fast Hodgkin–Huxley membrane conductances at all dendritic branch points (distal 40 μm segment of each distal apical branch and the distal 40 μm of each apical and basal dendritic tip) and NMDA synaptic input (Fig. 19.8F). These and many more permutations of parameters (different conductances with different voltage dependencies, different patterns of synaptic input, and so on) always yield the same robust sensitivity of the cell's response to the spatial patterning of synaptic input.

Mel, Ruderman, and Archie (1998) demonstrate in a convincing manner how this nonlinear synaptic interaction can implement a crucial operation found throughout sensory systems. The best characterized neurons in the mammalian visual cortex are *simple cells*, whose distinguishing feature is that they respond optimally to visual stimuli (such as a bar) having a particular orientation (e.g., horizontal) at some position within their narrow receptive field (Hubel and Wiesel, 1962). A *complex cell* differs in a number of ways from a simple cell, chiefly in that it shows orientation tuning at many locations throughout its receptive field, which is much larger than the optimal bar stimulus. The transition from simple to complex cells is common to many systems and occurs anytime some selectivity, be it for disparity, face detection, and so on, must be generalized across space.

The original Hubel and Wiesel (1962) model as well as subsequent models (Movshon, Thompson, and Tolhurst, 1978; Heeger, 1992) posit that this tuning across space is accomplished by pooling the output of a set of simple cells with different receptive field positions. However, much evidence has accumulated challenging the notion of a purely hierarchical processing. In particular, complex cells receive direct, monosynaptic input from the lateral geniculate nucleus (LeVay and Gilbert, 1976; Ferster and Lindstrom, 1983; see also Malpeli et al., 1981; Ghose, Freeman, and Ohzawa, 1994). Mel and colleagues (1998) show how the center-surround receptive field of geniculate cells can be transformed in a single step into oriented complex cells using intradendritic computations.

The principal idea is similar to Mel's (1993) study of pattern discrimination using synaptic clustering (Sec. 5.2.2 and Fig. 5.9). As before, the layer 5 pyramidal cell receives mixed AMPA/NMDA excitatory synaptic input from a number of geniculate cells (either 100 or 1000). The dendritic membrane of the cell incorporates the standard g_{Na} and g_K Hodgkin–Huxley conductances, whose density is one quarter of the somatic density, in qualitative agreement with Stuart and Sakmann's (1994) findings discussed above. Mel, Ruderman, and Archie (1998) mimic a developmental learning rule that preferentially encourages the growth of geniculo-cortical synapses into the neighborhood of those patches of dendritic membrane that already show strong depolarization (Shatz, 1990; Cline, 1991).

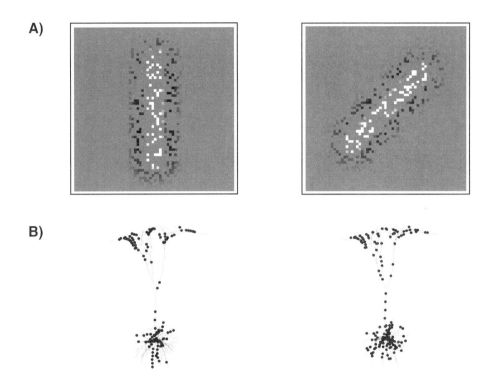

Fig. 19.9 ORIENTATION TUNING IN A COMPLEX CELL IN VISUAL CORTEX The sensitivity of the firing rate to the exact spatial arrangement of synapses in the dendritic tree is exploited by Mel, Ruderman, and Archie (1998) to implement nonlinear superposition. This is exemplified by replicating orientation selectivity at multiple locations in the receptive field of a complex cell in the primary visual cortex directly excited by neurons in the lateral geniculate nucleus. The dendritic tree of the pyramidal cell includes sodium- and potassium-dependent Hodgkin–Huxley conductances. **(A)** Input is provided by 1024 randomly sampled geniculate cells that activate mixed—NMDA and AMPA—synapses (white squares; the black squares indicate inhibited cells). **(B)** A developmental rule for mapping these inputs onto the cortical cell assures that geniculate input activated by a vertical bar anywhere in the receptive field is organized in spatial clusters along the dendrites, while the synapses activated by an obliquely oriented input activate an equal number of synapses that are spread homogeneously throughout the dendritic tree (the 150 most active synapses are shown here).

This local activity could be caused, for instance, by simultaneous active synapses on a neighboring patch of dendrite. An outcome of learning rules of this type is that strongly correlated inputs are more likely to form synapses at nearby sites in the dendritic tree. In the case of Fig. 19.9, geniculate input is correlated with respect to an ensemble of vertically oriented bars form neighboring synapses (within less than 100 μm of each other along the dendrite) while geniculate inputs triggered by diagonally oriented bars are spread throughout the dendritic tree. This results in the formation of clusters corresponding to a set of geniculate afferents aligned in such a way as to be maximally activated by a vertical bar and least by a horizontal one.

With many such clusters, the cell shows orientation selectivity throughout its receptive field (Fig. 19.9D). The bar stimulus will always activate roughly the same number of synapses; but for nonoptimal orientations these are spread diffusely throughout the cell and their postsynaptic response is reduced. That orientation tuning is mediated by cluster

C)

D)

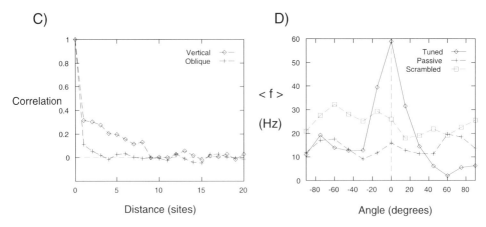

Distance (sites)

Angle (degrees)

Fig. 19.9 CONTINUED   (**C**) The spatial correlation function for synaptic input as a function of distance along the dendrite bears this out: for a vertical bar, synaptic input to the tree is correlated in space (up to approximately 100 μm). This spatial clustering leads to an optimal activation of NMDA and voltage-dependent inward currents. Very little such correlation is evident for the oblique input. (**D**) Averaging the spiking response of the cell for bars of different orientations presented at six different spatial positions yields a tuning curve very similar to those recorded in the cat visual cortex (Orban, 1984).   Spatially scrambling the synapses or blocking the NMDA input and the dendritic g_{Na} (in this latter passive scenario, the AMPA conductances were amplified thirtyfold to compensate for the loss of current) eliminates orientation tuning. Reprinted in modified form by permission from Mel, Ruderman, and Archie (1998).

sensitivity caused by voltage-dependent conductances can be demonstrated in two ways: scrambling the spatial location of all synapses (that is, the mapping from the geniculate to the cortical cell is randomized) or blocking both the NMDA and dendritic g_{Na} conductances, which eliminates orientation selectivity (Fig. 19.9D). Mel, Ruderman, and Archie (1998) conclude that the position independence of orientation tuning of complex cells in the primary visual cortex could originate within the dendritic tree of a single cell, rather than by the convergence of the output of many cortical simple cells as usually assumed. Functionally, such a neuron corresponds to a polynomial threshold unit, computationally more powerful than the standard linear threshold unit (Sec. 14.4.2).

Note that this sort of nonlinear superposition might well be commonplace. Take as an example neurons in the most anterior portion of the inferotemporal cortex that strongly fire to a unique view of a three-dimensional object, such as a paperclip or a face, almost anywhere within the monkey's visual field (Logothetis and Pauls, 1995). They could achieve this feat by exploiting the same biophysical mechanism: nonlinear interaction among adjacent synapses in an active dendritic tree.

19.3.5 Graded Amplification of Distal Synaptic Input

In the previous chapter, we derived an analytical relationship between synaptic input to one end of a single finite cable and the current necessary to clamp the membrane potential at the other terminal (Eq. 18.18). As illustrated in Fig. 18.8, this current saturates once the synaptic input is large enough to depolarize the local membrane potential close to the synaptic reversal potential. This insight was extended to synaptic input into the distal apical tuft of the layer 5 pyramidal cell (Fig. 18.10C). The total somatic current that can be delivered by 500 AMPA synapses distributed throughout layers 1, 2, and 3 (all the dendrites past the first major branch point along the apical tree; see Fig. 18.6) saturates at 0.65 nA, which is not enough to generate more than a handful of spikes per second at the soma. The culprit is the high axial resistance of the apical trunk, causing V_m in the distal dendrite to quickly approach the synaptic reversal potential. And this in a cell whose apical trunk is quite thick (4.4 μm diameter in the middle of layer 5; see Fig. 3.7) and has relatively low input impedances (see Table 3.1). The thinner dendrites in a rat pyramidal cell (Larkman and Mason, 1990) would give rise to even smaller values of $I_{syn,s}$.

It seems odd that a large fraction of the dendrite would be dedicated to synaptic input (in this case, originating to a large extent in higher cortical areas) that only has a minor effect on the output of the cell. Following earlier suggestions by Spencer and Kandel (1961) and more recent authors, Bernander, Douglas, and Koch (1994) investigated how voltage-dependent membrane conductances in the dendrites could be used to amplify distal synaptic input in a continuous manner.

Let us first consider the principle of the idea with the aid of a highly simplified model of a neuron (Fig. 19.10A). The three compartments crudely mimic the distal synaptic input (at right), the intervening apical tree (middle compartment), and the cell body, whose membrane potential is clamped to the average potential $\langle V_m \rangle$ during spiking (Fig. 17.10B). In the absence of any voltage-dependent membrane conductance, $I_{syn,s}$ will saturate (as in Eq. 18.18).

Because such saturation reduces the dynamic bandwidth of the input, Bernander, Douglas, and Koch (1994) argue that a voltage-dependent potassium conductance g_K needs to be introduced within the synaptic compartment in order to counteract saturation. The current does this by providing a hyperpolarizing current proportional to the degree of membrane depolarization. We can derive the voltage dependency of this current by demanding that

$$I_{syn,s} = \kappa_1 g_{syn} . \tag{19.1}$$

In the absence of g_K, $I_{syn,s}$ equals I_{syn}, the current flowing through the synapse (Fig. 19.10B). Kirchhoff's current conservation law demands that the current supplied by g_K must equal the difference between these two currents, that is, $I_K = I_{syn,s} - I_{syn}$.

From Fig. 19.10B this difference current has a parabolic shape: in the absence of synaptic input $g_K = 0$; likewise around $g_{syn} = 180$ nS, when the curves overlap again. At intermediate values of g_{syn}, activation of g_K makes up for the discrepancy between the two curves. Combining these equations, we find that the closed-form solution for the voltage dependency of g_K is a simple fractional polynomial (Fig. 19.10C),

$$g_K(V) = \frac{g}{2 \cdot \kappa_1} \frac{(V - V_{soma})(V - E_{syn} + \kappa_1)}{E_K - V} . \tag{19.2}$$

As the membrane potential depolarizes, g_K activates and opposes this increase. At large values of V, g_K inactivates, similar to the transient inactivating A type potassium current (Sec. 9.2.1), leading to an overall linear relationship between input and output, with $\kappa_1 = 5$ mV (Fig. 19.10B). Amplification of this "linearized" current, that is,

$$I_{syn,s} = \kappa_1 \kappa_2 g_{syn} \tag{19.3}$$

can be achieved by an inward current between the synaptic site and the soma. Depending on the local membrane potential, it will generate additional current across the membrane. The derivation of the voltage dependency of the associated noninactivating conductance g_{Ca} is tedious but straightforward (done in Fig. 19.10C for $\kappa_2 = 2$). Note that different

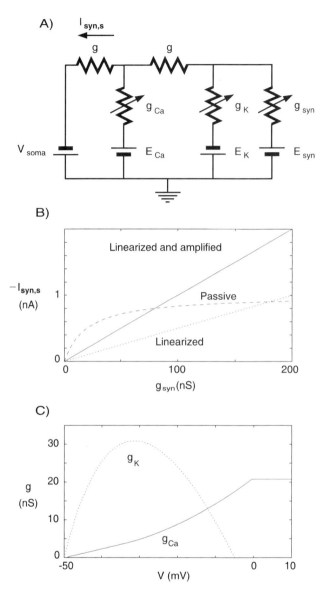

Fig. 19.10 PRINCIPLE OF DEN-
DRITIC AMPLIFICATION Lin-
earization and amplification in
the three-compartment model
shown in (A) with V_{soma} =
−50 mV. All capacitances have
been neglected. (B) "Somatic"
current I_{soma} flowing across the
conductance g in response to
a sustained synaptic input g_{syn}.
Due to synaptic saturation in the
rightmost compartment, the current is sublinear (dashed "Passive" curve). The introduction of
an inactivating potassium conductance g_K with the voltage dependency shown in (C) linearizes
$I_{syn,s}$ as a function of g_{syn} (dotted
"Linearized" curve). Positioning
an inward current (here g_{Ca}) between the synaptic input site and
the soma amplifies the synaptic response (thin "Linearized
and amplified" curve). Reprinted
by permission from Bernander,
Douglas, and Koch (1994).

combinations of κ_1 and κ_2 in Eq. 19.3 give rise to different voltage dependencies for the potassium and calcium currents, but to the same current as long as the product $\kappa_1\kappa_2$ remains constant.

By analogy with this simple model, the saturating behavior of synaptic input into the apical tuft for the pyramidal cell of Fig. 3.7 can be eliminated by evenly spreading a noninactivating potassium conductance throughout layers 1, 2, and 3. The determination of its exact voltage dependency is carried out using an iterative learning scheme that modifies $g_K(V_m)$, recomputes $I_{syn,s}$, and changes its activation curve proportional to the difference between the actual somatic current and the desired linear one (Bernander, 1993). For reasons of stability, potassium inactivation was not considered here (unlike the simple model of Fig. 19.10C). Subsequently, an inward current is inserted into the membrane around the first branch point of the apical dendrite (close to the layer 3 to layer 4 transition) and its activation curve is derived (Fig. 19.11B); it resembles a noninactivating high-threshold calcium conductance. Imaging experiments have visualized such a localized band of calcium activity 500 μm away from the cell body in neocortical pyramidal cells (Yuste, Delaney, and Tank, 1994).

In the presence of these two dendritic currents, distal synaptic input into the model cell delivers up to 1 nA of current to the soma (Fig. 19.11A), triggering 40 spikes per second in the absence of any other input (Fig. 19.11C). This sizable contribution is specific to apical tuft input (Bernander, Douglas, and Koch, 1994). Basal input will not be amplified since the calcium conductance is too far removed from the soma. The effect of the linearizing and amplifying currents remains hidden from the cell body and does not show up in the cell's f–I curve.

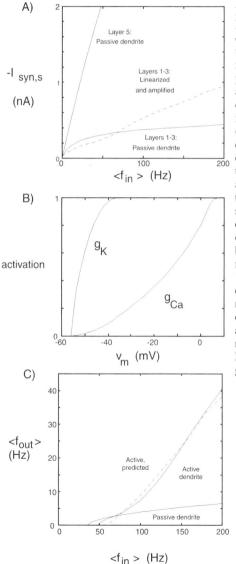

Fig. 19.11 Dendritic Amplification in a Pyramidal Cell The dynamic output range of the synaptic input into the apical tuft of the layer 5 pyramidal cell can be expanded via noninactivating voltage-dependent calcium and potassium conductances (Bernander, Douglas, and Koch, 1994). **(A)** The total somatic current delivered by 500 voltage-independent synaptic inputs into layers 1, 2, and 3 saturates quickly (see also Fig. 18.10). Spreading a potassium conductance g_K into all compartments in the apical tuft (beyond the major branching point of the apical dendrite in Figs. 3.7 and 18.6) and adding a hot spot of calcium conductances g_{Ca} at this branch point (with the voltage dependency shown in **B**) linearizes and amplifies the somatic current I_{soma} (dashed line in **A**). The final input-output relationship of the cell in **(C)** (computed by either clamping the soma to the mean somatic membrane potential or without restraining V_{soma} in any way) reveals that the distal input can modulate the cell's output firing rate in a significant manner. The function of these dendritic outward and inward currents is to linearize and amplify distal synaptic input. In this view, all-or-none calcium spikes are a mere epiphenomenon. Reprinted by permission from Bernander, Douglas, and Koch (1994).

The model treated here is quite simple and neglects many important aspects, such as calcium current inactivation, the detailed dynamics of the conductances, the effect of multiple calcium and potassium or other conductances, and so on. Furthermore, the exact distribution and voltage dependency of the two currents are only derived in a heuristic *ad hoc* manner. Yet the conclusion is similar to that of a biophysically more sophisticated simulation of active currents in cerebellar Purkinje cells (DeSchutter and Bower, 1994a,b): voltage-dependent inward dendritic currents can counteract the attenuation of distal synaptic inputs. In the Purkinje cell, this renders the somatic EPSP independent of the position of the synaptic input.

In this view, any all-or-none voltage events mediated by the amplifying inward current in the dendrites reflect a mere exuberance, an epiphenomenon. What is relevant is that the limited range of the current delivered to the soma is expanded by linearization and amplified in a graded manner.

A much more elegant and principled way of introducing a multitude of voltage-dependent dendritic currents is based on local, unsupervised learning rules that adjust the dynamics and the voltage dependency of the conductances. Such algorithms are related to the synaptic learning rules treated in Chap. 13 and achieve homeostasis by keeping the average calcium concentration at some setpoint (LeMasson, Marder, and Abbott, 1993) or by maximizing a quantity such as the mutual information between synaptic input and the output of the cell (Bell, 1992). Such learning schemes can operate continuously throughout the lifetime of the cell.

19.4 Recapitulation

It was held for a long time that voltage-dependent membrane conductances were restricted to the cell body and axon. Given the overwhelming evidence for the presence of a variety of sodium, calcium, and potassium conductances in the dendritic tree it is imperative that such active membranes be included into biophysically realistic compartmental models. We must also understand the functional role of such conductances.

Because of the great diversity of voltage-dependent currents found in so many different cell types throughout the animal kingdom, it is difficult to summarize the experimental data. The dendritic membrane of many neurons, in particular pyramidal cells, contains a relatively low but homogeneous distribution of ionic channels. This is enough to support the propagation of spikes from the axon and soma back into the dendritic arbor. That the spike is initiated in the axon and not locally in the dendrite can be explained by the fact that the difference in voltage threshold for spike initiation in the dendrite and V_{th} in the axon (due to much higher densities of active channels and/or intrinsic activation curves that are shifted toward more hyperpolarized values) is less than the voltage attenuation from the dendrite to the site of spike initiation

Under certain conditions, both rapid (1 msec) all-or-none sodium spikes and slower (10–50 msec) calcium-based action potentials can be observed in the dendritic tree. These may or may not propagate to the soma. The extent to which faster or slower spikes are initiated under physiological conditions in the dendrites and propagate to the soma remains unclear.

Numerous specific functions have been ascribed to active dendritic conductances, some dependent on dendritic spikes while others work in the absence of action potentials. Among the former resides the hypothesis that the timing of the back-propagated spike relative to the timing of any near-simultaneous presynaptic input is a crucial variable controlling whether the weight of the associated synapse decreases or increases. A number of researchers simulated the ability of locally generated spikes in dendrites and spines to instantiate various logical operations as well as temporal coincidence detection.

Two quite plausible proposals for the functional role of active inward currents in the dendrites see them as providing the biophysical substrate either to implement a multiplicative-like operation in a distributed fashion among spatially clustered groups of synapses (the neuron as a sigma-pi element) or to linearize and amplify distal synaptic input selectively so that this input can modulate the cell's discharge more effectively (dendritic amplifier). Any such proposal raises many more questions than it answers, one of the most pertinent one being the question of the necessary developmental and learning rules that can assemble all of this hardware with the required temporal and spatial specificity (that is, inserting a certain number of channels of a specific type into a specific location at some subneuronal site). While a passive dendritic tree would require much less specificity to "wire up," its computational abilities are far less than those of an active dendritic arbor.

20

UNCONVENTIONAL COMPUTING

As discussed in the introduction to this book, any (bio)physical mechanism that transforms some physical variable, such as the electrical potential across the membrane, in such a way that it can be mapped onto a meaningful formal mathematical operation, such as delay-and-correlate or convolution, can be treated as a computation. Traditionally only V_m, spike trains, and the firing rate $f(t)$ have been thought to play this role in the computations performed by the nervous system.

Due to the recent and widespread usage of high-resolution calcium-dependent fluorescent dyes, the concentration of free intracellular calcium $[Ca^{2+}]_i$ in presynaptic terminals, dendrites, and cell bodies has been promoted into the exalted rank of a variable that can act as a short-term memory and that can be manipulated using buffers, calcium-dependent enzymes, and diffusion in ways that can be said to instantiate specific computations.

But why stop here? Why not consider the vast number of signaling molecules that are localized to specific intra- or extracellular compartments to instantiate specific computations that can act over particular spatial and temporal time scales? And what about the peptides and hormones that are released into large areas of the brain or that circulate in the bloodstream? In this penultimate chapter, we will acquaint the reader with several examples of computations that use such unconventional means.

20.1 A Molecular Flip-Flop

The computation in question constitutes a *molecular switch* that stores a few bits of information at each of the thousands of synapses on a typical cortical cell. In order to describe its principle of operation, it will be necessary to introduce the reader to some basic concepts in biochemistry. The ability of individual synapses to potentially store analog variables is important enough that this modest intellectual investment will pay off. (For an introduction to biochemistry, consult Stryer, 1995).

20.1.1 Autophosphorylating Kinases

A group of proteins called *kinases* phosphorylate particular target proteins, that is they add a PO_4^{3-} group. The negative charge of the phosphate group changes the shape of the protein, altering its function in very specific ways. *Phosphorylation* is a very common mechanism to modulate ionic currents and synapses (Kennedy and Marder, 1992; Hille, 1992; Kennedy, 1994). Indeed, for the most part metabotropic synaptic receptors act by causing the final target channel to be phosphorylated. Many of these kinases are activated by elevated levels of intracellular calcium.

In principle such a kinase could be used as a trigger to a 1-bit storage device. Calcium rushes into the cell, directly or indirectly activating the kinase, which in turn switches protein molecules into their phosphorylated "on" state. This "on" state is then assumed to be read out by another protein or process.

The problem with this is twofold. Firstly, another class of enzymes, the *phosphatases*, undoes the work of the kinase by snipping off the PO_4^{3-} group. This returns the protein to its "off" state. The time constant of this degrading action is on the order of minutes. Secondly, all the molecules making up biological systems (with the exception of DNA) degrade over a period of days, weeks, or, at the most, within a few months. This relentless *molecular turnover* means that each and every protein, phosphorylated or not, is eventually replaced by newly synthesized, and thereby unphosphorylated, proteins.

How can such unstable molecules be used to construct memories that can last for a lifetime? One solution is to store information in the very stable DNA molecules in the nucleus at the cell body. While it appears plausible that such storage occurs for properties that affect the entire neuron, such as the level of expression of its receptors or the overall state of its firing rate adaptation (Kandel and Schwartz, 1982), it is unclear how the DNA in the nucleus could control the synaptic strength of each and every one of the thousands of synapses in the dendritic tree of a typical neuron. This has lead to the proposal (Crick 1984b; Lisman, 1985; Saitoh and Schwartz, 1985; Miller and Kennedy, 1986) that a bistable kinase stores the information locally at each synapse. (For older ideas on the regulation of cellular growth and differentiation with the help of enzymatic networks see Monod and Jacob, 1961.)

The principle uses something called *autophosphorylation*, in which a particular kinase, here generically termed kinase-1, can phosphorylate itself in an autocatalytic reaction. The basic switch, illustrated in Fig. 20.1, is built from two proteins, kinase-1 and a phosphatase. Some neuronal stimulus, such as a brief but strong calcium transient at a dendritic spine, triggers a local kinase, kinase-2. As long as it is present, it phosphorylates kinase-1, transforming it from its inactive K_1 to its active K_1^* form. The presence of a phosphatase turns off this form, by returning it to its native K_1 state. This system can only store information as long as the original kinase-2 is present, for in its absence, all of the kinase-1 eventually ends up in its inactive K_1 form.

Things become interesting if *intermolecular autophosphorylation* occurs. Here the phosphorylated kinase-1 molecules can phosphorylate other, not yet phosphorylated kinase-1 molecules. The more K_1^* is present, the larger the rate with which K_1 is transformed into its active counterpart. Conversely, the larger the amount of K_1^*, the smaller the pool of remaining K_1 that can be phosphorylated. The kinetic equation describing this first-order reaction can be formulated as (Lisman, 1985)

$$\frac{d[K_1^*]}{dt} = c_1 \frac{[K_1][K_1^*]}{K_{d1} + [K_1]} - c_2 \frac{[P][K_1^*]}{K_{d1}^* + [K_1^*]} \tag{20.1}$$

where K_{d1} and K_{d1}^* are the Michaelis constants for the kinase-1 and the phosphatase reactions, respectively, and $[P]$ is the concentration of the phosphatase. The *Michaelis constant* of an enzymatic reaction is defined as the concentration of the substrate (here

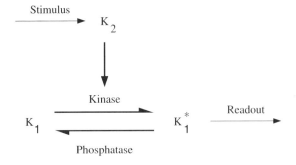

Stimulus

K_2

Kinase

K_1 K_1^* Readout

Phosphatase

Fig. 20.1 MOLECULAR FLIP-FLOP SWITCH Basic building block of a bistable molecular switch that can retain its memory indefinitely in the face of constant protein turnover. The switch is constructed from an inactive unphosphorylated and an active phosphorylated form of a protein kinase, labeled K_1 and K_1^* respectively. The transition from the inactive to the active form of the protein can be initiated from the outside by another kinase K_2 (which, in turn, is triggered by some neuronal stimulus such as an elevation in the local calcium concentration). The critical component of the switch is the ability of the active form of kinase-1 to facilitate the phosphorylation of its own inactive form: the higher the concentration of K_1^*, the larger the reaction rate. The amount of K_1^*, in turn, controls some other process. Some phosphatase molecule is assumed to return the phosphorylated kinase-1 to its native unphosphorylated state. Reprinted in modified form by permission from Lisman (1985).

either K_1 or K_1^*) at which the reaction rate is half of its maximal value. c_1 and c_2 are the turnover numbers of the kinase-1 and phosphatase.

The first term on the right-hand side can be thought of as the product of $[K_1]$ and a forward rate constant. Due to autophosphorylation, the rate constant increases with $[K_1^*]$, but saturates at high concentrations of $[K_1]$. The backward rate constant saturates as well and is proportional to the fixed concentration $[P]$. Equation 20.1 is supplemented by the requirement that the total amount of kinase-1 in its active and inactive forms be conserved,

$$T_K = [K_1] + [K_1^*]. \tag{20.2}$$

Equation 20.1 can be expressed in normalized units ($K = [K_1^*]/T_K$ with $0 \leq K \leq 1$; $K_{d1}' = K_{d1}/T_K$ and $K_{d1}^{*'} = K_{d1}^*/T_K$) as

$$T_K \frac{dK}{dt} = c_1 \frac{T_K(1-K)K}{K_{d1}' + 1 - K} - c_2 \frac{[P]K}{K_{d1}^{*'} + K}. \tag{20.3}$$

With our choice of parameters (see legend to Fig. 20.2), the first reaction essentially does not saturate and has the shape of a parabola, while the second reaction is a strongly saturating function of K (Fig. 20.2A). The difference between the two terms is shown in Fig. 20.2B and displays, in general, three zero crossings.

The stability of Eq. 20.3 can be investigated using the methods introduced in Chap. 7, except that here everything is simpler. The system is stable if its first temporal derivative is zero and its second derivative negative. It is clear that the origin, $K = 0$, is a stable point, since any small perturbation $K = \epsilon > 0$ will lead to a negative value of dK/dt, bringing the system back to the origin. However, if the calcium stimulus that initiates the entire reaction by activating kinase-2 is present for a long enough time, it can phosphorylate enough kinase-1 into its active form to move K past the second zero crossing in Fig. 20.2B.

The system will converge to the third zero crossing and will remain there, even once the initial kinase-2 stimulus has subsided. The basis of attraction of the second stable point is large enough for it to remain stable even in the face of the ubiquitous turnover of all proteins

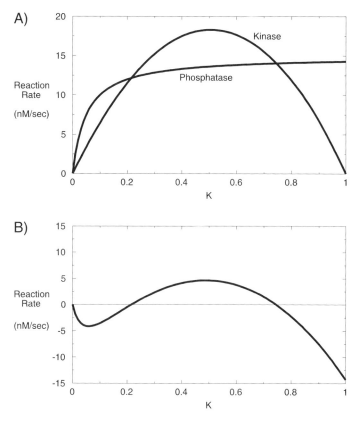

Fig. 20.2 BISTABILITY OF THE PHOSPHORYLATED FORM OF KINASE-I   (A) Turnover, that is, the rate of the two reactions, as a function of the relative fraction of kinase-1 in its phosphorylated form, $K = [K_1^*]/T_K$ (Eq. 20.3 and Fig. 20.1). "Kinase" refers to the first term in Eq. 20.3, expressing the rate at which kinase-1 is moved from its inactive to its active phosphorylated form, while the inverse reaction, here assumed to saturate, is labeled "Phosphatase." **(B)** Net rate of change in the phosphorylated form of kinase-1, that is, the difference between the two curves in the upper panel. The stable points of the systems are at $K = 0$ and 0.74. That is, either no active form is present or about three-quarters of kinase-1 is in its phosphorylated form. Parameters are from Lisman (1985) with $c_1 = 30/\text{sec}$, $c_2 = 3/\text{sec}$, $K_{d1} = 1\,\mu M$, $K_{d1}^* = 2.5\,\text{n}M$, $[P] = 5\,\text{n}M$ and $T_K = 50\,\text{n}M$. Reprinted in modified form by permission from Lisman (1985).

(unless this turnover is faster than the rate of autophosphorylation). In other words, 1 bit of information can be stored in this system, making it a molecular *flip-flop*, even though all the molecules that make up the switch can themselves slowly turn over.

Nothing is free in nature. In this case the cost of maintaining the switch in its "on" position is represented by the ATP molecules supplying the energy needed to rephosphorylate those kinase molecules that are being dephosphorylated (similar to the power requirements for maintaining dynamic RAM).

This switch can be turned off by increasing the amount of phosphatase such that the loss of the phosphorylated form of kinase-1 cannot be compensated any longer by the autophosphorylation.

The important concept that has emerged from this proposal (Crick 1984b; Lisman, 1985; Miller and Kennedy, 1986) is the feasibility of long-term information storage by a molecular switch that attains its stability using a biochemical reaction with positive feedback.

20.1.2 CaM Kinase II and Synaptic Information Storage

What is the experimental status of the molecular switch idea? Work on the biochemistry and biophysics of long-term potentiation (LTP) and long-term depression (LTD; see Chap. 13) has shown that a brain protein, called Ca^{2+}/calmodulin-dependent protein kinase II (*CaM kinase II*), has many of the properties of the hypothetical kinase-1 discussed above. It has also a number of differences that make the proposal even more attractive from a computational point of view (Lisman and Goldring, 1988; Lisman, 1989; Patton, Molloy, and Kennedy, 1993; Lisman, 1994; Hanson et al., 1994).

Firstly, CaM kinase II exists in high concentrations at the *postsynaptic density* at synapses in the central nervous system (Fig. 4.1; Kennedy, 1993). Secondly, the kinase, immobilized within the postsynaptic density, is activated by the calcium-calmodulin complex (Hanson et al., 1994; Fig. 11.7). Finally, once a particular concentration of calcium has been exceeded, the molecule switches into an "on" state that retains its activity even after the removal of the initial Ca^{2+} stimulus (Miller and Kennedy, 1986). And this calcium-independent form of the phosphorylated CaM kinase II has been found at least up to an hour after the induction of LTP (Fukunaga et al., 1993).

One important difference between the hypothetical kinase-1 and CaM kinase II is that the latter works by *intramolecular autophosphorylation*. Figure 20.3A illustrates a schematic version of the structure of this enzyme, which consists of 12 subunits. Each subunit, in turn, contains three to four phosphorylation sites. Miller and Kennedy (1986) determined experimentally that calcium is only necessary for the addition of the first two to four phosphate groups onto these sites (out of a possible 30 or so sites) on a single CaM kinase II molecule. Subsequently, even if Ca^{2+} is removed, the other sites phosphorylate on their own in an autocatalytic reaction. If subunits become dephosphorylated by the action of phosphatase or if a new—unphosphorylated—subunit is inserted into the molecule as part of the general protein turnover, the remaining phosphorylated sites are more than sufficient to rephosphorylate all sites.

The fact that the autocatalytic reaction only occurs within a single molecule of CaM kinase II (that is, one molecule cannot phosphorylate a subunit in another molecule of CaM kinase II) opens the door to the possibility that analog information, rather than 1 bit of information, can be stored at an individual synapse (Lisman and Goldring, 1988; Fig. 20.3B). This assumes that the calcium transient, initiated by ongoing synaptic activity, does not saturate the postsynaptic density for so long that all CaM kinase II molecules will be in their phosphorylated states. Each calcium event will cause some molecules to switch. The more calcium, the more molecules will be switched. The amount of information that can be encoded in this manner depends on the uncertainty in the fraction of kinase molecules that is switched into their "on" state. Given the stochastic nature of molecular binding, Lisman and Goldring (1988) estimate that about 80 of these molecules could effectively store 3 to 4 bits of information at each synapse.

It is very tempting to estimate the storage capacity of the brain by simply multiplying this number by the total synaptic density of the cortex. But the implicit assumption that all synapses are independent of each other is most certainly incorrect. Furthermore, we know almost nothing about the spatial and temporal specificity of the read-out mechanism of the autophosphorylating molecular switch. Even the best RAM memory in the world serves little purpose if its states cannot be independently accessed.

The specific hypothesis we discussed, although exciting and with some experimental support, needs to be worked out in all of its messy biochemical details. What it does show, at least in principle, is that individual molecules can be used to instantiate computations using positive and negative feedback loops.

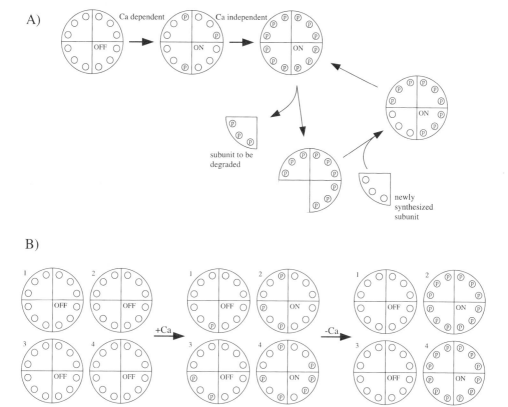

Fig. 20.3 SYNAPTIC INFORMATION STORAGE VIA CaM KINASE II Experimental observations implicate the brain type II Ca^{2+}/calmodulin-dependent protein kinase (CaM kinase II) as the crucial protein implementing the autocatalytic switching behavior proposed by Crick (1984b) and Lisman (1985). **(A)** One such molecule consists of a number of subunits with a total of about 30 phosphorylation sites, of which 12 are illustrated here. A rise in Ca^{2+} leads to formation of a calcium-calmodulin complex (Fig. 11.7) that induces phosphorylation of some of the sites. If a critical number of these sites (probably around three) has been phosphorylated, the molecule can itself phosphorylate the remaining sites in the absence of any further Ca^{2+}. This autocatalytic reaction also assures that the molecule remains completely phosphorylated in the face of the degrading action of phosphatase and the perpetually occurring protein turnover of individual subunits. **(B)** Because a phosphorylated molecule of this kinase cannot phosphorylate another molecule, an ensemble of molecules can encode graded information, as long as the initial calcium stimulus does not lead to phosphorylation occurring on all sites of all molecules. Due to the random nature of this process, the calcium transient causes a few sites on the four molecules to be phosphorylated (left and center). Here, it is assumed that if at least three sites per molecule are phosphorylated, the remaining sites on that molecule will be autophosphorylated, resulting in two completely activated and two inactivated molecules, even in the absence of elevated calcium (right). A longer calcium transient would have turned all four molecules on, allowing for the analog storage of information at each synapse (up to 4 bits). Reprinted in modified form by permission from Lisman and Goldring (1988).

In closing, let us recall the large number (on the order of 10^2–10^3) of regulatory proteins that can interact with calcium and other second messengers, giving rise to a complex web of dense interactions. We need to understand in what sense these molecules not only serve as metabolic intermediaries but also represent, store, and manipulate information. In principle, the computations involving V_m and $[Ca^{2+}]_i$ could be supplemented by molecular or protein computations. Their advantages are the minimal spatial requirements, usually operating at the submicrometer scale, and the speed of the reaction, limited only by the associated chemical rate constants (Koshland, 1987; Bray, 1995; Barkai and Leibler, 1997).

20.2 Extracellular Resources and Presynaptic Inhibition

The designer of analog very large scale integrated (VLSI) electronic circuits needs to be careful when considering the spatial placement of various circuit components onto the silicon chip. For instance, the wire carrying the digital clock cannot be placed too closely to the wire carrying analog information since capacitive coupling between the two can induce sufficient noise so that the possibly very small analog variable can become corrupted. If done cleverly, the parasitic capacitance of a standard metal-oxide-semiconductor transistor can be exploited for a time-derivative computation rather than being an undesirable feature of this particular type of hardware.

We would argue that it is one of the defining characteristics of any efficient information processing system that the algorithms implemented are carefully matched to the physics of the machine. If we are willing to spend enough resources, of which there are fundamentally three—space, time, and power—we can violate such design principles, but at a price. In today's digital computers we take enormous amounts of time and power to implement functions that a house fly, with a brain volume of less than 1 mm^3, can carry out in real time.

It is reasonable to assume that evolutionary pressures will have acted on the nervous system in such a way as to optimize the placement of all circuit elements using constraints that we are only now beginning to be dimly aware of.

A case in point is the resource limitation imposed by the extracellular space (Montague, 1996). The amount of space accessible to ions outside neurons and glia cells is very small. Correcting for shrinkage during histological preparation of biological material leads to a *volume fraction* α, which is the fraction of the volume that is extracellular space. It is around 20% for most tissues (that is, $\alpha = 0.2$), with peak values of 30% for the parallel fiber system in the cerebellum (Nicholson, 1995; Syková, 1997; Barbour and Häusser, 1997). Of course, $\alpha = 1$ for an unencumbered volume (e.g., a beaker of water).

The smallness of this space could have important functional consequences, as proposed by Montague (1996). As discussed already in Sec. 4.2, the arrival of an action potential at the presynaptic terminal causes voltage-dependent calcium channels to open. The resulting influx of extracellular calcium is the crucial signal that triggers exocytosis, when the vesicle containing neurotransmitter molecules fuses with the membrane and dumps its content into the synaptic cleft (Bennett, 1997). When not enough calcium is present outside the presynaptic terminal, the probability of release, and therefore the average postsynaptic response, drops. Reducing the extracellular concentration $[Ca^{2+}]_o$ by a factor of 2, from 2 to 1 mM, reduces the postsynaptic response to 30% of its original value (Mintz, Sabatini, and Regehr, 1995). Reducing the extracellular Ca^{2+} concentration fourfold attenuates the postsynaptic response more than tenfold.

It has been estimated that on the order of 13,000 Ca^{2+} ions enter the presynaptic terminal to trigger release of a vesicle (Borst and Sakmann, 1996). Given the tight extracellular space, filled with membranes and other organelles that impede the rapid diffusion of ions, this influx of Ca^{2+} into the presynaptic terminal will deplete the calcium concentration just outside the terminal. This reduction in $[Ca^{2+}]_o$ is compounded if the postsynaptic terminal also demands calcium, for instance, if it contains significant numbers of NMDA receptors or voltage-dependent calcium channels.

When considering the diffusion of calcium, potassium, or other ions outside the cell one must account for the fact that the diffusion coefficient in the tight tortuous space between the glia cells and neurons is not the same as the diffusion coefficient in aqueous milieu. The free diffusion of ions is hindered by these membrane obstructions, by macromolecules, by charged molecules and so on. The reduction of D is accounted for by the so-called *tortuosity* factor λ (not to be confused with the electrotonic space constant). The effective diffusion coefficient D_{eff} as measured in the tissue is related to the coefficient in aqueous solution, D, via

$$D_{\text{eff}} = \frac{D}{\lambda^2}. \tag{20.4}$$

In an idealized aqueous solution $\lambda = 1$; experimentally measured values for brain tissue fall in the 1.5–1.9 range (Nicholson and Phillips, 1981; Nicholson, 1995). In other words, the effective diffusion coefficient is reduced by a factor 0.3 to 0.4 (see Eq. 11.26).

Taking both the volume fraction and tortuosity into account modifies the one-dimensional diffusion equation (Eq. 11.18) as follows:

$$\frac{\partial C(x, t)}{\partial t} = \frac{D}{\lambda^2} \frac{\partial^2 C(x, t)}{\partial x^2} + \frac{\text{sources and sinks}}{\alpha}, \tag{20.5}$$

where the sources and sinks include membrane currents, pumps, buffers, and so on.

Egelman and Montague (1997) have carried out exploratory simulations of the diffusion of Ca^{2+} ions in the extracellular space around the synaptic terminals on the basis of Eq. 20.5 and estimate that $[Ca^{2+}]_o$ can drop by as much as a quarter in response to a single action potential invading the presynaptic terminal. A high-frequency burst can further deplete extracellular calcium. Given packing densities on the order of one billion synapses per cubic millimeter of neuronal tissue (Sec. 4.1), a neighboring synaptic terminal might now have difficulties to release a synaptic vesicle successfully, since the needed Ca^{2+} ions have been "stolen" by the first synapse.

Depending on how long it will take to replenish extracellular $[Ca^{2+}]_o$ by pumps and calcium-release mechanisms, the first synapse can inhibit synaptic transmission at all closely adjacent synapses. And this irrespective of whether its postsynaptic action is excitatory or inhibitory, since this form of inhibition relies on calcium stealing by the presynaptic terminal. This is an instance of *presynaptic inhibition* in the absence of any postsynaptic conductance change.

It is well possible, of course, that the temporary reduction in $[Ca^{2+}]_o$ is too brief or too minute to significantly affect any but directly apposed synapses, or that the effect exists but that the brain does not exploit it for computational purposes. Or the reduction of $[Ca^{2+}]_o$ might only be significant during calcium spikes in the dendrites. Another interesting possibility is that synaptic microcircuits (Sec. 5.3), such as the spine-triad arrangement so prevalent in the thalamus (Sec. 12.3.4), implement a special form of presynaptic inhibition that relies on the exact spatial placement of the synaptic terminals relative to each other.

20.3 Computing with Puffs of Gas

Throughout this book, we have lived with the convenient assumption that the three-dimensional arrangements of synapses, dendrites, axons, and cell bodies do not matter and that all neurons can be reduced to sets of one-dimensional cylinders. This simplification is a powerful one since it allows us to study the spatio-temporal distribution of the membrane potential and calcium with ease on the basis of one-dimensional cable and diffusion equations.

However, one of the most obvious features of almost any piece of nervous tissue is its high degree of structure: columns, layers, laminae, and other spatial organizational forms abound. It therefore behooves us to at least briefly consider the range of possible effects of three-dimensional geometry on neuronal computation.

As alluded to in Chap. 13, one or more retrograde messenger molecules have been postulated to mediate between the postsynaptic induction of LTP and its, at least partly, presynaptic expression. We also pointed out that the specificity of synapses during LTP and LTD might not be quite as high as frequently asserted. In the best explored model system for LTP, the synapses between the output fibers of CA3 pyramidal cells in the hippocampus and CA1 neurons, *spillover* exists. That is, synaptic plasticity is not only confined to the handful of synapses at the intersection of the single presynaptic axon and the postsynaptic cell recorded from, but is observed at "neighboring" synapses as well. LTP can be expressed at synapses from the same axon but made onto neighboring neurons that have not undergone the induction process (Bonhoeffer, Staiger, and Aertsen, 1989; Schuman and Madison 1994a). In a second form of spillover, excitatory synapses within 50–70 μm of the potentiated synapse onto the same neuron (but from nonstimulated axons) can be potentiated as well (Engert and Bonhoeffer, 1997; for a summary, see Murthy, 1997).

It has been hypothesized that these effects are mediated by a rather unusual class of neuronal messengers, of which the best known is the free radical gas *nitric oxide* (NO) (for reviews see Schuman and Madison, 1994b; Montague and Sejnowski, 1994; Schuman, 1995; Brenman and Bredt, 1997). Another possible candidate is the gas carbon monoxide (Zhuo et al., 1993).

While conventional neurotransmitters like glutamate, GABA, acetylcholine, or noradrenaline are packaged in synaptic vesicles and released from the presynaptic terminal in response to an invading action potential, nitric oxide is not released in vesicles but directly diffuses away from its site of production. Because it is gaseous and extremely membrane permeant, it readily moves *across cell membranes* irrespective of dendrites, axons, or other cellular processes. The second feature distinguishing it from a conventional neurotransmitter is that nitric oxide is "produced on demand" by a calcium-calmodulin-dependent enzyme.

Nitric oxide is limited in its spread by the fact that it is rapidly oxidized, with a half-life of 4 sec and possibly much less. Given its large diffusion coefficient (3.8 μm^2/msec; see Table 11.1) it can diffuse 160 μm in all directions in this time (Eq. 11.26), a sphere that encompasses around 20,000 synapses.

Nitric oxide has been implicated in the control of both LTP and LTD. In particular, inhibition of the enzyme responsible for producing NO, nitric oxide synthase, blocks the establishment of the NMDA-dependent form of LTP in the hippocampus. Furthermore, the correlation of presynaptic electrical activity and elevated levels of NO is sufficient to potentiate transmission at recently activated synaptic terminals (Zhuo et al., 1993).

A plausible scenario is based on conjoint pre- and postsynaptic electrical activity that causes the $[Ca^{2+}]_i$ in the spine to rise, triggering the production of NO by nitric oxide synthase (Montague et al., 1994). It immediately diffuses away from this site and into the local volume of tissue. Its sphere of influence includes its own presynaptic terminal as well as nearby synapses, those on the same postsynaptic neuron as well as those on other neurons (Fig. 20.4). At synapses that were active within some time window, for instance, where a vesicle was recently released, the NO leads to either LTP or LTD through a—as of yet—ill-characterized cascade of biochemical events (Kennedy, 1994).

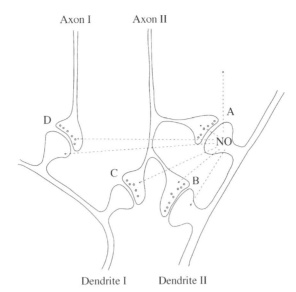

Fig. 20.4 Retrograde Messengers and Synaptic Specificity Schematic drawing illustrating the effect that the gas nitric oxide (NO) can have on synaptic weights (Gally et al., 1985). A sufficiently large calcium influx, reflecting an appropriate conjunction of pre- and postsynaptic activity, at spine A causes production of NO. Like a puff of gas, NO will freely diffuse from its site of production into the adjacent tissue volume. Presynaptic terminals that have recently been activated and that see an increase in the local concentration of NO (Eq. 20.6) upregulate their synaptic weight, leading to LTP. This *volume learning* (Gally et al., 1990) is less specific than classical associative Hebbian learning, since potentially thousands of synapses in the neighborhood of the primed one could be affected. Reprinted by permission from Gally et al., (1990).

As NO diffuses outward, its concentration drops, due to dilution, chemical degradation through oxidation to nitrates, and destruction by hemoglobin and other molecules. This imposes a limit on the volume throughout which the arrival of NO could trigger the biochemical cascade of events that eventually leads to the presynaptic expression of LTP.

Within this volume a large number of synapses could potentially change their synaptic weights, even those that lacked either pre- or postsynaptic activity to satisfy Hebb's rule (Eq. 13.7 or its variant Eq. 13.8). Experimental evidence suggests that synapses within 50 μm of a potentiated one do (Engert and Bonhoeffer, 1997).

The appropriate *volume learning* rule that needs to replace the covariance rule of Eq. 13.8 (Montague and Sejnowski, 1994) is

$$\Delta w_{kl} \propto (V_k - \langle V_k \rangle)([NO]_k - \langle [NO]_k \rangle) \tag{20.6}$$

where V_k is the presynaptic activity (and $\langle V_k \rangle$ its time average), $[NO]_k$ the instantaneous concentration of NO (or any other retrograde messenger molecule) at the presynaptic terminal, and $\langle [NO]_k \rangle$ its running average. The index l ranges over some neighborhood of the initially potentiated synapse (i, j) from which the NO diffuses. Of note is that Eq. 20.6 is independent of postsynaptic activity.

Conjoint increases in presynaptic activity and the concentration of NO (relative to their mean) cause LTP while a presynaptic increase in conjunction with a relative decrease in the concentration of the retrograde messenger (and vice versa) leads to LTD.

In other words, the unit of synaptic plasticity might not be individual synapses, as assumed by neural network learning algorithms, but groups of adjacent synapses, making for a more robust, albeit less specific learning rule. How specific depends, among other things, on the exact temporal relationship between the release of a presynaptic vesicle and the local change in $[NO]_i$. In any case, it is obvious that the detailed placement of axons and synapses in three dimensions will greatly affect their ability to locally store information.

It is important that plasticity rules, as in Eq. 20.6, be combined with realistic models of the three-dimensional configurations of axons and synapses in order to better understand developmental as well as ongoing learning processes in the brain (for an example, see Montague, Gally, and Edelman, 1991).

The picture that we are left with is one in which afferent patterns of activity are translated into local hot spots of calcium in spines and other postsynaptic terminals. These generate local puffs of gas that freely diffuse in three-dimensional space to up or down regulate synaptic weights at neighboring synapses.

20.4 Programming with Peptides

We mentioned in Chap. 4 the principle of synaptic *colocalization* of fast, classical neurotransmitters with much slower acting neuromodulators. In many synaptic terminals, neurotransmitters—small molecules such as ACh, GABA, or glutamate—are stored in small and clear vesicles while neuropeptides, short (typically 2–10) amino acid chains, are stored in *dense core vesicles* within the same terminal. The different vesicles can be released differentially, for instance, one preferentially at low stimulation frequencies and the other during high-frequency bursts. The number of neuroactive peptides in the brain, with very idiosyncratic names relating to the biological function they have historically first been identified with, totals 50 and keeps on rising. They are found throughout the animal kingdom and throughout all nervous structures (for reviews, see Kupfermann, 1991; Marder, Christie, and Kilman, 1995).

The study of their neuronal effects has progressed farthest in small nervous systems that are readily accessible to neurochemical methods. The model system of choice is the *stomatogastric ganglion* (STG) of the crab *Cancer borealis* (Selverston and Moulins, 1987). This ganglion consists of approximately 30 neurons and is responsible for controlling movement in the esophagus and stomach. The neurons are tightly coupled with both chemical and electrical synapses.

About 50 individually identifiable input fibers project from more forward located ganglia into the very dense central core of the STG. This neuroplex, consisting of nothing but axons, synapses, and dendrites, is about a quarter of a millimeter in diameter and is surrounded by the associated cell bodies (Fig. 20.5). Many of the input axons branch widely and throughout the ganglion while some have more restricted branching patterns. Some of the associated terminals resemble neurosecretory organs that are thought to release peptides into the hemolymph to act at distant sites while some resemble conventional point-to-point contacts. Colocalization of neuropeptides also fails to follow any simple rule, with any given peptide being associated with a different complement of cotransmitters in different neurons (for instance, all terminals staining immunocytochemically for peptide A may also stain for peptide B, but terminals positive for peptide B might be negative for A but positive for peptide C; Marder et al., 1997).

Hormones   Neuromodulators

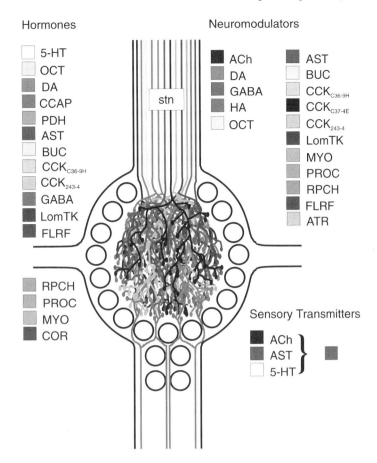

Hormones	Neuromodulators
5-HT	ACh AST
OCT	DA BUC
DA	GABA CCK_{C36-9H}
CCAP	HA CCK_{C37-4E}
PDH	OCT CCK_{243-4}
AST	LomTK
BUC	MYO
CCK_{C36-9H}	PROC
CCK_{243-4}	RPCH
GABA	FLRF
LomTK	ATR
FLRF	
RPCH	
PROC	Sensory Transmitters
MYO	ACh
COR	AST
	5-HT

stn

Fig. 20.5 Neuromodulators in a Small Nervous System   Summary of neuroactive substances present in the inputs to the stomatogastric ganglion (STG) of the crab. *Hormones* are released into the circulating blood or hemolymph and can thus act globally, while *neuromodulators*—usually peptides—are released by conventional synaptic terminals and can modulate neuronal properties over long time scales. *Sensory transmitters* act locally and rapidly. Some substances, such as acetylcholine (ACh), bind to fast nicotinic as well as to slow muscarinic receptors. The input fibers project into the dense, central core of the ganglion, indicated schematically, with the 30 or so cell bodies (open circles) located in the periphery. The neuromodulators are present in various and complex subsets of the input fibers and are colocalized with conventional neurotransmitters. The area of influence of these peptides ranges from local synapses to the entire ganglion. As illustrated in Fig. 20.6, their effects vary widely in scope, time scale, and sign. Unpublished data from E. Marder, A. E. Christie, and M. P. Nusbaum, printed with permission.

While the distribution of neuropeptides and their postsynaptic receptors ranges from the very local to the global and defies any simple classification, their effects on their targets are equally varied.

When food moves from the mouth into the esophagus of the crab, the neurons in the STG generate a variety of different rhythmic motor patterns. Accordingly, most of the STG cells can be identified on the basis of a characteristic oscillatory pattern involving EPSPs, IPSPs, bursts, plateau potentials, and the like. Some neurons display these regular sequences in isolation, that is, when all synaptic input has been removed, while others rely on network interactions. Such *central pattern generators* (CPG) are a common feature of vertebrate and invertebrate motor systems (Marder and Calabrese, 1996).

The action of peptides on these oscillatory discharges as well as on the 40 odd stomach muscles enervated by the STG neurons is complex and still ill-understood. We illustrate the possibilities in Fig. 20.6, involving the application of the peptide *proctolin*, released by fibers projecting into the neuropil. Applying proctolin to the entire ganglion (by adding it to the bath solution) changes the properties of the hardware at multiple organizational levels.

1. It affects specific motor patterns, here the so-called "pyloric rhythm" (Fig. 20.6A). Acting via a second-messenger cascade, proctolin modulates a voltage-dependent conductance in two identifiable neurons. The net effect is to increase the frequency and modulation depth of the oscillations (Hooper and Marder, 1987).

2. Even at very low concentrations, proctolin enhances the motor neuron evoked contractions of certain stomach muscles (Fig. 20.6B). The muscle may now be sensitive to firing rates it did not previously respond to (Marder et al., 1997).

3. Peptides can also affect the evoked amplitude of individual synaptic connections (Fig. 20.6C). Dual intracellular impalement in the IC and GM neurons in the absence of any spiking activity (by adding a sodium-channel blocking agent to the bath) fails to reveal any direct synaptic connection. Yet in the presence of proctolin a robust and profound inhibition can be observed.

A different substance, crustacean cardio-active peptide (CCAP), which is not present in any input fibers to the STG but is released into the hemolymph via neurosecretory structures, can initiate a switch from one pattern of neural activity into a qualitative different one (Weimann et al., 1997). The complex action of CCAP on various slow intrinsic membrane conductances changes the normal 1:1 alternation between two specific neurons to a 2:1, 3:1, or 4:1 alternation.

Peptides can be thought of as reprogramming the nervous system by changing its motor pattern, its synaptic gains, and its output. In an uncanny way this resembles loading a new program into an *application specific integrated circuit* (ASIC), with each peptide, or combination of different peptides, loading a slightly different motor program.

Because peptides rely on passive diffusion to influence an entire neural network, acting akin to a "global variable," the time scale of action is seconds, minutes, or longer (Jan and Jan, 1983; Kuffler and Sejnowski, 1983). An extreme form of this can be found in the mammalian *suprachiasmatic nucleus*, the central "clock" that transmits the circadian 24-hour rhythm to the rest of the brain and body. Grafting experiments have proven that circadian activity rhythms can be sustained via direct action of an as yet unidentified diffusible signal (Silver et al., 1996). The action of these modulatory signals must be viewed in constrast to the computational function of fast, synaptic input, which is quite local in time and space.

20.5 Routing Information Using Neuromodulators

We argued in Sec. 9.3 that the action of noradrenaline is contingent on excitatory synaptic input. By itself, its application causes only a small and long-lasting EPSP, which can be thought of as an epiphenomenon. Its real action is to close a potassium conductance and

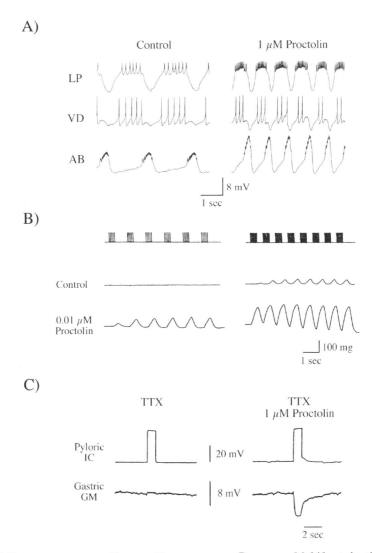

Fig. 20.6 REPROGRAMMING A NEURAL NETWORK VIA A PEPTIDE Multifaceted action of proctolin, one of the many peptides that are released from afferent fibers into the stomatogastric ganglion of the lobster (PROC in Fig. 20.5). **(A)** Intracellular recordings from three identifiable STG neurons reveal a complex oscillatory behavior in their membrane potential, characteristic of central pattern generators. The effect of 1 μM of proctolin on this "pyloric" rhythm is complex; among others, it increases the frequency of oscillation in the AB neuron. Reprinted by permission from Hooper and Marder (1987). **(B)** A hundredfold lower concentration of proctolin enhances contractions of two stomach muscles controlled by STG neurons. The motor nerve was stimulated repetitively (upper row); at low stimulation frequencies (left column) no muscle contraction is apparent while a weak one can be observed at high frequencies (right column). In the presence of proctolin, these contractions are much enhanced. Unpublished data from J. C. Jorge-Rivera, printed with permission. **(C)** Proctolin also affects synaptic gain. In this experiment, all spikes are blocked by the application of TTX. A depolarizing pulse in one identifiable neuron fails to cause any change in the membrane potential of another. Adding proctolin to the bath reveals a profound and long-lasting inhibitory synaptic connection from one to the other. Unpublished data from J. M. Weimann, printed with permission.

thereby increase the gain of the cell's discharge curve (as in Figs. 9.13 and 21.3) without affecting the cell's excitability. This mechanism might enable the nervous system to send information selectively this or that way in a dynamic manner.

Efforts to build massively parallel computers have lead to the realization that a major challenge facing the computer architect is the problem of routing information efficiently among the individual processors, that is, using the least amount of time and/or space (here, transistors). The cortex and similar structures, with upward of 10^{10} to 10^{11} processors operating in parallel, most likely face a similar conundrum. How is information routed among different neurons without quickly exceeding space by connecting every neuron with every other neuron? One possible mechanism to deal with this could be based on neuromodulators (Koch and Poggio, 1987).

Let us assume that a particular input neuron projects to a large number of neurons, that is, establishes conventional synapses with them. In the absence of any modulatory input, spikes in an afferent fiber evoke synaptic activity in all of its targets. Assume that in the presence of the neuromodulator A for which only a subset of neurons M_A, has receptors, the excitability of neurons in M_A is enhanced (and possibly suppressed in other neurons). Substance A could be released either from a local interneuron—as in Fig. 20.7—or from the axon terminal of a neuron far away (e.g., in the locus coeruleus). By itself A does not induce a signal. Yet in the presence of conventional synaptic input, a neuron that is part of M_A signals more vigorously than before while neurons that are not part of M_A respond in a much weaker manner. Conceptually, this can be thought of as routing synaptic input to the subset M_A of neurons. If another population M_B has receptors for neuromodulator B, the information can be routed to a different target.

In order for such an addressing scheme to work efficiently, a large number of neuromodulators is required to target specific subsets of neurons. Addressing works by selectively and temporarily (on the order of seconds) increasing the output gain of the class of neurons that are meant to be targeted without directly exciting them. This solution to the addressing problem is similar to the traditional telephone exchange system, in which connections are made and broken as required for exchanging information (in contrast to a dedicated-line solution).

20.6 Recapitulation

We here dealt with a number of mechanisms that are not conventionally thought of as subserving specific neuronal computations. None of them involve the membrane potential, the firing rate, or the intracellular calcium concentration.

The entire realm of biochemical computations has been neglected. Yet at present there are no solid arguments ruling out why specific molecular reactions might not subserve specific computations. As one instance we introduced a molecular flip-flop switch that relies on positive feedback—via autophosphorylation—to implement a long-term memory device with the storage capacity of a few bits that could reside at individual synapses. This is but one example of a realm of computation about which we know almost nothing. Given the extremely large and complex regulatory cascades and networks of proteins and enzymes, the possibilities for nested multilevel computations are staggering. The crucial question is whether the brain avails itself of these possibilities or whether such computations cannot be implemented for reasons having to do with lack of bandwidth, signal-to-noise ratio or specificity.

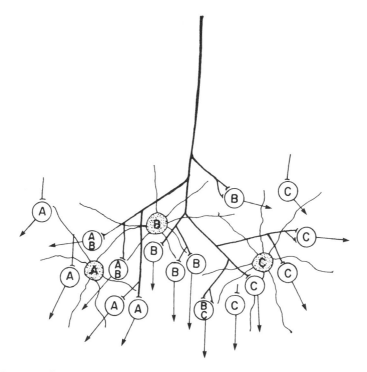

Fig. 20.7 ROUTING INFORMATION AMONG NEURONS USING NEUROMODULATORS Speculative addressing scheme based on neuromodulators (Koch and Poggio, 1987). The input axon (heavy line) is presynaptic to a variety of cells that project both within and outside the system. Neurons have receptors for many neuromodulators (here only three types are indicated schematically, A, B, and C). Neuromodulators can be released by an interneuron (as shown here) or by some external neuron, diffuse throughout the ganglion, and bind to their receptor sites on a specific subset of neurons (M_A and so on). The postsynaptic effect of a neuromodulatory substance is to change the gain of the firing response of the neuron, that is, the same neuron responds more or less vigorously to the same presynaptic input as before (e.g. Fig. 20.6C). Depending on which neuromodulatory substance has been released, action potentials coming in on the axon will only activate a subset of neurons. Functionally, this enables information to be selectively routed to neurons within the ganglion. Reprinted by permission from Koch and Poggio (1987).

Two other candidate mechanisms rely on the precise three-dimensional arrangements of neuronal components, either at the subcellular or at the cellular level. Whether or not the short-term depletion of extracellular Ca^{2+} ions implements a universal presynaptic inhibition that works without any conductance changes is pure speculation at the moment, but is too important to neglect from an experimental point of view.

That puffs of nitric oxide (and possibly carbon monoxide) are released in nervous tissue following local hot spots of synaptic-induced calcium activity opens up new avenues of spreading information in a retrograde manner, back across the synapse. Given the inexorable square-root law of diffusion and the aggressive chemical nature of nitric oxide, NO is unlikely to be effective beyond a small fraction of a millimeter from the site of its synthesis. This sphere of influence does include a potentially very large number of synapses. One of the "unfortunate" consequences of such a diffusing substance is that the specificity of Hebb's synaptic plasticity rule would be significantly reduced. The unit of learning would not be individual synapses, but groups of adjacent synapses.

The last two mechanisms exploit the very large laundry list of neuroactive substances (biogenic amines, neuropeptides, hormones) known to be present in any nervous system to implement global variables that act over a fraction of a millimeter and longer distances and on a seconds to minutes and longer time scale. We speculated on the role of neuromodulators in routing information, for reprogramming a particular neural network to change its mode of operation, for adapting the retina or other sensory surfaces to different operating conditions, and the like.

21

COMPUTING WITH NEURONS: A SUMMARY

We now have arrived at the end of the book. The first 16 chapters dealt with linear and nonlinear cable theory, voltage-dependent ionic currents, the biophysical origin of spike initiation and propagation, the statistical properties of spike trains and neural coding, bursting, dendritic spines, synaptic transmission and plasticity, the types of interactions that can occur among synaptic inputs in a passive or active dendritic arbor, and the diffusion and buffering of calcium and other ions. We attempted to weave these disparate threads into a single tapestry in Chaps. 17–19, demonstrating how these elements interact within a single neuron. The penultimate chapter dealt with various unconventional biophysical and biochemical mechanisms that could instantiate computations at the molecular and the network levels.

It is time to summarize. What have we learned about the way brains do or do not compute?

The brain has frequently been compared to a universal *Turing machine* (for a very lucid account of this, see Hofstadter, 1979). A Turing machine is a mathematical abstraction meant to clarify what is meant by algorithm, computation, and computable. Think of it as a machine with a finite number of internal states and an infinite tape that can read messages composed with a finite alphabet, write an output, and store intermediate results as memory. A universal Turing machine is one that can mimic any arbitrary Turing machine. We are here not interested in the renewed debate as to whether or not the brain can, *in principle*, be treated as such a machine (Lucas, 1964; Penrose, 1989), but whether this is a useful way to conceptualize nervous systems in this manner.

Because brains have limited precision, only finite amounts of memory and do not live forever, they cannot possibly be like "real" Turing machines. It is therefore more appropriate to ask: to what extent can brains be treated as *finite state machines* or *automata*? Such a machine only has finite computational and memory resources (Hopcroft and Ullman, 1979).

The answer has to be an ambiguous "it depends." If one seeks to understand the nature of computation in general and how operations carried out by nervous systems, such as associative memory or visual object recognition, can be implemented in machines, then Boolean circuits or finite state machines will be relevant (although not sufficient to

469

understand these operations, since the theory of computation includes no recipe to decide on the proper level of abstraction and representation).

Yet when the goal is to comprehend any one nervous system and why it has the features it does, for instance, probabilistic synaptic release, finite state machines and their relatives will not be very useful for a number of reasons. The most obvious is that these virtual machines are disembodied entities that fail to account for the fact that any information processing system existing in the real world has physical extent, takes some time before it comes to any sort of decision, and requires power to operate. And in the real world, space, time, and power are in short supply. Build a brain that is too big and the organism will be left behind in the struggle to survive; if the brain does not react in real time to an approaching threat, it will be quickly eaten by something that responds more rapidly. And an inefficient brain requires too much energy to sustain for long. None of these considerations are incorporated into our current notion of computation. It is only with the advent of portable computers, cellular phones, and the like that the power consumption associated with each elementary computation is beginning to be scrutinized (Sarpeshkar, 1998).

Brains differ in some rather obvious ways from present-day computers: memory and computation are not separated as they are in all of our current machines; the nervous system operates without any systemwide clock and is built from stochastic elements. Finally, developing as well as mature brains are constantly reprogramming themselves, up or down regulating synaptic weights, modulating the degree of adaptation, shifting the character and frequency of central pattern generators, changing the time constants of integration, and so on. Conceptually, this amounts to the input changing the transition function governing how the machine switches from one state to the next. Put differently, a nervous system will act like a machine that changes its instruction set as a function of its input.

On the positive side, what can we say about the manner in which the nervous system processes information (Koch, 1997)? Individual neurons convert the incoming streams of binary pulses into analog, spatially distributed variables, the postsynaptic voltage and calcium distribution throughout the dendritic tree, soma, and axon. These appear to be the two primary variables regulating and controlling information flow, each with a distinct spatio-temporal signature and dynamic bandwidth. The transformation between the input and the postsynaptic V_m and $[Ca^{2+}]_i$ involves highly dynamic synapses that adapt to different time scales, from milliseconds to seconds and longer. Information is processed in the analog domain, using a menu of linear and nonlinear operations. Second-order and higher multiplication, amplification, sharpening and thresholding, saturation, temporal and spatial filtering, and coincidence detection are the major operations readily available in the dendritic cable structure augmented by voltage-dependent membrane and synaptic conductances.

This is a large enough toolkit to satisfy the requirement of the most demanding analog circuit designer. To see this, take the case of implementing a spatial low-pass or band-pass filtering operation of the type described in Chap. 10 (e.g., Fig. 10.8). Much of early sensory processing, for instance, in the retina, involves such linear filtering operations. Convolving a two-dimensional $m \times m$ pixel image by an $n \times n$ pixel size filter requires on the order of $n^2 m^2$ operations on a digital computer (this can be reduced to $6m^2 \log m$ operations using the fast fourier transform). Implemented within a dendritic tree or across nonspiking retinal neurons connected by electrical gap junctions (Sec. 4.10), spatio-temporal filtering is carried out in parallel within one or two time constants, no matter the size of the image.

The outcome of these analog dendritic computations is once again converted into asynchronous pulses—possibly supplemented by bursts that could act as high-confidence symbols—and conveyed to the following neurons. The precision of this firing rate code is in the 5–10 msec range (and better in some cases). Synchrony or correlations across neurons are likely to contribute to coding, but the extent remains unclear. Reliability can be achieved by pooling the responses of a small number of neurons (on the order of 10–1000).

Individual nerve cells respond in a relatively slow and unreliable manner to sub- or peri-threshold input. However, for rapidly changing suprathreshold input, as expected during normal operations, neurons can respond in a rapid and reliable manner (Mainen and Sejnowski, 1995).

It is doubtful whether the effective resolution, that is, the ratio of minimal change in any one variable, such as V_m or $[Ca^{2+}]_i$, relative to the noise amplitude associated with this variable, exceeds a factor of 100. Functionally, this corresponds to between 6 and 7 bits of resolution, a puny number compared to a standard 32-bit machine architecture.

Memory is everywhere, intermixed with the computational elements, but it cannot be randomly accessed: in the concentration of free calcium in the presynaptic terminal and elsewhere, implementing various forms of short-term plasticity, in the density of particular molecules in the postsynaptic site for plasticity on a longer time scale, in the density and exact voltage dependency of the various ionic currents at the hour or day time scale, and ultimately, of course, in the genes at the cell's nucleus for lifetime memories.

21.1 Is a Biophysics of Computation Possible?

Multiplication is both the simplest and one of the most widespread of all nonlinear operations in the nervous system. It is closely related to the operations of correlation and of squaring. The biophysical implementation of these canonical operations has therefore received widespread scrutiny. Let us briefly summarize some of the more plausible scenarios.

21.1.1 The Many Ways to Multiply

1. *Nonlinear synaptic interaction via shunting inhibition* (Torre and Poggio, 1978; Koch, Poggio, and Torre, 1982; see Sec. 5.1.6).

2. *NMDA synapses in a passive or active dendritic tree* (Mel, 1992, 1993; see Secs. 5.2.2 and 19.3.4).

3. *Log-exp transform.* This exploits a technique common to analog electronic circuits. To multiply x with y take the logarithm of both, add them, and then invert by computing an exponential,

$$xy = e^{\log x + \log y}. \tag{21.1}$$

A possible example of this can be found in the locust's visual system (Hatsopoulos, Gabbiani, and Laurent, 1995). The instantaneous firing rate $f(t)$ of a single identified neuron, the lobula giant movement detector, can be modeled by the algebraic product of two terms,

$$f(t) = \alpha \dot{\theta}(t - \delta t) e^{-\beta \theta(t - \delta t)} \tag{21.2}$$

where θ is the angle subtended by an approaching object on the insect's retina, $\dot{\theta}$ is its angular velocity, α and β are appropriate gain factors, and δt is the latency (on the order of 40 msec) of the response. This system has the elegant property that the neuron's peak firing rate is reached when an approaching object has reached a critical angular size on the retina.

It has been hypothesized (F. Gabbiani, G. Laurent, and C. Koch, private communication) that the time-averaged membrane potential $\langle V_m \rangle$ (Eq. 17.12) at the spike initiation zone of this cell, when an object is rapidly moving toward the locust, is proportional to the sum of the synaptic contributions from two sources: the logarithm of $\dot{\theta}$ from the excitatory input, and $-\theta$ from an inhibitory input. Multiplication could be achieved if the firing rate is exponentially related to $\langle V_m \rangle$. Because this neuron is readily accessible to intracellular electrodes while the animal is visually stimulated, it represents an ideal model system to study the biophysical implementation of multiplication.

4. *Product of firing rates* (Srinivasan and Bernard, 1976). This mechanism relies on coincidence detection between two statistically independent spike trains by an integrate-and-fire unit. Imagine two input streams, carrying spikes at a mean rate of f_A and f_B, that converge onto a leaky integrate-and-fire unit (Fig. 21.1). The unitary EPSP of amplitude V_0 from either input is large but does not, by itself, exceed the threshold. For this, two EPSPs are required. In order for near simultaneous inputs from one neuron not to trigger a spike, their average minimal interspike interval must exceed Δt, as defined implicitly by $V_0 e^{-\Delta t / \tau} + V_{th} = V_{th}$. Because the probability of the joint occurrence of two statistically independent events is the product of the probabilities of the individual events, it can easily be shown that the average firing rate of the coincidence unit is

$$f_C = 2\Delta t f_A f_B . \tag{21.3}$$

This trick only works in the presence of sufficient jitter. If the input spikes are too regular, the response depends on the exact phase relationship between the two trains. (Srinivasan and Bernard, 1976, simulated this mechanism using a C_V of the input spike trains of around 30%.) Poisson distributed input spike trains give rise to a qualitatively similar result (Fig. 21.1).

5. *Linear summation of noisy linear threshold neurons* (Suarez and Koch, 1989). A squaring operation can be implemented using a population of spiking cells. The output of these neurons is zero up to a threshold V_{th} and linear in the input for larger values. An input x is applied to a population of such cells whose thresholds are drawn from a uniform probability distribution (the V_{th} value for any one cell remaining constant in time), and the output from all cells is summed. This sum approximates the area, between V_{th} and some upper limit, under the output curve using a Monte-Carlo sampling procedure. As the number of neurons goes to infinity (in simulations 50–100 cells suffice; Suarez and Koch, 1989), the sum converges to a quadratic function of the input (for a linear input-output curve). Computing xy requires two populations of such cells with jittered thresholds, one evaluating $(x + y)^2$ and one $(x - y)^2$. Subtracting these two yields the desired result.

Both this and the previous mechanism exploit random spike times for computational purposes. In other words, they do not work *in spite* but *because* of noise.

For further alternatives involving synaptic disinhibition, the synaptic transfer function at a graded dendro-dendritic synapse, or the cooperative nature of calcium binding to enzymes or other molecules (Sec. 11.4.2), see Koch and Poggio (1992).

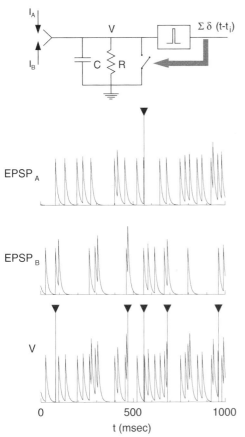

Fig. 21.1 MULTIPLYING FIRING RATES An elementary fact of probability theory, that independent probabilities multiply, is exploited here to achieve multiplication (Srinivasan and Bernard, 1976). A leaky integrate-and-fire unit is excited by two presynaptic axons, each with independent Poisson distributed spikes with $f_A = f_B = 20$ Hz giving rise to two associated synaptic current sources I_A and I_B. Each input gives rise to a 10-mV massive EPSP not sufficient to exceed threshold $V_{th} = 15$ mV ($\tau = 10$ msec). Occasionally, consecutive EPSPs from one fiber are close enough to each other to exceed the threshold (single arrow in upper trace). If two inputs from the two axons arrive "simultaneously," here within about 11 msec, V_{th} is exceeded and a postsynaptic spike is triggered (arrows in lower trace). The average output firing rate is proportional to $(f_A + f_B)^2$. This mechanism requires jitter in spike arrival times.

21.1.2 A Large Number of Biophysical Mechanisms for Computation

The integrated electronic circuits at the heart of today's microprocessors are fabricated using CMOS technology. The physics of these circuits makes logical inverters, NAND and NOR gates, particularly easy to implement. It is not that AND or AND-NOT gates cannot be designed in CMOS, but just that they use more transistors and are therefore less efficient (Fig. 0.1). We posed the question early on in this book whether the biophysics of membranes, synapses, and neurons equally constrain the types of mathematical operations that can be implemented efficiently in nervous tissue.

The original intent (Koch and Poggio, 1987) of the "Biophysics of Computation" program was motivated by the hope that a handful of biophysical computations would be universal, in the sense that they underlie all computations in the nervous system and that they can be implemented in a natural and "efficient" (see below) manner. In the earliest day of neural networks, the spiking threshold (or its continuous equivalent, the sigmoidal output function) was believed to be the only relevant nonlinear operation. Later on, it was argued (Torre and Poggio, 1978; Poggio and Torre, 1981) that the repertoire of primitive nonlinear operations should be expanded to include the nonlinear interactions occurring between conductance-increasing synapses in a passive dendritic tree.

As we saw in the preceding pages, the list of biophysical mechanisms that have been tied to specific computations has expanded considerably. Figure 21.2 tabulates these by borrowing from the summary diagram in Churchland and Sejnowski (1992) and ordering them along two dimensions, their approximate spatial extent and the time scales at which they act. And this listing is by no means exhaustive.

It should be kept in mind that many of the biophysical mechanisms discussed here are likely to have important noncomputational roles. Computers include circuits for managing power, for compensating parasitic capacities, for speeding up memory access, and so on, which make the system perform its task more efficiently. Yet these circuits are themselves neither strictly necessary nor directly related to specific computations. Likewise, neurons develop their characteristic properties over the lifetime of the animal, have metabolic requirements, and need to ensure stability in the face of relentless molecular turnover. Because natural selection acts to ensure the overall optimality of the organism given its particular ecological niche, it will frequently be difficult to isolate the exact role that any biophysical mechanism plays in brain-style computation.

21.1.3 Can Different Biophysical Mechanisms Be Selected For?

It is sobering to realize that to implemt a single operation—multiplication—the nervous system can choose from mechanisms that operate at the level of individual synapses and spines, to those that require small populations of cells. This raises the unsettling, but quite plausible scenario in which any one computation is carried out using a plurality of mechanisms at different spatial and temporal scales.

As just one example consider retinal direction selectivity, the differential response to the motion of a spot of light in the preferred direction compared to motion in the opposite direction (Sec. 5.1.7). This asymmetry could arise due to nonlinear synaptic interactions at the level of the starburst amacrine cells and is reinforced by exploiting temporal dispersion in the elongated asymmetrical dendrites of these cells and the nonlinear synaptic transfer function onto the dendrites of the ganglion cells. The response could be further accentuated by nonlinear synaptic interactions in the dendritic tree of the ganglion cell.

Numerous biophysical mechanisms acting in tandem is a robust way to achieve one's goal since failure of any of these operations would not imperil the entire computation. Furthermore, each mechanism most likely works optimally only under a narrow range of stimulus conditions (in the above example, contrast or speed of the spot of light), so that invariance to the vagaries of the stimulus requires several mechanisms operating in parallel.

Indeed, one can argue for an extreme version of this where the exact contributions that the mechanisms outlined in Fig. 21.2 make toward any one specific computation differ, depending on the developmental stage of the animal and its own idiosyncratic experience. In other words, any one computation would be implemented by the linear or nonlinear superposition of a host of biophysical mechanisms, where the coefficients specifying the contribution that each mechanism makes vary from one animal to the next.

Once we discovered all of these biophysical operations, it would remain for us to understand and characterize the learning rules that, given some input stream and various system level constraints, such as minimizing metabolic costs, would derive the exact contribution each mechanism makes toward the overall goal of achieving directional selectivity. Such a learning algorithm would need to take the nonlinear nature of the biophysical mechanisms into account and the way they interact with each other, whether

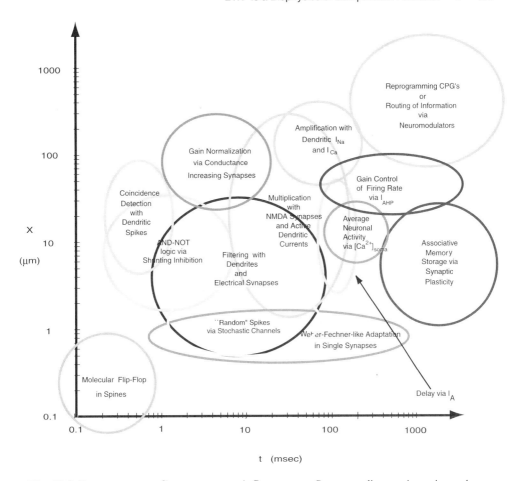

Fig. 21.2 BIOPHYSICS OF COMPUTATION: A SUMMARY Summary diagram imposing order on the biophysical mechanisms linked to specific computations by locating them in a space-time diagram. The degree of uncertainty about the extent to which any one mechanism actually exists and does what it has been proposed to do—admittedly a dicey and subjective matter—is indicated by the amount of shading: the darker, the more evidence in favor. The spatial extent of a particular mechanism that instantiates the computation and the time scale at which the operation occurs are indicated in logarithmic units (of μm and msec). This nonexhaustive list includes (with their section numbers in parentheses): autocatalytic molecular storage elements in single spines (20.1), synapses that signal deviation from their average input activity (13.5.4), low-, band-, or high-pass spatio-temporal filters implemented in single neurons or in neurons linked by electrical gap junctions (Chaps. 2 and 3, and Secs. 4.10 and 10.4), synaptic logic and multiplication on the basis of individual synapses (5.1.6 and 5.2.3) or groups of synapses in passive or active dendritic trees (5.2, 19.3.2, and 19.3.4), a random event generator via spontaneous generation of spikes by stochastic ionic channels (8.3.2), coincidence detection using fast dendritic spikes (19.3.3) or whole neurons (21.1.1), gain normalization and adaptation via the conductance increasing action of chemical synapses (1.5.3, 4.8.1, and 18.3), selective linearization and amplification of distal input via active dendritic currents (19.3.5), controlling the gain of a cell's discharge curve with I_{AHP} (9.3; see also Fig. 21.3), computing the average activity via intracellular calcium (9.1.6), implementing temporal delay elements using a potassium current (9.2.1), associative memory via Hebbian synapses (13.5), and reprogramming networks and the routing of information using peptides and hormones (Secs. 20.4 and 20.5).

cooperatively or competitively. The astute reader will have noticed that the selection scheme alluded to here has quite a few analogies with the way the vertebrate immune system works.

The conceptually simpler possibility is that by and large only a single biophysical operation is responsible for any one specific biological operation. It is to be hoped that ongoing research will help us decide this issue in the next decade.

It should not be forgotten that a concatenation of nonlinear mechanisms, operating at different levels in the nervous system, might compensate each other in such a way that the overall result is difficult to predict by contemplating each mechanism in isolation. A case in point is shunting inhibition. Within a passive cable, it can act in a very nonlinear, vetolike manner. Yet as we saw in Sec. 18.5, when treated within the context of a spiking neuron it affects its firing rate in a subtractive manner.

Sometimes neurons can act in a surprisingly linear fashion, even though we know that their components, synapses, voltage-dependent currents, and so on include substantial nonlinearities. A beautiful example of this are brainstem neurons in the medial vestibular nucleus that help mediate the vestibulo-ocular reflex. Their discharge curve is remarkably linear over the entire physiological firing range (Fig. 21.3). Reducing adaptation by pharmacological blockage of the underlying calcium-dependent potassium conductance fails to cause linearity to deteriorate but just leads to an increased gain (as in Fig. 9.13; du Lac, 1996). The linearity of the cell extends to the temporal domain (du Lac and Lisberger, 1995). The response to steps of current can be accurately predicted from the linear superposition of sinusoidal currents that are injected directly into the soma (in the 0.1–10 Hz frequency range). The high degree of linearity requires the active participation of voltage-gated currents to compensate for the low-pass filtering properties expected from a passive membrane.

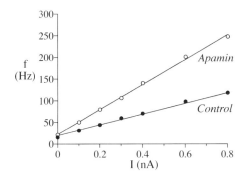

Fig. 21.3 A LINEAR NEURON We have come full circle. The last figure in the book is reminiscent of one of the first figures (1.4) and underlines that sometimes, despite all the nonlinear mechanisms known to exist in the nervous system, evolution conspires to construct neurons that act in a perfectly linear manner: the f–I curve of a medial vestibular nucleus (brainstem) neuron recorded in a slice with a slope of 124 Hz per nanoampere of injected current (du Lac, 1996). Circles are the data with the best linear fit indicated by the line. The neuron fires spontaneously. Apamin, one of the active components in bee venom, removes adaptation by blocking I_{AHP}. As in the model of bullfrog sympathetic ganglion cells (Fig. 9.13), this causes the gain of the f–I curve to increase (here by 1.67) without eliminating linearity. Linearity is also observed in the temporal domain and has been assessed using the superposition theorem. Reprinted by permission from du Lac (1996).

A further example is the linearization procedure thought to occur in distal dendrites (Sec. 19.3.5; see also Laurent, 1993). Because of the associated high input impedance, synaptic input will quickly saturate. This can be counteracted by an inactivating potassium current such that the transfer function between synaptic input and postsynaptic somatic current is linear.

It is almost as if there were a deeper principle of "compensating nonlinearities" at work in these systems. Yet what advantage would the nervous system derive from linear components that, at best, process no information and, at worst, lose information due to unavoidable contamination of the signal by noise? Is it possible that linear systems are easier to evolve and can be more readily incorporated into existing systems without causing stability problems?

21.2 Strategic Questions or How to Find a Topic for a Ph.D. Thesis

In the early phase of the scientific development of a field, progress occurs as much by asking the right sort of question as by answering them. Today, in the heat of the battle it is difficult, of course, to know whether we are asking the right sorts of questions. Undaunted by this challenge, let us finish the book by enumerating, in no particular order, a small set of strategic questions.

1. Given the canonical nature of *multiplication*, we need to understand in detail, in at least one neurobiological model system, how this nonlinear operation is implemented at the cellular level.

2. What are the principal, intrinsic noise sources (as compared to external noise sources, such as the shot nature of individual quanta of light impinging onto the retina) in the nervous system? Ultimately, they are what limits the accuracy with which signals in the environment can be detected and computations can be carried out. A catalogue of noise sources needs to include thermal noise, the ill-defined but ever present $1/f$ noise, the probabilistic nature of synaptic transmission, the flickering of stochastic ionic channels—in particular if they are present at low density in the dendrites—as well as the "spontaneous" neuronal background activity.[1] Noise sources need to be described in terms of their amplitudes, power spectra and temporal structure. How do they constrain particular computations?

3. To what extent do considerations involving the energy metabolism play a role in constraining the style of neuronal computation?

   "On" and "off" cells in the visual pathways are a case in point (Sterling, 1998). The retina communicates an increase in local spatial contrast (e.g., a bright spot on a dark background) to the rest of the brain by increasing the firing rate of so-called on ganglion cells (Kuffler, 1953). Since firing rates cannot be negative, how are decreases in local contrast coded? One solution would be for the on neurons to have a very high maintained discharge, say at 100 Hz, and then to signal a decrease in contrast by decreasing this rate. Yet a million neurons firing at 100 Hz is expensive in terms of the continuous

---

1. However, this background may just appear as noise from the limited point of view of the observer and may well turn out to represent a significant signal correlated across many cells.

upkeep of the ionic gradients across the membrane of the ganglion cells. Given that the retina is already one of the most metabolic active pieces of nervous tissue—as witnessed by its dense blood supply—this solution appears unfeasible. Instead, the visual system has adopted an alternative strategy by creating *off* cells whose discharge increases in response to a dark spot on a bright background (that is, a local decrease in contrast). By using two half-wave rectified channels the brain solves the problem of signalling negative numbers (Schiller, Sandell, and Maunsell, 1986) with the added benefit of doubling contrast sensitivity (albeit at the price of also doubling the number of neurons needed).

To what extent do metabolic considerations determine whether or not a neuron outputs graded potentials or spikes (van Hateren and Laughlin, 1990; Haag and Borst, 1997)? Knowing the amount of sodium and potassium ions that move across the membrane during the propagation of a spike and the metabolic cost in terms of the universal ATP energy currency of pumping them back via the Na^+-K^+ exchanger gives us an estimate of what a single action potential costs. Similar estimates could be made for the cost of synaptic signaling, although this calculation would be more extensive since the cost of maintaining and recycling vesicles would have to be factored in (Laughlin, 1994). And how expensive are dendritic computations?

Coding schemes have been proposed that minimize the firing rate of neurons on the assumption that the metabolic cost of spiking is substantial and must be kept to a minimum (Levy and Baxter, 1996). What fraction of the 13–15 watt power consumption of an adult human brain is necessary for maintaining homeostasis and what fraction directly supports signaling in general and spiking or dendritic computations in particular, needs to be determined (Kety, 1957; Sokoloff, 1989).[2]

The quiescent power consumption of CMOS integrated electronic circuits—due to leakage currents—is minuscule. The 30 watt or so required by a modern microprocessor can be attributed to charging and discharging the parasitic capacitances during every clock cycle. That is, most of the power goes to computing and switching and very little to maintenance. (For supercomputers the cost of air-conditioning must be factored into this estimate.) It would be interesting to "poison" an entire brain with TTX to block sodium action potentials, and to measure the associated energy metabolism in the absence of any spiking. It is possible that the fraction of the energy devoted to homeostasis versus computation/communication is heavily skewed toward the former while in our microprocessors it is toward the latter?

4. What is the function of the apical dendrite, the most prominent component of many pyramidal cells in the cortex? Is it a vehicle for rapidly communicating information perpendicular to the cortical surface, like a copper wire? Can input to the distal apical tree, frequently originating in higher cortical areas, only weakly modulate the firing rate of the pyramidal cell or can it, by itself, trigger vigorous spiking activity in the absence of other input (Sec. 19.3.5)?

Is the total length of the apical dendrite in a neocortical pyramidal cell—and hence the thickness of the cortical sheet—limited by the distance over which a synaptic signal from the distal apical tuft to the cell body can travel without being swamped by the

---

2. Given the 2.4×10^{14} synapses in a typical human cortex and an average firing rate of 10 Hz per neuron, we estimate that about 2.4×10^{15} synaptic switching events occur every second. With a power budget of 15 watts, this amounts to 1.6×10^{14} operations per joule of energy (neglecting all computations taking place in the dendrites). This compares very favorably to the cost of computing on a modern workstation (around 5 to 10 million operations per joule, Sarpeshkar, 1998).

distributed noise sources in the intervening dendritic membrane (e.g., channel noise)? In other words, if the apical dendrite were to be any longer, the signal arriving at the soma from the distal periphery could not be distinguished from the noisy background (this is an argument having to do with signal-to-noise ratio and not with signal attenuation as in Sec. 19.3.5).

5. We must derive unsupervised learning algorithms—and compare them against experimental data—that can explain how neurons develop, optimize, and maintain their complex properties in the face of constantly varying environmental conditions (Bell, 1992; Siegel, Marder, and Abbott, 1994). What type of learning rules could help pattern dendritic tree morpholgy and the spatial distribution of voltage-dependent membrane conductances? A simple feedback mechanism—acting via $[Ca^{2+}]_i$ at the cell body— has been shown to keep the time-averaged firing rate of a neuron pegged to some constant even though the maximal conductances are changing (LeMasson, Marder, and Abbott, 1993).

   We need to tackle the challenging problem of "learning" Hodgkin-Huxley-like spiking systems. Much would be achieved by the discovery of a local learning rule that starts out with the standard spiking conductances and their correct voltage and time dependencies but with either zero or random settings of the channel densities and then learns the associated maximal conductances $\overline{G}_{Na}$ and $\overline{G}_K$ so that the system generates action potentials (on the simplifying premise that the nature of the individual sodium and potassium channels is relatively fixed and that only the number of channels expressed per area needs to be adjusted to obtain spikes).

6. What is the function of dendritic trees? Why do different cell types possess so strikingly different morphologies? One possibility was discussed extensively in Chap. 5: the dendritic tree geometry, coupled with a unique synaptic architecture, implements specific computations. Although this hypothesis is by now almost two decades old, we still do not know to what extent individual synapses (as in the AND-NOT logic of Koch, Poggio, and Torre, 1982) or groups of them (as in Mel, 1993) are involved in such computations or whether the location of synapses in the tree is pretty much irrelevant.

   An alternate hypothesis is that dendrites are a means of maximizing the probability that diffuse axonal trees contact as many neurons as possible (within a fixed volume), subject to the constraint that neuronal processes have to have some minimal diameter to accommodate mitochondria and other intracellular organelles and to assure the transport of macromolecules. What sort of space-filling, fractal geometry will maximize the dendritic surface area for contacts with the largest number of axons under these conditions? Can a simple learning rule give rise to such a geometrical structure during development?

7. It is important to construct canonical single-cell models that are more faithful to biology than the linear threshold models in usage today. On the one hand, such models must be simple enough to be treatable using standard mathematical techniques. On the other hand, they must be rich enough to capture much of the passive and active dendritic processing occurring in a real dendritic tree. Polynomial threshold neurons (Sec. 14.4.2) that include temporal processing in the input and that retain spiking information in their output are a good starting point. The neural network community needs to come up with novel theoretical tools that can deal with signal and information processing in cascades of such digital-to-analog-to-digital computational elements.

Finding answers to any one of these questions, possibly in the form of a Ph.D. thesis, will ultimately help us better understand the most complex object in the known universe, the human brain. The author as well as the reader of these pages are fortunate enough to live in exciting times where much progress in the neurosciences occurs within a relatively short span of history. Let's get to it: *per aspera ad astra*!

Appendix A

Passive Membrane Parameters

A vexing and confusing question concerns the use of the all-important membrane parameters R_i, C_m, and R_m and their associated units.

A.1 Intracellular Resistivity R_i

In physics, the amount of *resistance* a piece of bulk material offers to electrical current flowing across the material is characterized by its *specific resistivity* ρ; its units are ohms-centimeter ($\Omega \cdot$cm). If the piece of material has a constant cross section A and length ℓ, its total resistance will be

$$R = \frac{\rho \cdot \ell}{A} . \tag{A.1}$$

In biophysics, the specific resistance of the intracellular medium, also known as the *intracellular resistivity*, is denoted by R_i, which can be thought of as the total resistance across a 1 cm cube of intracellular cytoplasm.

Given a cylindrical axon or dendrite of diameter d and length ℓ, its total resistance will be $4R_i \ell/\pi d^2$ and its resistance per unit length ℓ of cable,

$$r_a = \frac{4R_i}{\pi d^2} . \tag{A.2}$$

Conversely, given a neural process with axial resistance per unit length r_a, the total axial resistance will be ℓr_a. The units of r_a are ohms per centimeter (Ω/cm).

The resistivity of seawater, from where we all came half a billion years ago, is 20 $\Omega \cdot$cm, that of mammalian saline about 60, and that of physiological Ringer solution 80 $\Omega \cdot$cm. This is not surprising, since the resistivity decreases as salts are added to a solution. Computing the intracellular resistivity on the basis of the Nernst-Planck electrodiffusion equation yields 33.4 $\Omega \cdot$cm for K^+ ions and 267 $\Omega \cdot$cm for Na^+ ions (Eq. 11.33). Under the assumption that these are the dominant charge carriers, this leads to a final value of $R_i = 29.7\ \Omega \cdot$cm (see Sec. 11.3.1).

Because the intracellular environment contains many structures in addition to electrolyte, such as endoplasmic reticulum, the cytoskeleton, the Golgi apparatus, and so on, that restrict charge redistribution within neurons, the above cited values set a lower bound on the true value of R_i in neurons. The value of R_i conventionally used is in the 70–100 Ω·cm range, based on measurements in neurons and axons of marine invertebrates (Foster, Bidinger, and Carpenter, 1976; Gilbert, 1975) and on data taken from cat motoneurons (Barrett and Crill, 1974)

However, much higher values have been reported. Neher (1970) directly measured a value of 180 Ω·cm for the resistivity of the cell body of snail neurons, while Shelton (1985) and Segev et al., (1992) require values of 225–250 Ω·cm for cerebellar Purkinje cells in order to fit various experimentally observed records, such as pulse attenuation and input resistance, against detailed cable models.

In a careful study of this problem using *in vitro* cortical pyramidal cells recorded and stained with sharp intracellular electrodes and whole-cell patch pipettes, Major (1992; see also Major, Evans, and Jack, 1993a) concludes that a good match between his physiological records and a detailed cable model requires a R_i in the 200–300 Ω·cm range (see also Spruston and Johnston, 1992, and Spruston, Jaffe, and Johnston, 1994). Tissue shrinkage during the histological recovery of the cell's anatomy does, however, affect these numbers. After decreasing the diameters of all dendrites by 20% while simultaneously increasing all lengths by 20% to invert the effect of shrinkage, Major and his colleagues infer a reduced value of $R_i = 187$ Ω·cm.

A.2 Membrane Resistance R_m

The *passive membrane resistivity*, that is, the resistance associated with a unit area of membrane, of the electrical component of the neuronal membrane that does not depend on synaptic input or on the membrane potential is denoted by R_m, measured in units of ohms-square centimeter (Ω·cm^2). Its inverse is known as the passive conductance per unit area of dendritic membrane or, for short, as the *leak conductance* $G_m = 1/R_m$ and is measured in units of siemens per square centimeter (S/cm^2).

The molecular correlate of this leak conductance is not precisely known. A pure phospholipid in saline solution has an extremely high resistance of up to 10^{15} $\Omega \cdot$ cm^2 (Hille, 1992). Since measured membrane resistances are considerably lower, some mechanism has to permit ions to pass across the membrane.

The evidence for voltage-independent "leak" channels is not strong. Patch-clamp studies of frog sympathetic neurons reveal a nearly ohmic region between -70 and -110 mV (Jones, 1989). The underlying conductance is only weakly voltage dependent and is insensitive to blockers that block other known conductances in the cell. After adjusting for a 10 mV contribution of the sodium-potassium pump, the remainder of the "leak" current is carried by potassium ions and reverses around -65 mV. Evidence from hippocampal pyramidal cells suggests a very weakly voltage-dependent potassium current that is active at rest and that can be blocked by cholinergic input (Madison, Lancaster, and Nicoll, 1987).

In a cylindrical fiber of length ℓ and cross section $A = \pi d\ell$, the membrane resistance per unit length of fiber is defined as

$$r_m = \frac{R_m}{\pi d} \tag{A.3}$$

in units of Ω·cm. The total membrane resistance in a fiber of length ℓ is identical to r_m/ℓ. R_m varies widely from preparation to preparation, with the quality of the intracellular recording, with the amount of synaptic input to the cell, and other parameters. As intracellular recording techniques become more mature and sophisticated, the estimates of R_m increase.

Note that the "effective" specific resistivity associated with the membrane, on the order of 10^5 $\Omega\cdot$cm^2/40 Å$= 2.5 \times 10^{11}$ $\Omega\cdot$cm, is approximately one billion times larger than that of the intracellular fluid, explaining why by far the largest fraction of the cytoplasmic current flows within the dendrite or axon rather than across the membrane.

A.3 Membrane Capacitance C_m

The capacitance of the neuronal membrane is characterized by the *specific capacitance* per unit area C_m, measured in units of farads per square centimeter (F/cm^2). The generally agreed upon value for C_m is 1 μF/cm^2. This amounts to 10^{-7} coulomb (C) charge being distributed on both sides of a 1 cm^2 piece of neuronal membrane in the presence of a 100 mV voltage difference across the membrane. Assuming a parallel plate capacitor configuration and a dielectric constant of 2.1 for the hydrocarbon chains making up the bilipid layers, this implies a separation of 23 Å (Hille, 1992). Hence, the reason for the slow time constants in the millisecond time range is the molecular dimension of the neuronal membrane.

It is intriguing to compare the biological value of C_m against the capacitance in the analog CMOS circuit technology used to design neuromorphic electronic circuits (Mead, 1989; Mahowald and Douglas, 1991; Douglas, Mahowald, and Mead, 1995). Here, a capacitance is created by separating two layers of polysilicon with a 40-nm-thick layer of silicon oxide. The specific capacitance is about 0.5 μF/cm^2, about 20 times lower than its biological counterpart (compatible with the larger separation between the two layers). The higher value of C_m in biology is partly compensated for by the fact that the voltages across membranes (on the order of 0.1 V) are much smaller than the typical gate voltages of around 5 V in electronic circuits. Ultimately, this is due to the multiple gating charges of the voltage-dependent channels (see Chap. 8) allowing them to work on a much steeper exponential than transistors.

The 1 μF/cm^2 value of C_m appears to be somewhat of an overestimate. The membrane capacitance of a pure bilayer lipid membrane without proteins is between 0.6 and 0.8 μF/cm^2 (Cole, 1972; Fettiplace, Andrews, and Haydon, 1971; Benz et al., 1975), placing a lower bound on C_m. Mammalian red blood cells have a measured C_m of between 0.8 and 0.9 μF/cm^2. The "traditional" value of unity for C_m is based on the membrane of the squid giant axon and includes the nonlinear capacitances associated with the gating currents (for a discussion see Adrian, 1975). In a detailed investigation on dissociated hippocampal pyramidal cells, Sah, Gibb, and Gage (1988) report $C_m = 1.0 \pm 0.2$ μF/cm^2, while a more recent investigation of neocortical pyramidal cells finds values between 0.65 and 0.8 μF/cm^2 (Major, 1992).

A further complication are membrane invaginations and foldings which can multiply the effective membrane area—and therefore C_m—manyfold (as for instance in bullfrog sympathetic ganglion cells, where $C_m = 3$ μF/cm^2 has been inferred from the measured total cell capacitance and the area of these spherical cells (Yamada, Koch, and Adams, 1998; see also Segev et al., 1992).

Frequently, one defines

$$c_m = C_m \cdot \pi d \qquad (A.4)$$

as capacitance per unit fiber of diameter d (in units of F/cm), in analogy to r_m. The total capacitance of all of the membrane in a process of length ℓ is given by $c_m \ell$.

Appendix B

A Miniprimer on Linear Systems Analysis

For those readers who forgot linear systems analysis, crucial to this book, we here provide the briefest of reviews.

A system is always linear or nonlinear with respect to some particular input and output variable. Indeed, the same physical system can be linear when using one sort of input-output pairing and nonlinear when considering a different one. If we restrict ourselves to the case when the input and output variables are single-valued functions of time, termed $x(t)$ and $y(t)$, respectively, then for a system $\mathbf{L}$, defined as

$$y(t) = \mathbf{L}[x(t)] \tag{B.1}$$

to be linear, it must obey two constraints. Firstly, it must be *homogeneous*, that is,

$$\mathbf{L}[\alpha x(t)] = \alpha \mathbf{L}[x(t)]. \tag{B.2}$$

For instance, doubling the input should double the output. Secondly, the system must also be *additive*,

$$\mathbf{L}[x_1(t) + x_2(t)] = \mathbf{L}[x_1(t)] + \mathbf{L}[x_2(t)]. \tag{B.3}$$

The response of the system to the sum of two inputs is given by the sum of the responses to the individual inputs. These two properties are sometimes also summarized in the *superposition* principle, expressed as

$$\mathbf{L}[\alpha_1 x_1(t) + \alpha_2 x_2(t)] = \alpha_1 \mathbf{L}[x_1(t)] + \alpha_2 \mathbf{L}[x_2(t)]. \tag{B.4}$$

A further property that some (not all) linear systems possess is *shift* or *time invariance*; in other words, if the input is delayed by some Δt, the output will be delayed by the same interval. A linear system is *shift invariant* if and only if

$$y(t) = \mathbf{L}[x(t)] \quad \text{implies} \quad y(t - t_1) = \mathbf{L}[x(t - t_1)]. \tag{B.5}$$

If a system possesses these three properties, then its entire behavior can be summarized by its response to an impulse or *delta* function $\delta(t)$.[1] The impulse response or Green's function of the system is exactly what its name imples, namely, the response of the system to an impulse,

---

1. The delta, unit, or *Dirac* "function" is defined by being zero for all values of t except at the origin, where it diverges. It has the immensely useful property that $\int_{-\infty}^{+\infty} f(t)\delta(t)dt = f(0)$.

$$h(t) = \mathbf{L}[\delta(t)] \,.\tag{B.6}$$

Any input signal can be treated as an infinite sum of appropriately shifted and scaled impulses, or

$$x(t) = \int_{-\infty}^{+\infty} x(t_1)\delta(t - t_1)dt_1 \,.\tag{B.7}$$

The properties of homogeneity, additivity, and shift invariance ensure that the response to any arbitrary input $x(t)$ can be obtained by summing over appropriately shifted and scaled responses to an impulse function (or, equivalently, the output is a weighted sum of its inputs). In short,

$$y(t) = \int_{-\infty}^{+\infty} x(t_1)h(t - t_1)dt_1 \,.\tag{B.8}$$

The shorthand form of this integral operation, known as a *convolution*, is $*$, as in

$$y(t) = (x * h)(t) \,.\tag{B.9}$$

We conclude that once we know $h(t)$, the response of a linear system to an impulse, the response to an arbitrary input waveform can be obtained by the linear convolution operation.

Before we end this short digression, we briefly want to remind the reader of another way to analyze time-invariant linear systems, namely, by using sinusoidal inputs. If the input to a linear system is a sinusoidal wave of a particular frequency f (in hertz), the output is another sinusoidal of the same frequency but shifted in time and scaled,

$$\mathbf{L}[\sin(2\pi f t)] = A(f)\sin(2\pi f t + \phi(f)) \,.\tag{B.10}$$

The function $A(f)$ is known as the amplitude response and determines how much the output is scaled for an input at frequency f, while the phase $\phi(f)$ determines by how much the sinusoidal wave at the output is shifted in time with respect to the input.

Any input can always be represented as a sum of shifted and scaled sinusoidals. For a linear system, the impulse response function and the amplitude and phase functions are closely related by way of the Fourier transform. The Fourier transform of the impulse response function $h(t)$ is

$$\tilde{h}(f) = \mathbf{F}[h(t)] = \int_{-\infty}^{+\infty} e^{-i2\pi f t} h(t)dt \,.\tag{B.11}$$

The amplitude of this filter $\tilde{h}(f)$ corresponds to the amplitude of the Fourier transform of the impulse response function. Note that $\tilde{h}(f)$ (throughout the book, the $\tilde{h}$ symbol denotes the Fourier transform of some function h) is a complex function. We then have

$$\tilde{h}(f) = A(f)e^{-i\phi(f)} \,.\tag{B.12}$$

The reason we frequently talk in this book about the input being "filtered" by the filter function $\tilde{h}(f)$ is that formally the output can be obtained by convolving (another name for filtering) the input by the filter function. In the frequency space representation, convolution is turned into a straight multiplication, and Eq. B.8 can be rewritten as

$$\tilde{y}(f) = \tilde{h}(f) \cdot \tilde{x}(f) \,.\tag{B.13}$$

As emphasized before, when discussing linearity it is crucial to discuss with respect to what variable. There are a number of instances in which neurobiological systems can be treated, to some degree of approximation, by linear systems analysis. Prominent examples are linear cable theory, when the input is current and the output voltage (but not when the input is a conductance change; see Chap. 1), or receptive field analysis of retinal or cortical neurons in the visual system (Palmer, Jones, and Stepnoski, 1991). Here, the input is usually the stimulus contrast and the output the mean firing rate.

Certain nonlinear systems can frequently be treated as a linear system with the addition of a simple type of static nonlinearities, such as a threshold (Palm, 1978; French and Korenberg, 1989).

A perhaps surprising linear system is the one that relates a continuous input, call it $\mu(t)$, to a continuous firing rate $f(t)$ via a spiking process. The input, suitably scaled, can be thought of as input into an integrate-and-fire unit (Chap. 14) with no leak and no refractory period. The threshold for generating spikes V_{th} is not fixed but is some probability distribution $p_{th}(V)$ (Gestri, 1971; Gabbiani and Koch, 1997): every time a spike has been generated the threshold is set to a new value drawn from $p_{th}(V)$.[2]

For any input $\mu(t)$ this unit generates a particular random spike train sequence, abbreviated here as $\sum_i \delta(t - t_i)$. Obviously, the relationship between the continuous input $\mu(t)$ and the discrete spike train is highly nonlinear. However, let us assume a population of independent but otherwise identically integrate-and-fire units with identical distributed voltage thresholds. If all receive the same input $\mu(t)$ one can average, as discussed in Sec. 14.1, over this ensemble and define an instantaneous output rate $f(t)$.

It can be proven (Gestri, 1971) that varying the input by $\alpha\mu(t)$ changes the instantaneous firing rate of this fictive population of cells by $\alpha f(t)$. In other words, the relationship between the input and the instantaneous output rate is a linear one.

---

2. If $p_{th}(V)$ is exponentially distributed, the spikes generated by this process have the convenient property that they are Poisson distributed (Chap. 15).

Appendix C

Sparse Matrix Methods for Modeling Single Neurons

Barak A. Pearlmutter and Anthony Zador

In this appendix we describe numerical methods used in the efficient solution of the linear and nonlinear cable equations that describe single neuron dynamics. Our exposition is limited to the scale of a whole neuron; we will ignore both the simulation of circuits of neurons (see, for example, the monographs by Bower and Beeman, 1998 and by Koch and Segev, 1998), as well as the simulation of the stochastic equations governing single ion channel kinetics (Skaugen and Walloe, 1979; Chow and White, 1996).

The appendix is divided into two parts. The first deals with the solution of the *linear* component of the cable equation. Since the cable as well as the diffusion equation are linear second-order parabolic partial differential equations (PDE), this part draws on techniques and principles developed in the many other fields that deal with similar equations, though naturally the discussion will focus on those problems of particular interest in neurobiology. The theme common to this part is that for the purposes of numerical solution, the cable equation is best discretized into a system of ordinary differential equations coupled by sparse matrices. The main difficulty is that the resulting equations are *stiff*, that is, they display time scales of very different magnitude; even so it is possible to apply widely available and efficient techniques for sparse matrices. The second part deals with the *nonlinear* components of neurodynamics, particularly equations of the Hodgkin–Huxley type and those arriving from calcium dynamics. These nonlinearities are surprisingly benign, and can readily be handled with a few simple techniques, provided that the linear component is treated properly.

Many of the techniques described are implemented in widely used and freely available neural simulators (in particular GENESIS and NEURON; see DeSchutter, 1992; Hines, 1998; Bower and Beeman, 1998) in a manner that is relatively transparent to the user. Nevertheless, there are at least three good reasons for understanding the foundations of these numerical methods. Firstly, when such simulators produce surprising—and possibly spurious—results, an understanding of their internal workings can help determine whether the numerical method is to blame. Secondly, when conducting original research it is inevitable that some problem will arise for which the software must be customized. Finally, and most important, understanding these techniques provides insight into the underlying neurodynamics itself, and hence into the behavior of neurons.

C.1 Linear Cable Equation

C.1.1 Unbranched Cables and Tridiagonal Matrices

Discretization in Space

The passive cable equation (Eq. 2.7) describes the spread of potential in one dimension along an unbranched homogeneous cable,

$$C_m \frac{\partial V(x,t)}{\partial t} = \frac{d}{4R_i} \frac{\partial^2 V(x,t)}{\partial x^2} - G_m V(x,t) - J(x,t), \qquad \text{(C.1)}$$

where $G_m = 1/R_m$ and $J(x,t)$ is the injected current. (In this appendix, we use J rather than I to distinguish it from the identity matrix I below.) For the sake of convenience, we set the reversal potential in Eq. C.1 to zero; it can be thought of as being absorbed into the offset current $J(x,t)$. The membrane parameters C_m, R_i, G_m, and d are assumed to be independent of position along the cable. Note that the subscript on the intracellular resistivity R_i is not used here as a numerical index. We now discretize this partial differential equation in space by replacing the second spatial derivative with its simplest, second-order discrete approximation,

$$\frac{\partial^2 V(x,t)}{\partial x^2} \rightarrow \frac{V_{i+1} - 2V_i + V_{i-1}}{\Delta x^2}, \qquad \text{(C.2)}$$

resulting in the system of coupled, ordinary differential equations

$$C_m \frac{dV_i}{dt} = \frac{d}{4R_i} \frac{V_{i+1} - 2V_i + V_{i-1}}{\Delta x^2} - G_m V_i - J_i. \qquad \text{(C.3)}$$

Here $V_i = V(x_i, t)$ and $J_i = J(x_i, t)$ denote, respectively, the membrane potential and injected current, at some point x_i, where i is the discretization index, and the time variable t has been suppressed. Physically, these indices correspond to the locations along the cable—the *nodes* or *compartments*—at which the voltage is specified (see Fig. 2.2B). The resulting system of coupled ordinary differential equations can be written compactly in matrix form as

$$\frac{d\mathbf{V}}{dt} = \mathbf{C}^{-1}(\psi \mathbf{B'V} - \mathbf{GV}) - \mathbf{J}_{\text{inj}} = \mathbf{BV} - \mathbf{J}_{\text{inj}} \qquad \text{(C.4)}$$

where the matrix $\mathbf{B}$ is given by

$$\mathbf{B} = \mathbf{C}^{-1}(\psi \mathbf{B'} - \mathbf{G}), \qquad \text{(C.5)}$$

with $\psi = d/(4R_i \Delta x^2)$. Here $\mathbf{B'}$ is a tridiagonal second difference matrix with elements -2 along the diagonal and $+1$ off the diagonal

$$\mathbf{B'} = \begin{pmatrix} -2 & 1 & & & & \\ 1 & -2 & 1 & & & \\ & & \ddots & \ddots & \ddots & \\ & & & 1 & -2 & 1 \\ & & & & 1 & -2 \end{pmatrix} \qquad \text{(C.6)}$$

and $\mathbf{C}$ and $\mathbf{G}$ are diagonal matrices with C_m and G_m, respectively, along their diagonals (that is, scalar multiples of the identity matrix). Since the matrices $\mathbf{C}$ and $\mathbf{G}$ are diagonal, they do not contribute to the interaction between equations; coupling is exclusively through $\mathbf{B}'$. This is consistent with our physical expectations, since points along the cable interact only through the second spatial derivative (scaled by the effective axial R_i) captured in $\mathbf{B}'$. As discussed in detail below, the corner elements of this matrix determine the boundary conditions, which here correspond to $V(x) = 0$ at the boundaries—the *killed-end* or *Dirichlet* condition, which we have choosen here for the sake of convenience (Sec. C.1.3). For the *sealed-end* or *von Neumann* boundary condition, we have $dV/dx = 0$ at the boundary, and $\mathbf{B}'$ takes slightly different values in the upper and lower rows. Although the von Neumann condition arises more often, here we consider only the Dirichlet condition because it gives rise to a simpler form.

This set of coupled differential equations is (almost) the same set that arises from a network of discrete passive electrical components, for instance, the one shown in Figs. 2.3 and 3.4B. This is no coincidence. Another way to arrive at this set of equations is to first approximate the spatially continuous electrical structure specified by Eq. C.1 with a discrete electrical network that approximates its properties. As we shall see, however, the analogy with a discrete electrical circuit breaks down at the endpoints of the cable, so care must be taken to ensure the exact boundary conditions.

Eigenvalue Analysis

It is helpful to analyze the behavior of Eq. C.4 in terms of the eigenvalues of $\mathbf{B}$. For now we neglect the injected current term. We will first consider the eigenvalues of $\mathbf{B}'$—the second difference matrix from which $\mathbf{B}$ is derived—and then consider the eigenvalues of $\mathbf{B}$ itself. Recall that the eigenvalues m_z of any matrix $\mathbf{B}$ are those numbers that satisfy

$$\mathbf{B}\mathbf{V}^{(z)} = m_z\mathbf{V}^{(z)} \tag{C.7}$$

where $\mathbf{V}^{(z)}$ is the eigenvector corresponding to the z-th eigenvalue. For an $n \times n$ matrix $\mathbf{B}'$ there are n eigenvalues and n associated eigenvectors, and the index z ranges from 1 to n. We show below that the eigenvalues of the second difference matrix $\mathbf{B}'$ are given by

$$m_z' = -2 + 2\cos\frac{\pi z}{n+1} \tag{C.8}$$

and the corresponding eigenvectors are

$$V_i^{(z)} \propto \sin\frac{\pi zi}{n+1} . \tag{C.9}$$

The two sides are proportional rather than equal, since we are free to choose the magnitude of the right-hand side: any multiple of an eigenvector is itself and eigenvector. If the eigenvectors are scaled so that $||\mathbf{V}^{(z)}|| = 1$, then the solutions to Eq. C.4 can be expressed as a sum of exponentials,

$$\begin{pmatrix} V_1(t) \\ \vdots \\ V_i(t) \\ \vdots \\ V_n(t) \end{pmatrix} = c_1 \begin{pmatrix} V_1^{(1)} \\ \vdots \\ V_i^{(1)} \\ \vdots \\ V_n^{(1)} \end{pmatrix} e^{-t/\tau_1'} + \cdots + c_z \begin{pmatrix} V_1^{(z)} \\ \vdots \\ V_i^{(z)} \\ \vdots \\ V_n^{(z)} \end{pmatrix} e^{-t/\tau_z'} + \cdots + c_n \begin{pmatrix} V_1^{(n)} \\ \vdots \\ V_i^{(n)} \\ \vdots \\ V_n^{(n)} \end{pmatrix} e^{-t/\tau_n'}, \tag{C.10}$$

where $V_i^{(z)}$ is the i-th component of the z-th eigenvector, and τ_z' the reciprocal of the corresponding eigenvalue.[1] The constants $c_1, \ldots, c_n$ are determined by the projection of the initial conditions $V_i(0)$ onto the eigenvectors,

$$c_z = \sum_i V_i(0) V_i^{(z)}, \tag{C.11}$$

or $\mathbf{c} = \mathbf{M}^T \mathbf{V}_0$. This last expression holds since the eigenvectors of the symmetric matrix $\mathbf{B}$ are orthogonal. If we denote by $\mathbf{M}$ the matrix that has the eigenvectors in Eq. C.9 as its columns, then the vector solution in Eq. C.10 can be rewritten compactly in matrix form as

$$\mathbf{V}(t) = \mathbf{M}^T \mathbf{V}_0 \mathbf{M} e^{\mathbf{m}'t} \tag{C.12}$$

where $\mathbf{V}_0$ is the vector of initial conditions and $\mathbf{m}'$ is the vector of eigenvalues.

These equations tell us several things. (1) The eigenvalues appear as the time constants in a sum of exponentials. This sum represents the decay of voltage with time. It is precisely these time constants that are "peeled" in the classical Rall analysis of electrotonic structure (see Sec. 2.3.2). Second, the eigenvalues m_z' range from $m_1' = -2 + 2\cos[\pi/(n+1)] < 0$ to $m_n' = -2 + 2\cos[n\pi/(n+1)] > -4$. (2) The smallest eigenvalue m_1 of $\mathbf{B}$ is just less than 0, so the solution always decays to 0, even in the absence of a membrane leak term; this is due to the killed-end boundary condition we are considering. For the sealed-end case, the matrix is conservative for $G_m = 0$, charge only redistributes spatially without net loss or gain, so 0 is an eigenvalue (that is, the voltage decays to a constant). (3) Finally, note that the end elements V_1 and V_n do *not* equal zero, contrary to what we might expect from the boundary conditions. Only when we consider the "phantom" elements at $i = 0$ and $i = n + 1$ do we see that the boundary conditions, $V_0 = 0$ and $V_{n+1} = 0$, are satisfied.

How do the eigenvalues of $\mathbf{B}$—which is the operator that governs the physical behavior of the cable equation—relate to those of $\mathbf{B}'$? In general, if the eigenvalues of a matrix $\mathbf{B}$ are m_z, then the eigenvalues of $(\mathbf{B} - b\mathbf{I})/c$, where $\mathbf{I}$ is the identity matrix, are $(m_z - b)/c$; the eigenvectors remain unchanged. The dependence on b is called the *shifting property*.[2] The eigenvalues m_z can be obtained from those of the second difference matrix (Eq. C.8),

$$m_z = -\frac{1}{\tau_z} = \frac{\psi}{C_m}\left(2 - 2\cos\frac{\pi z}{n+1}\right) + \frac{G_m}{C_m}, \tag{C.13}$$

where as before z ranges from 1 to n. For $\mathbf{B}$, the smallest eigenvalue (associated with $z = 1$) is just larger than G_m/C_m (since $\cos[\pi z/(n+1)] \to 1$ for large n), which is just the inverse of the membrane time constant. The largest eigenvalue is just smaller than $4\psi/C_m + G_m/C_m$ (for large n). However, in a cable of fixed length as the number of compartments becomes larger and larger, the spatial discretization step Δx becomes smaller and smaller and this eigenvalue goes to infinity (and the corresponding time constant to 0), as expected from the corresponding eigenvalues of the continuous cable equation.

How do these expressions for the eigenvalues and eigenvectors arise? One way to understand their origin is by analogy to the continuous diffusion equation. The eigenfunctions of the continuous second derivative operator with a Dirichlet boundary condition (corresponding to a killed-end condition) on the interval [0, 1] are the sinusoidals

---

1. Following neuroscience convention, we express the solution in terms of $\tau_z' = -1/m_z'$ rather than m_z'.

2. This shifting property can be used in the change of variables $W = Ve^{t/\tau}$, with $\tau = C_m/G_m$, for reducing the cable equation (Eq. C.1) in V to the diffusion equation $C_m \partial W/\partial t = \frac{(dG_m)}{(4R_i)} \partial^2 W/\partial x^2$ in W. This shifting property is a special case of the property that for any polynomial (or more generally, any analytic function) $p(\cdot)$, the eigenvectors of $p(\mathbf{B})$ are $p(m_z)$, and the eigenvectors are unchanged.

$$\frac{\partial^2}{\partial x^2} \sin(\pi z x) = -(\pi z)^2 \sin(\pi z x), \tag{C.14}$$

where $-(\pi z)^2$ is the z-th eigenvalue. We need to find the corresponding expression for the discrete second difference operator $\mathbf{D}_2$:

$$\mathbf{D}_2 V_j = V_{j-1} - 2V_j + V_{j+1}. \tag{C.15}$$

We posit that the eigenvectors of $\mathbf{D}_2$ correspond to the eigenfunctions of the continuous operator, and we write the expression

$$\mathbf{D}_2 \mathbf{V}^{(z)} = \sin\frac{\pi(j-1)z}{n+1} - 2\sin\frac{\pi j z}{n+1} + \sin\frac{\pi(j+1)z}{n+1} = m'_z \sin\frac{\pi j z}{n+1} \tag{C.16}$$

and solve for m'_z. Application of the appropriate trigonometric identities confirms our expression Eq. C.8 for m'_z.

Another interpretation is in terms of the discrete Fourier transform (DFT) or rather its cousin, the discrete sine transform (DST). Although the DST is usually efficiently computed using the FFT algorithm, the DST is a linear transform, and can therefore be represented as a matrix $\mathbf{S}$. For an $n \times n$ operator $\mathbf{S}$—required to find the transform of a $1 \times n$ vector—the elements of $\mathbf{S}$ are given by

$$S_{jz} = \sin\frac{\pi j z}{n+1}. \tag{C.17}$$

The columns of this matrix are precisely the eigenvectors of $\mathbf{B}$. Therefore, $\mathbf{S}$ diagonalizes $\mathbf{B}$,

$$\mathbf{S}^{-1}\mathbf{B}\mathbf{S} = \text{diag}(m'_1, m'_2, \dots, m'_n). \tag{C.18}$$

This is entirely in analogy with the Fourier transform in the continuous case.

Explicit Discretization in Time

While the sum of exponentials suggests a possible solution to the equation, it does not generalize well to nonlinear problems and for many problems it is computationally inefficient (because n exponentials must be evaluated at each time point). In order to solve this system numerically, we transform the system of ordinary differential equations into a system of algebraic difference equations to be advanced by discrete time steps Δt. Just as before we replaced the spatial derivative by its finite difference approximation, we replace the temporal derivative by its first-order difference

$$\frac{d\mathbf{V}}{dt} \rightarrow \frac{\mathbf{V}^{t+1} - \mathbf{V}^t}{\Delta t}, \tag{C.19}$$

where the superscript refers to the index of discretized time. Now we have a choice for how to combine this with the right-hand side of Eq. C.4: do we use $\mathbf{V}^t$ or $\mathbf{V}^{t+1}$? The temptation is to use $\mathbf{V}^t$, since then $\mathbf{V}^{t+1}$ is an *explicit* function of $\mathbf{V}^t$,

$$C_m\frac{V_i^{t+1} - V_i^t}{\Delta t} = \frac{d}{4R_i}\frac{V_{i+1}^t - 2V_i^t + V_{i-1}^t}{\Delta x^2} - G_m V_i^t - J_i^t. \tag{C.20}$$

This choice for dV/dt is the basis of the *forward Euler* scheme. We can advance one time step at a very reasonable cost of three multiplications per node—one for the self-connection V_i and one each for the adjacent nodes $V_{i\pm 1}$.

We can rewrite a single iteration of the forward Euler scheme as

$$\mathbf{V}^{t+1} = \mathbf{B}_f \mathbf{V}^t - \mathbf{J}_i^t, \tag{C.21}$$

where $\mathbf{B}_f$ is a tridiagonal matrix and $\mathbf{V}^t$ is a vector. The matrix $\mathbf{B}_f$ is obtained by solving for $\mathbf{V}^{t+1}$ in the previous equation and is equal to $\Delta t\, \mathbf{B}$ with an additional 1 along the diagonal,

$$\mathbf{B}_f = \mathbf{I} + \Delta t\, \mathbf{B} \tag{C.22}$$

where $\mathbf{I}$ is the identity matrix. If we set $\mathbf{J} = 0$, then after p iterations, the solution can be written as a power of $\mathbf{B}_f$,

$$\mathbf{V}^{t+p} = (\mathbf{B}_f)^p \mathbf{V}^t$$

$$= c_1 (f_1)^p \mathbf{V}^{(1)} + \cdots + c_z (f_z)^p \mathbf{V}^{(z)} + \cdots + c_n (f_n)^p \mathbf{V}^{(n)},$$

where in the second line the solution is expressed in terms of the eigenvalues f_z of $\mathbf{B}_f$ and the c's are once again determined by the initial conditions. Note that while in the solution of the continuous differential equation (Eq. C.10) the eigenvalues appear in the exponent, here in the discrete difference equation they are raised to a power. This imposes a *stability* constraint on the eigenvalues. Stability refers to the behavior of the solution for $t \to \infty$, which for the discrete case translates into $p \to \infty$. For the discrete case the solution is stable if

$$|f_z| < 1 \tag{C.23}$$

for all z, since the eigenvalues are raised to a power. If any eigenvalue does not satisfy this condition, then for sufficiently large p it eventually diverges to infinity. By contrast, in the continuous case the eigenvalues appeared in the exponent, so the condition was that $m_z < 0$ for all z.

The forward Euler method is unstable if Δt is chosen too large. Consider the eigenvalues of the difference matrix $\mathbf{B}_f$, which we compute from Eq. C.13 using the shift property,

$$f_z = 1 + \Delta t\, m_z. \tag{C.24}$$

These eigenvalues are almost identical to those of the original differential system, but with the addition of the 1, which imposes a condition on the discretization parameters. Recall that we know only that $m_z < 0$, so there will exist discretization parameters such that $|f_z| > 1$. For $n \to \infty$ and $G_m = 0$, that is, no membrane leak, the stability condition simplifies to $|4\psi \Delta t / C_m - 1| < 1$, or $\Delta t \leq R_i C_m \Delta x^2 / d$. Thus if the spatial discretization Δx is made twice as small, the temporal discretization Δt must be made four times smaller to preserve stability. Since doubling the spatial discretization step also doubles the size of the matrix, computation time scales as the third power of n. This very rapidly becomes the limiting factor for large numbers of compartments.

Implicit Discretization in Time

One alternative is to use an *implicit* or *backward Euler* scheme, where the spatial derivative is evaluated at $t + 1$,

$$C_m \frac{V_i^{t+1} - V_i^t}{\Delta t} = \frac{d}{4 R_i} \frac{V_{i+1}^{t+1} - 2 V_i^{t+1} + V_{i-1}^{t+1}}{\Delta x^2} - G_m V_i^{t+1} - J_i^{t+1}. \tag{C.25}$$

Setting $J_i^{t+1} = 0$, we can rewrite (C.25) in matrix form as

$$\mathbf{V}^{t+1} = \mathbf{B}_b \mathbf{V}^t, \tag{C.26}$$

where

$$\mathbf{B}_b = [\mathbf{I} - \Delta t \; \mathbf{B}]^{-1}. \tag{C.27}$$

Using the result that the eigenvalues of the inverse of a matrix are just the inverse of the eigenvalues (and the eigenvectors unchanged), we write the eigenvalues b_z of $\mathbf{B}_b$,

$$b_z = (1 - \Delta t \; m_z)^{-1}. \tag{C.28}$$

The denominator is identical to Eq. C.24, except for the sign of the -1. This makes all the difference, since now $|b_z| < 1$ for all z, and the system is stable for all discretization steps. This does not guarantee, of course, that the solution to the discretized equation is close to the solution to the underlying continuous equations. This still requires small values of Δx and Δt.

The stability of the implicit scheme comes at a very high apparent cost: the inversion of an $n \times n$ matrix $\mathbf{B}_b$ at each time step. Now in general, the inversion of an $n \times n$ matrix requires $O(n^3)$ time; this is the same cost as the explicit scheme. Note that $\mathbf{B}_b$ is very sparse—in fact, it is tridiagonal. This sparseness is critical, since the solution to Eq. C.26 can be obtained by Gaussian elimination in $O(n)$ steps. (We defer the description of Gaussian elimination for the tridiagonal matrix until Sec. C.1.2, where it appears as a special case.) Thus, the implicit scheme is stable for all time steps and can be implemented in an efficient manner that scales linearly with the grain of spatial discretization.

Semi-Implicit Discretization in Time

The choice defining the implicit scheme, evaluating the spatial derivative at $\mathbf{V}^{t+1}$, is not the only one that leads to a stable scheme. We could also choose to evaluate the spatial derivative at the the midpoint between $\mathbf{V}^{t+1}$ and $\mathbf{V}^t$ (that is, by using the discretization scheme $\mathbf{V}^{t+1/2} = (\mathbf{V}^{t+1} + \mathbf{V}^t)/2$), to obtain

$$C \frac{V_i^{t+1} - V_i^t}{\Delta t} = \frac{d}{4R_i} \frac{\frac{1}{2}(V_{i+1}^t + V_{i+1}^{t+1}) - 2\frac{1}{2}(V_i^t + V_i^{t+1}) + \frac{1}{2}(V_{i-1}^t + V_{i-1}^{t+1})}{\Delta x^2}$$

$$- \frac{G_m}{2}(V_i^t + V_i^{t+1}) - \frac{J_i^t + J_i^{t+1}}{2}. \tag{C.29}$$

This choice is called the *semi-implicit* or *Crank-Nicolson* algorithm. Setting $J_i = 0$ as before, the resulting matrix equations can be expressed in terms of the implicit and explicit matrices,

$$\mathbf{B}_b \mathbf{V}^{t+1} = \mathbf{B}_f \mathbf{V}^t. \tag{C.30}$$

Stability is therefore determined by the eigenvalues e_z of $\mathbf{B}_b^{-1} \mathbf{B}_f$. Since $\mathbf{B}_b$ and $\mathbf{B}_f$ share eigenvectors, the eigenvalues of $\mathbf{B}_b^{-1} \mathbf{B}_f$ are simply the product of the eigenvalues of $\mathbf{B}_b^{-1}$ and $\mathbf{B}_f$,

$$e_z = \frac{1 + \Delta t \; m_z}{1 - \Delta t \; m_z}. \tag{C.31}$$

Observe that since $m_z < 0$ the numerator is always less than the denominator, so that the absolute value of this expression is less than unity. Hence the Crank-Nicolson method is stable for all choices of discretization parameters. This method is almost always preferable to the implicit method, because it is more accurate (as well as being stable).

The *accuracy* of any numerical method is measured by performing a Taylor expansion about $\Delta t = 0$. If there is a nonzero coefficient associated with Δt then the method is *first-order*, while if the first nonzero coefficient is associated with Δt^2 then the method is *second-order*. Both the explicit and the implicit discretizations are only first-order, while the semi-implicit discretization is correct up to second-order.

C.1.2 Branched Cables and Hines Matrices

So far we have developed techniques only for unbranched cables. Since many neurons have very extended and complex dendritic trees (see Fig. 3.1), these techniques must be extended if they are to be useful. The primary complication introduced by branching is that while the connectivity of an unbranched cable can be represented by a tridiagonal matrix, connectivity of a branched neuron can only be represented by a special matrix, a *Hines* matrix, of which the tridiagonal is a special case. This matrix has the same number of elements as a tridiagonal matrix, $3n - 2$ in the $n \times n$ case, but they can be scattered, rather than concentrated along the diagonals. Hines (1984) first introduced this transformation in the context of dendritic modeling and showed how to invert this matrix in $O(n)$ steps— the same as any tridiagonal matrix. This observation was key: until then implicit matrix techniques were believed inefficient for branched cables because the matrix inversion was thought to require $O(n^3)$ steps. We will see below that there are other cases where the sparseness of this matrix can be used to construct algorithms as efficient as those for the tridiagonal case.

The Hines matrix is defined by the structure of its sparseness. A matrix $\mathbf{B}$ is a Hines matrix if three conditions are satisfied: (1) The diagonal elements B_{jj} are nonzero; (2) B_{ij} is nonzero if and only if B_{ji} is nonzero; and (3) for any nonzero B_{ij} with $i < j$, there is no h such that $h > j$ and B_{ih} is nonzero. The second condition requires B to be structurally although not numerically symmetric, and the third requires that $\mathbf{B}$ have no more than one nonzero element to the right of the diagonal in any row, and by structural symmetry, no more than one nonzero element below the diagonal in any column. Notice that while there are only two tridiagonal matrices corresponding to an unbranched cable (since numbering can start from either end), there are many Hines matrices corresponding to a single tree structure.

In order to convert a graph into a matrix, the nodes must be numbered sequentially. There is a simple algorithm for appropriately labeling the nodes of a tree to construct a Hines matrix. The algorithm given here is a generalization of that proposed in Hines (1984).

Sequentially number nodes that have at most one unnumbered neighbor until there are no unnumbered nodes left.

At each step in the algorithm there may be many choices for which node to number next; although all choices are equally suitable for the algorithms discussed, the particular choice of numbering adjacent nodes successively wherever possible leads to the matrix elements maximally concentrated near the diagonal.

The advantage of this numbering scheme is that Gaussian elimination can be applied directly. At the abstract level, it amounts to solving for $\mathbf{V}^{t+1}$ in the matrix equation $\mathbf{B}\mathbf{V}^{t+1} = \mathbf{V}^t$ by

Consider the matrix $[\mathbf{B}|\mathbf{V}^t]$.

Using row operations, put it in the form $[\mathbf{I}|h]$, where $\mathbf{I}$ is the identity matrix.

Now $h = \mathbf{V}^{t+1} = \mathbf{B}^{-1}\mathbf{V}^t$.

The standard scheme for solving tridiagonal systems is a special case of this algorithm.

C.1.3 Boundary Conditions

The partial differential equation for a single unbranched cable (Eq. C.1) has a unique solution only if *boundary conditions* (BC) are specified at the endpoints. The BCs determine the behavior of the membrane potential or its derivative at the boundary points. The membrane potential along a branched cable is governed by a system of coupled partial differential equations, one for each branch, and BCs must be specified at the branch points. In matrix notation the BCs at the origin of the cable correspond to the first row of the matrix $\mathbf{B}$, and at the end of the cable to the last row of the matrix $\mathbf{B}$. Similarly the BCs at a branch point correspond to the matrix elements at that point. We shall see that the implementation of BCs at a branch point is just a natural extension of the elements at any point along the cable and so poses few problems. The BCs at endpoints is rather more subtle and requires careful consideration.

In general there are many BCs corresponding to different physical situations. For example, in the *killed-end* or *Dirichlet* case (see Sec. 2.2.2) the voltage is clamped to zero and the axial current "leaks" out to ground. The matrix $\mathbf{B}'$ we have been considering above in the eigenvalue analysis implements this condition. To see this, assume that the fictive point at $x = 0$, that is, V_0 is assumed to be zero. Following Eq. C.3, the difference equation for the point at $x = \Delta x$ is proportional to $V_2 - 2V_1 + V_0 = V_2 - 2V_1$. Thus, the top two entries in $\mathbf{B}'$ are -2 and 1, as are the bottom two entries. The killed-end solution is a special case of the voltage-clamp condition, where the voltage at the endpoint is held at some arbitrary value. This condition is simply $V_0 = V_{\text{clamp}}$.

Here we limit our attention to the most important class of boundary conditions for nerve equations, the so-called *sealed-end* or *von Neumann* condition (Sec. 2.2.2). The sealed-end condition requires that no current flow out of the end. Mathematically, this is expressed as $\partial V/\partial x = 0$ at the boundary (Eq. 2.19).

There are several ways this boundary condition associated with the continuous equation can be implemented in a discretized system. Perhaps the most intuitive is the one that arises from consideration of an equivalent electrical circuit model, as in Fig. 3.4B. Following Ohm's law, the current flow in the axial dircetion is proportional to $V_2 - V_1$. Thus, the diagonal element is -1 or half the size of the elements along the rest of the diagonal. Because the endpoint is connected only to a single neighbor, while all other nodes are connected to two neighbors, it seems reasonable that the loss to neighbors should be half of that at all other nodes. With these choices the resulting matrix $\mathbf{B}'$ turns out to be symmetric, which is in line with our intuition. For a single unbranched cable with a sealed end at both ends, we have

$$\mathbf{B}' = \begin{pmatrix} -1 & 1 & & & & \\ 1 & -2 & 1 & & & \\ & & \ddots & \ddots & \ddots & \\ & & & 1 & -2 & 1 \\ & & & & 1 & -1 \end{pmatrix}. \tag{C.32}$$

This implementation has been widely used in neural simulators (but not by Mascagni, 1989, or Hines, 1989).

Our intuition can be misleading, that is, the associated electrical circuit (as in Fig. 3.4B) does not approximate well a finite cable with a sealed-end boundary. While such a matrix does implement a form of the von Neumann condition, it is correct only to first order, that is, the error is $O(\Delta x)$. This implementation introduces systematic errors or "phantom currents" (Niebur and Niebur, 1991).

A more accurate scheme can be derived by considering the Taylor expansion of the potential around $x = 0$ for a "ficticious" element just beyond the end of the cable at $x = -\Delta x$,

$$V_{-1} = V_0 - \Delta x \left.\frac{\partial V}{\partial x}\right|_{x=0} + \frac{\Delta x^2}{2} \left.\frac{\partial^2 V}{\partial x^2}\right|_{x=0} + O((-\Delta x)^3). \quad (C.33)$$

We can now solve for V_{-1} in terms of V_0 and V_1 by setting the first derivative to 0 (the BC corresponding to no axial current across the membrane at $x = 0$) and setting the second derivative to the second difference approximation, $\partial^2 V/\partial x^2|_{x=0} = (V_{-1}+V_1-2V_0)/\Delta x^2$, to obtain

$$V_{-1} = V_1. \quad (C.34)$$

This gives us the matrix with the second-order correct von Neumann boundary conditions:

$$\mathbf{B}' = \begin{pmatrix} -2 & 2 & & & & \\ 1 & -2 & 1 & & & \\ & & \ddots & \ddots & \ddots & \\ & & & 1 & -2 & 1 \\ & & & & 2 & -2 \end{pmatrix}. \quad (C.35)$$

C.1.4 Eigensystems and Model Fitting

We have seen how the morphology of a neuron—its branching geometry and connectivity—together with the passive membrane parameters—R_m, C_m, and R_i—collectively determine the sparse matrix $\mathbf{B}$ which governs the dynamic behavior of the neuron. Equation C.10 shows how the eigensystem acts as the link between dynamics and morphology: the time course of membrane potentials can be expressed as the sum of exponentials whose decay constants are the eigenvalues of $\mathbf{B}$. Thus, starting from the physical description of the neuron, we can use the eigensystem to compute its response to stimuli, although in practice it is usually more efficient to use an implicit matrix scheme.

Suppose we wish to determine the eigensystem of a model neuron. For very simple morphologies, for instance, a single terminated cylinder, and homogeneous membrane parameters, Rall (1977) has used the underlying partial differential equation to derive some relatively simple expressions for the eigenvalues as a function of the membrane parameters (as in Eq. 2.38). These expressions provide some insight into the behavior of passive cables. For somewhat more complex morphologies (for instance, several cylinders attached to a single lumped soma) there are correspondingly more complex expressions involving the roots of transcendental equations. But as the morphologies become more complex the expressions become more difficult to evaluate and provide less insight. Analytic solutions are not generally useful for arbitrary dendritic trees.

For arbitrary trees there are at least two options for determining the eigensystem. First, we could compute a solution $V_i(t)$ from initial conditions, using for example an implicit matrix scheme as outlined above, and then attempt to extract the time constants τ_z and associated coefficients $c'_{iz} = V_i^{(z)} c_z$ by fitting them to the solution. One method of fitting is the so-called "exponential peeling" or just "peeling" method (Rall, 1977). In fact, exponential fitting is intrinsically very difficult. The reason is that exponentials are a poor set of basis functions, because they are so far from orthogonal. The inner product of two exponentials over an infinite interval is given by

$$\int_0^\infty e^{-t/\tau_1} e^{-t/\tau_2} dt = \frac{\tau_1 \tau_2}{\tau_1 + \tau_2}. \qquad \text{(C.36)}$$

Orthogonality would require that this expression be zero when $\tau_1 \neq \tau_2$, which is not the case. In fact, the general exponential fitting problem is equivalent (in the limit of a large number of time constants) to numerically inverting a Laplace transform, which is well known to be *ill-conditioned* (Bellman, Kalaba, and Locket, 1966). Since many combinations of time constants and coefficients fit almost equally well, no technique can reliably extract the time constants in the presence of noise. Typically only the first two or three eigenvalues can be extracted, as suggested by the characteristic decrease in the eigenvalues in Eq. C.8.

If we are working with a model of a neuron, a second and much more reasonable alternative is to extract the eigenvalues directly from the matrix **B**. (This option is not available to us if we are working with a real neuron, in which case we must fall back on some nonlinear fitting technique.) Typically we are interested only in the first few eigenvalues, so we exploit the sparseness of **B** by observing that multiplication of **B** with a vector is cheap—$O(n)$ for an $n \times n$ Hines matrix. We use the power method, which depends on the fact that the principal eigenpair dominates. Defining a new matrix $\mathbf{P} = \alpha \mathbf{I} - \mathbf{B}$ (where **I** is the identity matrix) with eigenvalues $p_z = \alpha - m_z$, the smallest eigenvalue of **B** (that is, the largest time constant) can be determined by the following algorithm, which finds the principal eigenpair of **P** for any initial **V**:

repeat

$$\mathbf{V} \leftarrow \frac{1}{\|\mathbf{V}\|} \mathbf{P} \mathbf{V} \qquad \text{(C.37)}$$

until **V** is stable.

Standard deflation can be used to extend this procedure directly to the computation of the next several smallest eigenvalues of **B** (Stoer and Bulirsch, 1980).

C.1.5 *Green's Functions and Matrix Inverses*

Prior to the ascent of sparse matrix techniques, methods based on Green's functions (also called *transfer impedances*) and Laplace transforms were widely used (Butz and Cowan, 1984; Koch and Poggio, 1985a; Holmes, 1986; see Sec. 3.4). The Green's function gives the response at any point i to a delta pulse of current applied at some other point j; their properties were discussed in Sec. 2.3.1 (e.g., Eq. 2.31). These methods have been largely abondoned because they are typically less efficient and more difficult to generalize to synaptic and nonlinear conductances. Nevertheless, for theoretical work they are often very convenient, and can sometimes offer a different perspective (see also Abbott, 1992). One interesting application of the Green's function is the morphoelectrotonic transform

(Zador, Agmon-Snir, and Segev, 1995; see Sec. 3.5.4). Another is in analyzing the effects of Hebbian learning (Pearlmutter, 1995). Here we reconsider the Green's functions in terms of sparse matrix methods.

One way to compute the Green's function is in terms of the eigensystem of the matrix **B**. First we expand the solution $V_j(t)$ at a point x_j in terms of the initial conditions $V_i(0)$ at $t = 0$ and the Green's function $K'_{ij}(t)$,

$$V_j(t) = \sum_i V_i(0) K'_{ij}(t). \tag{C.38}$$

Notice that K'_{ij} is identical to K_{ij} of Eq. 3.16 and in the following, except for a constant of proportionality with the dimensions of an impedance. Because the eigenvectors $V_i^{(z)}$ form a complete basis, we can use them to represent the initial conditions,

$$V_i(0) = \sum_z c_z V_i^{(z)}, \tag{C.39}$$

with c_z constants. From Eq. C.10 we can write $V_j(t)$ as

$$V_j(t) = \sum_z c_z V_j^{(z)} e^{-t/\tau_z}. \tag{C.40}$$

By orthogonality and completeness we can compute the constants as the projection of the initial condition vector on the eigenvectors,

$$c_z = \sum_i V_i(0) V_i^{(z)}. \tag{C.41}$$

Hence

$$V_j(t) = \sum_i V_i(0) \underbrace{\sum_z V_i^{(z)} V_j^{(z)} e^{-t/\tau_z}}_{\text{Green's function}} \tag{C.42}$$

that is,

$$K'_{ij}(t) = \sum_z V_i^{(z)} V_j^{(z)} e^{-t/\tau_z} . \tag{C.43}$$

Understanding the eigenvalue expansion of the Green's function can be useful in theoretical work and for small test problems. In practice, computing the Green's function from the eigensystem is often not the most efficient way. Rather, it is often better to exploit the fact that the Green's function can be considered an inverse operator. This reduces the problem to computing the inverse of a matrix. Once again, we can exploit the sparseness of **B** to compute the Green's function.

Extension to Two and Three Spatial Dimensions

The techniques we have described for solving the cable and diffusion equations in one dimension are readily generalized to two or even three spatial dimensions. Although the solution of electrical potential in several dimensions is not common (see, however, Chap. 2), the diffusion of second messengers such as Ca^{2+} often requires a consideration of two- or three-dimensional effects.

The separability of the three-dimensional diffusion operator suggests an easy and efficient method for solving multidimensional diffusion. Consider the three-dimensional Laplacian operator in Cartesian coordinates,

$$\nabla^2 = \frac{\partial^2}{\partial x^2} + \frac{\partial^2}{\partial y^2} + \frac{\partial^2}{\partial z^2}. \tag{C.44}$$

Similarly we can write the discrete approximation using matrix operators,

$$\mathbf{L}_{3d} = \mathbf{L}_x + \mathbf{L}_y + \mathbf{L}_z. \tag{C.45}$$

Here $\mathbf{L}_x$, $\mathbf{L}_y$, and $\mathbf{L}_z$ are very sparse matrices that approximate the second derivative in the x, y, and z directions, respectively. These operators can be applied sequentially, using for example a stable implicit method, to advance the solution from t to $t + 1$.

The main difficulty associated with solving three-dimensional diffusion is bookkeeping, although the bookkeeping with this approach is simpler than if $\mathbf{L}_{3d}$ is used directly. The matrices that comprise $\mathbf{L}_{3d}$ are sparse, but not tridiagonal or Hines. The locations of the nonzero entries depend on the precise spatial discretization.

C.2 Nonlinear Cable Equations

We have focused on the numerical solution of the linear cable equation for two reasons. First, it is impossible to solve the nonlinear equations efficiently without first understanding the techniques for linear solution. Second, the methods of numerical solution provide insight into the equations themselves. Our approach was to compare the eigensystems of the discrete and continuous systems, and the correspondences were strong.

Here we consider two important classes of nonlinear cable equations that arise frequently in the study of single neurons. The first class is a generalization of the Hodgkin-Huxley equations, which describe the propagation of the action potential. The second class arises from the effect of nonlinear saturable ionic buffering in diffusion. Unfortunately, we know of no approach to understanding the numerical solutions of these nonlinear equations that provides the same kind of insight. This may be because most numerical techniques begin by linearization (eliminating the interesting properties of these equations), while the behavior of the nonlinear equations is best understood using phase space analysis (see Chap. 7) or by numerical simulation. In the first section we provide an overview of the most widely used and efficient method for solving generalizations of the Hodgkin-Huxley equations. It is a direct method that is so easy to implement that since its introduction a decade ago (Hines, 1984) it seems largely to have supplanted the predictor-corrector method of Cooley and Dodge (1966) that reigned before. It is now standard on many widely distributed neural simulators, including GENESIS and NEURON (see www.klab.caltech/MNM for more information on these). In the next section we describe a simple scheme for handling nonlinear saturable buffers.

C.2.1 Generalized Hodgkin–Huxley Equations

The method we consider applies to a class of nonlinear cable equations generalized from the Hodgkin–Huxley equations. They differ from the linear cable equation only by the addition of a term, $I_{\text{HH}}(V(x,t),t)$, which gives the current contributed by voltage-dependent membrane channels,

$$C_m \frac{\partial V(x,t)}{\partial t} = \frac{d}{4R_i} \frac{\partial^2 V(x,t)}{\partial x^2} - G_m V(x,t) - I_{\text{HH}}(V(x,t),t). \tag{C.46}$$

The most general form we need to consider for $I_{\text{HH}}(V,x,t)$ includes k different ionic currents,

$$I_{HH}(V(x,t),t) = \overline{g_1}\, g_1(V,t)\,(V(x,t) - E_1) + \cdots + \overline{g_k}\, g_k(V,t)\,(V(x,t) - E_k) \quad\text{(C.47)}$$

(see Eq. 6.20). The total current is the sum of the contributions of the k individual currents in the membrane patch considered. Here $\overline{g_1}$ is the density of type 1 membrane channels at any point along the neuron; $V(x,t) - E_1$ is the driving force for this conductance; and $0 \le g_1(V,t) \le 1$ is the fraction of total conductance of the current, that is, the fraction of total channels of that type that is open in one patch of membrane. The interesting term in this expression is the function $g(V,t)$, because it includes the nonlinearity. It has the general form

$$g(V,t) = y(V,t)^r, \quad\text{(C.48)}$$

where y is a gating particle, and r is an integer corresponding to the number of identical gating particles that need to be simultaneously present in order for current to flow (see Sec. 6.2). In many cases, there are two or even three gating particles, usually *activating* and *inactivating* particles. In this case, a slightly more general product of the form $g(V,t) = y_m(V,t)^{r_m}\, y_h(V,t)^{r_h}$ must be used. This introduces no conceptual difficulties but does clutter the notation.

We have suppressed the spatial dependence of the currents, since this occurs only through $V(x,t)$. This fact simplifies the solution, since it means that the system of equations governing g in the spatially discretized system is diagonal. The variable $y(V,t)$ (or variables y_m and y_h) in turn obeys a first-order, nonlinear differential equation of the form

$$\frac{dy}{dt} = \frac{y_\infty(V) - y}{\tau_y(V)}. \quad\text{(C.49)}$$

Here the function $y_\infty(V)$ governs the steady-state behavior of y. It is monotonic and bounded between 0 and 1—it must be, for g to be guaranteed to always fall in the same range. The $\tau_y(V)$ governs the rate at which equilibrium is reached. For numerical solution an equivalent but more convenient form is

$$\frac{dy}{dt} = (1 - y)\,\alpha_y(V) - y\,\beta_y(V). \quad\text{(C.50)}$$

The method we describe depends on the fact that the generalized Hodgkin–Huxley system is *conditionally linear* (Mascagni, 1989), meaning that the system is linear in V^{t+1} if g_h^t is known, and likewise linear in g_h^{t+1} if V^t is known. Thus we can alternate between solving for V^t at the times V^t, V^{t+1}, V^{t+2}, ..., and solving for g_h at intermediate points $g^{t+\frac{1}{2}}$, $g^{t+\frac{3}{2}}$, ... We do this implicitly,

$$y^{t+\frac{1}{2}} = y^{t-\frac{1}{2}} + \Delta t[(1 - y^{t+\frac{1}{2}})\alpha_y(V^t) - y^{t+\frac{1}{2}}\beta_y(V^t)]. \quad\text{(C.51)}$$

We can solve this expression explicitly for $y^{t+\frac{1}{2}}$ in terms of $y^{t-\frac{1}{2}}$ and V^t, both of which are known. The algebra is straightforward and the details are well described in Hines (1984).

C.2.2 Calcium Buffering

Chapter 11 deals with how, under certain conditions, the equations describing calcium dynamics are formally equivalent to the cable equation. Furthermore, over a wide range of parameters, the linearized equations offer an adequate approximation to the nonlinear dynamics. Nevertheless, in some cases it is of interest to explore the behavior of the full nonlinear equations. (If linear two- or three-dimensional diffusional behavior is of interest, the methods of Sec. C.1.5 can be used.) For the most part the semi-implicit method described above generalizes directly to calcium dynamics. A nonlinear saturable buffer raises special issues that must be considered separately.

Let us treat the common case of a second-order buffer (Sec. 11.4.1),

$$B + Ca^{2+} \overset{f}{\underset{b}{\rightleftharpoons}} M \tag{C.52}$$

where the rate constants f and b govern the equilibrium of free and bound calcium and $[Ca^{2+}]_i$ is the concentration of free, intracellular calcium. (To simplify our exposition, we consider only the case of a second-order buffer, but higher-order buffers introduce no qualitative differences.)

Together with the basic diffusion equation, with a simple calcium extrusion process $P([Ca^{2+}]_i)$, and a saturable nondiffusible buffer (see also Eqs. 11.37, 11.43 and 11.48) we have,

$$\frac{\partial [Ca^{2+}]_i}{\partial t} = D \frac{\partial^2 [Ca^{2+}]_i}{\partial x^2} - P([Ca^{2+}]_i)$$

$$- f[Ca^{2+}]_i \, (T_B - [B \cdot Ca]) + b[B \cdot Ca] - \frac{4}{d} i(x, t) \tag{C.53}$$

$$\frac{\partial [B \cdot Ca]}{\partial t} = f[Ca^{2+}]_i \, (T_B - [B \cdot Ca]) - [B \cdot Ca] + D' \frac{\partial^2 [B \cdot Ca]}{\partial x^2}, \tag{C.54}$$

where the calcium concentration $[Ca^{2+}]_i(x, t)$ at time t and position x in response to the applied calcium current density $i(x, t)$ depends on the partial derivatives of $[Ca^{2+}]_i$, the diffusion constant D of Ca^{2+}, the diameter d of the cable, and the concentration of total buffer T_B (the sum of the free buffer and the bound buffer). The first equation specifies the rate of change in concentration of calcium as a function of diffusion, extrusion, buffer dynamics and influx. The second equation gives the rate of change in bound buffer concentration as a function of buffer diffusion and buffer binding and unbinding (see also Eq. 11.48).

The difficulty arises because $[Ca^{2+}]_i$ enters into Eq. C.53 in an essentially nonlinear way as $[Ca^{2+}]_i[B]$, and not in the conditionally linear way as in the case of the Hodgkin–Huxley equations. However, we can discretize Eq. C.53 in such a way that the system becomes conditionally linear. That is, we can rewrite the right-hand side of Eq. C.53 in terms of $[Ca^{2+}]_i^{t+1}$ and $[B \cdot Ca]^t$,

$$\frac{[Ca^{2+}]_i^{t+1} - [Ca^{2+}]_i^t}{\Delta t} = h([Ca^{2+}]_i^{t+1}, [B \cdot Ca]^t) \tag{C.55}$$

and then write Eq. C.54 in terms of $[Ca^{2+}]_i^t$ and $[B \cdot Ca]^{t+1}$,

$$\frac{[B \cdot Ca]^{t+1} - [B \cdot Ca]^t}{\Delta t} = g([B \cdot Ca]^{t+1}, [Ca^{2+}]_i^t) . \tag{C.56}$$

These implicit equations can now be advanced each time step by an $O(n)$ step (a single sparse matrix inversion for the Hines matrix) for Eq. C.55, and another $O(n)$ step for the diagonal matrix specified in Eq. C.56.

C.2.3 Conclusion

Although this appendix has ostensibly been about efficient numerical methods for single neuron simulation, the real aim has been to provide a deeper understanding of the equations that govern neurodynamics. In the case of the linear cable equation our approach has been to carefully analyze the properties of the discrete eigensystem corresponding to the continuous partial differential equations underlying the cable equation. This led to stable and efficient methods for simulating the cable equation, to an understanding of why fitting passive models to neurophysiological data is hard, and to a different way of looking at the Green's functions. For the nonlinear case our goals were more modest, namely, an overview of the best existing method for solving equations of the Hodgkin–Huxley class, and of some special problems that arise in the solution of calcium dynamics.

References

Abbott, L.F. "Realistic synaptic inputs for model neuronal networks," *Network* **2:** 245–258 (1991).

Abbott, L.F. "Simple diagrammatic rules for solving dendritic cable problems," *Physica A* **185:** 343–356 (1992).

Abbott, L.F. "Decoding neuronal firing and modeling neural networks," *Quart. Rev. Biophys.* **27:** 291–331 (1994).

Abbott, L.F., and Blum, K.I. "Functional significance of long-term potentiation for sequence learning and prediction," *Cerebral Cortex* **6:** 406–416 (1996).

Abbott, L.F., Farhi, E., and Gutmann, S. "The path integral for dendritic trees," *Biol. Cybern.* **66:** 49–60 (1991).

Abbott, L.F., and van Vreeswijk, C. "Asynchronous states in networks of pulse-coupled oscillators," *Physical Rev. E* **48:** 1483–1490 (1993).

Abbott, L.F., Varela, J.A., Sen, K., and Nelson, S.B. "Synaptic depression and cortical gain control," *Science* **275:** 220–224 (1997).

Abeles, M. *Local Cortical Circuits: An Electrophysiological Study.* Springer Verlag: New York, New York (1982a).

Abeles, M. "Role of the cortical neuron: Integrator or coincidence detector?" *Israel J. Medical Sci.* **18:** 83–92 (1982b).

Abeles, M. *Corticonics: Neural Circuits of the Cerebral Cortex.* Cambridge University Press: Cambridge, United Kingdom (1990).

Abeles, M., Bergman, H., Margalit, E., and Vaadia, E. "Spatiotemporal firing patterns in the frontal cortex of behaving monkeys," *J. Neurophysiol.* **70:** 1629–1638 (1993).

Abeles, M., Vaadia, E., Bergman, H., Prut, Y., Haalman, I., and Slovin, H. *Concepts Neurosci.* **4:** 131–158 (1994).

Adams, P.R., Jones, S.W., Pennefather, P., Brown, D.A., Koch, C., and Lancaster, B. "Slow synaptic transmission in frog sympathetic ganglia," *J. Exp. Biol.* **124:** 259–285 (1986).

Adams, D.J., and Nonner, W. "Voltage-dependent potassium channels: Gating, ion permeation and block." In: *Potassium Channels: Structure, Classification, Function and Therapeutic Potential.* Cook, N.S., editor, pp. 42-69. Ellis Horwood: Chichester, United Kingdom (1989).

Adelson, E.H., and Bergen, J.R. "Spatio-temporal energy models for the perception of motion," *J. Opt. Soc. Am. A* **2:** 284–299 (1985).

Adrian, E.D. *The Mechanism of Nervous Action: Electrical Studies of the Neurone.* University of Pennsylvania Press: Philadelphia, Pennsylvania (1932).

Adrian, E.D., and Zotterman, Y. "The impulses produced by sensory nerve endings: Part III. Impulses set up by touch and pressure," *J. Physiol.* **61:** 465–483 (1926).

Adrian, R.H. "Conduction velocity and gating current in the squid giant axon," *Proc. Roy. Soc. Lond. B* **189:** 81–86 (1975).

Agin, D. "Hodgkin-Huxley equations: Logarithmic relation between the membrane current and frequency of repetitive activity," *Nature* **201:** 625–626 (1964).

Agmon, A., and Connors, B.W. "Correlation between intrinsic firing patterns and thalamocortical synaptic responses of neurons in mouse barrel cortex," *J. Neurosci.* **12:** 319–329 (1992).

Agmon-Snir, H., Carr, C.E., and Rinzel, J. "The role of dendrites in auditory coincidence detection," *Nature*, **393:** 268–272 (1998).

Agmon-Snir, H., and Segev, I. "Signal delay and input synchronization in passive dendritic structures," *J. Neurophysiol.* **70:** 2066–2085 (1993).

Ahmed, B., Anderson, J.C., Douglas, R.J., Martin, K.A.C., and Nelson, J.C. "Polyneuronal innervation of spiny stellate neurons in cat visual cortex," *J. Comp. Neurol.* **341:** 39–49 (1994).

Ahmed, B., Anderson, J.C., Douglas, R.J., Martin, K.A.C., and Nelson, J.C. "Estimates of the major excitatory innervation on apical dendrites of layer 5 pyramids within layer 4 of cat visual cortex," *J. Physiol.* **494:** 15P–16P (1996).

Ahmed, B., Anderson, J.C., Douglas, R.J., Martin, K.A.C., and Whitteridge, D. "A method for estimating net somatic input current from the action potential discharge of neurons in the visual cortex of the anesthetized cat," *J. Physiol.* **459:** 134P (1993).

Aldrich, R.W., Corey, D.P., and Stevens, C.F. "A reinterpretation of mammalian sodium channel gating based on single channel recording," *Nature* **306:** 436–441 (1983).

Alkon, D.L. *Memory Traces in the Brain.* Cambridge University Press: Cambridge, United Kingdom (1987).

Alkon, D.L., Lederhendler, I., and Shoukimas, J.J. "Primary changes of membrane currents during retention of associative learning," *Science* **215:** 693–695 (1982).

Alkon, D.L., Sakakibara, M., Forman, R., Harrigan, J., Lederhendler, I., and Farley, J. "Reduction of two voltage-dependent K^+ currents mediates retention of a learned association," *Behav. Neural Biol.* **44:** 278–300 (1985).

Allbritton, N.L., Meyer, T., and Stryer, L. "Range of messenger action of calcium ion and inositol 1,4,5-trisphosphate," *Science* **258:** 1812–1815 (1992).

Allen, C., and Stevens, C.F. "An evaluation of causes for unreliability of synaptic transmission," *Proc. Natl. Acad. Sci. USA* **91:** 10380–10383 (1994).

Alonso, J.M., Usrey, W.M., and Reid, R.C. "Precisely correlated firing in cells of the lateral geniculate nucleus," *Nature* **383:** 815–819 (1996).

Amaral, D.G., Ishizuka, N., and Claiborne, B.J., "Neurons, numbers and the hippocampal network," *Prog. Brain Res.* **83:** 1–11 (1990).

Ames, A. "Energy requirements of brain functions: When is energy limiting?" In: *Mitochondria and Free Radicals in Neurodegenerative Disease.* Beal, M.F., Howell, N., and Bodis-Wollner, I., editors, pp. 12–27. John Wiley: New York, New York (1997).

Ames, A., Li, Y.Y., Heher, E.C., and Kimble, C.R. "Energy metabolism of rabbit retina as related to function—High cost of Na^+ transport," *J. Neurosci.* **12:** 840–853 (1992).

Amit, D.J., and Brunel, N. "Adequate input for learning in attractor neural networks," *Network* **4:** 177–194 (1993).

Amit, D.J., and Tsodyks, M.V. "Quantitative study of attractor neural network retrieving at low spike rates: I. Substrate—Spikes, rates and neuronal gain," *Network* **2:** 259–273 (1991).

Amitai, Y., Friedman, A., Connors, B.W., and Gutnick, M.J. "Regenerative electrical activity in apical dendrites of pyramidal cells in neocortex," *Cerebral Cortex* **3:** 26–38 (1993).

Amthor, F.R., and Grzywacz, N.M. "Direction selectivity in vertebrate retinal ganglion cells." In: *Visual Motion and its Role in the Stabilization of Gaze.* Miles, F.A., and Wallman, J., editors, pp. 79–100. Elsevier: Amsterdam, The Netherlands (1993).

Anderson, J.C., Douglas, R.J., Martin, K.A.C., and Nelson, J.C. "Map of the synapses formed with the dendrites of spiny stellate neurons of cat visual cortex," *J. Comp. Neurol.* **341:** 25–38 (1994).

Andersen, P., Raastad, M., and Storm, J.F. "Excitatory synaptic integration in hippocampal pyramids and dendate granule cells," *Cold Spring Harbor Symp. Quant. Biol.* **55:** 81–96 (1990).

Andersen, R.A., Essick, G.K., and Siegel, R.M. "Encoding of spatial location by posterior parietal neurons," *Science* **230**: 456–458 (1985).

Arbib, M., editor. *The Handbook of Brain Theory and Neural Networks.* MIT Press: Cambridge, Massachusetts (1995).

Armstrong, C.M., and Bezanilla, F. "Currents related to movement of the gating particles of the sodium channels," *Nature* **242**: 459–461 (1973).

Armstrong, C.M., and Bezanilla, F. "Inactivation of the sodium channel. II. Gating current experiments," *J. Gen. Physiol.* **70**: 567–590 (1977).

Art, J.J., and Fettiplace, R. "Variation of membrane properties in hair cells isolated from the turtle cochlea," *J. Physiol.* **385**: 207–242 (1987).

Artola, A., and Singer, W. "Long-term potentiation and NMDA receptors in rat visual cortex," *Nature* **330**: 649–652 (1987).

Artola, A., and Singer, W. "Long-term depression of excitatory synaptic transmission and its relationship to long-term potentiation," *Trends Neurosci.* **16**: 480–487 (1993).

Ascher, P., and Nowak, L. "The role of divalent cations in the N-methyl-D-aspartate responses of mouse central neurons in culture," *J. Physiol.* **399**: 247–266 (1988).

Attwell, D., and Wilson, W. "Behavior of the rod network in the tiger salamander retina mediated by membrane properties of individual rods," *J. Physiol.* **309**: 287–315 (1980).

Aziz, P.M., Sorensen, H.V., and van der Spiegel, J. "An overview of Sigma-Delta converters," *IEEE Signal Proc. Magazine* **13**: 61–84 (1996).

Azouz, R., Jensen, M.S., and Yaari, Y. "Ionic basis of spike after-depolarization and burst generation in adult rat hippocampal CA1 pyramidal cells," *J. Physiol.* **492**: 211–223 (1996).

Babu, Y.S., Sack, J.S., Greenhough, T.J., Bugg, C.E., Means, A.R., and Cook, W.J. "Three-dimensional structure of calmodulin," *Nature* **315**: 37–40 (1985).

Baer, S.M., and Rinzel, J. "Propagation of dendritic spikes mediated by excitable spines: A continuum theory," *J. Neurophysiol.* **65**: 874–890 (1991).

Bair, W. *Analysis of temporal structure in spike trains of visual cortical area MT.* PhD Thesis. California Institute of Technology: Pasadena, California (1995).

Bair, W., and Koch, C. "Temporal precision of spike trains in extrastriate cortex of the behaving macaque monkey," *Neural Comput.* **8**: 1185–1202 (1996).

Bair, W., Koch, C., Newsome, W., and Britten, K. "Power spectrum analysis of bursting cells in area MT in the behaving monkey," *J. Neurosci.* **14**: 2870–2892 (1994).

Baker, P.F., Hodgkin, A.L., and Ridgway, E.B. "Depolarization and calcium entry in squid giant axons," *J. Physiol.* **218**: 709–755 (1971).

Baker, P.F., Hodgkin, A.L., and Shaw, T.I. "Replacement of the axoplasm of giant nerve fibers with artificial solutions," *J. Physiol.* **164**: 330–354 (1962).

Baranyi, A., Szente, M.B., and Woody, C.D. "Electrophysiological characterization of different types of neurons recorded in vivo in the motor cortex of the cat. I. Patterns of firing activity and synaptic responses," *J. Neurophysiol.* **69**: 1850–1864 (1993).

Barbour, B., and Häusser, M. "Intersynaptic diffusion of neurotransmitter," *Trends Neurosci.* **20**: 377–384 (1997).

Barkai, N., and Leibler, S. "Robustness in simple biochemical networks," *Nature* **387**: 913–917 (1997).

Barlow, H.B. "Single units and sensation: A neuron doctrine for perceptual psychology?" *Perception* **1**: 371–394 (1972).

Barlow, H.B., Kaushal, T.P., Hawken, M., and Parker, A.J. "Human contrast discrimination and the threshold of cortical neurons," *J. Opt. Soc. Am. A* **4**: 2366–2371 (1987).

Barlow, H.B., and Levick, R.W. "The mechanism of directional selectivity in the rabbit's retina," *J. Physiol.* **173**: 477–504 (1965).

Barr, R.C., and Plonsey, R. "Electrophysiological interaction through the interstitial space between adjacent unmyelinated parallel fibers," *Biophys. J.* **61**: 1164–1175 (1992).

Barrett, E.F., and Barrett, J.N. "Intracellular recording from myelinated axons: Mechanism of the depolarizing afterpotential," *J. Physiol.* **323:** 117–144 (1982).

Barrett, E.F., and Stevens, C.F. "Quantal independence and uniformity of presynaptic release kinetics at the frog neuromuscular junction," *J. Physiol.* **227:** 665–689 (1972).

Barrett, J.N. "Motoneuron dendrites: Role in synaptic integration," *Fed. Proc. Am. Soc. Exp. Biol.* **34:** 1398–1407 (1975).

Barrett, J.N., and Crill, W. "Specific membrane properties of cat motoneurones," *J. Physiol.* **239:** 301–324 (1974).

Barrionuevo, G., and Brown, T.H. "Associative long-term potentiation in hippocampal slices," *Proc. Natl. Acad. Sci. USA* **80:** 7347–7351 (1983).

Barron, D.H., and Matthews, B.H.C. "Intermittent conduction in the spinal cord," *J. Physiol.* **85:** 73–103 (1935).

Bashir, Z.I., Alford, S., Davies, S.N., Randall, A.D., and Collingridge, G.L. "Long-term potentiation of NMDA receptor-mediated synaptic transmission in the hippocampus," *Nature* **349:** 156–158 (1991).

Bauerfeind, R., Galli, T., and De Camilli, P. "Molecular mechanisms in synaptic vesicle recycling," *J. Neurocytol.* **25:** 701–715 (1996).

Baylor, D.A., and Fettiplace, R. "Synaptic drive and impulse generation in ganglion cells of turtle retina," *J. Physiol.* **288:** 107–127 (1979).

Beam, K.G., and Donaldson, P.L. "A quantitative study of potassium channel kinetics in rat skeletal muscle from 1 to 37° C," *J. Gen. Physiol.* **81:** 485–512 (1983).

Bear, M.F., and Malenka, R.C. "Synaptic plasticity: LTP and LTD," *Current Opinion Neurobiol.* **4:** 389–399 (1994).

Bear, M.F. "Mechanism for a sliding synaptic modification threshold," *Neuron* **15:** 1–4 (1995).

Bear, M.F., Cooper, L.N., and Ebner, F.F. "A physiological basis for a theory of synapse modification," *Science* **237:** 42–48 (1987).

Beaulieu, C., and Colonnier, M. "A laminar analysis of the number of round-asymmetrical and flat-symmetrical synapses on spines, dendritic trunk and cell bodies in area 17 of the cat," *J. Comp. Neurol.* **231:** 180–189 (1985).

Bekkers, J.M. "Are autapses prodigal synapses?" *Current Biol.* **8:** R52-R55 (1998).

Bekkers, J. M., Richerson, G.B., and Stevens, C.F. "Origin of variability in quantal size in cultured hippocampal neurons and hippocampal slices," *Proc. Natl. Acad. Sci. USA* **87:** 5359–5362 (1990).

Bekkers, J.M., and Stevens, C.F. "NMDA and non-NMDA receptors are co-localized at individual excitatory synapses in cultured rat hippocampus," *Nature* **341:** 230–233 (1989).

Bell, A.J. "Self-Organization in real neurons: Anti-Hebb in channel space." In: *Advances in Neural Information Processing Systems*, Vol. 4. Moody, J.E., Hanson, S.J., and Lippman, R., editors, pp. 117–124. Morgan Kaufmann: San Mateo, California (1992).

Bell, A.J., Mainen, Z.F., Tsodyks, M., and Sejnowski, T.J. "Balanced conductances may explain irregular cortical spiking." *Technical Report INC–9502*, Institute for Neural Computation, La Jolla, California (1995).

Bellman, R., Kalaba, R.E., and Lockett, J.A. *Numerical Inversion of the Laplace Transform: Applications to Biology, Economics, Engineering and Physics*. Elsevier, North Holland: New York, New York (1966).

Bennett, M.K. "Ca^{2+} and the regulation of neurotransmitter secretion," *Current Opinion Neurobiol.* **7:** 316–322 (1997).

Bennett, M.V.L., and Spray, D.C. "Intercellular communication mediated by gap junctions can be controlled in many ways." In: *Synaptic Function*. Edelman, G.M., Gall, W.E., and Cowan, W.M., editors, pp. 109–136. John Wiley: New York, New York (1987).

Benz, R., Frölich, O., Läuger, P., and Montal, M. "Electrical capacity of black lipid films and of lipid bilayers made from monolayers," *Biochem. Biophys. Acta* **394:** 323–334 (1975).

Berardi, N., and Morrone, M.C. "The role of γ-aminobutyric acid mediated inhibition in the response properties of cat lateral geniculate nucleus neurones," *J. Physiol.* **357:** 505–524 (1984).

Berg, H.C. *Random Walks in Biology.* Princeton University Press: Princeton, New Jersey (1983).

Berman, N.J., Douglas, R.J., and Martin, K.A.C. "GABA-mediated inhibition in the neural networks of visual cortex." In: *Progress in Brain Research*, Vol. 90. Mize, R.R., Marc, R.E., and Sillito, A.M., editors, pp. 443–476. Elsevier: Amsterdam, The Netherlands (1992).

Bernander, Ö. *Synaptic integration and its control in neocortical pyramidal cells.* PhD Thesis. California Institute of Technology: Pasadena, California (1993).

Bernander, Ö., Douglas, R., and Koch, C. "A model of cortical pyramidal neurons," *CNS Memo No. 16*, California Institute of Technology: Pasadena, California (1992).

Bernander, Ö., Douglas, R.J., and Koch, C. "Amplification and linearization of synaptic input to the apical dendrites of cortical pyramidal neurons," *J. Neurophysiol.* **72:** 2743–2753 (1994).

Bernander, Ö., Douglas, R., Martin, K.A.C., and Koch, C. "Synaptic background activity influences spatiotemporal integration in single pyramidal cells," *Proc. Natl. Acad. Sci. USA* **88:** 11569–11573 (1991).

Berretta, B., and Jones, S.G. "A comparison of spontaneous EPSCs in layer II and layer IV–V neurons of the rat entorhinal cortex in vitro," *J. Neurophysiol.* **76:** 1089–1100 (1996).

Berridge, M. "Calcium oscillations," *J. Biol. Chem.* **265:** 9582–9586 (1990).

Berry, M.J., Warland, D.K., and Meister, M. "The structure and precision of retinal spike trains," *Proc. Natl. Acad. Sci. USA* 94: 5411–5416 (1997).

Bezanilla, F. "Single sodium channels from the squid giant axon," *Biophys. J.* **52:** 1087–1090 (1987).

Bhalla, U.S., Bilitch, D.H., and Bower, J.M. "Rallpacks: A set of benchmarks for neuronal simulators," *Trends Neurosci.* **15:** 453–458 (1992).

Bialek, W., Rieke, F., van Steveninck, R.R.D., and Warland, D. "Reading a neural code," *Science* **252:** 1854–1857 (1991).

Bienenstock, E.L., Cooper, L.N., and Munro, P.W. "Theory for the development of neuron selectivity: Orientation specificity and binocular interaction in visual cortex," *J. Neurosci.* **2:** 32–48 (1982).

Bindman, L., Meyer, T., and Prince, C. "Comparison of the electrical properties of neocortical neurones in slices *in vitro* and in the anesthetized cat," *Brain Res.* **68:** 489–496 (1988).

Bindman, L., and Prince, C. "Intracellular recordings from neurones in the cerebral cortex of the anaesthetized rat," *J. Physiol.* **341:** 7–8P (1983).

Bittner, G.D. "Differentiation of nerve terminals in the crayfish opener muscle and its functional significance," *J. Gen. Physiol.* **51:** 731–758 (1968).

Blaustein, M.P. "Calcium transport and buffering in neurons," *Trends Neurosci.* **11:** 438–443 (1988).

Blaustein, M.P., and Hodgkin, A.L. "The effect of cyanide on the efflux of calcium from squid axon," *J. Physiol.* **200:** 497–527 (1969).

Bliss, T.V.P., and Collingridge, G.L. "A synaptic model of memory: Long-term potentiation in the hippocampus," *Nature* **361:** 31–39 (1993).

Bliss, T.V.P., and Lomo, T. "Long-lasting potentiation of synaptic transmission in the dentate area of the anaesthetized rabbit following stimulation of the perforant path," *J. Physiol.* **232:** 331–356 (1973).

Bloomfield, S. "Arithmetical operations performed by nerve cells," *Brain Res.* **69:** 115–124 (1974).

Bloomfield, S.A., Hamos, J.E., and Sherman, S.M. "Passive cable properties and morphological correlates of neurons in the lateral geniculate nucleus of the cat," *J. Physiol.* **383:** 653–668 (1987).

Bloomfield, S.A., and Sherman, S.M. "Dendritic current flow in relay cells and interneurons of the cat's lateral geniculate nucleus," *Proc. Natl. Acad. Sci. USA* **86:** 3911–3914 (1989).

Bolshakov, V.Y., and Siegelbaum, S.A. "Postsynaptic induction and presynaptic expression of hippocampal long-term depression," *Science* **20:** 1148–1152 (1994).

Bolshakov, V.Y., and Siegelbaum, S.A. "Hippocampal long-term depression: Arachidonic acid as a potential retrograde messenger," *Neuropharmacology* **34:** 1581–1587 (1995).

Bonhoeffer, K.F. "Activation of passive iron as a model for the excitation of nerve," *J. Gen. Physiol.* **32:** 69–79 (1948).

Bonhoeffer, T., Staiger, V., and Aertsen, A. "Synaptic plasticity in rat hippocampal slice cultures: Local "Hebbian" conjunction of pre- and postsynaptic stimulation leads to distributed synaptic enhancement," *Proc. Natl. Acad. Sci. USA* **86:** 8113–8117 (1989).

Borg-Graham, L.J. *On directional selectivity in the vertebrate retina: An experimental and computational study*. PhD Thesis. Massachusetts Institute of Technology: Cambridge, Massachusetts (1991).

Borg-Graham, L.J. "Interpretations of data and mechanisms for hippocampal pyramidal cell models." In: *Cerebral Cortex*, Vol. 13. Ulinksi, P.S., Jones, E.G. and Peters, A., editors. Plenum Press: New York, New York, in press (1998).

Borg-Graham, L.J., and Grzywacz, N.M. "A model of the directional selectivity circuit in retina: Transformations by neurons singly and in concert." In: *Single Neuron Computation*. McKenna, T., Davis, J., and Zornetzer, S.F., editors, pp. 347–375. Academic Press: Boston, Massachusetts (1992).

Borg-Graham, L., Monier, C., and Frégnac, Y. "Visual input evokes transient and strong shunting inhibition in visual cortical neurons," *Nature* **393:** 369–373 (1998).

Bormann,J., Hamill, O.P., and Sakmann, B. "Mechanism of anion permeation through channels gated by glycine and γ-aminobutyric acid in mouse cultured spinal neurons," *J. Physiol.* **385:** 243–286 (1987).

Borst, A., and Egelhaaf, M. "Dendritic processing of synaptic information by sensory interneurons," *Trends Neurosci.* **17:** 257–263 (1994).

Borst, A., Egelhaaf, M., and Haag, J. "Mechanisms of dendritic integration underlying gain control in fly motion-sensitive interneurons," *J. Computational Neurosci.* **2:** 5–18 (1995).

Borst, J.G.G., and Sakmann, B. "Calcium influx and transmitter release in a fast CNS synapse," *Nature* **383:** 431–434 (1996).

Bose, A., and Jones, C.K. "Stability of the in-phase travelling wave solution in a pair of coupled nerve fibers," *Indiana Univ. Math.* **14:** 189–220 (1995).

Bourne, H.R., and Nicoll, R. "Molecular machines integrate coincident synaptic signals," *Neuron* **10 (Suppl.):** 65–76 (1993).

Bower, J.M., and Beeman, D., editors. *The Book of Genesis: Exploring Realistic Neural Models with the General Neural Simulation System*. Second edition. Springer Verlag: New York, New York (1998).

Boycott, B.B., and Wässle, H. "The morphological types of ganglion cells of the domestic cat's retina," *J. Physiol.* **240:** 397–419 (1974).

Bradley, P., and Horn, G. "Neuronal plasticity in the chick brain: Morphological effects of visual experience on neurons in hyperstriatum accessorium," *Brain Res.* **162:** 148–153 (1979).

Braitenberg, V. "Is the cerebellar cortex a biological clock in the millisecond range?" In: *Progress in Brain Research. The Cerebellum*. pp. 334–346. Elsevier: Amsterdam, The Netherlands (1967).

Braitenberg, V., and Atwood, R.P. "Morphological observations in the cerebellar cortex," *J. Comp. Neurol.* **109:** 1–34 (1958).

Braitenberg, V., and Schüz, A. *Anatomy of the Cortex*. Springer Verlag: Heidelberg, Germany (1991).

Brandon, J.G., and Coss, R.G. "Rapid dendritic spine stem shortening during one-trial learning: The honeybee's first orientation flight," *Brain Res.* **252:** 51–61 (1982).

Bray, D. "Protein molecules as computational elements in living cells," *Nature* **376:** 307–312 (1995).

Brenman, J.E., and Bredt, D.S. "Synaptic signaling by nitric oxide," *Current Opinion Neurobiol.* **7:** 374–378 (1997).

Britten, K.H., Shadlen, M.N., Newsome, W.T., and Movshon, A. "The analysis of visual motion: A comparison of neuronal and psychophysical performance," *J. Neurosci.* **12:** 4745–4765 (1992).

Britten, K.H., Shadlen, M.N., Newsome, W.T., and Movshon, A. "Responses of neurons in macaque MT to stochastic motion signals," *Visual Neurosci.* **10:** 1157–1169 (1993).

Brock, L.G., Coombs, J. S., and Eccles, J.C. "The recording of potential from motoneurones with an intracellular electrode," *J. Physiol.* **117:** 431–460 (1952).

Brotchie, P., Andersen, R.A., Snyder, L., and Goodman, S. "Head position signals used by parietal neurons to encode location of visual stimului," *Nature* **375:** 232–235 (1995).

Brotz, T.M., and Borst, A. "Cholinergic and GABAergic receptors on fly tangential cells and their role in visual motion detection," *J. Neurophysiol.* **76:** 1786–1799 (1996).

Brown, D. "M Currents: an update," *Trends Neurosci.* **11:** 294–299 (1988).

Brown, T.H., Chang, V.C., Ganong, A.H., Keenan, C.L., and Kelso, S.R. "Biophysical properties of dendrites and spines that may control the induction and expression of long-term synaptic potentiation." In: *Long Term Potentiation: From Biophysics to Behavior.* Landfield, P.W., and Deadwyler, S.A., editors, pp. 201–264. Alan Liss: New York, New York (1988).

Brown, T.H., Fricke, R.A., and Perkel, D.H. "Passive electrical constants in three classes of hippocampal neurons," *J. Neurophysiol.* **46:** 812–827 (1981).

Brown, T.H., and Johnston, D. "Voltage-clamp analysis of mossy fiber synaptic input to hippocampal neuron," *J. Neurophysiol.* **50:** 487–507 (1983).

Brown, T.H., Kairiss, E.W., and Keenan, C.L. "Hebbian synapses: Biophysical mechanisms and algorithms," *Ann. Rev. Neurosci.* **13:** 475–511 (1990).

Brown, T.H., Mainen, Z.F., Zador, A.M., and Claiborne, B.J. "Self-organization of Hebbian synapses in hippocampal neurons." In: *Advances in Neural Information Processing Systems.* Vol. 3. Lippmann, R., Moody, J., and Touretzky, D., editors, pp. 39–45. Morgan Kaufmann: Palo Alto, California (1991a).

Brown, T.H., and Zador, A.M. "Hippocampus." In: *The Synaptic Organization of the Brain.* Shepherd, G.M., editor, third edition, pp. 346–388. Oxford University Press: New York, New York (1990).

Brown, T.H., Zador, A.M., Mainen, Z.F., and Claiborne, J.B. "Hebbian modifications in hippocampal neurons." In: *Long Term Potentiation: A Debate of Current Issues.* Davis, J., and Baudry, M., editors, pp. 357–389. MIT Press: Cambridge, Massachusetts (1991b).

Brown, T.H., Zador, A.M., Mainen, Z.F., and Claiborne, B.J. "Hebbian computations in hippocampal dendrites and spines." In: *Single Neuron Computation.* McKenna, T., Davis, J., and Zornetzer, S.F., editors, pp. 81–116. Academic Press: Boston, Massachusetts (1992).

Bruck, J. "Harmonic analysis of polynomial threshold functions," *SIAM J. Disc. Math.* **3:** 168–177 (1990).

Bruck, J., and Smolensky, R. "Polynomial threshold functions," *SIAM J. Computing* **21:** 33–42 (1992).

Brühl, G., Jansen, W., and Vogt, H.-J. *Nachrichtenübertragungstechnik I.* Berliner Union: Stuttgart, Germany (1979).

Brunel, N. "Dynamics of an attractor neural network converting temporal into spatial correlations," *Network: Comp. Neural Sys.* **5:** 449–470 (1996).

Buchner, E. "Behavioral analysis of spatial vision in insects." In: *Photoreception and Vision in Invertebrates.* Ali, N.A., editor, pp. 561–621. Plenum Press: New York, New York (1984).

Buhl, E.H., Han, Z.S., Lorinczi, Z., Stezhka, V.V., Karnup, S.V., and Somogyi, P. "Physiological properties of anatomically identified axo-axonic cells in the rat hippocampus," *J. Neurophysiol.* **71:** 1289–1307 (1994).

Burgoyne, R.D., Gray, E.G., and Barron, J. "Cytochemical localization of calcium in the dendritic spine apparatus of the cerebral cortex and at synaptic sites in the cerebellar cortex," *J. Anat.* **136:** 634–635 (1983).

Burke, R.E. "Composite nature of the monosynaptic excitatory postsynaptic potential," *J. Neurophysiol.* **30:** 1114–1137 (1967).

Burke, R.E. "Spinal cord: Ventral horn." In: *Synaptic Organization of the Brain,* Shepherd, G.M., editor, fourth edition, pp. 77–120. Oxford University Press: New York, New York (1998).

Burke, R.E., and Ten Bruggencate, G. "Electrotonic characteristics of alpha motoneurones of varying size," *J. Physiol.* **212:** 1–20 (1971).

Burkitt, A.N. "Attractor neural networks with excitatory neurons and fast inhibitory interneurons at low spike rates," *Network: Comp. Neural Sys.* **4:** 437–448 (1994).

Burns, B.D., and Webb, A.C. "The spontaneous activity of neurones in the cat's visual cortex," *Proc. Roy. Soc. Lond. B* **194:** 211–223 (1976).

Busse, H., and Hess, B. "Information transmission in a diffusion-coupled oscillatory chemical system," *Nature* **244:** 203–205 (1973).

Butz, E.G., and Cowan, J.D. "Transient potentials in dendritic systems of arbitrary geometry," *Biophys. J.* **14:** 661–689 (1974).

Buzsaki, G., Penttonen, M., Nadasdy, Z., and Bragin, A. "Pattern and inhibition-dependent invasion of pyramidal cell dendrites by fast spikes in the hippocampus *in vivo*," *Proc. Natl. Acad. Sci. USA* **93:** 9921–9925 (1996).

Caceres, A., Payne, M.R., Binder, L.I., and Steward, O. "Immunocytochemical localization of actin, and microtubule-associated protein MAP2 in dendritic spines," *Proc. Natl. Acad. Sci. USA* **80:** 1738–1742 (1983).

Cain, D. P. "LTP, NMDA, genes, and learning," *Curr. Opinion Neurobiol.* **7:** 235–242 (1997).

Calkins, D.J., and Sterling, P. "Absence of spectrally specific lateral inputs to midget ganglion cells in primate retina," *Nature* **381:** 613–615 (1996).

Calverly, R.K.S., and Jones, D.G. "Contributions of dendritic spines, and perforated synapses to synaptic plasticity," *Brain Res. Rev.* **15:** 215–249 (1990).

Calvin, W., and Stevens, C.F. "Synaptic noise as a source of variability in the interval between action potentials," *Science* **155:** 842–844 (1967).

Calvin, W., and Stevens, C.F. "Synaptic noise and other sources of randomness in motoneuron interspike intervals," *J. Neurophysiol.* **31:** 574–587 (1968).

Cameron, H.A., Kaliszewski, C.K., and Greer, C.A. "Organization of mitochondria in olfactory bulb granule cell dendritic spines," *Synapse* **8:** 107–118 (1991).

Cao, B., and Abbott, L.F. "A new computational method for cable theory problems," *Biophys. J.* **64:** 303–313 (1993).

Carafoli, E. "Intracellular calcium homeostasis," *Ann. Rev. Biochem.* **56:** 395–433 (1987).

Carandini, M., and Heeger, D.J. "Summation and division by neurons in primate visual cortex," *Science* **264:** 1333–1336 (1994).

Carandini, M., Mechler, F., Leonard, C.S., and Movshon, J.A. "Spike train encoding by regular-spiking cells of the visual cortex," *J. Neurophysiol.* **76:** 3425–3441 (1996).

Carew, T.J. "Molecular enhancement of memory formation," *Neuron* **16:** 5–8 (1996).

Carlin, R.K., Bartlet, D.C., and Siekevitz, P. "Identification of fodrin as a major calmodulin-binding protein in postsynaptic density preparation," *J. Cell Biol.* **96:** 443–448 (1983).

Carnevale, N.T., and Johnston, D. "Electrophysiological characterization of remote chemical synapses," *J. Neurophysiol.* **47:** 606–621 (1982).

Carnevale, N.T., and Rosenthal, S. "Kinetics of diffusion in a spherical cell. I. No solute buffering," *J. Neurosci. Meth.* **41:** 205–216 (1992).

Carr, C.E., Heiligenberg, W., and Rose, G.J. "A time-comparison circuit in the electric fish midbrain. I. Behavior, and physiology," *J. Neurosci.* **6:** 107–119 (1986).

Carr, C.E., and Konishi, M. "A circuit for detection of interaural time differences in the brain stem of the barn owl," *J. Neurosci.* **10:** 3227–3248 (1990).

Catterall, W.A. "Structure and function of voltage-sensitive ion channels," *Science* **242:** 50–61 (1988).

Catterall, W.A. "Cellular and molecular biology of voltage-gated sodium channels," *Physiol. Rev.* **72:** 515–548 (1992).

Catterall, W.A. "Structure and function of voltage-gated ion channels," *Ann. Rev. Biochemistry* **64:** 493–531 (1995).

Catterall, W.A. "Molecular properties of sodium and calcium channels," *J. Bioenergetics & Biomembranes* **28:** 219–230 (1996).

Cauller, L.J., and Connors, B.W. "Functions of very distal dendrites: experimental and computational studies of layer I synapses on neocortical pyramidal cells." In: *Single Neuron Computation*, McKenna, T., Davis, J. and Zornetzer, S.F., editors, pp. 199–229. Academic Press: Boston, Massachusetts (1992).

Cauller, L.J., and Connors, B.W. "Synaptic physiology of horizontal afferents to layer I in slices of rat SI neocortex," *J. Neurosci.* **2:** 751–762 (1994).

Chandler, W.K., FitzHugh, R., and Cole, K.S. "Theoretical stability properties of a space-clamped axon," *Biophys. J.* **2:** 105–127 (1962).

Chang, F.F., and Greenough, W.T. "Transient and enduring morphological correlates of synaptic activity and efficacy change in the rat hippocampal slice," *Brain Res.* **309:** 35–46 (1984).

Chang, H.T. "Cortical neurons with particular reference to the apical dendrites," *Cold Spring Harbor Symp. Quant. Biol.* **17:** 189–202 (1952).

Chapman, R.A. "Prepulse and extra impulse experiments on a type of repetitive crab axon," *J. Physiol.* **168:** 17–18P (1963).

Chicurel, M.E., and Harris, K.M. "Three-dimensional analysis of the structure and composition of CA3 branched dendritic spines and their synaptic relationships with mossy fiber boutons in the rat hippocampus," *J. Comp. Neurol.* **325:** 169–182 (1992).

Chiu, S.Y., and Ritchie, J.M. "Potassium channels in nodal and intranodal axonal membrane of mammalian myelinated fibres," *Nature* **284:** 170–171 (1980).

Chiu, S.Y., Ritchie, J.M., Rogart, R.B., and Stagg, D. "A quantitative description of membrane currents in rabbit myelinated nerve," *J. Physiol.* **292:** 149–166 (1979).

Chow, C., and White, J. "Spontaneous action potentials due to channel fluctuations," *Biophys. J.* **71:** 3013–3021 (1996).

Christie, B.R., Eliot, L.S., Ito, K.-I., Miyakawa, H., and Johnston, D. "Different Ca^{2+} channels in soma and dendrites of hippocampal pyramidal neurons mediate spike-induced Ca^{2+} influx," *J. Neurophysiol.* **73:** 2553–2557 (1995).

Chua, L.O., Desoer, C.A., and Kuh, E.S. *Linear and Nonlinear Circuits.* Mc-Graw Hill: New York, New York (1987).

Chung S.H., Raymond S.A., and Lettvin J.Y. "Multiple meaning in single visual units," *Brain, Behavior Evol.* **3:** 72–101 (1970).

Churchland, P.S., and Sejnowski, T.J. *The Computational Brain.* MIT Press: Cambridge, Massachusetts (1992).

Clapham, D.E. "Calcium signaling," *Cell* **80:** 259–268 (1995).

Clapham, D.E. "The G-protein nanomachine," *Nature* **379:** 297–299 (1996).

Clapham, D.E., and DeFelice, L.J. "The theoretical small signal impedance of the frog node, *Rana pipiens*," *Pflügers Archiv* **366:** 273–276 (1976).

Clapham, D.E., and DeFelice, L.J. "Small signal impedance of the heart cell membrane," *J. Membrane Biol.* **67:** 63–71 (1982).

Clark, J.W., and Plonsey, R. "A mathematial evaluation of the core conductor model," *Biophys. J.* **6:** 95–112 (1966).

Clark, J.W., and Plonsey, R. "The extracellular potential field of the single active nerve fiber in a volume conductor," *Biophys. J.* **8:** 842–864 (1968).

Clark, J.W., and Plonsey, R. "Fiber interaction in a nerve trunk," *Biophys. J.* **11:** 281–294 (1971).

Clay, J., and DeFelice, L. "Relationship between membrane excitability and single channel open-close kinetics," *Biophys. J.* **42:** 151–157 (1983).

Cleland, B.G., Dubin, M.W., and Levick, W.R. "Sustained and transient neurones in the cat's retina and lateral geniculate nucleus," *J. Physiol.* **217:** 473–496 (1971).

Clements, J.D., and Redman, S.J. "Cable properties of cat spinal motoneurons measured by combining voltage clamp, current clamp and intracellular staining," *J. Physiol.* **409:** 63–87 (1989).

Clements, J.D., and Westbrook, G.L. "Activation kinetics reveal the number of glutamate and glycine binding sites on the NMDA receptor," *Neuron* **7:** 605–613 (1991).

Cline, H. "Activity-dependent plasticity in the visual systems of frogs and fish," *Trends Neurosci.* **14:** 104–111 (1991).

Cohen, M.S., and Grossberg, S. "Absolute stability of global pattern formation and parallel memory storage by competitive neural networks," *IEEE Trans. Systems, Man & Cybern.* **13:** 815–826 (1983).

Colbert, C.M., and Johnston, D. "Axonal action-potential initiation and Na$^+$ channel densities in the soma and axon initial segment of subicular neurons," *J. Neurosci.* **16:** 6676–6686 (1996).

Cole, K.S. "Rectification and inductance in the squid giant axon," *J. Gen. Physiol.* **25:** 29–47 (1941).

Cole, K.S. "Dynamic electrical characteristics of the squid axon membrane," *Arch. Sci. Physiol.* **3:** 253–258 (1949).

Cole, K.S. *Membranes, Ions and Impulses.* University of California Press: Berkeley, California (1972).

Cole, K.S., and Baker, R.F. "Transverse impedance of the squid giant axon during current flow," *J. Gen. Physiol.* **24:** 535–551 (1941).

Cole, K.S., and Curtis, H.J. "Electric impedance of nerve and muscle," *Cold Spring Harbor Symp. Quant. Biol.* **4:** 73–89 (1936).

Cole, K.S., Guttman, R., and Bezanilla, F. "Nerve excitation without threshold," *Proc. Natl. Acad. Sci. USA* **65:** 884–891 (1970).

Cole, A.E., and Nicoll, R.A. "Characterization of a slow cholinergic post-synaptic potential recorded *in vitro* from rat hippocampal pyramidal cells," *J. Physiol.* **352:** 173–188 (1984).

Colino, A., Huang, Y.Y., and Malenka, R.C. "Characterization of the integration time for the stabilization of long-term potentiation in area CA1 of the hippocampus," *J. Neurosci.* **12:** 180–187 (1992).

Collingridge, G.L., Kehl, S.J., and McLennan, H. "Excitatory amino acids in synaptic transmission in the Schaffer collateral-commissural pathway of the rat hippocampus," *J. Physiol.* **334:** 33–46 (1983).

Colquhoun, D., and Hawkes, A.G. "On the stochastic properties of single ion channels," *Phil. Trans. Roy. Soc. Lond. B* **211:** 205–235 (1981).

Connor, J.A., and Nikolakopoulou, G. "Calcium diffusion and buffering in nerve cytoplasm," *Lect. Math. Life Sci.* **15:** 79–101 (1982).

Connor, J.A., and Stevens, C. "Inward and delayed outward membrane currents in isolated neural somata under voltage clamp," *J. Physiol.* **213:** 31–53 (1971a).

Connor, J.A., and Stevens, C. "Voltage clamp studies of a transient outward membrane current in gastropod neural somata," *J. Physiol.* **213:** 21–30 (1971b).

Connor, J.A., and Stevens, C. "Prediction of repetitive firing behavior from voltage clamp data on an isolated neurone soma," *J. Physiol.* **213:** 31–53 (1971c).

Connors, B.W., and Gutnick, M.J. "Intrinsic firing patterns of diverse neocortical neurons," *Trends Neurosci.* **13:** 99–104 (1990).

Connors, B.W., Gutnick, M.J., and Prince, D.A. "Electrophysiological properties of neocortical neurons in vitro," *J. Neurophysiol.* **48:** 1302–1320 (1982).

Contreras, D., Destexhe, A., Sejnowski, T.J., and Steriade, M. "Spatiotemporal patterns of spindle oscillations in cortex and thalamus," *J. Neurosci.* **17:** 1179-1196 (1997).

Cooley, J., and Dodge, F.A. "Digital computer solutions for excitation and propagation of the nerve impulse," *Biophys. J.* **6:** 583–599 (1966).

Cooley, J., Dodge, F.A., and Cohen, H. "Digital computer solutions for excitable membrane models," *J. Cell. Comp. Physiol.* **66:** (Supp. 2) 99–109 (1965).

Coombs, J.S., Curtis, D.R., and Eccles, J.C. "The generation of impulses in motoneurones," *J. Physiol.* **139:** 232–249 (1957).

Coombs, J.S., Eccles, J. C., and Fatt, P. "The electrical properties of the motoneurone membrane," *J. Physiol.* **30:** 396–413 (1955).

Copenhagen, D.R., and Owen, W.G. "Functional characteristics of lateral interactions between rods in the retina of the snapping turtle," *J. Physiol.* **259:** 251–282 (1976).

Cornell-Bell, A.H., Finkbeiner, S.M., Cooper, M.S., and Smith, S.J. "Glutamate induces calcium waves in cultured astrocytes: Long-range glial signaling," *Science* **247:** 470–473 (1990).

Coss, R.G., and Globus, A., "Spine stems on tectal interneurons in jewel fish are shortened by social stimulation," *Science* **200:** 787–789 (1978).

Coss, R.G., and Perkel, D.H. "The function of dendritic spines—A review of theoretical issues," *Behav. Neural Bio.* **44:** 151–185 (1985).

Coulter, D.A., Huguenard, J.R., and Prince, D.A. "Calcium currents in rat thalamocortical relay neurons: Kinetic properties of the transient, low threshold current," *J. Physiol.* **414:** 587–604 (1989).

Cox, D.R. *Renewal Theory*. Meuthen: London, United Kingdom (1962).

Cox, D.R., and Isham, V. *Point Processes*. Chapman and Hall: London, United Kingdom (1980).

Crank, J. *The Mathematics of Diffusion*. Second edition. Oxford University Press: Oxford, United Kingdom (1975).

Crawford, A.C., and Fettiplace, R. "An electrical tuning mechanism in turtle cochlear hair cells," *J. Physiol.* **312:** 377–412 (1981).

Crick, F.C. "Do dendritic spines twitch?" *Trends Neurosci.* **5:** 44–46 (1982).

Crick, F.C. "The function of the thalamic reticular complex: The searchlight hypothesis," *Proc. Natl. Acad. Sci. USA* **81:** 4586–4590 (1984a).

Crick, F.C. "Memory and molecular turnover," *Nature* **312:** 101 (1984b).

Crick, F.C., and Koch, C. "Towards a neurobiological theory of consciousness," *Seminars Neurosci.* **2:** 263–275 (1990).

Cronin, J. *Mathematical Aspects of Hodgkin-Huxley Neural Theory*. Cambridge University Press: Cambridge, United Kingdom (1987).

Cullheim, S., Fleshman, J.W., and Burke, R.E. "Three-dimensional architecture of dendritic trees in type-identified alpha motoneurones," *J. Comp. Neurol.* **255:** 82–96 (1987).

Cummings, J.A., Mulkey, R.M., Nicoll, R.A., and Malenka, R.C. "Ca^{2+} signaling requirements for long-term depression in the hippocampus," *Neuron* **16:** 825–833 (1996).

Cussler, E.L. *Diffusion: Mass Transfer in Fluid Systems*. Cambridge University Press: Cambridge, United Kingdom (1984).

Dani, J.A. "Open channel structure and ion binding sites of the nicotinic acetylcholine receptor channel," *J. Neurosci.* **9:** 884–892 (1989).

Davidson, C.W. *Transmission Lines for Communication*. Macmillan Press: London, United Kingdom (1978).

Davies, S.N., Lester, R.A., Reymann, K.G., and Collingridge, G.L. "Temporally distinct pre- and post-synaptic mechanisms maintain long-term potentiation," *Nature* **338:** 500–503 (1989).

Davis, L., Jr., and Lorente de No, R. "Contribution to the mathematical theory of the electrotonus," *Stud. Rockefeller Inst. M. Res.* **131:** 442–496 (1947).

Davis, S., Butcher, S.P., and Morris, R.G. "The NMDA receptor antagonist D–2-amino–5-phosphono-pentanoate (D-AP5) impairs spatial learning and LTP *in vivo* at intracerebral concentrations comparable to those that block LTP *in vitro*," *J. Neurosci.* **12:** 21–34 (1992).

Daw, N.W., Stein, P.S.G., and Fox, K. "The role of NMDA receptors in information processing," *Ann. Rev. Neurosci.* **16:** 207–222 (1993).

Debanne, D., Gähwiler, B.H., and Thompson, S.M. "Asynchronous pre- and postsynaptic activity induces associative long-term depression in area CA1 of the rat hippocampus *in vitro*," *Proc. Natl. Acad. Sci. USA* **91:** 1148–1152 (1994).

Decharms, R.C., and Merzenich, M.M. "Primary cortical representation of sounds by the coordination of action potential timing," *Nature* **381:** 610–613 (1996).

DeFelice, L.J. *Introduction to Membrane Noise*. Plenum Press: New York, New York (1981).

DeFelice, L.J., and Clay, J. "Membrane current and membrane potential from single-channel kinetics." In: *Single Channel Recording*. Sakmann, B., and Neher, E., editors, pp. 323–342. Plenum Press: New York, New York (1983).

DeHaan, R.L., and Chen, Y.-H. "Development of gap junctions," *Ann. New York Acad. Sci.* **588:** 164–173 (1990).

Dehay, C., Douglas, R.J., Martin, K.A.C., and Nelson, C. "Excitation by geniculocortical synapses in not "vetoed" at the level of dendritic spines in cat visual cortex," *J. Physiol.* **440:** 723–734 (1991).

Delaney, K.R., and Tank, D.W. "A quantitative measurement of the dependence of short-term synaptic enhancement on presynaptic residual calcium," *J. Neurosci.* **14:** 5885–902 (1994).

Delaney, K.R., Zucker, R.S., and Tank, D.W. "Calcium in motor-nerve terminals associated with posttetanic potentiation," *J. Neurosci.* **9:** 3558–3567 (1989).

Denk, W., Strickler, J.H., and Webb, W.W. "Two-photon laser scanning fluorescence microscopy," *Science* **248:** 73–76 (1990).

Denk, W., Sugimore, M., and Llinás, R.R. "Two types of calcium response limited to single spines in cerebellar Purkinje cells," *Proc. Natl. Acad. Sci. USA* **92:** 8279–8282 (1995).

Denk, W., Yuste, R., Svoboda, K. and Tank, D.W. "Imaging calcium dynamics in dendritic spines," *Curr. Opinion Neurobiol.* **6:** 372–378 (1996).

Dermietzel, R., and Spray, D.C. "Gap junctions in the brain: Where, what type, how many and why?" *Trends Neurosci.* **16:** 186–191 (1993).

Derrington, A.M., and Lennie, P. "The influence of temporal frequency and adaptation level on receptive field organization of ganglion cells in cat," *J. Physiol.* **333:** 343–366 (1982).

DeSchutter, E. "A consumer guide to neuronal modeling software," *Trends Neurosci.* **15:** 462–464 (1992).

DeSchutter, E., and Bower, J.M. "Sensitivity of synaptic plasticity to the Ca^{2+} permeability of NMDA channels: A model of long-term potentiation in hippocampal neurons," *Neural Comput.* **5:** 681–694 (1993).

DeSchutter, E., and Bower, J.M. "An active membrane model of the cerebellar Purkinje cell: I. Simulation of current clamp in slice," *J. Neurophysiol.* **71:** 375–400 (1994a).

DeSchutter, E., and Bower, J.M. "An active membrane model of the cerebellar Purkinje cell: II Simulation of Synaptic Responses," *J. Neurophysiol.* **71:** 401–419 (1994b).

DeSchutter, E., and Smolen, P. "Calcium dynamics in large neuronal models." In: *Methods in Neuronal Modeling.* Koch, C., and Segev, I., editors, second edition, pp. 211–250. MIT Press: Cambridge, Massachusetts (1998).

Desmond, N.L., and Levy, W.B. "Anatomy of associative long-term synaptic modification." In: *Long Term Potentiation: From Biophysics to Behavior,* Landfield, P.W., and Deadwyler, S.A., editors, pp. 265–305. Alan Liss: New York, New York (1988).

Desmond, N.L., and Levy, W.B. "Morphological correlates of long-term potentiation imply the modification of existing synapses, not synaptogenesis, in the hippocampal dentate gyrus," *Synapse* **5:** 39–43 (1990).

Destexhe, A., Mainen, Z.F., and Sejnowski, T.J. "Synthesis of models for excitable membranes, synaptic transmission and neuromodulation using a common kinetic formalism," *J. Computational Neurosci.* **1:** 195–231 (1994a).

Destexhe, A., Mainen, Z.F., and Sejnowski, T.J. "An efficient method for computing synaptic conductances based on a kinetic model of receptor binding," *Neural Comput.* **6:** 14–18 (1994b).

Destexhe, A., McCormick, D.A., and Sejnowski, T.J. "A model for 8–10 Hz spindling in interconnected thalamic relay and reticular neurons," *Biophys. J.* **65:** 2473–2477 (1993).

Detwiler, P.B., Hodgkin, A.L., and McNaughton, P.A. "Temporal and spatial characteristics of the voltage response of rods in the retina of the snapping turtle," *J. Physiol.* **300:** 213–250 (1980).

Diamond, J., Gray, E.G., and Yasargil, G.M. "The function of the dendritic spine." In: *Excitatory Synaptic Mechanisms.* Andersen, P., and Jansen, J., editors, pp. 212–222. Universitetsforlag: Oslo, Norway (1970).

DiFrancesco, D., and Noble, D. "A model of cardiac electrical activity incorporating ionic pumps and concentration changes," *Phil. Trans. Roy. Soc. Lond. B* **307:** 353–398 (1985).

Dingledine, R.J., and Bennett, J.A. "Structure and function of ligand-gated channels." In: *The Cortical Neuron.* Gutnick, M.J. and Mody, I., editors, pp. 67–96. Oxford University Press: New York, New York (1995).

Diorio, C., Hasler, P., Minch, B. A., and Mead, C. "A single-transistor silicon synapse," *IEEE Trans. Electron Devices* **43:** 1972–1980 (1996).

DiPolo, R., and Beauge, L. "The calcium pump and sodium-calcium exchange in squid axons," *Ann. Rev. Neurosci.* **45:** 313–332 (1983).

Dobrunz, L.E., Huang, E.P., and Stevens, C.F. "Very short-term plasticity in hippocampal synapses," *Proc. Natl. Acad. Sci. USA* **94:** 14843–14847 (1997).

Dobrunz, L.E., and Stevens, C.F. "History of synaptic use determines strength of central synapses." In: *23rd Annual International Biophysics Congress*. Amsterdam, The Netherlands (1996).

Dobrunz, L. E. and Stevens, C.F. "Heterogeneity of release probability, facilitation and depletion at central synapses," *Neuron* **18:** 995–1008 (1997).

Dodge, F.A. *A study of ionic permeability changes underlying excitation in myelinated nerve fibers of the frog.* PhD Thesis. Rockefeller University: New York, New York (1963).

Dodge, F.A., Knight, B.W., and Toyoda, J. "Voltage noise in *Limulus* visual cells," *Science* **160:** 88–90 (1968).

Dodge, F.A., and Rahamimoff, R. "Cooperative action of calcium ions in transmitter release at the neuromuscular junction," *J. Physiol.* **193:** 419–432 (1967).

Dodt, H.-U., and Zieglgänsberger, W. "Infrared videomicroscopy: A new look at neuronal structures and function," *Trends Neurosci.* **17:** 453–458 (1994).

Douglas, R., Koch, C., Mahowald, M., Martin, K., and Suarez, H. "Recurrent excitation in neocortical circuits," *Science* **269:** 981–985 (1995).

Douglas, R., Mahowald, M., and Mead, C. "Neuromorphic analog VLSI," *Ann. Rev. Neurosci.* **18:** 255–281 (1995).

Douglas, R.J., and Martin, K.A.C. "Neocortex." In: *The Synaptic Organization of the Brain*. Shepherd, G.M., editor, fourth edition, pp. 459–510. Oxford University Press: New York, New York (1998).

Douglas, R.J., Martin, K.A.C., and Whitteridge, D. "Selective responses of visual cortical cells do not depend on shunting inhibition," *Nature* **332:** 642–644 (1988).

Douglas, R.J., Martin, K.A.C., and Whitteridge, D. "An intracellular analysis of the visual responses of neurons in cat visual cortex," *J. Physiol.* **440:** 659–696 (1991).

Doupnik, C.A., Davidson, N., and Lester, H.A. "The inward rectifier potassium channel family," *Current Opinion Neurobiol.* **5:** 268–277 (1995).

Dowling, J.E. "Information processing by local circuits: The vertebrate retina as a model system." In: *The Neurosciences: Fourth Study Program*. Schmitt, F.O., and Worden, F.G., editors, pp. 163–181. MIT Press: Cambridge, Massachusetts (1979).

Dowling, J.E. *The Retina: An Approachable Part of the Brain*. Harvard University Press: Cambridge, Massachusetts (1987).

Doyle, D.A., Cabral, J.M., Pfuetzner, R.A., Kuo, A., Gulbis, J.M., Cohen, S.L., Chait, B.T., and MacKinnon, R. "The structure of the potassium channel: Molecular basis of K^+ conduction and selectivity," *Science* **280:** 69-77 (1998).

Drenckhahn, D., and Kaiser, H.W. "Evidence for the concentration of F-actin and myosin in synapses and in the plasmalemmal zone of axons," *Eur. J. Cell. Biol.* **31:** 235–240 (1983).

Dudek, S. M., and Bear, M.F. "Homosynaptic long-term depression in area CA1 of hippocampus and effects of *N*-methyl-D-aspartate receptor blockade," *Proc. Natl. Acad. Sci. USA* **89:** 4363-4367 (1992).

du Lac, S. "Candidate cellular mechanisms of vestibulo-ocular reflex plasticity," *Ann. New York Acad. Sci.* **781:** 489–498 (1996).

du Lac, S., and Lisberger, S.G. "Cellular processing of temporal information in medial vestibular nucleus neurons," *J. Neurosci.* **15:** 8000–8010 (1995).

Dupont, G., and Goldbeter, A. "Oscillations and waves of cytosolic calcium: insights from theoretical models," *Bioessays* **14:** 485–493 (1992).

Durand, D., Carlen, P.L., Gurevich, N., Ho, A., and Kunov, H. "Electrotonic parameters of rat dendate granule cells measured using short current pulses and HRP staining," *J. Neurophysiol.* **50:** 1080–1097 (1983).

Eccles, J.C. *The Physiology of Synapses.* Springer Verlag: Berlin, Germany (1964).

Eccles, J.C. "Developing concepts of the synapses," *J. Neurosci.* **10:** 3769–3781 (1990).

Eccles, J.C., Libet, B., and Young, R.R. "The behavior of chromatolysed motoneurones studied by intracellular recording," *J. Physiol.* **143:** 11–40 (1958).

Eckhorn R., Bauer, R., Jordan, W, Brosch, M., Kruse, W., Munk, M., and Reitboeck, H.J. "Coherent oscillations: A mechanism of feature linking in the visual cortex?" *Biol. Cybern.* **60:** 121–130 (1988).

Edwards, F.A. "LTP is a long term problem," *Nature* **350:** 271–272 (1991).

Edwards, F.A. "LTP—A structural model to explain the inconsistencies," *Trends Neurosci.* **18:** 250–255 (1995).

Edwards, F.A., Konnerth, A., and Sakmann, B. "Quantal analysis of inhibitory synaptic transmission in the dendate gyrus of rat hippocampal slices: A patch-clamp study," *J. Physiol.* **430:** 213–249 (1990).

Edwards, F.A., Konnerth, A., Sakmann, B., and Takahashi, T. "A thin slice preparation for patch clamp recordings from neurones of the mammalian central nervous systems," *Pflügers Archiv* **414:** 600–612 (1989).

Egelhaaf, M., Borst, A., and Pilz, B. "The role of GABA in detecting visual motion," *Brain Res.* **509:** 156–160 (1990).

Egelhaaf, M., Borst, A., and Reichardt, W. "Computational structure of a biological motion-detection system as revealed by local detector analysis in the fly's nervous system," *J. Opt. Soc. Am. A* **6:** 1070–1087 (1989).

Egelman, D.M., and Montague, P.R. "An example of external calcium fluctuations due to presynaptic spike arrival," *Center for Theoretical Neuroscience Technical Report No. 42*. Baylor College of Medicine: Houston, Texas (1997).

Eigen, M. "Molecules, information and memory: From molecules to neural networks." In: *The Neurosciences: Third Study Program*. Schmitt, F.O., and Worden, F.G., editors, pp. 19–27. MIT Press: Cambridge, Massachusetts (1974).

Eilers, J., Augustine, G.J., and Konnerth, A. "Subthreshold synaptic Ca^{2+} signaling in fine dendrites and spines of cerebellar Purkinje neurons," *Nature* **373:** 155–158 (1995).

Einstein, A. "Über die von der molekularkinetischen Theorie der Wärme geforderte Bewegung von in ruhenden Flüssigkeiten suspendierten Teilchen," *Ann. Physik* **17:** 549–560 (1905).

Eisenberg, R.S., and Johnson, E.A. "Three-dimensional electrical field problems in physiology," *Prog. Biophys. Mol. Biol.* **20:** 1–65 (1970).

Emerson, R.C., Bergen, J.R., and Adelson, E.H. "Directionally selective complex cells and the computation of motion energy in cat visual cortex," *Vision Res.* **32:** 203–218 (1992).

Emerson, R.C., Citron, M.C., Felleman, D.J., and Kaas, J.H. "A proposed mechanism and role for cortical directional selectivity." In: *Models of the Visual Cortex*. Rose, D., and Dobson, V.G., editors, pp. 420–431. John Wiley: New York, New York (1985).

Engert, F., and Bonhoeffer, T. "Synapse specificity of long-term potentiation breaks down at short distances," *Nature* **388:** 279–284 (1997).

Enroth-Cugell, C., Robson, J.G., Schweitzer-Tong, D.E., and Watson, A.B. "Spatio-temporal interactions in cat retinal ganglion cells showing linear spatial summation," *J. Physiol.* **341:** 279–307 (1983).

Eskandar, E.N., Richmond, B.J., and Optican, L.M. "Role of inferior temporal neurons in visual memory: I. Temporal encoding of information about visual images, recalled images and behavioral context," *J. Neurophysiol.* **68:** 1277–1295 (1992).

Falk, G., and Fatt, P. "Linear electrical properties of striatal muscle fibers observed with intracellular electrodes," *Proc. Roy. Soc. Lond. B* **160:** 69–123 (1964).

Falke, J.J., Drake, S.K., Hazard, A.L., and Peersen, O.B. "Molecular tuning of ion binding to calcium signaling proteins, " *Quart. Rev. Biophys.* **27:** 219–290 (1994).

Fano, U. "Ionization yield of radiations. II The fluctuations of the number of ions," *Phys. Rev.* **72:** 26–29 (1947).

Farinas, I., and DeFelipe, J. "Patterns of synaptic input on corticocortical and corticothalamic cells in the cat visual cortex. II. The axon initial segment," *J. Comp. Neurol.* **304:** 70–77 (1991).

Fatt, P. "Sequence of events in synaptic activation of a motoneurone," *J. Neurophysiol.* **20:** 61–80(1957).

Fatt, P., and Katz, B. "Spontaneous subthreshold activity at motor nerve endings," *J. Physiol.* **117:** 109–128 (1952).

Feldman, J.A., and Ballard, D.H. "Connectionist models and their properties," *Cognitive Sci.* **6:** 205–254 (1982).

Feller, W. *An Introduction to Probability Theory and Its Applications.* John Wiley: New York, New York (1966).

Feng, T.P. "Studies on the neuromuscular junction. XXVI. The changes of the end-plate potential during and after prolonged stimulation," *Chin. J. Physiol.* **16:** 341–372 (1941).

Fenwick, E.M., Marty, A., and Neher, E. "A patch-clamp study of bovine chromaffin cells and of their sensitivity to acetylcholine," *J. Physiol.* **331:** 577–597 (1982).

Ferster, D., and Jagadeesh, B. "EPSP-IPSP interactions in cat visual cortex studied with *in vivo* whole-cell patch recording," *J. Neurosci.* **12:** 1262–1274 (1992).

Ferster, D., and Lindstrom, S. "An intracellular analysis of geniculocortical connectivity in area 17 of the cat," *J. Physiol.* **342:** 181–215 (1983).

Fettiplace, R., Andrews, D.M., and Haydon, D.A. "The thickness, composition and structure of some lipid bilayers and natural membranes," *J. Mem. Biol.* **5:** 277–296 (1971).

Fetz, E., Toyama, K., and Smith, W. "Synaptic interactions between cortical neurons." In: *Cerebral Cortex*, Vol. 9. Peters, A., and Jones, E.G., editors, pp. 1–48. Plenum Press: New York, New York (1991).

Feynman, R.P., and Hibbs, A.R. *Quantum Mechanics and Path Integrals.* McGraw-Hill: New York, New York (1965).

Feynman, R.P., Leighton, R.B., and Sands, M. *The Feynman Lectures on Physics.* Addison-Wesley: Reading, Massachusetts (1964).

Fick, A. "Über Diffusion," *Ann. Physik & Chemie* **4:** 59–86 (1855).

Fifková, E. "A possible mechanism of morphometric changes in dendritic spines induced by stimulation," *Cell. Molec. Neurobiol.* **5:** 47–63 (1985).

Fifková, E., Eason, H., and Schaner, P. "Inhibitory contacts on dendritic spines of the dendate fascia," *Brain Res.* **577:** 331–336 (1992).

Fifková, E., Markham, J.A., and Delay, R.J. "Calcium in the spine apparatus of dendritic spines in the dendate molecular layer," *Brain Res.* **266:** 163–168 (1983).

Fine, A., Amos, W.B., Durbin, R.M., and McNaughton, P.A. "Confocal microscopy: Applications in neurobiology," *Trends Neurosci.* **11:** 345–351 (1988).

Fischer, M., Kalch, S., Knulti, D., and Matus, A. "Rapid actin-based plasticity in dendritic spines," *Neuron* **20:** 1–20 (1998).

Fisher, R.E., Gray, R., and Johnston, D. "Properties and distribution of single voltage-gated calcium channels in adult hippocampal neuron," *J. Neurophysiol.* **64:** 91–104 (1990).

FitzHugh, R. "Impulses and physiological states in theoretical models of nerve membrane," *Biophys. J.* **1:** 445–466 (1961).

FitzHugh, R. "Computation of impulse initiation and saltatory conduction in a myelinated nerve fiber," *Biophys. J.* **2:** 11–21 (1962).

FitzHugh, R. "Mathematical models for excitation and propagation in nerve." In: *Biological Engineering.* Schwan, H.P., editor, pp. 1–85. McGraw Hill: New York, New York (1969).

Fleshman, J., Segev, I., and Burke, R. "Electrotonic architecture of type identified α-motoneurons in the cat spinal cord," *J. Neurophysiol.* **60:** 60–85 (1988).

Foote, S.L., Aston-Jones, G., and Bloom, F.E. "Impulse activity of locus coeruleus neurons in awake rats and monkeys is a function of sensory stimulation and arousal," *Proc. Natl. Acad. Sci. USA* **77:** 3033–3037 (1980).

Foote, S.L., and Morrison, J.H. "Extrathalamic modulation of cortical function," *Ann. Rev. Neurosci.* **10:** 67–95 (1987).

Foster, K.R., Bidinger, J.M., and Carpenter, D.D. "The electrical resistivity of cytoplasm," *Biophys. J.* **16:** 991–1001 (1976).

Fox, R.F., and Lu, Y. "Emergent collective behavior in large numbers of globally coupled independently stochastic ion channels," *Phys. Rev. E* **49:** 3421–3431 (1994).

Fox, K., Sato, H., and Daw, N. "The effect of varying stimulus intensity on NMDA receptor activity in cat visual cortex," *J. Neurophysiol.* **64:** 1413–1428 (1990).

Franceschetti, S., Guatteo, E., Panzica, F., Sancini, G., Wanke, E., and Avanzini, G. "Ionic mechanisms underlying bursts firing in pyramidal neurons: Intracellular study in rat sensorimotor cortex," *Brain Res.* **696:** 127 139 (1995).

Frankenhaeuser, D., and Huxley, A.F. "The action potential in the myelinated nerve fibre of *Xenopus laevis* as computed on the basis of voltage clamp data," *J. Physiol.* **171:** 302–315 (1964).

Franklin, J., and Bair, W. "The effect of a refractory period on the power spectrum of neuronal discharge," *SIAM J. Appl. Math.* **55:** 1074–1093 (1995).

Freed, M.A., Smith, R.G., and Sterling, P. "Computational model of the ON alpha ganglion cell receptive field based on bipolar cell circuitry," *Proc. Natl. Acad. Sci. USA* **89:** 236–240 (1992).

Freed, M.A., and Sterling, P. "The ON alpha ganglion cell of the cat retina and its presynaptic cell types," *J. Neurosci.* **8:** 2303–2320 (1988).

Freeman, W.J. *Mass Action in the Nervous System.* Academic Press: New York, New York (1975).

French, A.S., and Korenberg, M.J. "A nonlinear cascade model for action potential encoding in an insect sensory neuron," *Biophys. J.* **55:** 655–661 (1989).

French, C.R., Sah, P., Buckett, K.J., and Gage, P.W. "A voltage-dependent persistent sodium current in mammalian hippocampal neurons," *J. Gen. Physiol.* **95:** 1139–1157 (1990).

Friedman, D.P. "Laminar patterns of terminations of cortico-cortical afferents in the somatosensory system," *Brain Res.* **273:** 147–151 (1983).

Fromherz, P., and Gaede, V. "Exclusive-OR function of single arborized neuron," *Biol. Cybern.* **69:** 337–344 (1993).

Fukunaga, K., Stoppini, L., Miyamoto, E., and Muller, D. "Long-term potentiation is associated with an increased activity of Ca^{2+} calmodulin-dependent protein kinase-II," *J. Biol. Chem.* **268:** 7863–7867 (1993).

Fuortes, M.G.F., Frank, K., and Becker, M.C. "Steps in the production of motoneuron spikes," *J. Gen. Physiol.* **40:** 725–752 (1957).

Furshpan, E.J., and Potter, D.D. "Transmission at the giant motor synapses of the crayfish," *J. Physiol.* **145:** 289–325 (1959).

Gabbiani, F. and Koch, C. "Principles of spike train analysis." In: *Methods in Neuronal Modeling.* Koch, C., and Segev, I., editors, second editon, pp. 313–360. MIT Press: Cambridge, Massachusetts (1998).

Gabbiani, F., Metzner, W., Wessel, R., and Koch, C. "From stimulus encoding to feature extraction in weakly electric fish," *Nature* **384:** 564–567 (1996).

Gabbiani, F., Midtgaard, J., and Knöpfel, T. "Synaptic integration in a model of cerebellar granule cells," *J. Neurophysiol.* **72:** 999–1009 (1994).

Gabso, M., Neher, E., and Spira, M.E. "Low mobility of the Ca^{2+} buffers in axons of cultured Aplysia neurons," *Neuron* **18:** 473–481 (1997).

Gally, J.A., Montague, P.R., Reeke, G.N. Jr., and Edelman, G.M. "The NO hypothesis: Possible effects of a short-lived, rapidly diffusible signal in the development and function of the nervous system," *Proc. Natl. Acad. Sci. USA* **87:** 3547–3551 (1990).

Galvan, M., Constanti, A., Franz, P., Sedlmeir, C., and Endres, W. "Calcium spikes and inward current in mammalian peripheral and central neurons," *Exp. Brain Res.* **14:** 61–70 (1986).

Gamble, E., and Koch, C. "The dynamics of free calcium in dendritic spines in response to repetitive synaptic input," *Science* **236:** 1311–1315 (1987).

Gammaitoni, L. "Stochastic resonance and the dithering effect in threshold physical systems," *Phys. Rev. E* **52:** 4691–4698 (1995).

Garrahan, P.J., and Rega, A.F. "Plasma membrane calcium pump." In: *Intracellular Calcium Regulation*. Bronner, F., editor, pp. 271–303. Alan R. Liss: New York, New York (1990).

Gasser, H.S. "Unmedullated fibers originating in dorsal root ganglia," *J. Gen. Physiol.* **33:** 651–690 (1950).

Gennis, R.B. *Biomembranes: Molecular Structure and Function*. Springer Verlag: Berlin, Germany (1989).

Gerstein, G., and Mandelbrot, B. "Random walk models for the spike activity of a single neuron," *Biophys. J.* **4:** 41–68 (1964).

Gerstner, W., Kempter, R., van Hemmen, J.L., and Wagner, H. "A neuronal learning rule for submillisecond temporal coding," *Nature* **383:** 76–78 (1996).

Gestri, G. "Pulse frequency modulation in neural systems: A random model," *Biophys. J.* **11:** 98–109 (1971).

Gestri, G., Masterbroek, H.A.K., and Zaagman, W.H. "Stochastic constancy, variability and adaptation of spike generation. Performance of a giant neuron in the visual system of the fly," *Biol. Cybern.* **38:** 31–40 (1980).

Getting, P. "Mechanisms of pattern generation underlying swimming in *Tritonia*. II. Intrinsic and cellular mechanisms of delayed excitation," *J. Neurophysiol.* **49:** 1036–1050 (1983).

Getting, P. "Reconstruction of small neural networks." In: *Methods in Neuronal Modeling*. Koch, C. and Segev, I., editors, pp. 171–194. MIT Press: Cambridge, Massachusetts (1989).

Ghose, G.M., Freeman, R.D., and Ohzawa, I. "Local intracortical connections in the cat's visual cortex: Postnatal development and plasticity," *J. Neurophysiol.* **72:** 1290–1303 (1994).

Ghosh, A., and Greenberg, M.E. "Calcium signaling in neurons: Molecular mechanisms and cellular consequences," *Science* **268:** 239–247 (1995).

Giaume, C., and McCarthy, K.D. "Control of gap-junctional communication in astrocytic networks," *Trends Neurosci.* **19:** 319–325 (1996).

Gilbert, C. "Laminar differences in receptive field properties in cat primary visual cortex," *J. Physiol.* **268:** 391–421 (1977).

Gilbert, D.S. "Axoplasm architecture and physical properties as seen in the *Myxicola* giant axon," *J. Physiol.* **253:** 257–301 (1975).

Glass, L., and Mackey, M.C. *From Clocks to Chaos: The Rhythms of Life*. Princeton University Press: Princeton, New Jersey (1988).

Goda, Y., and Stevens, C.F. "Long-term depression properties in a simple system," *Neuron* **16:** 103–111 (1996).

Gold, J.I., and Bear, M.F. "A model of dendritic spine Ca^{2+} concentration exploring possible bases for a sliding synaptic modification threshold," *Proc. Natl. Acad. Sci. USA* **91:** 3941–3945 (1994).

Goldman, D.E. "Potential, impedance and rectification in membranes," *J. Gen. Physiol.* **27:** 37–60 (1943).

Goldstein, S.S., and Rall, W. "Changes in action potential shape and velocity for changing core conductor geometry," *Biophys. J.* **14:** 731–757 (1974).

Gorman, A.L.F., and Thomas, M.V. "Intracellular calcium accumulation during depolarization in a molluscan neurone," *J. Physiol.* **308:** 259–285 (1980).

Graham, J., and Gerard, R.W. "Membrane potentials and excitation of impaled single muscle fiber," *J. Cellular Comp. Physiol.* **28:** 99–117 (1946).

Granit, R., Kernell, D., and Shortess, G.K. "Quantitative aspects of repetitive firing of mammalian motoneurons, caused by injected currents," *J. Physiol.* **168:** 911–931 (1963).

Gray, E.G. "Axo-somatic and axo-dendritic synapses of the cerebral cortex: An electron-microscopic study," *J. Anat.* **93:** 420–433 (1959).

Gray, C.M., König, P., Engel, A.K., and Singer, W. "Oscillatory responses in cat visual cortex exhibit inter-columnar synchronization which reflects global stimulus properties," *Nature* **338:** 334–337 (1989).

Gray, C.M., and McCormick, D.A. "Chattering cells: Superficial pyramidal neurons contributing to the generation of synchronous oscillations in the visual cortex," *Science* **274:** 109–113 (1996).

Greenough, W.T., and Chang, F.-L.F. "Synaptic structural correlates of information storage in the mammalian nervous system." In: *Synaptic Plasticity.* Cotman, C.W., editor, pp. 335–372. Guilford Press: New York, New York (1985).

Greer, C.A. "Golgi analyses of dendritic organization among denervated olfactory bulb granule cells," *J. Comp. Neurol.* **257:** 442–452 (1987).

Griniasty, M., Tsodyks, M.V., and Amit, D.J. "Conversion of temporal correlations between stimuli to spatial correlations between attractors," *Neural Comput.* **5:** 1–17 (1993).

Grossman, Y., Parnas, I., and Spira, M.E. "Differential conduction block in branches of a bifurcating axon," *J. Physiol.* **295:** 283–305 (1979a).

Grossman, Y., Parnas, I., and Spira, M.E. "Ionic mechanisms involved in differential conduction of action potentials at high frequency in a branching axon," *J. Physiol.* **295:** 307–322 (1979b).

Grzywacz, N.M., and Amthor, F.R. "Facilitation in ON-OFF directionally selective ganglion cells of the rabbit retina," *J. Neurophysiol.* **69:** 2188–2199 (1993).

Guido, W., Lu, S.M., Vaughan, J.W., Godwin, D.W., and Sherman, S.M. "Receiver operating characteristic (ROC) analysis of neurons in the cat's lateral geniculate nucleus during tonic and burst response-mode," *Visual Neurosci.* **12:** 723–741 (1995).

Guido, W., and Weyand, T. "Burst responses in thalamic relay cells of the awake behaving cat," *J. Neurophysiol.* **74:** 1782–1786 (1995).

Gulyás, A.I., Miles, R., Sik, A., Tóth, K., Tamamaki, N., and Freund, T.F. "Hippocampal pyramidal cells excite inhibitory neurons through a single release site," *Nature* **366:** 683–687 (1993).

Gustafsson, B., Wigstrom, H., Abraham, W.C., and Huang, Y. -Y. "Long-term potentiation in the hippocampus using depolarizing current pulses as the conditioning stimulus to single volley synaptic potentials," *J. Neurosci.* **7:** 774–780 (1987).

Guthrie, P.B., Segal, M., and Kater, S.B. "Independent regulation of calcium revealed by imaging dendritic spines" *Nature* **354:** 76–80 (1991).

Gutnick, M.J., and Crill, W.E. "The cortical neuron as an electrophysiological unit." In: *The Cortical Neuron.* Gutnick, M.J., and Mody, I., editors, pp. 33–51. Oxford University Press: New York, New York (1995).

Guttman, R., and Barnhill, R. "Oscillation and repetitive firing in squid axons. Comparison of experiments with computations," *J. Gen. Physiol.* **55:** 104–118 (1970).

Haag, J., and Borst, A. "Encoding of visual motion information and reliability in spiking and graded potential neurons," *J. Neurosci.* **17:** 4809–4819 (1997).

Hagiwara, S. *Membrane Potential-Dependent Ion Channels in Cell Membrane. Phylogenetic and Developmental Approaches.* Raven Press: New York, New York (1983).

Hagiwara, S., and Byerly, L. "Calcium channels," *Ann. Rev. Neurosci.* **4:** 69–125 (1981).

Hagiwara, S., and Oomura, Y. "The critical depolarization for the spike in the squid giant axon," *Jap. J. Physiol.* **8:** 234–245 (1958).

Hall, Z.W. *Introduction to Molecular Neurobiology.* Sinauer Associates: Sunderland, Massachusetts (1992).

Halliwell, J.V., and Adams, P.R. "Voltage-clamp analysis of muscarinic excitation in hippocampal neurons," *Brain Res.* **250:** 71–92 (1982).

Hamori, J., Pasik, T., Pasik, P., and Szentagothai, J. "Triadic synaptic arrangement and their possible significance in the lateral geniculate nucleus of the monkey," *Brain Res.* **80:** 379–393 (1974).

Hamos, J.E., Van Horn, S.C., Raczkowski, D., Uhlrich, D.J., and Sherman, S.M. "Synaptic connectivity of a local circuit neurone in lateral geniculate nucleus of the cat," *Nature* **317:** 618–621 (1985).

Hansel, D., and Mato, G. "Patterns of synchrony in a heterogeneous Hodgkin-Huxley neural network with weak coupling," *Physica A* **200:** 662–669 (1993).

Hanson, P.I., Meyer, T., Stryer, L., and Schulman, H. "Dual role of calmodulin in autophosphorylation of multifunctional CAM kinase may underlie decoding of calcium signals," *Neuron* **12:** 943–956 (1994).

Harris, R.W. "Morphology of physiologically identified thalamocortical relay neurons in the rat ventrobasal thalamus," *J. Comp. Neurol.* **254:** 382–402 (1986).

Harris, E.W., and Cotman, C.W. "Long-term potentiation of guinea pig mossy fiber responses is not blocked by N-methyl D-aspartate antagonists," *Neurosci. Lett.* **70:** 132–137 (1986).

Harris, K.M., Jensen, F.E., and Tsao, B. "Three-dimensional structure of dendritic spines and synapses in rat hippocampus (CA1) at postnatal day 15 and adult ages: Implications for the maturation of synaptic physiology and long-term potentiation," *J. Neurosci.* **12:** 2685–2705 (1992).

Harris, K.M., and Kater, S.B. "Dendritic spines: Cellular specializations imparting both stability and flexibility to synaptic function," *Ann. Rev. Neurosci.* **17:** 341–371 (1994).

Harris, K.M., and Stevens, J.K. "Study of dendritic spines by serial electron microscopy and three-dimensional reconstruction." In: *Intrinsic Determinants of Neuronal Form and Function.* Lasek, R.J. and Black, M.M., editors, pp. 179–199. Alan Liss: New York, New York (1988a).

Harris, K.M., and Stevens, J.K. "Dendritic spines of rat cerebellar Purkinje cells: Serial electron microscopy with reference to their biophysical characteristics," *J. Neurosci.* **8:** 4455–4469 (1988b).

Harris, K.M., and Stevens, J.K. "Dendritic spines of CA1 pyramidal cells in the rat hippocampus: Serial electron microscopy with reference to their biophysical characteristics," *J. Neurosci.* **9:** 2982–2997 (1989).

Hassenstein, B., and Reichardt, W. "Systemtheoretische Analyse der Zeit-, Reihenfolgen- und Vorzeichenauswertung bei der Bewegungsperzeption des Rüsselkäfers *Chlorophanus*," *Z. Naturforsch. B* **11:** 513–524 (1956).

Hatsopoulos, N., Gabbiani, F., and Laurent, G. "Elementary computation of object approach by a wide field visual neuron," *Science* **270:** 1000–1003 (1995).

Hawkins, R.D., Kandel, E.R., and Siegelbaum, S.A. "Learning to modulate transmitter release: Themes and variations in synaptic plasticity," *Ann. Rev. Neurosci.* **16:** 16:625–665 (1993).

Hebb, D. O. *The Organization of Behavior: A Neuropsychological Theory.* John Wiley: New York, New York (1949).

Heeger, D.J. "Normalization of cell responses in cat striate cortex," *Visual Neurosci.* **9:** 181–197 (1992).

Heeger, D.J., Simoncelli, E.P., and Movshon, J.A. "Computational models of cortical visual processing," *Proc. Natl. Acad. Sci. USA* **93:** 623–627 (1996).

Heiligenberg, W.F. *Neural Nets in Electric Fish.* MIT Press: Cambridge, Massachusetts (1991).

Heinemann, S.H., Teriau, H., Stühmer, W., Imoto, K., and Numa, S. "Calcium-channel characteristics conferred on the sodium-channel by single mutations," *Nature* **356:** 441–443 (1992).

Helmchen, F., Borst, J.G., and Sakmann, B. "Calcium dynamics associated with a single action potential in a CNS presynaptic terminal," *Biophys. J.* **72:** 1458–1471 (1997).

Helmchen, F., Imoto, K., and Sakmann, B. "Ca^{2+} buffering and action potential-evoked Ca^{2+} signaling in dendrites of pyramidal neurons," *Biophys. J.* **70:** 1069–1081 (1996).

Hermann, L. "Zur Theorie der Erregungsleitung und der elektrischen Erregung," *Pflügers Archiv* **75:** 574–623 (1899).

Hermann, M., Hertz, J.A., and Prügel-Bennett, A. "Analysis of synfire chains," *Networks: Comput. Neural Sys.* **6:** 403–414 (1995).

Hernández-Cruz, A., and Pape, H.-C. "Identification of two calcium currents in acutely dissociated neurons from the rat lateral geniculate nucleus," *J. Neurophysiol.* **61:** 1270–1283 (1989).

Hernández-Cruz, A., Sala, F., and Adams, P.R. "Subcellular calcium transients visualized by confocal microscopy in a voltage-clamped vertebrate neuron," *Science* **247:** 858–862 (1990).

Hertz, J., Krogh, A., and Palmer, R.G. *Introduction to the Theory of Neural Computation.* Addison-Wesley: Redwood City, California (1991).

Hessler, N.A., Shirke, A.M., and Malinow, R. "The probability of transmitter release at a mammalian central synapse," *Nature* **366:** 569–572 (1993).

Hestrin, S. "Activation and desensitization of glutamate-activated channels mediating fast excitatory synaptic currents in the visual cortex," *Neuron* **9**: 991–999 (1992).

Hestrin, S., Nicoll, R.A., Perkel, D.J., and Sah, P. "Analysis of excitatory synaptic action in pyramidal cells using whole-cell recording from rat hippocampal slices," *J. Physiol.* **422**: 203–225 (1990a).

Hestrin, S., Sah, P., and Nicoll, R.A. "Mechanisms generating the time course of dual component excitatory synaptic currents recorded in hippocampal slices," *Neuron* **5**: 247–253 (1990b).

Hildreth, E., and Koch, C. "The analysis of visual motion: From computational theory to neuronal mechanisms," *Ann. Rev. Neurosci.* **10**: 477–533 (1987).

Hille, B. "Potassium channels in myelinated nerve. Selective permeability to small cations," *J. Gen. Physiol.* **61**: 669-686 (1973).

Hille, B., *Ionic Channels of Excitable Membranes.* Second edition. Sinauer Associates: Sunderland, Massachusetts (1992).

Hillman, D.E. "Neuronal shape parameters and substructures as a basis of neuronal form." In: *The Neurosciences: Fourth Study Program.* Schmitt, F.O., and Worden, F.G., editors, pp. 477–498. MIT Press: Cambridge, Massachusetts (1979).

Hines, M. "Efficient computation of branched nerve equations," *Intl. J. Biomed. Comp.* **15**: 69–76 (1984).

Hines, M. "A program for simulation of nerve equations with branching geometries," *Intl. J. Biomed. Comp.* **24**: 55–68 (1989).

Hines, M. "The Neurosimulator NEURON." In: *Methods in Neuronal Modeling.* Koch, C. and Segev, I., editors, second edition, pp. 129–136. MIT Press: Cambridge, Massachusetts (1998).

Hirsch, J., and Gilbert, C. "Synaptic physiology of horizontal connections in the cat's visual cortex," *J. Neurosci.* **11**: 1800–1809 (1991).

Hirsch, M.W., and Smale, S. *Differential Equations, Dynamical Systems and Linear Algebra.* Academic Press: New York, New York (1974).

Hirst, G.D.S., Redman, S.J., and Wong, K. "Post-tetanic potentiation and facilitation of synaptic potentials evoked in cat spinal motoneurones," *J. Physiol.* **321**: 97–109 (1981).

Hjelmfelt, A., and Ross, J. "Chemical implementation and thermodynamics of collective neural networks," *Proc. Natl. Acad. Sci. USA* **89**: 388–391 (1992).

Hodgkin, A.L. "Evidence for electrical transmission in nerve," *J. Physiol.* **90**: 183–232 (1937).

Hodgkin, A.L. "Ionic movements and electrical activity in giant nerve fibres," *Proc. R. Soc. Lond. B* **148**: 1–37 (1958).

Hodgkin, A.L "Chance and design in electrophysiology: An informal account of certain experiments on nerve carried out between 1934 and 1952," *J. Physiol.* **263**: 1–21 (1976).

Hodgkin, A.L., and Huxley, A.F. "Action potentials recorded from inside a nerve fibre," *Nature* **144**: 710–711 (1939).

Hodgkin, A.L., and Huxley, A.F. "Currents carried by sodium and potassium ions through the membrane of the giant axon of *Loligo*," *J. Physiol.* **116**: 449–472 (1952a).

Hodgkin, A.L., and Huxley, A.F. "The components of membrane conductance in the giant axon of *Loligo*," *J. Physiol.* **116**: 473–496 (1952b).

Hodgkin, A.L., and Huxley, A.F. "The dual effect of membrane potential on sodium conductance in the giant axon of *Loligo*," *J. Physiol.* **116**: 497–506 (1952c).

Hodgkin, A.L., and Huxley, A.F. "A quantitative description of membrane current and its application to conduction and excitation in nerve," *J. Physiol.* **117**: 500–544 (1952d).

Hodgkin, A.L., Huxley, A.F., and Katz, B. "Ionic currents underlying activity in the giant axon of the squid," *Arch. Sci. Physiol.* **3**: 129–150 (1949).

Hodgkin, A.L., Huxley, A.F., and Katz, B. "Measurements of current-voltage relations in the membrane of the giant axon of *Loligo*," *J. Physiol.* **116**: 424–448 (1952).

Hodgkin, A.L., and Katz, B. "The effect of temperature on the electrical activity of the giant axon of the squid," *J. Physiol.* **109**: 240–249 (1949).

Hodgkin, A.L., and Keynes, R.D. "Movements of labeled calcium in squid giant axons," *J. Physiol.* **138:** 253–281 (1957).

Hodgkin, A.L., and Rushton, W.A.H. "The electrical constants of a crustacean nerve fibre," *Proc. Roy. Soc. Lond. B* **133:** 444–479 (1946).

Hoffman, D.A., Magee, J.C., Colbert, C.M., and Johnston, D. "K^+ channel regulation of signal propagation in dendrites of hippocampal pyramidal neurons," *Nature* **387:** 869–875 (1997).

Hofmann, F., Biel, M., and Flockerzi, V. "Molecular basis for Ca^{2+} channel diversity," *Ann. Rev. Neurosci.* **17:** 399–418 (1994).

Hofstadter, D.R. *Gödel, Escher, Bach: An Eternal Golden Braid.* Basic Books: New York, New York (1979).

Holden, A.V. *Models of the Stochastic Activity of Neurones.* Springer Verlag: New York, New York (1976).

Holmes, W.R. "A continuous cable method for determining the transient potential in passive dendritic trees of known geometry," *Biol. Cybern.* **55:** 115–124 (1986).

Holmes, W.R. "The role of dendritic diameter in maximizing the effectiveness of synaptic inputs," *Brain Res.* **478:** 127–137 (1989).

Holmes, W.R., and Levy, W. "Insights into associative long-term potentiation from computational models of NMDA receptor-mediated calcium influx and intracellular calcium concentration changes," *J. Neurophysiol.* **63:** 1148–1168 (1990).

Holmes, W.R., Segev, I., and Rall, W. "Interpretation of time constant and electrotonic length estimates in multicylinder or branched neuronal structures," *J. Neurophysiol.* **68:** 1401–1420 (1992).

Holmes, W.R., and Woody, C.D. "Effects of uniform and non-uniform synaptic activation-distributions on the cable properties of modeled cortical pyramidal neurons," *Brain Res.* **505:** 12–22 (1989).

Holt, G. *A critical reexamination of some assumptions and implications of cable theory in neurobiology.* PhD Thesis. California Institute of Technology: Pasadena, California (1998).

Holt, G., and Koch, C. "Shunting inhibition does not have a divisive effect on firing rates," *Neural Comput.* **9:** 1001–1013 (1997).

Holt, G., Softky, W., Koch, C., and Douglas, R.J. "A comparison of discharge variability *in vitro* and *in vivo* in cat visual cortex neurons," *J. Neurophysiol.* **75:** 1806–1814 (1996).

Hooper, S.L., and Marder, E. "Modulation of the lobster pyloric rhythm by the peptide proctolin," *J. Neurosci.* **7:** 2097–2112 (1987).

Hopcroft, J.E., and Ullman, J.D. *Introduction to Automata Theory, Languages and Computation.* Addison-Wesley: Reading, Massachusetts (1979).

Hopfield, J.J. "Neural networks and physical systems with emergent collective computational abilities," *Proc. Natl. Acad. Sci. USA* **79:** 2554–2558 (1982).

Hopfield, J.J. "Neurons with graded responses have collective computational properties like those of two-state neurons," *Proc. Natl. Acad. Sci. USA* **81:** 3088–3092 (1984).

Hopfield, J.J. "Pattern-recognition computation using action potential timing for stimulus representation," *Nature* **376:** 33–36 (1995).

Horwitz, B. "An analytical method for investigating transient potentials in branched neurons with branching dendritic trees," *Biophys. J.* **36:** 155–192 (1981).

Hoshi, T., Zagotta, W.N., and Aldrich, R.W. "Biophysical and molecular mechanisms of *Shaker* potassium channel inactivation," *Science* **250:** 533-538 (1990).

Hubbard, J.L., Llinás, R.R., and Quastel, D.M.J. *Electrophysiological Analysis of Synaptic Transmission.* Williams and Wilkins: Baltimore, Maryland (1969).

Hubel, D.H., and Wiesel, T.N. "Receptive fields, binocular interaction and functional architecture in the cat's visual cortex," *J. Physiol.* **160:** 106–154 (1962).

Hudspeth, A.J. "The cellular basis of hearing: The biophysics of hair cells," *Science* **230:** 745–752 (1985).

Hudspeth, A.J., and Lewis, R.S. "Kinetic analysis of voltage- and ion-dependent conductances in saccular hair cells of the bullfrog *Rana catesbeiana*," *J. Physiol.* **400:** 237–274 (1988a).

Hudspeth, A.J., and Lewis, R.S. "A model for electrical resonance and frequency tuning in saccular hair cells of the bullfrog *Rana catesbeiana*," *J. Physiol.* **400:** 275–297 (1988b).

Huguenard, J.R., Hamill, O.P., and Prince, D.A. "Developmental changes in Na^+ conductances in rat neocortical neurons: Appearance of a slowly inactivating component," *J. Neurophysiol.* **59:** 778–794 (1988).

Huguenard, J.R., and McCormick, D.A. "Simulation of the currents involved in rhythmic oscillations in thalamic relay neurons," *J. Neurophysiol.* **68:** 1373–1383 (1992).

Hursh, J.B. "Conduction velocity and diameter of nerve fibers," *Am. J. Physiol.* **127:** 131–139 (1939).

Huxley, A.F. "Can a nerve propagate a subthreshold disturbance?" *J. Physiol.* **148:** 80–81 (1959).

Huxley, A.F., and Stämpfli, R. "Evidence for saltatory conduction in peripheral myelinated nerve fibres," *J. Physiol.* **108:** 315–339 (1949).

Iansek, R., and Redman, S.J. "An analysis of the cable properties of spinal motoneurons using a brief intracellular current pulse," *J. Physiol.* **234:** 613–636 (1973).

Ito, M. "The cellular basis of cerebellar plasticity," *Current Opinion Neurobiol.* **1:** 616–620 (1991).

Ito, M., and Oshima, T. "Electrical behavior of the motoneurone membrane during intracellular applied current steps," *J. Physiol.* **180:** 607–635 (1965).

Jack, J.J.B., Noble, D., and Tsien, R. *Electric Current Flow in Excitable Cells.* Oxford University Press: Oxford, United Kingdom. (1975). Revised paperback edition (1983).

Jaeger, D., De Schutter, E., and Bower, J.M. "The role of synaptic and voltage-gated currents in the control of Purkinje cell spiking: A modeling study," *J. Neurosci.* **17:** 91–106 (1997).

Jaffe, D.B., Fisher, S.A., and Brown, T.H. "Confocal laser scanning microscopy reveals voltage-gated calcium signals within hippocampal dendritic spines," *J. Neurobiol.* **25:** 220–233 (1994).

Jagadeesh, B., Gray, C.M., and Ferster, D. "Visually evoked oscillations of membrane potential in cells of cat visual cortex," *Science* **257:** 552–554 (1992).

Jahnsen, H., and Llinás, R. "Electrophysiological properties of guinea-pig thalamic neurons: An *in vitro* study," *J. Physiol.* **349:** 205–226 (1984a).

Jahnsen, H., and Llinás, R. "Ionic basis for the electroresponsivenes and oscillatory properties of guinea-pig thalamic neurons *in vitro*," *J. Physiol.* **349:** 227–247 (1984b).

Jahr, C., and Stevens, C. "A quantitative description of NMDA receptor-channel kinetic behavior," *J. Neurosci.* **10:** 1830–1837 (1990).

Jan, Y.N., and Jan, L.Y. "A LHRH-like peptidergic neurotransmitter capable of 'action at a distance' in autonomic ganglia," *Trends Neurosci.* **6:** 320–325 (1983).

Jaslove, S.W. "The integrative properties of spiny distal dendrites," *Neurosci.* **47:** 495–519 (1992).

Jefferys, J.G.R. "Nonsynaptic modulation of neuronal activity in the brain: Electric currents and extracellular ions," *Physiol. Rev.* **75:** 689–723 (1995).

Jessell, T.M., and Kandel, E.R. "Synaptic transmission: A bidirectional and self-modifiable form of cell-cell communication," *Neuron* **10:** (Suppl.) 1–30 (1993).

Johansson, S., and Arhem, P. "Single-channel currents trigger action potentials in small cultured hippocampal neurons," *Proc. Natl. Acad. Sci. USA* **91:** 1761–1765 (1994).

Johnson, E.M., Jr., Koike, T., and Franklin, J. "A 'calcium set-point hypothesis' of neuronal dependence on neurotrophic factor," *Experim. Neurol.* **115:** 163–166 (1992).

Johnston, D. "Passive cable properties of hippocampal CA3 pyramidal neurons," *Cell. Mol. Neurobiol.* **1:** 41–55 (1981).

Johnston, D., Magee, J.C., Colbert, C.M., and Christie, B.R. "Active properties of neuronal dendrites," *Ann. Rev. Neurosci.* **19:** 165–186 (1996).

Johnston, D., Williams, S., Jaffe, D., and Gray, R. "NMDA receptor-independent long-term potentiation," *Ann. Rev. Neurosci.* **54:** 489–505 (1992).

Johnston, D., and Wu, S.M.-S. *Foundations of Cellular Neurophysiology.* MIT Press: Cambridge, Massachusetts (1995).

Jonas, P., Major, G., and Sakmann, B. "Quantal components of unitary EPSCs at the mossy fibre synapse on CA3 pyramidal cells of rat hippocampus," *J. Physiol.* **472:** 615–663 (1993).

Jonas, P., and Spruston, N. "Mechanisms shaping glutamate-mediated excitatory postsynaptic currents in the CNS," *Current Opinion Neurobiol.* **4:** 366–372 (1994).

Jones, S. "On the resting potential of isolated frog sympathetic neurons," *Neuron* **3:** 153–161 (1989).

Jones, E.G., and Powell, T.P.S. "Morphological variations in the dendritic spines of the neocortex," *J. Cell. Sci.* **5:** 509–529 (1969).

Kandel, E.R., and Schwartz, J.H. "Molecular biology of learning: modulation of transmitter release," *Science* **218:** 433–443 (1982).

Kandel, E.R., Schwartz, J.H., and Jessell, T.M., editors. *Principles of Neural Science.* Third edition. Elsevier, North Holland: New York, New York (1991).

Kaneko, A. "Electrical connections between horizontal cells in the dogfish retina," *J. Physiol.* **213:** 95–105 (1976).

Karabelas, A.B., and Purpura, D.P. "Evidence for autapses in the substantia nigra," *Brain Res.* **200:** 467–473 (1980).

Kasai, H., and Petersen, O.H. "Spatial dynamics of second messengers: IP$_3$ and cAMP as long-range and associative messengers," *Trends Neurosci.* **17:** 95–101 (1994).

Kasper, E.M., Larkman, A.U., Lübke, J., and Blakemore, C. "Pyramidal neurons in layer 5 of rat visual cortex. I. Correlation among cell morphology, intrinsic electrophysiological properties and axon targets," *J. Comp. Neurol.* **339:** 459–474 (1994).

Katz, B. "Depolarization of sensory terminals anad the initiation of impulses in the muscle spindle," *J. Physiol.* **111:** 261–282 (1950).

Katz, B. *Nerve, Muscle and Synapse.* McGraw-Hill: New York, New York (1966).

Katz, B. *The Release of Neural Transmitter Substances.* Liverpool University Press: Liverpool, United Kingdom (1969).

Katz, B., and Miledi, R. "The measurements of synaptic delay, and the time course of acetylcholine release at the neuromuscular junction," *Proc. Roy. Soc. Lond. B* **161:** 483–495 (1965).

Katz, B., and Miledi, R. "The role of calcium in neuromuscular facilitation," *J. Physiol.* **195:** 481–492 (1968).

Kauer, J.A., Malenka, R.C., and Nicoll, R.A. "A persistent postsynaptic modification mediates long-term potentiation in the hippocampus," *Neuron* **1:** 911–917 (1988).

Kauffman, S.A. *The Origins of Order: Self-Organization and Selection in Evolution.* Oxford University Press: New York, New York (1993).

Kawaguchi, Y., Wilson, C.J., and Emson, P.C. "Projection subtypes of rat neostriatal matrix cells revealed by intracellular injection of biocytin," *J. Neurosci.* **10:** 3421–3438 (1990).

Kawasaki, M., Rose, G., and Heiligenberg, W. "Temporal hyperacuity in single neurons of electric fish," *Nature* **336:** 173–176 (1988).

Kawato, M., and Tsukahara, N. "Theoretical study on electrical properties of dendritic spines." *J. theor. Biol.* **103:** 507-522 (1983).

Kay, A.R., and Wong, R.K.S. "Calcium current activation kinetics in isolated pyramidal neurons of the CA1 region of the mature guinea-pig hippocampus," *J. Physiol.* **392:** 603–616 (1987).

Keener, J. *Principles of Applied Mathematics: Transformation and Approximation.* Addison-Wesley: Reading, Massachusetts (1988).

Kelso, S.R., Ganong, A.H., and Brown, T.H. "Hebbian synapses in hippocampus," *Proc. Natl. Acad. Sci.* **83:** 5326–5330 (1986).

Kennedy, M.B. "Regulation of synaptic transmission in the central nervous system," *Cell* **59:** 777–787 (1989).

Kennedy, M.B. "Second messengers and neuronal function." In: *An Introduction to Molecular Neurobiology.* Hall, Z., editor, pp. 207–246. Sinauer: Sunderland, Massachusetts (1992).

Kennedy, M.B. "The postsynaptic density," *Current Opinion Neurobiol.* **3:** 732–737 (1993).

Kennedy, M.B. "The biochemistry of synaptic regulation in the central nervous system," *Ann. Rev. Biochem.* **63:** 571–600 (1994).

Kennedy, M.B., and Marder, E. "Cellular and molecular mechanisms of neuronal plasticity." In: *An Introduction to Molecular Neurobiology*. Hall, Z., editor, pp. 463–494. Sinauer: Sunderland, Massachusetts (1992).

Kepler, T.H., Abbott, L.F., and Marder, E. "Reduction of conductance-based neuron models," *Biol. Cybern.* **66:** 381–387 (1992).

Kernell, D. "Synaptic conductance changes and the repetitive impulse discharge of spinal motoneurones," *Brain Res.* **15:** 291–294 (1969).

Kernell, D. "Effects of synapses on dendrites and soma on the repetitive impulse firing of a compartmental neuron model," *Brain Res.* **35:** 551–555 (1971).

Kety, S.S. "The general metabolism of the brain *in vivo*." In: *Metabolism of the Nervous System*. Richter, D., editor, pp. 221–237. Pergamon: London, United Kingdom (1957).

Keyes, R.W. "What makes a good computer device?" *Science* **230:** 138-144 (1985).

Khodorov, B.I., and Timin, E.N. "Nerve impulse propagation along nonuniform fibres (investigations using mathematical models)," *Prog. Biophys. Mol. Biol.* **30:** 145–184 (1975).

Kim, H.G., and Connors, B.W. "Apical dendrites of the neocortex: Correlation between sodium- and calcium-dependent spiking and pyramidal cell morphology," *J. Neurosci.* **13:** 5301–5311 (1993).

Klee, C.B. "Interaction of calmodulin with Ca^{2+} and target proteins." In: *Calmodulin*, Cohen, P., and Klee, C.B., editors, pp. 35–56. Elsevier: Amsterdam, The Netherlands (1988).

Knight, B. "Dynamics of encoding in a population of neurons," *J. Gen. Physiol.* **59:** 734–766 (1972a).

Knight, B. "The relationship between the firing rate of a single neuron and the level of activity in a population of neurons," *J. Gen. Physiol.* **59:** 767–778 (1972b).

Knight, B., Toyoda, J.-I., and Dodge, F.A. "A quantitative description of the dynamics of excitation and inhibition in the eye of *Lumulus*," *J. Gen. Physiol.* **56:** 421–437 (1970).

Knudsen, E.I., du Lac, S., and Esterly, S.D. "Computational maps in the brain," *Ann. Rev. Neurosci.* **10:** 41–65 (1987).

Koch, C. *Nonlinear information processing in dendritic trees of arbitrary geometry*. PhD Thesis. University of Tübingen: Tübingen, Germany (1982).

Koch, C. "Cable theory in neurons with active, linearized membranes," *Biol. Cybern.* **50:** 15–33 (1984).

Koch, C. "Understanding the intrinsic circuitry of the cat's lateral geniculate nucleus: Electrical properties of the spine-triad arrangement," *Proc. Roy. Soc. Lond. B* **225:** 365–390 (1985).

Koch, C. "The action of the corticofugal pathway on sensory thalamic nuclei: A hypothesis," *Neuroscience* **23:** 399–406 (1987).

Koch, C. "Computational approaches to cognition: The bottom-up view," *Curr. Opinion Neurobiol.* **3:** 203–208 (1993).

Koch, C. "Computation and the single neuron," *Nature* **385:** 207–210 (1997).

Koch, C., Bernander, Ö., and Douglas, R.J. "Do neurons have a voltage or a current threshold for action potential initiation," *J. Comp. Neurosci.* **2:** 63–82 (1995).

Koch, C., and Crick, F. "Some Further Ideas Regarding the Neuronal Basis of Awareness." In: *Large-Scale Neuronal Theories of the Brain*. Koch, C., and Davis, J., editors, pp. 93–110. MIT Press: Cambridge, Massachusetts (1994).

Koch, C., Douglas, R., and Wehmeier, U. "Visibility of synaptically induced conductance changes: Theory and simulations of anatomically characterized cortical pyramidal cells," *J. Neurosci.* **10:** 1728–1744 (1990).

Koch, C., and Mathur, B. "Neuromorphic vision chips," *IEEE Spectrum* **33 (5):** 38–46 (1996).

Koch, C., and Poggio, T. "Electrical properties of dendritic spines," *Trends Neurosci.* **6:** 80–83 (1983a).

Koch, C., and Poggio, T. "A theoretical analysis of the electrical properties of spines," *Proc. Roy. Soc. Lond. B* **218:** 455–477 (1983b).

Koch, C., and Poggio, T. "A simple algorithm for solving the cable equation in dendritic trees of arbitrary geometry," *J. Neurosci. Meth.* **12:** 303–315 (1985a).

Koch, C. and Poggio, T. "The synaptic veto mechanism: Does it underlie direction and orientation selectivity in the visual cortex?" In: *Models of the Visual Cortex*. Rose, D., and Dobson, V.G., editors, pp. 408–419. John Wiley: New York, New York (1985b).

Koch, C., and Poggio, T. "The biophysical properties of spines as a basis for their electrical function: A comment of Kawato & Tsukahara (1983)," *J. theor. Biol.* **113:** 225–229 (1985c).

Koch, C., and Poggio, T. "Biophysics of computation: Neurons, synapses and membranes." In: *Synaptic Function*. Edelman, G.M., Gall, W.E., and Cowan, W.M., editors, pp. 637–698. John Wiley: New York, New York (1987).

Koch, C., and Poggio, T. "Multiplying with synapses and neurons." In: *Single Neuron Computation*. McKenna, T., Davis, J., and Zornetzer, S.F., editors, pp. 315–345. Academic Press: Boston, Massachusetts (1992).

Koch, C., Poggio, T., and Torre, V. "Retinal ganglion cells: A functional interpretation of dendritic morphology," *Phil. Trans. Roy. Soc. B* **298:** 227–264 (1982).

Koch, C., Poggio, T., and Torre, V. "Nonlinear interaction in a dendritic tree: Localization timing and role in information processing," *Proc. Natl. Acad. Sci. USA* **80:** 2799–2802 (1983).

Koch, C., Poggio, T., and Torre, V. "Computations in the vertebrate retina: Motion discrimination, gain enhancement and differentiation," *Trends Neurosci.* **9:** 204–211 (1986).

Koch, C., Rapp, M., and Segev, I. "A short history of time (constants)," *Cerebral Cortex* **6:** 93–101 (1996).

Koch, C., and Schuster, H. "A simple network showing burst synchronization without frequency-locking," *Neural Comput.* **4:** 211–223 (1992).

Koch, C., and Segev, I., editors. *Methods in Neuronal Modeling*, second edition. Editors. MIT Press: Cambridge, Massachusetts (1998).

Koch, C., Stemmler, M., Suarez, H., and Douglas, R. "Adapting recurrent cortical excitation." In: *Long Term Potentiation*, Vol. 3. Baudry, M., and Davis, J., editors, pp. 351–377. MIT Press: Cambridge, Massachusetts (1996).

Koch, C., and Zador, A. "The function of dendritic spines: Devices subserving biochemical rather than electrical compartmentalization," *J. Neurosci.* **13:** 413–422 (1993).

Koch, C., Zador, A., and Brown, T.H. "Dendritic spines: Convergence of theory and experiment," *Science* **256:** 973–974 (1992).

König, P., Engel, A.K., and Singer, W. "Integrator or coincidence detector: The role of the cortical neuron revisited," *Trends Neurosci.* **19:** 130–137 (1996).

Kondo, S., and Asai, R. "A reaction-diffusion wave on the skin of the marine angelfish *Pomacanthus*," *Nature* **376:** 765–768 (1995).

Konishi, M. "The neural algorithm for sound localization in the owl," *Harvey Lectures* **86:** 47–64 (1992).

Korn, H., and Faber, D.S. "Quantal analysis and synaptic efficacy in the CNS," *Trends Neurosci.* **14:** 439–445 (1991).

Korn, H., and Faber, D.S. "Transmitter release mechanism: Relevance for neuronal network functions." In: *Exploring Brain Functions: Models in Neuroscience*. Poggio, T., and Glaser, D.A., editors, pp. 5–17. John Wiley: Chichester, United Kingdom (1993).

Korn, H., Faber, D.S., and Triller, A. "Probabilistic determination of synaptic strength," *J. Neurophysiol.* **55:** 402–421 (1986).

Kosaka, T. "Axon initial segments of the granule cell in the rat dentate gyrus: Synaptic contacts on bundles of axon initial segments," *Brain Res.* **274:** 129–134 (1983).

Koshland, D.E. Jr., "Switches, thresholds and ultrasensitivity," *Trends Biochemistry* **12:** 225–229 (1987).

Kozyrakis, C.E., Perissakis, S., Patterson, D., Anderson, T., Asanovic, K., Cardwell, N., Fromm, R., Golbus, J., and Gribstad, B. "Scalable processors in the billion transistor era: IRAM," *IEEE Computer* **30:** 75–78 (1997).

Kuffler, S.W. "Discharge patterns and functional organization of mammalian retina," *J. Neurophysiol.* **16:** 37–68 (1953).

Kuffler, S.W., and Sejnowski, T.J. "Peptidergic and muscarinic excitation at amphibian sympathetic synapses," *J. Physiol.* **341:** 257–278 (1983).

Kullmann, D.M., Perkel D.J., Manabe, T., and Nicoll R.A. "Ca^{2+} entry via postsynaptic voltage-sensitive Ca^{2+} channels can transiently potentiate excitatory synaptic transmission in the hippocampus," *Nature* **9:** 1175–1183 (1992).

Kullmann, D.M., and Siegelbaum, S.A. "The site of expression of NMDA receptor-dependent LTP: New fuel for an old fire," *Neuron* **15:** 997–1002 (1995).

Kuno, M., and Miyahara, J.T. "Non-linear summation of unit synaptic potentials in spinal motoneurones of the cat," *J. Physiol.* **201:** 465–477 (1969).

Kupfermann, I. "Functional studies of cotransmission," *Physiol. Rev.* **71:** 683–732 (1991).

LaCaille, J.C. "Postsynaptic potentials mediated by excitatory and inhibitory amino acids in interneurons of stratum pyramidale of the CA1 region of rat hippocampal slices *in vitro*," *J. Neurophysiol.* **66:** 1441–1454 (1991).

Lancaster, B., and Adams, P.R. "Calcium-dependent current generating the afterhyperpolarization of hippocampal neurons," *J. Neurophysiol.* **55:** 1268–1282 (1986).

Lando, L., and Zucker, R.S. "Ca^{2+} cooperativity in neurosecretion measured using photolabile Ca^{2+} chelators," *J. Neurophysiol.* **72:** 825–830 (1994).

Langdon, R.B., Johnson, J.W., and Barrionuevo, G. "Posttetanic potentiation and presynaptically induced long-term potentiation at the mossy fiber synapse in rat hippocampus," *J. Neurophysiol.* **26:** 370–385 (1995).

Langmoen, I.A., and Andersen, P. "Summation of excitatory postsynaptic potentials in hippocampal pyramidal neurons," *J. Neurophysiol.* **50:** 1320–1329 (1983).

Lapicque, L. "Recherches quantitatifs sur l'excitation electrique des nerfs traitée comme une polarisation," *J. Physiol. Paris* **9:** 620–635 (1907).

Lapicque, L. *L'excitabilité en fonction du temps.* Presses Universitaires de France: Paris, France (1926).

Larkman, A.U. "Dendritic morphology of pyramidal neurones of the visual cortex of the rat," *J. Comp. Neurol.* **306:** 332–343 (1991).

Larkman, A.U, Stratford, K., and Jack, J. "Quantal analysis of excitatory synaptic action and depression in hippocampal slices," *Nature* **350:** 344–347 (1991).

Larkman, A.U., Major, G., Stratford, K.J., and Jack, J.J.B. "Dendritic morphology of pyramidal neurones of the visual cortex of the rat. IV: Electrical geometry," *J. Comp. Neurol.* **323:** 137–152 (1992).

Larkman, A.U., and Mason, A. "Correlations between morphology and electrophysiology of pyramidal neurons in slices of rat visual cortex. I: Establishment of cell classes," *J. Neurosci.* **10:** 1407–1414 (1990).

Larsson, H.P., Baker, O.S., Dhillon, D.S., and Isacoff, E.Y. "Transmembrane movement of the Shaker K$^+$ channel S4," *Neuron* **16:** 387–397 (1996).

Latorre, R., and Miller, C. "Conduction and selectivity in potassium channels," *J. Membr. Biol.* **71:** 11–30 (1983).

Latorre, R., Oberhauser, A., Labarca, P., and Alvarez, O. "Varieties of calcium-activated potassium channels," *Ann. Rev. Physiol.* **51:** 385–399 (1989).

Laughlin, S.B. "Matching coding, circuits, cells and molecules to signals—General principles of retinal design in the fly eye," *Prog. Ret. Eye Res.* **13:** 165–196 (1994).

Laurent, G. "Voltage-dependent nonlinearities in the membrane of locust nonspiking local interneurons and their significance for synaptic integration," *J. Neurosci.* **10:** 2268–2280 (1990).

Laurent, G. "A dendritic gain control mechanism in axonless neurons of the locust, *Schistocerca americana*," *J. Physiol.* **470:** 45–54 (1993).

Laurent, G. "Dynamical representation of odors by oscillating and evolving neural assemblies," *Trends Neurosci.* **19:** 489–496 (1996).

Laurent, G., and Burrows, M. "Distribution of intersegmental inputs to nonspiking local interneurons and motor neurons in the locust," *J. Neurosci.* **9:** 3019–3029 (1989).

Laurent, G., and Sivaramakrishnan, A. "Single local interneurons in the locust make central synapses with different properties of transmitter release on distinct postsynaptic neurons," *J. Neurosci.* **12:** 2370–2380 (1992).

Lecar, H., and Nossal, R. "Theory of threshold fluctuations in nerves. I. Relationships between electrical noise and fluctuations in axon firing," *Biophys. J.* **11:** 1048–1067 (1971a).

Lecar, H., and Nossal, R. "Theory of threshold fluctuations in nerves. II. Analysis of various sources of membrane noise," *Biophys. J.* **11:** 1068–1084 (1971b).

Lee, C., Rohrer, W.H., and Sparks, D.L. "Population coding of saccadic eye movements by neurons in the superior colliculus," *Nature* **332:** 357–360 (1988).

Lee, K.S., Schottler, F., Oliver, M., and Lynch, G. "Brief bursts of high-frequency stimulation produce two types of structural changes in the rat hippocampus," *J. Neurophsyiol.* **44:** 247–258 (1980).

Lee, K.S., and Tsien, R.W. "Reversal of current through calcium channels in dialysed single heart cells," *Nature* **297:** 498–501 (1982).

LeMasson, G., Marder, E., and Abbott, L.F. "Activity-dependent regulation of conductances in model neurons," *Science* **259:** 1915–1917 (1993).

Lestienne, R., and Strehler, B.L. "Time structure and stimulus dependence of precisely replicating patterns present in monkey cortical neuronal spike trains," *Brain Res.* **437:** 214–238 (1987).

Lettvin, J.P., Maturana, H.R., McCulloch, W.S., and Pitts, W.H. "What the frog's eye tells the frog's brain," *Proc. Inst. Rad. Eng.* **47:** 1940–1951 (1959).

LeVay, S. "Synaptic organization of claustral and geniculate afferents to the visual cortex of the cat," *J. Neurosci.* **6:** 3564–3575 (1986).

LeVay, S., and Gilbert, C. "Laminar patterns of geniculocortical projection in the cat," *Brain Res.* **113:** 1–19 (1976).

Leventhal, A., and Hirsch, H. "Receptive-field properties of neurons in different laminae of visual cortex of the cat," *J. Neurophysiol.* **41:** 948–962 (1978).

Levy, W.B., and Baxter, R.A. "Energy efficient neural codes," *Neural Comput.* **8:** 531–544 (1996).

Levy, W.B., and Steward, O. "Temporal contiguity requirements for long-term associative potentiation/depression in the hippocampus," *Neurosci.* **8:** 791–797 (1983).

Lewis, R.S., and Hudspeth, A.J. "Voltage- and ion-dependent conductances in solitary vertebrate hair cell," *Nature* **304:** 538–541 (1983).

Liao, D., Hessler, N.A., and Malinow, R. "Activation of postsynaptically silent synapses during pairing-induced LTP in CA1 region of hippocampal slice," *Nature* **375:** 400–404 (1995).

Linden, D.J., and Connor, J.A. "Long-term synaptic depression," *Ann. Rev. Neurosci.* **18:** 319–357 (1995).

Linsker, R. "Self-organization in a perceptual network," *Computer* **21:** 105–117 (1988).

Lisman, J.E. "A mechanism for memory storage insensitive to molecular turnover: A bistable autophosphorylating kinase," *Proc. Natl. Acad. Sci. USA* **82:** 3055–3057 (1985).

Lisman, J.E. "A mechanism for the Hebb and the anti-Hebb processes underlying learning and memory," *Proc. Natl. Acad. Sci. USA* **86:** 9574–9578 (1989).

Lisman, J.E. "The CaM kinase II hypothesis for the storage of synaptic memory," *Trends Neurosci.* **17:** 406–412 (1994).

Lisman, J.E. "Bursts as a unit of neural information: Making unreliable synapses reliable," *Trends Neurosci.* **20:** 38–43 (1997).

Lisman, J.E., and Goldring, M.A. "Feasibility of long-term storage of graded information by the Ca^{2+}/calmodulin-dependent protein kinase molecules of the postsynaptic density," *Proc. Natl. Acad. Sci. USA* **85:** 5320–5324 (1988).

Llano, I., Webb, C.K., and Bezanilla, F. "Potassium conductance of the squid giant axon," *J. Gen. Physiol.* **92:** 179–196 (1988).

Llinás, R. "The intrinsic electrophysiological properties of mammalian neurons: Insights into central nervous system function," *Science* **242:** 1654–1664 (1988).

Llinás, R., Grace, A.A., and Yarom, Y. "*In vitro* neurons in mammalian cortical layer 4 exhibit intrinsic oscillatory activity in the 10 to 50 Hz frequency range," *Proc. Natl. Acad. Sci. USA* **88:** 987–991 (1991).

Llinás, R., and Nicholson, C. "Electroresponsive properties of dendrites and somata in alligator Purkinje cells," *J. Neurophysiol.* **34:** 532–551 (1971).

Llinás, R., Nicholson, C., Freeman, J.A., and Hillman, D.E. "Dendritic spikes and their inhibition in alligator Purkinje cells," *Science* **160:** 1132–1135 (1968).

Llinás, R., Steinberg, I.Z., and Walton, K. "Presynaptic calcium currents in squid giant synapse," *Biophys. J.* **33:** 289–322 (1981a).

Llinás, R., Steinberg, I.Z., and Walton, K. "Relationship between presynaptic calcium current and postsynaptic potential in squid giant synapse," *Biophys. J.* **33:** 323–351 (1981b).

Llinás, R., and Sugimori, M. "Electrophysiological properties of *in vitro* Purkinje cell dendrites in mammalian cerebellar slices," *J. Physiol.* **305:** 197–213 (1980).

Llinás, R., and Sugimori, M. "The electrophysiology of the cerebellar Purkinje cells revisited." In: *The Cerebellum Revisited.* Llinás, R., and Sotelo, C., editors, pp. 167–181. Springer Verlag: New York, New York (1992).

Llinás, R., Sugimori, M., and Silver, R.B. "The concept of calcium concentration microdomains in synaptic transmission," *Neuropharmacol.* **34:** 1443–1451 (1995).

Llinás, R., and Walton, K.D. "Cerebellum." In: *The Synaptic Organization of the Brain.* Shepherd, G.M., editors, fourth edition, pp. 255–288. Oxford University Press: New York, New York (1998).

Logothetis, N.K., and Pauls, J. "Psychophysical and physiological evidence of viewer-centered object representations in the primate," *Cerebral Cortex* **5:** 270–288 (1995).

Lorente de Nó, R. "Analysis of the distribution of action currents of nerve in volume conductors," *Studies Rockefeller Institute Medical Res.* **132:** 384–477 (1947).

Lorente de Nó, R. "Conduction of impulses in the neurons of the oculomotor nucleus." In: *The Spinal Cord*, Malcolm, J.L., and Gray, J.A.B., editors, pp. 132–173. CIBA Foundation Synmposium (1953).

Lowndes, M., and Stewart, M.G. "Dendritic spine density in the lobus parolfactorius of the domestic chick is increased 24 h after one-trial passive avoidance training," *Brain Res.* **654:** 129–136 (1994).

Lucas, J.R. "Minds, Machines and Gödel." In: *Minds and Machines*. Anderson, A.R., editor. Prentice-Hall: Englewood Cliffs, New Jersey (1964).

Lübke, J., Markram, H., Frotscher, M., and Sakmann, B. "Frequency and dendritic distribution of autapses established by layer 5 pyramidal neurons in the developing rat neocortex–comparisons with synaptic innervation of adjacent neurons of the same class," *J. Neurosci.* **16:** 3209–3218 (1996).

Lüscher, H.R., and Shiner, J.S. "Computation of action potential propagation and presynaptic bouton activation in terminal arborization of different geometries," *Biophys. J.* **58:** 1377–1388 (1990a).

Lüscher, H.R., and Shiner, J.S. "Simulation of action potential propagation in complex terminal arborization," *Biophys. J.* **58:** 1389–1399 (1990b).

Lynch, J.W., and Barry, P.H. "Action potentials initiated by single channels opening in a small neuron (rat olfactory receptor)," *Biophys. J.* **55:** 755–768 (1989).

Lynch, G., and Baudry, M. "Brain spectrin, calpain and long-term changes in synaptic efficacy," *Brain Research Bulletin* **18:** 809–815 (1987).

Lynch, G., Larson, J., Kelso, S., Barrionuevo, G., and Schottler, F. "Intracellular injections of EGTA block induction of hippocampal long-term potentiation," *Nature* **305:** 719–721 (1983).

Lytton, W.W., and Sejnowski, T.J. "Simulations of cortical pyramidal neurons synchronized by inhibitory interneurons," *J. Neurophysiol.* **66:** 1059–1079 (1991).

Maass, W. "Lower bounds for the computational power of networks of spiking neurons," *Neural Comput.* **8:** 1–40 (1996).

Maass, W., and Zador, A.M. " Dynamic stochastic synapses as computational units." In: *Advances in*

Neural Information Processing Systems, Vol. 10. Mozer, M.C., Jordan, M.I., and Petsche, T., editors, pp. 194-200. MIT Press: Cambridge, Massachusetts (1998).

Madison, D.V., and Nicoll, R.A. "Control of the repetitive discharge of rat CA1 pyramidal neurones *in vitro*," *J. Physiol.* **354:** 319–331 (1984).

Madison, D.V., and Nicoll, R.A. "Actions of noradrenaline recorded intracellularly in rat hippocampal CA1 pyramidal neurons *in vitro*," *J. Physiol.* **372:** 221–244 (1986).

Madison, D., Lancaster, V., and Nicoll, R. "Voltage clamp analysis of cholinergic action in the hippocampus," *J. Neurosci.* **7:** 733–741 (1987).

Madison, D., Malenka, R.C., and Nicoll, R. "Mechanisms underlying long-term potentiation of synaptic transmission," *Ann. Rev. Neurosci.* **14:** 379–397 (1991).

Magee, J.C., and Johnston, D. "Synaptic activation of voltage-gated channels in the dendrites of hippocampal pyramidal neurons," *Science* **268:** 301–304 (1995a).

Magee, J.C., and Johnston, D. "Characterization of single voltage-gated Na^+ and Ca^{2+} channels in apical dendrites of rat CA1 pyramidal neurons," *J. Physiol.* **487:** 67–90 (1995b).

Magleby, K.L. "Short-term changes in synaptic efficacy." In: *Synaptic Function*, Edelman, G.M., Gall, W.E., and Cowan, W.M., editors, pp. 21–56. John Wiley: New York, New York (1987).

Mahowald, M., and Douglas, R. "A silicon neuron," *Nature* **354:** 515–518 (1991).

Mainen, Z.F., Joerges, J., Huguenard, J.R., and Sejnowski, T.J. "A model of spike initiation in neocortical pyramidal cells," *Neuron* **15:** 1427–1439 (1995).

Mainen, Z.F., and Sejnowski, T.J. "Reliability of spike timing in neocortical neurons," *Science* **268:** 1503–1506 (1995).

Mainen, Z.F., and Sejnowski, T.J. "Influence of dendritic structure on firing pattern in model neocortical neurons," *Nature* **382:** 363–366 (1996).

Mainen, Z.F. and Sejnowski, T.J. "Modeling active dendritic processes in pyramidal neurons." In: *Methods in Neuronal Modeling*. Koch, C., and Segev, I., editors, second edition, pp. 171–210. MIT Press: Cambridge, Massachusetts (1998).

Major, G. *The physiology, morphology and modeling of cortical pyramidal neurons*. PhD Thesis. Oxford University: Oxford, United Kingdom (1992).

Major, G. "Solutions for transients in arbitrarily branching cables: III. Voltage clamp problems," *Biophys. J.* **65:** 469–491 (1993).

Major, G., Evans, J.D., and Jack, J.J.B. "Solutions for transients in arbitrarily branching cables: I. Voltage recording with a somatic shunt," *Biophys. J.* **65:** 423–449 (1993a).

Major, G., Evans, J.D., and Jack, J.J.B. "Solutions for transients in arbitrarily branching cables: II. Voltage clamp theory," *Biophys. J.* **65:** 450–468 (1993b).

Major, G., Larkman, A.U., Jonas, P., Sakmann, B., and Jack, J.J.B. "Detailed passive cable properties of whole-cell recorded CA3 pyramidal neurons in rat hippocampal slices," *J. Neurosci.* **14:** 4613–4638 (1994).

Malenka, R.C. "Synaptic plasticity in the hippocampus: LTP and LTD," *Cell* **26:** 535–538 (1994).

Malenka, R.C. "Synaptic plasticity in hippocampus and neocortex: A comparison." In: *The Cortical Neuron*. Gutnick, M.J., and Mody, I., editors, pp. 98–110. Oxford University Press: New York, New York (1995).

Malenka, R.C., Kauer, J.A., Zucker, R.S., and Nicoll, R.A. "Postsynaptic calcium is sufficient for potentiation of hippocampal synaptic transmission," *Science* **242:** 81–84 (1988).

Malenka, R.C., and Nicoll, R.A. "NMDA receptor-dependent synaptic plasticity: Multiple forms and mechanisms," *Trends Neurosci.* **16:** 521–527 (1993).

Malinow, R., and Miller, J.P. "Postsynaptic hyperpolarization during conditioning reversibly blocks induction of long-term potentiation," *Nature* **320:** 529–530 (1986).

Malinow, R., and Tsien, R.W. "Presynaptic enhancement shown by whole-cell recordings of long-term potentiation in hippocampal slices," *Nature* **346:** 177–180 (1990).

Mallart, A., and Martin, A.R. "The relation between quantum content and facilitation at the neuromuscular junction of the frog," *J. Physiol.* **196:** 593–604 (1968).

Malpeli, J.G., Lee, C., Schwark, H.D., and Weyand, T.G. "Cat area 17. I. Pattern of thalamic control of cortical layers," *J. Neurophysiol.* **46:** 1102–1119 (1981).

Manalan, A.S., and Klee, C.B. "Calmodulin," *Adv. Cyclic Nucleotide Protein Phosphorylation Res.* **18:** 227–278 (1984).

Mandelbrot, B. *The Fractal Geometry of Nature.* Freeman: New York, New York (1977).

Manor, Y., Koch, C., and Segev, I. "Effect of geometrical irregularities on propagation delay in axonal trees," *Biophys. J.* **60:** 1424–1437 (1991).

Marchiafava, P.L. "The responses of retinal ganglion cells to stationary and moving visual stimuli," *Vis. Res.* **19:** 1203–1211 (1979).

Marder, E., and Calabrese, R.L. "Principles of rhythmic motor pattern generation," *Physiol. Rev.* **76:** 687–717 (1996).

Marder, E., Christie, A.E., and Kilman, V.L. "Functional organization of cotransmission systems: Lessons from small nervous systems," *Invertebrate Neurosci.* **1:** 105–112 (1995).

Marder, E., Jorge-Rivera, J.C., Kilman, V., and Weimann, J.M. "Peptidergic modulation of synaptic transmission in a rhythmic motor system," *Adv. Organ Biology* **2:** 213–233 (1997).

Markram, H., Helm, P.J., and Sakmann, B. "Dendritic calcium transients evoked by single back-propagating action potentials in rat neocortical pyramidal neurons," *J. Physiol.* **485:** 1–20 (1995).

Markram, H., Lübke, J., Frotscher, M., and Sakmann, B. "Regulation of synaptic efficacy by coincidence of postsynaptic APs and EPSPs," *Science* **275:** 213–215 (1997).

Markram, H., and Tsodyks, M. "Redistribution of synaptic efficacy between neocortical pyramidal neurons," *Nature* **382:** 807–810 (1996).

Marks, W.B., and Loeb, G.E. "Action currents, internodal potentials and extracellular records of myelinated mammalian nerve fibers derived from node potentials," *Biophys. J.* **16:** 655–668 (1976).

Marmont, G. "Studies on the axon membrane. I. A new method," *J. Cell. Comp. Physiol.* **34:** 351–382 (1949).

Marr, D. *Vision.* Freeman: New York, New York (1982).

Marr, D., and Poggio, T. "Cooperative computation of stereo disparity," *Science* **194:** 283–287 (1976).

Marsalek, P., Koch, C., and Maunsell, J. "On the relationship between synaptic input and spike output jitter in individual neurons," *Proc. Natl. Acad. Sci. USA* **94:** 735–740 (1997).

Mascagni, M. "Numerical methods for neuronal modeling." In: *Methods in Neuronal Modeling.* Koch, C., and Segev, I., editors, pp. 439–484. MIT Press: Cambridge, Massachusetts (1989).

Masland, R.H., Mills, W., and Cassidy, C. "The functions of acetylcholine in the rabbit retina," *Proc. Roy. Soc. Lond. B* **223:** 121–139 (1984).

Maslim, J., Webster, M., and Stone, J. "Stages in the structural differentiation of retinal ganglion cells," *J. Comp. Neurol.* **254:** 382–402 (1986).

Mason, A., and Larkman, A. "Correlations between morphology and electrophysiology of pyramidal neurons in slices of rat visual cortex. II. Electrophysiology," *J. Neurosci.* **10:** 1415–1428 (1990).

Mason, A., Nicoll, A., and Stratford, K. "Synaptic transmission between individual pyramidal neurons of the rat visual cortex *in vitro*," *J. Neurosci.* **11:** 72–84 (1991).

Matthews, G. "Neurotransmitter release," *Ann. Rev. Neurosci.* **19:** 219–233 (1996).

Mauro, A., Conti, F., Dodge, F., and Schor, R. "Subthreshold behavior and phenomenological impedance of the squid giant axon," *J. Gen. Physiol.* **55:** 497–523 (1970).

Mayer, M.L. "A calcium-activated chloride current generates the after-depolarization of rat sensory neurones in culture," *J. Physiol.* **364:** 217–239 (1985).

McBurney, R.N., and Neering, I.R. "Neuronal calcium homeostasis," *Trends Neurosci.* **10:** 164–169 (1987).

McCleskey, E.W. "Calcium channels: Cellular roles and molecular mechanisms," *Current Opinion Neurobiol.* **4:** 304–312 (1994).

McCormick, D.A. "Membrane properties and neurotransmitter actions." In: *The Synaptic Organiza-*

tion of the Brain. Shepherd, G.M., editor, fourth edition, pp. 37–77. Oxford University Press: New York, New York (1998).

McCormick, D.A. "Functional properties of a slowly inactivating potassium current in guinea pig dorsal lateral geniculate relay neurons," *J. Neurophysiol.* **66:** 1176–1189 (1991).

McCormick, D.A. "Neurotransmitter actions in the thalamus and cerebral cortex and their role in neuromodulation of thalamocortical activity," *Prog. Neurobiol.* **39:** 337–388 (1992).

McCormick, D.A., Connors, B.W., Lighthall, J.W., and Prince, D.A. "Comparative electrophysiology of pyramidal and sparsely spiny stellate nerons of the neocortex," *J. Neurophysiol.* **54:** 782-806 (1985).

McCormick, D.A., and Huguenard, J.R. "A model of the electrophysiological properties of thalamo-cortical relay neurons," *J. Neurophysiol.* **68:** 1384–1400 (1992).

McCormick, D.A., Huguenard, J.R., and Strowbridge, B.W. "Determination of state-dependent processing in thalamus by single neuron properties and neuromodulators." In: *Single Neuron Computation.* McKenna, T., Davis, J., and Zornetzer, S.F., editors, pp. 259–290. Academic Press: Boston, Massachusetts (1992).

McCormick, D.A., and Prince, D.A. "Acetylcholine induces burst firing in thalamic reticular neurones by activating a potassium conductance," *Nature* **319:** 402–405 (1986).

McCulloch, W.S., and Pitts, W. "A logical calculus of the ideas immanent in nervous activity," *Bull. Math. Biophys.* **5:** 115–133 (1943).

McHugh, T.J., Blum, K.I., Tsien, J.Z., Tonegawa, S., and Wilson, M.A. "Impaired hippocampal representation of space in CA1-specific NMDAR1 knockout mice," *Cell* **87:** 1339–1349 (1996).

McKenna, T., Davis, J., and Zornetzer, S.F., editors. *Single Neuron Computation.* Academic Press: Boston, Massachusetts (1992).

McManus, O.B., and Magleby, K.L. "Kinetic states and modes of single large-conductance calcium-activated potassium channels in cultured rat skeletal muscle," *J. Physiol.* **402:** 79–129 (1988).

Mead, C. *Analog VLSI and Neural Systems.* Addison-Wesley: Reading, Massachusetts (1989).

Mead, C. "Neuromorphic electronic systems," *Proc. IEEE* **78:** 1629–1636 (1990).

Mead, C., and Conway, L. *Introduction to VLSI Systems.* Addison-Wesley: Reading, Massachusetts (1980).

Meek, J., and Nieuwenhuys, R. "Palisade pattern of mormyrid Purkinje cells—A correlated light and electron-microscopic study," *J. Comp. Neurol.* **306:** 156–192 (1991).

Meinhardt, H. "Biological pattern formation: New observations provide support for theoretical predictions," *BioEssays* **16:** 627–632 (1994).

Mel, B.W. "NMDA-based pattern discrimination in a modeled cortical neuron," *Neural Comput.* **4:** 502–517 (1992).

Mel, B.W. "Synaptic integration in an excitable dendritic tree," *J. Neurophysiol.* **70:** 1086–1101 (1993).

Mel, B.W. "Information processing in dendritic trees," *Neural Comput.* **6:** 1031–1085 (1994).

Mel, B.W., and Koch, C. "Sigma-pi learning: On radial basis functions and cortical associative learning." In: *Advances in Neural Information Processing Systems*, Vol. 2. Touretzsky, D.S., editor, pp. 474–481. Morgan Kaufmann: San Mateo, California (1990).

Mel, B.W., Ruderman, D.L., and Archie, K.A. "Translation-invariant orientation-tuning in visual complex cells could derive from intradendritic computations," *J. Neurosci.,* **18:** 4325–4334 (1998).

Meulemans, A. "Diffusion coefficients and half-lives of nitric oxide and N-nitroso-L-arginine in rat cortex," *Neurosci. Lett.* **171:** 89–93 (1994).

Meyer, T., and Stryer, L. "Calcium spiking," *Ann. Rev. Biophysics Biophysical Chem.* **20:** 153–174 (1991).

Migliore, M. "Modeling the attenuation and failure of action potentials in the dendrites of hippocampal neurons," *Biophys. J.* **71:** 2394–2403 (1996).

Miller, K.D. "Models of activity-dependent neural development," *Prog. Brain Res.* **102:** 303–18 (1994).

Miller, K.D., Chapman, B., and Stryker, M.P. "Visual responses in adult cat visual cortex depend on *N*-methyl-D-aspartate receptors," *Proc. Natl. Acad. Sci. USA* **86:** 5183–5187 (1989).

Miller, S.G., and Kennedy, M.B. "Regulation of brain type-II Ca^{2+} calmodulin-dependent protein-kinase by autophosphorylation: A Ca^{2+} triggered molecular switch," *Cell* **44:** 861–870 (1986).

Miller, J.P., Rall, W., and Rinzel, J. "Synaptic amplification by active membrane in dendritic spines," *Brain Res.* **325:** 325–330 (1985).

Milner, P.M. "A model for visual shape recognition," *Psychol. Rev.* **81:** 521–535 (1974).

Milner, B., Squire, L.R., and Kandel, E.R. "Cognitive neuroscience and the study of memory," *Neuron* **20:** 445-468 (1998).

Minai, A.A., and Levy, W.B. "Sequence learning in a single trial," *Intl. Neural Network Society World Congress of Neural Networks*, Vol. 2. 505–508 (1993).

Mintz, I.M., Adams, M.E., and Bean, B.P. "P-type calcium channels in rat central and peripheral neurons," *Neuron* **9:** 85–95 (1992).

Mintz, I.M., Sabatini, B.L., and Regehr, W.G. "Calcium control of transmitter release at a cerebellar synapse," *Neuron* **15:** 675–688 (1995).

Moiseff, A., and Konishi, M. "Neuronal and behavioral sensitivity to binaural time differences in the owl," *J. Neurosci.* **1:** 40–48 (1981).

Monod, J., and Jacob, F. "General conclusions: Teleonomic mechanisms in cellular metabolism, growth and differentiation," *Cold Spring Harbor Symp. Quant. Biol.* **26:** 389–401 (1961).

Montague, P.R. "The resource consumption principle: Attention and memory in volumes of neural tissue," *Proc. Natl. Acad. Sci. USA* **93:** 3619–3623 (1996).

Montague, P.R., Dayan, P., Person, C., and Sejnowski, T.J. "Bee foraging in uncertain environments using predictive Hebbian learning," *Nature* **377:** 725–728 (1995).

Montague, P.R., Gally, J.A., and Edelman, G.M. "Spatial signaling in the development and function of neural connections," *Cerebral Cortex* **1:** 199–220 (1991).

Montague, P.R., Gancayco, C.D., Winn, M.J., Marchase, R.B., and Friedlander, M.J. "Role of NO production in NMDA receptor-mediated neurotransmitter release in cerebral cortex," *Science* **263:** 973–977 (1994).

Montague, P.R., Dayan, P., and Sejnowski, T.J. "A framework for mesencephalic dopamine systems based on predictive Hebbian learning," *J. Neurosci.* **16:** 1936–1947 (1996).

Montague, P.R., and Sejnowski, T.J. "The predictive brain: Temporal coincidence and temporal order in synaptic learning mechanisms," *Learning & Memory* **1:** 1–33 (1994).

Moore, E.F., and Shannon, C.E. "Reliable circuits using less reliable relays," *J. Franklin Inst.* **262:** 191–208 (1956).

Moore, J.W., Stockbridge, N., and Westerfield, M. "On the site of impulse generation in a neuron," *J. Physiol.* **336:** 301–311 (1983).

Morris, C., and Lecar, H. "Voltage oscillations in the barnacle giant muscle fiber," *Biophys. J.* **193:** 193–213 (1981).

Moser, M.-B., Trommald, M., and Andersen, P. "An increase in dendritic spine density on hippocampal CA1 pyramidal cells following spatial learning in adult rats suggests the formation of new synapses," *Proc. Natl. Acad. Sci. USA* **91:** 12673–12675 (1994).

Movshon, J.A., Thompson, I.D., and Tolhurst, D.J. "Receptive field organization of complex cells in the cat's striate cortex," *J. Physiol.* **283:** 79–99 (1978).

Muller, D., Joly, M., and Lynch, G. "Contributions of quisqualate and NMDA receptors to the induction and expression of LTP," *Science* **242:** 1694–1697 (1988).

Müller, W., and Connor, J.A. "Dendritic spines as individual neuronal compartments for synaptic Ca^{2+} responses," *Nature* **354:** 73–76 (1991).

Murthy, V.N. "Synaptic plasticity: Neighborhood influences," *Current Biol.* **7:** R612-R615 (1997).

Nagumo, J.S., Arimoto, S., and Yoshizawa, S. "An active pulse transmission line simulating nerve axon," *Proc. IRE* **50:** 2061–2070 (1962).

Neher, E. *Dynamische Eigenschaften von Nervsoma Membranen*. PhD Thesis. University of München: München, Germany (1970).

Neher, E. "The use of fura–2 for estimating Ca buffers and Ca fluxes," *Neuropharmacology* **34:** 1423–1442 (1995).

Neher, E., and Sakmann, B. "Single-channel currents recorded from membrane of denervated frog muscle fibres," *Nature* **260:** 779–802 (1976).

Neher, E., and Stevens, C.F. "Conductance fluctuations and ionic pores in membranes," *Ann. Rev. Biophys. Bioeng.* **6:** 345–381 (1977).

Nelson, M.E. "A mechanism for neuronal gain control by descending pathways," *Neural Comput.* **6:** 242–254 (1994).

Nernst, W. "Zur Kinetik der in Lösung befindlichen Körper: Theorie der Diffusion," *Z. Phys. Chemie* **3:** 613–637 (1888).

Nernst, W. "Die elektromotorische Wirksamkeit der Ionen," *Z. Phys. Chemie* **4:** 129–181 (1889).

Neumcke, B., and Stämpfli, R. "Sodium currents and sodium-current fluctuations in rat myelinated nerve fibers," *J. Physiol.* **329:** 163–184 (1982).

Neveu, D., and Zucker, R.S. "Postsynaptic levels of $[Ca^{2+}]_i$ needed to trigger LTD and LTP," *Neuron* **16:** 619–629 (1996).

Newsome, W.T., Britten, K.H., and Movshon, J.A. "Neuronal correlates of a perceptual decision," *Nature* **341:** 52–54 (1989).

Nicholson, C. "Extracellular space as the pathway for neuron-glial cell interaction." In: *Neuroglia*. Kettenmann, H., and Ransom, B.R., editors, pp. 387–397. Oxford University Press: Oxford, United Kingdom (1995).

Nicholson, C. and Phillips, J.M. "Ion diffusion modified by tortuosity and volume fraction in the extracellular microenvironment of the rat cerebellum," *J. Physiol.* **321:** 225–257 (1981).

Nicoll, R.A. "The coupling of neurotransmitter receptors to ion channels in the brain," *Science* **241:** 545–551 (1988).

Nicoll, R.A., and Malenka, R.C. "Contrasting properties of two forms of long-term potentiation in the hippocampus," *Nature* **377:** 115–118 (1995).

Niebur, E., and Niebur, D. "Numerical implementation of sealed-end boundary conditions in cable theory," *IEEE Trans. Biomed. Eng.* **38:** 1266–1272 (1991).

Niebur, E., and Koch, C. "A Model for the neuronal implementation of selective visual attention based on temporal correlation among neurons," *J. Comput. Neurosci.* **1:** 141–158 (1994).

Nitzan, R., Segev, I., and Yarom, Y. "Voltage behavior along the irregular dendritic structure of morphologically and physiologically characterized vagal motoneurons in the guinea pig," *J. Neurophysiol.* **63:** 333–346 (1990).

Noble, D. *The Initiation of the Heart Beat*. Clarendon: Oxford, United Kingdom (1979).

Noble, D., and Stein, R.B. "The threshold conditions for initiation of action potentials by excitable cells," *J. Physiol.* **187:** 129–162 (1966).

Noda, H., and Adey, R. "Firing variability in cat asociation cortex during sleep and wakefulness," *Brain Res.* **18:** 513–526 (1970).

Norsworthy, S.J., Schreier, R., and Temes, G.C. *Delta-Sigma Data Converters: Theory, Design, and Simulation*. Editors. IEEE Press (1996).

Nowak, L., Bregestovski, P., Ascher, A.H., and Prochiantz, A. "Magnesium gates glutamate-activated channels in mouse central neurones," *Nature* **307:** 462–465 (1984).

Nuñez, A., Amzica, F., and Steriade, M. "Electrophysiology of cat association cortical cells in vivo: Intrinsic properties and synaptic responses," *J. Neurophysiol.* **70:** 418–430 (1993).

Ogden, D. "Intracellular calcium release in central neurons," *Sem. Neurosci.* **8:** 281–291 (1996).

Optican, L.M., and Richmond, B.J. "Temporal encoding of two-dimensional patterns by single units in primate inferior temporal cortex. III Information theoretic analysis," *J. Neurophysiol.* **57:** 162–178 (1987).

Orban, G. *Neuronal Operations in the Visual Cortex*. Springer Verlag: New York, New York (1984).

Oster, G.F., Perelson, A., and Katchalsky, A. "Network thermodynamics," *Nature* **234:** 393–399 (1971).

Oyster, C.W., Amthor, F.R., and Takahashi, E.S. "Dendritic architecture of ON-OFF direction-selective ganglion cells in the rabbit retina," *Vision Res.* **33:** 579–608 (1993).

Palm, G. "On representation and approximation of nonlinear systems," *Biol. Cybern.* **31:** 119–124 (1978).

Palm, G. *Neural Assemblies.* Springer Verlag: Berlin, Germany (1982).

Palmer, L.A., Jones, J.P., and Stepnoski, R.A. "Striate receptive fields as linear filters: Characterization in two spatial dimensions of space." In: *The Neural Basis of Visual Function.* Leventhal, A.G., editor, pp. 246–265. CRC Press: Boca Raton, Florida (1991).

Panzeri, S., Biella, G., Rolls, E.T., Skaggs, W.E., and Treves, A. "Speed, noise, information and the graded nature of neuronal responses," *Network: Comp. Neural. Sys.* **7:** 365–370 (1996).

Papa, M., Bundman, M.C., Greenberger, V., and Segal, M. "Morphological analysis of dendritic spine development in primary cultures of hippocampal neurons," *J. Neurosci.* **15:** 1–11 (1995).

Papoulis, A. *Probability, Random Variables and Stochastic Processes.* Third edition. McGraw-Hill: New York, New York (1984).

Parnas, I. "Differential block at high frequency of branches of a single axon innervating two muscles," *J. Neurophysiol.* **35:** 903–914 (1972).

Parnas, H., Hovav, G., and Parnas, I. "Effect of Ca^{2+} diffusion on the time course of neurotransmitter release," *Biophys. J.* **55:** 859–874 (1989).

Parnas, I., and Segev, I. "A mathematical model for conduction of action potentials along bifurcating axons," *J. Physiol.* **295:** 323–343 (1979).

Patlak, J.B., and Ortiz, M. "Two modes of gating during late Na^+ channel currents in frog sartorius muscle," *J. Gen. Physiol.* **87:** 305–326 (1986).

Patton, B.L., Molloy, S.S., and Kennedy, M.B. "Autophosphorylation of type II CAM kinase in hippocampal neurons: Localization of phosphokinase and dephosphokinase with complementary phosphorylation site-specific antibodies," *Mol. Biol. Cell* **4:** 159–172 (1993).

Pauling, L., and Pauling, P. *Chemistry.* W.H. Freeman: San Francisco, California (1975).

Pearlmutter, B.A., "Time-skew Hebb rule in a nonisopotential neuron," *Neural Comp.* **7:** 706–712 (1995).

Pei, X., Volgushev, M., Vidyasagar, T.R., and Creutzfeldt, O.D. "Whole cell recording and conductance measurements in cat visual cortex *in vivo*," *NeuroReport* **2:** 485–488 (1991).

Peichl, L., and Wässle, H. "The structural correlate of the receptive field centre of α ganglion cells in the cat retina," *J. Physiol.* **341:** 309–324 (1983).

Penfield, P., Spence, R., and Duinker, S. *Tellegen's Theorem and Electrical Networks.* MIT Press: Cambridge, Massachusetts (1970).

Pennefather, P., Lancaster, B., Adams, P.R., and Nicoll, R.A. "Two distinct Ca-dependent K currents in bullfrog sympathetic ganglion cells," *Proc. Natl. Acad. Sci. USA* **82:** 3040–3044 (1985).

Penny, G.R., Wilson, C.J., and Kitai, S.T. "Relationship of the axonal and dendritic geometry of spiny projection neurons to the compartmental organization of the neostriatum," *J. Comp. Neurol.* **269:** 275–289 (1988).

Penrose, R. *The Emperor's New Mind.* Oxford University Press: New York, New York (1989).

Perkel, D.H., and Bullock, T.H. "Neural coding," *Neurosci. Res. Prog. Sum.* **3:** 405–527 (1968).

Perkel, D.H., Gerstein, G.L., and Moore, G.P. "Neuronal spike trains and stochastic point processes: I. The single spike train," *Biophys. J.* **7:** 391–418 (1967).

Perkel, D.H., Mulloney, B., and Budelli, R.W. "Quantitative methods for predicting neuronal behavior," *Neurosci.* **6:** 823–837 (1981).

Perkel, D.H., and Perkel, D.J. "Dendritic spines: Role of active membrane in modulating synaptic efficiency," *Brain Res.* **325:** 331–335 (1985).

Peters, A., and Kaiserman-Abramof, I.R. "The small pyramidal neuron of the rat cerebral cortex. The perikaryon, dendrites and spines," *Am. J. Anat.* **127:** 321–356 (1970).

Peters, A., Palay, S.L., and Webster, H. deF. *The Fine Structure of the Nervous System.* Saunders Company: Philadelphia, Pennsylvania (1976).

Peters, A., and Payne, B.R. "Numerical relationships between geniculocortical cell modules in cat primary visual cortex," *Cereb. Cortex* **3:** 69–78 (1993).

Peters, A., Payne, B.R., and Rudd, J. "A numerical analysis of the geniculocortical input to striate cortex in the monkey," *Cereb. Cortex* **4:** 215–229 (1994).

Planck, M. "Über die Potentialdifferenz zwischen zwei verdünnten Lösungen binärer Elektrolyte," *Ann. Phys. Chemie, Neue Folge* **40:** 561–576 (1890).

Plonsey, R. *Bioelectric Phenomena.* McGraw-Hill: New York, New York (1969).

Pockberger, H. "Electrophysiological and morphological properties of rat motor cortex neurons *in vivo*," *Brain Res.* **539:** 181–190 (1991).

Poggio, T., and Koch, C. "Ill-posed problems in early vision: From computational theory to analog networks," *Proc. Roy. Soc. Lond. B* **226:** 303–323 (1985).

Poggio, T., and Torre, V. "A new approach to synaptic interactions." In: *Lecture Notes in Biomathematics: Theoretical Approaches to Complex Systems*, Vol. 21. Heim, R., and Palm, G., editors, pp. 89–115. Springer Verlag: Berlin, Germany (1978).

Poggio, T., and Torre, V. "A theory of synaptic interactions." In: *Theoretical Approaches in Neurobiology.* Reichardt, W.E., and Poggio, T., editors, pp. 28–46. MIT Press: Cambridge, Massachusetts (1981).

Poggio, G.F., and Viernstein, L.J. "Time series analysis of impulse sequences of thalamic somatic sensory neurons," *J. Neurophysiol.* **27:** 517–545 (1964).

Poggio, T., Yang, W., and Torre, V. "Optical flow: Computational properties and networks, biological and analog." In: *The Computing Neuron.* Durbin, R., Miall, C., and Mitchison, G., editors, pp. 355–370. Addison-Wesley: Wokingham, United Kingdom (1989).

Pongracz, F. "The function of dendritic spines: A theoretical study," *Neurosci.* **15:** 933–946 (1985).

Porter, R., and Lemon, R. *Corticospinal Function and Voluntary Movement.* Clarendon Press: Oxford, United Kingdom (1993).

Pouget, A., and Sejnowski, T.J. "Spatial transformations in the parietal cortex using basis functions," *J. Cog. Neurosci.* **9:** 222–237 (1997).

Powers, P., Tobinson, R., and Konodi, M. "Effective synaptic current can be estimated from measurements of neuronal discharge," *J. Neurophysiol.* **68:** 964–698 (1992).

Purpura, D., "Dendritic spines 'dysgenesis' and mental retardation," *Science* **186:** 1126–1128 (1974).

Qian, N., and Sejnowski, T.J. "Electrodiffusion model of electrical conduction in neuronal processes." In: *Cellular Mechanisms of Conditioning and Behavioral Plasticity.* Woody, C.D., Alkon, D.L., and McGaugh, J.L., editors, pp. 237–244. Plenum Press: New York, New York (1988).

Qian, N., and Sejnowski, T.J. "An electro-diffusion model for computing membrane potentials and ionic concentrations in branching dendrites, spines, and axons," *Biol. Cybern.* **62:** 1–15 (1989).

Qian, N., and Sejnowski, T.J. "When is an inhibitory synapse effective?," *Proc. Natl. Acad. Sci. USA* **87:** 8145–8149 (1990).

Raastad, M., Storm, J.F., and Andersen, P. "Putative single quantum and single fibre excitatory postsynaptic currents show similar amplitude range and variability in rat hippocampal slices," *Eur. J. Neurosci.* **4:** 113–117 (1992).

Rall, W. "Membrane time constant of motoneurons," *Science* **126:** 454–455 (1957).

Rall, W. "Branching dendritic trees and motoneuron membrane resistivity," *Exp. Neurol.* **1:** 491–527 (1959).

Rall, W. "Theory of physiological properties of dendrites," *Ann. N. Y. Acad. Sci.* **96:** 1071–1092 (1962).

Rall, W. "Theoretical significance of dendritic tree for input-output relation." In: *Neural Theory and Modeling.* Reiss, R.F., editor, pp. 73–97. Stanford University Press: Stanford, California (1964).

Rall, W. "Distinguishing theoretical synaptic potentials computed for different soma-dendritic distributions of synaptic input," *J. Neurophysiol.* **30:** 1138–1168 (1967).

Rall, W. "Time constants and electrotonic length of membrane cylinders and neurons," *Biophys. J.* **9**: 1483–1508 (1969a).

Rall, W. "Distribution of potential in cylindrical coordinates and time constants for a membrane cylinder," *Biophys. J.* **9**: 1509–1541 (1969b).

Rall, W. "Cable properties of dendrites and effects of synaptic location." In: *Excitatory Synaptic Mechanisms*. Anderson, P., and Jansen, J.K.S., editors, pp. 175–187. Universitetsforlag: Oslo, Norway (1970).

Rall, W. "Dendritic spines, synaptic potency and neuronal plasticity." In: *Cellular Mechanisms Subserving Changes in Neuronal Activity*. Woody, C.D., Brown, K.A., Crow, T.J., and Knispel, J.D., editors, pp. 13–21. University of California Press: Los Angeles, California (1974).

Rall, W. "Core conductor theory and cable properties of neurons." In: *Handbook of Physiology, The Nervous System. Cellular Biology of Neurons*. Vol. I, pp. 39–97. American Physiological Society: Bethesda, Maryland (1977).

Rall, W. "Dendritic spines and synaptic potency." In: *Studies in Neurophysiology*. Porter, R., editor, pp. 203–209. Cambridge University Press: Cambridge, United Kingdom (1978).

Rall, W. "Cable theory for dendritic neurons." In: *Methods in Neuronal Modeling*. Koch, C., and Segev, I., editors, pp. 9–62. MIT Press: Cambridge, Massachusetts (1989).

Rall, W., Burke, R.E., Holmes, W.R., Jack, J.J.B., Redman, S.J., and Segev, I. "Matching dendritic neuron models to experimental data," *Physiol. Rev.* **72**: S159-S186 (1992).

Rall, W., Burke, R.E., Smith, T.G., Nelson, P.G., and Frank, K. "Dendritic location of synapses and possible mechanism for the monosynaptic EPSP in motoneurons," *J. Neurophysiol.* **30**: 1169–1193 (1967).

Rall, W., and Rinzel, J. "Branch input resistance and steady attenuation for input to one branch of a dendritic neuron model," *Biophys. J.* **13**: 648–688 (1973).

Rall, W., and Segev, I. "Space clamp problems when voltage clamping branched neurons with intracellular microelectrodes." In: *Voltage and Patch Clamping with Microelectrodes*. Smith, T.G., Lecar, H., Redman, S.J., and Gage, P.W., editors, pp. 191–215. American Physiological Society: Bethesda, Maryland (1985).

Rall, W., and Segev, I. "Functional possibilities for synapses on dendrites and dendritic spines." In: *Synaptic Function*. Edelman, G.M., Gall, W.E., and Cowan, W.M., editors, pp. 605–636. John Wiley: New York, New York (1987).

Rall, W., and Shepherd, G.M. "Theoretical reconstruction of field potentials and dendrodendritic synaptic interactions in olfactory bulb," *J. Neurophysiol.* **31**: 884–915 (1968).

Rall, W., Shepherd, G.M., Reese, T.S., and Brightman, M.W. "Dendrodendritic synaptic pathway for inhibition in the olfactory bulb," *Expl. Neurol.* **14**: 44–56 (1966).

Ramón y Cajal, S. *Histologie du système nerveux de l'homme et des vertébrés*. Azouly, L., transl. Malaine: Paris, France (1909).

Ramón y Cajal, S. "New ideas on the structure of the nervous system of man and vertebrates." Translated by Swanson, N. and Swanson, L.M. from *Les nouvelles idées sur la structure du système nerveux chez l'homme et chez les vertébrés*. MIT Press: Cambridge, Massachusetts (1991).

Randall, A.D., and Tsien, R.W. "Pharmacological dissection of multiple types of Ca^{2+} channel currents in fat cerebellar granule neurons," *J. Neurosci.* **15**: 2995–3012 (1995).

Rapp, M. *The computational role of excitable dendrites*. PhD Thesis. Hebrew University. Jerusalem: Israel (1997).

Rapp, M., Segev, I., and Yarom, Y. "Physiology, morphology and detailed passive models of cerebellar Purkinje cells," *J. Physiol.* **474**: 101–118 (1994).

Rapp, M., Yarom, Y., and Segev, I. "The impact of parallel fiber background activity on the cable properties of cerebellar Purkinje cells," *Neural Comput.* **4**: 518–533 (1992).

Rapp, M., Yarom, Y., and Segev, I. "Modeling back propagating action potential in weakly excitable dendrites of neocrotical pyramidal cells," *Proc. Natl. Acad. Sci. USA* **93**: 11985–11990 (1996).

Rasgado-Flores, H., and Blaustein, M.P. "ATP-dependent regulation of cytoplasmic free calcium in nerve terminals," *Am. J. Physiol. (Cell Physiol.)* **21:** C588-C594 (1987).

Rashevsky, N. *Mathematical Biophysics.* University of Chicago Press: Chicago, Illinois (1938).

Rausch, G., and Scheich, H. "Dendritic spine loss and enlargement during maturation of the speech control system in the mynah bird (*Gracula religiosa*)," *Neurosci. Lett.* **29:** 129–133 (1982).

Raymond, S.A. "Effects of nerve impulses on threshold of frog sciatic nerve fibres," *J. Phyiol.* **290:** 273–303 (1979).

Redman, S. "Quantal analysis of synaptic potentials in neurons of the central nervous system," *Physiol. Rev.* **70:** 165–198 (1990).

Redman, S., and Walmsley, B. "The time course of synaptic potentials evoked in cat spinal motoneurones at identified group Ia synapses," *J. Physiol.* **343:** 117–133 (1983).

Regehr, W.G., and Armstrong, C.M. "Where does it all begin?" *Current Biol.* **4:** 436–439 (1994).

Regehr, W.G., Kehoe, J.S., Ascher, P., and Armstrong, C. "Synaptically triggered action potentials in dendrites," *Neuron* **11:** 145–151 (1993).

Regehr, W.G., Konnerth, A., and Armstrong C.M. "Sodium action potentials in the dendrites of cerebellar Purkinje cells," *Proc. Natl. Acad. Sci. USA* **89:** 181–190 (1992).

Regehr, W.G., and Tank, D.W. "Calcium concentration dynamics produced by synaptic activation of CA1 hippocampal pyramidal cells," *J. Neurosci.* **12:** 4202–4223 (1992).

Regehr, W.G., Delaney, K.R., and Tank, D.W. "The role of presynaptic calcium in short-term enhancement at the hippocampal mossy fiber synapse," *J. Neurosci.* **14:** 523–537 (1994).

Reichardt, W. "Autocorrelation, a principle for evaluation of sensory information by the central nervous system." In: *Principles of Sensory Communication.* Rosenblith, W.A., editor, pp. 303–317. John Wiley: New York, New York (1961).

Reyes, A.D., Rubel, E.W., and Spain, W.J. "Membrane properties underlying the firing of neurons in the avian cochelar nucleus," *J. Neurosci.* **14:** 5352–5364 (1994).

Reza, F.M. "On passivity, reciprocity and linear operators." In: *Aspects of Network and System Theory.* Kalman, R.E., and Declaris, N., editors. Holt, Rinehart & Winston: New York, New York (1971).

Rhodes, P.A., and Gray, C.M. "Simulations of intrinsically bursting neocortical pyramidal neurons," *Neural Comput.* **6:** 1086–1110 (1994).

Richmond, B.J., and Optican, L.M. "The structure and interpretation of neuronal codes in the visual system." In: *Neural Networks for Perception.* Wechsler, H., editor, pp. 104–119. Academic Press: Boston, Massachusetts (1992).

Rieke, F., Warland, D., and Bialek, W. "Coding efficiency and information rates in sensory neurons," *Europhys. Lett.* **22:** 151–156 (1993).

Rieke, F., Warland, D., van Steveninck, R.R.D., and Bialek, W. *Spikes: Exploring the Neural Code.* MIT Press: Cambridge, Massachusetts (1996).

Rinzel, J. "Integration and propagation of neuroelectric signals." In: *Studies in Mathematical Biology*, Vol. 15. Levin, S.A., editor, pp. 1–66. Mathematical Association of America: Washington, DC (1978).

Rinzel, J. "A formal classification of bursting mechanisms in excitable systems." In: *Proc. Intl. Congress Mathematicians.* Gleason, A.M., editor, pp. 1578–1594. American Mathematical Society: Providence, Rhode Island (1987).

Rinzel, J., and Ermentrout, B.B. "Analysis of neural excitability and oscillations." In: *Methods in Neuronal Modeling.* Koch, C., and Segev, I., editors, second edition, pp. 251–292. MIT Press: Cambridge, Massachusetts (1998).

Rinzel, J., and Miller, R.N. "Numerical calculation of stable and unstable periodic solutions to the Hodgkin–Huxley equations," *Math. Biosci.* **49:** 27-59 (1980).

Rinzel, J., and Rall, W. "Transient response in a dendritic neuron model for current injected at one branch," *Biophys. J.* **14:** 759–790 (1974).

Ritchie, J.M. "On the relation between fibre diameter and conduction velocity in myelinated nerve fibres," *Proc. Roy. Soc. Lond. B* **217:** 29–35 (1982).

Ritchie, J.M. "Physiology of axons." In: *The Axon: Structure, Function and Pathophysiology.* Waxman, S.G., Kocsis, J.D., and Stys, P.K., editors, pp. 68–96. Oxford University Press: New York, New York (1995).

Ritchie, J.M., Rang, H.P., and Pellegrino, R. "Sodium and potassium channels in demyelinated and remyelinated mammalian nerve," *Nature* **294:** 257–259 (1981).

Roberts, A., and Bush, B.M.H., editors. *Neurones without Impulses: Their Significance for Vertebrate and Invertebrate Nervous Systems.* Cambridge University Press: Cambridge, United Kingdom (1981).

Robinson, H.P.C., and Koch, C., "Calcium, spines and memory: A specific proposal," *Artif. Intell. Memo 779*, MIT, Artificial Intelligence Laboratory, Cambridge, Massachusetts (1984).

Rockel, A.J., Hiorns, R.W., and Powell, T.P.S. "The basic uniformity in structure of the neocortex," *Brain* **103:** 221–244 (1980).

Rockland, K.S. "The organization of feedback connections from area V2 (18) to V1 (17)." In: *Cerebral Cortex*, Vol. 10. Peters, A., and Rockland, K.S., editors, pp. 261–299. Plenum Press: New York, New York (1994).

Rogan, M.T., Stäubli, U.V., and LeDoux, J.E. "Fear conditioning induces associative long-term potentiation in the amygdala," *Nature* **390:** 604-607 (1997).

Rogart, R.B., and Ritchie, J.M. "Physiological basis of conduction in myelinated nerve fibers." In: *Myelin.* Morell, P., editor, pp. 117–159. Plenum Press: New York, New York (1977).

Rose, R.M., and Hindmarsh, J.L. "A model of a thalamic neuron," *Proc. Roy. Soc. Lond. B* **225:** 161–193 (1985).

Rosenfalck, P. "Intra- and extracellular potential fields of active nerve and muscle fibers," *Acta Physiol. Scand. (Suppl.)* **321:** 1–168 (1969).

Ross, E.M. "Signal sorting and amplification through G protein-coupled receptors," *Neuron* **3:** 141–152 (1989).

Rothman, J.E., and Wieland, F.T. "Protein sorting by transport vesicles," *Science* **272:** 227–234 (1996).

Rudy, B. "Diversity and ubiquity of K channels," *Neurosci.* **25:** 729–749 (1988).

Rushton, W.A.H. "Initiation of the propagated disturbance," *Proc. Roy. Soc. Lond. B* **124:** 210 (1937).

Rushton, W.A.H. "A theory of the effects of fibre size in medullated nerve," *J. Physiol.* **115:** 101–122 (1951).

Sabah, N.H., and Leibovic, K.N. "Subthreshold oscillatory responses of the Hodgkin-Huxley cable model for the squid giant axon," *Biophys. J.* **9:** 1206–1222 (1969).

Sabatini, B.L., and Regehr, W.G. "Timing of neurotransmission at fast synapses in the mammalian brain," *Nature* **384:** 170–173 (1996).

Sah, P., Gibb, A.J., and Gage, P.W. "The sodium current underlying action potentials in guinea pig hippocampal CA1 neurons," *J. Gen. Physiol.* **91:** 373–398 (1988).

Saito, H. "Morphology of physiologically identified X-, Y- and W-type retina ganglion cells of the cat," *J. Comp. Neurol.* **221:** 279–288 (1983).

Saitoh, T., and Schwartz, J.H. "Phosphorylation-dependent subcellular translocation of a Ca^{2+}/calmodulin-dependent protein kinase produces an autonomous enzyme in Aplysia neurons," *J. Cell Biol.* **100:** 835–842 (1985).

Sakmann, B., and Neher, E., editors. *Single Channel Recording.* Plenum Press: New York, New York (1983).

Sala, F., and Hernández-Cruz, A. "Calcium diffusion modeling in a spherical neuron," *Biophys. J.* **57:** 313–324 (1990).

Saleh, B.E.A. *Photoelectron Statistics.* Springer Verlag: New York, New York (1978).

Saleh, B.E.A., and Teich, M.C. "Multiplication and refractoriness in the cat's retinal ganglion cell discharge at low light levels," *Biol. Cybern.* **52:** 101–107 (1985).

Salin, P.-A., and Bullier, J. "Corticocortical connections in the visual system: Structure and function," *Physiol. Rev.* **75:** 107–154 (1995).

Salpeter, M.M., editor. *The Vertebrate Neuromuscular Junction.* A.R. Liss: New York, New York (1987).

Sanderson, M.J. "Intercellular waves of communication," *New Physiol. Sci.* **11:** 262–269 (1996).

Sarpeshkar, R. "Analog versus digital: Extrapolating from electronics to neurobiology," *Neural Computation,* **10:** 1601–1638 (1998).

Saucier, D., and Cain, D.P. "Spatial learning without NMDA receptor-dependent long-term potentiation," *Nature* **378:** 186–189 (1995).

Saul, A.B., and Humphrey, A.L. "Spatial and temporal response properties of lagged and nonlagged cells in cat lateral geniculate nucleus," *J. Neurophysiol.* **64:** 206–224 (1990).

Schiller, P.H., Sandell, J.H., and Maunsell, J.H.R. "Functions of the ON and OFF channels of the visual system," *Nature* **322:** 824–825 (1986).

Schneggenburger, R., Zhou, Z., Konnerth, A., and Neher, E. "Fractional contribution of calcium to the cation current through glutamate receptor channels," *Neuron* **11:** 133–143 (1993).

Schuman, E.M. "Nitric oxide signalling, long-term potentiation and long-term depression." In: *Nitric Oxide in the Nervous System.* Vincent, S., editor, pp. 125–150. Academic Press: London, United Kingdom (1995).

Schuman, E.M. "Synapse specificity and long-term information storage," *Neuron* **18:** 339–342 (1997).

Schuman, E.M., and Madison, D.V. "Locally distributed synaptic potentiation in the hippocampus," *Science* **263:** 532–534 (1994a).

Schuman, E.M., and Madison, D.V. "Nitric oxide and synaptic function," *Ann. Rev. Neurosci.* **17:** 153–183 (1994b).

Schüz, A. "Comparison between the dimensions of dendritic spines in the cerebral cortex of newborn and adult guinea pigs," *J. Comp. Neurol.* **244:** 277–285 (1986).

Schüz, A., and Dortenmann, M. "Synaptic density on non-spiny dendrites in the cerebral cortex of the house mouse. A phosphotungstic acid study," *J. Hirnforschung* **28:** 633–639 (1987).

Schwark, H.D., and Jones, E.G. "The distribution of intrinsic cortical axons in area 3b of cat primary somatosensory cortex," *Exp. Brain Res.* **78:** 501–513 (1989).

Schwindt, P.C., Spain, W.J., Foehring, R.C., Stafstrom, C.E., Chubb, M.C., and Crill, W.E. "Multiple potassium conductances and their functions in neurons from cat sensorimotor cortex in vitro," *J. Neurophysiol.* **59:** 424–449 (1988).

Scott, A.C. "Effect of the series inductance of a nerve axon upon its conduction velocity," *Math. Biosci.* **11:** 277–290 (1971).

Scott, A.C. "The electrophysics of a nerve fiber," *Rev. Mod. Phys.* **47:** 487–533 (1975).

Scott, A.C., Chu, F.Y.F., and McLaughlin, D.W. "The soliton: A new concept in applied science," *Proc. IEEE* **61:** 1443–1483 (1973).

Scott, A.C., and Luzader, S.D. "Coupled solitary waves in neurophysics," *Physica Scripta* **20:** 395–401 (1979).

Segal, M. "Fast imaging of $[Ca^{2+}]_i$ reveals presence of voltage-gated calcium channels in dendritic spines of cultured hippocampal neurons," *J. Neurophysiol.* **74:** 484–488 (1995a).

Segal, M. "Dendritic spines for neuroprotection: A hypothesis," *Trends Neurosci.* **18:** 468–471 (1995b).

Segal, M., and Barker, J.L. "Rat hippocampal neurons in culture: Potassium conductances," *J. Neurophysiol.* **51:** 1409–1433 (1984).

Segev, I., Friedman, A., White, E.L., and Gutnick, M.J. "Electrical consequences of spine dimensions in a model of a cortical spiny stellate cell completely reconstructed from serial thin sections," *J. Comp. Neurosci.* **2:** 117–130 (1995).

Segev, I., and Parnas, I. "Synaptic integration mechanisms: theoretical and experimental investigation of temporal postsynaptic interactions between excitatory and inhibitory inputs," *Biophys. J.* **41:** 41–50 (1983).

Segev, I., and Rall, W. "Computational study of an excitable dendritic spine," *J. Neurophysiol.* **60:** 499–523 (1988).

Segev, I., Rapp, M., Manor, Y., and Yarom, Y. "Analog and digital processing in single nerve cells." In: *Single Neuron Computation.* McKenna, T., Davis, J., and Zornetzer, S., editors, pp. 173–198. Academic Press: Boston, Massachusetts (1992).

Segev, I., Rinzel, J. and Shepherd, G.M., editors. *The Theoretical Foundation of Dendritic Function: Selected Papers of Wilfrid Rall with Commentaries.* MIT Press: Cambridge, Massachusetts (1995).

Segundo, J.P., Moore, G.P., Stensaas, L.J., and Bullock, T.H. "Sensitivity of neurones in *Aplysia* to temporal patterns of arriving impulses," *J. Exp. Biol.* **40:** 643–667 (1963).

Sejnowski, T.J. "Storing covariance with nonlinearly interacting neurons," *J. Math. Biol.* **4:** 303–321 (1977a).

Sejnowski, T.J. "Statistical constraints on synaptic plasticity," *J. theor. Biol.* **69:** 385–389 (1977b).

Sejnowski, T.J., Koch, C., and Churchland, P.S. "Computational neuroscience," *Science* **241:** 1299–1306 (1988).

Selverston, A.I., and Moulins, M. *The Crustacean Stomatogastric System.* Springer Verlag: Berlin, Germany (1987).

Sereno, M.I., and Ulinski, P.S. 'Caudal topographic nucleus isthmi and the rostral nontopographic nucleus isthimi in the turtle, *Pseudemys scripta*," *J. Comp. Neurol.* **261:** 319–346 (1987).

Sestokas, A.K., and Lehmkuhle, S. "Visual response latency of X- and Y-cells in the dorsal lateral geniculate nucleus of the cat," *Vision Res.* **26:** 1041–1054 (1986).

Shadlen, M.N., and Newsome, W.T. "Noise, neural codes and cortical organization," *Current Opinion Neurobiol.* **4:** 569–579 (1994).

Shatz, C.J. "Impulse activity and the patterning of connections during CNS development," *Neuron* **5:** 745–756 (1990).

Shelton, D. "Membrane resistivity estimated for the Purkinje neuron by means of a passive computer model," *Neurosci.* **14:** 111–131 (1985).

Shepherd, G.M. "The neuron doctrine: A revision of functional concepts," *Yale J. Biol. Med.* **45:** 584–599 (1972).

Shepherd, G.M. "Microcircuits in the nervous system," *Sci. Am.* **238:** 93–103 (1978).

Shepherd, G.M. "Canonical neurons and their computational organization." In: *Single Neuron Computation.* McKenna, T., Davis, J., and Zornetzer, S.F., editors, pp. 27–60. Academic Press: Boston, Massachusetts (1992).

Shepherd, G.M. "The dendritic spine: A multifunctional unit," *J. Neurophysiol.* **75:** 2197–2210 (1996).

Shepherd, G.M., editor. *The Synaptic Organization of the Brain.* Fourth edition. Oxford University Press: New York, New York (1998).

Shepherd, G.M., and Brayton, R.K. "Logic operations are properties of computer simulated interactions between excitable dendritic spines," *Neurosci.* **21:** 151–166 (1987).

Shepherd, G.M., Brayton, R.K., Miller, J.F., Segev, I., Rinzel, J., and Rall, W. "Signal enhancement in distal cortical dendrites by means of interactions between active dendritic spines," *Proc. Natl. Acad. Sci. USA* **82:** 2192–2195 (1985).

Shepherd, G.M., and Erulkar, S.D. "Centenary of the synapse: From Sherrington to the molecular biology of the synapse and beyond," *Trends Neurosci.* **20:** 385–392 (1997).

Shepherd, G.M., and Koch, C. "Introduction to synaptic circuits." In: *The Synaptic Organization of the Brain.* Shepherd, G.M., editor, fourth edition, pp. 1–36. Oxford University Press: New York, New York (1998).

Shepherd, G.M., Woolf, T.B., and Carnevale, N.T. "Comparisons between active properties of distal dendritic branches and spines: Implications for neuronal computations," *J. Cognitive Neurosci.* **1:** 273–286 (1989).

Sherman, S.M., and Guillery, R.W. "Functional organization of thalamocortical relays," *J. Neurophysiol.* **76:** 1367–1395 (1996).

Sherman, S.M., and Koch, C. "The control of retinogeniculate transmission in the mammalian lateral geniculate nucleus," *Exp. Brain Res.* **63:** 1–20 (1986).

Sherman, S.M., and Koch, C. "Thalamus." In: *The Synaptic Organization of the Brain.* Shepherd, G.M., editor, fourth edition, pp. 289–328. Oxford University Press: New York, New York (1998).

Siegel, M., Marder, E., and Abbott, L.F. "Activity-dependent current distribution in model neurons," *Proc. Natl. Acad. Sci. USA* **91:** 11308–11312 (1994).

Sigworth, F.J. "An example of analysis." In: *Single Channel Recording.* Sakmann, B., and Neher, E., editors, pp. 301–321. Plenum Press: New York, New York (1983).

Silva, R.L., Amitai, Y., and Connors, B.W. "Intrinsic oscillations of neocortex generated by layer 5 pyramidal neurons," *Science* **251:** 432–435 (1991).

Silver, R., LeSauter, J., Tresco, P.A., and Lehman, M.N. "A diffusible coupling signal from the transplanted suprachiasmatic nucleus controlling circadian locomotor rhythms," *Nature* **382:** 810–813 (1996).

Simon, S., and Llinás, R. "Compartmentalization of submembrane calcium activity during calcium influx and its significance in transmitter release," *Biophys J.* **48:** 559–569 (1985).

Singer, W., and Gray, C.M. "Visual feature integration and the temporal correlation hypothesis," *Ann. Rev. Neurosci.* **18:** 555–586 (1995).

Skaugen, E. "Firing behaviour in stochastic nerve membrane models with different pore densities," *Acta Physiol. Scand.* **108:** 49–60 (1980a).

Skaugen, E. "Firing behaviour in nerve cell models with a two-state pore system," *Acta Physiol. Scand.* **109:** 377–392 (1980b).

Skaugen, E., and Walloe, L. "Firing behaviour in a stochastic nerve membrane model based upon the Hodgkin-Huxley equations," *Acta Physiol. Scand.* **107:** 343–363 (1979).

Skydsgaard, M., and Hounsgaard, J. "Spatial integration of local transmitter responses in motoneurones of the turtle spinal cord *in vitro*," *J. Physiol.* **479:** 233–246 (1994).

Smetters, D.K.S. *Electrotonic structure and synaptic integration in cortical neurons.* PhD Thesis. Massachusetts Institute of Technology: Cambridge, Massachusetts (1995).

Smetters, D.K.S., and Nelson, S.B. "Estimates of functional synaptic convergence in rat and cat visual cortical neurons," *Soc. Neurosci. Abst.* **19:** 628 (1993).

Smetters, D.K.S., and Zador, A.M. "Synaptic transmission: Noisy synapses and noisy neurons," *Current Biology* **6 :** 1217–1218 (1996).

Smith, D.O. "Axon conduction failure under *in vivo* conditions in crayfish," *J. Physiol.* **344:** 327–333 (1983).

Smith, C.E. "A heuristic approach to stochastic models of single neurons." In: *Single Neuron Computation.* McKenna, T., Davis, J., and Zornetzer, S., editors, pp. 561–588. Academic Press: Boston, Massachusetts (1992).

Smith, T.G., Lecar, H., Redman, S.J., and Gage, P.W., editors. *Voltage and Patch Clamping with Microelectrodes.* American Physiological Society: Bethesda, Maryland (1985).

Smith, J.S., MacDermott, A.B., and Weight, F.F. "Detection of intracellular Ca^{2+} transients in sympathetic neurones using arsenazo III," *Nature* **304:** 350–352 (1983).

Smith, D.R., and Smith, D.K. "A statistical analysis of the continual activity of single cortical neurones in the cat unanaesthetized isolated forebrain," *Biophys.J.* **5:** 47–74 (1965).

Smith, T.G., Wuerker, R.B., and Frank, K. "Membrane impedance changes during synaptic transmission in cat spinal motoneurons," *J. Neurophysiol.* **30:** 1072–1096 (1967).

Snowden, R.J., Treue, S., and Andersen, R.A. "The response of neurons in areas V1 and MT of the alert rhesus monkey to moving random dot patterns," *Exp. Brain Res.* **88:** 389–400 (1992).

Sobel, E.S., and Tank, D.W. "In vivo Ca^{2+} dynamics in a cricket auditory neuron: An example of chemical computation," *Science* **263:** 823–826 (1994).

Softky, W.R. "Sub-millisecond coincidence detection in active dendritic trees," *Neurosci.* **58:** 15–41 (1994).

Softky, W.R. "Simple codes versus efficient codes," *Curr. Opin. Neurobiol.* **5:** 239–247 (1995).

Softky, W.R., and Koch, C. "Cortical cells should fire regularly, but do not," *Neural Comput.* **4:** 643–646 (1992).

Softky, W.R., and Koch, C. "The highly irregular firing of cortical cells is inconsistent with temporal integration of random EPSP's," *J. Neurosci.* **13:** 334–350 (1993).

Sokoloff, L. "Circulation and energy metabolism of the brain." In: *Basic Neurochemistry: Molecular, Cellular and Medical Aspects.* Siegel, G.J., Agranoff, B.W., Albers, R.W., and Molinoff, P.B., editors, pp. 565–590. Raven Press: New York, New York (1989).

Sommer, B., and Seeburg, P.H. "Glutamate receptor channels: novel properties and new clones," *Trends Pharmacol. Sci.* **13:** 291–296 (1992).

Soriano, E., and Frotscher, M. "A GABAergic axo-axonic cell in the fascia dentata controls the main excitatory hippocampal pathway," *Brain Res.* **503:** 170–174 (1989).

Sorra, K.E., and Harris, K.M. "Occurrence and three-dimensional structure of multiple synapses between individual radiatum axons and their target pyramidal cells in hippocampal area CA1," *J. Neurosci.* **13:** 3736–3748 (1993).

Spain, W.J., Schwindt, P.C., and Crill, W.E. "Anomalous rectification in neurons from cat sensorimotor cortex *in vitro*," *J. Neurophysiol.* **57:** 1555–1576 (1987).

Spain, W., Schwindt, P., and Crill, W. "Post-inhibitory excitation and inhibition in layer V pyramidal neurons from cat sensorimotor cortex," *J. Physiol.* **434:** 609–626 (1990).

Spain, W., Schwindt, P., and Crill, W.W. "Two transient potassium currents in layer V pyramidal neurons from cat sensorimotor cortex," *J. Physiol.* **434:** 591–607 (1991).

Spencer, W.A., and Kandel, E.R. "Electrophysiology of hippocampal neurons. IV. Fast prepotentials," *J. Neurophysiol.* **24:** 272–285 (1961).

Spitzer, N.C. "Development of voltage-dependent and ligand-gated channels in excitable membranes," *Progress Brain Res.* **102:** 169–179 (1994).

Spruston, N., and Johnston, D. "Perforated patch-clamp analysis of the passive membrane properties of three classes of hippocampal neurons," *J. Neurophysiol.* **67:** 508–529 (1992).

Spruston, N., Jaffe, D.B., and Johnston, D. "Dendritic attenuation of synaptic potential and currents: The role of passive membrane properties," *Trends Neurosci.* **17:** 161–166 (1994).

Spruston, N., Jaffe, D.B., Williams, S.H., and Johnston, D. "Voltage- and space-clamp errors associated with the measurement of electrotonically remote synaptic events," *J. Neurophysiol.* **70:** 781–802 (1993).

Spruston, N., Schiller, Y., Stuart, G., and Sakmann, B. "Activity-dependent action potential invasion and calcium influx into hippocampal CA1 dendrites," *Science* **268:** 297–300 (1995).

Squire, L.R. *Memory and Brain.* Oxford University Press: New York, New York (1987).

Srinivasan, M.V., and Bernard, G.D. "A proposed mechanism for multiplication of neural signals," *Biol. Cybern.* **21:** 227–236 (1976).

Stafstrom, C.E., Schwindt, P.C., Chubb, M.C., and Crill, W.E. "Properties of persistent sodium conductance and calcium conductance of layer V neurons from cat sensorimotor cortex *in vitro*," *J. Neurophysiol.* **53:** 153–170 (1985).

Staley, K.J., Soldo, B.L., and Proctor, W.R. "Ionic mechanisms of neuronal excitation by inhibitory $GABA_A$ receptors," *Science* **269:** 977–981 (1995).

Stämpfli, R., and Hille, B. "Electrophysiology of the peripheral myelinated nerve." In: *Frog Neurobiology.* Llinás, R., and Precht, W., editors, pp. 3–32. Springer Verlag: Berlin, Germany (1976).

Stanford, L.R., and Sherman, S.M. "Structure/function relationships of retinal ganglion cells in the cat," *Brain Res.* **297:** 381–386 (1984).

Stein, R.B. "A theoretical analysis of neuronal variability," *Biophys. J.* **5:** 173–194 (1965).

Stein, R.B. "The frequency of nerve action potentials generated by applied currents," *Proc. Roy. Soc. Lond. B* **167:** 64–86 (1967a).

Stein, R.B. "Some models of neuronal variability," *Biophys. J.* **7**: 37–68 (1967b).

Steinbuch, K. "Die Lernmatrix," *Kybernetik* **1**: 36–45 (1961).

Stemmler, M. "A single spike suffices: The simplest form of stochastic resonance in model neurons," *Network: Comput. Neural Systems* **7**: 1–29 (1996).

Stent, G. "A physiological mechanism for Hebb's postulate of learning," *Proc. Natl. Acad. Sci. USA* **70**: 997–1001 (1973).

Steriade, M., and McCarley, R.W. *Brainstem Control of Wakefulness and Sleep* Plenum Press: New York, New York (1990).

Steriade, M., McCormick, D.A., and Sejnowski, T.J. "Thalamocortical oscillations in the sleeping and aroused brain," *Science* **262**: 679–685 (1993).

Sterling, P. "Retina." In: *The Synaptic Organization of the Brain*. Shepherd, G.M., editor, fourth edition, pp. 205–254. Oxford University Press: New York, New York (1998).

Stern, P., Edwards, F.A., and Sakmann, B. "Fast and slow components of unitary EPSCs on stellate cells elicited by focal stimulation in slices of rat visual cortex," *J. Physiol.* **449**: 247–278 (1992).

Stevens, C.F. "Quantal release of neurotransmitter and long-term potentiation," *Neuron* **10**: (Suppl.) 55–64 (1993).

Stevens, C.F. "What form should a cortical theory take." In: *Large-Scale Neuronal Theories of the Brain*. Koch, C., and Davis, J.L., editors, pp. 239–256. MIT Press: Cambridge, Massachusetts (1994).

Stevens, C.F., and Tsujimoto, T. "Estimates for the pool size of releasable quanta at a single central synapse and for the time required to refill the pool," *Proc. Natl. Acad. Sci. USA* **92**: 846–849 (1995).

Stevens, C.F., and Wang, Y. "Changes in reliability of synaptic function as a mechanism for plasticity," *Nature* **371**: 704–707 (1994).

Stevens, C.F., and Wang, Y. "Facilitation and depression at single central synapses," *Neuron* **14**: 795–802 (1995).

Stockbridge, N., and Moore, J.W. "Dynamics of intracellular calcium and its possible relationship to phasic transmitter release and facilitation at the frog neuromuscular junction," *J. Neurosci.* **4**: 803–811 (1984).

Stoer, J., and Bulirsch, R. *Introduction to Numerical Analysis*. Springer Verlag: New York, New York (1980).

Stopfer, M., Bhagavan, S., Smith, B.H., and Laurent, G. "Impaired odor discrimination on desynchronization of odor-encoding neural assemblies," *Nature* **390**: 70–74 (1997).

Storm, J.F. "Temporal integration by a slowly inactivating K^+ current in hippocampal neurons," *Nature* **336**: 379–381 (1988).

Strassberg, A.F., and DeFelice, L.J. "Limitations of the Hodgkin-Huxley formalism: Effects of single channel kinetics on transmembrane voltage dynamics," *Neural Comput.* **5**: 843–855 (1993).

Stratford, K., Mason, A., Larkman, A., Major, G., and Jack, J. "The modeling of pyramidal neurons in the visual cortex." In: *The Computing Neuron*. Durbin, R., Miall, C., and Mitchison, G., editors, pp. 296–321. Addison-Wesley: London, United Kingdom (1989).

Stratford, K.J., Tarczy-Hornoch, K., Martin, K.A.C., Bannister, N.J., and Jack, J.J.B. "Excitatory synaptic inputs to spiny stellate cells in cat visual cortex," *Nature* **382**: 258–261 (1996).

Strausfeld, N.J. *Atlas of an Insect Brain*. Springer Verlag: Berlin, Germany (1976).

Strehler, B.L., and Lestienne, R. "Evidence on precise time-coded symbols and memory of patterns in monkey cortical neuronal spike trains," *Proc. Natl. Acad. Sci. USA* **83**: 9812–9816 (1986).

Strogatz, S.H. *Nonlinear Dynamics and Chaos*. Addison-Wesley: Reading, Massachusetts (1994).

Struijk, J.J. "The extracellular potential of a myelinated nerve fiber in an unbounded medium and in nerve cuff models," *Biophys. J.* **72**: 2457–2469 (1997).

Stryer, L. *Biochemistry*. Fourth edition. W.H. Freeman: San Francisco, California (1995).

Stuart, G.J., and Häusser, M. "Initiation and spread of sodium action potentials in cerebellar Purkinje cells," *Neuron* **13:** 703–712 (1994).

Stuart, G., and Spruston, N. "Probing dendritic function with patch pipettes," *Current Opinion Neurobiol.* **5:** 389–394 (1995).

Stuart, G.J., and Sakmann, B. "Active propagation of somatic action potentials into neocortical pyramidal cell dendrites," *Nature* **367:** 69–72 (1994).

Stuart, G., Spruston, N., Sakmann, B., and Häusser, M. "Action potential initiation and backpropagation in neurons of the mammalian CNS," *Trends Neurosci.* **20:** 125–131 (1997).

Suarez, H., and Koch, C. "Linking linear threshold units with quadratic models of motion perception," *Neural Comput.* **1:** 318–320 (1989).

Suarez, H., Koch, C., and Douglas, R.J. "Modeling direction selectivity of simple cells in striate visual cortex using the canonical microcircuit," *J. Neurosci.* **15:** 6700–6719 (1995).

Sudhof, T.C. "The synaptic vesicle cycle: A cascade of protein-protein interactions," *Nature* **375:** 645–653 (1995).

Sutor, B., and Zieglgänsberger, W. "A low-voltage activated, transient calcium current is responsible for the time-dependent depolarizing inward rectification of rat neocortical neurons in vitro," *Eur. J. Physiol.* **410:** 102–111 (1987).

Svoboda, K., Tank, D., and Denk, W. "Direct measurement of coupling between dendritic spines and shafts," *Science* **272:** 716–719 (1996).

Svoboda, K., Denk, W., Kleinfeld, D., and Tank, D.W. "In-vivo dendritic calcium dynamics in neocortical pyramidal neurons," *Nature* **385:** 161–165 (1997).

Swindale, N.V. "Dendritic spines only connect," *Trends Neurosci.* **4:** 240–241 (1981).

Syková, E. "The extracellular space in the CNS: Its regulation, volume and geometry in normal and pathological function," *Neuroscientist* **3:** 28–41 (1997).

Tamás, G., Buhl, E.H., and Somogyi, P. "Massive autaptic self-innervation of GABAergic neurons in cat visual cortex," *J. Neurosci.* **17:** 6352–6364 (1997).

Tank, D.W, Sugimori, M., Connor, J., and Llinás, R. "Spatially resolved calcium dynamics of mammalian Purkinje cells in cerebellar slice," *Science* **242:** 773–777 (1988).

Tanzi, E. "I fatti e la induzione nell'odierna istologia del sistema nervoso," *Riv. Sper. Freniatr. Med. Leg. Alienzioni Ment Soc. Ital. Psichiatria* **19:** 419–472 (1893).

Tasaki, I. *Nervous Transmission.* Charles C. Thomas: Springfield, Illinois (1953).

Tauc, L., and Hughes, G.M. "Modes of initiation and propagation of spikes in the branching axons of molluscan central neurons," *J. Gen. Physiol.* **46:** 533–549 (1963).

Teich, M.C. "Fractal neuronal firing patterns." In: *Single Neuron Computation.* McKenna, T., Davis, J., and Zornetzer, S.F., editors, pp. 289–625. Academic Press: Boston, Massachusetts (1992).

Teich, M.C., Heneghan, C., Lowen, S.B., Ozaki, T., and Kaplan, E. "Fractal character of the neural spike train in the visual system of the cat," *J. Opt. Soc. Am. A* **14:** 529–546 (1997).

Teich, M.C., Turcott, R.G., and Siegel, R.M. "Temporal correlation in cat striate-cortex neural spike trains," *IEEE Eng. Medicine & Biol.* **15:** 79–87 (1996).

Terashima, T., Inoue, K., Inoue, Y., Yokoyama, M., and Mikoshiba, K. "Observations on the cerebellum of normal-reeler mutant mouse chimera," *J. Comp. Neurol.* **252:** 264–278 (1986).

Theunissen, F., and Miller, J.P. "Representation of sensory information in the cricket cercal sensory system. II: Information theoretic calculations of system accuracy and optimal tuning curve widths of four primary interneurons," *J. Neurophysiol.* **66:** 1690–1703 (1991).

Thomson, A.M., Deuchars, J., and West, D.C. "Single axon excitatory postsynaptic potentials in neocortical interneurons exhibit pronounced paired pulse facilitation," *Neurosci.* **54:** 347–360 (1993a).

Thomson, A.M., Deuchars, J., and West, D.C. "Large, deep layer pyramid-pyramid single axon EPSPs in slices of rat motor cortex display paired pulse and frequency-dependent depression, mediated presynaptically and self-facilitation mediated postsynaptically," *J. Neurophysiol.* **70:** 2354–2369 (1993b).

Thomson, A.M., Girdlestone, D., and West, D. "Voltage-dependent currents prolong single-axon postsynaptic potentials in layer III pyramidal neurons in rat neocortical slices," *J. Neurophysiol.* **60:** 1896–1907 (1988).

Tolhurst, D., Movshon, J.A., and Dean, A. "The statistical reliability of signals in single neurons in cat and monkey visual cortex," *Vision Res.* **23:** 775–785 (1983).

Tonegawa, S. "Mammalian learning and memory studied by gene targeting," *Ann. N.Y. Acad. Sci.* **758:** 213–217 (1995).

Torre, V., and Owen, W.G. "High-pass filtering of small signals by the rod network in the retina of the toad, *Bufo Marinus,*" *Biophys. J.* **41:** 305–324 (1983).

Torre, V., Owen, W.G., and Sandini, G. "The dynamics of electrically interacting cells," *IEEE Trans. Systems, Man & Cybern.* **13:** 757–765 (1983).

Torre, V., and Poggio, T. "A synaptic mechanism possibly underlying directional selectivity to motion," *Proc. Roy. Soc. Lond. B* **202:** 409–416 (1978).

Toyoshima, C., and Unwin, N. "Ion channel of acetylcholine receptor reconstructed from images of postsynaptic membranes," *Nature* **336:** 247–250 (1988).

Traub, R. D. "Neocortical pyramidal cells: A model with dendritic calcium conductances reproduces repetitive firing and epileptic behavior," *Brain Res.* **173:** 243–257 (1979).

Traub, R.D. "Simulation of intrinsic bursting in CA3 hippocampal neurons," *Neurosci.* **7:** 1233–1242 (1982).

Traub, R.D., and Llinás, R. "Hippocampal pyramidal cells: Significance of dendritic ionic conductances for neuronal function and epileptogenesis," *J. Neurophysiol.* **42:** 476–496 (1979).

Traub, R.D., and Miles, R. *Neuronal Networks of the Hippocampus.* Cambridge University Press: Cambridge, United Kingdom (1991).

Traub, R.D., Whittington, M.A., Stanford, I.M., and Jefferys, J.G.R. "A mechanism for generation of long-range synchronous fast oscillations in the cortex," *Nature* **383:** 621–624 (1996).

Traub, R. D., Wong, R., Miles, R., and Michelson, H. "A model of a CA3 hippocampal pyramidal neuron incorporating voltage-clamp data on intrinsic conductances," *J. Neurophysiol.* **66:** 635–650 (1991).

Troy, W.C. "The bifurcation of periodic solutions in the Hodgkins–Huxley equations," *Quart. Appl. Math.* **36:** 73-83 (1978).

Troyer, T.W., and Miller, K.D. "Physiological gain leads to high ISI variability in a simple model of a cortical regular spiking cell," *Neural Comput.* **9:** 971–983 (1997).

Trussell, L.O., Zhang, S., and Raman, I.M. "Desensitization of AMPA receptors upon multiquantal neurotranmsitter release," *Neuron* **10:** 1185–1196 (1993).

Tseng, G.F., and Haberly, L.B. "Characterization of synaptically mediated fast and slow inhibitory processes in piriform cortex in an *in vitro* slice preparation," *J. Neurophysiol.* **59:** 1352–1376 (1988).

Tseng, G.F., and Prince, D.A. "Heterogeneity of rat corticospinal neurons," *J. Comp. Neurol.* **335:** 92–108 (1993).

Tsien, R.Y. "Fluoresence measurement and photochemical manipulation of cytosolic free calcium," *Trends Neurosci.* **11:** 419–424 (1988).

Tsien, R.Y. "Fluorescent probes of cell signaling," *Ann. Rev. Neurosci.* **12:** 227–253 (1989).

Tsien, R.W., Lipscombe, D., Madison, D.V., Bley, K.R., and Fox, A.P. "Multiple types of neuronal calcium channels and their selective modulation," *Trends Neurosci.* **11:** 431–438 (1988).

Tsodyks, M., and Markram, H. "The neural code between neocortical pyramidal neurons depends on neurotransmitter release probability," *Proc. Natl. Acad. Sci. USA* **94:** 719–723 (1997).

Tsodyks, M., and Sejnowski, T.J. "Rapid switching in balanced cortical network models," *Network Comput. Neural Sytems* **6:** 111–124 (1995).

Tsubokawa, H., and Ross, W.N. "IPSPs modulate spike backpropagation and associated $[Ca^{2+}]_i$ changes in the dendrites of hippocampal CA1 pyramidal neurons," *J. Neurophysiol.* **76:** 2896–2906 (1996).

Tsumoto, T. "Long-term potentiation and long-term depression in the necortex," *Prog. Neurobiol.* **39:** 209–328 (1992).

Tuckwell, H.C. "On stochastic models of the activity of single neurons," *J. Theor. Biol.* **65:** 783–785 (1977).

Tuckwell, H.C. *Introduction of Theoretical Neurobiology.* Vol. 1: *Linear Cable Theory and Dendritic Structure.* Cambridge University Press: Cambridge, United Kingdom (1988a).

Tuckwell, H.C. *Introduction of Theoretical Neurobiology.* Vol. 2: *Nonlinear and Stochastic Theories.* Cambridge University Press: Cambridge, United Kingdom (1988b).

Tuckwell, H.C. *Stochastic Processes in the Neurosciences.* SIAM: Philadelphia, Pennsylvania (1988c).

Tuckwell, H.C., and Wan, F.Y.M. "The response of a nerve cylinder to spatially distributed white noise inputs," *J. Theor. Biol.* **87:** 275–295 (1980).

Turcott, R.G., Barker, P.D.R., and Teich, M.C. "Long-duration correlation in the sequence of action potentials in an insect visual interneuron," *J. Stat. Comput. Simul.* **52:** 253–271 (1995).

Turing, A.M. "The chemical basis of morphogenesis," *Phil. Trans. Roy. Soc. Lond. B* **237:** 37–57 (1952).

Turner, D.A. "Conductance transients onto dendritic spines in a segmental cable model of hippocampal neurons," *Biophys. J.* **46:** 85–96 (1984).

Turner, D.A., Li, X.-G., Pyapali, K., Ylinen, A., and Buzsaki, G. "Morphometric and electrical properties of reconstructed hippocampal CA3 neurons recorded in vivo," *J. Comp. Neurol.* **356:** 580–594 (1995).

Turner, R.W., Maler, L., Deerinck, T., Levinson, S.R., and Ellisman, M.H. "TTX-sensitive dendritic sodium-channels underlie oscillatory discharge in a vertebrate sensory neuron," *J. Neurosci.* **14:** 6453–6471 (1994).

Turner, R.W., Meyers, D.E.R., and Barker, J.L. "Fast pre-potentials generation in rat hippocampal CA1 pyramidal neurons," *Neurosci.* **53:** 949–959 (1993).

Turner, R.W., Meyers, D.E.R., Richardson, T.L., and Barker, J.L. "The site for initiation of action potential discharge over the somatodendritic axis of rat hippocampal CA1 pyramidal neurons," *J. Neurosci.* **11:** 2270–2280 (1991).

Turrigiano, G., Abbott, L.F., and Marder, E. "Activity-dependent changes in the intrinsic properties of cultured neurons," *Science* **264:** 974–977 (1994).

Usher, M., Stemmler, M., Koch, C. and Olami, Z. "Network amplification of local fluctuations causes high spike rate variability, fractal firing patterns and oscillatory local field potentials," *Neural Comput.* **6:** 795–836 (1994).

van der Loos, H., and Glaser, E.M. "Autapses in neocortex cerebri: Synapses between a pyramidal cells axon and its own dendrites," *Brain Res.* **48:** 355–360 (1972).

van der Pol, B. "On relaxation oscillations," *Phil. Mag.* **2:** 978–992 (1926).

van der Pol, B., and van der Mark, J. "The heartbeat considered as a relaxation oscillation and an electrical model of the heart," *Phil. Mag. Suppl.* **6:** 763–775 (1928).

Vaney, D.I., Collin, S.P., and Young, H.M. "Dendritic relationships between cholinergic amacrine cells and direction-selective retinal ganglion cells." In: *Neurobiology of the Inner Retina.* Weiler, R., and Osborne, N.N., editors, pp. 157–168. Springer Verlag: Berlin, Germany (1989).

Van Harrefeld, A., and Fifková, E. "Swelling of dendritic spines in the fascia dentata after stimulation of the perforant fibers as a mechanism of post- tetanic potentiation," *Exp. Neurol.* **49:** 736–749 (1975).

van Hateren, J.H., and Laughlin, S.B. "Membrane parameters, signal transmission and the design of a graded potential neuron," *J. Comp. Physiol. A* **166:** 437–448 (1990).

van Opstal, A., and Hepp, K. "A novel interpretation for the collicular role in saccade generation," *Biol. Cybern.* **73:** 431–445 (1995).

van Steveninck, R.R.R., Lewen, G.D., Strong, S.R., Koberly, R., and Bialek, W. "Reproducibility and variability in neural spike trains," *Science* **275:** 1805–1808 (1997).

van Vreeswijk, C., and Abbott, L.F. "Self-sustained firing in populations of integrate-and-fire neurons," *SIAM J. Appl. Math.* **53**: 253–264 (1993).

van Vreeswijk, C., and Sompolinsky, H. "Chaos in neuronal networks with balanced excitatory and inhibitory activity," *Science* **274**: 1724–1726 (1996).

Verveen, A.A., and Derksen, H.E. "Fluctuation phenomena in nerve membrane," *Proc. IEEE* **56**: 906–916 (1968).

Vogels, R., and Orban, G. A. "How well do response changes of striate neurons signal differences in orientation—A study in the discriminating monkey," *J. Neurosci.* **10**: 3543–3558 (1990).

Vogels, R., Spileers, W., and Orban, G.A. "The response variability of striate cortical neurons in the behaving monkey," *Exp. Brain Res.* **77**: 432–436 (1989).

Volper, D.J., and Hampson, S.E. "Learning using specific instances," *Biol. Cybern.* **57**: 57–71 (1987).

von der Malsburg, C., "The correlation theory of brain function." *Internal Report 81-2*. Max-Planck-Institute for Biophysical Chemistry: Göttingen, Germany (1981).

von der Malsburg, C., and Schneider, W. "A neural cocktail-party processor," *Biol. Cybern.* **54**: 29–40 (1986).

von Neumann, J. *The Computer and the Brain*. Yale University Press: New Haven, Connetticut (1956).

Vu, E.T., and Krasne, F.B. "Evidence for a computational distinction between proximal and distal neuronal inhibition," *Science* **255**: 1710–1712 (1992).

Vu, E.T., Lee, S.C., and Krasne, F.B. "The mechanism of tonic inhibition of crayfish escape behavior: Distal inhibition and its functional significance," *J. Neurosci.* **13**: 4379–4393 (1993).

Wagner, J., and Keizer, J. "Effects of rapid buffers on Ca^{2+} diffusion and Ca^{2+} oscillations," *Biophys. J.* **67**: 447–456 (1994).

Walmsley, B., Edwards, F.R., and Tracey, D.J. "The probabilistic nature of synaptic transmission at a mammalian excitatory central synapse," *J. Neurosci.* **7**: 1027–1048 (1987).

Wandell, B. A. *Foundations of Vision*. Sinauer: Sunderland, Massachusetts (1995).

Wang, X.-J. "Multiple dynamical modes of thalamic relay neurons: Rhythmic bursting and intermittent phase-locking," *Neurosci.* **59**: 21–31 (1994).

Wang, Z., and McCormick, D.A. "Control of firing mode of corticotectal and corticopontine layer V burst-generating neurons by norepinephrine, acetylcholine and 1S,3R-ACPD," *J. Neurosci.* **13**: 2199–2216 (1993).

Wang, X.-J., and Rinzel, J. "Oscillatory and bursting properties of neurons." In: *The Handbook of Brain Theory and Neural Networks*. Arbib, M., editor, pp. 686–691. MIT Press: Cambridge, Massachusetts (1996).

Wang, X.-J., Rinzel, J., and Rogawski, M.A. "A model of the T-type calcium current and the low-threshold spike in thalamic neurons," *J. Neurophysiol.* **66**: 839–850 (1991).

Wässle, H., Illing, R.-B., and Peichl, L. "Morphologische Klassen und zentrale Projektion von Ganglienzellen in der Retina der Katze," *Verh. Dtsch. Zool. Ges.* **1979**: 180–193 (1979).

Watanabe, A., and Grundfest, H. "Impulse propagation at the septal and commissural junctions of crayfish lateral giant axons," *J. Gen. Physiol.* **45**: 267–308 (1961).

Watanabe, S., and Murakami, M. "Synaptic mechanisms of directional selectivity in ganglion cells of frog retina as revealed by intracellular recordings," *Jap. J. Physiol.* **34**: 497–511 (1984).

Waxman, S.G., Kocsis, J.D., and Stys, P.K., editors. *The Axon: Structure, Function and Pathophysiology*. Oxford University Press: New York, New York (1995).

Waxman, S.G., and Ritchie, J.M. "Organization of ion channels in the myelinated nerve fiber," *Science* **228**: 1502–1507 (1985).

Wehmeier, U., Dong, D., Koch, C., and Van Essen, D. "Modeling the mammalian visual system." In: *Methods in Neuronal Modeling*. Koch, C., and Segev, I., editors, pp. 335–360. MIT Press: Cambridge, Massachusetts (1989).

Wehr, M., and Laurent, G. "Odour encoding by temporal sequences of firing in oscillating neural assemblies," *Nature* **384**: 162–166 (1996).

Weidmann, S. *Electrophysiologie der Herzmuskelfaser*. Verlag Hans Huber: Bern, Switzerland (1956).

Weimann, J.M., Skiebe, P., Heinzel, H.-G., Soto, C., Kopell, N., Jorge-Rivera, J.C., and Marder, E. "Modulation of oscillator interactions in the crab stomatogastric ganglion by crustacean cardioactive peptide," *J. Neurosci.* **17:** 1748–1760 (1997).

Weiss, T.F. *Cellular Biophysics.* Vol. 2: *Electrical Properties.* MIT Press: Cambridge, Massachusetts (1996).

Werner, G., and Mountcastle, V.B. "The variability of central neural activity in a sensory system and its implications for the central reflection of sensory events," *J. Neurophysiol.* **26:** 958–977 (1963).

Wessel, R., Koch, C., and Gabbiani, F. "Coding of time-varying electric field amplitude modulations in a wave-type electric fish," *J. Neurophysiol.* **75:** 2280–2293 (1996).

Westbrook, G.L. "Glutamate receptor update," *Curr. Opinion Neurobiol.* **4:** 337–346 (1994).

Westenbroek, R.E., Ahlijanian, M.K., and Catterall, W.A. "Clustering of L-type Ca^{2+} channels at the base of major dendrites in hippocampal pyramidal neurones," *Nature* **347:** 281–284 (1990).

Westenbroek, R.E., Hell, J.W., Warner, C., Dubel, S., Snutch, T.P., and Catterall, W.A. "Biochemical properties and subcellular distribution of an N-type calcium channel $\alpha 1$ subunit," *Neuron* **9:** 1099–1115 (1992).

Westenbroek, R.E., Merrick, D.K., and Catterall, W.A. "Differential subcellular localization of the R_I and R_{II} Na^+ channel subtypes in central neurons," *Neuron* **3:** 695–704 (1989).

White, E.L. *Cortical Circuits.* Birkhäuser: Boston, Massachusetts (1989).

White, E.L., and Rock, M.P. "Three-dimensional aspects and synaptic relationships of a Golgi-impregnated spiny stellate cell reconstructed from serial thin sections," *J. Neurocytol.* **9:** 615–636 (1980).

Wickens, J. "Electrically coupled but chemically isolated synapses: Dendritic spines and calcium in a rule for synaptic modification," *Prog. Neurobiol.* **31:** 507–528 (1988).

Wigstrom, H., Gustafsson, B., Huang, Y.Y., and Abraham, W.C. "Hippocampal long-term potentiation is induced by pairing single afferent volleys with intracellularly injected depolarizing current pulses," *Acta Physiol. Scand.* **126:** 317–319 (1986).

Wilbur, W.J., and Rinzel, J. "A theoretical basis for large coefficients of variation and bimodality in neuronal interspike interval distributions," *J. Theor. Biol.* **105:** 345–368 (1983).

Williams, J.H., Errington, M.L., Lynch, M.A., and Bliss, T.V.P. "Arachidonic acid induces a long-term activity-dependent enhancement of synaptic transmission in the hippocampus," *Nature* **341:** 739–742 (1989).

Williams, S., and Johnston, D. "Long-term potentiation of hippocampal mossy fiber synapses is blocked by postsynaptic injection of calcium chelators," *Neuron* **3:** 583–588 (1989).

Williams, S., and Johnston, D. "Kinetic properties of two anatomically distinct excitatory synapses in hippocampal CA3 neurons," *J. Neurophysiol.* **66:** 1010–1020 (1991).

Wilson, C.J. "Passive cable properties of dendritic spines and spiny neurons," *J. Neurosci.* **4:** 281–297 (1984).

Wilson, C.J. "Cellular mechanisms controlling the strength of synapses," *J. Electron Microscopic Tech.* **10:** 293–313 (1988).

Wilson, C.J. "Basal Ganglia." In: *The Synaptic Organization of the Brain.* Shepherd, G.M., editor, fourth edition, pp. 329–376. Oxford University Press: New York, New York (1998).

Wilson, H.R., and Cowan, J.D. "Excitatory and inhibitory interactions in localized populations of model neurons," *Biophys. J.* **12:** 1–24 (1972).

Wilson, J.R., Friedlander, M.J., and Sherman, S.M. "Ultrastructural morphology of identified X- and Y-cells in the cat's lateral geniculate nucleus," *Proc. Roy. Soc. Lond. B* **221:** 411–436 (1984).

Wilson, C.J., Groves, P.M., Kitai, S.T., and Linder, J.C. "Three-dimensional structure of dendritic spines in the rat neostriatum," *J. Neurosci.* **3:** 383–398 (1983).

Wilson, C.J., Mastronarde, D.N., McEwen, B., and Frank, J. "Measurement of neuronal surface area using high-voltage electron microscope tomography," *Neuroimage* **1:** 11–22 (1992).

Wise, D.L., and Hougton, G. "Diffusion coefficients of neon, krypton, xenon, carbon monoxide and nitric oxide in water at 10–60° C," *Chem. Eng. Sci.* **23:** 1211–1216 (1968).

Wong, P.W., and Gray, R.M. "Sigma-Delta modulation with i.i.d. Gaussian inputs," *IEEE Trans. Inf. Theory* **36**: 784–798 (1990).

Woodbury, J.W., and Patton, H.D. "Electrical activity of single spinal cord element," *Cold Spring Harbor Symp. Quant. Biol.* **17**: 185–188 (1952).

Wood-Jones, F., and Porteus, S.D. *The Matrix of the Mind*. Mercantile: Honolulu, Hawaii (1928).

Woody, C.D., Gruen, E., and McCarley, K. "Intradendritic recordings from neurons of the motor cortex of cats," *J. Neurophysiol.* **51**: 925–938 (1984).

Woolf, T.B., and Greer, C.A. "Local communication within dendritic spines: Models of second messenger diffusion in granule cell spines of the mammalian olfactory bulb," *Synapse* **17**: 247–267 (1994).

Woolf, T.B., Shepherd, G.M., and Greer, C.A. "Serial reconstructions of granule cell spines in the mammalian olfactory bulb," *Synapse* **7**: 181–192 (1991a).

Woolf, T.B., Shepherd, G.M., and Greer, C.A. "Local information processing in dendritic trees: Subsets of spines in granule cells of the mammalian olfactory bulb," *J. Neurosci.* **11**: 1837–1854 (1991b).

Woolley, C.S., Gould, E., Frankfurt, M., and McEwen, B.S. "Naturally occurring fluctuations in dendritic spine density on adult hippocampal pyramidal neurons," *J. Neurosci.* **10**: 4035–4039 (1990).

Wu, L.G., and Saggau, P. "Block of multiple presynaptic calcium channel types by omega-conotoxin-MVIIC at hippocampal CA3 to CA1 synapses," *J. Neurophysiol.* **73**: 1965–1970 (1995).

Wyatt, J.L. Jr. *Lectures on Nonlinear Circuit Theory. Microsystems Technology Laboratories Memo 92-685*. Massachusetts Institute of Technology: Cambridge, Massachusetts (1992).

Yamada, W., Koch, C., and Adams, P. "Multiple channels and calcium dynamics." In: *Methods in Neuronal Modeling: From Synapses to Networks*. Koch, C., and Segev, I., editors, second edition, pp. 137–170. MIT Press: Cambridge, Massachusetts (1998).

Yamada, W., and Zucker, R. "Time course of transmitter release calculated from simulations of a calcium diffusion model," *Biophys. J.* **61**: 671–682 (1992).

Yang, J., Ellinor, P.T., Sather, W.A., Zhang, J.-F., and Tsien, R.W. "Molecular determinants of Ca^{2+} selectivity and ion permeation in L-type Ca^{2+} channels," *Nature* **366**: 158–161 (1993).

Yang, C.R., Seamans, J.K., and Gorelova, N. "Electrophysiological and morphological properties of layers V-VI pyramidal cells in rat prefrontal cortex *in vitro*," *J. Neurosci.* **16**: 1904–1921 (1996).

Yang, C.-Y., and Yazulla, S. "Neuropeptide-like immunoreactive cells in the retina of the larval tiger salamander: Attention to the symmetry of dendritic projections," *J. Comp. Neurol.* **248**: 105–118 (1986).

Yellen, G. "Ionic permeation and blockage in Ca^{2+}-activated K^+ channels of bovine chromaffin cells," *J. Gen. Physiol.* **84**: 157–186 (1984).

Young, A.B., and MacDonald, R.L. "Glycine as a spinal cord neurotransmitter." In: *Handbook of the Spinal Cord*. Vol. 1: *Pharmacology*. Davidoff, R.E., editor, 1–43. Marcel Dekker: New York, New York (1983).

Yuste, R., Delaney, K.R., and Tank, D.W. "Ca^{2+} accumulations in dendrites of neocortical pyramidal neurons—An apical band and evidence for two functional compartments," *Neuron* **13**: 23–43 (1994).

Yuste, R., and Denk, W. "Dendritic spines as a basic functional units of neuronal integration," *Nature* **375**: 682–684 (1995).

Zador, A.M. *Biophysics of computation in single hippocampal neurons*. PhD Thesis. Yale University: New Haven, Connecticut (1993).

Zador, A.M., Agmon-Snir, H., and Segev, I. "The morphoelectronic transform: A graphical approach to dendritic function, " *J. Neurosci.* **15**: 1668–1682 (1995).

Zador, A.M., Claiborne, B.J., and Brown, T.J. "Nonlinear pattern separation in single hippocampal neurons with active dendritic membrane." In: *Advances in Neural Information Processing*

Systems, Vol. 4. Moody, J., Hanson, S.J., and Lippmann, R., editors, pp. 51–58. Morgan Kaufmann: San Mateo, California (1992).

Zador, A.M., and Dobrunz, L.E. "Dynamic synapses in the cortex," *Neuron* **19**: 1–4 (1997).

Zador, A.M., and Koch, C. "Linearized Models of Calcium Dynamics: Formal Equivalence to the Cable Equation," *J. Neurosci.* **14**: 4705–4715 (1994).

Zador, A.M., Koch, C., and Brown, T. "Biophysical model of a Hebbian Synapse," *Proc. Natl. Acad. Sci. USA* **87**: 6718–6722 (1990).

Zagotta, W.N., Hoshi, T., and Aldrich, R.W. "Restoration of inactivation in mutants of *Shaker* potassium channels by a peptide derived from ShB," *Science* **250**: 568–571 (1990).

Zhuo, M., Small, S.A., Kandel, E.R., and Hawkins, R.D. "Nitric oxide and carbon monoxide produce activity-dependent long-term synaptic enhancement in hippocampus," *Science* **260**: 1946–1950 (1993).

Zipser, D., and Andersen, R.A. "A back-propagating programmed network that simulates response properties of a subset of posterior perietal neurons," *Nature* **331**: 679–684 (1988).

Zipser, D., Kehoe, B., Littlewort, G., and Fuster, J. "A spiking network model of short-term active memory," *J. Neurosci.* **13**: 3406–3420 (1993).

Zohary, E., Hillman, P., and Hochstein, S. "Time course of perceptual discrimination and single neuron reliability," *Biol. Cybern* **62**: 475–486 (1990).

Zola-Morgan, S., and Squire, L.R. "Neuroanatomy of memory," *Ann. Rev. Neurosci.* **16**: 547–563 (1993).

Zona, C., Pirrone, G., Avoli, M., and Dichter, M. "Delayed and fast transient potassium currents in rat neocortical neurons in cell culture," *Neurosci. Lett.* **94**: 285–290 (1988).

Zucker, R.S. "Exocytosis: a molecular and physiological perspective," *Neuron* **17**: 1049–1055 (1996).

Zucker, R., and Fogelson, A. "Relationship between transmitter release and presynaptic calcium influx when calcium enters through discrete channels," *Proc. Natl. Acad. Sci. USA* **83**: 3032–3036 (1986).

Zucker, R., and Stockbridge, N. "Presynaptic calcium diffusion and the time courses of transmitter release and synaptic facilitation at the squid giant synapse," *J. Neurosci.* **3**: 1236–1269 (1983).

Index

Note: Page numbers followed by *f* indicate figures.